Fluid Flow in Fractured Porous Media

Fluid Flow in Fractured Porous Media

Special Issue Editors

Richeng Liu
Yujing Jiang

MDPI • Basel • Beijing • Wuhan • Barcelona • Belgrade

Special Issue Editors

Richeng Liu
China University of Mining and Technology
China

Yujing Jiang
Nagasaki University
Japan

Editorial Office
MDPI
St. Alban-Anlage 66
4052 Basel, Switzerland

This is a reprint of articles from the Special Issue published online in the open access journal *Processes* (ISSN 2227-9717) from 2018 to 2019 (available at: https://www.mdpi.com/journal/processes/special_issues/porous_media).

For citation purposes, cite each article independently as indicated on the article page online and as indicated below:

LastName, A.A.; LastName, B.B.; LastName, C.C. Article Title. *Journal Name* **Year**, *Article Number*, Page Range.

Volume 2
ISBN 978-3-03921-473-0 (Pbk)
ISBN 978-3-03921-474-7 (PDF)

Volume 1-2
ISBN 978-3-03921-475-4 (Pbk)
ISBN 978-3-03921-476-1 (PDF)

Contents

About the Special Issue Editors

Richeng Liu is Research Associate of Rock Mechanics and Rock Engineering at China University of Mining and Technology, China. His research is focused on fractal and nonlinear flow properties of complex rock fracture networks which are strongly connected with projects covering underground assessments, such as CO_2 sequestration, enhanced oil recovery, and geothermal energy development. He has authored over 60 refereed journal publications. He has received numerous awards for his outstanding contributions to Engineering, including the Best Research Paper and Best Doctoral Thesis awards by the Japanese Society for Rock Mechanics (JSRM). He is also a JSPS Postdoctoral Research Fellow in Japan, supported by JSPS (Japan Society for the Promotion of Science) and has been awarded with the Young Elite Scientist Sponsorship Program by CAST (China Association for Science and Technology). He is the leading Guest Editor of *Processes*, *Water*, and *Computer Modeling in Engineering and Sciences*.

Yujing Jiang is Professor of Civil Engineering at Nagasaki University, Japan. His main research activities involve the experimental characterization and numerical modeling of fluid flow through fractured porous media during shearing. He has co-authored around 200 refereed journal papers. He is a member of the Engineering Academy of Japan, Japan.

Article

Temporal Mixing Behavior of Conservative Solute Transport through 2D Self-Affine Fractures

Zhi Dou [1,2,*], Brent Sleep [2,*] , Pulin Mondal [2] , Qiaona Guo [1], Jingou Wang [1] and Zhifang Zhou [1]

1   School of Earth Science and Engineering, Hohai University, 8 Fochengxi Rd., Nanjing 210098, China; guoqiaona2010@hhu.edu.cn (Q.G.); wang_jinguo@hhu.edu.cn (J.W.); zhouzf@hhu.edu.cn (Z.Z.)
2   Department of Civil Engineering, University of Toronto, 35 St. George Street, Toronto, ON M5S 1A4, Canada; pulin.mondal@utoronto.ca
*   Correspondence: douz@hhu.edu.cn (Z.D.); Sleep@ecf.utoronto.ca (B.S.); Tel.: +86-25-8378-7234 (Z.D.); +1-416-978-3005 (B.S.)

Received: 21 August 2018; Accepted: 4 September 2018; Published: 5 September 2018

Abstract: In this work, the influence of the Hurst exponent and Peclet number (Pe) on the temporal mixing behavior of a conservative solute in the self-affine fractures with variable-aperture fracture and constant-aperture distributions were investigated. The mixing was quantified by the scalar dissipation rate (SDR) in fractures. The investigation shows that the variable-aperture distribution leads to local fluctuation of the temporal evolution of the SDR, whereas the temporal evolution of the SDR in the constant-aperture fractures is smoothly decreasing as a power-law function of time. The Peclet number plays a dominant role in the temporal evolution of mixing in both variable-aperture and constant-aperture fractures. In the constant-aperture fracture, the influence of Hurst exponent on the temporal evolution of the SDR becomes negligible when the Peclet number is relatively small. The longitudinal SDR can be related to the global SDR in the constant-aperture fracture when the Peclet number is relatively small. As the Peclet number increases the longitudinal SDR overpredicts the global SDR. In the variable-aperture fractures, predicting the global SDR from the longitudinal SDR is inappropriate due to the non-monotonic increase of the longitudinal concentration second moment, which results in a physically meaningless SDR.

Keywords: mixing; conservative solute; fractal; roughness; fracture

---

1. Introduction

It has been widely recognized that fractures can play an important role in the transport and fate of contaminants. Characterizing the spreading and mixing processes of conservative solute through the fractures is very important for the understanding of reaction rates and mass transport rates associated with nuclear waste disposal, enhanced oil recovery, and bioremediation [1–6]. Although, in recent decades, many studies have provided new insights into the mechanisms and properties of mixing processes in homogeneous and heterogeneous porous media [7–17], to date, little attention has been focused on mixing behavior in fractures.

Since the heterogeneity of geological formations is ubiquitous, a fundamental issue about the difference between spreading and mixing processes of conservative solute needs to be understood. Several authors [9,12,18] emphasized the difference between spreading and mixing. Spreading indicates the change of the spatial extent of a solute plume whereas mixing describes the process that uniformizes the concentration distribution of solute inside the plume. In other words, spreading leads to the stretching and deformation of a solute plume while mixing gives rise to dilution of a conservative solute with time. Thus, spreading and mixing are not the same, but complete conservative solute transport can be thought of as being composed of both spreading and mixing. For a conservative solute transport in a spatially variable velocity field, the spreading of the solute plume is driven by the differences in advection that deform and stretch out the plume along the streamlines, whereas at the same time molecular diffusion causes the mixing that smooths out the concentration gradients within the solute plume. There is a complex interaction between the spreading and mixing processes, especially in heterogeneous flow fields [19,20]. Due to the naturally-coupled property of spreading and mixing, separating the spreading and mixing process is challenging, but studying the temporal evolution of mixing is still useful and important for improving predictions of reactive transport and mixing. Describing conservative solute transport only by the spreading is valid for some applications (for example, risk analysis). However, due to the influence of the mixing behavior of reactants on rate of reaction, describing the transport with the mixing-controlled chemical reactions only by spreading is insufficient [3,21,22].

To this end, many efforts have been made to develop methods for quantifying mixing and spreading. Following Aris's method of moments [23], the second central spatial moment of a conservative solute is a measure of spreading. This is because the spatial extent of the plume can be easily estimated based on the temporal evolution of spatial moments of the plume and is related to an apparent dispersion coefficient even for pre-asymptotic times [7,9,24,25]. In a homogeneous flow field, the second central spatial moment increases monotonically with time, which can be expected as a good measure for both spreading and mixing. This approach is invalid for a heterogeneous flow field where the second central spatial moment could decrease due to the convergence of streamlines [14,16,26]. Various metrics for quantifying mixing have been proposed.

Since the dilution caused by mixing is an irreversible process, Kitanidis [12] proposed the dilution index that measures the volume occupied by the solute plume. The dilution index is obtained from the statistical entropy (Shannon entropy) of the solute distribution. As opposed to the second central spatial moment, the dilution index is capable of quantifying true mixing and a mixing rate can be calculated by the rate of change of the entropy. The dilution index is useful not only for conservative solute transport but also for reactive transport [27]. The ratio of actual to theoretical maximum dilution index can be an indicator of the influence of incomplete mixing on reactive transport [15,28]. Moreover, based on the original concept of the dilution index, the flux-related dilution index that describes dilution as "act of distributing a given solute mass flux over a larger water flux" was proposed by Rolle et al. [14] for steady-state transport with continuous injection mode.

From the view of stochastic hydrogeology, the concentration distribution of a solute plume can be decomposed into a cross-sectional mean and a fluctuation about that mean. The concentration variance method has been proposed as a measure of mixing [11,29,30]. In addition to the dilution index and concentration variance, the mixing can be alternatively quantified by the scalar dissipation rate (SDR) which is determined from the time derivative of the integral of squared concentration within the solute plume [22,31]. Although the SDR was proposed for the study of turbulent flow and combustion, several studies [32] have shown that the SDR can be also applied to a variety of problems of subsurface contaminant transport (e.g., compound-specific transport, conservative solute transport, and multicomponent reactive transport). Le Borgne et al. [31] investigated the temporal evolution of the SDR in heterogeneous porous media and demonstrated the occurrence of a non-Fickian scaling of mixing. Bolster et al. [7] used the SDR to decompose the global mixing state into a dispersive mixing state and a local mixing state. Jha et al. [33] applied the SDR to quantify the mixing in a viscously unstable flow. Dreuzy et al. [21] considered that mixing resulted from competition between velocity fluctuations and local scale diffusion, and they proposed a new decomposition of mixing into potential mixing and departure rate. This new decomposition of mixing showed a generic characterization and could offer new ways to establish a transport equation with consideration of both advection, spreading, and mixing. A comparison of different transport models can be found in [10]. A series of analytical solutions for the SDR was derived in non-conservative transport systems by Engdahl, Ginn, and Fogg [32]. Furthermore, previous studies [7,21,31] in porous media showed that the transverse mixing generating the concentration gradients in the transverse direction influenced longitudinal mixing. If the global mixing was dominated by the spreading at asymptotical time, the transverse mixing could be negligible and the global mixing could be predicted by the longitudinal mixing. However, these current studies on the SDR are limited to specific homogeneous or heterogeneous porous media. Since anomalous (non-Fickian) transport has been observed in single rough fractures [34–38] and mixing and spreading could play an important role in fractures, the study of the performance, characteristic, and evolution of the SDR in other important subsurface geological formations (e.g., single rough fractures) needs further investigation.

The primary objective of this work is to investigate the effects of the Hurst exponent (which can indicate the roughness features of the fracture walls) and Peclet number on the temporal behavior of mixing. The validity of using longitudinal mixing to predict global mixing was evaluated in self-affine fractures. Two groups of different self-affine fractures were considered and denoted as the constant-aperture fracture and the variable-aperture fracture. The computational fluid dynamics (CFD) simulations of the flow field and solute transport in fractures were implemented. There are three major contributions here relative to previous work. The first is to show the capability of the SDR for characterizing the mixing in self-affine fractures. The second is to quantify the influence of the Hurst exponent and the Peclet number on the SDR scaling in constant-aperture fractures and the variable-aperture fractures. The third is to test and evaluate the validity of using the longitudinal mixing to predict the global mixing in self-affine fractures.

2. Methodology

2.1. Fracture Generation

Previous studies [39] on the morphology of natural fracture walls indicated that the walls of natural fractures could be characterized as statistical self-affine distributions. The mathematical characterization of the self-affine rough fracture wall was briefly reviewed here. A two-dimensional single fracture was considered, whose height is defined by a single-value function $Z(x)$ and the statistical self-affine property of the height can be expressed as:

$$\lambda^H Z(x) = Z(\lambda x) \tag{1}$$

where H indicates the magnitude of the roughness or the so-called Hurst exponent varying from 0 to 1 and λ is a scaling factor. $Z(x)$ can be thought of as a function of an independent spatial or temporal variable x. For a self-affine fracture wall, the stationary increment $[Z(x + h_x) - Z(x)]$ over the distance h_x follows a Gaussian distribution with mean zero. Thus, for the arbitrary λ, the mean and variance of the increments can be expressed as:

$$\langle Z(x + \lambda h_x) - Z(x) \rangle = 0 \tag{2}$$

$$\sigma^2(\lambda) = \lambda^{2H} \sigma^2(1) \tag{3}$$

where $\langle \cdot \rangle$ represents the mathematical expectation. For the different distances, the variance $\sigma^2(\lambda)$ is defined as a function of λ:

$$\sigma^2(\lambda) = \langle [Z(x + \lambda) - Z(x)]^2 \rangle \tag{4}$$

$$\sigma^2(\lambda h_x) = \langle [Z(x + \lambda h_x) - Z(x)]^2 \rangle \tag{5}$$

Then, depending on Equation (3):

$$\sigma^2_{\lambda h_x} = \lambda^{2H} \sigma^2_{h_x} \tag{6}$$

$$\sigma_{\lambda h_x} = \lambda^H \sigma_{h_x} \tag{7}$$

where $\sigma^2_{\lambda h_x}$ and $\sigma^2_{h_x}$ represent the variances of increments with different distances λh_x and λ, respectively.

Based on the self-affine scaling law given by Equations (1)–(7), a number of algorithms (e.g., the successive random additions, the randomization of the Weierstrass-Mandelbrot function, and the Fourier transformation) have been developed for synthetic self-affine fracture generation. In the present study, the successive random addition algorithm [40] was used to generate the synthetic self-affine fracture wall. It should be noted that to generate the self-affine fracture wall, the desired Hurst exponent must be selected. For this, the rough surface morphology of the bottom fracture wall of a single-fracture dolomite rock block (of 280 × 210 × 70 mm in size) was measured by using a 3D stereo-topometric measurement system (ATOS II from GOM mbH, Braunschweig, Germany). Preparation of this single-fracture dolomite rock block has been described in [41]. A 3D model of the fracture wall surface was generated by ATOS II using non-contact optical scanning technique [42]. A raw dataset of fracture surface heights from the dolomite rock fracture sample was obtained with a spatial resolution of ~250 μm (See Figure 1). From variogram analysis [43], the self-affinity of dolomite rock surface was evaluated. The distribution of Hurst exponent values was examined over 200 profiles along the longitudinal direction (for example, A-B profile in Figure 1). The calculated Hurst exponent values were between 0.55 and 0.91, where Hurst exponent between 0.6 and 0.8 covers ~90% of profiles. Thus, due to the computational cost, three different Hurst exponents were selected (i.e., $H = 0.6$, $H = 0.7$, and $H = 0.8$).

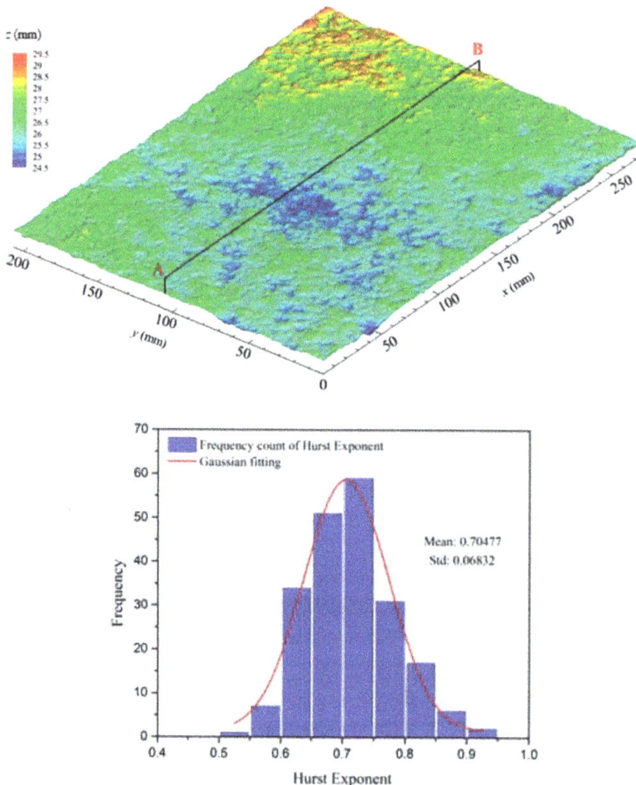

Figure 1. Dolomite rock fracture surface 3D profile and distribution of Hurst exponent for the different 2D profiles along the longitudinal direction.

Once the desired Hurst exponent was selected, the self-affine fracture wall was generated by the successive random addition algorithm. Figure 2 shows the generated self-affine fracture walls with $H = 0.8$, $H = 0.7$, and $H = 0.6$, respectively. It can be seen from Figure 2 that the larger Hurst exponent leads to a higher spatial correlation and a smoother wall.

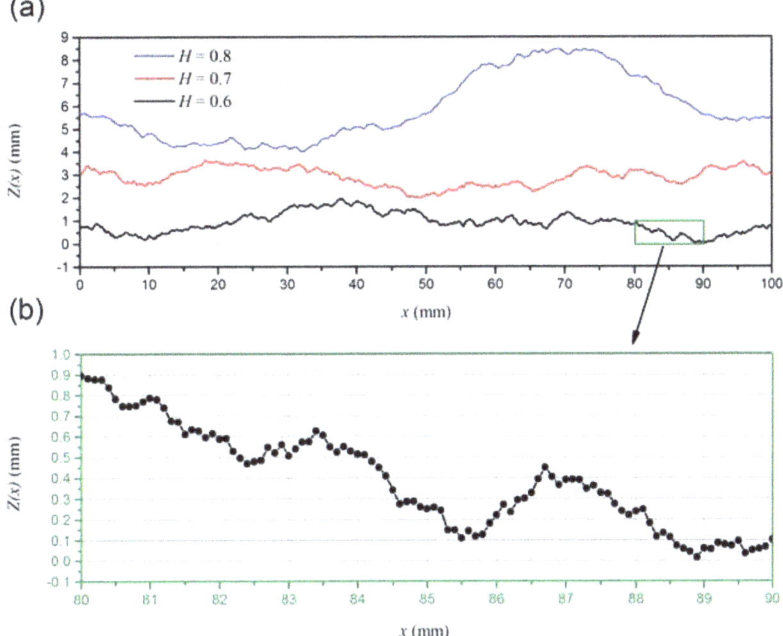

Figure 2. Self-affine fracture walls with the different Hurst exponents. (**a**) The self-affine fracture walls with $H = 0.6$, $H = 0.7$, and $H = 0.8$, respectively. (**b**) The zoom-in self-affine fracture wall with $H = 0.6$ between $x = 80$ mm and $x = 90$ mm.

In self-affine fractures, the aperture distribution can have a significant influence on the spreading and mixing processes. The reconstruction of the aperture field from a pair of generated fracture walls is dependent upon the way in which the walls are oriented relative to each other. Note that the Hurst exponent could be different for the top and bottom fracture walls. However, our aim here is not to perform an exhaustive investigation for all possibilities. It is assumed that the reconstructed fracture is two-dimensional, uniform mineral component, there is no contact area (no zero-aperture region), and the Hurst exponent for both of the top and bottom fracture walls is the same. Two possibilities to reconstruct the rough fracture aperture field from the specific self-affine fracture walls were considered. First, a constant-aperture rough fracture was introduced, where the top fracture wall was a replica of the bottom fracture wall translated a distance $\bar{b}$ normal to the mean plane (See Figure 3a). In this case, the fracture walls are rough and self-affine, but the local aperture $b(x)$ is constant and equal to $\bar{b}$. Alternatively, the variable-aperture rough fracture was studied, where the top fracture wall was a replica of the bottom wall. The bottom wall was sheared along the horizontal direction by a displacement d_0 and then translated a distance $\bar{b}$ normal to the mean plane (See Figure 3b). Obviously, the two walls of the reconstructed fracture with shear displacement d_0 do not overlap and the local aperture is a function of the horizontal location x. Based on the self-affine scaling law, Wang, et al. [44] developed the shear displacement model to obtain the aperture field with Gaussian distribution. The local aperture $b(x)$ is given by:

$$S_2(x) = S_1(x + d_0) + \bar{b} \tag{8}$$

$$b(x) = \begin{cases} S_2(x) - S_1(x) & \text{if } S_2(x) > S_1(x) \\ 0 & \text{otherwise} \end{cases} \tag{9}$$

where $S_1(x)$ and $S_2(x)$ are the top and bottom fracture walls, respectively. Figure 3c shows that when the Hurst exponent of the self-affine fracture wall is $H = 0.7$, the aperture field of the variable-aperture rough fracture follows a Gaussian distribution with the mean aperture $\bar{b} = 0.5$ mm and the standard deviation of the aperture $\sigma_b = 0.17$ mm.

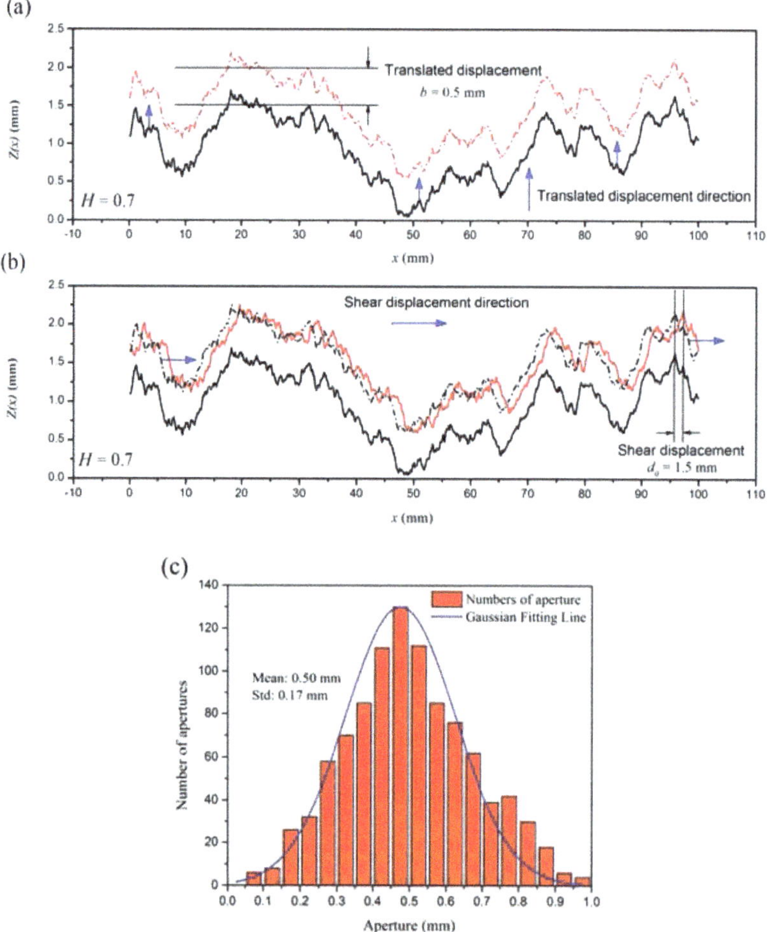

Figure 3. Reconstruction of aperture field in the self-affine rough fracture with $H = 0.7$. (**a**) Reconstruction of constant-aperture rough fracture. (**b**) Reconstruction of variable-aperture rough fracture. (**c**) Gaussian distribution of aperture field for the variable-aperture rough fracture.

In this study, the total length of the generated self-affine fracture wall is set as 100 mm and the horizontal distances between two adjacent points in the self-affine fracture wall were equal to 0.1 mm (see Figure 2b). Three self-affine fracture walls with $H = 0.6$, $H = 0.7$, and $H = 0.8$ were generated by the successive random addition algorithm. Each self-affine fracture wall was used to reconstruct the constant-aperture fracture and the variable-aperture fracture, respectively. Thus, there were six fractures in this study. The coefficient of variation (COV) was set to 0.35 for the variable-aperture fracture.

2.2. Computational Fluid Dynamics (CFD) Simulations of the Flow Field and Solute Transport in Single Rough Fractures

Since the mixing behavior is highly dependent on the flow field [8,9,16,45], the flow field in a single rough fracture was solved directly by using the Navier-Stokes and continuity equations for isothermal, incompressible, and homogenous single Newtonian steady flow:

$$\nabla \cdot \boldsymbol{u} = 0 \tag{10}$$

$$\rho(\boldsymbol{u} \cdot \nabla \boldsymbol{u}) - \nabla(\mu \nabla \boldsymbol{u}) = -\nabla p \tag{11}$$

where ρ is the density of fluid, $\boldsymbol{u} = [u, w]$ is the velocity vector, p is the fluid pressure, and μ is the dynamic viscosity of fluid. Two given pressure values were implemented at the inlet and outlet boundary. The steady-state flow field was solved by the pressure drop over the entire fracture.

Transient solute transport in a single self-affine fracture was described by the advection-diffusion equation for conservative non-sorbing solute transport:

$$\frac{\partial c}{\partial t} = -\nabla \cdot (\boldsymbol{u} c) + D_m \nabla^2 c \tag{12}$$

where c is the solute concentration, t is time, and D_m is the molecular diffusion coefficient. The velocity vector in Equation (12) is from the flow field based on solution of Equations (10) and (11). It is assumed that the initial concentrations were given by:

$$c(x, t = 0) = \begin{cases} \frac{m_0}{b(x)*\Delta L*W} & \text{if } x_L^* < x < x_L^* + \Delta L \\ 0 & \text{otherwise} \end{cases} \tag{13}$$

where m_0 is the mass of injected solute, $b(x)$ is the local aperture, W is the width of fracture in the out of plane direction (equal to 1 m in the 2D problem) and ΔL is the width of injected solute. The ΔL is constant for all of simulations and assumed as $\Delta L/L = 0.001$, where the L is the length of the whole fracture. To avoid boundary effects, the initial injection location of the solute e mass is shifted downstream from the fracture inlet by a distance of $x_L^* = 0.01L$. The inlet and outlet boundary conditions for transient solute transport were specified as:

$$c(0, t) = 0 \ \ t \geq 0 \tag{14}$$

$$\partial c(L, t)/\partial n = 0 \ \ t \geq 0 \tag{15}$$

where n represents the normal direction to the outlet boundary.

2.3. Mixing: Scalar Dissipation Rate (SDR)

Mixing can be described by the SDR that is a global mixing measure based on the integral of concentration gradients. Recently, several studies have focused on the SDR evolution and scaling properties during solute transport in porous media [7,10,13,21,27,31,32,46]. However, all of those studies on SDR were restricted to porous media. The study on the SDR evolution in rough fractures is still limited, which motivates our investigation. The SDR of a conservative scalar is given by:

$$\chi(t) = \int_\Omega D_m \nabla c(x,t) \cdot \nabla c(x,t) dx \tag{16}$$

To obtain the SDR by using the Equation (16), the local concentration gradients needs to be determined. However, due to the occurrence of the sharp concentration gradient over small distances in the relatively heterogeneous flow field, a very fine numerical discretization for both of the flow and the concentration field is required to obtain an accurate quantification of concentration gradients, which results in the huge computational cost. Le Borgne et al. [31] showed that the SDR can be approximated from the concentration second moment (the integral of the squared concentrations). After multiplying Equation (12) by $c(x,t)$ and integrating over the entire domain:

$$\frac{1}{2}\frac{\partial}{\partial t}\int_\Omega c(x,t)^2 d\Omega + \frac{1}{2}\int_\Omega \nabla\cdot[uc(x,t)^2]d\Omega = \frac{1}{2}D_m\int_\Omega \nabla\cdot\nabla c(x,t)^2 d\Omega \\ - \int_\Omega D_m\nabla c(x,t)\cdot\nabla c(x,t)d\Omega \tag{17}$$

Assuming that the fractured domain is infinite, there is no mass flux out of the domain, and the flow field is divergence-free, the terms involving a divergence operator in Equation (17) are zero. Then it can obtain:

$$\chi(t) = -\frac{1}{2}\frac{dM^2(t)}{dt} = D_m\int_\Omega \nabla\cdot\nabla c(x,t)^2 d\Omega \tag{18}$$

where $M^2(t)$ is the concentration second moment and can be expressed as:

$$M^2(t) = \int_\Omega c(x,t)^2 d\Omega \tag{19}$$

Le Borgne et al. (2010) reported that the results from using Equation (18) instead of Equation (16) are more accurate than from using Equation (16) directly and the calculation for Equation (18) is computationally more efficient. In this study, the temporal mixing state is obtained from Equation (18).

For an infinite 1D homogeneous domain with zero velocity Fick's Law of diffusion can be used to describe the spatial distribution of solute concentration. The corresponding analytical solution for the concentration distribution in the absence of reaction is given by:

$$c_0(x,t) = \frac{m_0}{\sqrt{4\pi D_m t}}\exp\left(-\frac{x^2}{4\pi D_m}\right) \tag{20}$$

By integrating the square of Equation (20) over all domains, the corresponding concentration second moment can be expressed as:

$$\int_\Omega c_0(x,t)^2 d\Omega = \frac{m_0^2}{\sqrt{8\pi D_m t}} = M_0(t) \tag{21}$$

From Equation (18), the analytical 1D SDR solution can thus be expressed as:

$$\chi_0(t) = -\frac{1}{2}\frac{dM_0(t)}{dt} = \frac{1}{8}\frac{m_0^2}{\sqrt{2\pi D_m}}t^{-\frac{3}{2}} \tag{22}$$

3. Results and Discussion

3.1. Model Setup

In this study, the water with standard properties at 20 °C (e.g., ρ = 998.2 kg/m^3 and μ = 1.002 × 10^{-3} Pa·s) was used to saturate the void space in the fractures. The typical conservative solute transport (e.g., Cl$^-$ in water) and the corresponding D_m = 2.03 × 10^{-9} m^2/s were assumed depending on the reference of [47]. The matrix of the fracture was assumed impermeable and the rough fracture walls were considered as non-slip boundaries. As background flow, the steady-state flow was induced by a given pressure drop over the entire fracture. The solved flow field serves as the input for the transient solute transport model. The flow field and transient solute transport models based on Equations (10)–(16) were implemented in the COMSOL Multiphysics package version 5.2

(COMSOL Inc., Burlington, MA, USA) using the Galerkin finite-element method [48]. In order to ensure numerical stability and accuracy, the fracture domain was discretized into ~152,000 triangular elements. The number of triangular elements was determined by the mesh independence analysis. Under the same pressure gradient ($-\nabla p = 185\ \mathrm{Pa/m}$), the steady-state flow rate changes about 0.95% (from $9.550 \times 10^{-4}\mathrm{m}^3/\mathrm{s}$ to $9.641 \times 10^{-4}\mathrm{m}^3/\mathrm{s}$) as the number of triangular elements increases by about 104% (from 152,000 to 310,000). This indicates that 152,000 triangular elements are sufficient to provide stable and accurate numerical results.

The Peclet number, $Pe = \tau_D/\tau_a = \overline{u}\overline{b}/D_m$, was defined by the ratio of the characteristic diffusion time ($\tau_D = \overline{b}^2/D_m$) to the characteristic advection time ($\tau_a = \overline{b}/\overline{u}$) where the $\overline{u}$ is the mean flow velocity in the fractures. In each simulation, three different Pe values ($Pe = 10$, $Pe = 100$, and $Pe = 1000$) were considered. Without loss of generality, Figure 4 shows the flow fields in variable-aperture and constant-aperture fractures for a self-affine fracture wall with $H = 0.6$ and $Pe = 1000$. Figure 5 shows the results for both constant-aperture and variable-aperture fractures when the Pe is set as 1000 and Hurst exponent is equal to 0.6.

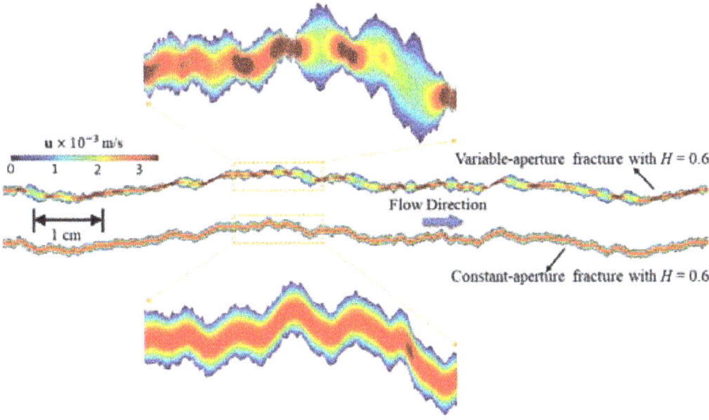

Figure 4. The flow fields in variable-aperture and constant-aperture fractures with $H = 0.6$ for $Pe = 1000$.

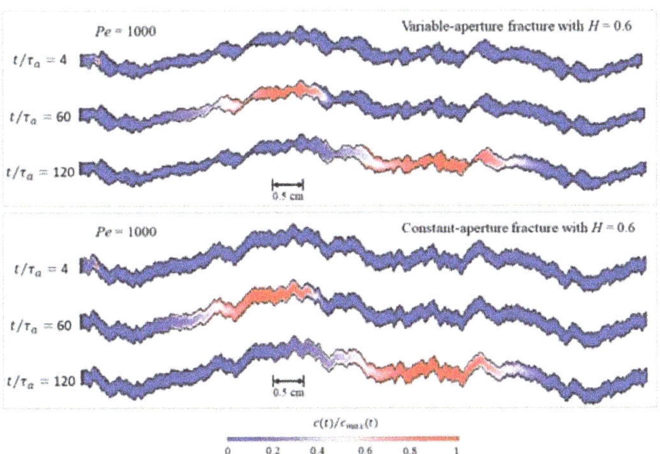

Figure 5. The solute transport in variable-aperture and constant-aperture fractures with $H = 0.6$ for $Pe = 1000$.

3.2. Influence of the Roughness of Fracture Walls on the Temporal Behavior of the Global SDR

To investigate the influence of the roughness of fracture walls on the temporal behavior of the global SDR, the conservative solute transport in self-affine rough fractures with different Hurst exponents and Peclet number values were simulated. Figure 6 shows the evolution of SDR in the constant-aperture fracture and variable-aperture fracture for Hurst exponents of 0.6, 0.7, and 0.8 and Peclet number values of 10, 100, and 1000. In Figure 6, the time is normalized by the characteristic advection time τ_a and the temporal SDR $\chi(t)$ is normalized by $\chi_0(\tau_a)$. It can be seen from Figure 6 that the SDR in general decreases with time for all conditions considered. The temporal evolution of the SDR in the variable-aperture fracture fluctuates more than that in the constant-aperture fracture. This indicates that the spatial distribution of local apertures and the corresponding spatial changes in velocity have a significant influence on the temporal evolution of the SDR. Since mixing is the only process of solute mass exchange between the different streamlines, the distribution of streamlines significantly impacts the mixing [8]. In the constant-aperture fracture, the streamlines are uniform over the entire fracture, whereas the streamlines in the variable-aperture fracture are deformed and bent due to the spatial distribution of local apertures (see Figure 4). The deformed and bent streamlines lead to changes in the transverse concentration gradients during transport along the fracture. In contrast, in the constant-aperture fractures, the transverse concentration gradients are smoothed out by diffusion after the characteristic diffusion time (i.e., $> \bar{b}^2/D_m$). In addition, the trend in the temporal evolution of the SDR for the variable-aperture fractures reveals that the magnitude of this fluctuation is affected by the Hurst exponent. For the smaller Hurst exponents ($H = 0.6$ and $H = 0.7$), the fluctuation of the SDR around the corresponding power-law fitting line exists over the entire interval of the transport time. However, the fluctuation is much less when $H = 0.8$ for $t < 10\tau_a$.

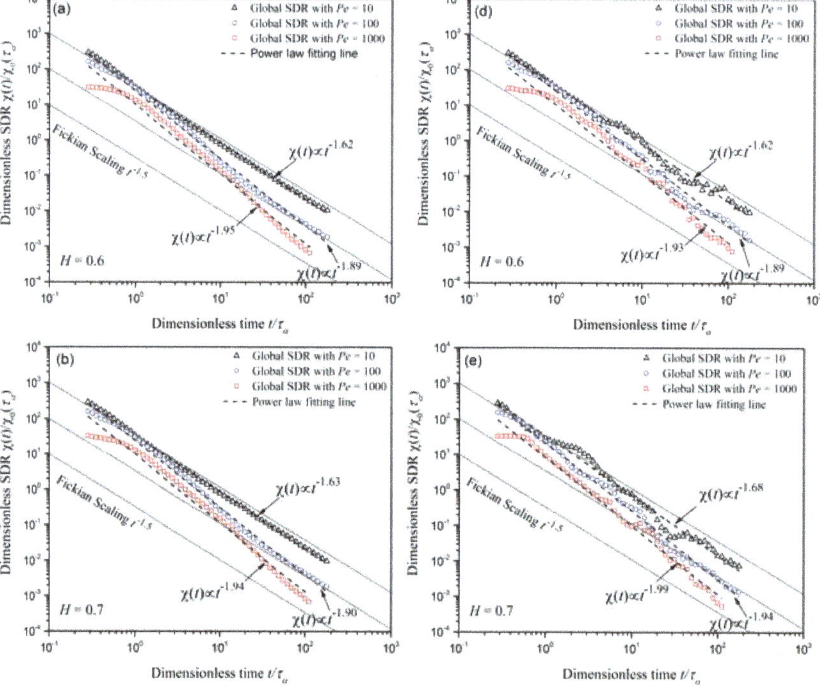

Figure 6. *Cont.*

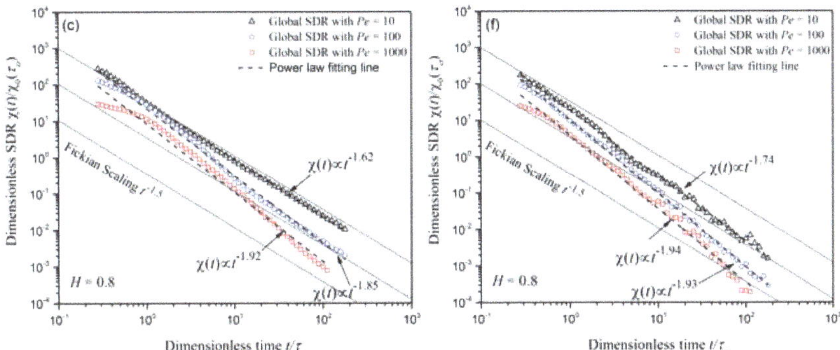

Figure 6. Scalar dissipation rate estimated in the constant-aperture fracture (**a–c**) and variable-aperture fracture (**d–f**) with $H = 0.6$, 0.7, and 0.8 for $Pe = 10$, 100, and 1000, respectively.

Each SDR in Figure 6 was fitted with a power-law function of the dimensionless time over the entire temporal range ($10^{-1} < t/\tau_a < 10^3$). For both constant-aperture and variable-aperture fractures, the SDR scaling for a given Peclet number is generally independent of the Hurst exponent of the fracture wall, while the Peclet number has a significant influence on the SDR scaling. When the Hurst exponent ranges from $H = 0.6$ to $H = 0.8$ the exponents in the power law fit are very similar for a particular Peclet number value in both constant-aperture and variable-aperture fractures. For one-dimensional fractures, the analytical SDR scaling follows the Fickian scaling $\chi(t) \propto t^{-1.5}$, as shown in Equation (22). Thus, as per the definition by Le Borgne et al. [31], the results in Figure 6 indicate that the SDR scaling is non-Fickian and the exponent of the best-fit power law equation for the SDR increases as the Peclet number decreases. This indicates that predicting the temporal evolution of the SDR over the entire temporal range in both constant-aperture and variable-aperture fractures using the one-dimensional analytical SDR scaling is inappropriate as this underpredicts the SDR, especially for the cases with the high Peclet number. Furthermore, it can be seen from Figure 6a–c that in constant-aperture fractures, for a given Peclet number, the results are similar for $H = 0.6$, $H = 0.7$, and $H = 0.8$. This implies that the roughness of the self-affine fracture walls, as characterized by Hurst exponent, is not a dominant influence on the temporal evolution of the SDR in the constant-aperture fractures. In contrast, as shown in Figure 6d–f, for a given Peclet number, the exponent of the best-fitting power law decreases as the Hurst exponent increases and spatial correlation increases.

The characteristic of the mixing can be seen from the SDR scaling over the different temporal ranges. For the cases with $Pe = 10$ and $Pe = 100$ in both constant-aperture and variable-aperture fractures, the SDR scaling is approximately Fickian at late time ($t > 10\tau_a$), but becomes non-Fickian for the temporal range examined for $Pe = 1000$. The transition time between the Fickian and non-Fickian temporal evolution of the SDR is important for predicting the mixing process. Before the transition time, the temporal evolution of the SDR shows non-Fickian mixing behavior [31,49]. In constant-aperture fractures (see Figure 6a–c) the transition time increases as the Peclet number increases while the transition time is insensitive to the Hurst exponent. For the cases with $Pe = 100$ the corresponding transition time is equal to $30\tau_a$, whereas for the cases with $Pe = 1000$ it is more than $100\tau_a$. In the variable-aperture fractures, both the Hurst exponent and the Peclet number influence the transition time. It fails to capture the exact transition time for the cases with $Pe = 1000$ due to the limitation of fracture length scale. For a given Peclet number and Hurst exponent, the occurrence of non-Fickian mixing in fractures depends on both the aperture distribution and the fracture length scale.

3.3. Validity of Predicting Global SDR from the Longitudinal SDR

Since Taylor's work [23,50] demonstrated that solute transport in an asymmetrical shear flow field can be reduced to a one-dimensional dispersion process by using a longitudinal effective dispersion coefficient, this idea has been proven useful and effective among a wide range of fields and applications [37,51–54]. The spatially variable velocity field dominates the behavior of solute spreading process, while the local diffusion process can smooth concentration gradients at the same time. These two coupled processes lead to the incomplete mixing inside the plume and reflect the fact that the mixing can be considered as a result of both the local diffusion and solute spreading. Le Borgne et al. [31] pointed out that the incomplete mixing inside the plume, which generates the concentration gradients in the transverse direction, results in the non-Fickian scaling of mixing. Bolster et al. [7] distinguished the two coupled process of mixing by expressing the concentration as the sum of the average of the concentration over the transverse cross-section and the deviation about it. In order to investigate the validity of predicting the global SDR by the longitudinal SDR in both constant-aperture and variable-aperture fractures, the method in [31] was used to obtain the mean longitudinal concentration projected in the transverse direction $\bar{c}(x,t)$ and then calculate the longitudinal SDR:

$$\bar{c}(x,t) = \frac{\int_0^{b(x)} c(x,t)dx}{b(x)} \tag{23}$$

The validity of predicting the global SDR by the longitudinal SDR can be evaluated by comparing the longitudinal SDR to the global SDR. The longitudinal SDR can also be considered as the contribution of the transverse concentration gradients to the global SDR. For the cases in the constant-aperture fractures, Figure 7 shows the global SDR estimated from the full concentration field and the longitudinal SDR estimated from Equation (23). In general, regardless of the influence of the Hurst exponent and Peclet number, the temporal evolution of the longitudinal and global SDR both decrease as power-law functions of time. For the cases with $Pe = 10$, the longitudinal SDR is very close to the global SDR, which indicates that the longitudinal SDR is capable of predicting the global SDR in the constant-aperture fractures. This also reveals that the transverse concentration gradients, in this case, do not play a dominant role in global mixing. Thus, both of the longitudinal and global SDR follow Fickian scaling ($\chi(t) \propto t^{-1.5}$).

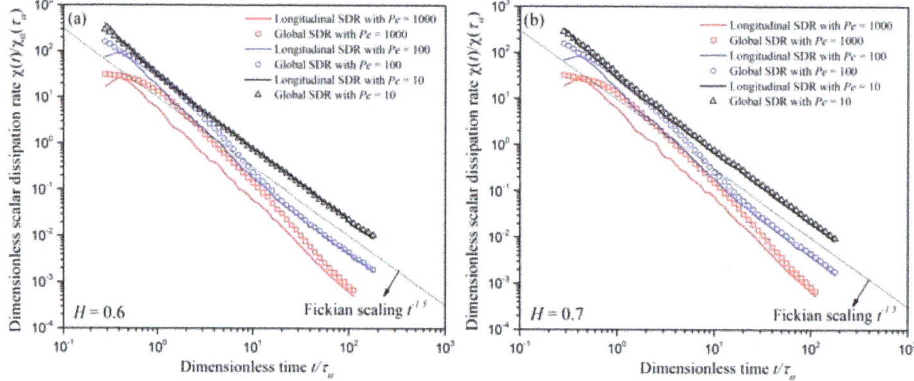

Figure 7. *Cont.*

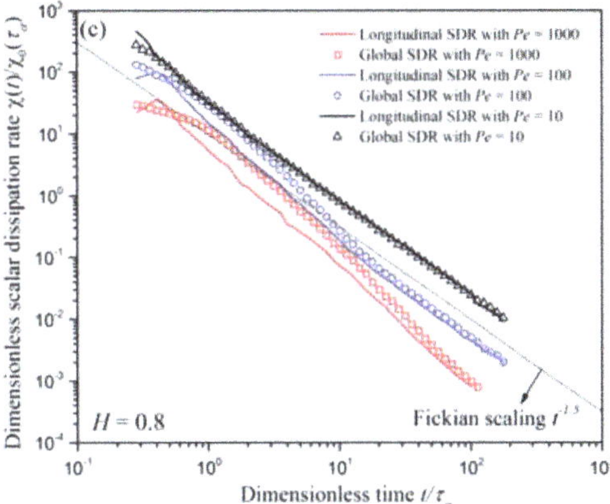

Figure 7. The global SDR estimated from the full concentration field and the longitudinal SDR estimated from the average of the concentration over the transverse cross-section in the constant-aperture fractures with the Hurst exponent $H = 0.6$ (**a**), $H = 0.7$ (**b**), and $H = 0.8$ (**c**) for the $Pe = 10$, 100, and 1000.

For the cases with $Pe = 100$ and $Pe = 1000$, the longitudinal SDR is smaller than the global SDR for most of the temporal range, indicating that the longitudinal SDR underpredicts the global SDR at that time. In contrast to the cases with $Pe = 10$, the transverse concentration gradients are relatively large and make significant contributions to the temporal evolution of the global SDR. As the transverse concentration gradients decrease due to diffusion, the concentration distribution inside the plume becomes more homogeneous. The characteristic time for diffusion smoothing the transverse concentration gradients depends on the Peclet number. For the case with $Pe = 100$, the longitudinal SDR gradually converges towards the global SDR around $t/\tau_a = 100$, which implies that the longitudinal SDR appears to be valid for predicting the global SDR when the influence of transverse concentration gradients becomes negligible. Moreover, Figure 7 shows that the Hurst exponent has little influence on the relationship between the longitudinal and the global SDR. These results indicate that in the constant-aperture fractures, the Hurst exponent is not a dominant factor in the development of transverse concentration gradients.

To predict the global SDR from the longitudinal SDR, it is essential that the temporal evolution of the longitudinal SDR has the same tendency as the global SDR, which implies that the longitudinal concentration second moment should decrease as a power-law function of time. This would be true for the cases in the constant-aperture fracture, however, it is not necessary in the variable-aperture fractures where the abruptly changing aperture leads to the convergence of streamlines. The temporal evolutions of both the longitudinal and global concentration second moments in the variable-aperture fractures were investigated. The longitudinal and global concentration second moments is associated with the temporal evolutions of longitudinal and global SDR, respectively, as shown in Figure 8. Unlike the temporal evolution of the global concentration second moment, the temporal evolution of the longitudinal concentration second moment does not follow a strictly monotonic decrease with time for all conditions considered. Consequently, the longitudinal SDR in the variable-aperture fractures has a negative value (corresponding to the discontinuities in the lines in Figure 8b,d,f) though this is not possible as per the definition of Equation (16). This negative longitudinal SDR results from the fact that the longitudinal concentration second moment increases when the plume is transported in the small-aperture regions. For conservative solute transport, the mixing is an irreversible process and associates with the entropy of the mixing system towards its maximum. Although the longitudinal SDR generally decrease as a power-law function of time in the variable-aperture fractures, the negative value in SDR is physically meaningless. Therefore, predicting the global SDR by the longitudinal SDR is inappropriate in the variable-aperture fractures.

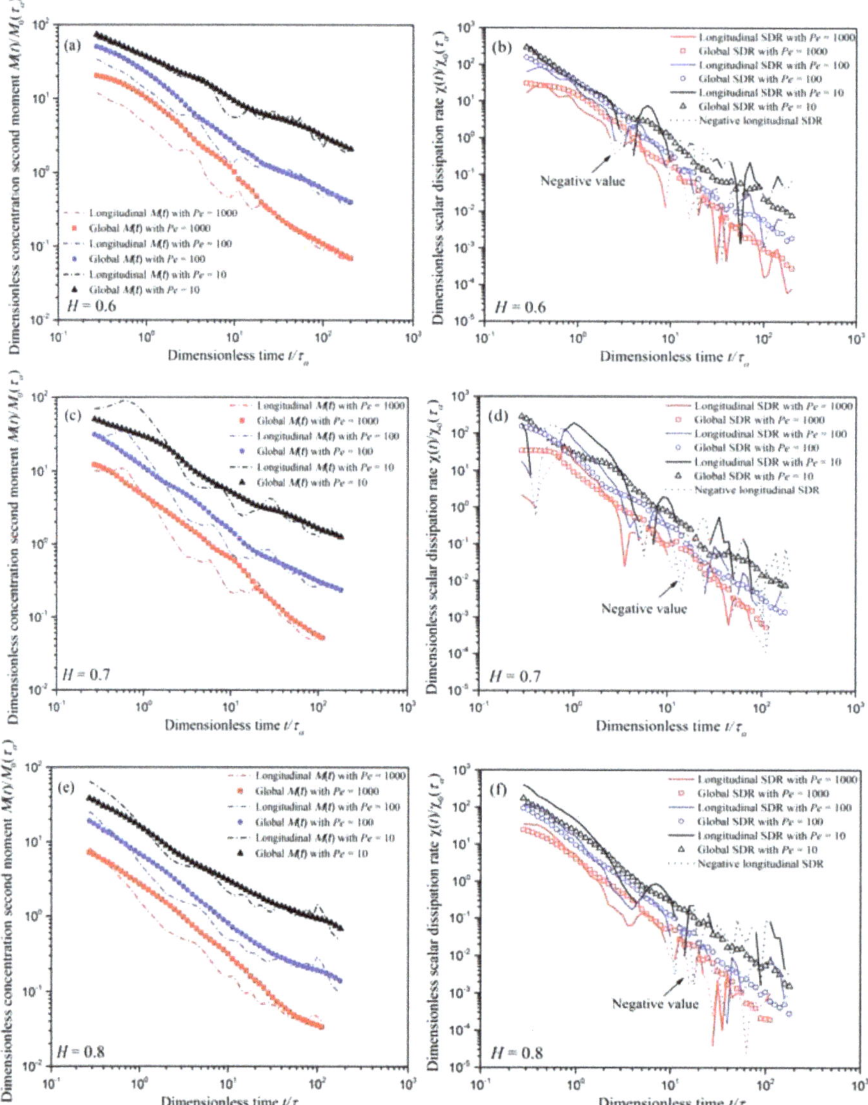

Figure 8. (**a**,**c**,**e**) represent the global concentration second moment estimated from the full concentration field and the longitudinal concentration second moment estimated from the average of the concentration over the transverse cross-section in the variable-aperture fractures with $H = 0.6$, $H = 0.7$, and $H = 0.8$, respectively. (**b**,**d**,**f**) represent the global SDR estimated from the full concentration field and the longitudinal SDR estimated from the average of the concentration over the transverse cross-section in the variable-aperture fractures with the Hurst exponent $H = 0.6$, $H = 0.7$, and $H = 0.8$ for the $Pe = 10$, 100, and 1000, respectively.

4. Summary and Conclusions

In this work, the successive random additions technique was applied to generate the 2D self-affine fracture wall for both the variable-aperture and constant-aperture fractures. The Hurst exponent of the generated 2D fracture wall was determined by the statistical analysis of the surface heights of the dolomite rock. Although there was no contact area (zero-aperture zone) in the generated 2D fractures, the simulations of conservative solute through variable-aperture and constant-aperture fractures highlighted the influence of the Hurst exponent and Peclet number on the temporal mixing behavior. It was found that, as in porous media, the SDR, decreased as a power-law function of time, and was characteristic of the mixing in both variable-aperture and constant-aperture fractures. The Peclet number had a significant influence on the temporal evolution of the SDR. The variable-aperture distribution led to the local fluctuation of the temporal evolution of the SDR, even for the small *Pe*. For both the variable-aperture and constant-aperture fractures, as the Peclet number increased, the exponent of the best-fitting power law for SDR scaling decreased significantly, indicating that the relatively large Peclet number enhanced the mixing process.

The influence of Hurst exponent on the temporal evolution of the SDR was dependent on the Peclet number and the aperture distribution. For the constant-aperture fracture, the influence of Hurst exponent on the temporal evolution of the SDR become negligible when the Peclet number was relatively small. The transition time between Fickian and non-Fickian mixing process was sensitive to the Hurst exponent, Peclet number, aperture distribution, and fracture lengh scale. For a given Peclet number and Hurst exponent, the occurrence of the non-Fickian mixing in fractures was dependent on the aperture distribution and the fracture length scale.

Comparisons of the longitudinal and global SDR showed that the global SDR could be predicted from the longitudinal SDR for Peclet number values of 100 or less in the constant-aperture fracture, independent of the value of the Hurst exponent for the self-affine fracture wall. However, for Peclet number of 1000, the longitudinal SDR would overpredict the global SDR. For the variable-aperture fracture, predicting the global SDR from the longitudinal SDR was inappropriate for all Peclet number and Hurst exponents investigated due to the non-monotonic increases of the longitudinal concentration second moment, which resulted in the physically meaningless SDR (i.e., negative SDR).

This study provides new insights into the temporal mixing behavior in self-affine fractures. However, the results for the validity of predicting the global SDR from the longitudinal SDR needs to be extended and tested in 3D real fractures with a comparison with experimental data, particularly to determine the influence of contact area (zero-aperture zone) and preferential flow on the SDR. In addition, the complex flow field (i.e., turbulent flow) in fractures would have a significant influence on the temporal mixing behavior, which also needs to be considered.

Author Contributions: Conceptualization, Z.D.; Formal analysis, Z.D.; Investigation, Z.D.; Methodology, Z.D.; Project administration, J.W.; Supervision, Brent Sleep and Z.Z.; Validation, Z.D.; Visualization, Q.G.; Writing—original draft, Z.D.; Writing—review & editing, Z.D., Brent Sleep, P.M., J.W. and Z.Z.

Funding: This research was funded by the National Natural Science Foundation of China (grant nos. 41602239, 41877171, 41402197, and 41572209), the Natural Science Foundation of Jiangsu (grant nos. BK20160861 and BK20140843), the International Postdoctoral Exchange Fellowship Program (grant no. 20150048), and the Fundamental Research Funds of the Central Universities (grant no. 2016B05514).

Conflicts of Interest: The authors declare no conflict of interest.

Nomenclature

$b(x)$	The local aperture of the fracture
$\bar{b}$	The distance normal to the mean plane
c	The solute concentration
$\bar{c}(x,t)$	The mean longitudinal concentration projected in the transverse direction
d_0	The shear displacement distance along the horizontal direction
D_m	The molecular diffusion coefficient
H	Hurst exponent
h_x	The horizontal distance
L	The length of the whole fracture
ΔL	The width of injected solute
m_0	The mass of injected solute
$M^2(t)$	The concentration second moment
n	The normal direction to the outlet boundary
Pe	Peclet number
p	The fluid pressure
$S_1(x)$	The top fracture wall
$S_2(x)$	The bottom fracture wall
t	Time
$\mathbf{u}$	The velocity vector
W	The width of fracture in the out of plane direction
x_L^*	The initial injection location of the solute mass
$Z(x)$	A function of independent variable x
λ	Scaling factor
μ	The dynamic viscosity
ρ	The density of fluid
$\sigma^2(\lambda)$	The variance
$\sigma^2_{\lambda h_x}$	The variance of increments with the distances λh_x
$\sigma^2_{h_x}$	The variance of increments with the distances λ
σ_b	The standard deviation of the aperture
τ_D	The CHARACTERISTIC diffusion time
τ_a	The characteristic advection time
$\chi(t)$	The scalar dissipation rate
$\chi_0(t)$	The analytical 1D SDR solution

References

1. Berkowitz, B.; Cortis, A.; Dentz, M.; Scher, H. Modeling non-fickian transport in geological formations as a continuous time random walk. *Rev. Geophys.* **2006**, *44*. [CrossRef]
2. Dentz, M.; Le Borgne, T.; Englert, A.; Bijeljic, B. Mixing, spreading and reaction in heterogeneous media: A brief review. *J. Contam. Hydrol.* **2011**, *120–121*, 1–17. [CrossRef] [PubMed]
3. Rolle, M.; Kitanidis, P.K. Effects of compound-specific dilution on transient transport and solute breakthrough: A pore-scale analysis. *Adv. Water Resour.* **2014**, *71*, 186–199. [CrossRef]
4. Cai, J.; Hu, X.; Xiao, B.; Zhou, Y.; Wei, W. Recent developments on fractal-based approaches to nanofluids and nanoparticle aggregation. *Int. J. Heat Mass Transf.* **2017**, *105*, 623–637. [CrossRef]
5. Dou, Z.; Zhou, Z.; Wang, J.; Liu, J. Pore-scale modeling of mixing-induced reaction transport through a single self-affine fracture. *Geofluids* **2018**, *2018*, 9095143. [CrossRef]
6. Liu, R.; Li, B.; Jiang, Y.; Yu, L. A numerical approach for assessing effects of shear on equivalent permeability and nonlinear flow characteristics of 2-d fracture networks. *Adv. Water Resour.* **2018**, *111*, 289–300. [CrossRef]
7. Bolster, D.; Valdés-Parada, F.J.; LeBorgne, T.; Dentz, M.; Carrera, J. Mixing in confined stratified aquifers. *J. Contam. Hydrol.* **2011**, *120–121*, 198–212. [CrossRef] [PubMed]
8. Chiogna, G.; Cirpka, O.A.; Herrera, P.A. Helical flow and transient solute dilution in porous media. *Trans. Porous Media* **2016**, *111*, 591–603. [CrossRef]

9. Dentz, M.; Carrera, J. Mixing and spreading in stratified flow. *Phys. Fluids* **2007**, *19*, 017107. [CrossRef]
10. Dreuzy, J.-R.D.; Carrera, J. On the validity of effective formulations for transport through heterogeneous porous media. *Hydrol. Earth Syst. Sci.* **2016**, *20*, 1319–1330. [CrossRef]
11. Kapoor, V.; Kitanidis, P.K. Concentration fluctuations and dilution in aquifers. *Water Resour. Res.* **1998**, *34*, 1181–1193. [CrossRef]
12. Kitanidis, P.K. The concept of the dilution index. *Water Resour. Res.* **1994**, *30*, 2011–2026. [CrossRef]
13. Le Borgne, T.; Dentz, M.; Villermaux, E. Stretching, coalescence, and mixing in porous media. *Phys. Rev. Lett.* **2013**, *110*, 204501. [CrossRef] [PubMed]
14. Rolle, M.; Eberhardt, C.; Chiogna, G.; Cirpka, O.A.; Grathwohl, P. Enhancement of dilution and transverse reactive mixing in porous media: Experiments and model-based interpretation. *J. Contam. Hydrol.* **2009**, *110*, 130–142. [CrossRef] [PubMed]
15. Tartakovsky, A.M.; Tartakovsky, G.D.; Scheibe, T.D. Effects of incomplete mixing on multicomponent reactive transport. *Adv. Water Resour.* **2009**, *32*, 1674–1679. [CrossRef]
16. Cirpka, O.A.; Kitanidis, P.K. Characterization of mixing and dilution in heterogeneous aquifers by means of local temporal moments. *Water Resour. Res.* **2000**, *36*, 1221–1236. [CrossRef]
17. Dou, Z.; Zhou, Z.-F.; Wang, J.-G. Three-dimensional analysis of spreading and mixing of miscible compound in heterogeneous variable-aperture fracture. *Water Sci. Eng.* **2016**, *9*, 293–299. [CrossRef]
18. Cirpka, O.A. Choice of dispersion coefficients in reactive transport calculations on smoothed fields. *J. Contam. Hydrol.* **2002**, *58*, 261–282. [CrossRef]
19. Dou, Z.; Chen, Z.; Zhou, Z.; Wang, J.; Huang, Y. Influence of eddies on conservative solute transport through a 2d single self-affine fracture. *Int. J. Heat Mass Transf.* **2018**, *121*, 597–606. [CrossRef]
20. Dou, Z.; Zhou, Z.; Wang, J.; Huang, Y. Roughness scale dependence of the relationship between tracer longitudinal dispersion and peclet number in variable-aperture fractures. *Hydrol. Process.* **2018**, *32*, 1461–1475. [CrossRef]
21. Dreuzy, J.R.; Carrera, J.; Dentz, M.; Le Borgne, T. Time evolution of mixing in heterogeneous porous media. *Water Resour. Res.* **2012**, *48*, W06511. [CrossRef]
22. De Simoni, M.; Carrera, J.; Sánchez-Vila, X.; Guadagnini, A. A procedure for the solution of multicomponent reactive transport problems. *Water Resour. Res.* **2005**, *41*, 1–17. [CrossRef]
23. Aris, R. On the dispersion of a solute in a fluid flowing through a tube. *Proc. R. Soc. Lond. A Math. Phys. Eng. Sci.* **1956**, *235*, 67–77. [CrossRef]
24. Detwiler, R.L.; Rajaram, H.; Glass, R.J. Solute transport in variable-aperture fractures: An investigation of the relative importance of Taylor dispersion and macrodispersion. *Water Resour. Res.* **2000**, *36*, 1611–1625. [CrossRef]
25. Freyberg, D.L. A natural gradient experiment on solute transport in a sand aquifer: 2. Spatial moments and the advection and dispersion of nonreactive tracers. *Water Resour. Res.* **1986**, *22*, 2031–2046. [CrossRef]
26. Rolle, M.; Hochstetler, D.; Chiogna, G.; Kitanidis, P.K.; Grathwohl, P. Experimental investigation and pore-scale modeling interpretation of compound-specific transverse dispersion in porous media. *Trans. Porous Media* **2012**, *93*, 347–362. [CrossRef]
27. Chiogna, G.; Hochstetler, D.L.; Bellin, A.; Kitanidis, P.K.; Rolle, M. Mixing, entropy and reactive solute transport. *Geophys. Res. Lett.* **2012**, *39*. [CrossRef]
28. Rolle, M.; Chiogna, G.; Hochstetler, D.L.; Kitanidis, P.K. On the importance of diffusion and compound-specific mixing for groundwater transport: An investigation from pore to field scale. *J. Contam. Hydrol.* **2013**, *153*, 51–68. [CrossRef] [PubMed]
29. Miralles-Wilhelm, F.; Gelhar, L.W. Stochastic analysis of oxygen-limited biodegradation in heterogeneous aquifers with transient microbial dynamics. *J. Contam. Hydrol.* **2000**, *42*, 69–97. [CrossRef]
30. Kapoor, V.; Gelhar, L.W. Transport in three-dimensionally heterogeneous aquifers: 1. Dynamics of concentration fluctuations. *Water Resour. Res.* **1994**, *30*, 1775–1788. [CrossRef]
31. Le Borgne, T.; Dentz, M.; Bolster, D.; Carrera, J.; de Dreuzy, J.-R.; Davy, P. Non-fickian mixing: Temporal evolution of the scalar dissipation rate in heterogeneous porous media. *Adv. Water Resour.* **2010**, *33*, 1468–1475. [CrossRef]
32. Engdahl, N.B.; Ginn, T.R.; Fogg, G.E. Scalar dissipation rates in non-conservative transport systems. *J. Contam. Hydrol.* **2013**, *149*, 46–60. [CrossRef] [PubMed]

33. Jha, B.; Cueto-Felgueroso, L.; Juanes, R. Quantifying mixing in viscously unstable porous media flows. *Phys. Rev. E* **2011**, *84*, 066312. [CrossRef] [PubMed]

34. Cardenas, M.B.; Slottke, D.T.; Ketcham, R.A.; Sharp, J.M. Navier-stokes flow and transport simulations using real fractures shows heavy tailing due to eddies. *Geophys. Res. Lett.* **2007**, *34*. [CrossRef]

35. Cardenas, M.B.; Slottke, D.T.; Ketcham, R.A.; Sharp, J.M. Effects of inertia and directionality on flow and transport in a rough asymmetric fracture. *J. Geophys. Res.* **2009**, *114*. [CrossRef]

36. Wang, L.; Cardenas, M.B. Non-fickian transport through two-dimensional rough fractures: Assessment and prediction. *Water Resour. Res.* **2014**, *50*, 871–884. [CrossRef]

37. Wang, L.; Cardenas, M.B. An efficient quasi-3d particle tracking-based approach for transport through fractures with application to dynamic dispersion calculation. *J. Contam. Hydrol.* **2015**, *179*, 47–54. [CrossRef] [PubMed]

38. Kang, P.K.; Brown, S.; Juanes, R. Emergence of anomalous transport in stressed rough fractures. *Earth Planet. Sci. Lett.* **2016**, *454*, 46–54. [CrossRef]

39. Mandelbrot, B.B. *The Fractal Geometry of Nature*; Macmillan: San Francisco, CA, USA, 1983; Volume 173.

40. Dou, Z.; Zhou, Z.; Sleep, B.E. Influence of wettability on interfacial area during immiscible liquid invasion into a 3d self-affine rough fracture: Lattice boltzmann simulations. *Adv. Water Resour.* **2013**, *61*, 1–11. [CrossRef]

41. Mondal, P.K.; Sleep, B.E. Colloid transport in dolomite rock fractures: Effects of fracture characteristics, specific discharge, and ionic strength. *Environ. Sci. Technol.* **2012**, *46*, 9987–9994. [CrossRef] [PubMed]

42. Tatone, B.S.; Grasselli, G. A method to evaluate the three-dimensional roughness of fracture surfaces in brittle geomaterials. *Rev. Sci. Instrum.* **2009**, *80*, 125110. [CrossRef] [PubMed]

43. Develi, K.; Babadagli, T. Quantification of natural fracture surfaces using fractal geometry. *Math. Geol.* **1998**, *30*, 971–998. [CrossRef]

44. Wang, J.S.Y.; Narasimhan, T.N.; Scholz, C.H. Aperture correlation of a fractal fracture. *J. Geophys. Res. Solid Earth* **1988**, *93*, 2216–2224. [CrossRef]

45. Liu, R.; Jiang, Y.; Jing, H.; Yu, L. Nonlinear flow characteristics of a system of two intersecting fractures with different apertures. *Processes* **2018**, *6*, 94. [CrossRef]

46. Bolster, D.; Dentz, M.; Le Borgne, T. Hypermixing in linear shear flow. *Water Resour. Res.* **2011**, *47*. [CrossRef]

47. Li, Y.-H.; Gregory, S. Diffusion of ions in sea water and in deep-sea sediments. *Geochim. Cosmochim. Acta* **1974**, *38*, 703–714.

48. Comsol, A. *Comsol Multiphysics User's Guide*; Version 5.2; COMSOL Inc.: Stockholm, Sweden, 2015; Volume 10, p. 333.

49. Werth, C.J.; Cirpka, O.A.; Grathwohl, P. Enhanced mixing and reaction through flow focusing in heterogeneous porous media. *Water Resour. Res.* **2006**, *42*, 1165–1173. [CrossRef]

50. Taylor, G. Dispersion of soluble matter in solvent flowing slowly through a tube. *Proc. R. Soc. Lond. A Math. Phys. Eng. Sci.* **1953**, *219*, 186–203. [CrossRef]

51. Bouquain, J.; Méheust, Y.; Bolster, D.; Davy, P. The impact of inertial effects on solute dispersion in a channel with periodically varying aperture. *Phys. Fluids* **2012**, *24*, 083602. [CrossRef]

52. Briggs, S.; Karney, B.W.; Sleep, B.E. Numerical modeling of the effects of roughness on flow and eddy formation in fractures. *J. Rock Mech. Geotech. Eng.* **2017**, *9*, 105–115. [CrossRef]

53. Jin, Y.; Dong, J.; Zhang, X.; Li, X.; Wu, Y. Scale and size effects on fluid flow through self-affine rough fractures. *Int. J. Heat Mass Transf.* **2017**, *105*, 443–451. [CrossRef]

54. Dou, Z.; Zhou, Z.-F. Lattice boltzmann simulation of solute transport in a single rough fracture. *Water Sci. Eng.* **2014**, *7*, 277–287.

Article

Shear-Flow Coupled Behavior of Artificial Joints with Sawtooth Asperities

Cheng Zhao [1,2,3], Rui Zhang [1], Qingzhao Zhang [1,2,*], Zhenming Shi [1,2] and Songbo Yu [1,2]

[1] Department of Geotechnical Engineering, Tongji University, Shanghai 200092, China; zhaocheng@tongji.edu.cn (C.Z.); zhangrui724@tongji.edu.cn (R.Z.); shi_tongji@tongji.edu.cn (Z.S.); yusongbo@tongji.edu.cn (S.Y.)

[2] Key Laboratory of Geotechnical and Underground Engineering of Ministry of Education, Department of Geotechnical Engineering, Tongji University, Shanghai 200092, China

[3] College of Engineering, Tibet University, Lhasa 850011, China

* Correspondence: zqz0726@163.com; Tel.: +86-138-1687-3098

Received: 20 July 2018; Accepted: 21 August 2018; Published: 1 September 2018

Abstract: The coupling between hydraulic and mechanical processes in rock joints has significantly influenced the properties and applications of rock mass in many engineering fields. In this study, a series of regular shear tests and shear-flow coupled tests were conducted on artificial joints with sawtooth asperities. Shear deformation, strength, and seepage properties were comprehensively analyzed to reveal the influence of joint roughness, normal stress, and seepage pressure on shear-flow coupled behavior. The results indicate that the shear failure mode, which can be divided into sliding and cutting, is dominated by joint roughness and affected by the other two factors under certain conditions. The seepage process makes a negative impact on shear strength as a result of the mutual reinforcing of offsetting and softening effects. The evolution of hydraulic aperture during the shear-flow coupled tests embodies a consistent pattern of four stages: shear contraction, shear dilation, re-contraction, and stability. The permeability of joint sample is considerably enlarged with the increase of joint roughness, but decreases with the addition of normal stress.

Keywords: artificial joint rock; shear-flow coupled test; hydraulic aperture; roughness; seepage pressure

1. Introduction

The strength and deformability of rock joints have been the subjects of numerous investigations, in design and analysis of underground structures, foundation, slope stability, and risk assessment of underground disposal [1–3]. The coupling between hydraulic and mechanical processes in rock joints, as one of these subjects, has received wide attention, since a series of events, including dam failures, landslides, and injection-induced earthquakes, were believed to result from it [4–6].

Performing laboratory coupled stress-flow tests is an effective way to study the coupled stress-flow characteristics of rock joint [7]. Iwai [8] presented a one-dimensional model where a rough fracture consists of a series of wedge-shaped increments. Raven and Gale [9] studied the effect of changes in sample size on the normal stress-permeability properties of natural fractures. Durham and Bonner [10] characterized the hydraulic behavior of joints under conditions where fluid migrates. These early studies were limited to investigating the effect of normal loading on fluid flow through rock fractures.

The mutual effects of normal and shear stress on fractures with fluid flow were considered in further studies. Singh [11] measured the permeability of sandstone using triaxial tests, proposing that permeability changed as a function of differential stress. Some studies have been dedicated to jointed rock behavior under triaxial undrained conditions, including pore water pressure behavior, with or without infilled joints [12,13]. Chen [14] investigated the effects of normal stress on the mechanical and hydraulic behavior of fractures. Liu [15] analyzed the relationships between aperture and triaxial stresses, and developed coupled models of seepage and triaxial stresses. Although these investigations involved seepage of rock fractures subjected to three-dimensional stresses, which is more appropriate, the effect of shear stress on the jointed rock has not been involved.

In the past decades, numerous studies have been performed to investigate the shear behavior of rock joints under different conditions: constant normal load (CNL) and constant normal stiffness (CNS) [16–20]. However, the seepage behavior coupling with stress was not comprehensively involved in these studies. A series of shear-flow coupling tests were carried out by Jiang et al., with some empirical relations proposed [21–24], but they did not investigate the influence of water pressure on permeability characteristics. Olsson and Barton [25] carried out a shear-flow coupled test under low seepage pressure, and studied the influence of joint dilatancy on the joint width. Shi [26] introduced a new sealing system of shear box to achieve higher hydraulic pressure shear-flow coupled tests on jointed rock samples. They developed an empirical equation for obtaining the shear strength of jointed rocks experiencing a higher seepage pressure.

Studies on the interaction of high water pressure, joint roughness coefficients, and normal stress in the shear-flow coupled tests are still few, due to various reasons. In this paper, direct shear tests and shear-flow coupled tests, with different joint roughness, normal stresses, and seepage pressures, were performed to investigate the shear behavior of joint rock, choosing sawtooth surface as a simplification of the natural structure planes. The mechanisms of shear strength, seepage process, and shear failure modes under different conditions are compared and evaluated. The results could improve the basis work for studying the mechanics and hydraulic characteristics of jointed rock mass, thus providing certain engineering application value for problems in underground cavern excavation, reservoir, mining, dam, and other fields.

2. Materials and Methods

2.1. Test Equipment

The regular shear tests and shear-flow coupled tests were both carried out on the digit-control shear-flow coupled test system, as shown in Figure 1, which was developed by Tongji University and manufactured by Digit-Control Technology Company of Changsha YaXing [27]. The test system mainly includes four parts: servo control unit, normal and tangential loading system, sealed shear box, and seepage loading system. The servo control unit is characterized by rapid speed, high measurement precision, and short feedback time, which can realize the control mode of constant normal load (CNL). The largest normal and tangential stress is 30 MPa, with the measuring accuracy of 1%; the largest deformation measurement range is 25 mm, with the measuring accuracy of $\pm 0.5\%$.

The maximum shear deformation along the shear flow direction is 15 mm and the maximum seepage pressure is 3 MPa. The joint sample was placed in the middle of the sealed shear box. During the test, the water was pumped from a water tank and flowed into the joint through water inlet under the set pressure. The outflow of the water was collected and weighted. Three sensors were set on the water outlet for averaging to reduce errors.

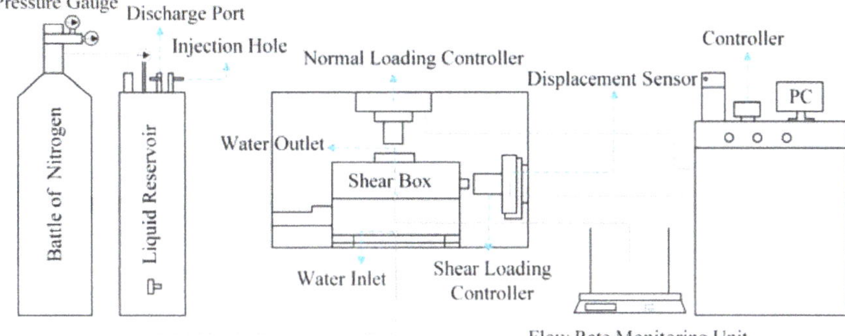

Figure 1. Structure diagram of shear-flow coupled test system.

2.2. Sample Preparation

Considering that this study was a fundamental research study, artificial joint samples made of plaster were used to simulate the natural rock structure. Plaster and water were mixed in the ratio of 4:1 by weight, which was based on a previous study [28]. The plaster material could be easily molded into any shape and size, and was ideal for modelling rock samples with low to medium strength [18,19].

To meet the requirement of the shear-flow coupled test system, cylindrical samples with a circular shear surface (ϕ200 mm × 150 mm) were needed. Some previous studies investigated the hydromechanical properties of joint rock, using cylindrical samples, but with different scales or structures [11,13,14,26,29]. Before the preparation of samples, the molds, as shown in Figure 2, were cleaned and smeared with Vaseline, to ensure successful knockout. The mix of plaster and water was poured into the molds to be solidified for about 30 min, and then left undisturbed for 8 h after being taken out. Then, a hole (ϕ5 mm) was drilled in the center of the samples as the injection hole of the radiation flow.

Figure 2. Molds for sample preparation.

Four kinds of samples were prepared: intact samples, samples with smooth profiles, and rough profiles consisting of triangular asperities with either 15° or 30° inclination angles. Details of these profiles are shown in Figure 3. It can be seen that the results of angles greater than 45° are symmetrical with the results of angles less than 45°. Additionally, a previous similar study suggested that the shear properties of the joint with this morphology showed uniqueness when the angle is around 20°, while the cases of 30° and 45° did not show much difference [28]. Therefore, the characteristics of artificial joints samples of joint roughness of 0°, 15°, 30° were studied. In an earlier study, Dove and Frost [30] introduced a roughness parameter, $R = A_s / A_0$, where A_s is the actual area of the surface, and A_0 is the projected area of the surface. In our experiments, this roughness parameter can be calculated as $R = \sin \theta + \cos \theta$, where θ is the sawtooth angle and $0° < \theta < 90°$. Therefore, the angle of the asperity was used to represent the roughness of joint surface in this paper, and the roughness increased as this angle increased. The uniaxial compressive strength of these samples was about 25 MPa, obtained though uniaxial compression tests. The regular shear tests were performed to obtain the shear strength parameters of intact samples, which were shown in Table 1.

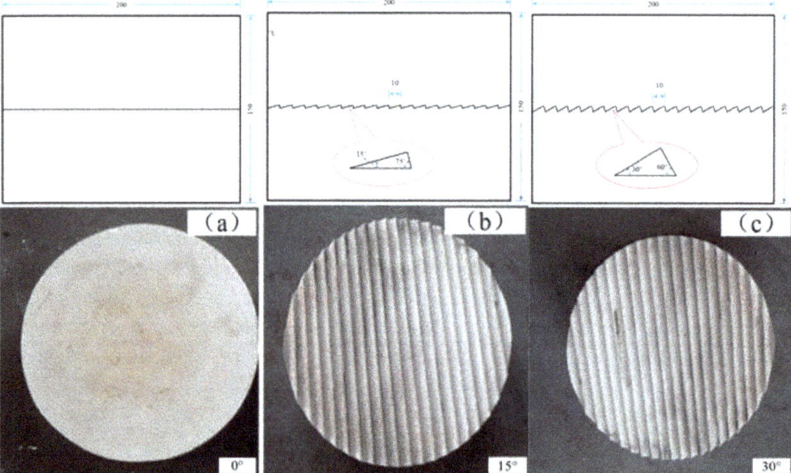

Figure 3. Diagram of artificial joint samples: (**a**) joint roughness is 0°; (**b**) joint roughness is 15°; (**c**) joint roughness is 30° (unit: mm).

Table 1. Shear strength parameters of intact cylindrical specimens.

Normal Stress (MPa)	1.5	2.5	3.5	4.5	C = 1.62 MPa
Shear Strength (MPa)	2.23	2.63	3.03	3.44	$\varphi_0 = 32°$

2.3. Testing Scheme

In both shear tests and shear-flow coupled tests, the normal stresses were chosen to be 1.5 MPa and 2.5 MPa, which were approximately the overburden pressures of the rock masses with burial depths of 60 m and 100 m, respectively. All the tests were performed with the maximum shear displacement of 20 mm, and the shear rate of 2 mm/s.

Three groups of samples with different artificial joint surface were subjected to regular shear testing, and each group has two samples for the two levels of normal stress. In regular shear tests, the normal stress was applied on the sample to the predefined value, and remained for 5 min to ensure the normal stress was stable, and then the shear stress was applied until the sample failed. As for shear-flow coupled tests, the samples were divided into three groups according to different joint roughness, and six samples in each group. When the normal stress was stable, the seepage pressure was applied to the predefined value, which was 0.5 MPa, 1.0 MPa, and 1.5 MPa. Being set through the pressure gauge, as shown in Figure 1, this pressure represented the pressure of the water inlet, namely, the pressure of radiant point on the joint surface. Since the joint surface pressure was impossible to measure directly, the water pressure of radiant point was considered as the mean pressure of joint surface. Although the flow form was radiation, theoretically, the water mostly flowed along the joint surface during the experiments, especially when the seepage pressure is small. Basically, the direction of fluid velocity is parallel to the asperities. As the outlet radiant flow was stable, the shear stress was then applied until the sample failed.

In both shear tests and shear-flow coupled tests, the changing values of shear stress with the increase of shear displacement were documented for further analysis. In order to calculate the hydraulic aperture, it was necessary to measure the seepage amount during the shear-flow coupled tests. The method for obtaining the seepage volume was to stop the test every displacement of 1 mm and measure the volume of water flowing out within 30 s at the outlet of the test system.

3. Results of Shear Tests

3.1. Shear Deformation Curves

The regular shear tests under two levels of normal stress were carried out on the joint samples with different sawtooth angles. Figure 4 shows the shear deformation curves of these tests. The failure mode of shearing can be obtained by analyzing the shear deformation curves. As shown in Figure 4a, when the joint roughness is 0°, the shear displacement curve is a typical sliding failure mode, showing obvious sliding characteristics, which includes two regions: the shear dilatancy area and the sliding area. As the normal stress increases, the deformation curves keep a similar shape, while the peak shear stress and the residual shear stress increase. Take Figure 4c, for example: the peak stress varies from 0.93 MPa to 1.60 MPa when the normal stress increases to 2.5 MPa; meanwhile, the residual shear stress varies from 0.54 MPa to 0.79 MPa.

In Figure 4b, when the joint roughness is 15°, the shear deformation curves of the joint surface under the normal stress of 1.5 MPa and 2.5 MPa show multiple cutting types. It can be seen that a sliding stage occurs after the initial shear dilatancy area, and then the shear stress rapidly decreases with the increase of deformation. After a fluctuant valley, the shear stress grows slowly until the second peak stress appears. Then, it swiftly falls without a sliding stage and begins to repeat this process. The second peak stress is even larger than the first one in the condition of smaller normal stress. This phenomenon illustrates that the shear failure is sliding failure mode before the first batch of sawteeth are sheared, and the subsequent shear is cutting type, until the wear and destruction of joint surface reach a threshold.

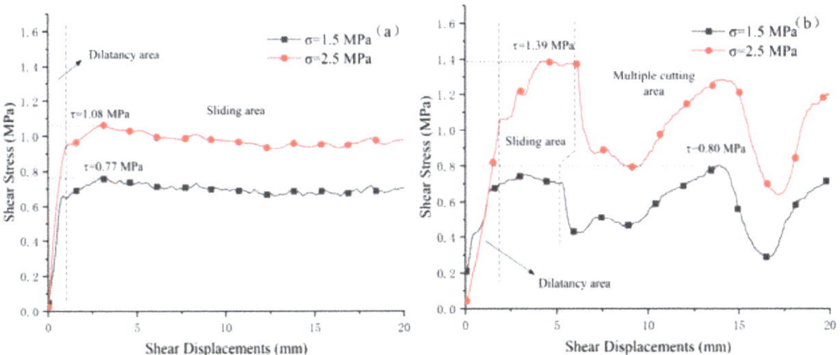

Figure 4. *Cont.*

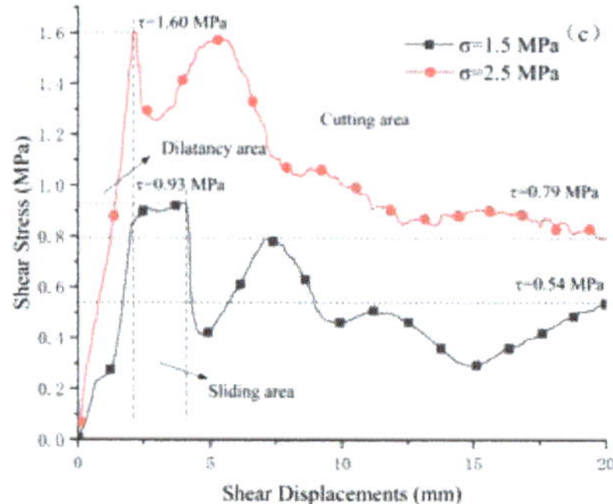

Figure 4. Shear deformation curves of joint surfaces with different roughness under different normal stresses: (**a**) joint roughness is 0°; (**b**) joint roughness is 15°; (**c**) joint roughness is 30°.

As for the condition of larger sawtooth angle, 30°, it can be observed from Figure 4c that the shear deformation curve also exhibits multiple cutting mode with a short sliding stage when the normal stress is 1.5 MPa. However, the one with larger normal stress of 2.5 MPa does not display a sliding stage after the dilatancy area, but decreases quickly. This indicates that the cutting type failure will be more significant with the increase of normal stress.

It is obvious that the roughness of the joint surface and the normal stress mutually affect the shear failure mode. When the joint surface is smooth, the shear failure is totally in sliding mode. However, when it is rough with a number of high sawteeth, the failure mode tends to change to cutting type. The larger the normal stress is, the bigger this kind of tendency is. The increase of the roughness, meaning the enlargement of the sawtooth, results in a more remarkably interlaced occlusion effect of the contact area, leading to higher strength and more possibility for cutting failure.

3.2. Shear Strength

Shear strength parameters are listed in Table 2, using peak shear strengths obtained from direct shear tests and corresponding normal stresses. In this study, two levels of normal stress are selected. Previous studies have been performed to investigate the shear behavior of joint samples made of similar materials, so the strength parameters in our study are obtained, comprehensively, based on shear tests and other similar studies [26,28].

Table 2. Shear strength parameters of joint samples with different roughness.

Roughness	Normal Stress (MPa)	Shear Strength (MPa)	Cohesion (MPa)	Friction Angle (°)
0°	1.5	0.77	0.30	17.22
	2.5	1.08		
15°	1.5	0.80	0.08	30.54
	2.5	1.39		
30°	1.5	0.93	0.07	33.82
	2.5	1.60		

In Table 2, it is notable that the friction angle increases remarkably as the roughness of joint surface increases. This result is consistent with the observations from shear deformation curves (Figure 4). As the roughness of joint surface increases, the shear failure mode is inclined to cutting type, which allows the interlaced occlusion effect of contact area increasing. In the plaster—plaster interfaces, the ductile and destructive behavior of asperities is proven to have an important influence on the shear strength [19,31]. As a result of the enlargement of the sawtooth, the friction angle reflected from tests is becoming larger. Therefore, the increasing occurrence of cutting mode will result in a larger shear strength, with the addition of joint sawtooth angle.

4. Results of Shear-Flow Coupled Tests

4.1. Shear Deformation Curves of Shear-Flow Coupled Tests

The shear-flow coupled tests were performed on joints samples under two levels of normal stress and three levels of seepage pressure. Figure 5 shows the shear deformation curves with different roughness under different seepage pressures when the normal stress is 1.5 MPa. The curves with the same roughness exhibit the similar pattern. In Figure 5a, under different seepage pressures, the shear stress stays stable after an initial increase, showing the typical characters of sliding failure mode. When the sawtooth angle is 15°, a short stage of sliding mode occurs before the cutting process in Figure 5b. However, the sliding process disappears with the increase of joint roughness, as shown in Figure 5c. After the dilatancy area, the shear stresses increase rapidly until the cutting failure when the sawtooth angle is 30°. Comparing with the results of the regular shear tests, it can be found that the seepage pressure does not have obvious effect on the shear failure mode when the normal stress is small.

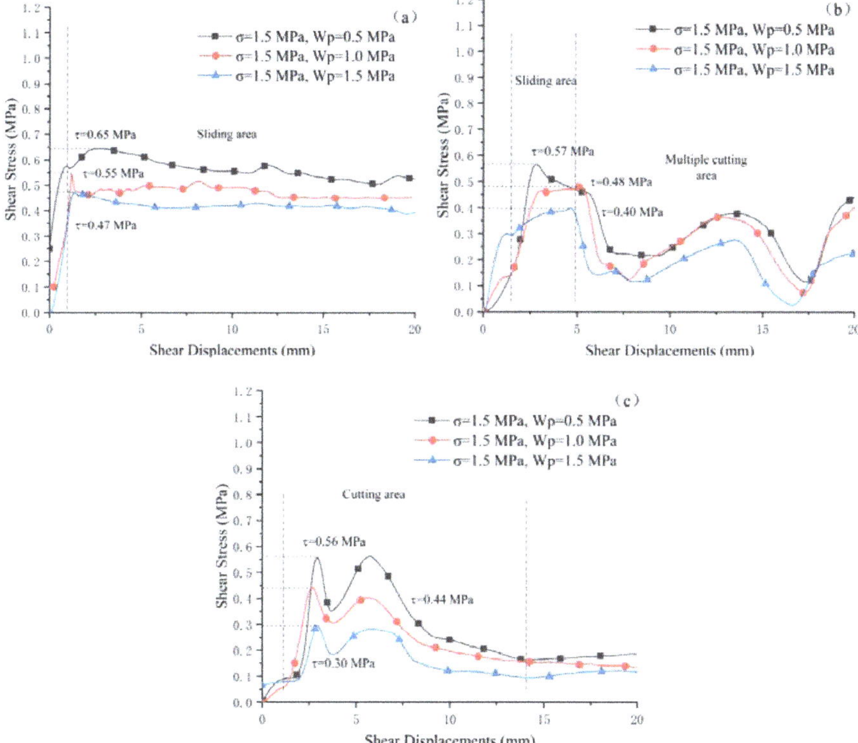

Figure 5. Shear deformation curves of joint surfaces under different seepage pressures when normal stress is 1.5 MPa: (**a**) joint roughness is 0°; (**b**) joint roughness is 15°; (**c**) joint roughness is 30°.

Figure 6 gives the shear deformation curves of joint surfaces with different roughness values under different seepage pressures when normal stress is 2.5 MPa. The shear deformation curves with the same roughness show very similar characteristics. The exception appears after the dilatancy area when the joint sawtooth angle is 15°. When the seepage pressure is 0.5 MPa or 1.0 MPa, a small cutting area occurs after the dilatancy (Figure 6b); as the seepage pressure increases to 1.5 MPa, the failure mode changes from cutting to sliding mode at the first peak strength. However, the curves in Figure 6c do not display this feature. It seems that larger seepage pressure contributes to the occurrence of sliding failure when the normal stress is higher, while the increase of roughness is helpful for the occurrence of cutting failure. The impact of roughness exceeds the effect of seepage pressure in Figure 6c, so that cutting failure is prevailing, even with the high seepage pressure.

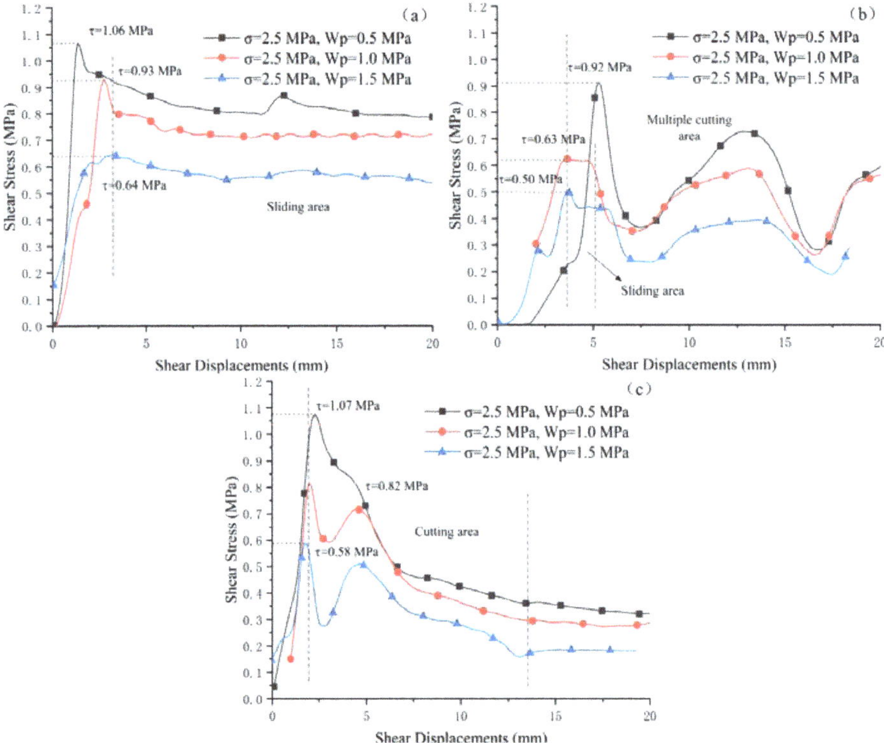

Figure 6. Shear deformation curves of joint surfaces under different seepage pressures when normal stress is 2.5 MPa: (**a**) joint roughness is 0°; (**b**) joint roughness is 15°; (**c**) joint roughness is 30°.

In summary, it can be found that the increase of seepage pressure results in the reduction of peak shear strength and residual shear strength, in Figures 5 and 6. The shear failure mode changes from sliding to cutting type with the increase of joint roughness, while the increasing seepage pressure has an opposite effect when the normal stress is 2.5 MPa. Comparing Figure 5 with Figure 6, it can be summarized that the normal stress only contributes to the change of peak shear strength, and has no effect on the shear failure mode.

In shear-flow coupled tests, the displacements at the peak strength are extracted and plotted in Figure 7. Figure 7a shows the displacements corresponding to the peak shear strength under the normal stress of 1.5 MPa. When the seepage pressure is 0.5 MPa, it can be seen that the displacements corresponding to the peak shear strength are almost same for different joint roughness. However, for the larger seepage pressure of 1.0 MPa and 1.5 MPa, the peak shear strengthes with the roughness of 15° are obtained at the largest displacements. This phenomenon indicates that the sliding or cutting failure is less likely to happen with the sawtooth angle of 15° when the seepage pressure is large, while this influence of joint roughness can be neglected when the seepage pressure is small (0.5 MPa).

Figure 7b shows the displacement values corresponding to the peak shear stresses under the normal stress of 2.5 MPa. Like those in Figure 7a, the displacements corresponding to the peak stresses for the sawtooth angle of 15° are the largest, even including the one under the seepage pressure of 0.5 MPa. This may suggest that the increase of normal stress highlights the role of joint roughness. Further study is needed to explain why the angle of 15° is special in the shear-flow coupled tests.

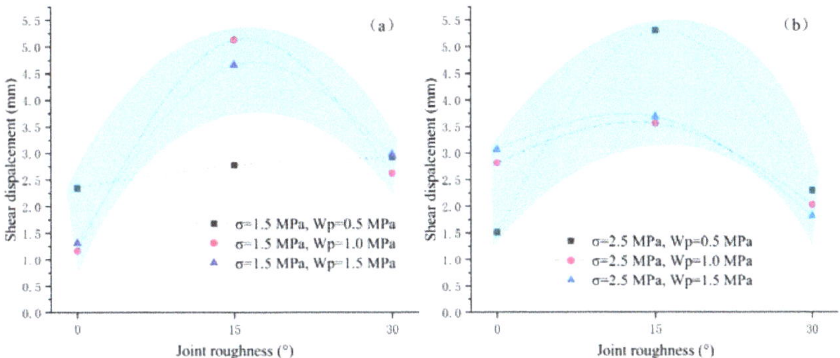

Figure 7. Displacements corresponding to the peak shear stresses under different seepage pressures: (**a**) normal stress is 1.5 MPa; (**b**) normal stress is 2.5 MPa.

4.2. Shear Strength of Shear-Flow Coupled Tests

The shear strength in the shear-flow coupled tests is generally controlled by the roughness, normal stress, and seepage pressure. The shear strength parameters with different roughness and seepage pressures are listed in Table 3. According to the results of regular shear tests, the friction angle increases as the roughness increase. However, in shear-flow coupled tests, the friction angle decreases rapidly with the increase of roughness and seepage pressure, as seen in Table 3.

Table 3. Shear strength parameters of joint samples with different roughness and seepage pressures.

Roughness	Seepage Pressure (MPa)	Shear Stress (MPa) σ = 1.5 MPa	σ = 2.5 MPa	Cohesion (MPa)	Friction Angle (°)
	0.5	0.65	1.06	0.03	22.29
0°	1.0	0.55	0.93	0.02	20.81
	1.5	0.47	0.64	0.21	9.65
	0.5	0.57	0.92	0.04	19.29
15°	1.0	0.48	0.63	0.25	8.53
	1.5	0.40	0.50	0.25	5.71
	0.5	0.56	1.07	0.20	27.02
30°	1.0	0.44	0.82	0.13	20.80
	1.5	0.30	0.58	0.12	15.64

It can be speculated that the reason for the decrease of shear strength caused by seepage pressure is that this pressure counteracts the effect of normal stress. Furthermore, the seepage process can apparently soften the plaster–plaster interface to damage the original sawtooth structure. This softening effect, which will enhance with the increase of seepage pressure, has been reported by previous studies [26,32]. The mutually reinforcing of offsetting and softening eventually ruins the structure of the joint surface, leading to the quick decrease of shear strength.

The photos of failed joint in the shear-flow coupled tests when normal stress is 2.5 MPa and joint roughness is 15° are selected to illustrate the failure morphology, as shown in Figure 8. It can be observed that when the seepage pressure is 0.5 MPa (Figure 8a), the joint surface with the traces of cutting shear is almost intact, while obvious characteristics of sliding shear can be found under larger seepage pressure in Figure 8b,c. In Figure 8c, some sludge regions can be clearly observed under seepage pressure of 1.5 MPa, which reflects the existence of the softening effect. An intact area in the center of each surface occurs because the contact region will become smaller as the shear test proceeds, due to the cylindrical shape of samples.

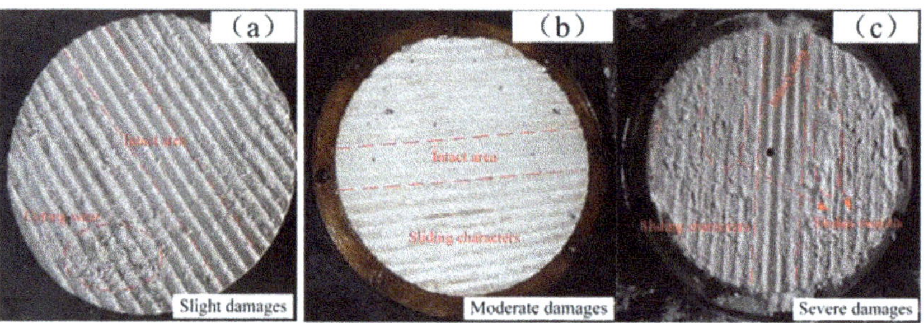

Figure 8. Photos of joint surface after the shear-flow coupled tests when normal stress is 2.5 MPa and joint roughness is 15°: (**a**) W_p = 0.5 MPa; (**b**) W_p = 1.0 MPa; (**c**) W_p = 1.5 MPa.

4.3. Analysis of Seepage Properties

In the shear-flow coupled field, the permeability of the joint samples can be represented by the value of hydraulic aperture, which can be calculated using the so-called cubic law [33], with the results obtained from the shear-flow coupled test. The cubic law is expressed as follows:

$$Q = \frac{g}{v}\frac{we^3}{12}i,$$
(1)

where Q is the flow rate; g is the acceleration of gravity; e is the hydraulic aperture; v is the dynamic viscosity coefficient; w is the width of the flow region; and i is the unit hydraulic gradient.

The hydraulic apertures of joint surfaces under following conditions are calculated: sawtooth angle of 15°/30°, seepage pressure of 1.0 MPa, and normal stress of 1.5/2.5 MPa. Figure 9 displays the hydraulic aperture curves under different normal stresses. It can be observed that these curves have four stages as the shear deformation increases: shear contraction, shear dilation, re-contraction, and stability. In the initial stage of shearing (Stage I), the shear displacement causes joint shrinkage, and the hydraulic aperture of the joint decreases. The reason for this phenomenon is that the joint surfaces are not completely meshed with cracks in the middle, and their initial hydraulic apertures are not zero. Immediately after the adding of normal load, the joint surface has a tendency of compression and closing. As the shearing progresses (Stage II), the dilations of the joint surfaces make the hydraulic aperture increase rapidly [34]. Due to the sliding and cutting damage during the shearing process of the 15° joint surfaces, their hydraulic apertures have some fluctuations with a second peak value in stage III (Figure 9a). However, the shearing of 30° joint surfaces are dominated by the cutting failure mode so that their hydraulic apertures decrease quickly as a result of the complete occlusal, even destruction of sawteeth in stage III. As the tests progress, the contact surfaces become smooth, and all the hydraulic apertures are almost no longer changed, reaching the residual hydraulic aperture (stage IV).

The comparison of experimental results in Figure 9 reflects the influence of the surface morphology on the mechanical aperture evolution. For a roughness of 15° (Figure 9a), a relatively larger void space will be generated between the surfaces after the partial shearing. This will result in smaller values of hydraulic apertures compared to those of 30°. In general, when the shear displacement increases to 6–8 mm, the hydraulic aperture of the joints reaches the maximum value due to the effect of dilatancy.

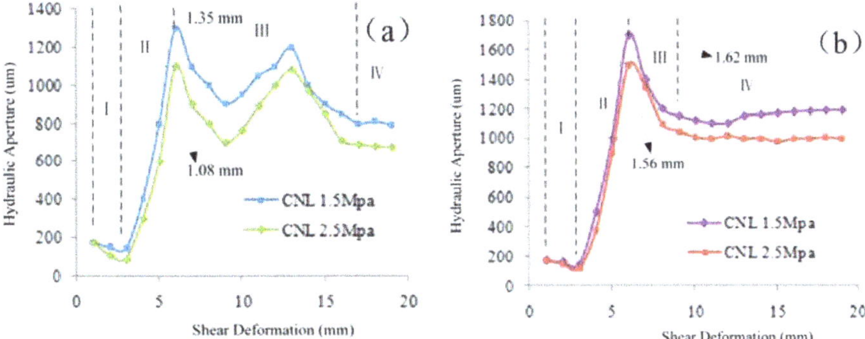

Figure 9. Hydraulic aperture curves under different normal stresses when seepage pressure is 1.0 MPa: (**a**) joint roughness is 15°; (**b**) joint roughness is 30°.

The increasing normal stress leads to the decrease of hydraulic aperture, which represents the decrease of permeability. This conclusion was demonstrated by many previous studies [11,15,35]. In our tests, the comparison between Figures 9a and 9b illustrates that the permeability of joint sample is considerably enlarged by the increasing surface sawtooth angle. However, some previous studies proved that roughness increases would decrease the fracture transmissivity, thus resulting in greater resistance to flow, and finally, in lower conductivity of rough joints [22,36,37]. This discrepancy can be explained by the significant softening effect of the seepage process, which brings severe damage to the joint surfaces and makes the fracture transmissivity larger.

5. Conclusions

This study performed the regular shear tests and shear-flow coupled tests to comprehensively consider the effect of roughness, normal stress, and seepage pressure on the shear process of artificial joint rock samples. The major conclusions are as follows:

(1) In a dry environment, the roughness of joint surface has a significant influence on the shear failure mode and strength. With the increase of the roughness, the failure mode of the joint changes from sliding failure to cutting failure, and the shear strength becomes larger. The normal stress has little influence on the failure mode of the joint surface when it is small, but has a notable effect when large. With the increase of normal stress, it is less likely that sliding failure would occur.

(2) The seepage pressure has no obvious effect on the failure mode of the joint sample when the normal stress is 1.5 MPa. However, the failure mode under the normal stress of 2.5 MPa varies from cutting to sliding type as the seepage pressure increases. Under the normal stress of 2.5 MPa, the mode of macroscopic failure depends on the mutual competition between roughness and seepage pressure. The normal stress only contributes to the change of peak shear strength, and has no effect on the shear failure mode in the shear-flow coupled test.

(3) The seepage process can clearly lower the shear strength of the joint sample. Moreover, this effect will be more significant as the pressure increases. The internal mechanism for this phenomenon is the mutual reinforcing of offsetting and softening effects.

(4) The evolution of hydraulic aperture during the shear-flow coupled test can be divided into four stages: shear contraction, shear dilation, re-contraction, and stability. The permeability of joint sample is considerably enlarging with the increasing surface sawtooth angle, but decreasing with the addition of normal stress.

Author Contributions: Conceptualization, C.Z. and Z.S.; Data curation, R.Z.; Formal analysis, R.Z.; Funding acquisition, C.Z., Q.Z. and S.Y.; Investigation, R.Z. and S.Y.; Methodology, Q.Z. and Z.S.; Project administration,

Z.S.; Supervision, C.Z., Q.Z., Z.S. and S.Y.; Validation, C.Z.; Writing—original draft, R.Z.; Writing—review & editing, C.Z. and Q.Z.

Funding: This research was funded by (the National Key Research and Development Plan) grant number (2017YFC0806000), (the National Natural Science Foundation of China) grant number (41572262, 41502275, 41602287), and (Shanghai Rising-Star Program) grant number (17QC1400600).

Conflicts of Interest: The authors declare no conflict of interest. The funders had no role in the design of the study; in the collection, analyses, or interpretation of data; in the writing of the manuscript, and in the decision to publish the results.

References

1. Niktabar, S.M.M.; Rao, K.S.; Shrivastava, A.K. Effect of rock joint roughness on its cyclic shear behavior. *J. Rock Mech. Geotech. Eng.* **2017**, *9*, 1071–1084. [CrossRef]

2. Barton, N.; Bandis, S.; Bakhtar, K. Strength, deformation and conductivity coupling of rock joints. *Int. J. Rock Mech. Min. Sci. Geomech. Abstr.* **1985**, *22*, 121–140. [CrossRef]

3. Chappell, B.A. Rock bolts and shear stiffness in jointed rock masses. *J. Geotech. Eng. ASCE* **1989**, *115*, 179–197. [CrossRef]

4. Rutqvist, J.; Stephansson, O. The role of hydromechanical coupling in fractured rock engineering. *Hydrogeol. J.* **2003**, *11*, 7–40. [CrossRef]

5. Gangi, A.F. Variation of whole and fractured porous rock permeability with confining pressure. *Int. J. Rock Mech. Min. Sci. Geomech. Abstr.* **1978**, *15*, 249–257. [CrossRef]

6. Tsang, Y.W.; Tsang, C.F. Channel model of flow through fractured media. *Water Resour. Res.* **1987**, *23*, 467–479. [CrossRef]

7. Xiao, W.M.; Xia, C.C.; Deng, R.G. Advances in development of coupled stress-flow test system for rock joins. *Chin. J. Rock Mech. Eng.* **2014**, *33*, 3456–3465.

8. Iwai, K. Fundamental Studies of the Fluid Flow through a Single Fracture. Ph.D. Thesis, University of California, Berkeley, CA, USA, 1976.

9. Raven, K.G.; Gale, J.E. Water flow in a natural rock fracture as a function of stress and sample size. *Int. J. Rock Mech. Min. Sci. Geomech. Abstr.* **1985**, *22*, 251–261. [CrossRef]

10. Durham, W.B.; Bonner, B.P. Self-propping and fluid flow in slightly offset joints at high effective pressures. *J. Geophys. Res. Solid Earth* **1994**, *99*, 9391–9399. [CrossRef]

11. Singh, A.B. Study of rock fracture by permeability method. *J. Geotech. Geoenviron. Eng.* **1997**, *123*, 601–608. [CrossRef]

12. Archambault, G.; Poirier, S.; Rouleau, A.; Gentier, S.; Riss, J. The behavior of induced pore fluid pressure in undrained triaxial shear tests on fractured porous analog rock material specimens. In *Mechanics of Jointed and Faulted Rock*; CRC Press: Boca Raton, FL, USA, 1998; pp. 555–560.

13. Indraratna, B.; Jayanathan, M. Measurement of pore water pressure of clay-infilled rock joints during triaxial shearing. *Geotechnique* **2005**, *55*, 759–764. [CrossRef]

14. Chen, Z.; Narayan, S.P.; Yang, Z.; Rahman, S.S. An experimental investigation of hydraulic behaviour of fractures and joints in granitic rock. *Int. J. Rock Mech. Min. Sci.* **2000**, *37*, 1061–1071. [CrossRef]

15. Liu, C.H.; Chen, C.X.; Jaksa, M.B. Seepage properties of a single rock fracture subjected to triaxial stresses. *Prog. Nat. Sci.* **2007**, *17*, 1482–1486.

16. Crawford, A.M.; Curran, J.H. The influence of rate- and displacement-dependent shear resistance on the response of rock slopes to seismic loads. *Int. J. Rock Mech. Min. Sci. Geomech. Abstr.* **1982**, *19*, 1–8. [CrossRef]

17. Barbero, M.; Barla, G.; Zaninetti, A. Dynamic shear strength of rock joints subjected to impulse loading. *Int. J. Rock Mech. Min. Sci. Geomech. Abstr.* **1996**, *33*, 141–151. [CrossRef]

18. Jafari, M.K.; Pellet, F.; Boulon, M.; Hosseini, K.A. Experimental study of mechanical behaviour of rock joints under cyclic loading. *Rock Mech. Rock Eng.* **2004**, *37*, 3–23. [CrossRef]

19. Atapour, H.; Moosavi, M. The influence of shearing velocity on shear behavior of artificial joints. *Rock Mech. Rock Eng.* **2014**, *47*, 1745–1761. [CrossRef]

20. Niktabar, S.M.M.; Rao, K.S.; Shrivastava, A.K. Automatic static and cyclic shear testing machine under constant normal stiffness boundary conditions. *Geotech. Test. J.* **2018**, *41*, 508–525. [CrossRef]

21. Li, B.; Jiang, Y.J. Experimental study and numerical analysis of shear and flow behaviors of rock with single joint. *Chin. J. Rock Mech. Eng.* **2008**, *27*, 2431–2439.

22. Li, B.; Jiang, Y.J.; Koyama, T.; Jing, L.R.; Tanabashi, Y. Experimental study of the hydro-mechanical behavior of rock joints using a parallel-plate model containing contact areas and artificial fractures. *Int. J. Rock Mech. Min. Sci.* **2008**, *45*, 362–375. [CrossRef]

23. Jiang, Y.J.; Wang, G.; Li, B.; Zhao, X.D. Experimental study and analysis of shear-flow coupling behaviors of rock joints. *Chin. J. Rock Mech. Eng.* **2007**, *26*, 2253–2259.

24. Jiang, Y.J.; Li, B.; Tanabashi, Y. Estimating the relation between surface roughness and mechanical properties of rock joints. *Int. J. Rock Mech. Min. Sci.* **2006**, *43*, 837–846. [CrossRef]

25. Olsson, R.; Barton, N. An improved model for hydromechanical coupling during shearing of rock joints. *Int. J. Rock Mech. Min. Sci.* **2001**, *38*, 317–329. [CrossRef]

26. Shi, Z.M.; Shen, D.Y.; Zhang, Q.Z.; Peng, M.; Li, Q.D. Experimental study on the coupled shear flow behavior of jointed rock samples. *Eur. J. Environ. Civ. Eng.* **2017**, 1–18. [CrossRef]

27. Xia, C.C.; Wang, W.; Wang, X.R. Development of coupling shear-seepage test system for rock joints. *Chin. J. Rock Mech. Eng.* **2008**, *27*, 1285–1291.

28. Zhang, Q.Z.; Shen, M.R.; Ding, W.Q.; Clark, C. Shearing creep properties of cements with different irregularities on two surfaces. *J. Geophys. Eng.* **2012**, *9*, 210–217. [CrossRef]

29. Wang, G.; Zhang, Y.Z.; Jiang, Y.J.; Liu, P.X.; Guo, Y.S.; Liu, J.K.; Ma, M.; Wang, K.; Wang, S.G. Shear behaviour and acoustic emission characteristics of bolted rock joints with different roughnesses. *Rock Mech. Rock Eng.* **2018**, *51*, 1885–1906. [CrossRef]

30. Dove, J.E.; Frost, J.D. A method for measuring geomembrane surface roughness. *Geosynth. Int.* **1996**, *3*, 369–392. [CrossRef]

31. Wang, W.B.; Scholz, C.H. Micromechanics of the velocity and normal stress dependence of rock friction. *Pure Appl. Geophys.* **1994**, *143*, 303–315. [CrossRef]

32. Su, H.J. Seepage Evolution Mechanism of Deep Buried Jointed Rock Mass and Its Engineering Application. Ph.D. Thesis, China University of Mining & Technology, Xu Zhou, China, 2015.

33. Witherspoon, P.A.; Wang, J.S.Y.; Iwai, K.; Gale, J.E. Validity of cubic law for fluid-flow in a deformable rock fracture. *Water Resour. Res.* **1980**, *16*, 1016–1024. [CrossRef]

34. Tang, Z.C.; Xia, C.C.; Xiao, S.G. Constitutive model for joint shear stress-displacement and analysis of dilation. *Chin. J. Rock Mech. Eng.* **2011**, *30*, 917–925.

35. Lee, D.; Juang, C.H. Use of permeability as an index to characterize internal structural changes and fracture mechanism. *Geotech. Test. J.* **1988**, *11*, 63–67. [CrossRef]

36. Oron, A.P.; Berkowitz, B. Flow in rock fractures: The local cubic law assumption reexamined. *Water Resour. Res.* **1998**, *34*, 2811–2825. [CrossRef]

37. Dimadis, G.; Dimadi, A.; Bacasis, I. Influence of fracture roughness on aperture fracture surface and in fluid flow on coarse-grained marble, experimental results. *J. Geosci. Environ. Prot.* **2014**, *2*, 59–67. [CrossRef]

Article

Analysis of Overlying Strata Movement and Disaster-Causing Effects of Coal Mining Face under the Action of Hard Thick Magmatic Rock

Quanlin Wu [1,†], Quansen Wu [1,2,*,†], Yanchao Xue [2,*], Peng Kong [2] and Bin Gong [3]

[1] Department of Chemistry and Chemical Engineering, Jining University, Qufu 273100, China; jnxywql@163.com
[2] State Key Laboratory of Mining Disaster Prevention and Control Co-Founded by Shandong Province and the Ministry of Science and Technology, Shandong University of Science and Technology, Qingdao 266590, China; wangxiao900105@126.com
[3] Graduate School of Engineering, Nagasaki University, Nagasaki 852-8521, Japan; gongbin0412@gmail.com
* Correspondence: wuquansen1989@126.com (Q.W.); stxycgs@126.com (Y.X.);
  Tel.: +86-155-508-39826 (Q.W.); +86-173-191-57131 (Y.X.)
† These authors contributed equally to this work.

Received: 9 August 2018; Accepted: 24 August 2018; Published: 1 September 2018

Abstract: When the hard and thick key strata are located above the working face, the bed separation structure is easy to be formed after mining because of the high strength and integrity of the hard and thick key strata and the initial breaking step is large. After the hard, thick strata are broken, the overburden will be largely collapsed and unstable in a large area and the dynamic disaster is easily induced. In this study, considering the fundamental deformation and failure effect of coal seam, the development law of the bed separation and the fractures under hard and thick magmatic rocks and the mechanism of breaking induced disaster of hard and thick magmatic rocks are studied by similar simulation tests. The results of the study are as follows: (1) The similar material ratio of coal seam is obtained by low-strength orthogonal ratio test of similar materials of coal seam, that is, cement:sand:water:activated carbon:coal = 6:6:7:1.1:79.9. (2) The magmatic rocks play a role in shielding the development of the bed separation, which makes the bed separation beneath the magmatic rock in an unclosed state for a long time, providing space for the accumulation of gas and water. (3) The distribution pattern of the fracture zone shows different shapes as the advancing of working face and the fracture zone width of the rear of working face coal wall is larger than that of the front of the open-off. (4) The breaking of magmatic rocks will press the gas and water accumulated in the bed separation space below to rush towards the working face along the fracture zone at both ends of the goaf. The above results are verified through the drainage borehole gas jet accident in the Yangliu coal mine. The research results are of great significance for revealing the occurrence process of dynamic disasters and adopting scientific and reasonable preventive measures.

Keywords: hard and thick magmatic rocks; orthogonal ratio test; similar simulation; fracture; bed separation; disaster-causing mechanism

---

1. Introduction

There are often magmatic rock intrusions in coal measures strata and the variation process and occurrence state have great influence on mining safety [1–3]. When the overburden of the working face contains tens of meters or even hundreds of meters of magmatic rock, due to the large thickness, high strength, good integrity and large initial breaking step of the magmatic rock, the hard and thick strata are suspended in the large area after the working face is mined and the lower bed separation space is large, which is easy to cause the stress of the stope to be concentrated and its breaking and movement easily induces strong dynamic phenomena and even leads to dynamic disasters such as mine shock, large surface subsidence, bed separation water, gas outburst, rock burst and so on, which poses a serious threat to mine safety production, causing casualties and economic losses [4–7]. Magmatic rocks are prevalent in most of Chinese mining areas, such as the Huaibei, Yanzhou and Datong coal fields [8–10]. During the mining process, the hard and thick strata break down, resulting in the occurrence of dynamic disasters. For example, in China, Huafeng coal mine is covered with 400–800 m thick magmatic rocks and the abutment pressure during the mining process is abnormal [11]. Since 1992, rockbursts have occurred tens of thousands of times and the maximum magnitude is up to 2.9. The number of damage to the working face is 108 times, forcing the working face to stop production 12 times, resulting in a total of 43 serious injuries and countless number of roadway maintenance, adversely affecting social and economic benefits. During the mining process of Huafeng mine, the surface subsidence has a jump growth after the breaking of the thick magmatic rock. The collapse and cracks on the ground have severely damaged farmland and buildings. On 17 July 2011, a gas jet occurred at the ground #2 drainage borehole from the 10,414 working face of the Yangliu mine in China. During the gas jet period, the water outlet of 87# rack (borehole location) in the working face was abnormal and the water inflow reached heights of up to 46 m^3/h. The entire process lasted 33 h, with gas discharge of 166,383 m^3 and the water inflow in the working face reached 7845.6 m^3. The analysis shows that the dynamic phenomena in the 10,414 working face, such as the gas jet from drainage borehole and the roof water burst, are caused by high igneous rock collapse, which results in the rapid closure of the bottom separation space and induces water-gas outburst. Therefore, it is of great theoretical significance and application value for the control of rock strata and prevention and control of dynamic disasters to study the evolution law of the mining induced bed separation and fracture under the hard and thick magmatic rocks and to reveal the mechanism of the corresponding dynamic disasters.

At present, many scholars have applied numerical simulation, theoretical analysis and on-site observation methods to study the occurrence of dynamic disasters and overburden structures under hard and thick strata for mining [12–16]. Guo studied the mechanism of the rock burst of hard and thick upper strata and proposed the method of overburden bed separation grouting for controlling rock bursts [17]. Taking the No.103$_{up}$02 working face of Baodian coal mine as an engineering example, according to the field measured microseismic data, Wu studied the law of the overlying hard and thick sandstone breaking caused by the mining face and explained the relationship between the microseismic data and the rock movement [18]. In view of the two layers of hard and thick strata overlying the working face, microseismic monitoring was carried out on the site by Ning et al. and the relationship between the fracture of hard thick rock strata and microseismic data was studied by using the law of microseismic distribution and the prevention measures were proposed [19]. Jiang used numerical simulation method to analyze the stress distribution characteristics, energy evolution law of the working face and surface subsidence under hard and thick strata and found that the vertical stress and energy reaches the maximum before the hard and thick strata break and the vertical stress and energy decrease after the hard and thick strata break but the subsidence of the ground surface increases sharply [11]. Based on the special geological conditions of two layers of hard and thick strata overlying the working face of Yangliu coal mine, Xu used theoretical analysis to derive the analytical formula of the fracture of hard and thick strata and verified the theoretical model by numerical simulation and field data measurement. Based on the key strata theory and numerical simulation method [20]. Wang et al. analyzed the fracture of hard sandstone roof and found that the coal and gas dynamic disaster in the working face is caused by the fracture of hard and thick sandstone [21]. Based on the key strata theoretical analysis of strata movement, He et al. believed that the mine pressure reaches the maximum when the key strata of the overburden fracture, which easily leads to the occurrence of strong mine earthquake and rock burst [22]. Wang et al. analyzed gas disaster characteristics and evolution rules under the igneous sill in the Haizi coal mine, and gas control methods were proposed and tested [23]. Xuan et al. used UDEC numerical simulation to study the mining stress evolution law under magmatic rocks and explained the causes of coal and gas outburst disaster from the perspective of stress [24]. In summary, the existing research uses field measurement, theoretical analysis and numerical simulation to study and analyze the disaster mechanism and breaking law of hard and thick strata in working face mining. Although the field measured data are reliable, they cannot reflect the characteristics of bed separation development and fracture distribution in the whole strata. Theoretical analysis is often based on field measurements and experimental results, it is difficult to systematically analyze the development of the bed separation and the distribution of fractures. Numerical simulation technology is currently the most widely used research technique and some scholars have introduced stochastic modelling of crack propagation, discrete element method, fracture toughness into numerical calculations and achieved good results [25–29]. But it also has some shortcomings. For example, it is difficult to achieve true restoration in the simulation of overlying rock movement process. The physical similar material simulation test can approach the real simulation of the collapse movement, the bed separation development and the fracture formation process of the overburden. However, in the physical similar simulation material test, the existing research rarely considers the influence of the fundamental deformation and damage of the coal seam on the overburden movement.

Therefore, in view of previous studies, the similar simulation test method is adopted in this study. The ratio of similar materials in coal seam is firstly determined by the low-strength orthogonal ratio test of the coal seam similar materials and then the law of the bed separation and fracture development under the hard and thick magmatic rocks during the mining process is studied and analyzed and finally the breaking mechanism of the hard and thick magmatic rocks is proposed, which is of great significance to predict and prevent dynamic disasters of coal mining under the hard and thick strata.

2. Experimental Study on Low-Strength Orthogonal Ratio of Similar Materials in Coal Seam

2.1. Selection of Ratio Test Scheme

After investigating the existing similar material test ratios for simulating coal seams, the aggregates are mainly coal fines or sand and the cement is mostly made of gypsum and cement [30]. Comparing the comprehensive investigation data, in order to choose a more reasonable ratio scheme, coal fines, sand, activated carbon were chosen as aggregate, cement or gypsum as cement. First of all, a preliminary strength test was performed on the cement and the gypsum and after the cement was preferably used, an orthogonal test was performed. The test piece size was ϕ 50 mm $\times$ 100 mm. The test piece making mold and some test pieces are shown in Figure 1. The test was carried out using the Shimadzu electronic universal testing machine.

Figure 1. Test piece making mold and some test pieces.

2.2. Low-Strength Orthogonal Ratio Test and Scheme Optimization

2.2.1. Cement Optimization

The test selected gypsum and cement as cement, respectively. After the test piece was finished, the test piece was tested for strength. The test results are shown in Table 1. According to the similarity theory, similarity ratio and mechanical properties of coal seams in similar simulation tests [31], combined with the test results of Table 1, the low-strength of gypsum and cement can meet the needs of the test. For different proportions of gypsum, uniaxial compressive strength was 85–118 kPa, for different ratios of cement, uniaxial compressive strength was 114–456 kPa, which means that when coal fines are used as the main aggregate, the strength of the test piece is stronger when cement is used as cement. In order to meet the needs of the test coal body strength, this test used cement as a cement for similar simulated test coal.

Table 1. Optimum test results of cement.

Test Number	Cement (%)	Sand (%)	Water (%)	Activated Carbon (%)	Coal (%)	Sand to Rubber Ratio (%)	Density (g·cm^{-3})	Uniaxial Compressive Strength (kPa)
Gypsum #1	8	7	11	0.9	73.1	10.13	0.972	85
Gypsum #2	12	6	11	0.5	70.5	6.42	0.961	97
Gypsum #3	15	6	11	0.5	67.5	4.93	0.885	118
Gypsum #1	8	4	7	0.5	80.5	10.63	0.866	114
Gypsum #2	10	4	11	1.1	73.9	7.9	0.935	228
Gypsum #3	12	7	13	1.1	66.9	6.25	1.089	456

2.2.2. Orthogonal Ratio Test Scheme

Select the orthogonal test table of 5 factors and 4 levels (factors are cement, sand, water and activated carbon, the fifth factor is empty, the proportion of coal is determined by the total mass ratio of the above four factors) for orthogonal test. The ratio scheme and results are shown in Table 2.

Table 2. Low-strength orthogonal ratio test scheme for coal seam similar materials.

Factor Number	Cement (%)	Sand (%)	Water (%)	Activated Carbon (%)	Coal (%)	Density (g·cm^{-3})	Uniaxial Compressive Strength (kPa)	Sand to Rubber Ratio
1	6	5	11	0.7	77.3	0.917	96	13.83
2	10	7	7	0.7	75.3	0.986	121	8.30
3	8	7	11	0.9	73.1	0.958	204	10.13
4	12	5	7	0.9	75.1	0.958	187	6.75
5	6	6	7	1.1	79.9	0.913	76	14.50
6	10	4	11	1.1	73.9	0.935	228	7.90
7	8	4	7	0.5	80.5	0.866	114	10.63
8	12	6	11	0.5	70.5	0.949	319	6.42
9	6	4	13	0.9	76.1	0.953	210	13.50
10	10	6	9	0.9	74.1	1.012	156	8.10
11	8	6	13	0.7	72.3	1.046	147	9.88
12	12	4	9	0.7	74.3	1.013	264	6.58
13	6	7	9	0.5	77.5	1.008	146	14.17
14	10	5	13	0.5	71.5	1.011	160	7.70
15	8	5	9	1.1	76.5	0.959	167	10.38
16	12	7	13	1.1	66.9	1.089	456	6.25

2.2.3. Optimization of Ratio Scheme for Similar Materials in Coal Seam

According to the geological data of the simulated coal seam and the mine, the average density of the coal seam is 1.2 g/cm^3 and the uniaxial compressive strength is 16 MPa. According to the low-strength ratio test results and the similarity theory requirements, the uniaxial compressive strength of the coal seam in the similar material simulation test should be 0.07 MPa and the ratio #5, that is, cement:sand:water:activated carbon:coal = 6:6:7:1.1:79.9, was selected as the ratio of the coal seam in the similar material simulation, the density of the coal seam produced was 0.913 g/cm^3 and the uniaxial compressive strength was 0.076 MPa.

3. Similar Simulation Material Test Design

3.1. An Overview of Similar Simulation Test

Similar simulation test is a test method of making model in laboratory using similar materials, according to the actual strata data of coal mines, following similar theory and similarity criterion. After the model is completed, the model is mined and analyzed to observe the movement of the overlying strata. The test results and the laws of overburden movement are analyzed. The similar material simulation method is based on the similar simulation conditions and the formation prototype can be simplified appropriately. If there are many rock layers laid, the upper strata can be properly simplified and converted into corresponding loads as compensation to the top of the model.

3.2. Similar Conditions for Similar Simulation Experiments

The similar material simulation test is based on the similarity theory [32–34]. The geometrical similarity, time similarity, bulk density similarity, elastic modulus similarity and Poisson's ratio similarity must be followed between the laying model and the original model.

1. Geometrical similarity. Assuming that the three vertical dimensions of the original model are X_P, Y_P, Z_P and the corresponding dimensions of the similar model are X_m, Y_m, Z_m, then, the geometric similarity ratio is: $C_l = \frac{X_m}{X_p} = \frac{Y_m}{Y_p} = \frac{Z_m}{Z_p}$.

2. Time similarity. The time similarity ratio is: $C_t = \frac{T_m}{T_p} = \sqrt{C_l}$.

3. Bulk density similarity. Assuming that the bulk density of the i layer of the original model is γ_{pi} and the bulk density of the rock is γ_{mi} in the similar model, then the similarity ratio of the bulk density is: $C_\gamma = \frac{\gamma_{mi}}{\gamma_{pi}}$.

4. Strength similarity. Assuming that the uniaxial compressive strength of each rock layer of the original model is σ_{cpi}, σ_{cmi} and the uniaxial compressive strength similarity coefficient of each layer material is $C_{\sigma c}$, then the strength similarity ratio is: $C_{\sigma c} = \sigma_{cmi} \cdot \sigma_{cpi} = C_l \cdot C_\gamma$.

5. Elastic modulus similarity. Assuming that the elastic modulus of each rock layer in the original model is E_{pi}, E_{mi}, the elastic modulus similarity ratio is $C_E = \frac{E_{mi}}{E_{pi}} = C_l \cdot C_r$.

6. Poisson's ratio similarity. Assuming that the Poisson's ratio of each rock layer in the original model is μ_{mi}, μ_{pi}, then the Poisson's ratio similarity coefficient is $C_u = \frac{\mu_{mi}}{\mu_{pi}}$.

In this study, the similar simulation experiment was based on the coal seam geological comprehensive histogram and rock mechanics parameters of the 104 mining area of Yangliu coal mine in Huaibei Mining Group. The thickness of the magmatic rock in the model was 60 m, which was 80 m away from the mining coal seam and the designed coal seam was 8 m thick. There was 8# coal overlying the mining coal seam, but it could not stoped. Select model geometric similarity constant $C_l = 200$, bulk density similarity constant $C_\gamma = 1.5$, then, other similar constants can be obtained: time similarity constant $C_t = \sqrt{C_l} \approx 14$, elastic modulus similarity constant $C_E = C_l \cdot C_\gamma = 300$, strength similarity constant $C_{\sigma c} = C_l \cdot C_\gamma = 300$. A 500 kg iron block was applied on the model to simulate some unpaved rock formations.

3.3. Test Material Selection and Model Making

In this simulation test, except for the coal seam, other lithology selected fine river sand as aggregate, calcium carbonate and gypsum as cementing materials and mica powder was used as the sandwich material to simulate interlayer joints. The different lithologies of each stratum layer can be achieved through different ratios of materials in the test. According to the determined similarity constant, the ratiometric parameters of the similar material test were determined by the proportioning test, as shown in Table 3.

Table 3. Ratio of materials and materials for different strata in similar simulation tests.

Rock Number	Rock Name	Thickness (cm)	Cumulative Thickness (cm)	Proportioning Number	Bulk Density (g/cm^3)
R-29	Siltstone	6	162.7	755	1.6
R-28	Mudstone	6	156.7	864	1.5
R-27	Fine sandstone	6	150.7	782	1.6
R-26	Siltstone	5.2	144.7	755	1.6
R-25	Mudstone	5.2	139.5	864	1.5
R-24	Siltstone	5	134.3	755	1.6
R-23	Mudstone	12	129.3	864	1.5
R-22	Fine sandstone	7	117.3	782	1.6
R-21	Sandy mudstone	4.4	110.3	864	1.5
R-20	Fine sandstone	4.8	105.9	782	1.6
R-19	Mudstone	4.6	101.1	864	1.5
R-18	Siltstone	5.4	96.5	755	1.6
R-17	Mudstone	3.6	91.1	864	1.5
R-16	**Magmatic rock**	30	87.5	737	1.5
R-15	Mudstone	1.5	57.5	864	1.5
R-14	Fine sandstone	2.8	56	782	1.6
R-13	Sandy mudstone	3	53.2	864	1.5
R-12	Siltstone	3.2	50.2	755	1.6
R-11	8# coal	1.6	47	864	1.5
R-10	Siltstone	1.5	45.4	755	1.5
R-9	Mudstone	3.2	43.9	864	1.5
R-8	Siltstone	3.2	40.7	755	1.6
R-7	Sandy mudstone	4	37.5	864	1.5
R-6	Siltstone	3	33.5	755	1.6
R-5	Mudstone	2.8	30.5	864	1.5
R-4	Siltstone	3	27.7	755	1.6
R-3	Piebald mudstone	3	24.7	864	1.5
R-2	Siltstone	3	21.7	755	1.6
R-1	Fine sandstone	1.2	18.7	782	1.6
Coal	**Coal**	4	17.5		
F-1	Coarse sandstone	13.5	13.5	773	1.6

3.4. Test Model Laying

According to the design, the aggregate and cement materials required for each layer were calculated and uniformly mixed together. The stirred materials were evenly laid on the model support, rammed with iron blocks and then a layer of mica powder was evenly spread on the layered surface to simulate interlayer joints. Lay the other strata in turn as described above until the model design height. The weight of overburden which cannot be simulated above the model is implemented by adding weights.

3.5. Monitoring Arrangement and Mining Method

Grids of 100 mm (length) × 100 mm (width) were arranged on the surface of the test model. A total of five monitoring lines were arranged on the model and the distance from the coal seam was 12.5 cm, 32.5 cm, 37.5 cm, 42.5 cm and 112.5 cm, respectively. Each monitoring line was equipped with 29 measuring points, numbering #1 to #29 respectively. The test model was excavated 50 mm each time and the interval between two excavations was generally 1.5 to 2 h. Record the state of the overburden structure with a camera before each excavation and record the test phenomenon in time. The total station was used to record the displacement change state of the monitoring points during the simulation. The model monitoring line and point layout are shown in Figure 2.

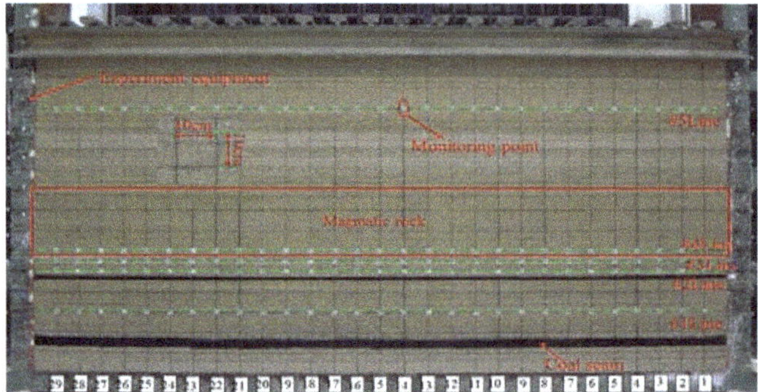

Figure 2. Similar simulation test model and monitoring line and point layout.

4. Evolution Process of Overburden Bed Separation under Hard Thick Magmatic Rock Conditions

4.1. Overburden Bed Separation Morphology and Developmental Height Evolution

The whole process of the overburden bed separation evolution can be visually displayed by the method of similar simulation test, as shown in Figure 3.

Figure 3a shows that when the working face advances 40 m, the falling immediate roof cannot fully enrich the goaf, causing the lower part of the unfallen roof to lose support and bending and sinking. Due to the difference in lithology, the roof strata are not synchronously bent. The sinking results in the bed separation of the siltstone-mudstone interface at 26 m above the coal seam, with a span of 19.6 m. When the working face advances 70 m, Figure 3b shows that the main roof of the working face is pressed and the bed separation position is moved up to 32 m above the coal seam and the bed separation span is slightly increased but the shape of the bed separation remains the same. As the mining area continues to increase, when the working face advances 100 m, the working face is periodically pressed. Figure 3c shows that the roof of the working face is broken and moved in a large area. The height (the maximum displacement difference of the upper and lower strata at the largest bed separation) and the shape of the bed separation space change rapidly, the height of the bed separation space is instantaneously increased to 3 m, the span reaches 40 m and the bed separation develops to 40 m above the coal seam and the distribution pattern of the bed separation changes. When the working face advances 130 m and 160 m, as shown in Figure 3d,e, the overburden fracture movement is carried out in groups and the shape of the bed separation gradually develops to the higher strata and develops to the bottom of the magmatic rock. The vertical height of the bed separation is 4.6 m and the span reaches 66 m. When the working face advances 200 m and 240 m, the overlying rock falls in a large area and the fallen strata in the middle of the goaf is gradually compacted. As shown in Figure 3f,g, the height of the bed separation space below the magmatic rock increases to 5 m and the span reaches 104 m, 134 m respectively, the development pattern of bed separation increases in both horizontal and vertical directions. When the working face advances 340 m, the magmatic rock breaks and loses stability and the magmatic rock and its upper strata sink and the bed separation under the magmatic rock is closed.

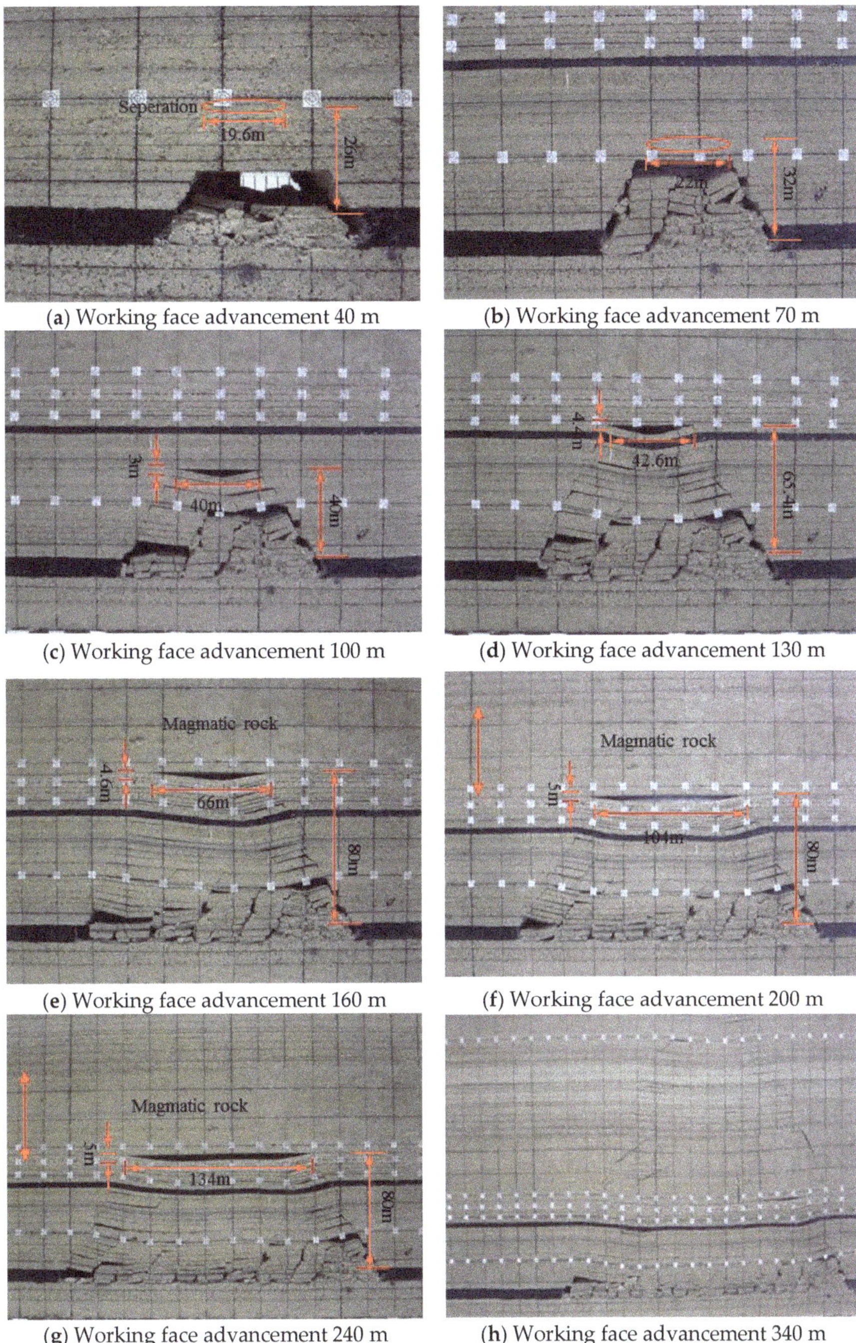

Figure 3. Evolution law of overburden bed separation.

In summary, under the condition that the working face is covered with hard and thick magmatic rocks, the evolution characteristics of the overburden bed separation are mainly characterized by the change of the morphology and developmental height. For the bed separation morphology, as shown in Figure 4, as the working face advances, the falling immediate roof cannot completely fill the goaf and the upper strata of the goaf does not move synchronously, resulting in the bed separation. With the advance of the working face, the goaf falls and the range of the fractures increases continuously in both the height and the length. The middle part of the fractured rock mass is hinged, forming a "triangular" shaped bed separation zone with the upper unbroken strata and compared with the previous bed separation, the volume of the bed separation is significantly increased. As the working face continues to advance, the "triangular" shaped bed separation zone is developed to the bottom of the hard and thick key strata. Under the shielding of the hard and thick key strata, the bed separation does not continue to develop upwards but is developed into a "basin" at the bottom of the hard and thick key strata. The bed separation has experienced the entire process of appearance-expansion-closing at the bottom of the hard and thick key strata.

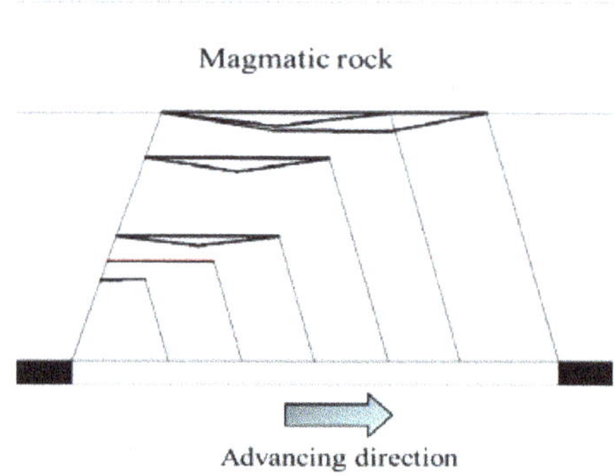

Figure 4. Evolution process of the bed separation under the hard and thick key strata.

4.2. Dynamic Evolution Law of Mining-Induced Overburden Bed Separation

The bed separation value (bed separation height between adjacent strata, m) and the bed separation rate (bed separation value in unit thickness, mm/m or ‰) can quantitatively describe the dynamic process of overburden bed separation during coal mining. According to the sinking amount of each monitoring point in the mining process, the dynamic change law of the bed separation value and the bed separation rate between the monitoring lines is calculated. Figures 5–7 shows the dynamic changes of the bed separation value and the bed separation rate between monitoring lines #1 and #2, #2 and #3, #3 and #4 respectively under different advancing lengths.

As shown in Figure 5, when the working face advances 130 m, the main roof of the working face is periodically collapsed and the falling height of the roof has not touched the strata where the measuring line 2 is located and the maximum bed separation value between the monitoring lines #1 and #2 reaches 6.8 m. The bed separation rate is as high as 170‰ and the maximum bed separation value is located in the middle of the goaf. As the working face continues to advance, the high strata break down and the bed separation value and the bed separation rate between the monitoring lines will decrease. When the working face advances 200 m, the bed separation value and the bed separation rate between the monitoring lines are greatly reduced. The maximum bed separation value is only 4 m, the maximum bed separation rate is 100‰ and the bed separation value and the bed separation rate is reduced by 41%. The fallen strata in the middle of the goaf are gradually compacted and the bed separation between the monitoring lines is closed or reduced, resulting in a significant change in the distribution of the bed separation value and the bed separation rate. The peaks are located at 50 m in front of the open-off cut and behind the coal wall respectively. When the working face advances 290 m and 350 m, the distribution of the bed separation value and the bed separation rate are similar to that of 200 m. The peaks of the bed separation value and the bed separation rate before the open-off cut are slightly reduced compared with 200 m but the peaks position is stable at 50 m in front of the open-off cut and the peaks distribution in front of the open-off cut are approximately equal.

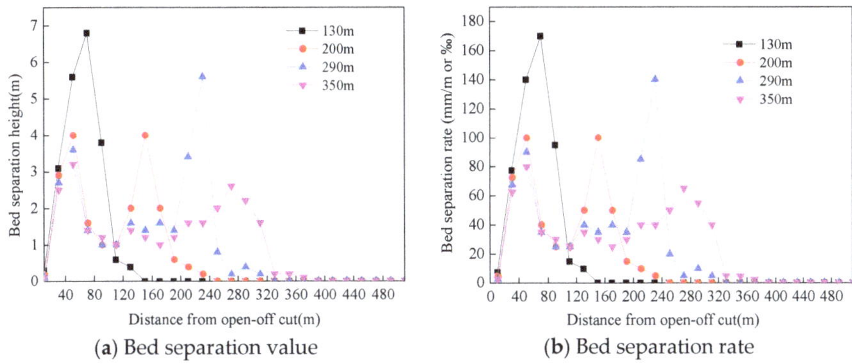

(a) Bed separation value

(b) Bed separation rate

Figure 5. Dynamic evolution of the bed separation between the monitoring lines #1 and #2.

As shown in Figure 6, the variation of the distribution pattern of the bed separation value or the bed separation rate between the monitoring lines #2 and #3 is basically the same as that of the monitoring lines #1 and #2. The #2 line was arranged in the sub-critical strata. The #3 line follows closely the #2 line movement. The bed separation between the monitoring lines is quickly compacted, so that the maximum bed separation value in the "Camel Valley" position is only 4 m before the magmatic rock is broken (working face advance 340 m). When the working face advances 350 m, the magmatic rock has been broken and the bed separation between the monitoring lines is basically compacted, leaving only the 0.2 m remnant bed separation. In addition, when the working face advancement length is greater than 130 m, the peak value of bed separation in front of the open-off cut is gradually decreased. Figures 5 and 6 show that the bed separation value between the #2 and the #3 is much smaller than that of the #1 and the #2 but the distance between the #2 and the #3 is only 10 m, so the bed separation does not change significantly.

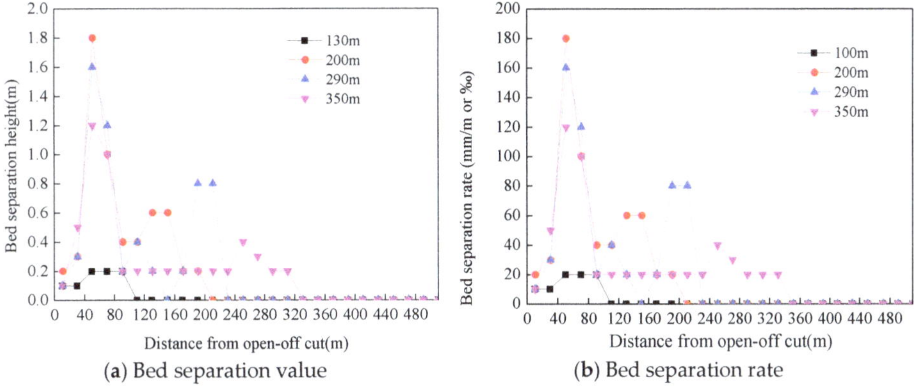

(a) Bed separation value

(b) Bed separation rate

Figure 6. Dynamic evolution of the bed separation between the monitoring lines #2 and #3.

The bed separation value and the bed separation rate between the magmatic rock and the lower rock are in a dynamic process, as shown in Figure 7. When the working face advances 130 m, the overburden movement does not affect the strata where the #3 is located. There is almost no bed separation between the #3 and the #4. As the working face continues to advance, the lower strata of the magmatic rock begin to bend, break and sink and the bed separation between the magmatic rock and the lower strata continues to expand. When the working face advances 200 m, the maximum bed separation value between the monitoring lines reaches 5 m and the bed separation rate is as high as 500‰. At this time, the bed separation value between the monitoring lines reaches the maximum. As the working face continues to advance, the bed separation value between the monitoring lines undergoes a stabilization-reduction-closing process. When the working face advances 290 m, the maximum bed separation value between the monitoring lines is reduced to 4 m and the bed separation value in the middle of the goaf is reduced to 3.2 m. The central part of the suspended magmatic rock has been obviously bent and sunk. When the working face advances 350 m, the magmatic rock breaks and loses stability and the bed separation is basically closed except for the rear of the coal wall and the front of the open-off cut. Compared with before breaking, the maximum bed separation rate in front of the open-off cut is reduced by 120‰ and the maximum bed separation rate behind the coal wall is reduced by 320‰.

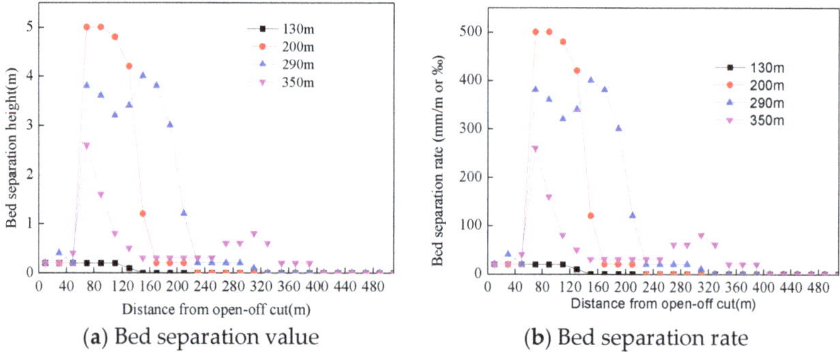

(a) Bed separation value   (b) Bed separation rate

Figure 7. Dynamic evolution of the bed separation between the monitoring lines #3 and #4.

5. Analysis of Evolution Law of Mining-Induced Overburden Fractures

The similar simulation test results show that the fracture distribution in the overburden has obvious zoning characteristics, as shown in Figure 8. Figure 8 shows that after mining of coal seam, the overburden breaks to produce fractures and as the working face advances, the fracture range expands in height and width. As the mining activity progresses, new fractures continue to occur. Under the action of the overburden, the original part of the fractures is gradually compacted, forming a compacted and fractured areas. The boundary shape of the compaction zone and the fracture zone is approximately trapezoidal. When the fracture development does not reach the magmatic rock, as shown in Figure 8a, the upper boundary of the compaction zone is lower than the upper boundary of the fracture zone. When the fractures develop to the lower surface of the magmatic rock, due to the shielding effect of the magmatic rock on the fracture development, the height of the fracture zone before the magmatic rock breaks no longer develops upward, while the middle part of the goaf is compacted, resulting in the upper boundary of the compaction zone coinciding with the upper boundary of the fracture zone, as shown in Figure 8b,c. After the magmatic rock is broken, the height of the fracture zone changes rapidly and the fracture zone height develops to the top of the model. The distribution of the fracture zone also undergoes singularity change. The boundary morphology of the fracture zone evolves into a double trapezoidal distribution with the lower surface of the magmatic rock as the boundary. As shown in Figure 8d. During the development of the fractures, the weak and small thickness strata can only remain stable in a short period of time, which has little effect on the development speed of the fractures, while the strata with large hardness can maintain stability and integrity for a long time and play a vital role in the formation and development of the fractures. Under the shielding effect of the key strata, the height of the fracture zone changes rapidly, especially in the case of overlying the hard and thick key strata. At the early stage of working face mining, the height of the fracture zone is continuously expanded and stops at the magmatic rock. The boundary area of the fracture zone advances with the advancement of the working face. During the process of fracture zone distribution evolution, the width of the largest fracture zone behind the coal wall is larger than the front of the open-off cut and the width of the fracture zone behind the coal wall is constantly changing. The maximum width is about 40–60 m and the width of the fracture zone in front of the open-off cut is stable at around 30 m.

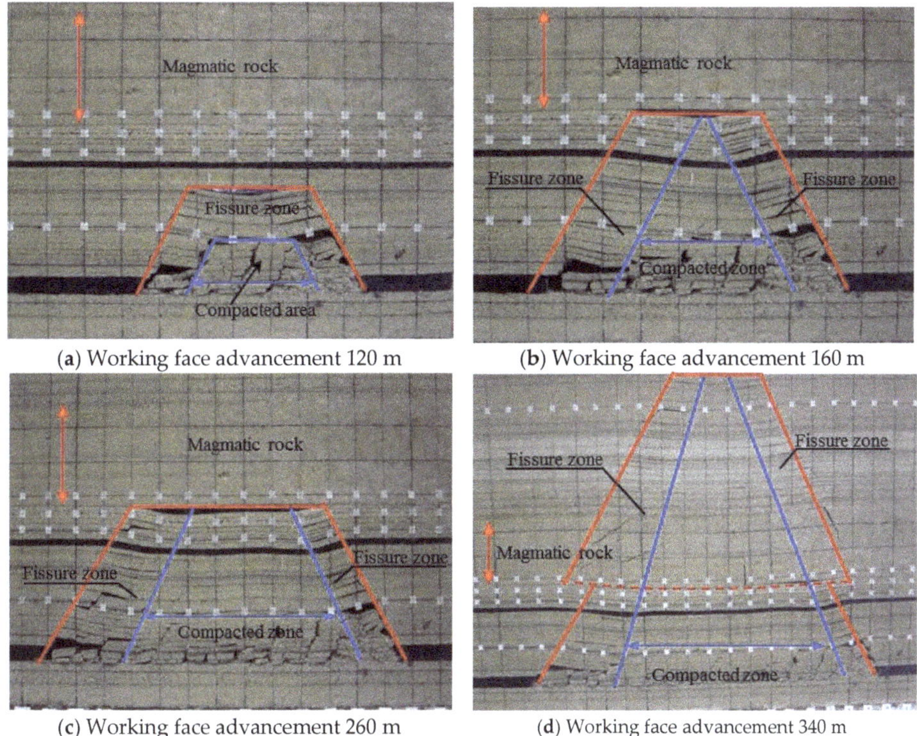

(**a**) Working face advancement 120 m

(**b**) Working face advancement 160 m

(**c**) Working face advancement 260 m

(**d**) Working face advancement 340 m

Figure 8. The distribution characteristics of fractures in overburden.

When the working face is covered with hard thick magmatic rocks, the development of overburden fractures has obvious characteristics in the direction of the working face height and advancement. In the height direction, in the early stage of mining, with the working face advancing, the range of fracture development under the magmatic rock expands continuously but the height of the fractures is only developed to the bottom of the magmatic rock under the shielding of hard and thick magmatic rocks. As the working face continues to advance, the hard thick magmatic rock breaks and loses stability and the development range of the fractures above the goaf changes rapidly and the fractures develop instantly to the top of the model. In the advancing direction, the maximum width of the fracture areas behind the coal wall of the working face is larger than the front of the open-off cut.

When the hard and thick magmatic rock is broken, the width distribution of the fracture zone remains the same. Under this influence, the gas in the bed separation space below the hard and thick key strata will flood into the working face through the fracture areas behind the coal wall after the hard and thick key strata is broken, resulting in a safety accident.

6. Mechanism of Breaking Induced Disaster of Hard and Thick Magmatic Rocks

The thermal metamorphism near coal seams under the hard thick magmatic rocks is high and the coal gas content is high, which is the main source of the bed separation gas. Through the law of the bed separation and fracture development of the overburden before and after the breaking of the hard and thick magmatic rock, it is known that before the magmatic rock breaks, there is a huge bed separation space between the magmatic rock and its lower strata and it does not communicate with the fractures and the evolution of the developmental morphology of the bed separation and the height of the bed separation space provides a good space carrier for gas accumulation, which plays a good role in trap accumulation. If the bed separation beneath the magmatic rock communicates with the perforating fractures of the lower strata of the magmatic rock, this will provide a natural passage and accumulation space for the accumulation of gas. Under the combined action of vacuum negative pressure and physical properties, the gas released in the coal seam mining process and the gas remaining in the goaf are collected into the bed separation space of the lower part of the magmatic rock through the surrounding fractures, forming a bed separation gas, as shown in Figure 9. The direction of the arrow is the direction of gas flow.

The results of similar simulation test show that the bed separation space at the bottom of the magmatic rock changes dynamically with the mining of the working face. Therefore, the gas in the bed separation space is also in a dynamic process. Magmatic rocks control the development height of the bed separation and fractures and limit the migration height of the gas. When the hard thick magmatic rock breaks, it exerts a strong dynamic impact on the gas in the bed separation, causing the secondary development of the vertical fractures. The bed separation gas is subjected to the strong dynamic pressure and the accumulated gas is partially poured into the working face along the fractures around the goaf, resulting in the gas outburst accident of the working face and the other part is poured out the ground to form drainage borehole gas jet, as shown in Figure 10.

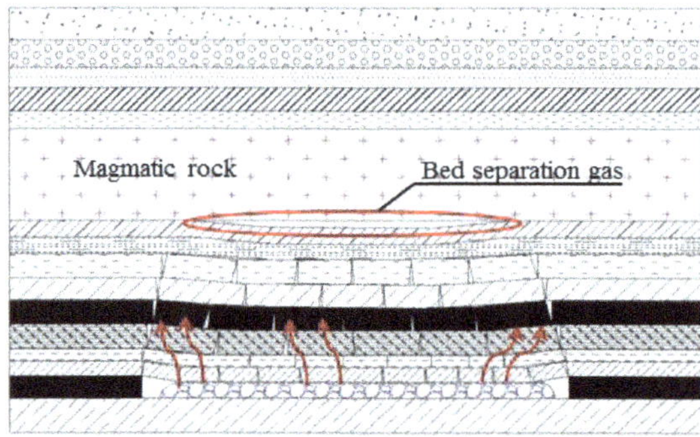

Figure 9. Schematic diagram of the formation of bed separation gas.

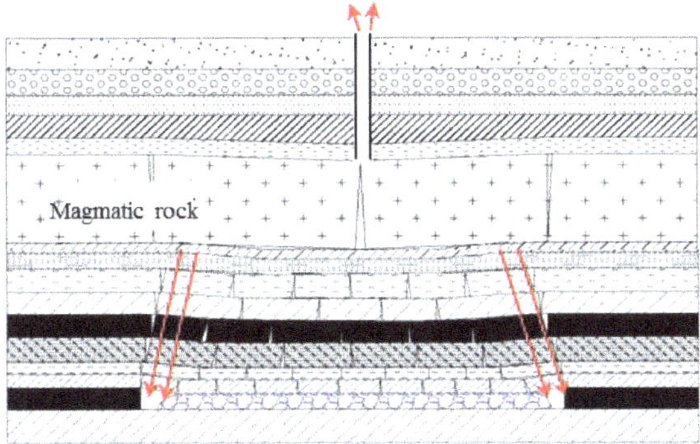

Figure 10. Schematic diagram of bed separation gas inrushing.

7. Engineering Case Analysis

The 10,414 working face of Yangliu coal mine is located in Suixi County, Huaibei City, Anhui Province, China. The average coal seam is 3.05 m thick and the buried depth is about 600 m. The 10,414 working face mining area is covered with a layer of magmatic rock with a thickness of 60 m, with an average spacing of 80 m from the coal seam. In order to effectively control the gas content of the working face and ensure safe production, four gas drainage holes of diameter 300 mm, such as 1#, 2#, 3-1#, 3# and so forth, were arranged in the working face range, which were 316 m, 515 m, 654 m and 721 m respectively from the open-off cut. The gas drilling arrangement is shown in Figure 11.

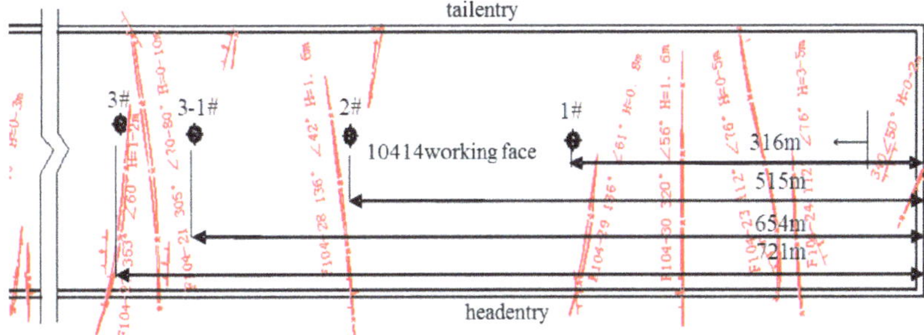

Figure 11. Schematic diagram of the gas drainage drilling arrangement of the 10,414 working face.

At 14 o'clock on the 16 July 2011, the extraction parameters of the 80-type ground pump began to change. The extraction concentration continued to drop from about 90% under normal conditions to about 20% of the lowest point. The extraction flow had a sharp fluctuation from the normal value of 55 m^3/min, while negative pressure and temperature fluctuations were not large. At 17:20 on the 17 July, the extraction concentration rose sharply from 20% to 100% in just 32 min and the negative pressure dropped sharply to 0. At the same time, a gas jet occurred at 2# drainage borehole, as shown in Figure 12. The ejecta was a water-gas mixture and the gas jet lasted 21 h and 30 min, with a total amount of 77,400 m^3.

Figure 12. Gas jet at the ground 2# drainage borehole.

Figure 13 shows the relationship between the gas extraction amount and the mining distance in the gas drainage borehole during the mining of the 10,414 working face. As shown in Figure 13 that when the working face advances 500 m, the amount of extraction of the 2# gas drainage hole begins to increase sharply. When the working face advances 525 m, the extraction volume of the 2# gas drainage hole reaches 0.15 m^3/min and the gas jet begin to occur. When the working face advances 543 m, the gas extraction volume in the 2# gas drainage hole reaches a peak value of 45.22 m^3/min. When the working face advancing distance is greater than 600 m, the gas extraction volume of the wells in all areas is greatly reduced.

The magmatic rock structure area contains a large amount of gas and water. Because the magmatic rock is dense and hard, the gas is not easily diffused and lost. In addition, No. 10 coal is rich in gas. After the working face is mined, the gas adsorbed in the coal seam is liberated and spreads along the fractures in the goaf to the top of the coal seam. According to the results of similar simulation, it is known that in the process of the working face advancing, a large amount of negative pressure separation space is formed between the magmatic rock and the lower rock strata and a large amount of gas and water are drawn into the space due to the caving and sinking of No. 10 coal overburden are not synchronized with magmatic rocks above.

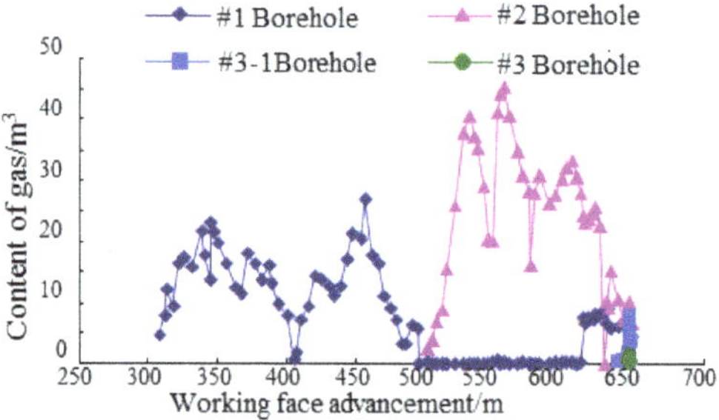

Figure 13. Relationship between the gas extraction amount and the mining distance in the gas drainage borehole.

When the magmatic rock sinks quickly, the gas in the bed separation space is squeezed. Since the ground #2 borehole penetrates the bed separation space through the fractures, causing the gas to flow through the borehole to the ground, resulting in the gas jet. According to the site geological and mine pressure data, the reason for the rapid movement of magmatic rocks is probably the fracture of the magmatic rocks. According to the records of the field engineering technicians, when the working face advances 480 m, the microseismic frequency and the microseismic energy level in front of the working face and above the goaf begin to increase. When the working face advances 500 m, the microseismic frequency and energy level increase rapidly and the large energy microseisms occur in front of the work face. As the working face continues to advance, the vibration energy continues to strengthen and large energy events with energy greater than 50,000 J are concentrated in front of the working face, accompanied by the occurrence of drainage borehole gas jet events. According to the on-site analysis, the gas jet of the #2 borehole is caused by the breaking of the overlying hard and thick magmatic rock. The occurrence of the gas jet event in the 10,414 working face confirmed the law of the bed separation and fracture development under the hard thick magmatic rock in mining.

8. Conclusions

1. Through the low-strength mechanical test, the cement with a wider range of strength is selected as the cementing material to carry out the low strength orthogonal ratio test of the similar simulation material of coal seam and the ratio of coal seam which conforms to the actual field and the similar theory is determined, namely cement:sand:water:activated carbon:coal = 6:6:7:1.1:79.9, the density of the coal seam produced is 0.913 g/cm^3 and the uniaxial compressive strength is 0.076 MPa.

2. Before the hard thick magmatic rock breaks, the developmental height of the overburden bed separation develops upward in a nonlinear way with the advance of the working face. The developmental height of the bed separation is blocked by the hard thick magmatic rock and the height is stopped at the bottom of the magmatic rock and the bed separation is not closed for a long time. As the working face advances, the bed separation changes only in the lateral direction. The bed separation space at the bottom of the hard thick magmatic rock is expanded for a long time, providing a breeding space for the bed separation gas and bed separation water.

3. When the working face is covered with hard thick magmatic rock, the development of overburden fracture has obvious characteristics in the direction of the working face height and advancement. In the height direction, in the early stage of mining, with the working face advancing, the range of fractures development under the magmatic rock expands continuously but the height of the fractures is only developed to the bottom of the magmatic rock under the shielding of hard and thick magmatic rock. As the working face continues to advance, the hard thick magmatic rock breaks and loses stability and the development range of the fracture above the goaf changes rapidly and the fractures develop instantly to the top of the model. In the advancing direction, the maximum width of the fracture area behind the coal wall of the working face is larger than the front of the open-off cut. When the hard and thick magmatic rock is broken, the width distribution of the fracture zone remains the same.

4. Before the magmatic rock breaks, there is a huge bed separation space between the magmatic rock and its lower strata and it does not communicate with the fractures and the evolution of the developmental morphology of the bed separation and the height of the bed separation space provides a good space carrier for gas accumulation, which plays a good role in trap accumulation. If the bed separation beneath the magmatic rock communicates with the perforating fractures of the lower strata of the magmatic rock, this will provide a natural passage and accumulation space for the accumulation of gas. When the hard thick magmatic rock breaks, it exerts a strong dynamic impact on the gas in the bed separation, causing the secondary development of the vertical fractures. The bed separation gas is subjected to the strong dynamic pressure and the accumulated gas is partially poured into the working face along the fractures around the goaf, resulting in the gas outburst of the working face and ground drainage borehole gas jet accidents.

Author Contributions: Conceptualization & Methodology, Q.W. (Quanlin Wu) and Q.W. (Quansen Wu); Writing-Original Draft Preparation, Y.X. and P.K.; Writing-Review & Editing, B.G.

Funding: This research was funded by the Natural Science Foundation of Shandong Province (Grant No. ZR2018PEE007).

Conflicts of Interest: The authors declare that there is no conflict of interests regarding the publication of this paper.

References

1. Cao, A.; Zhu, L.; Li, F. Characteristics of T-type overburden structure and tremor activity in isolated face mining under thick-hard strata. *J. China Coal Sci.* **2014**, *39*, 328–335.
2. Lu, C.P.; Liu, Y.; Wang, H. Microseismic signals of double-layer hard and thick igneous strata separation and fracturing. *Int. J. Coal Geol.* **2016**, *160–161*, 28–41. [CrossRef]
3. Jiang, J.; Zhang, P.; Nie, L. Fracturing and dynamic response of high and thick stratas of hard rocks. *Chin. J. Rock Mech. Eng.* **2014**, *33*, 1366–1374.
4. Jiang, J.; Xu, B. Study on the development laws of bed-separation under the hard-thick magmatic rock and its fracture disaster-causing mechanism. *Geotech. Geol. Eng.* **2018**, *36*, 1525–1543. [CrossRef]
5. Wu, L.; Qian, M.; Wang, J. The influence of a thick hard rock stratum on underground mining subsidence. *Int. J. Rock Mech. Min. Sci.* **1997**, *34*, 341–344. [CrossRef]
6. Xu, C.; Cheng, Y.; Ren, T. Gas ejection accident analysis in bed splitting under igneous sills and the associated control technologies: A case study in the Yangliu Mine, Huaibei Coalfield, China. *Nat. Hazards* **2014**, *71*, 109–134. [CrossRef]
7. Song, D.; Wang, E.; Liu, Z.; Liu, X. Numerical simulation of rock-burst relief and prevention by water-jet cutting. *Int. J. Rock. Mech. Min. Sci.* **2014**, *70*, 318–331. [CrossRef]
8. Wang, W.; Cheng, Y.; Wang, H. Coupled disaster-causing mechanisms of strata pressure behavior and abnormal gas emissions in underground coal extraction. *Environ. Earth Sci.* **2015**, *74*, 6717–6735. [CrossRef]
9. Wang, L.; Cheng, Y.; Xu, C.; An, F.; Jin, K.; Zhang, X. Thecontrolling effect of thick-hard igneous rock on pressure relief gas drainage and dynamic disasters in outburst coal seams. *Nat. Hazards* **2013**, *66*, 1221–1241. [CrossRef]
10. Wang, J.; Jiang, F.; Meng, X.; Wang, X.; Zhu, S.; Feng, Y. Mechanism of rock burst occurrence in specially thick coal seam with rock parting. *Rock Mech. Rock Eng.* **2016**, *49*, 1953–1965. [CrossRef]
11. Jiang, J.; Wu, Q.; Wu, Q.; Wang, P.; Zhang, C.; Gong, B. Study on distribution characteristics of mining stress and elastic energy under hard and thick igneous rocks. *Geotech. Geol. Eng.* **2018**, 1–16. [CrossRef]
12. Jiang, L.; Sainoki, A.; Mitri, H. Influence of fracture-induced weakening on coal mine gateroad stability. *Int. J. Rock Mech. Min. Sci.* **2016**, *88*, 307–317. [CrossRef]
13. Jiang, H.; Cao, S.; Zhang, Y. Analytical solutions of hard roof's bending moment, deflection and energy under the front abutment pressure before periodic weighting. *Int. J. Min. Sci. Technol.* **2016**, *26*, 175–181. [CrossRef]
14. Wang, P.; Jiang, J.; Zhang, P. Breaking process and mining stress evolution characteristics of a high-position hard and thick stratum. *Int. J. Min. Sci. Technol.* **2016**, *26*, 563–569. [CrossRef]
15. Qian, M.; Miao, X.; Xu, J. Theoretical study of key stratum in ground control. *J. China Coal Soc.* **1996**, *21*, 225–230.
16. Dou, L.; Hu, H. Study of OX-F-T spatial structure evolution of overlying strata in coal mines. *Chin. J. Rock Mech. Eng.* **2012**, *31*, 453–460.
17. Guo, W.; Li, Y.; Yin, D. Mechanisms of rock burst in hard and thick upper strata and rock-burst controlling technology. *Arab. J. Geosci.* **2016**, *9*, 561. [CrossRef]
18. Wu, Q.; Jiang, J.; Wu, Q. Study on the fracture of hard and thick sandstone and the distribution characteristics of microseismic activity. *Geotech. Geol. Eng.* **2018**, *2*, 1–17. [CrossRef]
19. Ning, J.; Wang, J.; Jiang, L. Fracture analysis of double-layer hard and thick roof and the controlling effect on strata behavior: A case study. *Eng. Fail. Anal.* **2017**, *81*, 117–134. [CrossRef]
20. Xu, C.; Yuan, L.; Cheng, Y.; Wang, K.; Zhou, A.; Shu, L. Square-form structure failure model of mining-affected hard rock strata: Theoretical derivation, application and verification. *Environ. Earth Sci.* **2016**, *75*, 1180. [CrossRef]
21. Wang, W.; Cheng, Y.; Wang, H. Fracture failure analysis of hard–thick sandstone roof and its controlling effect on gas emission in underground ultra-thick coal extraction. *Eng. Fail. Anal.* **2015**, *54*, 150–162. [CrossRef]
22. He, H.; Dou, L.; Gong, S.; Zhou, P.; Xue, Z. Rock burst rules induced by cracking of overlying key stratum. *Chin. J. Geotech. Eng.* **2010**, *32*, 1260–1265.

23. Wang, L.; Cheng, L.; Cheng, Y. Characteristics and evolutions of gas dynamic disaster under igneous intrusions and its control technologies. *J. Nat. Gas Sci. Eng.* **2014**, *18*, 164–174. [CrossRef]

24. Mousavi Nezhad, M.; Gironacci, E.; Rezania, M.; Khalili, N. Stochastic modelling of crack propagation in materials with random properties using isometric mapping for dimensionality reduction of nonlinear data sets. *Int. J. Numer. Methods Eng.* **2018**, *113*, 656–680. [CrossRef]

25. Mousavi Nezhad, M.; Fisher, Q.J.; Gironacci, E.; Rezania, M. Experimental study and numerical modeling of fracture propagation in shale rocks during brazilian disk test. *Rock Mech. Rock Eng.* **2018**, *51*, 1755–1775. [CrossRef]

26. Guo, L.; Latham, J.; Xiang, J. A numerical study of fracture spacing and through-going fracture formation in layered rocks. *Structures* **2017**, *110–111*, 44–57. [CrossRef]

27. Park, B.; Min, K. Bonded-particle discrete element modeling of mechanical behavior of transversely isotropic rock. *Int. J. Rock Mech. Min. Sci.* **2015**, *76*, 243–255. [CrossRef]

28. Chong, K.P.; Kuruppu, M.D. Fracture toughness determination of layered materials. *Eng. Fract. Mech.* **1987**, *28*, 43–54. [CrossRef]

29. Xuan, D.; Xu, J.; Feng, J.; Zhu, J.; Zhu, W. Disaster and evolvement law of mining-induced stress under extremely thick igneous rock. *J. China Coal Soc.* **2011**, *36*, 1252–1256.

30. Chen, S.; Wang, H.; Zhang, J.; Xing, H.; Wang, H. Experimental study on low-strength similar-material proportioning and properties for coal mining. *Adv. Mater. Sci. Eng.* **2015**, *3*, 1–6. [CrossRef]

31. Dai, H.; Lian, X.; Liu, J. Model study of deformation induced by fully mechanized caving below a thick loess layer. *Int. J. Rock Mech. Min. Sci.* **2010**, *47*, 1027–1033.

32. Xiao, T. Study on Surrounding Rock Stability and Control of Deep Roadway in Thick Coal Seam under the Action of Tectonic Stress. Ph.D. Thesis, China University of Mining and Technology, Beijing, China, 2011.

33. Lin, Y. *Experimental Rock Mechanics-Simulation*; China Coal Industry Publishing House: Beijing, China, 1984. (In Chinese)

34. Kai, W.; Gong, P.; Zhang, X.; Lian, Q.; Li, J.; Duan, D. Characteristics and control of roof fracture in caving zone for residual coal mining face. *Chin. J. Rock Mech. Eng.* **2016**, *35*, 2080–2088.

 *processes*

Article

A Numerical Study of Stress Distribution and Fracture Development above a Protective Coal Seam in Longwall Mining

Chunlei Zhang [1,2,3,*], Lei Yu [1], Ruimin Feng [4], Yong Zhang [2] and Guojun Zhang [2]

[1]   International Engineering Company of China Coal Technology and Engineering Group,
     Beijing 100013, China; lngdyulei@163.com
[2]   College of Resources and Safety Engineering, China University of Mining and Technology (Beijing),
     Beijing 100083, China; Johnzy68@163.com (Y.Z.); cherish-guojun@hotmail.com (G.Z.)
[3]   Department of Mining and Mineral Resources Engineering, Southern Illinois University,
     Carbondale, IL 62901, USA
[4]   Department of Chemical and Petroleum Engineering, University of Calgary, Calgary, AB T2N 1N4, Canada;
     fengruimin@126.com
*   Correspondence: chunleizhang@student.cumtb.edu.cn; Tel.: +86-10-156-5294-0306

Received: 7 August 2018; Accepted: 17 August 2018; Published: 1 September 2018

Abstract: Coal and gas outbursts are serious safety concerns in the Chinese coal industry. Mining of the upper or lower protective coal seams has been widely used to minimize this problem. This paper presents new findings from longwall mining-induced fractures, stress distribution changes in roof strata, strata movement and gas flow dynamics after the lower protective coal seam is extracted in a deep underground coal mine in Jincheng, China. Two Flac^{3D} models with varying gob loading characteristics as a function of face advance were analyzed to assess the effect of gob behavior on stress relief in the protected coal seam. The gob behavior in the models is incorporated by applying variable force to the floor and roof behind the longwall face to simulate gob loading characteristics in the field. The influence of mining height on the stress-relief in protected coal seam is also incorporated. The stress relief coefficient and relief angle were introduced as two essential parameters to evaluate the stress relief effect in different regions of protected coal seam. The results showed that the rock mass above the protective coal seam can be divided into five zones in the horizontal direction, i.e. pre-mining zone, compression zone, expansion zone, recovery zone and re-compacted zone. The volume expansion or the dilation zone with high gas concentration is the best location to drill boreholes for gas drainage in both the protected coal seam and the protective coal seam. The research results are helpful to understand the gas flow mechanism around the coal seam and guide industry people to optimize borehole layouts in order to eliminate the coal and gas outburst hazard. The gas drainage programs are provided in the final section.

Keywords: longwall mining; gob behaviors; stress relief; permeability; gas drainage

---

1. Introduction and Background

Coal as a major energy source accounted for 62% of Chinese energy consumption in 2016. It has remained the most important fuel in China's energy mix for the last few decades [1]. In recent years, the safety situation of Chinese coal industry has improved, but gas and coal outbursts still seriously threaten coal mine safety due to the fact that the mining depth increases at a rate of 30–50 m/year in China [2]. According to statistical data, the death toll of coal miners reached up to 608 in 76 gas accidents from 2012 to 2015 [3]. Additional effective measures and research, hence, should be carried out to reduce and avoid gas disasters in China.

Stress redistribution and rock mass fracture are two dominant factors influencing rock mass permeability [4–6]. Coal and rock mass permeability increases due to stress relief and mining induced fracturing [5]. It has been reported by several researchers at both laboratory and field scales [1–4,7]. A great deal of previous work showed that changes in coal permeability is induced by changes in the confining stress, and found an exponential relationship between permeability and stress [8]. A confining stress change of 10 MPa may result in changes in permeability with approximately one order of magnitude. Yang et al. [9] provides a visible relationship between dimensionless permeability and multiples of the initial normal stress (see Figure 1, σ_{n0} is initial normal stress, σ_n/σ_{n0} is multiples of the initial normal stress, and k_f is dimensionless permeability). It shows permeability decreases with increasing normal stress. Therefore, the mining of the protective coal seam is wildly used in the multiple coal seams in China (Figure 2). As shown in Figure 2, the coal seam which overlies the target seam is designated as the upper protective seam, whereas the underlying coal layer is called the lower protective seam. Extraction of a protective seam results in redistribution and relieving of some of the stress around underlying or overlying rock mass, thereby establishing new stress equilibrium [10]. Reestablishment of a new stress state will inevitably lead to the changes of structure and properties of rock masses, which will eventually promote the desorption rate of gas from coal matrix and considerably increase permeability in the target coal seam or strata with high gas content.

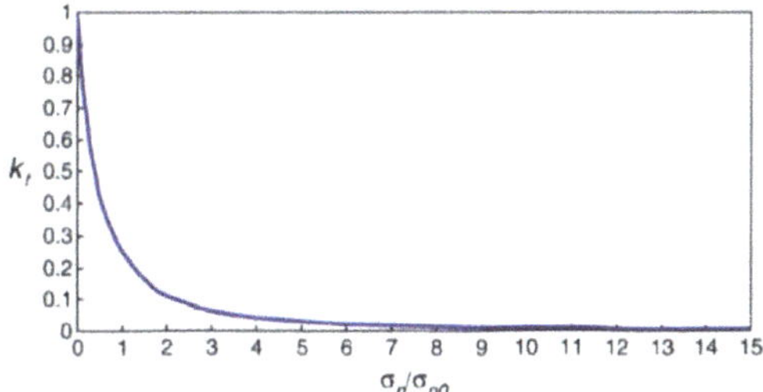

Figure 1. Relationship between permeability and normal stress [9] (Reproduced with permission from [9], Copyright Elsevier, 2011).

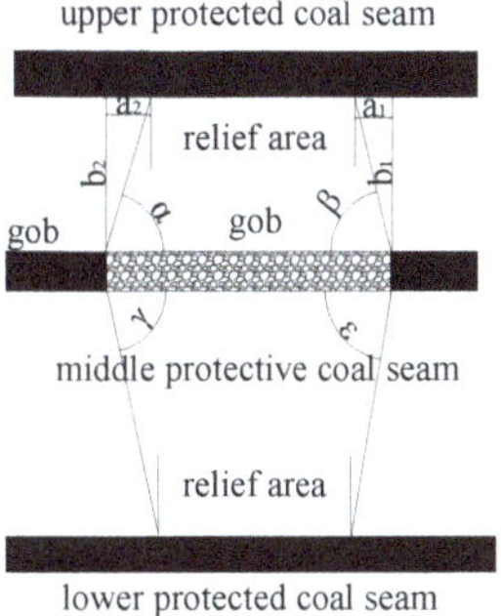

Figure 2. Illustration of protective coal seam mining.

2. Objective of Research

Stress relief coefficient and pressure relief angles are introduced by authors of this paper to evaluate the influence of lower protective excavation on its upper protected coal seam. The location of the protective layer, the horizontal butt entry and the mining height are key factors affecting the safety of protected coal seam mining [11]. This is because the relief area in the protected coal seam depends on a variety of parameters such as: The thickness and strength of the strata between the protected coal seam and the protective coal seam, the mining thickness, gob loading behavior, face advance rate, panel width, and pillar width of the extracted protective coal seam [11]. All of these factors should be taken into consideration to insure the safety of mining and the protected coal seam. In this paper, the mining height and gob behavior characteristics of protective coal seam are chosen as the two main factors affecting the results from protected coal seam excavation. As the effect of these two factors on the safety of mining the protected coal seam is still not clear. Therefore, an in-depth investigation on these two factors is necessitated in order to acquire the relief area in the protected coal seam and insure engineering safety. The final goal of this paper is to determine the stress relief coefficient and pressure relief angles along the strike and incline of the protective coal seam in order to locate the protection region (which is used for drilling boreholes) in the protected coal seam.

In this paper, authors attempt to develop a 3D model of two adjacent longwall faces in a physically realistic way with a consideration of gob loading behavior behind the face to improve understanding of stress distribution and stress relief not only around the protective coal seam but also in the protected coal seam after the excavation of protective coal seam (Figure 3). The stress distribution in the gob depends on mining depth, mining speed, excavation height, bulking factor and compressive strength. Since increasing the mining speed may delay the time for the stress recovery in the gob, it is worthwhile to do some research with a consideration of different gob loading behaviors behind the face. The longwall mining method has been widely used around the world in recent years because of its most important advantage of high efficiency [12]. A mine in Shanxi province (China) with longwall mining method is chosen for the research in this paper.

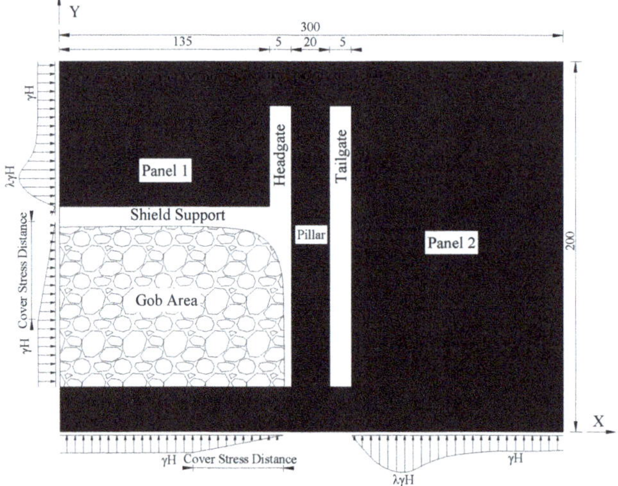

Figure 3. A schematic plan view of two adjacent longwall faces in the numerical model, showing the geometry of the longwall faces. Both panel 1 and panel 2 represent half of original panel width in the studied mine site.

3. Literature Review on Gob Behaviors in Longwall Mining

Numerical modeling of protective coal seam mining's influence on the protected coal seam has been extensively performed in the previous research [1,11]. However, gob loading characteristics have been widely ignored in their models. The authors believe that incorporating the gob loading characteristics into the model is essential for obtaining meaningful and realistic results. This is because the waste in the gob area can transfer the load from overburden to floor after the excavation of the coal seam. Ignoring load carried by the gob material will tend to increase stress concentrations in the face area and cause different stress distribution around excavation.

As the gob is inaccessible and direct measurements cannot be easily taken, it is very difficult to determine rock mass properties and stress distribution in the gob. A few techniques and research had been used and reported by previous researchers [13–16]. Figure 4 shows the strata pressure redistribution in the plane view of the seam around a longwall face after extraction. It shows that the abutment stress in coal ribs gradually recovers to the original stress state with the increasing distance from the rib-edge. The loading in the waste area gradually starts to pick up the overburden loading, and recovers its original stress at a distance from the face after the gob material can take the load from the overburden. Peng et al. [13] studied the supporting role of the gob material by performing three-dimensional element analysis. They divided the gob into three different zones: Loosely packed zone, packed zone, and well packed zone. The side and front abutment stresses will decrease considerably because of the support offered by the compacted gob. Whittaker [14] claimed that the cover pressure re-establishment distance is 0.3–0.4 times the coal seam depth (H) from the solid abutment (Figure 4). Wilson et al. [15] found that the vertical stress in the gob changes linearly, increasing from 0 at the rib to the original stress at a distance of 0.2–0.3 times the overburden depth. Whittaker and Singh [16] assumed that the gob began to take the load at a distance of 45 m from the face, and whether the gob can recover its original stress only depended on the width of the panel. Campoli et al. [17] investigated the longwall gob behavior in No. 3 coal seam of Pocahontas by field measurements under varying sets of geological conditions, and they found that the stress recovery distance in the gob was approximately 0.2 times of coal seam depth (360 m).

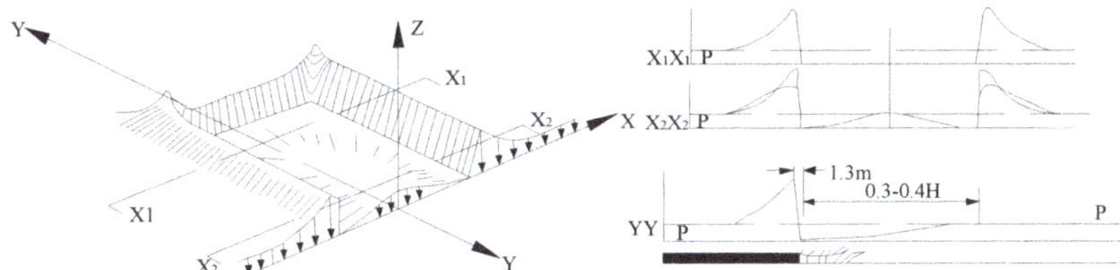

Figure 4. Strata pressure redistribution in the plane of the seam around a longwall face [16] (Reproduced with permission from [16], Copyright Elsevier, 1979).

Modeling of Longwall mining with consideration of gob behaviors has also been documented in previous studies [18–20]. In these studies, the "double yield" elements were incorporated in their simulation model and the gob was regarded as a strain-hardening material. The stress-strain response of gob material was obtained through uniaxial compression tests and displacement measurements in a laboratory test [17]. Li et al. [18] evaluated the stress distribution in the yield pillar and the other rib of the entry in order to find the principle for yield pillar design by Flac^{3D} numerical modeling. Esterhuizen et al. [19] examined the interaction between the surrounding rock mass and typical pillar systems for different geological conditions at various spans and depths of cover by establishing Flac^{2D} numerical models. A method was proposed by H. Yavuz [20] to estimate the pressure distribution in the gob of flat-lying longwall panels and cover pressure distance, and this method was verified with curves obtained through numerical models. However, all these numerical longwall models with the gob behaviors were about pillar design and stability of entries, but not for the protective coal seam mining.

Generally, gob modeling can follow two approaches, explicit model and implicit model [21]. The explicit model needs to explicitly model the gob formation process based on the study of roof fracturing, roof caving and gob development in response to coal mining, while the implicit simulates the effect of the gob on the stability of surrounding coal mine entries and pillars, making sure that the both load redistribution to the surrounding rock and the large-scale overburden defection and subsidence are correct. Both the double-yield model and gob loading model belong to the second approach [4]. This paper addresses the second method. Alternatively, setting up a large-scale model is capable of avoiding simulating the complex material behavior that corresponds the response curve, and equivalent gob stress distribution can be created by simply following the gob response curve. The gob load model has been used in Abbasi et al. [22] work with Flac^{3D}. They estimated the overall gob mechanical behavior by comparing model outputs with field measurements at several points around a longwall face in a coal mine of Illinois. The gob achieved pre-mining vertical stress at approximately 55 m behind the back end of the shields, and the gob load installed in the mine-out area was estimated by field measurements. The gob load varied both along and cross the face advance directions. The gob load model not only characterizes the effect of the gob on the surrounding entries and coal pillars but also simplifies the process of calculation, thus the model size was reduced with finer mesh at the area of interest. As such, in this paper, the gob load model was adopted.

4. Materials and Methods

4.1. Description of Case Coal Mine

The Changping coal mine, with the annual production capability of 5 Mt, is located in the northwestern of Gaoping City, Shanxi, China. The thickness of the coal-bearing geological section ranges from 125 m to 174 m. It contains about sixteen coal seams. The four key minable seams include seams 2#, 3#, 8# and 15#, only portions of 2# and 8# coal seams can be mined. The position relation of Changping coal mine is shown in Figure 5. The thickness of 2# and 3# coal seams are 0–2.95 m and 4.67–6.58 m, respectively. Currently the mine is extracting seam 3# which is classified as a low permeability seam with a methane content of 6.5–15.09 m^3/t. The gas content affected by the geological structure and mining depth is extremely unbalanced; it increases gradually from the eastern part to the western part of the coal field. As the coal mining turns to the west, the gas content and emission quantity increases constantly. The gas dynamic phenomenon and gas outbursts were found in the roadway of Faces 4304 and 4306, which largely decreased the roadway excavation efficiency and brought potential dangers to the production. To avoid the coal and gas outbursts, 8# coal seam was taken as the lower protective coal seam for 3# coal seam.

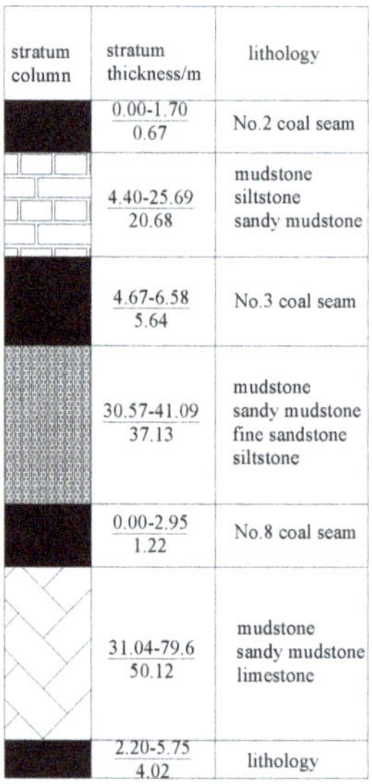

stratum column	stratum thickness/m	lithology
	0.00-1.70 / 0.67	No.2 coal seam
	4.40-25.69 / 20.68	mudstone siltstone sandy mudstone
	4.67-6.58 / 5.64	No.3 coal seam
	30.57-41.09 / 37.13	mudstone sandy mudstone fine sandstone siltstone
	0.00-2.95 / 1.22	No.8 coal seam
	31.04-79.6 / 50.12	mudstone sandy mudstone limestone
	2.20-5.75 / 4.02	lithology

Figure 5. Position relation of coal seams.

The basic parameters of coal seam 3# (No. 3 coal seam) are listed in Table 1. It shows the characteristics of coal seam 3#: Low permeability, high gas content and low hardness. The 4306 panel is in seam 3#, its dip angle ranges from 0–6°. Average dip angle can be regarded as 0°. The coal seam has an average thickness of 5.23 m, with overburden depth ranging between 500–550 m. The panel advance length is about 1100 m and the panel width is about 270 m. The 84306 panel is in seam 8#, its gas content is 4–6 m^3/t, there is no danger of coal and gas outburst in 8# coal seam, and the 84306 panel is mined first as a protective seam for 4306 panel.

Table 1. Basic parameters of No. 3 coal seam.

Coal Seam	Coal Face	Mining Thickness/m	Gas Content/m^3/t	Original Gas Pressure/MPa	Coal Stiffness	Outburst Risk
No. 3	4306 protected longwall face	5.23	3.5–15.09	0.38–0.55	0.44–0.56	Yes
No. 8	84306 protective longwall face	1–3	2–3	None	0.51–1.12	No

4.2. Description of Numerical Models

Flac^{3D}, an explicit finite-difference program for engineering mechanics computation, is used to model the excavations of protective coal seam in this paper, which has been widely used in mining engineering [23]. In this paper, Flac^{3D} was employed to investigate the displacement and stress distribution in the overlying strata (including the No. 3 coal seam) of coal seam 8# (No. 8 coal seam—see Figure 6). In order to eliminate the effect of boundary conditions to the simulation results, both the downside boundaries and the upper coal pillar were confirmed by taking the movement and deformation of the cover rock into account [24]. Therefore, the size of the model is 300 m × 100 m × 200 m. The working face advances in the Y direction, and the surrounding displacement boundary is restricted horizontally; the top of the model is free in Z direction, while the vertical displacement of bottom is restricted. A load of 11.25 MPa was added in the upside boundary of the model for representing an overlying strata of 450 m. The lateral pressure stress was added according to the measured data of the original in-situ stress of this mine. The horizontal stress in the X and Y direction was approximately confirmed as 14.56 MPa and 13.26 MPa, respectively in the numerical models by comparing calculated values with measured values. The dimension of the 8# coal seam is 1 m × 1 m × 1 m, extraction of the coalbed was modeled by removing elements over the height of the coalbed by 1 element height, 2 element heights, 3 element heights, respectively.

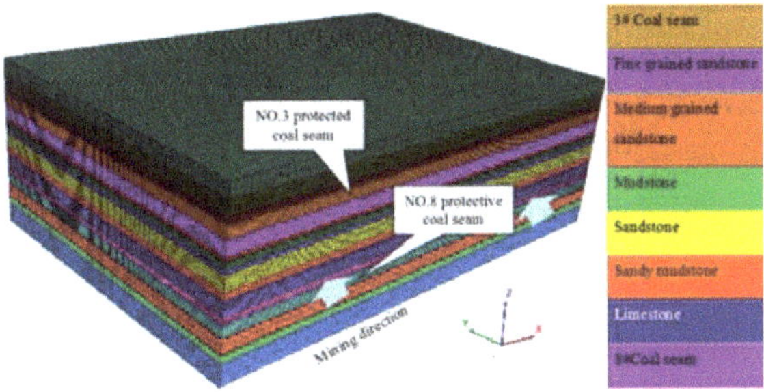

Figure 6. Flac^{3D} model.

There are 2 monitoring lines that are totally installed in the model: (1 m above the No. 8 coal seam along mining direction), (2 m above No. 3 coal seam floor along mining direction).

Rock mass engineering properties for different lithologies were described in Table 2. These values were calculated through estimated the geological strength index (GSI) rock mass system and

Hoek-Brown failure criterion [25]. (Equations (1)–(7)). The generalized Hoek-Brown criterion is expressed as:

$$\sigma_1' = \sigma_3' + \sigma_{ci}\left(m_b\frac{\sigma_3'}{\sigma_{ci}} + s\right)^a \tag{1}$$

where σ_1' and σ_3' are the major and minor effective principle stresses at failure, σ_{ci} is the uniaxial compressive strength (UCS) of the intact rock material, m_b, s and a are material constants. where m_b is a reduced value of the material constant m_i and is given by

$$m_b = m_i\exp\left(\frac{GSI - 100}{28 - 14D}\right) \tag{2}$$

s and a are constants for the rock mass, and we have the following relationships:

$$s = \exp\left(\frac{GSI - 100}{9 - 3D}\right) \quad a = \frac{1}{2} + \left(e^{-GSI/15} - e^{-20/3}\right) \tag{3}$$

where D is a factor depending upon the degree of disturbance to which the rock mass has been subjected by blast damage and stress relaxation. It varies from 0 for undisturbed in situ rock mass to 1 for very disturbed rock masses. $D = 0$ is assumed in this paper.

The estimation of rock masses strength are as follows:

The UCS is:

$$\sigma_c = \sigma_{ci}s^a \text{ when } \sigma_3' = 0 \tag{4}$$

The tensile strength is:

$$\sigma_t = -\frac{s\sigma_{ci}}{m_b} \text{ by setting } \sigma_1' = \sigma_3' = \sigma_t \tag{5}$$

The rock mass modulus of deformation (GPa) is:

$$E_m = \left(1 - \frac{D}{2}\right)\sqrt{\frac{\sigma_{ci}}{100}}\cdot 10^{((GSI-10)/40)} \text{ for } \sigma_{ci} \leq 100 \text{ MPa} \tag{6}$$

$$E_m = \left(1 - \frac{D}{2}\right)\cdot 10^{((GSI-10)/40)} \text{ for } \sigma_{ci} > 100 \text{ MPa} \tag{7}$$

Table 2. Rock mass properties, *GSI* and Hoek-Brown parameters used in the numerical modeling.

Lithology	ν	σ_{ci}/MPa	GSI	m_i	m_b	s	a	E_m/MPa
3# Coal seam	0.33	6	75	11	4.504	0.0622	0.501	2938.86
Fine grained sandstone	0.19	90	90	16	11.195	0.3292	0.500	6566.53
Medium grained sandstone	0.20	73	88	15	9.772	0.2636	0.500	5020.52
Mudstone	0.28	16	80	12	5.874	0.1084	0.501	3433.36
Sandstone	0.24	40	86	13	7.885	0.2111	0.500	4203.05
Sandy mudstone	0.26	35	85	13	7.084	0.1512	0.500	4095.07
Limestone	0.19	75	90	10	6.997	0.3292	0.500	9682.03
8# Coal seam	0.29	6.4	75	11	4.504	0.0622	0.201	3020.50

4.3. Gob Behaviors in the Model

Numerical modeling of gob behavior in longwall mining is a challenge. In order to make a better understanding of the stress recover distance in the gob, some researchers use double yield model to simulate the gob behavior [18–20], H. Yavuz et al. [20] proposes a methodology for estimating the re-establishment distance of the cover pressure and stress distribution in the gob by analyzing a large quantity of field data from British longwall coal mine over several years. He introduced a three-parameter power function, in which the independent variables are excavation height, mining depth, bulking factor and compressive strength of the rock fragments. The cover pressure re-establishment distance in the gob is estimated via Equation (8). His conclusion was finally verified with curves obtained through numerical models. As the bulking factor of the rock pile increases with mining height increasing, hence this results in increase in cover pressure re-establishment distance. According to the mining condition of the case coal mine, the parameters obtained for the case mine are given as: b = 1.2–1.4, the bulking factor of the caved roof; σ_c = 30 MPa, the compressive strength of the rock pieces; H = 480–530 m, the mining depth; h = 1–3 m, coal thickness; γ = 0.025 MN/m^3, the unite weight of overburden. As a result, the cover pressure re-establishment distance for the example working face were found to be 120 m–140 m from Equation (8).

$$X_{cd} = 0.2H^{0.96}\gamma Hb^{8.7} / [10.39\sigma_c^{1.042}(b-1)+\gamma Hb^{8.7}] \tag{8}$$

In this paper, gob behaviors were regarded as linear recovery, two gob behaviors (cover pressure re-establishment distance for Gob 1 and Gob 2 at 120 m and 140 m respectively) were incorporated in the 4 Flac^{3D} numerical models with different mining height (Table 3).

Table 3. Parameters of each Flac^{3D} model.

Model Parameters	Cover Pressure Re-Establishment Distance/m	Mining Height/m
Model 1	Gob1-120	1
Model 2	Gob1-120	2
Model 3	Gob1-120	3
Model 4	Gob2-140	3

5. Results and Discussion

5.1. Stress Distribution around the Protective Coal Seam

5.1.1. Stress Distribution in the Gob

The lower roof will move downward and collapse after the coal seam is extracted. The weight of the upper strata will be supported by both sides of the panel as the lower strata fall into the extracted space. The stress in the gob area will be redistributed due to coal seam extraction (Figure 7), as the overburden strata broken condition likes "O" shape, Lin et al. [26] and Qian et al. [27] characterized it as "Ring" circle and "O" circle, respectively. The pressure arch will develop across the solid coal and the destressed zone will be formed above the gob. As the longwall face retreats and the caving process continues, the caved material comes into contact with the roof and takes load from the upper strata due to the combined influence of floor heave, roof sag, and waste rock bulking. And then the stress in the gob will make a further redistribution.

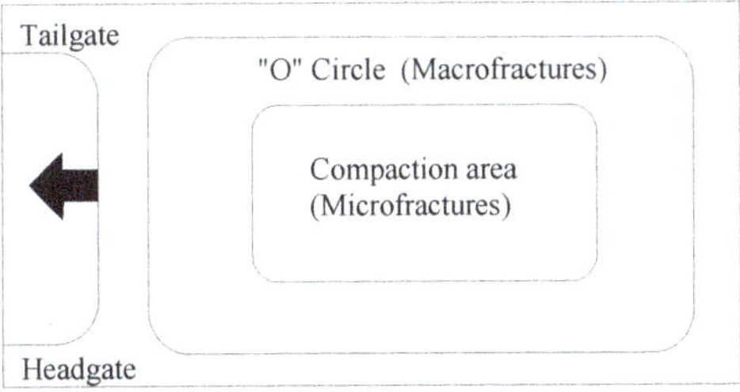

Figure 7. The plane schematic diagram of goaf mining fissure.

Figure 8 shows the vertical stress distribution in the roof of gob area after the excavation of No. 8 protective coal seam (results of model 3). It can be seen from the figure that the vertical stress is small in the area close to the face and pillar, but the stress values increase when moving toward the center of gob. The results are consistent with the "O" circle theory mentioned above. This is because there is less support near the face and pillar area, the roof therefore gets stress relief. These areas typically make up the main gas flow fissure in protective coal seam.

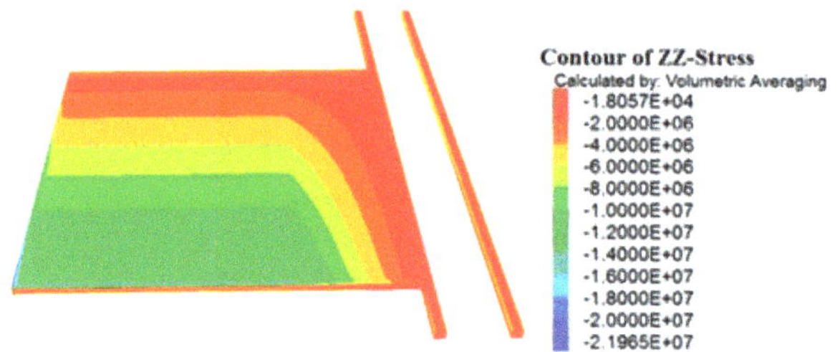

Figure 8. Vertical stress distribution in the roof of gob area.

5.1.2. The Zones in the Horizontal Direction of Protective Coal Seam

Figure 9 shows the stress and displacement distribution in the immediate roof of protective coal seam after 130 m of face advance (results from Model 3), the face position is at 0 m in the horizontal coordinate, as can be seen from Figure 9, and the overburden can be divided into 5 zones according to the stress and displacement change: The original zone, compression zone, expansion zone, recovering zone and re-compacted zone [9].

The original zone was 40 m beyond the coal face, this area was far away from the coal face, the stress was almost initial stress and the displacement decreased to 0, very little deformation occurred, the pr-existing fracture was isolated, showing that the mining activity had little effect on this area, the gas extraction therefore is difficult in this area.

The compression zone was from 5 m to 40 m ahead of the coal face, it can also be called the abutment stress area, the pre-existing cracks in the coal and rock mass of this area was closed due to the high abutment stress, new cracks developed and expanded. The cracks were gradually connected to the network especially from the 5 m to the peak abutment stress and the gas emission and transportation fissure was gradually formed.

The expansion zone extended from 5 m ahead of the coal face to −60 m behind the coal face, including the stress relief area ahead of the coal face and part of the gob area. As plastic failure occurred in the stress relief area of the coal, it reduced the ability to support the overburden. In the working face area, the overburden was broken due to the shield supporting, and more and more new cracks developed and connected in this area. The roof displacement increased gradually from coal face to gob, which means the expansion of the rock mass. The cracks in this area were open and connected into the network, which made the rock mass permeability increasing and the methane much easier to be extracted. This area was the best place for gas drainage of the protected coal seam.

The recovering zone extended from −60 m to −120 m within the gob. The vertical stress gradually increased with the increasing distance from the working face, and the rock mass in the gob was gradually compacted. The gob material could take some load from overburden, which resulted in the gradual closure of cracks in the roof and gob area. The rock mass permeability and the gas extraction quantity in this area decreased slowly.

The re-compacted zone was located at the back of the gob starting from −70 m behind the coal face. The vertical stress gradually recovered to its original stress. At the same time, the rock mass deformation, roof displacement and permeability became stable in this area. The expansion cracks shrunk and most of the cracks were closed. Gas extraction in this area was therefore very difficult.

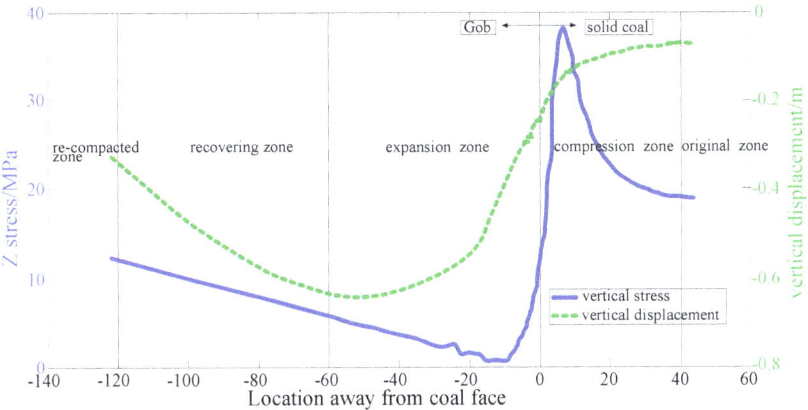

Figure 9. Five zones above the protective coal seam along mining direction.

5.2. Factors Affecting Pressure Relief and Deformation of Protected Coal Seam

5.2.1. Stress Relief Coefficient and Relief Angle of Protected Coal Seam

After the extraction of the protective coal seam, the stress around the opening will redistribute. Figure 10 shows that the stress added to the surrounding rib of the gob and the working face is concentrated at both ends of protective coal seam. At the same time, the roof and floor at the back of the gob were relieved. This is due to the influence of the abutment pressure of two-way rib of the gob, which lead to a certain pressure relief of the protected coal seam. The extent of the pressure relief can be described with pressure relief coefficient [28], which is defined as Equation (9).

$$k = \frac{\sigma_0 - \sigma}{\sigma_0} \tag{9}$$

where k is stress relief coefficient, σ_0 is the initial stress of the coal seam (MPa), σ is the stress in the protected coal seam after extraction of the protective coal seam (MPa). When $k > 0$, $\sigma_0 > \sigma$, it means pressure relief occurred in the protected coal seam; when $k < 0$, $\sigma_0 < \sigma$, it means pressure concentration occurred in the protected coal seam; when $k = 0$, $\sigma_0 = \sigma$, it means there is no stress change in the protected coal seam.

Figure 10. Contour of vertical stress along incline in the gob for 120 m face advance.

The protection criteria taking mining stresses into account is adopted in our study [29]. The relief angle can be then evaluated by the following critical stress relief value, as given in Equation (10).

$$|\sigma_{zc}| \leq \left(cos^2\alpha + \lambda sin^2\alpha\right)\gamma H \tag{10}$$

where σ_{zc} is the vertical stress when the coal seam angle is $0°$. σ_{zc} is the Z-stress. α is the coal seam angle. λ is the lateral pressure coefficient. γ is the bulk density and H is the initial depth of the outburst.

According to the conditions of Jincheng coal mine, the followings are taken: $\gamma = 25$ kN/m^3, $H = 470$ m, $\alpha = 0$, then $|\sigma_{zc}| \leq 2500 \times 470 = 11.7$ MPa. Therefore, when the stress decreases to 11.7 MPa, a gas outburst will not occur. Therefore, 11.7 MPa is taken as the critical value, the cross point of the critical value line and the stress curves are critical points. By substitution of value 11.7 MPa into Equation (9), the critical value of stress relief coefficient (0.1) could be obtained.

5.2.2. Abutment Stress Changes for Different Gob Behaviors and Mining Height in Protective Coal Seam

Figure 11 shows the vertical stress concentration factor (or VSCF) in the immediate roof of 84306 working face with different gob behaviors after 120 m face advance at the mining height of 3 m. The coal face is at 0 point on the horizontal coordinate. As can be seen from Figure 11, the vertical stress concentration factor varies in the gob and ahead of the coal face. In the back of the coal face, the value gradually decreases from 1 at the coal face to almost 0 at approximately 10 m back of the coal face, then it gradually increases as it is far away from the coal face. While in front of the coal face, the VSCF gradually increases until it reaches its peak value at around 7 m ahead of the coal face. Comparing the cures of Gob 1 with Gob 2, the VSCF value is almost the same from 7 m behind the coal face to 4m ahead of the coal face, and this means the length of yield zones in ahead of the coal face is the same, while the peak value of the VSCF in ahead of coal face and affecting range is different, the affecting range of Gob 2 is larger than Gob 1, and the peak value of VSCF of gob 2 is 3.1 compared with 2.8 for Gob 1 model. This is because the gob 2 has larger recovery distance to pre-mining stress value and the coal face need to take more overlying strata load than gob 1.

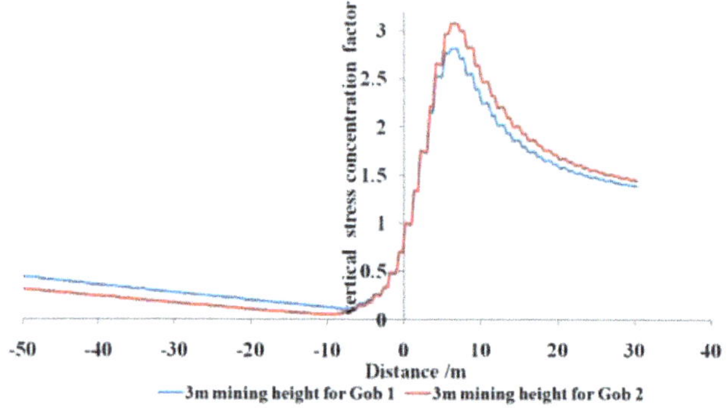

Figure 11. Vertical stress concentration factor in the face area of protected coal seam.

Figure 12 shows the VSCF curves with gob 1 for different mining height of 1 m, 2 m, and 3 m, respectively. It shows that the VSCF value is equal at the back of the coal face for different mining heights with the same gob behaviors. While the peak value of VSCF ahead of the coal face and their affecting range is different with mining height changing. The peak values of VSCF for 1 m, 2 m, and 3 m mining height are 3 m, 2.8 m and 2.7 m, respectively; the peak value of VSCF is located in 4 m, 6 m, 8 m respectively ahead of the coal face. And the influence area of abutment stress increases with the mining height. According to the above analysis, a negative relationship is found between seam height and the VSCF values ahead of the coal face. This is because the stiffness of the coal seam decreases with increasing coal seam height. The coal seam deforms more both laterally and vertically. Larger deformation on the face or even face falls in the return, distresses the solid coal and decreases the VSCF values. On the other hand, thinner seams can sustain more load because of higher stiffness.

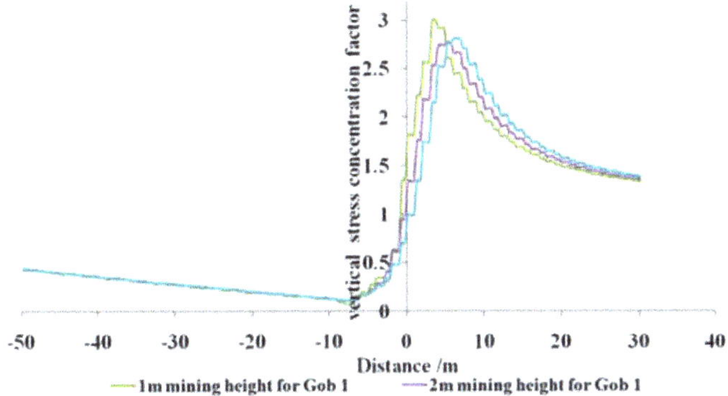

Figure 12. Vertical stress concentration factor with different mining height for Gob 1.

5.2.3. The Effect of Gob Behaviors to Protected Coal Seam

Figure 13 shows the pressure relief coefficient of the protected coal seam with different gob behaviors for different face advance. The starting line lies 20 m of the horizontal coordinate. As shown in Figure 13, overall, the pressure relief coefficient of the protected coal seam has same changing trends under the condition of two different gob behaviors in the protective coal seam, stress concentration occurs both ahead of the coal face and behind the set-up room. The pressure relief changes regularly during the mining process. Taking the gob 1 for example, it increases gradually at the beginning of the mining process. It then reaches the peak value at approximately 0.3 when the protective coal seam is mined 50 m; and later, the peak value gradually decreases and levels off at approximately 0.26. While it also shows some difference in terms of magnitude and scope for different face advance with different gob behaviors. It does not show the difference until the face advance is larger than 50 m, this shows these two kinds of gob behaviors have the same effect to the protected coal seam when the protective coal seam is mined within 50 m. When the protective coal seam is mined 100 m, the magnitude and scope of protected coal seam begins to show difference with gob 1 and gob 2. The peak value of pressure relief coefficient of gob 1 is about 0.28, which is approximately 34 m behind the coal face, and the pressure relief scope is about 74 m, while the corresponding values of gob 2 are 0.33 m, 37 m and 79 m, respectively, increasing by 18%, 8.8% and 6.8%, respectively.

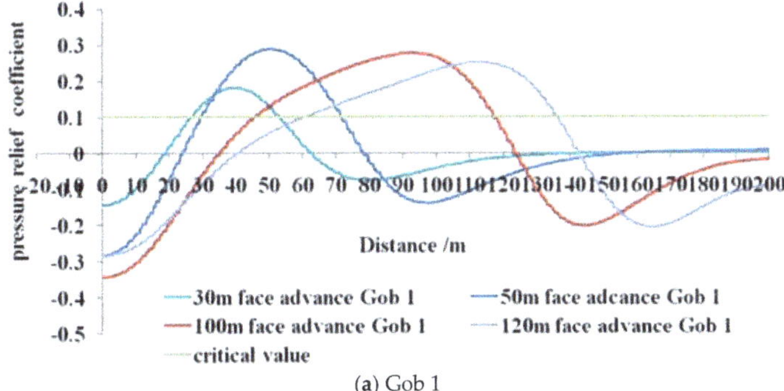

(a) Gob 1

Figure 13. *Cont.*

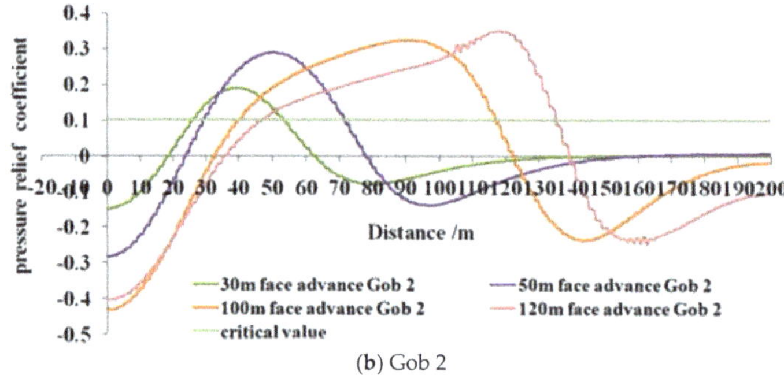

(b) Gob 2

Figure 13. Pressure relief coefficient of protected coal seam with different gob behaviors for different face advance.

The relief angle in Figure 2, is one of the major parameters used to describe the relief effect after mining a protective coal seam. Relief angles α and β can be expressed as $\alpha = arctan(b_1/a_1)$ and $\beta = arctan(b_1/a_1)$ [1]. The relief area of the protected coal seam presents the range where the coal and gas outburst danger is eliminated. A greater the relief angle corresponds to a larger relief area. As stress distribution in the gob is different, it will result in a different relief angle. Table 4 shows the relief angles change along strike with different gob behaviors for different face advance. Overall, the relief angles in the coal face are larger than those in the set-up room side. this is because the broken rock mass in the gob can pick up some load from overlying strata while there is only a shield support in the face area which can only take a little load, and the pressure of overlying strata above the face area transfers to the solid coal, which result in larger pressure relief above face area than gob side. In the starting line side, the relief angles gradually decrease during the mining process, decreasing from 82° for 30 m face advance to 43.1° (gob 1) and 55.2° (gob 2) for 120 m face advance, and the relief angle in the set up room side become different for gob 1 and gob 2 after the coal seam is mined 100 m, gob 2 has larger relief angle than gob 1 in gob side, this is because gob 2 has larger stress recover distance than gob 1, the relief angle for gob 1 and gob 2 is 55.2° and 61.0°, respectively, when the coal seam is mined 100 m. In the coal face side, the relief angles experience the increase, decrease, and stabilization process; it decreases from 82° for 30 m face advance to 76° for 100 m face advance and stabilizes at approximately 76°. The relief angle is almost the same in the face area for different gob behaviors; it means the gob behaviors have little effect on the pressure relief of the face area.

Table 4. Relief angles change along strike with different gob behaviors for different face advance.

Face Advance	30 m		50 m		100 m		120 m	
Gob	1	2	1	2	1	2	1	2
Set up room/°	82.0	82.0	74.5	74.5	55.2	61.0	41.3	55.2
Coal face/°	82.0	82.0	80.6	80.6	76.0	76.0	76.0	76.0

5.2.4. The Effect of Mining Height to Protected Coal Seam

Figure 14 shows the pressure relief coefficient of protected coal seam with different mining heights as a function of face advance. It is the results of Models 1 and 3. The peak value of pressure relief coefficient is 0.12 and 0.14 respectively for 1 m mining height and 3 m when the coal seam is mined 30 m from the start position. The value of 3 m mining height increases by 17% than that of 1 m mining height, and both of the peak values are at 17 m behind the coal face. The peak values are 0.21 and 0.28 respectively for 1 m and 3 m mining height when the coal seam is mined 50 m, the value of 3 m mining height increases by 33%, while these values become 0.23%, 0.24%, and 4.3% respectively when the coal seam is mined 70 m, and the peak values are equal for 1 m and 3 m mining height when the coal face in mined 120 m. According to the above data, we can draw the following conclusion: the mining height of protective coal seam mainly has an effect to the protected coal seam at the beginning of the mining process, the effect to protected coal seam decreases and even becomes very little when the coal seam is mined a certain distance.

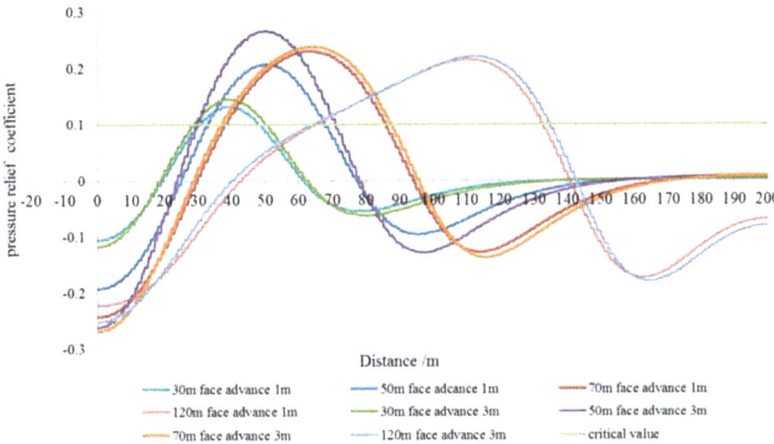

Figure 14. Pressure relief coefficient of protected coal seam with different mining height for different face advance.

In order to further quantitatively determine the regularities of mining height of protective coal seam to the stress relief and range of protected coal seam, the relief angles along strike with different mining heights for different face advances are obtained by Model 1 and Model 3 (Table 3). As can be seen from Table 5, the relief angles decrease with face advance. In the set-up room side, the relief angles decrease from 82° for 30 m face advance to 41.3° for 120 m face advance when the mining height is 3 m, and the relief angles are a little larger compared with 1 m mining height at the same condition. The difference becomes less and less with further face advance. The relief angles of two different mining heights are the same when the coal seam is mined 120 m; while in the coal face side, the relief angle of 3 m mining height is always larger than 1 m mining height with face advancing.

Table 5. Relief angles along strike with different mining height for different face advance.

Face Advance	30 m		50 m		70 m		120 m	
Ming height/m	3	1	3	1	3	1	3	1
Set-up room/°	82.0	74.5	74.5	71.6	64.7	63.5	41.3	41.3
Coal face/°	82.0	77.5	80.6	76.0	77.5	74.5	76.0	68.1

5.2.5. Pressure Relief Scope in Protected Coal Seam

As the lower protective coal seam (No. 8 seam) mining will generate stress relief in low permeability protected coal seams (No. 3 seam), the adsorbed gas will desorb and flow into the fracture network as the effective stress decreases and new fracture system growth occurs. The cross-measure boreholes are constructed as an artificial channel because the desorbed gas cannot flow out of the stress relief coal seams independently. A good understanding of stress-relief area with protective coal seam mining therefore is the key point for high efficiency gas drainage. The simulation results are capable of helping us identify the stress relief area of target coal seam prior to mining operation. According to the results of the four Flac^{3D} models, the pressure relief scope in protected coal seam can be calculated (Table 6), both the pressure relief scope and location in protected coal seam changes with the excavation of protective coal seam, so the boreholes can be dynamically connected and disconnected from the gas drainage system to get high-concentration gas.

Table 6. Pressure relief scope in protected coal seam.

Face Advance	30 m		50 m		70 m		100 m		120 m	
Ming height/m	3	1	3	1	3	1	3	1	3	1
Gob 1/m	27	19	41	36	52	49	73	–	77	72
Gob 2/m	27	–	41	–	–	–	78	–	93	–

5.3. Gas Drainage Program

A field study was carried in protected longwall face 4306 and protective longwall face 84306 for the drainage of desorbed gas and elimination of coal and gas outburst risks. 4036 working face has three roadways (Figure 15): 43061, 43062 and 43063 (original 43042). As 43063 is the original roadway of 4304 working face, 43061 and the 43042 roadway still need to be excavated. According to the field data, 4306 working face has a methane content of 7.66–13.25 m^3/t. The methane emission content during the excavation of 43061 and 43062 roadways is 2.25–6.32 m^3/min and 2.35–6.34 m^3/min, respectively. The prediction of methane emission content in 4306 face area is about 21.2–37.38 m^3/min. A combination gas drainage program of gas drainage while driving the roadway, drilling boreholes along coal seam and gas drainage by floor Tunnel 2 after stress-relief are used to eliminate outburst risk in 4306 working face. The roadway layout and boreholes parameters are shown in Figure 15.

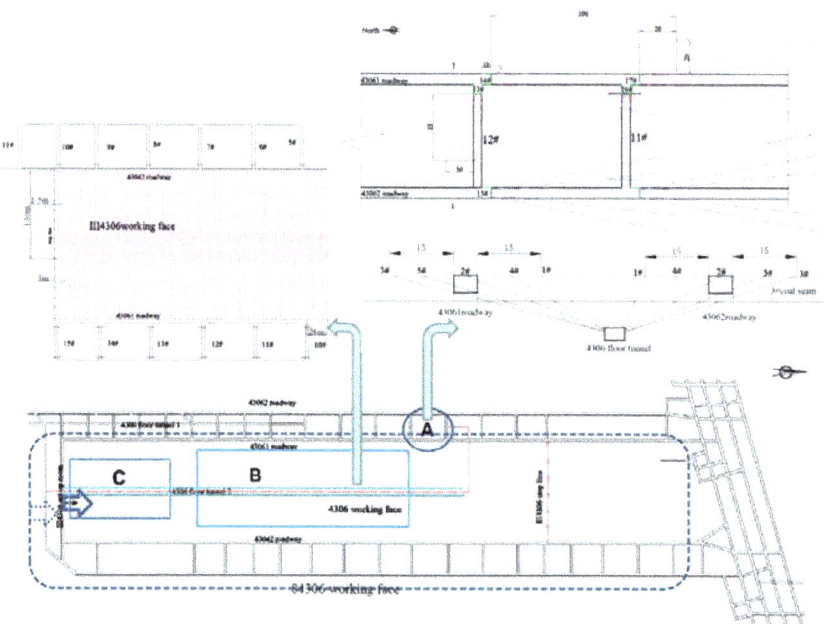

Figure 15. Illustration of gas drainage during and after roadway excavation.

5.3.1. Gas Drainage While Driving the Roadway and Drilling Boreholes along Coal Seam

Areas A and B in the Figure 15 represent the process of gas drainage while driving the roadway and drilling boreholes along the coal seam, respectively. The construction of floor Tunnel 1 and cross-measure boreholes was completed prior to gas drainage. The drilling of the borehole can increase plasticity of coal mass, reduce outburst prevention indexes effectively, and make sure the roadway tunnel safely and effectively.

The results of gas drainage by floor Tunnel 1 are shown in Table 7. It is proven through field application that draining methane while driving cannot only effectively prevent gas and coal outburst from occurring and reduce methane emission in the roadway, but also greatly reduce the frequency to adopt technical measures against outburst and raise the advancing speed.

Table 7. The results after gas drainage by floor tunnel 1.

No.	Parameters		Before Gas Drainage	After Gas Drainage
1	Average gas content/m^3/t		12.41	7.68
2	Maximum gas emission during roadway excavation/m^3/min		7.5	3.48
3	Gas outburst prevention index K1	Average value	0.4–0.42	0.32–0.33
		Exceed or not	Y	N
4	Roadway excavation speed	Excavation footage/m	2.4	5–7
		Weather gas affect roadway excavation or not	Y	N

5.3.2. Gas Drainage by Floor Tunnel 2 after Stress-Relief

As the high initial stress, low permeability and low hardness condition of protected coal seam, the gas drainage by borehole along coal seam cannot meet the safety mining requirements of protected coal seam. Further gas drainage must be done before excavating the protected coal seam. The floor tunnel is therefore driven for further gas drainage during the excavation of the lower protective coal seam (longwall face 84306). The floor tunnel is located in the K7 fine sandstone of No. 3 coal seam floor at a depth of 6 m. The cross-measure boreholes are drilled through the full thickness of the No. 3 coal seam. The horizontal and vertical cross-sectional views of the protected longwall face are shown in Figure 16. The spacing of adjacent holes and rows is 10 m × 5 m in length and width. Each borehole is attached to the gas drainage system immediately after drilling. The range of boreholes is 140 m × 900 m in width and length. There are a total of 18 gas drainage units in 4306 longwall face, every unit has 10 groups of boreholes, the quantity of boreholes is 15 and 14 respectively for single array and double array (Figure 16a). The borehole parameters of 4306 floor tunnel 2 are listed in Table 8.

According to the results of Table 6, for different mining height and face advance of lower protective coal seam, the pressure relief scope in protected coal seam is different. As such, the results of Table 6 can be used to guide the gas drainage scope along the strike. This will highly improve gas drainage efficiency and reduce unnecessary work.

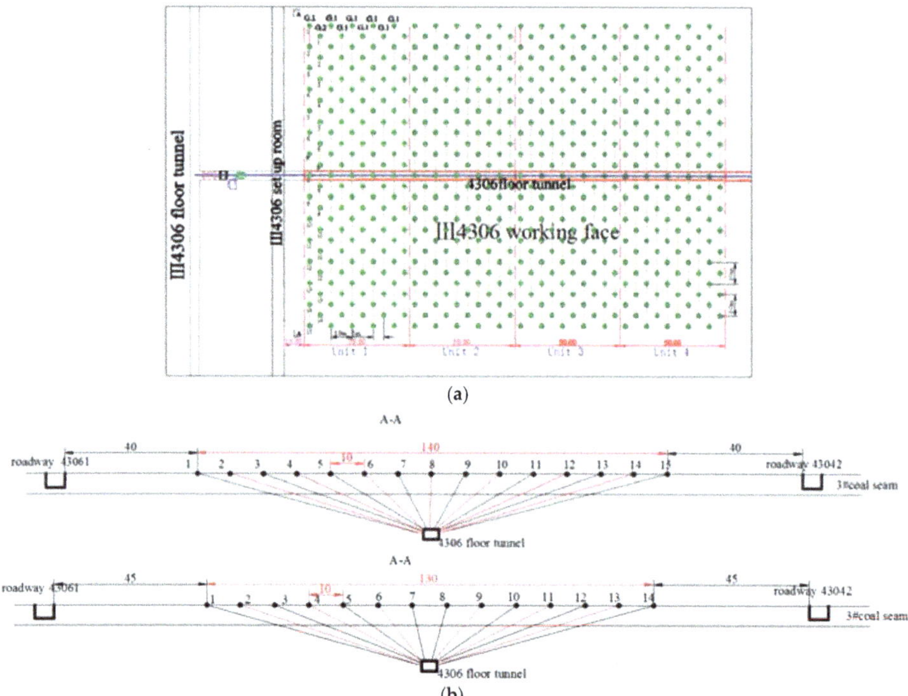

Figure 16. Diagram of the spatial configuration of the protected longwall and stress-relief gas drainage. (**a**) Horizontal plane view; (**b**) vertical section view.

Table 8. The borehole parameters of 4306 floor tunnel 2.

Singular Group				Even Group			
No.	Dip/i	Azimuth/z	Depth/m	No.	Dip/i	Azimuth/z	Depth/m
1	14	270	70.5	1	15	270	65.5
2	16	270	60.5	2	18	270	56
3	19	270	51	3	21	270	46.5
4	23	270	41.5	4	26	270	37
5	29	270	32.5	5	34	270	28
6	41	270	24.5	6	51	270	20.5
7	61	270	18.5	7	56	200	19.5
8	57	177	19	8	55	156	19.5
9	59	90	18.5	9	48	90	21.5
10	40	90	25	10	34	90	28.5
11	29	90	32.5	11	26	90	37
12	23	90	41.5	12	20	90	46.5
13	18	90	51	13	17	90	56
14	16	90	60.5	14	15	90	65.5
15	14	90	70.5				

6. Conclusions

The mining of protective coal seam is wildly used in the multiple coal seams in China to eliminate or at least mitigate the coal and gas outburst. In this study, a 3D numerical model has been established with a consideration of the gob loading characteristics, to make a better understanding of the mining influence of lower protective coal seam to the protected coal seam. The gob characteristics are developed by H. Yavuz's method and calibrated by the realistic VSCF value ahead of coal face. The model can be easily changed for specific coal mine gob loading behaviors based on field measurements and has the potential to realistically model longwall mining. The stress recovering distance in the gob and mining height of protected coal seam were selected as the important variables to assess their influence on the stress relief of the protected coal seam. The stress relief coefficient and relief angle were introduced as two important parameters to evaluate the stress relief effect in different regions of the protected coal seam. Finally, the reasonable gas drainage programs were provided. The main results of this study are summarized below.

(1) "O" circle was found in the gob after the mining of protective coal seam. The area closes to pillar and coal face made up of the main macro-fractures for gas flow.

(2) The overburden of protective coal seam can be divided into 5 zones according to the distribution of stress and displacement in the mining horizon: the original zone, compression zone, expansion zone, recovering zone, and re-compacted zone; the expansion zone had largest permeability compared with other areas and was the best place for gas drainage.

(3) The peak value of VSCF ahead of coal face increased with the increase of the stress recover distance; it decreased with mining height.

(4) For the same mining height with different gob behaviors of protective coal seam, the relief angles experienced the increase, decrease and stabilization process with face advancing in the coal face side; the relief angle was almost the same in the face area for different gob behaviors, indicating that it has little effect to the pressure relief of the face area.

(5) For the same gob behaviors with different mining height of protective coal seam, the mining height of protective coal seam mainly has an effect to the protected coal seam at the beginning of the mining process. This effect decreases and becomes very little when the coal seam is mined a certain distance. And the relief angle of 3 m mining height is always larger than 1m mining height with face advancing on the face side.

(6) Pressure relief scope in protected coal seam was acquired under specific mining condition. A combination gas drainage programs were provided for the case working face. The gas content reduced from 12.41 m3/t to 7.68 m3/t in 15 m range of roadway. It effectively prevented gas and coal outburst from occurring and reduced methane emission in roadway, and raised the advancing speed at the same time. And the research results can be used to guide the scope of gas drainage and highly improve gas drainage efficiency and reduce unnecessary work. More field data need to be collected for further research.

Author Contributions: C.Z. and Y.Z. conceived the main idea of the paper and designed the numerical model; C.Z. performed the numerical model; C.Z. and L.Y. analyzed the data; Y.Z. and R.F. contributed analysis tools and theoretical analysis; C.Z. wrote the paper; R.F. and G.Z. did a lot work to modify figures and proofread the revised version.

Funding: This paper was supported by the State Key Research Development Program of China (Grant No. 2016YFC0600708), CUMTB fund for creative PHD Student Project (00-800015z685), Tiandi Technology Co., Ltd. Technology Innovation Fund, China Scholarship Council (CSC), Zhongguancun Ten Hundred Thousand Special Project-7m High Mining Support Technology and Equipment (KJ-2017-GJGC-01). The authors would like to cordially acknowledge excellent guidance of Y.P. Chugh in Southern Illinois University Carbondale and G. Song in North China University of technology. The authors would also like to express special thanks to the editor and anonymous reviewers for their professional and constructive suggestion. In addition, C.Z. wants to thank, in particular, the patience, support and love from Y.Z. over the past years.

Conflicts of Interest: The authors declare no conflict of interest.

References

1. Yang, W.; Lin, B.Q.; Qu, Y.A.; Zhao, S.; Zhai, C.; Jia, L.L.; Zhao, W.Q. Stress evolution with time and space during mining of a coal seam. *Int. J. Rock Mech. Min. Sci.* **2011**, *48*, 1145–1152. [CrossRef]

2.	Li, W.; Cheng, Y.P.; Guo, P.K.; An, F.H.; Chen, M.Y. The evolution of permeability and gas composition during remote protective longwall mining and stress-relief gas drainage: A case study of the underground Haishiwan Coal Mine. *Geosci. J.* **2014**, *18*, 427–437. [CrossRef]
3.	State Administration of Coal Mine Safety. Available online: http://english.gov.cn/state_council/2014/10/06/content_281474992926692.htm (accessed on 6 October 2014).
4.	Esterhuizen, G.S.; Karacan, C.O. Development of Numerical Models to InvestigatePermeability Changes and Gas Emission around Longwall Mining Panel. In Proceedings of the 40th U.S. Rock Mechanics Symposium (USRMS), Anchorage, AK, USA, 25–29 June 2005.
5.	Liu, R.C.; Li, B.; Jiang, Y.J. A fractal model based on a new governing equation of fluid flow in fractures for characterizing hydraulic properties of rock fracture networks. *Comput. Geotech.* **2016**, *75*, 57–68. [CrossRef]
6.	Li, B.; Liu, R.C.; Jiang, Y.J. Influences of hydraulic gradient, surface roughness, intersecting angle, and scale effect on nonlinear flow behavior at single fracture intersections. *J. Hydrol.* **2016**, *538*, 440–453. [CrossRef]
7.	Zhang, C.L.; Zhang, Y. Stress and Fracture Evolution Based on Abutment Change in Thick Coal Seam—A Case Study in China Colliery. *Electron. J. Geotech. Eng.* **2016**, *21*, 4369–4386.
8.	Lowndes, I.S.; Reddish, D.J.; Ren, T.X.; Whittles, D.N.; Hargreaves, D.M. *Mine Ventilation: Proceedings of the North American/Ninth US Mine Ventilation Symposium*; CRC Press: Boca Raton, FL, USA, 2002; pp. 267–272.
9.	Yang, W.; Lin, B.Q.; Qu, Y.A.; Zhao, S.; Zhai, C.; Jia, L.L.; Zhao, W.Q. Mechanism of strata deformation under protective seam and its application for relieved methane control. *Int. J. Coal Geol.* **2011**, *85*, 300–306. [CrossRef]
10.	Liu, Y.K.; Zhou, F.B.; Liu, L.; Liu, C.; Hu, S.Y. An experimental and numerical investigation on the deformation of overlying coal seams above double-seam extraction for controlling coal mine methane emissions. *Int. J. Coal Geol.* **2011**, *87*, 139–149.
11.	Hu, G.Z.; Wang, H.T.; Li, X.H.; Fan, X.G.; Yuan, Z.G. Numerical simulation of protection range in exploiting the upper protective layer with a bow pseudo-incline technique. *Min. Sci. Technol.* **2009**, *19*, 58–64. [CrossRef]
12.	Suchowerska, A.M.; Merifield, R.S.; Carter, J.P. Vertical stress changes in multi-seam mining under supercritical longwall panels. *Int. J. Rock Mech. Min. Sci.* **2013**, *61*, 306–320. [CrossRef]
13.	Peng, S.S.; Matsuki, K.; Su, W. 3-D Structural Analysis of Longwall Panels. In Proceedings of the 21st U.S. Symposium on Rock Mechanics (USRMS), Rolla, MO, USA, 27–30 May 1980; pp. 44–56.
14.	Whittaker, B.N. An Appraisal of Strata Control Practice: 4F, 4T, 30R. *Int. J. Rock Mech. Min. Sci. Geomech. Abstr.* **1974**, *11*, 9–24.
15.	Wilson, A.H.; Carr, F. A new approach to the design of multi-entry developments for retreat longwall Mining. In Proceedings of the Second Conference on Ground Control in Mining, Morgantown, WV, USA, 5–7 August 1982; pp. 1–21.
16.	Whittaker, B.N.; Singh, R.N. Evaluation of the design requirements and performance of gate roadways. *Min. Eng.* **1979**, *138*, 535–548.
17.	Campoli, A.A.; Timothy, B.; Dyke, F.C.V. *Gob and Gate Road Reaction to Longwall Mining in Bump-Prone Strata*; U.S. Dept. of the Interior, Bureau of Mines Bureau of Mines: Michigan, MI, USA, 1993; p. 48.
18.	Li, W.F.; Bai, J.B.; Peng, S.S.; Wang, X.Y.; Xu, Y. Numerical modeling for yield pillar design: A case study. *Rock Mech. Rock Eng.* **2015**, *48*, 305–318. [CrossRef]
19.	Esterhuizen, E.; Mark, C.; Murphy, M.M. The ground response curve, pillar loading and pillar failure in coal mines. In Proceedings of the 29th International Conference on Ground Control in Mining, Morgantown, WV, USA, 27–29 July 2010; pp. 19–27.
20.	Yavuz, H. An estimation method for cover pressure re-establishment distance and pressure distribution in the goaf of longwall coal mines. *Int. J. Rock Mech. Min. Sci.* **2004**, *41*, 193–205. [CrossRef]
21.	Esterhuizen, E.; Mark, C.; Murphy, M.M. Numerical model calibration for simulating coal pillars, gob and overburden response. In Proceedings of the 29th International Conference on Ground Control in Mining, Morgantown, WV, USA, 27–29 July 2010; pp. 46–57.
22.	Abbasi, B.; Chugh, Y.P.; Gurley, H. An Analysis of the Possible Fault Displacements Associated with a Retreating Longwall Face in Illinois. In Proceedings of the 48th US Rock Mechanics/Geomechanics Symposium, Minneapolis, MN, USA, 1–4 June 2014.
23.	Flac[3D] User's Guide. Itasca Consulting Group, Inc. Available online: http://www.civil.utah.edu/~bartlett/CVEEN6920/FLAC%20manual.pdf (accessed on 5 June 2017).

24. Jing, L. A review of techniques, advances and outstanding issues in numerical modelling for rock mechanics and rock engineering. *Int. J. Rock Mech. Min. Sci.* **2003**, *40*, 283–353. [CrossRef]
25. Hoek, E.; Carranza-Torres, C.; Corkum, B. Hoek-Brown failure criterion-2002 edition. In Proceedings of the 5th North American Rock Mechanics Symposium and the 17th Tunnelling Association of Canada Conference (NARMS-TAC), Toronto, ON, Canada, 7–10 July 2002; pp. 267–273.
26. Zhao, B.T.; Lin, B.Q. *Methane Emission and Control Technology of "Three Soft" Unstable and Low Permeability Coal Seam Mining*; China University of Mining and Technology Press: Xuzhou, China, 2007. (In Chinese)
27. Qian, M.G.; Xu, J.L. Study on the "O" shape circle distribution characteristics of mining induced fractures in the overlaying strata. *J. Chin. Coal Soc.* **1998**, *23*, 466–468. (In Chinese)
28. Yuan, Z.G.; Wang, H.T.; Hu, G.Z.; Li, X.H.; Fan, X.G.; Hong, S. Numerical simulation for protection scope of upper protective seam in steeply inclined multi-coal seam. *J. Chin. Coal Soc.* **2009**, *34*, 594–598.
29. Малышеl, Ю.Н.; Фоменко, А.Т.; Далецкий, Ю.Л. *Forecast Method and Preventing Measure of Coal and Gas Burst*; China Coal Industry Publishing House: Beijing, China, 2004. (In Chinese)

Article

Experimental Study on the Reinforcement Mechanism of Segmented Split Grouting in a Soft Filling Medium

Zhipeng Li [1,2,*], Shucai Li [2], Haojie Liu [2], Qingsong Zhang [2] and Yanan Liu [2]

[1] School of Transportation and Civil Engineering, Shandong Jiaotong University, 5001 Haitang Road, Ji'nan 250023, China

[2] Geotechnical and Structural Engineering Research Center, Shandong University, Ji'nan 250061, China; lishucai@sdu.edu.cn (S.L.); rlhaojie@163.com (H.L.); zhangqingsong@sdu.edu.cn (Q.Z.); lyn_zhm@126.com (Y.L.)

* Correspondence: lizhipengsdu@163.com; Tel.: +86-18663791467

Received: 25 July 2018; Accepted: 13 August 2018; Published: 17 August 2018

Abstract: Subsection split grouting technology can effectively improve the grouting efficiency and homogeneity of grouting in a target reinforcement area. It is therefore necessary to clarify the reinforcement mechanism and characteristics of the soft filling medium under the condition of split grouting. A three-dimensional grouting simulation test of segmented split grouting in a soft filling medium was conducted. The distribution characteristics and thicknesses of the grouting veins were obtained under the condition of segmented grouting. The mechanical mechanism of segmented split grouting reinforcement, based on the distribution characteristics of different grouting veins, was revealed. After grouting, a uniaxial compression test and an indoor permeation test were conducted. Based on the method of the region-weighted average, the corresponding permeability coefficient and the elastic modulus of each splitting-compaction region were obtained. The quantitative relationship between the mechanical properties and the impermeability of the soft filling medium before and after grouting was established. The results revealed that three different types of veins were formed as the distance from the grouting holes increased; namely, skeleton veins, cross-grid grouting veins, and parallel dispersed grouting veins. The thicknesses of the grouting veins decreased gradually, whereas the number of grouting veins increased. Moreover, the strikes of the grouting vein exhibited increased randomness. The reinforcement effect of segmental split grouting on soft filling media was mainly confirmed by the skeleton support and compaction. The elastic modulus of the grouting reinforcement solid increased on average by a factor that was greater than 100, and the permeability coefficient decreased on average by a factor that was greater than 40 in the direction of the parallel grouting vein with the most impermeable solid. The research results may be helpful in the investigation of the split grouting reinforcement mechanism under the condition of segmented grouting.

Keywords: rock-soil mechanics; soft filling medium; segmented grouting; split grouting; model experiment; reinforcement mechanism

---

1. Introduction

Grouting [1–3] is increasingly and widely being used in underground engineering, to effectively strengthen soft rock and soil. The reinforcement mode of the fault-fracture zone and the quaternary topsoil layer, which are composed of a soft filling medium [4], are usually dominated by fracturing grouting [5]. Moreover, the mechanical properties and impenetrability performance of the soil can be significantly improved by grouting [3]. In the grouting processes, split grouting technology [6] can effectively improve the grouting efficiency when the method of single drilling and multiple grouting is adopted. In addition, it can significantly improve the homogeneity of grouting in a target reinforcement area [7]. However, the grouting reinforcement mechanism and simulation test research of the segmented grouting do not meet the requirements of engineering.

At present, many scholars have conducted numerous studies on the split grouting of soft filling mediums. In a theoretical study, Zhang et al. [8] considered the effect of an asymmetrical load, and obtained the analytical solution of splitting-opening pressure. Based on the flat slit model, Li Shucai and Sun Feng et al. [9–12] established a split grouting diffusion equation for different constitutive models, and a relationship between the grouting pressure or slurry viscosity and the diffusion radius of the slurry. Based on a hypothesis that the semi-infinite space of the injected medium on both sides of the grouting vein was subjected to a uniform force, Zhang Qingsong et al. [13] considered the coupling effect between the slurry flow field and the stress field of the injected medium, and established a spatial distribution equation for the width of the splitting channel. In experimental research, Zhang Lianzhen [14] developed a two-dimensional visualization split grouting simulation test device to study the split grouting mechanism of water-rich sands. The spatial attenuation characteristics of the thickness of the split grouting vein and the approximate influence range of the split grouting were obtained. Li Peng et al. [15,16] reported the improvement of the mechanical properties of a fractured mud medium before and after split grouting. However, the focus of the study was on the splitting-opening pressure [17], the relationship between the diffusion distance [18] and pressure, and the effect of single split grouting. There are few discussions on the spatial distribution and the thickness variation of the grouting veins [19], and the reinforcement that is caused by segmental split grouting.

To obtain the reinforcement characteristics of the soft filling medium under the condition of segmented split grouting, a three-dimensional grouting simulation test system was developed, which could realize segmented split grouting. A simulation test was conducted with a typical soft filling medium of the F2 fault shale filling of the Yonglian Tunnel in the Jilian Expressway, Jiangxi Province, China The reinforcement mechanism under the condition of segmented split grouting was studied. The distribution characteristics and the thickness variation of the grouting veins were obtained. The mechanism of segmental split grouting reinforcement was revealed. A uniaxial compression test and indoor penetration test were carried out by the grouting reinforcement solid in different splitting and compacting areas, and the corresponding permeability coefficient and elastic modulus of different splitting and compacting areas were obtained. Based on the methods of the regional weighted average [20], a quantitative relationship between the mechanical properties and the impermeability of the soft filling media before and after grouting was established.

2. Three-Dimensional Grouting Simulation Test System

The segmental split grouting simulation test system mainly consisted of a simulation test frame and a grouting module. The test system could realize segmental split grouting in a soft filling medium. Moreover, it could also be used to excavate, to obtain the distribution characteristics and the thickness of the grouting veins, and sampling to analyze the grouting reinforcement effect.

2.1. The Simulation Test Frame System

In order to expediently install, disassemble, observe, and sample, a simulation test frame was designed by stratified combination. It consisted of four rings with inner diameters of 1.5 m and heights of 30 cm, as shown in Figure 1. The material of the rings was high-strength steel, and they were connected to each other by high-strength bolts.

Figure 1. Model test equipment.

2.2. Grouting Module

The grouting module included an internal preset device and an external grouting support system. As presented in Figure 2a, the internal preset device included grouting holes, a preset grouting pipe, and a pressure bearing separator. The diameter of the preset grouting pipe was 20 mm, and it was composed of a galvanized steel pipe with overflow holes arranged at intervals. The grout was injected into the ground layer through the overflow hole under the grouting pressure. As presented in Figure 2b, the pressure-bearing separator included a diffuser-bearing body and a waterproof sand bag. The waterproof sand bag was filled with sand to prevent the passage of the slurry during the grouting process.

The external grouting supporting system included a grouting pump, a grouting pipe, and a slurry mixer. To obtain medium pressure and high pressure grouting, a pneumatic grouting pump and a manual grouting pump were used for the experiment.

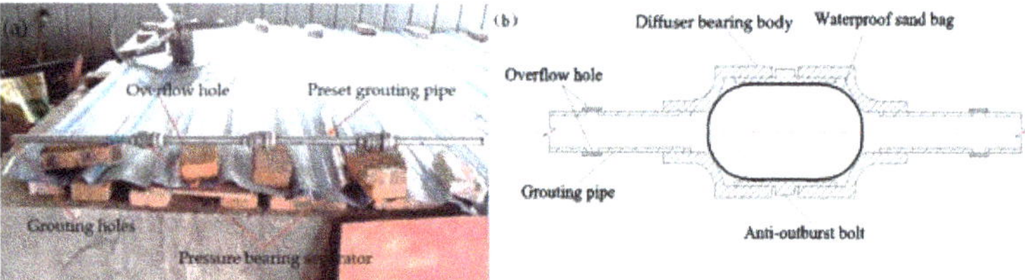

Figure 2. The internal preset device: (**a**) the physical picture of the internal preset device, (**b**) the principle of pressure bearing separator.

3. The Experiment Design for Grouting

3.1. Test Target

1. The thickness variation characteristics and the spatial distribution of the grouting veins along the direction of grout migration were obtained.
2. The mechanical parameters and permeability coefficient of the injected media before and after grouting were tested, and the quantitative relationship between the mechanical properties and permeability coefficient of the soft filling medium before and after the grouting was established.

3.2. Segmented Split Grouting Process

The segmented split grouting process was as follows. Firstly, the grouting pipe was vertically arranged in the simulation test frame, as shown in Figure 3. The grouting work was divided into three segmental grouting works by the pressure-bearing separator, and the grouting was begun from the bottom up. Then, before a segmental grouting work, the separator above the grouting pipe of this segmental grouting work was closed. When the segmental grouting work was finished and the slurry was consolidated, the waterproof sand bag in the pressure bearing separator was destroyed manually to allow the slurry to pass through. Next, the next segment grouting was carried out, so the grout could enter the injected medium from the overflow hole of the next section of the grouting pipe. Finally, three segmental grouting works were finished, the experiment was set for 14 days, and then it was excavated.

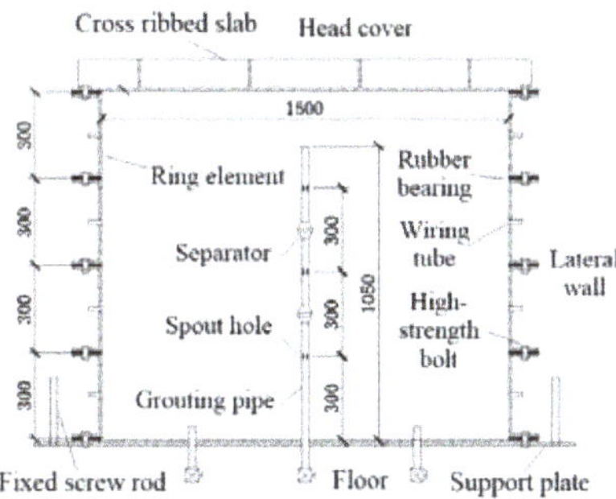

Figure 3. Internal structure of the simulated test.

3.3. Design of Filling Scheme

The fault muddy filling was used as the filling medium in this experiment, which was sourced from the F2 fault of the Yonglian Tunnel in the Jilian Expressway, Jiangxi Province, China [21]. The basic parameters are presented in Table 1.

Table 1. Basic physical parameters of the soil [21].

	Initial Water Content	Dry Density	Permeability Coefficient	Liquid Limit	Fine Cement	Porosity
Material type	ω (%)	ρ_d (g·cm^{-1})	k (cm·s^{-1})	W_l (%)	M (%)	P (%)
Soft clay	37.8	1.17	5.2×10^{-4}	45.9	36.7	31.5

First, before filling, the side walls of the model test stand were smoothed using glass glue. To simulate the real conditions of the tunnel fault, the soft filling medium was divided into eight layers into the mold, and then these were artificially compacted. Finally, when the filling was completed, glass glue was used to fill the gap between the inside and the filling of the model frame, to prevent the gap from becoming the dominant channel in the grouting.

3.4. Selection of Grouting Parameters

3.4.1. Selection of Grouting Pressure

The grouting pressure range was 0–1 MPa, and the grouting rate was 0–5 L/min. The grouting end standard were as follows: the grouting volume of each section was 20–40 kg, and the pressure of the external pressure gauge was a maximum of 1 MPa.

3.4.2. Selection of Grouting Material

The grouting material in this experiment consisted of Ordinary Portland cement (P.O 42.5) and a new special grouting material GT-1 [22,23]. GT-1 was self-developed by Shandong University, and it is composed of cement, polymer, dispersant, accelerator, early strength agent, pumping aid, and a defoamer. GT-1 has a quick setting effect and a higher strength in the early stages. In the grouting process, GT-1 was mixed with P.O 42.5 slurry at a certain volume ratio to form a new C-GT slurry, and it was injected into the stratum. In this experiment, the ratio of C-GT was 1:1. The basic parameters of the C-GT slurry are presented in Table 2.

Table 2. Parameters of the C-GT slurry.

W/C	$V_C{:}V_{GT}$	Stone Rate (%)	Gel Time (s)		Age Strength (MPa)					
			Initial Setting Time	Final Setting Time	1 h	5 h	1 Day	3 Days	14 Days	28 Days
1:1	1:1	85	45	80	0.5	1.8	3.0	7.0	8.0	8.9

3.4.3. Design of Slurry Color

To clearly determine the process of slurry migration and to compare the changes in the thicknesses of the grouting veins in different segmented split grouts, three kinds of color slurries were developed; namely, original, yellow, and red. The grouting parameters are shown in Table 3.

Table 3. Selection of grouting parameters.

Serial Number	Order	Design Grouting Rate	Design Grouting Pressure (MPa)	Slurry Color	Bottom Height of Overflow Hole (cm)
1	First	0–5 L/min	1	original	30
2	Second	0–5 L/min	1	yellow	60
3	Third	0–5 L/min	1	red	90

4. Experiment Result Analysis

The experimental results were analyzed from two aspects of distribution characteristics and thickness, as well the grouting reinforcement effect. The distribution characteristics and the thicknesses of the grouting veins of the soft filling medium were observed and measured, obtaining the distribution characteristics and the thickness variation of the grouting veins under the condition of segmented grouting. In the aspects of the grouting reinforcement effect, the soft filling medium was sampled before and after grouting, and the uniaxial compression test and the indoor penetration test were carried out to measure the mechanical properties and the permeabilities of the samples. The changes of the elastic modulus and the permeability coefficient before and after grouting were compared. The test results were as follows.

4.1. Distribution Characteristics and Thickness Analysis of Grouting Vein

Figure 4 shows the location of excavation section in the simulated test, and the excavation sections were A, B and C, respectively.

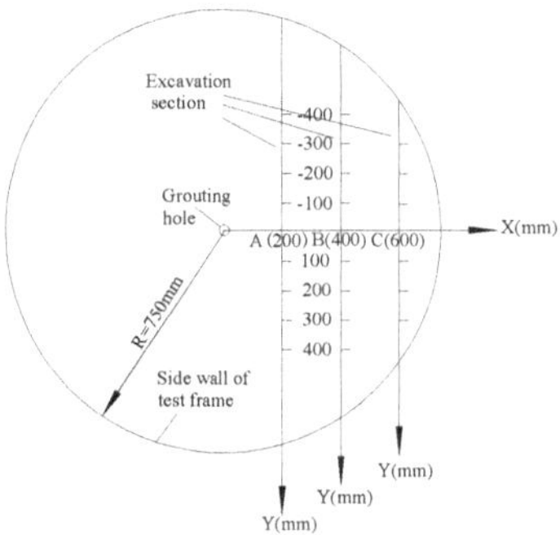

Figure 4. Location of excavation section.

As shown in Figure 4, the excavation sections were perpendicular to the bottom plate of the simulated test frame, and they were excavated from the simulated test frame to the grouting hole. The line that crossed the grouting hole and that was perpendicular to the excavation section was taken as the x-axis. The intersection line between the excavation section and the floor of the simulated test rack was taken as the y-axis. In the excavation process, the section excavation method was adopted. The excavation step length was 200 mm, and the excavation plane was smooth. The position coordinates of the grouting veins and thickness variation of the grouting veins were recorded in detail.

Figure 5 presents the veins of the whole excavation section of the three segmental grouting works.

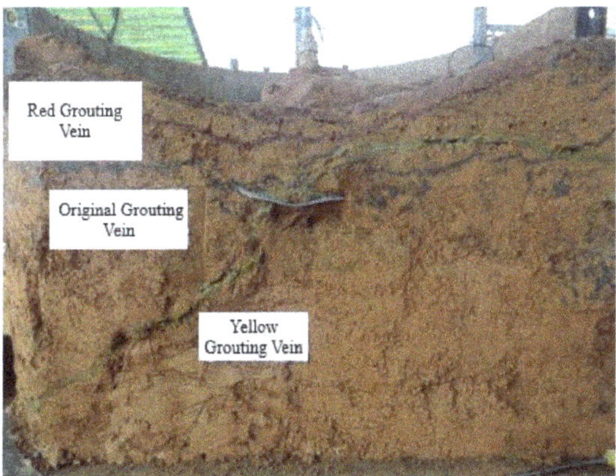

Figure 5. The veins of the whole excavation section.

With respect to the grouting vein that was exposed by the excavation, it was found that when the segmented split grouting was compared with the traditional full grouting, the segmented split grouting had a complex network of veins in the soft filling medium, and the spatial distribution of the veins was significantly different. Moreover, based on the thickness distribution and the morphological characteristics of the grouting veins, it could be divided into skeleton veins, cross-grid grouting veins, and parallel dispersed grouting veins.

4.1.1. Skeleton Vein

Figure 6 presents the skeleton veins of the three segmental grouting work. Figure 6a–c present the original skeleton vein in the first segmental grouting work, the yellow grouting vein in the second segmental grouting work, and the red grouting vein in the third segmental grouting work.

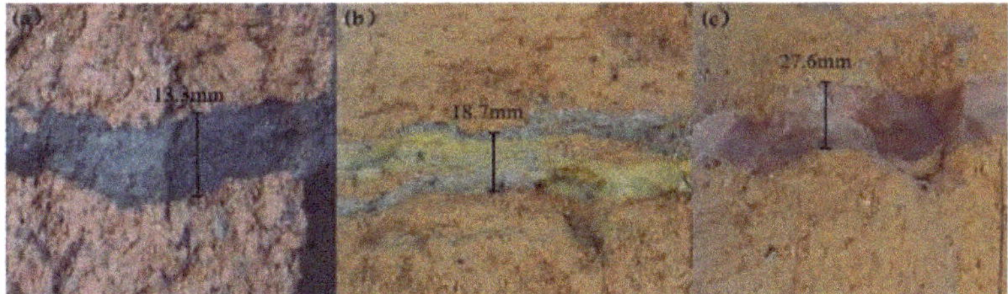

Figure 6. Framework grouting vein: (**a**) original grouting vein, (**b**) yellow grouting vein, and (**c**) red grouting vein.

Considering the grouting hole as the origin, the skeleton vein extended outward along the direction of grout migration. Moreover, the thickness of the grouting vein was larger near the grouting hole. From the excavation results, it was found that the thickness of the grouting veins of each segmental grouting work was different along the grouting pipe. The thickness of the grouting veins was larger near the grouting hole, and the injected medium compressed by the grouting vein was denser. This indicated that the supporting role of the skeleton vein was more significant when the grouting vein was closer to the location of the grouting hole.

Figure 7 presents the thickness distribution of the skeleton veins in the A (200 mm) section. Figure 7a presents the thickness distribution of the skeleton veins along the direction of grout migration. Figure 7b presents the thickness distribution proportions of the different skeleton veins.

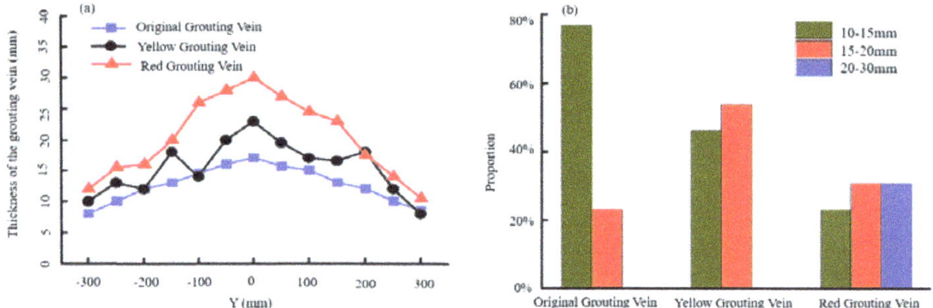

Figure 7. Thickness distribution of the grouting veins in A (200 mm) section: (**a**) thickness distribution of the grouting veins along the direction of grout migration; and (**b**) thickness distribution proportions of the different grouting veins.

As presented in Figure 7, the skeleton veins were mainly distributed near the A (200 mm) section, and the thicknesses of the skeleton veins were mainly concentrated within the range of 10–20 mm. Moreover, the skeleton veins were close to the overflow hole, and the thicknesses of the skeleton veins decreased significantly along the direction of grout migration.

4.1.2. Cross-Grid Grouting Vein

Figure 8 presents the cross-grid grouting veins after the three segmental grouting works.

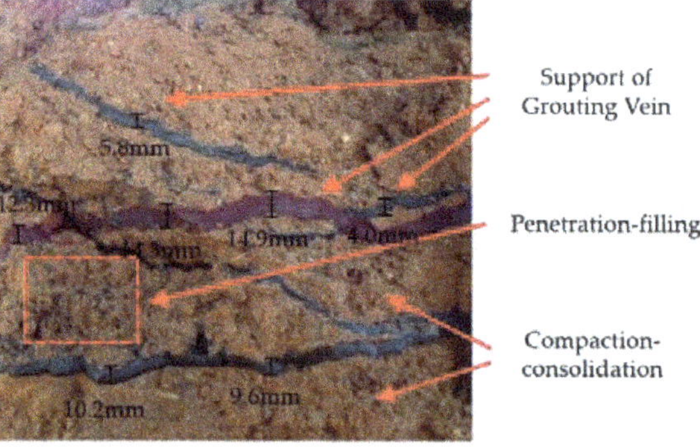

Figure 8. Cross-grid grouting vein.

As presented in Figure 8, the grouting veins of each segmental grouting work intersected each other to form a more distinct cross-grid grouting vein structure. Under the condition of subsection split grouting, the formation stress was the same as the original formation stress in the first segmental grouting work, whereas the formation stress was changed in the residual grouting segments. The change of the formation stress resulted in a change in split direction; thus, the splitting channel developed in a zigzag manner along the direction of grout migration. The grouting veins of each segmental grouting work converged and intersected in the section of the lower formation stress to form cross-grid grouting veins in the region. There were three different reinforcement areas in the area affected by the cross-grid grouting vein, namely, the grouting vein skeleton area, the penetration-filling area, and the compacted area. Therefore, the support of the grouting vein, penetration-filling, and compaction-consolidation were the three main reinforcement modes of the cross-grid grouting vein, and they were synergistic and complementary.

Figure 9 presents the thickness distribution of the grouting veins in the B (400 mm) section. Figure 9a–c present the thickness distribution of the original grouting vein, the yellow grouting vein, and the red grouting vein along the direction of grout migration, respectively. Figure 9d presents the thickness distribution proportions of the different grouting veins.

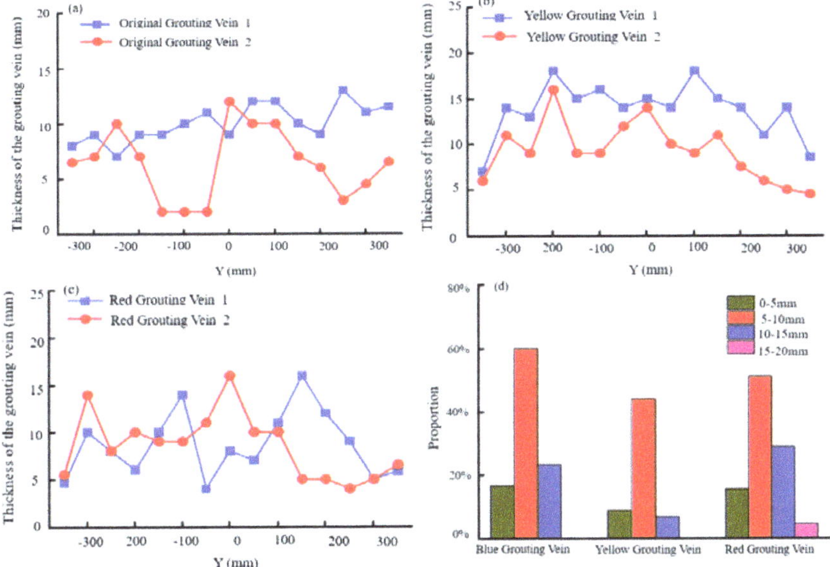

Figure 9. Thickness distribution of the grouting veins in the B (400 mm) section: (**a**) thickness distribution of the original grouting vein; (**b**) thickness distribution of the yellow grouting vein; (**c**) thickness distribution of the red grouting vein; and (**d**) thickness distribution proportions of the different grouting veins.

From the results of the excavation, the cross-grid grouting veins were concentrated near the B (400 mm) section. As presented in Figure 9, it was found that the thicknesses of the grouting veins were within the range of 5–15 mm, and an increase in the distance from the grouting hole resulted in a decrease in the thickness of the grouting veins. Due to the mutual influence of each segmental grouting work, the veins exhibited irregular variations at the local location.

4.1.3. Parallel Dispersed Grouting Vein

Figure 10 presents the parallel dispersed grouting veins. Figure 10a presents the original-red parallel dispersed grouting veins, and Figure 10b presents the blue-yellow parallel dispersed grouting veins.

Figure 10. Parallel dispersed grouting veins: (**a**) original-red parallel dispersed grouting veins and (**b**) blue-yellow parallel dispersed grouting veins.

The parallel dispersed grouting vein was mostly present at a position that was far away from the grouting hole and close to the grout frontal surface. With an increase in the distance of the grout migration from the grouting hole, the grout pressure gradually decreased, resulting in a smaller opening of the split channel. The grouting veins of this type were approximately parallel. There were many permeable diffusion regions near the parallel dispersed grouting vein. In the region that was affected by the parallel dispersed grouting vein, the reinforcement modes were the same as that of the cross-grid grouting vein, including the support of the grouting vein, penetration-filling, and compaction-consolidation. Compared with the skeleton vein and the cross-grid grouting vein, the thicknesses of most of the parallel dispersed grouting veins were relatively small, and the reinforcement effect of the grouting veins was significantly limited to the injected medium.

Figure 11 presents the thickness distribution of the grouting veins in the C (600 mm) section. Figure 11a–c presents the thickness distributions of the original grouting vein, the yellow grouting vein, and the red grouting vein along the direction of grout migration, respectively. Figure 10d presents the thickness distribution proportions of the different grouting veins.

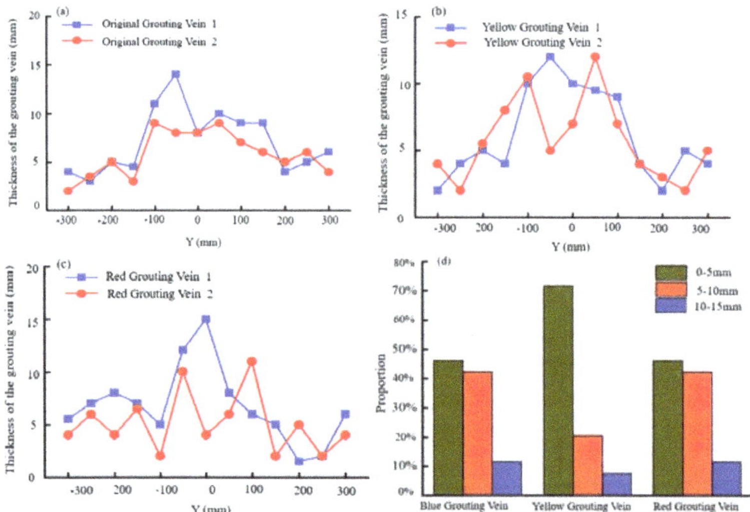

Figure 11. Thickness distribution of the grouting veins in the C′ (600 mm) section: (**a**) thickness distribution of the original grouting veins; (**b**) thickness distribution of the original grouting veins; (**c**) thickness distribution of the red grouting veins; (**d**) thickness distribution proportions of the different grouting veins.

From the results of the excavation, more parallel grouting veins were mainly distributed near the C (600 mm) section. As presented in Figure 11, it was found that the thicknesses of the parallel grouting veins were concentrated to within a range of 0–10 mm. The grouting veins were distributed near the grout frontal surface, and a decrease in the grouting vein thickness was clearly observed.

Therefore, when the soft filling medium was reinforced by segmented grouting, skeleton grouting veins, cross-grid grouting veins, and parallel dispersed grouting veins were formed in an orderly manner, in accordance with the increase in the distance from the grouting hole. The reinforcement modes of the subsection split grouting were the main supports of the grouting vein skeleton, permeation-filling, and compaction consolidation.

The thicknesses of the grouting veins in the grouting area decreased with an increase in the distance from the grouting hole. For example, the thicknesses of the skeleton grouting veins were in the range of 10–20 mm, whereas the thicknesses for the parallel grouting veins were in the range of 0–10 mm. The number of grouting veins increased from the grouting hole to the grout frontal surface, and the strikes of the grouting veins exhibited increased randomness.

4.2. Analysis of Grouting Reinforcement Effect

From the results of the distribution characteristics of the grouting veins, the reinforcement effect of the soft filling medium by the split grouting was mainly reflected by two phenomena. On the one hand, the soft filling medium at the corresponding position was replaced by the grouting vein skeleton, which could act as a support for the injected medium. On the other hand, the grouting vein skeleton had a significant compaction-consolidation effect and a permeation-filling effect on the soft filling medium. After the permeation-filling and the compaction-consolidation of the grouting veins, the mechanical properties and the impermeability of the soft filling medium were significantly improved.

To quantitatively analyze the reinforcement effect, and to determine the quantitative relationship between the mechanical properties and the permeability before and after grouting, the grouting reinforcement area was divided, based on the spatial distribution positions and the reinforcement methods of the grouting veins. It could be divided into three regions: the replacement region of the grouting veins, the grouting split zones, and the grouting compaction zone. The replacement region of the grouting veins was the grouting vein skeleton area; the grouting split zone was the area that was significantly affected by permeation-filling and compaction-consolidation; and the grouting compaction zone was the edge of the influence area of the segmental grouting area, which was dominated by compaction.

Figure 12 presents the schematic diagram of the injected medium zoning. As presented in Figure 12, ① and ⑨ indicate the compact area; ③, ⑤, and ⑦ indicate the replacement region of grouting vein; and ②, ④, ⑥, and ⑧ indicate the grouting split zone formed by the slurry split.

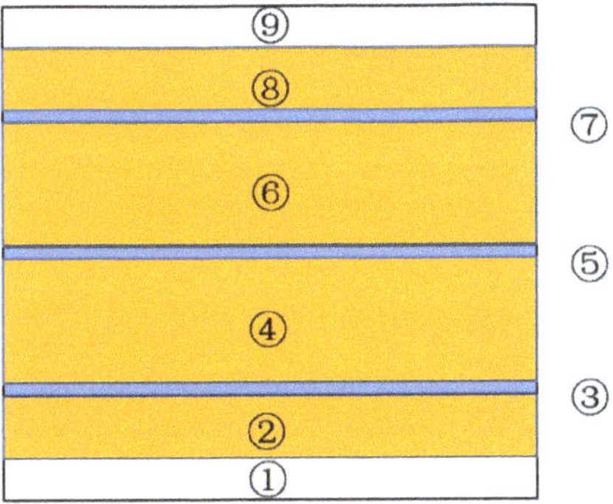

Figure 12. Schematic diagram of the injected medium zoning.

As presented in Figure 13, the cross-section samples of the typical grouting reinforcement area (non-grouting vein replacement area) were selected along the direction of grout diffusion, and the space of each sample was a distance of 20 cm.

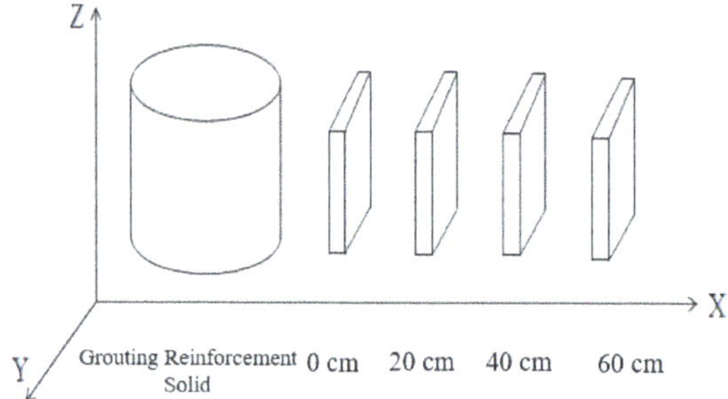

Figure 13. Location of cross-section sampling.

4.2.1. Compression Strength

Figures 14 and 15 present the cross section sampling and single cycle compression test in the grouting reinforcement area, respectively.

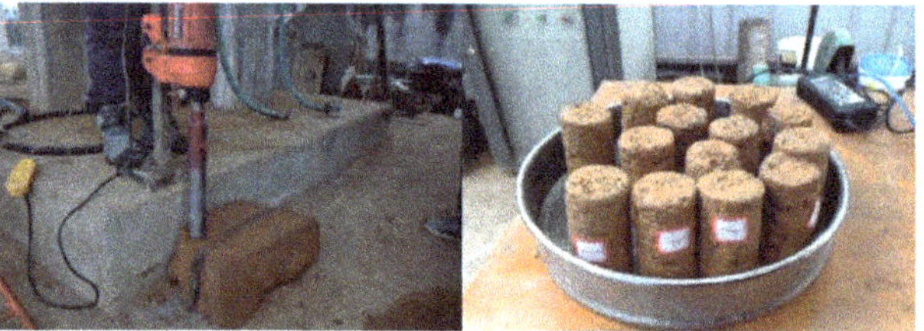

Figure 14. Cross section sampling.

Figure 15. Compression test.

An electric rotary coring machine was used to sample the grouting reinforcement solid area. The 144 samples were selected, and the diameter and height of the sample were 50 mm and 100 mm, respectively. The compressive strength of the sample was tested according to GB/T 50081-2002, and the elastic modulus of the sample was calculated from the compression stress-strain curve.

The replacement region of the grouting veins was mainly filled with grout stone; therefore, the elastic modulus (14 days) of the grout stone was used as the elastic modulus of the replacement region of the grouting veins. The elastic modulus and the standard deviations of other regions are presented in Table 4.

Table 4. Data sheet of the specimens in the splitting and compacting zone.

Location of Sample Cross-Section (cm)	Elastic Modulus(E)/MPa and Standard Deviations(S)											
	Compacting Zone 1		Splitting Zone 2		Splitting Zone 4		Splitting Zone 6		Splitting Zone 8		Compacting Zone 9	
	E	S	E	S	E	S	E	S	E	S	E	S
x = 0	96	1.56	182	1.45	284	2.23	235	2.21	168	1.10	94	3.21
	124	1.36	205	2.11	296	3.12	312	5.36	212	3.65	126	2.10
	178	1.23	285	4.21	340	4.96	325	4.25	271	4.23	182	0.86
	152	2.65	247	3.95	388	6.02	363	6.21	243	5.21	148	0.79
	110	3.21	226	4.23	312	2.70	318	3.65	184	1.89	120	1.89
	85	2.21	196	1.87	275	1.14	304	1.58	150	5.11	87	2.36
x = 20	80	1.26	127	4.56	226	3.21	185	3.65	106	2.68	82	0.68
	104	1.03	196	2.36	288	4.05	302	5.65	172	2.13	94	1.65
	148	2.01	264	3.29	307	3.04	325	4.96	203	0.98	138	2.31
	94	0.98	225	1.95	334	5.69	296	3.68	245	3.69	124	1.15
	87	1.68	158	0.98	272	4.12	252	1.56	164	5.45	91	2.35
	74	0.45	135	1.14	215	3.14	201	1.02	124	2.12	80	1.89
x = 40	65	0.65	105	1.68	202	4.56	196	3.68	94	1.56	52	1.22
	92	1.56	150	1.89	243	3.68	230	2.65	144	1.68	68	2.36
	106	1.89	226	2.68	271	2.68	264	2.31	191	3.21	93	1.10
	85	0.96	210	3.87	238	1.59	256	1.58	244	2.68	102	0.65
	72	1.45	154	1.56	222	3.68	228	2.10	141	1.36	74	0.98
	54	1.54	96	1.68	195	4.65	184	1.22	116	1.26	65	0.48
x = 60	45	0.98	67	0.98	93	0.65	97	0.98	54	0.79	52	2.56
	57	1.32	75	1.35	114	1.56	110	1.65	67	0.86	66	5.65
	93	1.89	101	1.68	146	1.75	158	2.65	96	1.12	80	2.65
	86	1.68	93	2.68	134	2.31	146	2.31	114	2.01	64	3.57
	78	2.12	78	0.98	102	1.62	112	1.12	84	1.10	48	2.32
	64	0.97	66	1.10	88	1.59	103	2.02	65	0.88	35	0.68

In Figure 12, the area weighting coefficients of areas ①–⑨ were calculated from the partition characteristics of the injected media, and the equivalent elastic modulus of the grouting reinforcement solid area could then be calculated using the method of the weighted mean value.

The average elastic modulus of a grouting reinforcement area was calculated by the equation [24] as follows:

$$E_{j-i} = \frac{1}{6}\sum_{m=1}^{6} E_{j-i-m} \tag{1}$$

where j is the section number ranging from 1–4, which represents the sections corresponding to $x = 60$ cm, $x = 40$ cm, $x = 20$ cm, and $x = 0$ cm, respectively; i is the area number of the section, which ranges from 1–9; m is the sample number in a certain area, which ranges from 1–6; E_{j-i} is the average elastic modulus of the excavation section j and area i; and E_{j-i-m} is the average elastic modulus of the excavation section j, area i, and sample m.

The average elastic modulus of a section can be expressed by the equation [24] as follows:

$$E_j = \sum_{i=1}^{9} \frac{S_{j-i}}{S_j} E_{j-i} \tag{2}$$

where S_j is the total section area of the excavation section j; S_{j-i} is the total section area of the excavation section j and area i; and E_j is the average elastic modulus of the excavation section j.

The equivalent elastic modulus of the grouting reinforcement solid is:

$$Ee = \frac{1}{4}\sum_{j=1}^{4} E_j \tag{3}$$

where E_e is the equivalent elastic modulus of the grouting reinforcement solid.

According to the calculation Formula (1)–(3) and the elastic modulus data in Table 4, the equivalent elastic modulus of the grouting reinforcement solid E_e was 205.2 MPa. Before the split grouting, the elastic modulus E_S of the soft filling medium sample without grouting was 2 MPa. Thus:

$$E_e = 102.6E_S \tag{4}$$

Under the experimental conditions used in this study, the elastic modulus of the grouting reinforcement solid increased on average by a factor greater than 100 after the grouting reinforcement, thus indicating that the support of the grouting vein, permeation-filling, and the compaction-consolidation formed by the split grouting could significantly improve the mechanical properties of the injected medium. Segmental split grouting was therefore effective in improving the mechanical properties of the weak filling media.

4.2.2. Anti-Permeability

Figure 16 presents the specimen sampling of the grouting reinforcement solid area and the permeability test. The permeability coefficient of the sample was tested using a TST-55 permeameter, and the diameter and height of the sample were 61.8 mm and 40 mm, respectively. The sampling distance of each reinforced area was 30 cm.

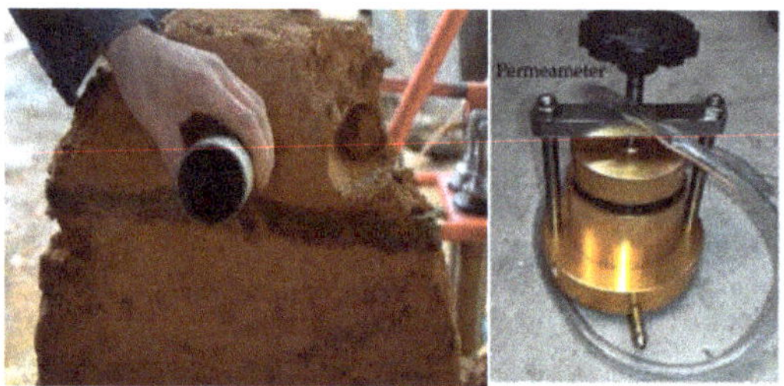

Figure 16. Specimen sampling and permeability test.

As shown in the Table 5, the permeability coefficients of each section were calculated using Darcy's law. The permeability coefficient of the grouting reinforcement solid was measured separately, and was 5.5×10^{-8} cm/s.

Table 5. Permeability coefficients.

Location of Sample Cross-Section (cm)	Permeability Coefficient (10^{-6} cm/s)					
	Compacting Zone		Splitting Zone			
	1	9	2	4	6	8
$x = 0$	8.6	27.4	0.7	0.2	0.4	1.7
	16.3	32.6	1.6	0.4	0.6	2.2
	10.7	44.8	2.2	0.6	0.8	3.4
$x = 20$	22.5	38.5	1.2	0.4	0.8	3.4
	27.4	45.4	2.3	0.4	0.6	4.4
	38.6	56.8	3.3	0.8	1.0	5.8
$x = 40$	48.1	67.4	4.6	0.7	0.7	5.7
	52.7	71.5	5.2	0.8	0.8	6.3
	64.2	82.8	6.7	0.9	1.2	7.8
$x = 60$	68.8	78.5	5.5	0.9	1.3	7.3
	74.5	86.2	6.6	1.12	1.4	7.9
	82.4	95.6	7.8	1.4	1.8	9.1

Figure 17 presents the schematic diagram of the stratified seepage in the parallel and vertical grouting vein directions.

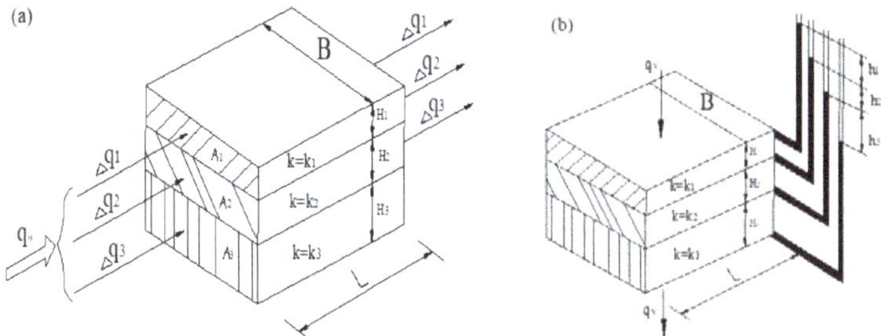

Figure 17. Schematic diagram of stratified seepage: (**a**) parallel grouting vein direction and (**b**) vertical grouting vein direction.

As presented in Figure 17, the soft filling medium exhibited significant stratification characteristics after grouting. The grouting reinforcement solid area was composed of lamellar regions with different permeability coefficients. According to the principle of seepage mechanics, the equivalent permeability coefficient in the parallel grouting vein direction was calculated by the equation [25] as follows:

$$K_h = \sum_{i=1}^{n} \frac{H_i}{H} K_i \tag{5}$$

where K_h is the equivalent permeability coefficient in the parallel grouting vein direction; H_i is the thickness of region i; H is the total thickness of the injected medium; K_i is the average permeability coefficient of region i.

The equivalent permeability coefficient in the vertical grouting vein direction was calculated by the equation [25] as follows:

$$K_v = \frac{1}{\sum_{i=1}^{n} \frac{H_i}{H} \frac{1}{K_i}} \tag{6}$$

where K_h is the equivalent permeability coefficient in the vertical grouting vein direction.

The permeability coefficient of the C-GT grouting reinforcement solid was small, thus $K_h \gg K_v$. The equivalent permeability coefficient K_h, which was most unfavorable in the parallel slurry direction, was calculated to represent the permeability coefficient of the grouting reinforcement solid. According the permeability coefficient of the soft filling medium in Table 5, the permeability coefficient of the grouting vein, and formula (5), the equivalent permeability coefficient of the grouting reinforcement solid K_h was calculated as 12.95 $\times$ 10^{-6} cm/s, whereas the initial permeability coefficient of the soft filling medium K_S was 5.2 $\times$ 10^{-4} cm/s. Therefore:

$$K_h = 40.15^{-1} KS \tag{7}$$

After grouting, the permeability coefficient was the biggest in the parallel grouting direction, but the permeability coefficient of the soft filling medium was reduced by a factor greater than 40. This indicated that the compaction effect of the grouting vein formed by the split grouting reduced the permeability of the soft filling medium. Thus, the subsection split grouting could effectively improve the anti-seepage performance of the soft filling medium.

5. Conclusions

Segmental splitting grouting is widely used in soft ground grouting reinforcement projects, such as fault fracture zones and the quaternary topsoil layers. The study of the section splitting grouting mechanism is helpful to realize the quantification of grouting design. The simulation test formed the following important conclusions:

(1) A three-dimensional grouting simulation test system was developed, which could realize the subsection split grouting in a soft filling medium. Moreover, it could be used to conduct the excavation observation and sampling analysis after grouting.

(2) There was no permeable grouting in the soft filling medium with poor permeability, but the form of grouting in this medium was mainly split grouting. Three types of veins were formed when the soft filling medium was consolidated by subsection split grouting; namely, skeleton support veins, cross-grid pulp veins, and parallel dispersed pulp veins. The thicknesses of the grouting veins decreased gradually as the distance from the grouting holes increased, and the strikes of grouting veins exhibited increased randomness.

(3) The special grouting reinforcement mechanism under the section split grouting was determined, and it consisted mainly of the reinforcement mode of the grout vein skeleton support, permeation-filling, and compaction-consolidation.

(4) The elastic modulus of the grouting reinforcement solid increased on average by a factor greater than 100, and the permeability coefficient decreased on average by a factor greater than 40 in the direction of the parallel grouting vein with the most impermeable solid. This confirmed that segmental split grouting is an effective method for the improvement of the mechanical properties and the impermeability of the weak filling media.

The soil used is sourced from the F2 fault of Yonglian Tunnel in Jilian Expressway, Jiangxi Province, China in this paper. The test results obtained are aimed at the injected medium that is used in this experiment. It is not clear whether the test results are applicable to other media. A plan has been drawn up to conduct experimental research in this field.

Author Contributions: Z.L., S.L. and Q.Z. conceived of and designed the study. Z.L. and Q.Z. developed the three-dimensional grouting simulation test system. Z.L., H.L. and Y.L. performed the experiments. Z.L., S.L., Q.Z., H.L. and Y.L. wrote and modified the paper.

Funding: This work was supported by the National Key Research and Development Project (grant number 2016YFC0801604), and the National Natural Science Foundation of China (grant number 51779133).

Conflicts of Interest: The authors declare no conflict of interest.

References

1. Marchi, M.; Gottardi, G.; Soga, A.K. Fracturing Pressure in Clay. *J. Geotech. Geoenviron. Eng.* **2014**, *140*, 04013008. [CrossRef]

2. Bezuijen, A.; te Grotenhuis, R.; van Tol, A.F.; Bosch, J.W.; Haasnoot, J.K. Analytical Model for Fracture Grouting in Sand. *J. Geotech. Geoenviron. Eng.* **2011**, *137*, 611–620. [CrossRef]

3. Hernqvist, L.; Gustafson, G.; Fransson, Å.; Norberg, T. A statistical grouting decision method based on water pressure tests for the tunnel construction stage—A case study. *Tunn. Undergr. Space Technol.* **2013**, *33*, 54–62.

4. Tan, Y.H. Evolutionary Mechanism of Mud Bursting through Water-Inrich Fault in Tunnels and Engineering Applications. Ph.D. Thesis, Shan Dong University, Jinan, China, 2017. (In Chinese)

5. Yu, W.S. Grouting Diffusion Mechanism in Mud Filled Fault Fracture Zone of Tunnel and It's Engineering Application. Ph.D. Thesis, Changsha University of Science & Technology, Changsha, China, 2017. (In Chinese)

6. Wang, Z.C.; Li, F.Q. Slurry water curtain grouting and its mechanism in complex coal mine area. *J. Cent. South Univ. (Sci. Technol.)* **2013**, *44*, 778–784. (In Chinese)

7. Zhou, K.F.; Li, Y.Z. Influence of grouting on the mechanical characteristic of stratified rock slope. *J. Cent. South Univ. (Sci. Technol.)* **2012**, *43*, 724–729. (In Chinese)

8. Zhang, M.; Zou, J.; Chen, J.; Li, X.; Li, Z. Analysis of soil fracturing grouting pressure under asymmetric loads. *Rock Soil Mech.* **2013**, *34*, 2255–2263. (In Chinese)
9. Li, S.C.; Zhang, W.J.; Zhang, Q.; Zhang, X.; Liu, R.; Pan, G.; Li, Z.; Chen, Z. Research on advantage-fracture grouting mechanism and controlled grouting method in water-rich fault zone. *Rock Soil Mech.* **2014**, *3*, 745–751. (In Chinese)
10. Sun, F.; Chen, T.; Zhang, D.; Zhang, Z.; Li, P. Study on fracture grouting mechanism in subsea tunnel based on Bingham fluids. *J. Beijing Jiaotong Univ.* **2009**, *33*, 1–6. (In Chinese)
11. Zhang, Z.; Zou, J. Penetration radius and grouting pressure in fracture grouting. *Chin. J. Geotech. Eng.* **2008**, *30*, 181–184. (In Chinese)
12. Sun, F.; Zhang, D.; Chen, T. Fracture grouting mechanism in tunnels based on time-dependent behaviors of grout. *Chin. J. Geotech. Eng.* **2011**, *33*, 88–93. (In Chinese)
13. Zhang, Q.; Zhang, L.; Liu, R.; Yu, W.; Zheng, Z.; Wang, H.; Zhu, G. Split grouting theory based on slurry-soil coupling effects. *Chin. J. Geotech. Eng.* **2016**, *38*, 323–330. (In Chinese)
14. Zhang, L. Study on Penetration and Reinforcement Mechanism of Grouting in Sand Layer Disclosed by Subway Tunnel and Its Application. Ph.D. Thesis, Shandong University, Jinan, China, 2017. (In Chinese)
15. Li, P.; Zhang, X.; Zhang, Q. Analysis on three-dimensional diffusion mechanism of multi-sequence grouting in fault. *Geotech. Spec. Publ.* **2016**, *2016*, 121–129. (In Chinese)
16. Li, P.; Zhang, Q.; Zhang, X. Grouting diffusion characteristics in faults considering the interaction of multi-sequence grouting. *Int. J. Geomech.* **2016**, *17*, 04014117. (In Chinese)
17. Zhang, B.W. Experimental Study on Compaction Grouting and Fracture Grouting of Soft Clay in the High-Speed Rail Way Foundation. Ph.D. Thesis, Southwest Jiaotong University, Chengdu, China, 2013. (In Chinese)
18. Zheng, C.C. Estimation on Penetrating Radius of Stable Cement Grouts in Rock Fractures. *J. Water Resour. Arch. Eng.* **2006**, *4*, 1–5. (In Chinese)
19. Li, P.; Zhang, Q.S.; Zhang, X.; Li, S.; Zhang, W.; Li, M.; Wang, Q. Analysis of fracture grouting mechanism based on model test. *Chin. J. Rock Mech. Eng.* **2014**, *35*, 3221–3230. (In Chinese)
20. Darling, W.G.; Bath, A.H.A. Stable isotope study of recharge processes in the English Chalk. *J. Hydrol.* **1988**, *101*, 31–46. [CrossRef]
21. Li, Z. Mechanism of Grouting Spread and Reinforcement on Soft Medium in Fault and Its Application. Ph.D. Thesis, Shan Dong University, Jinan, China, 2015. (In Chinese)
22. Liu, R.; Li, S.; Zhang, Q.; Yuan, X.; Han, W. Experiment and application research on a new type of dynamic water grouting material. *Chin. J. Rock Mech. Eng.* **2011**, *30*, 1454–1459. (In Chinese)
23. Sha, F.; Liu, R.; Li, S.; Lin, C.; Li, Z.; Liu, B.; Bai, J. Application on different types of cementitious grouts for water-leakage operational tunnels. *J. Cent. South Univ. (Sci. Technol.)* **2016**, *47*, 4163–4172. (In Chinese)
24. Xu, Z.L. *Elasticity*; Higher Education Press: Beijing, China, 2013. (In Chinese)
25. Liu, J.J.; Zhang, B.H. *Fluid Mechanic*; Peking University Press: Beijing, China, 2016. (In Chinese)

Article

A Strain-Based Percolation Model and Triaxial Tests to Investigate the Evolution of Permeability and Critical Dilatancy Behavior of Coal

Dongjie Xue [1,2,3,]*, Jie Zhou [1], Yintong Liu [1] and Sishuai Zhang [1]

[1] School of Mechanics and Civil Engineering, China University of Mining and Technology, Beijing 100083, China; cumtb.zhoujie2018@gmail.com (J.Z.); cumtb.liuyintong2018@gmail.com (Y.L.); cumtb.zhangsishuai2018@gmail.com (S.Z.)

[2] State Key Laboratory of Coal Mine Disaster Dynamics and Control, Chongqing University, Chongqing 400030, China

[3] Key Laboratory of Safety and High-efficiency Coal Mining, Anhui University of Science and Technology, Huainan 232001, China

* Correspondence: xuedongjie@163.com; Tel.: +86-1-15101127335

Received: 5 July 2018; Accepted: 6 August 2018; Published: 13 August 2018

Abstract: Modeling the coupled evolution of strain and CH_4 seepage under conventional triaxial compression is the key to understanding enhanced permeability in coal. An abrupt transition of gas-stress coupled behavior at the dilatancy boundary is studied by the strain-based percolation model. Based on orthogonal experiments of triaxial stress with CH_4 seepage, a complete stress-strain relationship and the corresponding evolution of volumetric strain and permeability are obtained. At the dilatant boundary of volumetric strain, modeling of stress-dependent permeability is ineffective when considering the effective deviatoric stress influenced by confining pressure and pore pressure. The computed tomography (CT) analysis shows that coal can be a continuous medium of pore-based structure before the dilatant boundary, but a discontinuous medium of fracture-based structure. The multiscale pore structure geometry dominates the mechanical behavior transition and the sudden change in CH_4 seepage. By the volume-covering method proposed, the linear relationship between the fractal dimension and porosity indicates that the multiscale network can be a fractal percolation structure. A percolation model of connectivity by the axial strain-permeability relationship is proposed to explain the transition behavior of volumetric strain and CH_4 seepage. The volumetric strain on permeability is illustrated by axial strain controlling the trend of transition behavior and radical strain controlling the shift of behavior. A good correlation between the theoretical and experimental results shows that the strain-based percolation model is effective in describing the transition behavior of CH_4 seepage in coal.

Keywords: enhanced permeability; deviatoric stress; mechanical behavior transition; CH_4 seepage; volumetric strain; strain-based percolation model

1. Introduction

Enhanced coal permeability after failure under triaxial stress can be determined quantitatively by the stress-dependent model. However, the sudden change in permeability from porous to fractured coal is more important in order to understand the enhanced mechanism and improve the stress-relieving technology and hydromechanical (HM) fracturing method. On the other hand, the production of coalbed methane (CBM) as green energy has been increasingly important, particularly in winter 2017, with more determination to fight air pollution in the face of the gas shortage initiative in China [1]. Most of China's coal has high density and low permeability, typically less than 1 mD [2,3]. In coal seams, to improve permeability for easy extraction, premining the protective coal is often carried out to relieve the stress [4,5]. At the working face, the hydraulic fracturing method is often employed to achieve local stress adjustment. The huge demand in CH_4 promotes a deep understanding of sudden changes in permeability, especially induced by the coupled effect of gas and stress. The key to all these problems is to accurately understand the transition of permeability. We need a new model that can illustrate the whole change in permeability, even sudden transitions.

Under triaxial compression, the permeability of CH_4 mainly depends on the shearing-induced dilatancy of the coal. Modeling enhanced permeability with deviatoric stress is useful for understanding the coupled mechanism between the dilatant deformation and the seepage or percolation channel. Many outstanding achievements focus on the study of stress-dependent permeability without considering transition effect. Many models [6–11], as shown in Table 1, have been made based on the excavation damage zone (EDZ), which depends on continuous media to explain the coupled process [12]. To our knowledge, the stress-dependent model can effectively explain the coupled effect before the dilatant boundary, taking coal as a continuous porous medium, and after, taking coal as discontinuous fractured medium. According to percolation theory, there must be a state transition of the continuous-discontinuous behavior controlling the coal permeability by connected clusters of multiscale networks [10]. The multiscale can be referred as nano, mirco, meso, macro, and even mega scale. The corresponding structure of discontinuities changes from pore to fracture. Here, we only focus on the study of cylindrical coal sample in lab. The resolution for pore structure by CT is about 60 μm, which can be the lower limit of size. There will be fracture generation connecting the top and bottom of the coal sample, which is about 12 cm long. The study is based on the multiscale structure from 60 μm to 12 cm, and the upper limit is 2000 times the lower limit. Indeed, the multiscale network of pore-crack-fracture is mainly generated at the dilatant boundary, and the focus of study is on the multiscale behavior and sudden transition of permeability. It is difficult to determine the boundary of two kinds of adjacent pore structures and the scale invariance is often used to understand the multiscale behavior based on the percolation theory. So, the sudden transition in permeability needs a new understanding of phase transition for the continuous-discontinuous behavior.

Considering the transition behavior, the local stress redistribution around isolated fractures changes dramatically, and the deviatoric stress cannot control the sudden increase of volumetric strain and permeability. Eventually, open fractures provide the main path for gas seepage until there is a steady flow. So, the initial permeability k_0 can be determined by Darcy's law [6,7,13] or the Carman-Kozeny equation [14,15]. The final permeability k_f is often determined by the cubic law [11,13,16] through fissure flow in discontinuous fractured coal. For the transition behavior, there is always a sudden increase of volumetric strain or permeability [17,18]. Meanwhile, the coal changes from continuous to discontinuous, corresponding to the generation of a multi scale network of void structures. In mathematical terms, we can understand such behavior as a continuous and nonderivative relationship. In physical terms, we propose the sudden changes of mechanical behavior from pore-dominated structure to fracture-dominated structure as the phase transition according to percolation theory [19,20]. The initial phase dominated by the continuous behavior moves into the final phase dominated by discontinuous structural behavior. The pore-based material behavior is continuous for Darcy's flow and the fracture-based structural behavior is discontinuous for the cubic law.

Table 1. Stress-dependent models.

Model	Classification	Gas	Lithological Type	Related Theory
$k_f/k_0 = e^{-3\times10^{-3}\overline{\sigma}k_0^{-0.1}} + 2 \times 10^{-4}\overline{\sigma}^{1/3}k_0^{1/3}$	Theory and measured data (Somerton et al., 1975)	N_2 CH_4	Coal	Darcy's law
$k_f = (1.12 - 0.03\sigma_3)K \times e^{-(1.12-0.03\sigma_3)C_f\sigma_3}$	Measured data (Durucan and Edwards, 1986)	N_2	Coal	Darcy's law
$k_f/k_0 = e^{-3\times C_f\times\Delta\sigma} / \left[1 - \phi_0\left(1 - e^{-C_f\times\Delta\sigma}\right)\right]$	Theory and measured data (McKee et al., 1988)	CH_4 H_2O	Coal	Carman-Kozeny equation
$k_f/k_0 = e^{-3\times C_f\times\Delta\sigma}$	Theoretical solution (Seidle et al., 1992)	H_2O	Coal	Carman-Kozeny equation
$k_f/k_0 = \left[\left(\eta + e^{-C_f\sigma_f}\right) / \left(\eta + e^{-C_f\sigma_0}\right)\right]^3$	Theoretical solution (Liu et al., 2009, 2010)	CH_4	Coal	Cubic law

Note. k_0 and k_f mean initial and final permeability corresponding to the initial effect stress σ_0 and final effect stress σ_f. $\Delta\sigma = \sigma_f - \sigma_0$. $\overline{\sigma}$ is the mean stress and σ_3 is the confining pressure. K defines the relative coefficient of existing fractures, C_f represents the compressibility of coal, and η defines the influence of the effective stress induced closure of aperture.

Therefore, it is necessary to establish a new model that will help us understand transition behavior in permeability. Percolation theory [21–24] is very effective at explaining such phase transitions in disordered media. For coal under conventional triaxial compression, the continuous-discontinuous transition depends on the cluster generation of multiscale pore-crack-fracture, as well as the structural phase transition behavior of the connected clusters. Percolation models have been used to analyze the fracture-induced transitions of various rocks [23–26]. However, due to our limited knowledge, there has been almost no analysis of CH_4 flow considering the mechanical-structural transition in coal.

Another focus is the multiscale geometric topology of the network that influences percolation depending on the generation, distribution, and opening of fractures. It is impossible to find the exact solution of local stress near all fractures, which dominate local opening and closing. The topology evolution can be considered by the equivalent strain of coal considering the discontinuities. We mainly focus on the percolation model of strain-dependent permeability by designing a series of orthogonal experiments of conventional triaxial compression for coupled stress-strain permeability.

2. Materials and Experimental Methods

2.1. Coal Sample Preparation for Orthogonal Experiments

The coal blocks in China were selected at a depth of 690 m, and the coal seam has an average thickness of 3.5 m with an inclination of 23°. The roof and floor of the roadway are fine-grained sandstone. Then, the coal blocks were processed into cylinder-shaped samples with a standard size of 50 mm diameter × 100 mm height [27]. The 9 samples were numbered C01, C02, C03, C04, C05, C06, C07, C08, and C09.

Figure 1 illustrates the designed cycle of experimental procedures for coupled research including the determined physical parameters, the multiscale structures by computed tomography (CT) before and after failure, and triaxial compression coupled with CH_4 seepage. Triaxial servo-controlled seepage equipment for coupled thermos-fluid-solid of coal was adopted for the coupled behavior. The system collected data at 1 s intervals, automatically recording axial force, axial and radial displacement, confining pressure, temperature, and gas flow. After the samples were dried in an oven for 12 h, their height, area, and porosity were measured, as shown in Table 2.

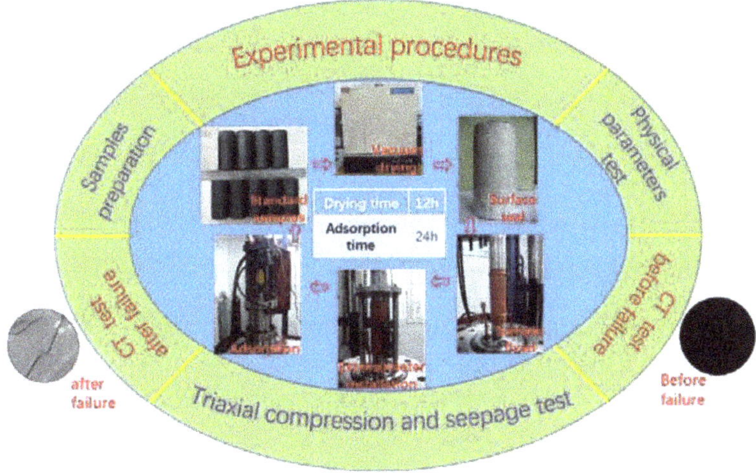

Figure 1. Experimental design.

Table 2. Physical parameter determination. Porosity is measured by computed tomography (CT) after failure.

Samples	Height	Cross-Section Area	Volume	Dry Weight	Density	Porosity
	/cm	/cm^2	/cm^3	/g	g/cm^3	%
C01	10.40	19.63	204.15	279.0	1.37	4.83
C02	10.56	19.63	207.29	281.5	1.36	3.32
C03	10.50	19.63	206.12	276.7	1.34	3.13
C04	10.50	19.63	206.12	269.8	1.31	1.44
C05	10.27	19.63	201.60	299.0	1.48	4.87
C06	10.56	19.63	207.29	279.3	1.35	1.17
C07	9.44	19.63	185.31	250.2	1.35	6.02
C08	10.40	19.63	204.15	269.2	1.32	8.25
C09	10.52	19.63	206.51	278.9	1.35	4.32

Then, the coupled tests were carried out. To prevent gas leakage through the sidewalls, they were covered with a 1 mm thick layer of silica gel. Sliding the heat-shrink tubing over the surface keeps the gel in close contact with the sidewalls, preventing the hydraulic oil from penetrating into the samples. After that, the extensometer for radial displacement was installed and the flow meter was connected. The outlet valve was closed after removing the air from the chamber with a vacuum pump, filling it with CH_4 and allowing 24 h for adsorption equilibrium. We then started the system and imposed displacement controlled axial stress load at a rate of 0.1 mm/min. Table 3 shows the designed orthogonal experiments with confining pressures of 3 MPa, 6 MPa, and 9 MPa, and gas pressures of 1 MPa, 1.5 MPa, and 2 MPa.

Table 3. Experimental design by orthogonal stress.

Sample	Confining Pressure	Inlet Gas Pressure	Dilatant Stress and Strain		Peak Stress and Strain		Residual Stress and Strain		Permeability		Stress Ratio
	σ_3	p_1	σ_D	ε_D	σ_P	ε_P	σ_R	ε_R	k_0	k_f	σ_D/σ_P
C01	9	1	22.77	0.45	43.12	−0.31	21.58	−6.45	3.93	56.96	0.53
C02	6	1.5	17.30	0.38	35.62	−1.05	26.77	−9.01	16.75	31.73	0.49
C03	3	2	18.08	0.39	25.83	0.17	19.04	−4.56	0	2434	0.70
C04	9	1.5	25.25	0.67	39.39	−1.14	31.59	−4.78	65.75	27.17	0.64
C05	6	2	22.01	0.44	43.12	−4.03	21.58	−6.44	11.07	1086	0.51
C06	3	1	24.37	0.63	39.20	−0.14	29.53	−4.50	187.5	51.87	0.62
C07	9	2	27.38	0.45	50.78	0.19	33.43	−2.51	0	Failed	0.54
C08	6	1	27.36	0.51	40.53	−0.59	26.15	−8.25	375.2	3653	0.68
C09	3	1.5	35.90	0.80	42.06	0.25	39.60	−4.20	0	4343	0.85

Note. σ_3, p_1, σ_D, σ_P and σ_R have the same unit of MPa. ε_D, ε_P and ε_R have the same unit of %. k_0 and k_f have the same unit of 10^{-18} m^2.

2.2. CT Observation of Multiscale Structures

Before the coupled test for permeability, the samples were scanned using an industrial CT scanner. A random zone of 30 mm to 35 mm from the bottom was determined as the interesting area, and the scan interval was set at 0.1 mm, with a total of 51 layers, shown in Figure 2. After sample failure, the same zone of all 9 samples was scanned again for CT images of 1024 × 1024 pixels and 57 μm resolution.

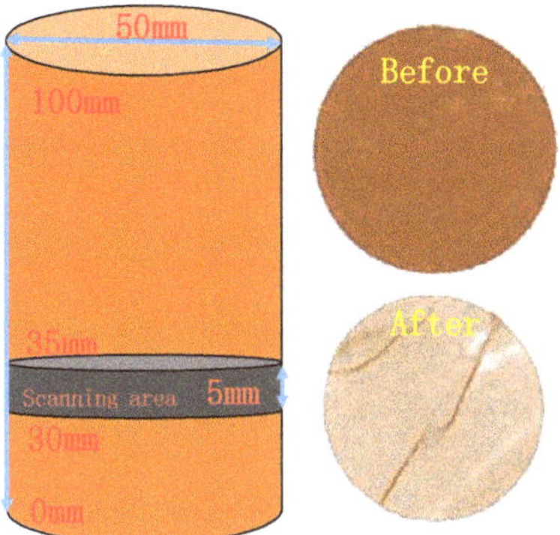

Figure 2. CT scan of interesting area.

2.3. Theory of CH_4 Seepage in Coal

Coal is often described as a dual-porosity medium considering the multiscale effect. When there is CH_4 flow in the multiscale network, an assumption of isothermal seepage in coal is often given as the ideal gas. The gas flow rate is automatically recorded and the permeability at different points can be determined by Darcy's law [28,29]:

$$k = \frac{2qp_0\mu L}{A(p_1^2 - p_2^2)} \tag{1}$$

where q is the flow rate (m/s), k is the permeability (m^2), p_0 is the atmospheric pressure (0.1 MPa), A is the cross-sectional area (m^2), μ is the gas viscosity coefficient (1.087×10^{-5} Pa $\cdot$ s) at 20 °C, L is the height of the cylinder (m), and p_1 and p_2 are gas pressure at the air inlet and outlet, respectively (MPa).

3. Experimental Results and Analysis

3.1. Effect of Volumetric Deformation on CH_4 Seepage

Figure 3 shows the coupled evolution of permeability with volumetric strain. The nine samples experienced the same behavior, including linear elastic compression, nonlinear compression with local damage, plastic deformation, and post-peak softening before failure. Accordingly, the volumetric strain underwent the two main stages, compression and dilatancy. There are two kinds of transition point: the first is the dilatant point, indicating the deformation transition from compression to dilatancy, and the second is the continuous-discontinuous point, indicating the phase from pore-based to fracture-based structure. The confining pressure promotes strain-softening behavior in the post-peak deformation. The higher the confining pressure, the more likely that volumetric strain will increase nonlinearly with stress.

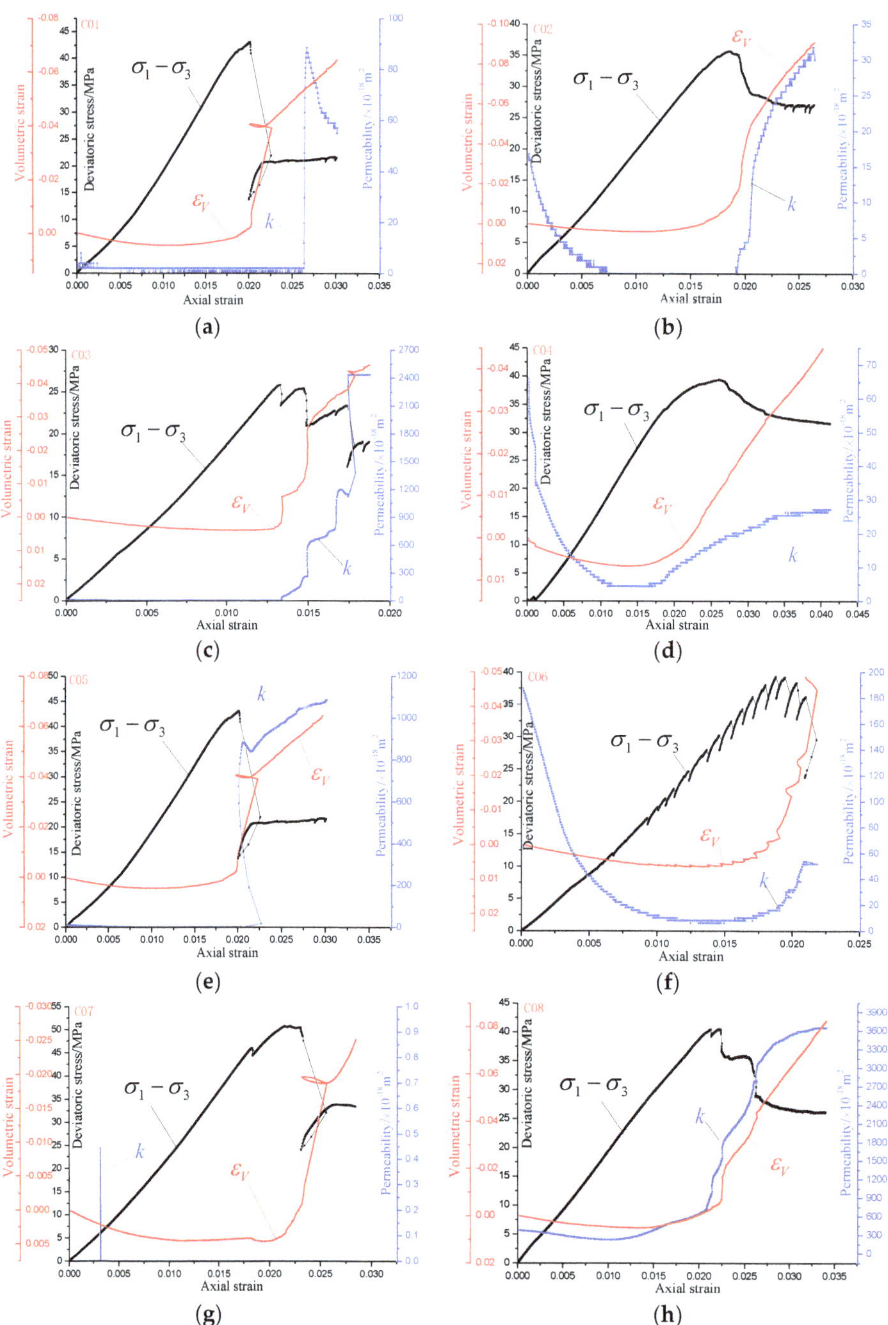

Figure 3. *Cont.*

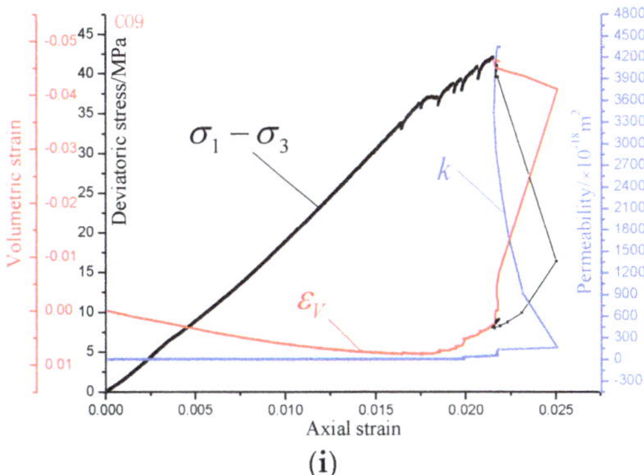

(i)

Figure 3. Evolution of volumetric strain and permeability on samples (**a**) C01, (**b**) C02, (**c**) C03, (**d**) C04, (**e**) C05, (**f**) C06, (**g**) C07, (**h**) C08, and (**i**) C09. Black lines indicate the stress–strain relationship, red lines indicate the volumetric–axial strain relationship, and blue lines indicate permeability with axial strain.

The permeability evolution with deviatoric stress can be determined according to Equation (1), and it shows high consistency with volumetric strain. The initial permeability k_0 for different samples varies, as shown in Table 2. For C01, C03, C05, C07, and C09, k_0 is close to 0. For C02 and C04, it is measured at 10^{-17} m^2, and for C06 and C08, it is at 10^{-16} m^2, which can be considered low-permeability coal. Increasing deviatoric stress often causes pores to open and fractures to form. Since the volumetric change of a solid matrix is much smaller than the discontinuous geometric structures (e.g., pores, cracks, and fractures), the volumetric strain can be a multiscale variable to illustrate the discontinuous geometries.

When the deviatoric stress is low, low-permeability coal can be considered as a porous medium with preexisting pores under compression. The corresponding volumetric strain decreases, seepage channels close, and permeability decreases. When it reaches the dilatant point, an increase in deviatoric stress will cause crack generation, maintaining volumetric compression, and the channels of the multiscale network are not completely connected. Until the linear-nonlinear transition in volumetric strain, permeability starts to increase significantly. Nonlinear deviatoric stress can be considered as a mechanical phase in percolation theory, in which geometric channels tend to interconnect. During the post-peak stage, fractures connect through the coal sample with a typical stress drop corresponding to strain softening. Fracture geometries, which determine the change of volumetric strain, show a sharp and dramatic increase in dilatancy and seepage. At the end, with the stabilization of residual stress, permeability remains constant. Meanwhile, the confining pressure still has a constraint effect on fracture channels. Therefore, the ultimate permeability is not necessarily greater than the initial value for all samples, such as C04 and C06. There is another possible reason that shear-induced fractures often have an off-axial fracture angle and the connected fracture cannot effectively connect the upper and lower ends of the surface.

3.2. Effect of Confining Pressure on Deformation and CH$_4$ Seepage

During the complete stress-strain evolution, dilatant stress represents nonlinear dilatancy, peak stress indicates coal strength, and residual stress indicates strain-softening behavior. Corresponding to volumetric deformation, there are dilatant strain, peak strain, and residual strain, respectively. Furthermore, permeability changes consistently with nonlinear deformation, and the corresponding points are dilatant permeability, peak permeability, and residual permeability.

The change rates of stress, deformation, and permeability are influenced by the coupled effect of confining pressure and gas pressure. Thus, the stress drop rate can be illustrated by the percentage of stress drop relative to the peak value, the volumetric strain change rate by the percentage of initial to final volumetric strain, and the permeability change rate by the percentage of initial to final permeability.

With increasing confining pressure, peak stress increases accordingly, as shown in Figure 4a,c. However, both the randomness in sampling and the inhomogeneity influence the peak stress; for example, the peak stress of sample C02 was the lowest when gas pressure was 1.5 MPa, shown in Figure 4b. There is no significant correlation between dilatant stress or residual stress and confining pressure. According to the Griffith crack theory, dilatant stress is mainly influenced by nonlinear deformation and damage accumulation. After the peak strength, the stress drop with discontinuous fractures dominates the residual stress. The increasing confining pressure has an inhibiting effect on dilatancy, while the heterogeneity of fractures determines that the effect under a confining pressure of 9 MPa is not significant.

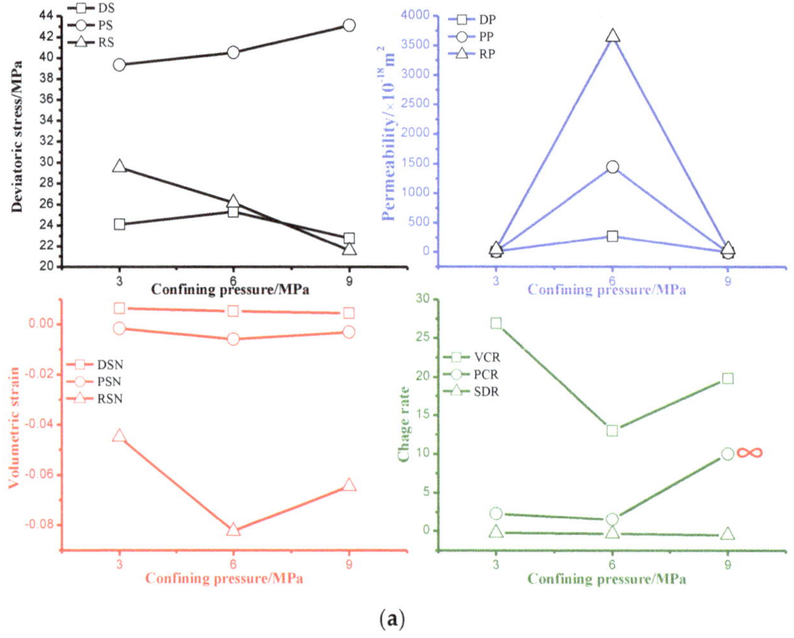

(**a**)

Figure 4. *Cont.*

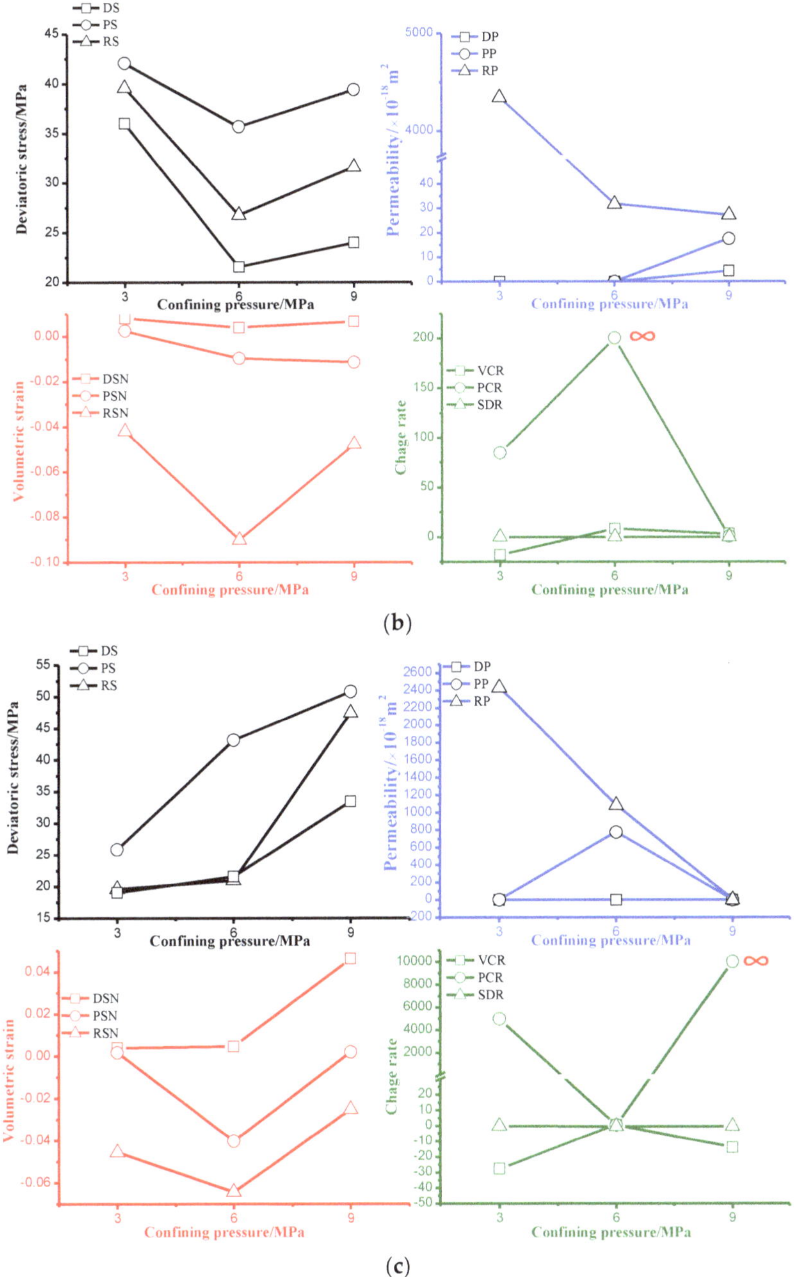

Figure 4. Deviatoric stress, volumetric strain, permeability, and permeability change rate at dilatant point, peak point, and residual point under confining pressures of (**a**) 1 MPa, (**b**) 1.5 MPa, and (**c**) 2 MPa. ∞ indicates an infinite value because the denominator may be 0, represented by a larger value such as 200 or 10,000.

Positive dilatant strain means a state of compression, and confining pressure does not increase significantly. Peak strain changes slightly with an increase in confining pressure. Residual strain largely changes in the same trend as peak strain. Residual strain is much greater than peak strain, and post-peak volumetric strain increases rapidly, as shown in Figure 4.

Change in volumetric deformation is mainly due to the interconnection of the multiscale network, and the nonlinear volumetric deformation leads to a change in permeability, not the linear change. In spite of the heterogeneity of the different samples, overall, confining pressure has a significant inhibiting effect on permeability, with residual permeability being the highest, peak permeability next, and dilatant permeability the lowest, even as low as 0. During the post-peak stage, volumetric strain and permeability change significantly. The change rates of volumetric strain and permeability are highly consistent, but have low correlation with confining pressure.

3.3. Effect of Gas Pressure on Deformation and CH$_4$ Seepage

According to the principle of effective stress, the enclosed gas pressure can reduce the net value of external pressure on the coal matrix. Overall, with an increase in gas pressure, dilatant stress, peak stress, and residual stress all increase accordingly. With an increase in confining pressure, the enclosed space tends to be ideal and the three types of stress show a much more significant increase, as shown in Figure 5. In most cases, residual stress is greater than dilatant stress. After peak stress, the enclosed space is broken and the static pressure in the porous network does not play a determining role, while under dilatant stress, the sample is still in a nonlinear enclosed space and pore pressure does not effectively cause an increase. Such difference indicates that pore pressure plays a less significant role than confining pressure. For the applicable conditions of the principle of effective stress, there is an available-to-unusable transition corresponding to the enclosed space existing and breaking after failure.

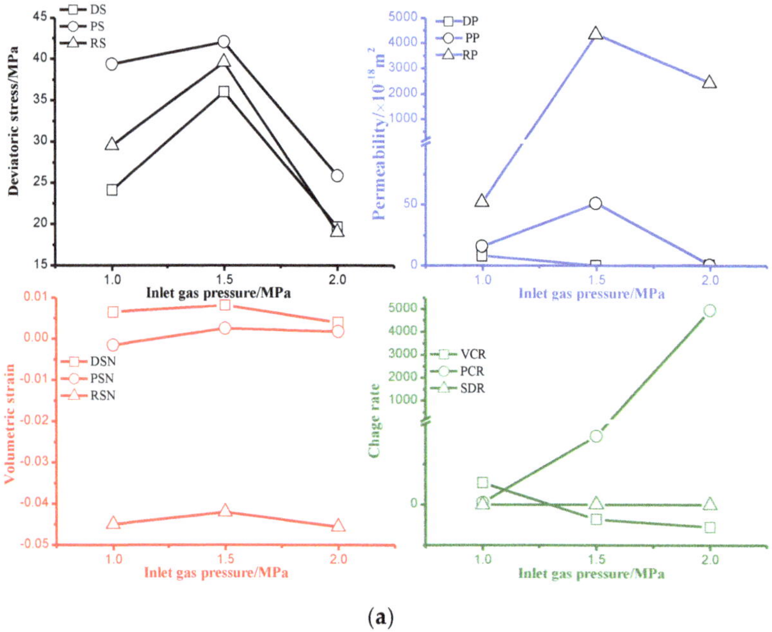

(a)

Figure 5. *Cont.*

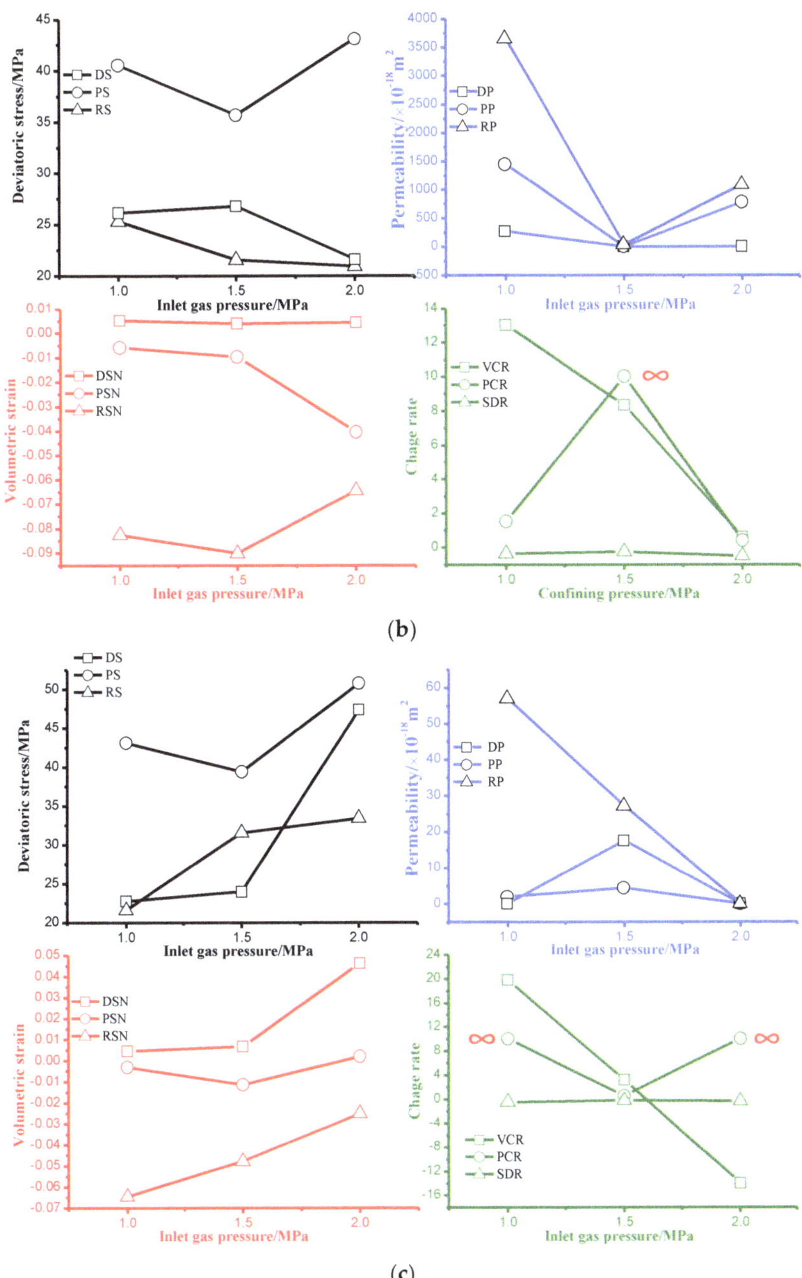

Figure 5. Deviatoric stress, volumetric strain, permeability, and permeability change rate at dilatant point, peak point, and residual point under confining pressures of (**a**) 3 MPa, (**b**) 6 MPa, and (**c**) 9 MPa ∞ indicates an infinite value because the denominator may be 0, represented by a larger value such as 10.

Volumetric change under different pore pressures is not significant, and dilatant strain experiences no change. Therefore, the change in peak strain and residual strain is mainly due to deviatoric stress. Permeability changes in a similar way to volumetric change. The sensitivity of pore pressure on the rate of volumetric strain and permeability is greater than the stress drop.

Coal deformation includes continuous deformation of matrix influenced by effective stress, pore deformation influenced by pore pressure, and discontinuous deformation of fracture influenced by deviatoric stress. Based on triaxial compression, before the dilatant strain, coal can be considered as a porous medium, dominated by matrix deformation and pore deformation, and the principle of effective stress is met in the enclosed pore space. After dilatant strain, nonlinear deformation promotes the development of a multiscale network, including pore interconnection and fracture generation. After peak strain, the deformation is mainly due to discontinuous fracture connection, and the principle of effective stress is no longer applicable in the broken enclosed space. Before peak strain, deviatoric stress and effective stress dominate the deformation continuously. After, deviatoric stress dominates the structural deformation by discontinuous fractures. Confining pressure and pore pressure can be external stress factors, and the complicated distribution and formation of discontinuities of the multiscale network are internal factors. External stress controls coal deformation by the local stress influenced by internal factors.

4. Percolation Model for Critical Behavior of CH_4 Seepage

4.1. CT Reconstruction of Network and Determination of Fractal Dimension

The CT topography of fractures for all nine samples is shown in Figure 6. After failure, the discontinuities can be divided into micropore network and macro-shear fractures. The complicated distribution of discontinuities shows the strong multiscale effect, and the multiscale connection forms a shearing block. Under triaxial compression, shear stress dominates the connected fracture generation and the shear in 3D space can occur in multiple fracture planes resulting in single network (C01 and C04), multiple network (C02 and C03), bending network (C06), crossed network (C07 and C08), pore-dominated network (C05), and complex network (C09).

Fractal dimension is often used to measure the roughness of a fracture surface, and spatial roughness reflects the complexity of the distribution of shear planes. In this study, we propose a self-similar fractal method [30], i.e., the volume-covering method, to characterize the roughness of a fracture surface. The fundamental concept here is to replace specific grid-covered shape with volume.

For a fractal set F, most researchers would use the n-dimensional ball or another collection method such as measuring fractal box scales. A collection of box dimension, often with an easy-to-use and less restrictive concept of s-dimensional tolerance degree, has a close relationship to volume. To illustrate this point, F_δ of δ parallel to the body is defined by:

$$A_\delta = \{x \in R^n : |x - y| \le \delta, \forall y \in A\} \tag{2}$$

So, the definition of F_δ of δ parallel to the body includes all points set at less than distance δ. For example, in the space R^3, assuming a single point of collection $F = \{x_0\}$, there is a sphere of radius δ and center coordinates x_0; assuming a line segment of collection $F = \{x_0\}$, there is a cylinder with the center line F. The n-dimensional volume of F_δ can be expressed by $vol^n(F_\delta)$ considering the speed of convergence under the condition $\delta \to 0$ with

$$vol^n(F_\delta) \sim c\delta^{n-s} \ (\delta \to 0) \tag{3}$$

where s denotes the dimension of F and c can be taken as the s-dimensional tolerance in set F.

The three-dimensional volume of F_δ could be expressed by $vol^3(F_\delta)$ considering the speed of convergence under the condition $\delta \to 0$ with form

$$\frac{vol^3(F_\delta)}{\delta^3} \sim c\delta^{-s}, (\delta \to 0) \tag{4}$$

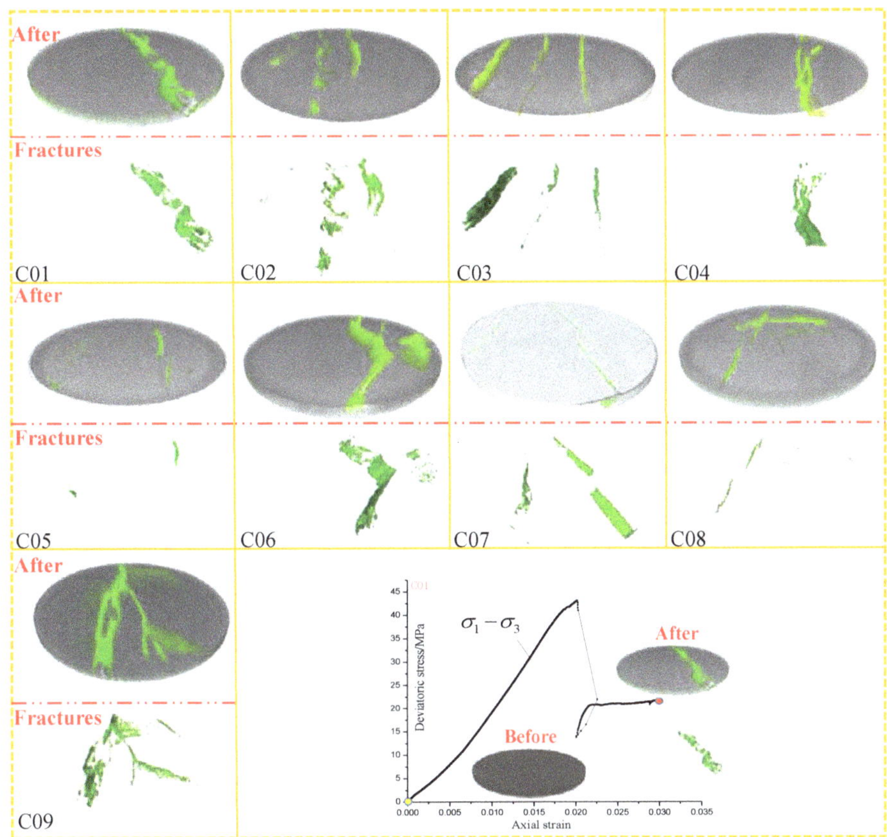

Figure 6. CT topography of multiscale network. Yellow and red points indicate CT before and after failure, respectively. C01 and C04 are single network, C02 and C03 are multiple network, C06 is bending network, C07 and C08 are crossed network, C05 is pore-dominated network, and C09 is complex network.

To extend the meaning of the cubic volume δ^3, there is the definition of $v(\delta)$ with the condition $\lim\limits_{\delta \to 0} v(\delta) = 0$. On the whole, the prism-like volume, including the rough surface, could be covered by the volume of $v(\delta)$ instead of the cubic shape to calculate all the numbers. The total number $N_{i,j}$ of volume covering one prism-like volume within grid (i, j) could be expressed by

$$N_{i,j} = INT\left(\frac{vol^3(F_\delta)}{v(\delta)} + 1\right) \tag{5}$$

So, the total number N of volume covering all prism-like volume could be calculated by

$$N(\delta) = \sum_{i,j=1}^{n-1} N_{i,j} \tag{6}$$

By changing the scale of observation, the total number N could be calculated repeatedly. Based on fractal theory, the rough surface has fractal properties that strictly follow the relationship

$$N(\delta) \sim c\delta^{-s} \tag{7}$$

According to the volume-covering method, the relationship between fractal dimension D and porosity ϕ (Table 2) of rough fractures after failure is obtained, shown in Figure 7. The relationship shows good linear characterization with fractal dimension and porosity, verifying that the fractal dimensions can be used to quantitatively characterize the complexity of the spatial distribution of rough fractures.

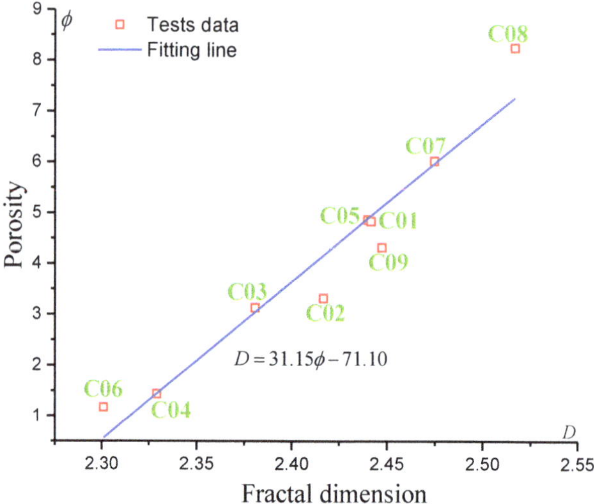

Figure 7. Relationship between fractal dimension and porosity of rough fractures.

4.2. Probability of Multiscale Connectivity Denfined by Strain-Based Damage

Under triaxial compression, spherical stress and deviatoric stress both increase. Before the dilatancy boundary, compression is mainly induced by spherical stress with the multiscale network closed. After, deviatoric stress starts to play a role in the open network. In addition, the multiscale behavior influenced by the network connection promotes increasing dilatancy. The deviatoric stress-induced dilatancy and contact surface of asperity decreasing is the main reason for enhanced permeability. Such explanation cannot quantitatively reveal the rapid increase until the network is connected. Indeed, the new interconnection channels by the deviatoric stress-induced deterioration of coal, which connect isolated pores into clusters, are the basis for the sudden transition behavior in permeability.

Moreover, volumetric strain can be used to describe the interconnectivity of the multiscale network indirectly, as shown in Figure 3. There is an increase in damage with shear stress, as well as the growing multiscale network in coal. Based on axial strain, damage D induced by deviatoric stress can be defined as:

$$D = \frac{\varepsilon_1}{\varepsilon_{1D}} \tag{8}$$

where ε_1 is the axial strain corresponding to axial stress σ_1 and ε_{1D} is the critical axial strain at the dilatant boundary.

When the damage reaches the value of 1, the stress–strain relationship turns from increasing into decreasing, indicating that strengthen behavior changes into strain-softening. The volumetric strain with axial strain also shows the sudden increase at the boundary as well as permeability. We can understand that the transition behavior is caused by the sudden connection of multiscale networks corresponding to the volumetric behavior shifting from compression to dilatancy. With the complete connection of multiscale networks in coal, the probability of CH_4 flow through the coal increases. Without micropores in the linear deformation, we mainly focus on the permeability evolution influenced by nonlinear deformation.

Clusters are formed by connecting adjacent pores, with an increase in local damage. Then, the clusters keep connecting to each other to form a bigger multiscale network. The connectivity of two pores can be defined as the number of paths allowing CH_4 to flow as well as the two clusters, even two networks. On a different scale, the number is different and its evolution with the damage is scale invariance [21–24]. Based on the percolation model, we can understand the phenomenon of phase transition caused by the multiscale network [20,26].

When the percolation threshold of network connectivity is reached, permeability increases drastically. The transition of connectivity means the phase of coal changes from porous to fractured medium. By taking the void generation and connection between them as sites and bonds in the percolation modeling, the probability of connectivity P can be defined as:

$$P \propto (p - p_c)^\beta \tag{9}$$

where p and p_c mean the cause of phase transition of connectivity, which is defined as the probability of the occupied lattice of a detailed parameter, i.e., void density ρ or volumetric strain, and critical probability, respectively. β is a conductivity index of 2 considering the 3D effect [31–33].

Void density is defined as the number of sites in a covering element, and it is also influenced by the multiscale network. As for the sharply increasing transition, crack density instantaneously changes with axial stress, confining pressure, and gas pressure. Its evolution depends on the spatial topology of the network's distribution, generation, and connection. Determination of density can be obtained quantitatively by CT technology. CT is often used to describe the micropore structure in coal, and the phase transition of multiscale structure may adapt the same way.

The final CT topography of the multiscale network can be easily obtained, as shown in Figure 6. The multiscale effect indicates that the scale-dependent variables should be taken into consideration. The critical behavior often shows self-similarity and scale invariance, and fractal theory is often used to describe the similarity of the disordered structures. The linear relationship of the fractal dimension by the volume-covering method and porosity, shown in Figure 7, indicates the self-similarity of the network. The fractal dimension has the same scale invariance as the conductivity index β in percolation theory. Therefore, we can define the multiscale structure of this nature as fractal percolation.

4.3. Percolation Model for Transition Behavior of CH_4 Seepage

Multiscale and quantitative descriptions of conductivity are normally based on permeability and correlate with connected clusters. The geometric opening and closing of pore-crack-fracture are determined by the competition behavior between normal stress-induced compression and shear stress-induced dilatancy. Indeed, shear stress can cause slipping behavior with opening of the contact at the aperture scale, and the improved permeability enhances conductivity.

The evolution of stress-strain-connectivity-conductivity-permeability is stress-dependent permeability, or stress-permeability for short [34]. Many results of stress-permeability show a good fitting relation before failure without strain-connectivity-conductivity. When coal as a porous medium changes into a fracture-dominated structure, few results of stress-permeability can illustrate the sudden transition of the geometric phase. Darcy's law can describe the seepage behavior in porous media, and the cubic law is the theoretical basis for analyzing stress-permeability for fractures [35–37]. For steady flow, the two laws can well describe before and after the dilatant boundary. Based on the steady state of stress, stress-permeability can have good applicability. When at the dilatant boundary, brittle failure with decreasing stress and unchanged strain indicates that stress-permeability is invalid to describe the abrupt change in permeability. Indeed, using stress-induced strain for the direct variable as a geometric factor for permeability and strain-permeability needs more consideration.

It is difficult, if not impossible, to precisely measure and quantify the aperture of multiscale networks. Alternatively, strain can be measured directly and has an advantage of taking into consideration all complicated factors. Establishing strain-permeability may be a more effective method. The relationship between the probability of connectivity and permeability can be defined as:

$$P \propto \frac{k}{k_f} \tag{10}$$

The percolation model for a multiscale network depends on volumetric strain and can be established by triaxial compression of coal. Volumetric strain at the dilatant boundary, shown in Table 3, is determined by axial strain and radical strain. Therefore, critical probability p_c can be expressed as:

$$p_c \propto \varepsilon_{1D} + \varsigma(\varepsilon_{3D}) \tag{11}$$

where ε_{1D} is critical axial strain at the dilatant boundary and determines the trend of the percolation curve and $\varsigma(\varepsilon_{3D})$ represents the effect of radial stain and determines the shift of the trend.

The probability density function of axial strain can be used for the multiscale analysis of percolation structures. Therefore, the corresponding occupied probability is:

$$p = 1 - \exp\left(-\frac{\varepsilon_1}{\varepsilon_{1D}}\right) \tag{12}$$

When the multiscale structure reaches the critical state corresponding to ε_{1D}, then the critical percolation can be defined as:

$$p_c = 1 - e^{-1} = 0.63 \tag{13}$$

With Equations (9), (10), (12), and (13), the permeability relationship under percolation theory can be obtained as:

$$k = Ck_f\left[1 - e^{-(\varepsilon_1/\varepsilon_{1D})} - 0.63\right]^2 \tag{14}$$

where C is a constant with a value of 40, which is obtained from the fitting curve in Figure 8.

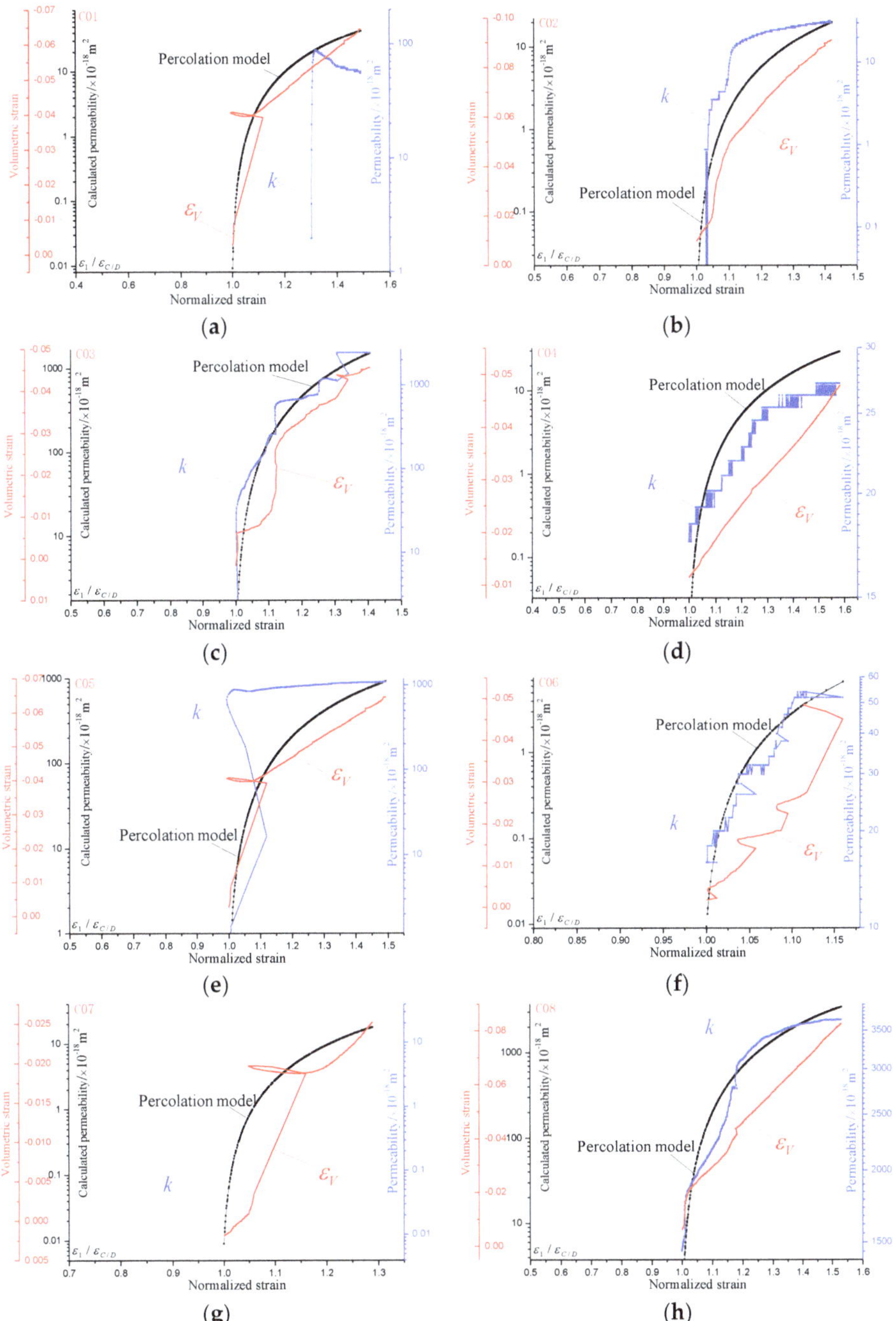

Figure 8. *Cont.*

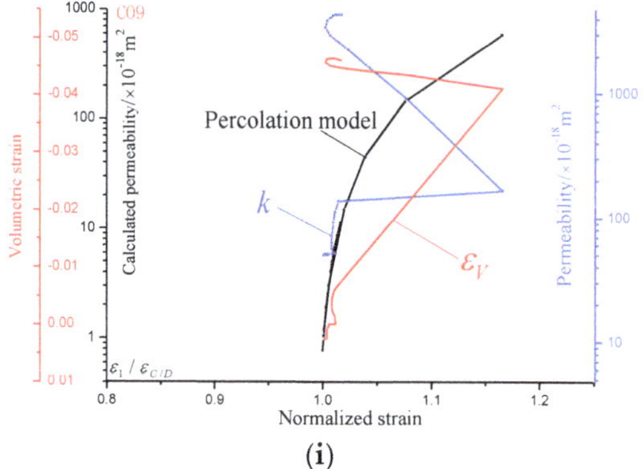

(**i**)

Figure 8. Percolation model validation and comparison with the experimental data for samples (**a**) C01, (**b**) C02, (**c**) C03, (**d**) C04, (**e**) C05, (**f**) C06, (**g**) C07, (**h**) C08, and (**i**) C09. Black lines indicate the percolation model, red lines indicate the volumetric-normalized strain relationship, and blue lines indicate permeability with normalized strain.

Using Equation (14), we compare the theoretical model with the experimental data for the permeability of all nine samples under triaxial compression. In Figure 8, the effective permeability of C07 is not obtained experimentally. The results for the remaining samples show acceptable matches, and even satisfying matches for C02, C03, C04, C06, and C08. Although confining pressure and gas pressure boundaries are different for each sample in Figures 4 and 5, the percolation model for CH$_4$ in coal shows excellent effectiveness.

5. Conclusions

In this study, the multiscale network is reconstructed by CT technology and determined quantitatively by fractal theory according to the volume-covering method proposed. The coupled tests are carried out by orthogonal experiments of confining pressure and gas pressure. The evolution of CH$_4$ seepage is modeled by percolation theory for the coupled strain-permeability. The main conclusions are drawn as follows:

1. The gas-stress coupled experiment was carried out for ineffectiveness of the stress-dependent model of permeability at the dilatant boundary. The results show that before failure, coal, as a porous medium, can be coupled with CH$_4$ seepage by the Darcy's flow, and after failure, coal as a fractured medium can be solved by the cubic law. The closed space before and after failure is the key controlling the assumptions for the law. The stress-dependent model is only effective based on the representative volume element as the continuous medium. The results also show that breakage of the closed space at the dilatant boundary will cause the multiscale behavior, with a sudden increase in volumetric strain and permeability.

2. The coupled strain-permeability of orthogonal experiments on confining pressure and gas pressure was investigated. The results show that there is not good agreement between deviatoric stress, volumetric strain, permeability, change rate, and confining pressure and pore pressure. Almost all the samples have a highly consistent relationship between volumetric strain and permeability at the dilatant point, peak point, and residual point of the stress-strain curve.

3. CT-based reconstruction of multiscale networks in coal at the dilatant boundary is proposed. The results of all samples show various kinds of multiscale networks, and the isolated micropores can be neglected in the linear deformation for the transition of permeability. The shear stress induced by nonlinear deformation of the network should receive more attention.

4. The fractal percolation model of the multiscale network by shear stress is verified by the volume-covering method. The linear relationship between porosity and fractal dimension by the volume-covering method shows the multiscale effect and effective description by fractal theory. The scale invariance of void density, connectivity, and conductivity suggests the percolation model's understanding of the coupled behavior of gas-stress-induced behavior.

5. The strain-based percolation model of permeability evolution under triaxial compression is proposed. The volumetric strain of connected clusters in triaxial compression dominates the transition behavior. The evolution of permeability with volumetric strain shows a good correlation. The results show that the percolation model is very effective at illustrating and predicting the sudden transition behavior of CH_4 seepage in coal by high agreement with the experimental data.

Author Contributions: Conceptualization and writing, D.X.; Methodology and software, J.Z. and Y.L.; Visualization, S.Z.

Funding: This study was supported by the National Natural Science Foundation of China (grant no. 51504257), the State Key Research Development Program of China (grant no. 2016YFC0600704), the Fund of Yueqi Outstanding Scholars from China University of Mining and Technology (Beijing) (grant no. 2017A03 and 2018A06), and Open Fund of the State Key Laboratory of Coal Mine Disaster Dynamics and Control at Chongqing University (2011DA105287-FW201604).

Acknowledgments: We are indebted to Liu at Sichuan University and Jiao at Nanyang Technological University for their critical and constructive review of a preliminary version of this manuscript. We would also like to cordially acknowledge excellent guidance of Zhou and technical work of Wang.

Conflicts of Interest: The authors declare no conflict of interest.

Abbreviations

DS	Dilatant stress (deviatoric stress)
PS	Peak stress (deviatoric stress)
RS	Residual stress (deviatoric stress)
DP	Dilatant permeability
PP	Peak permeability
RP	Residual permeability
DSN	Dilatant strain
PSN	Peak strain
RSN	Residual strain
SDR	Stress drop rate
PCR	Permeability change rate
VCR	Volumetric strain change rate
EDZ	Excavation damaged zone
CT	X-ray computed tomography

Symbol List

k, k_0, k_f	permeability, initial permeability, final permeability
p_0	atmospheric pressure, 0.1 MPa
p_1, p_2	inlet and outlet gas pressure
q	flow rate
A	cross-sectional area
μ	gas viscosity coefficient
L	length
F	fractal set
A_F^δ	parallel body with distance δ

δ	distance
x_0	center coordinates
x, y	Cartesian coordinates
$vol^n(F_\delta)$	volume in n-dimensional space
s	dimension of F
c	s-dimensional tolerance in set F
$v(\delta)$	prism-like volume
i, j	grid coordinate
$N_{i,j}$	total amount of volume covering one prism-like volume
N	total amount of volume covering all prism-like volume
D	damage
ϕ	porosity
P	probability of connectivity
p	probability of occupied lattice
p_c	critical probability
β	conductivity index
σ_1, σ_3	maximum and minimum principal stress
$\varepsilon_1, \varepsilon_3$	maximum and minimum principal strain
$\varepsilon_{1D}, \varepsilon_{3D}$	maximum and minimum principal strain at critical dilatant boundary
C	a constant obtained from the fitting curve

References

1. Qin, Y.; Tong, F.; Yang, G.; Mauzerall, D.L. Challenges of using natural gas as a carbon mitigation option in China. *Energy Policy* **2018**, *117*, 457–462. [CrossRef]
2. Li, H.; Lau, H.C.; Huang, S. China's coalbed methane development: A review of the challenges and opportunities in subsurface and surface engineering. *J. Pet. Sci. Technol.* **2018**, *166*, 621–635. [CrossRef]
3. Li, Q.; Lin, B.; Zhai, C. A new technique for preventing and controlling coal and gas outburst hazard with pulse hydraulic fracturing: A case study in Yuwu coal mine, China. *Nat. Hazards* **2015**, *75*, 2931–2946. [CrossRef]
4. Yuan, L. Theory of pressure-relieved gas extraction and technique system of integrated coal production and gas extraction. *J. China Coal Soc.* **2009**, *34*, 1–8.
5. Xue, D.J.; Zhou, H.W.; Chen, C.F.; Jiang, D.Y. A combined method for evaluation and prediction on permeability in coal seams during enhanced methane recovery by pressure-relieved method. *Environ. Earth Sci.* **2015**, *73*, 5963–5974. [CrossRef]
6. Somerton, W.H.; Söylemezoğlu, I.M.; Dudley, R.C. Effect of stress on permeability of coal. *Int. J. Rock. Mech. Min. Sci. Geomech. Abstr.* **1975**, *12*, 129–145. [CrossRef]
7. Durucan, S.; Edwards, J.S. The effects of stress and fracturing on permeability of coal. *Min. Sci. Technol.* **1986**, *3*, 205–216. [CrossRef]
8. McKee, C.R.; Bumb, A.C.; Koenig, R.A. Stress-dependent permeability and porosity of coal and other geologic formations. *SPE Form. Eval.* **1988**, *3*, 81–91. [CrossRef]
9. Seidle, J.P.; Jeansonne, M.W.; Erickson, D.J. Application of matchstick geometry to stress dependent permeability in coals. *Soc. Petrol. Eng. J.* **1992**, 1–11. [CrossRef]
10. Liu, H.H.; Rutqvist, J.; Berryman, J.G. On the relationship between stress and elastic strain for porous and fractured rock. *Int. J. Rock. Mech. Min. Sci.* **2009**, *46*, 289–296. [CrossRef]
11. Liu, H.H.; Rutqvist, J. A new coal-permeability model: Internal swelling stress and fracture—Matrix interaction. *Transp. Porous Media* **2010**, *82*, 157–171. [CrossRef]
12. Shi, J.Q.; Durucan, S. Drawdown induced changes in permeability of coalbeds: A new interpretation of the reservoir response to primary recovery. *Transp. Porous Media* **2004**, *56*, 1–16. [CrossRef]
13. Witherspoon, P.A.; Wang, J.S.Y.; Iwai, K.; Gale, J.E. Validity of Cubic Law for fluid flow in a deformable rock fracture. *Water Resour. Res.* **1980**, *16*, 1016–1024. [CrossRef]

14. Valdes-Parada, F.J.; Ochoa-Tapia, J.A.; Alvarez-Ramirez, J. Validity of the permeability Carman–Kozeny equation: A volume averaging approach. *Phys. A Stat. Mech. Appl.* **2009**, *388*, 789–798. [CrossRef]

15. Kruczek, B. Carman–Kozeny Equation. In *Encyclopedia of Membranes*; Springer: Berlin, Germany, 2014; pp. 1–3.

16. Brush, D.J.; Thomson, N.R. Fluid flow in synthetic rough-walled fractures: Navier-Stokes, Stokes, and local cubic law simulations. *Water Resour. Res.* **2003**, *39*, 1–5. [CrossRef]

17. Zhu, W.; Wong, T.F. The transition from brittle faulting to cataclastic flow: Permeability evolution. *J. Geophys. Res. Solid Earth* **1997**, *102*, 3027–3041. [CrossRef]

18. Zhu, W.; Wong, T.F. Network modeling of the evolution of permeability and dilatancy in compact rock. *J. Geophys. Res. Solid Earth* **1999**, *104*, 2963–2971. [CrossRef]

19. Isichenko, M.B. Percolation, statistical topography, and transport in random media. *Rev. Mod. Phys.* **1992**, *64*, 961–1043. [CrossRef]

20. Hunt, A.; Ewing, R.; Ghanbarian, B. *Percolation Theory for Flow in Porous Media*; Springer: Berlin, Germany, 2014.

21. Broadbent, S.R.; Hammersley, J.M. *Percolation Processes: I. Crystals and Mazes*; Cambridge University Press: London, UK, 1957; pp. 629–641.

22. Shante, V.K.S.; Kirkpatrick, S. An introduction to percolation theory. *Adv. Phys.* **1971**, *20*, 325–357. [CrossRef]

23. Chelidze, T.L. Percolation and fracture. *Phys. Earth Planet. Inter.* **1982**, *28*, 93–101. [CrossRef]

24. Bebbington, M.; Vere-Jones, D.; Zheng, X. Percolation Theory: A model for rock fracture? *Geophys. J. Int.* **1990**, *100*, 215–220. [CrossRef]

25. Sahimi, M. Flow phenomena in rocks: From continuum models to fractals, percolation, cellular automata, and simulated annealing. *Rev. Mod. Phys.* **1993**, *65*, 1393–1534. [CrossRef]

26. Alkan, H. Percolation model for dilatancy-induced permeability of the excavation damaged zone in rock salt. *Int. J. Rock Mech. Min. Sci.* **2009**, *46*, 716–724. [CrossRef]

27. ASTM D2664. Standard Test Method for Triaxial Compressive Strength of Undrained Rock Core Specimens without Pore Pressure Measurement. 1986. Available online: https://www.astm.org/DATABASE.CART/HISTORICAL/D2664-95A.htm (accessed on 10 August 2018).

28. Hsieh, P.A.; Tracy, J.V.; Neuzil, C.E.; Bredehoeft, J.D.; Silliman, S.E. A transient laboratory method for determining the hydraulic properties of "tight" rocks—I. Theory. *Int. J. Rock Mech. Min. Sci. Geomech. Abstr.* **1981**, *18*, 245–252. [CrossRef]

29. Neuzil, C.E.; Cooley, C.; Silliman, S.E.; Bredehoeft, J.D.; Hsieh, P.A. A transient laboratory method for determining the hydraulic properties of "tight" rocks—II. Application. *Int. J. Rock Mech. Min. Sci. Geomech. Abstr.* **1981**, *18*, 253–258. [CrossRef]

30. Zhou, H.W.; Xue, D.J.; Jiang, D.Y. On fractal dimension of a fracture surface by volume covering method. *Surf. Rev. Lett.* **2014**, *21*, 1–11. [CrossRef]

31. Stauffer, D.; Aharony, A. *Introduction to Percolation Theory: Revised Second Edition*; CRC Press: London, UK, 2014.

32. Sahini, M. *Applications of Percolation Theory*; CRC Press: London, UK, 2014; pp. 23–40.

33. Hestir, K.; Long, J.C.S. Analytical expressions for the permeability of random two-dimensional Poisson fracture networks based on regular lattice percolation and equivalent media theories. *J. Geophys. Res. Solid Earth* **1990**, *95*, 21565–21581. [CrossRef]

34. Sakhaee-Pour, A.; Agrawal, A. Integrating acoustic emission into percolation theory to predict permeability enhancement. *J. Pet. Sci. Eng.* **2018**, *160*, 152–159. [CrossRef]

35. Louis, C. A study of groundwater flow in jointed rock and its influence on the stability of rock masses. *Rock Mech. Res. Rep.* **1969**, *10*, 1–90.

36. Lomize, G.M. Flow in fractured rocks. *Gosenergoizdat* **1951**, *127*, 496. (In Russian)

37. Xue, D.J.; Zhou, H.W.; Liu, Y.T.; Deng, L.S.; Zhang, L. Study of drainage and percolation of nitrogen-water flooding tight coal by NMR Imaging. *Rock Mech. Rock Eng.* **2018**, 1–17. [CrossRef]

Article

Deformation and Hydraulic Conductivity of Compacted Clay under Waste Differential Settlement

Sifa Xu [1], Cuifeng Li [1], Jizhuang Liu [2], Mengdan Bian [1], Weiwei Wei [1], Hao Zhang [1] and Zhe Wang [1,*]

[1]   Institute of Geotechnical Engineering, Zhejiang University of Technology, Hangzhou 310014, China; xusifa@zjut.edu.cn (S.X.); licuifeng93@163.com (C.L.); mengdanbian@163.com (M.B.); 18451390922@163.com (W.W.); zhanghao@zjut.edu.cn (H.Z.)

[2]   Zhejiang College of Construction, Hangzhou 310014, China; liujizhuang137@sohu.com

*   Correspondence: wangzsd@163.com; Tel.: +86-139-6811-0665; Fax: +86-0571-8832-0460

Received: 20 July 2018; Accepted: 6 August 2018; Published: 8 August 2018

Abstract: Landfill is still the most important process to dispose of municipal solid waste in China, while landfill closure aims for pollution control, security control, and better land reuse. However, uneven settlement of landfill cover system is very likely to cause deformation and cracking. The objective of this paper is to examine the effects of geogrid reinforcement on the deformation behaviour and hydraulic conductivity of the bentonite-sand mixtures that are subjected to differential settlement. The laboratory model tests were performed on bentonite-sand mixtures with and without the inclusion of geogrid reinforcement. By maintaining the type and location of the geogrid within the liner systems as constant, the thickness of the bentonite-sand mixtures is varied. The performation of the liner systems with and without the inclusion of geogrid reinforcement was assessed by using jack to control differential settlement. Un-reinforced bentonite-sand mixtures of 100 mm and 200 mm thickness were observed to begin cracking at settlement levels of 2.5 mm and 7 mm, respectively. When settlement reached 25 and 42.5 mm, cracks for 100 mm and 200 mm thick bentonite-sand mixtures without geogrid penetrated completely. The settlement levels for bentonite-sand mixtures of 100 mm thickness with and without geogrid reinforcement was found to be 10 mm and 15 mm, respectively, when its hydraulic conductivity was around $5 * 10^{-7}$ cm/s. In comparison, geogrid reinforced bentonite-sand mixtures was found to sustain large deformation with an enhanced imperviousness. The results from the present study can provide theory evidence of predicting deformation and hydraulic conductivity of the landfill cover system.

Keywords: bentonite-sand mixtures; differential settlement; deformation; hydraulic conductivity; crack; geogrid

---

1. Introduction

Apart from exhaust and drainage layers, an impervious layer is set in landfill terminal cover systems to prevent methane entry into the atmosphere or explosion that is caused by agglomeration following landfill closure. Meanwhile, impermeable layer can prevent rain from flowing into landfill to form leachate, thus avoiding pressure to treatment, and provide space for land reuse after landfill closure. Of the two structures for landfill closure cover systems in China, the more commonly adopted one is to use natural clay barrier materials to collect air and prevent rain from entering the garbage, with a required clay layer hydraulic conductivity of lower than 10^{-7} cm/s and a soil thickness of larger than 300 mm [1].

In this paper, municipal solid waste is defined as high compressibility with large pores. Actual observation finds that a lot of compression remains in landfills after the closure. When the fill height reaches the designed level and the landfill is closed, with compression settlement exceeding fill height by 30%, bending deformation appears at the top of landfill impervious layer. When subjected to shear and tension in the cover system, cracks occur in imperious materials and such fissures become channels for surface water and landfill gas.

It is suggested that clay has very low cracking resistance performance and would crack with slightly bending deformation, meanwhile change of soil moisture content is likely to cause cracking and thus directly affect permeability [2]. Cracking behavior of compacted clay beams with different water content was studied, with crack propagation analyzed, revealing the strain localization of crack tip and obtaining the cracking strain of compacted clay with various water content [3,4]. The tension cracking process of clay was studied through a three-point bending test of compacted clay beam, along with the effects of water content on cracking properties [5]. In addition, saturated/unsaturated flow models for compacted clay layer with single slit were established, while the water migration process in the fractured clay layer was analyzed. As a result, it was thought that crack development has significant effect on seepage properties of the compacted clay layer [6]. Deformation characteristics of clay under tension are mostly demonstrated through uniaxial tensile test, triaxial tensile test, bending test, and splitting test of soil beam [7,8]. Uniaxial tensile test of compacted clay reveals a different tensile strength of compacted clay with different tensile stress and strain curve profile [9]. Some tests studied the triaxial tensile and compressive properties of dam clay, established clay joint strength criteria, and extended the Duncan-Chang model [10].

Since the solid waste is highly heterogeneous material and can settle either due to biodegradation of waste, or by its own weight or by overlying pressure applied above the barrier, development of differential settlement within the landfill area is common. The excessive differential settlements can result in the development of tension cracks in the soil barrier [11,12]. Bending tests were also conducted to investigate deformation characteristics of compacted clay cover systems under differential settlements, while considering that clay was cracking when its distortion reached 2~3% [13]. Four points bending test and on-site bending test were carried out on the surface of soil layer to measure soil strain, showing that cracking strain was about 0.1~2% [14,15]. Deformation characteristics of clay liner under local settlement were analyzed by centrifugal model, finding that cracking occurred in a clay liner under no overburden pressure [16]. The Mohr-Coulomb (M.C) Elasto-plastic model for soil beam bending tests was carried out by Plé et al. [17] to interpret data from both laboratory and field tests. Viswanadham and Jessberger and Viswanadham and Muthukumaran [18,19] introduced a biaxial geogrid layer within soil barrier to suppress cracks in the soil barrier, while being subjected to differential settlements.

This simulation was then used to study the crack formation in the clay cover barrier caused by differential settlement and predict the initiation cracks. Test site of compacted clay cover system was monitored for many years, while detecting the integrity of compacted clay layer by excavation, which showed that factors, including desiccation cracks and plant root holes, would have a significant impact on the impermeability of clay layer [20–23]. In effect, field measurements and researches of existing landfills have revealed that many traditional covers are not obtaining anticipative performances in semi-arid and arid areas [24]. Cracks in clay liner and complicated composition of landfill leachate may have effect on hydraulic conductivity of compacted clay liner, while desiccation cracks and bentonite have more crucial influence on permeability than analog leachates [25]. Desiccation cracking is a common phenomenon in soils [26], with significant factors evaluated, proving that soil shear strength and tensile strength and soil thickness had a dominant effect on desiccation cracking [27,28]. It is worth noting that wet and dry cycles are considered to affect soil permeability by changing joint soil parameters of the entire non-saturated permeability, as well as to reveal seepage control of water content on compacted clay cover system [29,30].

As mentioned above, existing researches mainly focus on tensile strength and crack characteristics of soil. However, there is no adequate investigation on development of cracks in soil regarding the change of differential settlement, change of hydraulic conductivity during crack development, and effects of geogrid on deformation and hydraulic conductivity of bentonite-sand mixtures.

2. Material and Methods

2.1. Materials

2.1.1. Low-Permeability Soil

As imperious material in our test, low-permeability soil is made up of sandy soil and bentonite due to that clay is difficult to obtain in many places. Sandy soil particle size distribution is shown in Figure 1, showing that the maximum particle size is less than 5 mm, with 2.52% fine grain fraction, 5.1 inhomogeneity coefficient, and 1.15 curvature coefficient, which belongs to graded soil.

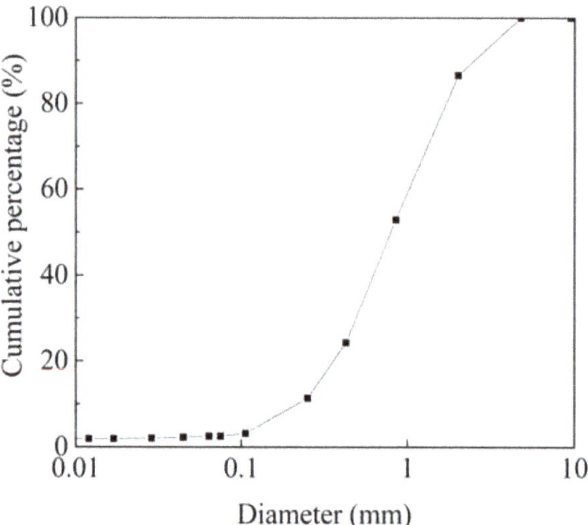

Figure 1. Particle-size distribution of sand.

Bentonite is Na-bentoite obtained from the United States, which has low-permeability, good expansibility, high ion exchange capacity, as well as sufficient mechanical cushioning properties. Specific gravity is 2.76, plastic limit 38, liquid limit 581, plastic index 543, and cation exchange capacity 0.77 meq/g.

The content of bentonite is 10% (kg/kg, dry basis) of dry sand. Results of Proctor compaction tests (compaction energy = 592.2 kJ/m^3) on bentonite-sand mixtures shows that the maximum dry density of sand is 1.60 g/cm^3 with an optimal moisture content of 17.5%. The hydraulic conductivity is 8.99 * 10^{-7} cm/s at steady state by penetration test.

2.1.2. Geogrid

The geogrid with the thickness of 1.2 mm is made of high density polyethylene (Figure 2). It has a hole size of 15 mm and tensile strength of 7.6 kN/m.

Figure 2. Geogrid.

2.2. Apparatus

Experimental apparatus is shown in Figure 3. Low-permeability clay is used for laying in the model box, with length of 800 mm and width of 200 mm. A plastic pipe with holes is embedded just beneath the top of the bentonite-sand mixtures. The distilled water after infiltration is collected in the lower part of the apparatus. A rubber air bag is embedded on the top of soil, and hydraulic conductivity of low-permeability soil after local subsidence is obtained by adopting the Darcy's law under constant head condition. The middle 400 mm is settlement area, which is controlled by Jack, with the displacement meter set on both sides. A rubber bag is placed on low-permeability soil, and uniform load is applied by filling air pressure. Tests are conducted to record settlement, deformation, load, and soil cracks.

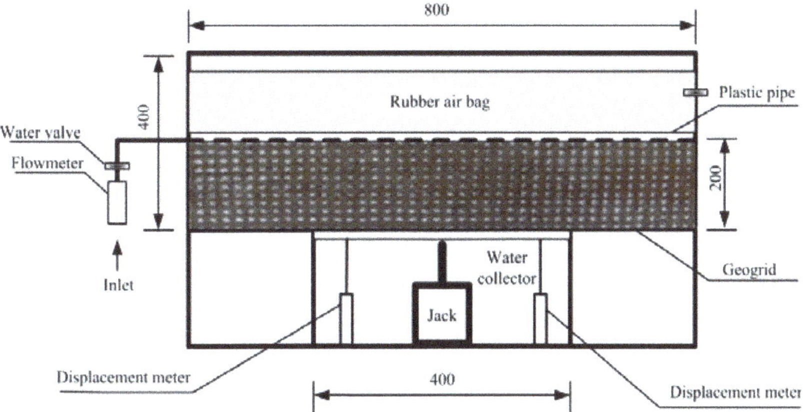

Figure 3. Sketch of apparatus (all dimensions are in mm).

2.3. Test Procedure

2.3.1. Experimental Scheme

To observe the influence of soil thickness, geogrid, and upper load on the cracking and hydraulic conductivity of low-permeability soil, upper load is set constant at 147 KPa, with four schemes, as shown in Table 1.

Table 1. Details of model tests.

S.No.	Study Focus	d (mm)	Geogrid
1	Deformation and cracks	100	No
2	Deformation and cracks	200	No
3	Deformation and cracks	100	Yes
4	Deformation and cracks	200	Yes
5	Hydraulic conductivity	100	No
6	Hydraulic conductivity	200	No
7	Hydraulic conductivity	100	Yes
8	Hydraulic conductivity	200	Yes

d—thickness of the bentonite-sand mixtures; Yes/No—with or without the inclusion of geogrid reinforcement.

2.3.2. Methods

Mixed soil is prepared with a bentonite mixing amount of 10% and water content 17.5%, and then compacted in three layers with more than 95% of compaction degree. As shown in Figure 4, the dimensions of specimens were: length 500 mm, height 100 (or 200), width 200 mm. Water pipes were placed on the soil, then with waterproof membrane laid above, and finally pressure rubber bags. A layer of bentonite is expanded on the contact surface between soil and model box wall, which swells with water and thus can seal the gap between the soil and model box wall to prevent leakage.

At the start of test, water valve was first opened, with water pressure maintained at 9.8 kPa, to observe water level and flow changes every 10 min. When the flow became stable and was then pressurized to the set load, Jack was falling at the speed of 0.5 mm/min until the set settlement. The designed settlements of bentonite-sand mixtures with thickness of 100 and 200 mm were 30 and 45 mm, respectively. Load, settlement, crack, and water flow were recorded during settlement. Deformation and cracking was analyzed through photographing marks that were set on the sides of soil.

Figure 4. Sample.

3. Results and Discussion

3.1. Settlement and Crack Development

Soil cracked with lower settlement and first occurred on the settlement edge. The relationship between settlement and crack development is presented in Figure 5, which shows crack development on the right side. It can be seen that cracks of soil layer gradually develop from bottom to top with the increase of settlement. With soil thickness of 100 mm, cracks started when settlement was 2.5 and 3.5 mm, respectively, for bottom not laid and laid with geogrid. As settlement continued to increase, cracks became more serious from bottom to top along the height, accounting for 25%, 50%, 75% of total height, and even crack transfixion. Settlement was, respectively, 5, 12.5, 22.5, and 25 mm when not laid with geogrid, while 6.5, 14.5, 23, and 30 mm when otherwise. Test results for thick soil layer of 200 mm and 100 mm are similar, i.e., cracking accounted for 25%, 50% of total height, and even crack transfixion. Settlement was, respectively, 7, 18, 34.5, and 42.5 mm when not laid with geogrid, while 10.5, 20.5, 38, and 45 mm when otherwise.

According to experimental results, with equal soil thickness, the relationship between crack depth ratio and settlement fluctuates, but on the whole under the same ratio settlement is slightly larger when laid with geogrid than otherwise. The difference is likely due to that geogrid laid has an inhibiting effect on the bottom layer and thus results in delayed cracking. When compared under the same crack depth ratio, settlement value for soil layer of 200 mm is larger than that for 100 mm.

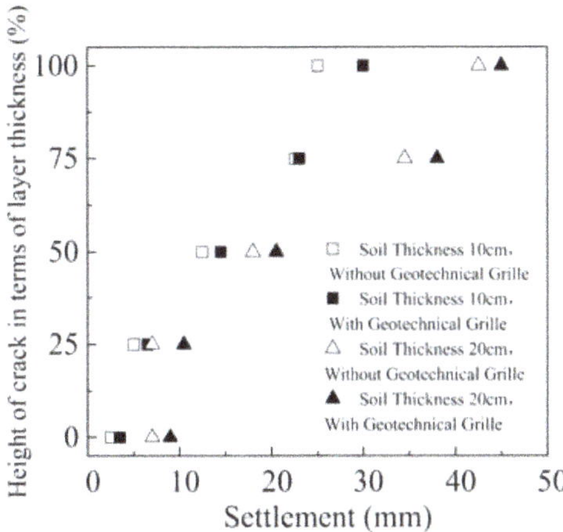

Figure 5. Relationship between settlement and crack development.

3.2. Crack Angle and Size

Soil cracks occur with the increase of settlement in the lower section, and the condition for soil thickness of 200 mm upon crack transfixion is shown in Figure 6. It can be seen that cracks occur on both sides of the subsidence area, which are in splayed shape, big on the top, while small on the bottom, with basically the same angle on the left and right side. Cracks development and angle are shown in Table 2.

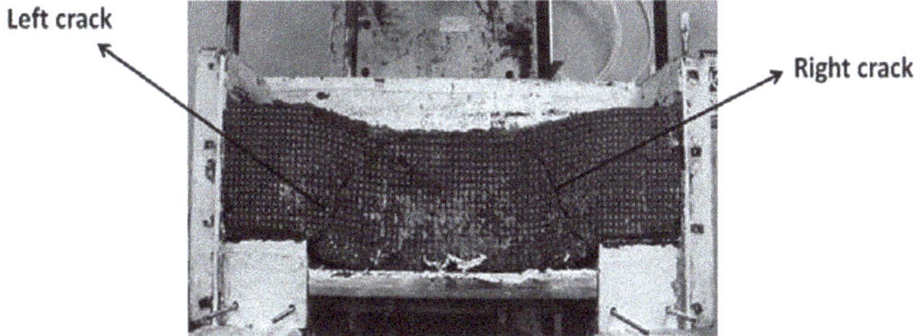

Figure 6. Picture of sample after test.

Table 2. Crack trend.

Crack Development	Angle (°)			
	Soil Thickness 100 mm		Soil Thickness 200 mm	
	No	Yes	No	Yes
Initiation	40	45	45	40
25% of thickness	47	45	50	48
50% of thickness	50	50	55	53
75% of thickness	53	57	60	59
Penetration	62	59	58	60

Yes/No—with or without the inclusion of geogrid reinforcement.

Table 2 records the angle between the crack and the height direction, showing that the angle is as small as about 40–45 degrees when cracks first occur whether or not being laid with geogrid and regardless of soil layer thickness. The angle increases with cracks development and reaches 60 degrees upon crack transfixion. When compared with a soil layer thickness of 100 mm, crack angle for soil layer thickness of 200 mm is slightly larger when the crack height is small. However, the results are basically the same when crack height becomes greater than 50% of soil thickness, indicating that soil thickness has little effect on fracture angle.

Similarly, soil bottom not laid with geogrid also has little effect on crack angle. It can be analyzed from digital camera images that with soil thickness of 100 mm and not laid with geogrid on the bottom, crack width is smaller than 1 mm at the start of cracking. With the increase of settlement, cracks in the lower section become larger, with crack width reaching 5 mm and 15 mm, respectively, when crack height is 50% of soil thickness and upon cracks transfixion. Crack development for soil thickness of 200 mm is similar to that for 200 mm, with a maximum width of 20 mm upon cracks transfixion.

With geogrid laid on the bottom, crack width remains smaller than 1mm from the occurrence of cracks to cracks developing upward along the height till crack development height reaching 50% of soil thickness. When settlement reaches 45 mm, crack width keeps around 3 mm, for soil thickness of both 100 mm and 200 mm.

3.3. Settlement and Hydraulic Conductivity

Settlement and Hydraulic Conductivity

The relationship between settlement and hydraulic conductivity is presented in Figures 7–10. As shown in Figures 7 and 8, with soil thickness of 100 mm, whether or not being laid with geogrid on the bottom, hydraulic conductivity is as big as 10^{-5}–10^{-6} cm/s at the start of watering and before the occurrence of cracking. With infiltration time prolonging, bentonite swelling in the soil filled the space between particles and hydraulic conductivity decreased. After watering for 100 h, hydraulic conductivity of soil was reduced to 10^{-7} cm/s level, or to even lower level of 10^{-8} cm/s, which meets the penetration requirements as impermeable layer. As the foundation settled, bentonite-sand mixtures started to deform and cracks appeared and exceeded 25% of the bentonite-sand mixtures thickness when the bentonite-sand mixtures without geogrid reinforcement settlement was 10 mm. The hydraulic conductivity increased from 10^{-8} cm/s to about 5×10^{-7} cm/s, because cracks were formed at the bottom of the bentonite-sand mixtures without geogrid reinforcement and negatively impacted permeability. With settlement over 20 mm, bottom cracks developed upward rapidly and formed crack transfixion, thus losing impermeability. As shown in Figure 8, relationship between variation of settlement and hydraulic conductivity was similar whether or not laid with geogrid. With settlement of 15 mm, crack height was about 50%, with hydraulic conductivity ranging between 5×10^{-7} cm/s and 10^{-6} cm/s. However, when the settlement was over 30 mm, hydraulic conductivity could still reach 10^{-6} cm/s level, and crack transfixion occurred after 150 min, with the breaking of soil layer and the loss of impermeability.

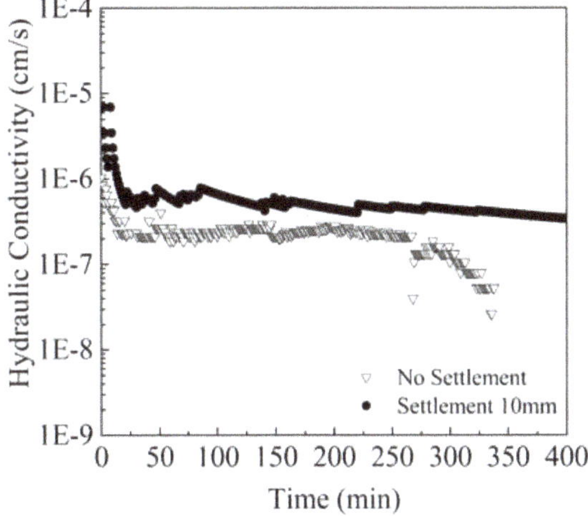

Figure 7. Time and hydraulic conductivity (soil thickness 100 mm; without geogrid).

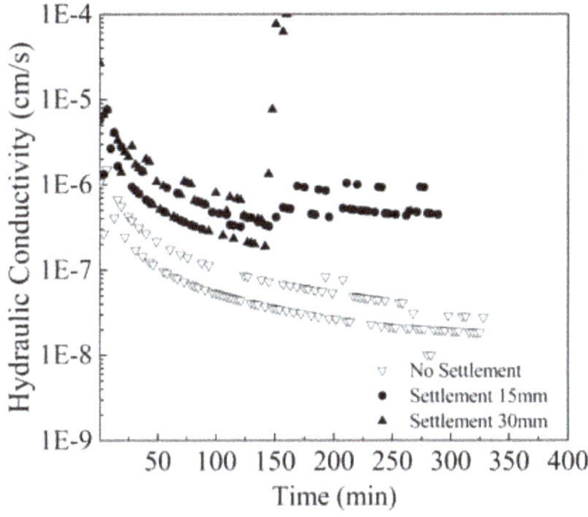

Figure 8. Time and hydraulic conductivity (soil thickness 100 mm; with geogrid).

With soil thickness of 200 mm and whether or not being laid with geogrid, relationship between settlement and hydraulic conductivity was similar, as shown in Figures 9 and 10. After 400 min of infiltration and without foundation settlement, hydraulic conductivity was kept at about 5×10^{-8} cm/s. With 15 mm of settlement, hydraulic conductivity was about 8×10^{-8} cm/s and 7×10^{-8} cm/s, respectively, when being laid with and without geogrid, which was very close mainly due to that crack height did not reach 50% of soil sickness with crack width less than 1 mm, thus having little effect on hydraulic conductivity. With further increase of settlement, cracks extended upward with increased width. When settlement reached 30 mm, crack height was close to 70% of soil thickness, with only about 60 mm uncracked and a maximum width of over 5 mm, which led to decreased hydraulic conductivity, being 2×10^{-7} cm/s and 7×10^{-7} cm/s, respectively, when being laid with and without geogrid. With settlement exceeding 40 mm, crack transfixion gradually occurred, and thus lost impermeability.

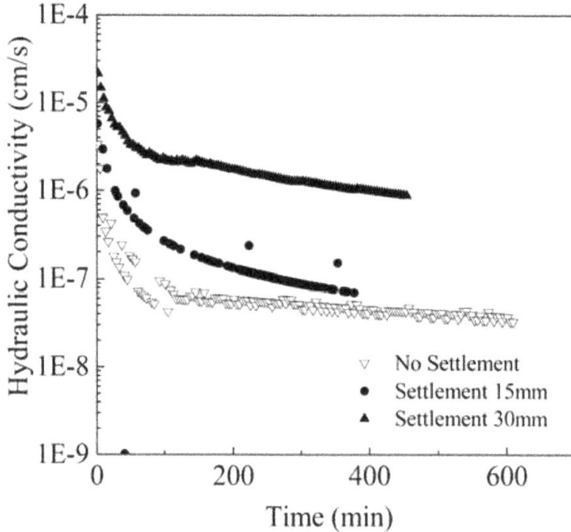

Figure 9. Time and hydraulic conductivity (soil thickness 200 mm; without geogrid).

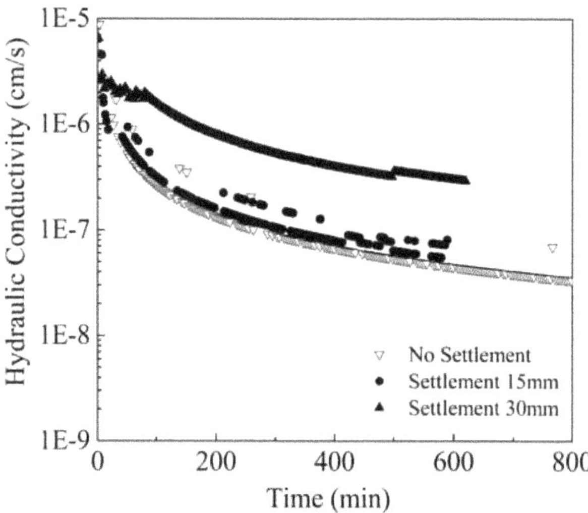

Figure 10. Time and hydraulic conductivity (soil thickness 200 mm; with geogrid).

4. Conclusions

Based on the analysis and interpretation of the model test results, the following conclusions can be drawn when the length of the settlement area was is relatively close to the thickness of the bentonite-sand mixtures:

(1)	Bentonite-sand mixtures reinforced with geogrid restrained carcking better. Un-reinforced bentonite-sand mixtures of 100 mm and 200 mm thickness were observed to begin cracking at settlement levels of 2.5 mm and 7 mm, respectively. The bentonite-sand mixtures of 100 mm and 200 mm thickness with geogrid reinforcement were observed to begin cracking at settlement levels of 3.5 mm and 9 mm, respectively. When settlement reached 25 and 42.5 mm, cracks for 100 mm and 200 mm thick bentonite-sand mixtures without geogrid being penetrated completely. Moreover, cracks for 100 mm and 200 mm thick bentonite-sand mixtures without geogrid penetrated completely when settlement reached 30 and 45 mm.

(2) Bentonite-sand mixtures reinforced with geogrid restrained carck width better. The maximum crack width for 100 mm and 200 mm thick bentonite-sand mixtures unreinforced with geogrid was found be 15 mm and 20 mm. The cracks of bentonite-sand mixtures reinforced with geogrid were <1 mm in width when the crack approached 50% of the layer thickness. When settlement reached maximum settlement, crack width kept around 3mm no matter whether the crack thickness was 100 mm or 200 mm.

(3) The settlement levels for bentonite-sand mixtures of 100 mm thickness with and without geogrid reinforcement was found to be 10 mm and 15 mm, respectively, when its hydraulic conductivity was around 5×10^{-7} cm/s. With the crack approaching 70% of the layer thickness, the hydraulic conductivity of bentonite-sand mixtures with and without geogrid reinforcement increased to 2×10^{-7} cm/s and 7×10^{-7} cm/s, respectively.

(4) The effect of geogrid on development of crack angle of bentonite-sand mixtures is little. Angle of the cracks just initiated was relatively small, at around 40–45°. The angle increased gradually as the crack continued to grow, with angle of the penetrating crack at about 60°.

Author Contributions: S.X., J.L. and H.Z. conceptualized the study and designed the experiments. C.L., M.B. and W.W. performed the experiments. C.L., S.X. and Z.W. acquired, collected and analyzed the data in the experiments. C.L. and S.X. co-wrote the paper.

Funding: The authors gratefully acknowledge the support from Natural Science Foundation of Zhejiang Province (grant numbers: Y14E080049, LY15E080021), Postdoctoral Scholarships of Zhejiang Province (grant number: BSH13020) and Project 51778585 supported by Natrual Science Foundation of China.

Conflicts of Interest: The authors declare no conflict of interest.

References

1. China Architecture & Building. *2012 Technical Code for Geotechnical Engineering of Municipal Solid Waste Sanitary Landfill*; CJJ 176–2012; China Architecture & Building Press: Beijing, China, 2012.
2. Albrecht, B.; Benson, C. Effect of desiccation on compacted natural clays. *J. Geotech. Geoenviron. Eng.* **2001**, *127*, 67–75. [CrossRef]
3. Krisnanto, S.; Rahardjo, H.; Fredlund, D.G.; Leong, E.C. Water content of soil matrix during lateral water flow through cracked soil. *Eng. Geol.* **2016**, *210*, 168–179. [CrossRef]
4. Ling, D.; Xu, Z.; Cai, W.; Wang, Y. Experimental study on characteristics of bending cracks of compacted soil beams. *Chin. J. Geotech. Eng.* **2015**, *37*, 1165–1172. [CrossRef]
5. Cai, W.; Ling, D.; Xu, Z.; Chen, Y. Influence of preferential flow induced by a single crack on anti-seepage performance of clay barrier. *Rock Soil Mech.* **2014**, *7598*, 2838–2845.
6. Vo, T.D.; Pouya, A.; Hemmati, S.; Tang, A.M. Numerical modelling of desiccation cracking of clayey soil using a cohesive fracture method. *Comput. Geotech.* **2017**, *85*, 15–27. [CrossRef]
7. Zhang, B.; Li, Q.; Yuan, H.; Sun, X. Tensile Fracture Characteristics of Compacted Soils under Uniaxial Tension. *J. Mater. Civ. Eng.* **2015**, *27*. [CrossRef]
8. Chen, X.; Zhang, J.; Li, Z. Shear behaviour of a geogrid-reinforced coarse-grained soil based on large-scale triaxial tests. *Geotext. Geomembr.* **2014**, *4*, 312–328. [CrossRef]
9. Zhang, Y.; Wang, H.M.; Yan, L.F. Test research on tensile properties of compacted clay. *Rock Soil Mech.* **2013**, *34*, 2151–2157.
10. Zhang, Y.; Zhang, B.Y.; Li, G.X.; Sun, X. Combined tension-compression triaxial tests and extended Duncan-Chang model of compacted clay. *Chin. J. Geotech. Eng.* **2010**, *32*, 999–1004. [CrossRef]
11. Dickinson, S.; Brachman, R.W.I. Deformations of a geosynthetic clay liner beneath a geomembrane wrinkle and coarse gravel. *Geotext. Geomembr.* **2006**, *24*, 285–298. [CrossRef]

12. Sharma, H.D.; De, A. Municipal Solid Waste Landfill Settlement: Postclosure Perspectives. *J. Geotech. Geoenviron. Eng.* **2007**, *133*. [CrossRef]

13. Gourc, J.P.; Camp, S.; Viswanadham, B.V.S.; Rajesh, S. Deformation behavior of clay cap barriers of hazardous waste containment systems: Full-scale and centrifuge tests. *Geotext. Geomembr.* **2010**, *28*, 281–291. [CrossRef]

14. Viswanadham, B.V.S.; Rajesh, S. Centrifuge model tests on clay based engineered barriers subjected to differential settlements. *Appl. Clay Sci.* **2009**, *42*, 460–472. [CrossRef]

15. Camp, S.; Gourc, J.P.; Ple, O. Landfill clay barrier subjected to cracking: Multi-scale analysis of bending tests. *Appl. Clay Sci.* **2010**, *48*, 384–392. [CrossRef]

16. Viswanadham, B.V.; Jha, B.K.; Sengupta, S.S. Centrifuge Testing of Fiber-Reinforced Soil Liners for Waste Containment Systems. *Pract. Period. Hazard. Toxic Radioact. Waste Manag.* **2009**, *13*, 45–58. [CrossRef]

17. Plé, O.; Manicacci, A.; Gourc, J.P.; Camp, S. Flexural behaviour of a clay layer: Experimental and numerical study. *Can. Geotech. J.* **2012**, *49*, 485–493. [CrossRef]

18. Viswanadham, B.V.S.; Jessberger, H.L. Centrifuge Modeling of Geosynthetic Reinforced Clay Liners of Landfills. *J. Geotech. Geoenviron. Eng.* **2005**, *131*, 564–574. [CrossRef]

19. Viswanadham, B.V.S.; Muthukumaran, A.E. Influence of geogrid layer on theintegrity of compacted clay liners of landfills. *Soils Found* **2007**, *47*, 519–534. [CrossRef]

20. Melchior, S.; Sokollek, V.; Berger, K.; Vielhaber, B.; Steinert, B. Results from 18 Years of In Situ Performance Testing of Landfill Cover Systems in Germany. *J. Environ. Eng.* **2010**, *136*, 815–823. [CrossRef]

21. Hettiarachchi, H.; Meegoda, J.; Hettiaratchi, P. Effects of gas and moisture on modeling of bioreactor landfill settlement. *Waste Manag.* **2009**, *29*, 1018–1025. [CrossRef] [PubMed]

22. Hirobe, S.; Oguni, K. Coupling analysis of pattern formation in desiccation cracks. Comput. *Methods Appl. Mech. Eng.* **2016**, *307*, 470–488. [CrossRef]

23. Zhu, H.; He, X.; Wang, K.; Su, Y.; Wu, J. Interactions of vegetation succession, soil bio-chemical properties and microbial communities in a Karst ecosystem. *Eur. J. Soil Biol.* **2012**, *51*, 1–7. [CrossRef]

24. Albright, W.H.; Benson, C.H.; Gee, G.W.; Abichou, T.; Tyler, S.W.; Rock, S. Field performance of a compacted clay langfill final cover at a humid site. *J. Geotech. Geoenviron. Eng.* **2006**, *132*, 1393–1403. [CrossRef]

25. He, J.; Wang, Y.; Li, Y.; Ruan, X. Effects of leachate infiltration and desiccation cracks on hydraulic conductivity of compacted clay. *Water Sci. Eng.* **2015**, *8*, 151–157. [CrossRef]

26. Kodikara, J.; Costa, S. *Desiccation Cracking in Clayey Soils: Mechanisms and Modelling*; Multiphysical Testing of Soils and Shales; Laloui, L., Ferrari, A., Eds.; Springer: Berlin/Heidelberg, Germany, 2013; pp. 21–32.

27. Gui, Y.L.; Zhao, Z.Y.; Kodikara, J.; Bui, H.H.; Yang, S.Q. Numerical modelling of laboratory soil desiccation cracking using UDEC with a mix-mode cohesive fracture model. *Eng. Geol.* **2016**, *202*, 14–23. [CrossRef]

28. Sima, J.; Jiang, M.; Zhou, C. Numerical simulation of desiccation cracking in a thin clay layer using 3D discrete element modeling. *Comput. Geotech.* **2014**, *56*, 168–180. [CrossRef]

29. Wang, D.Y.; Tang, C.S.; Cui, Y.J.; Shi, B.; Li, J. Effects of wetting-drying cycles on soil strength profile of a silty clay in micro-penetrometer tests. *Eng. Geol.* **2016**, *206*, 60–70. [CrossRef]

30. Tang, C.S.; Wang, D.Y.; Shi, B.; Li, J. Effect of wetting-drying cycles on profile mechanical behavior of soils with different initial conditions. *Catena* **2016**, *139*, 105–116. [CrossRef]

Article

A Coupled Thermal-Hydraulic-Mechanical Nonlinear Model for Fault Water Inrush

Weitao Liu, Jiyuan Zhao * , Ruiai Nie, Yuben Liu and Yanhui Du

College of Mining and Safety Engineering, Shandong University of Science and Technology, Qingdao 266590, China; wtliu@sdust.edu.cn (W.L.); nrasdust@163.com (R.N.); sdust_lyb@163.com (Y.L.); dyhsdust@163.com (Y.D.)
* Correspondence: zjysdust@126.com; Tel.: +86-157-6425-0061

Received: 1 July 2018; Accepted: 2 August 2018; Published: 7 August 2018

Abstract: A coupled thermal-nonlinear hydraulic-mechanical (THM) model for fault water inrush was carried out in this paper to study the water-rock-temperature interactions and predict the fault water inrush. First, the governing equations of the coupled THM model were established by coupling the particle transport equation, nonlinear flow equation, mechanical equation, and the heat transfer equation. Second, by setting different boundary conditions, the mechanical model, nonlinear hydraulic-mechanical (HM) coupling model, and the thermal-nonlinear hydraulic-mechanical (THM) coupling model were established, respectively. Finally, a numerical simulation of these models was established by using COMSOL Multiphysics. Results indicate that the nonlinear water flow equation could describe the nonlinear water flow process in the fractured zone of the fault. The mining stress and the water velocity had a great influence on the temperature of the fault zone. The temperature change of the fault zone can reflect the change of the seepage field in the fault and confined aquifer. This coupled THM model can provide a numerical simulation method to describe the coupled process of complex geological systems, which can be used to predict the fault water inrush induced by coal mining activities.

Keywords: fault water inrush; coupled THM model; nonlinear flow in fractured porous media; numerical model; warning levels of fault water inrush

---

1. Introduction

Fault water inrush frequently occurs in China's coal mines, which is a major threat to mine safety and production. The complexity of geological systems in deep mines, including high ground temperature, high ground water pressure, and high ground stress, has motivated researchers to consider the temperature (T), hydraulic flow (H), and mechanical deformation (M) coupling model for fault water inrush. The coupled THM process has been widely studied in recent years, including the TH [1,2], TM [3], and THM [4,5] coupling process. The international project DECOVALEX has promoted the research on the coupled THM process. Tsang et al. [6] summarized some studies of the project DECOVALEX III. These studies included two field experiments on coupled THM processes in a crystalline rock-bentonite system and in unsaturated tuff, three benchmark tests to evaluate the impact of coupled THM processes under different scenarios and geometries, and different approaches and computer codes for coupled THM processes. Bond et al. [7] presented a study that included 2D and 3D high-resolution coupled thermo-hydro-mechanical-chemical models to study the impact of complex physical and chemical processes on the geological formation surrounding the nuclear waste disposal facility. Graupner et al. [8] introduced and compared different THM coupling models developed by eight modelling teams to study the impact of THM processes on the properties of bentonite where all models were able to reproduce the coupled THM processes of the experiment. Sheng et al. [9] proposed a mechanical model of coupled THM processes for saturated porous media to study the impact of the coupled THM processes on the stresses of borehole wall and wellbore stability, and this model was translated into a set of partial differential equations by using COMSOL (COMSOL Co., Ltd., Shanghai, China). Zhu et al. [10] proposed a THM coupling model to study the effects of the coupled THM processes on the rock damage. Hence, the recent research in this area has mainly focused on the impact of coupled THM processes on the properties of geological environment, while very few of them have studied the hydraulic impact on temperature under the coupled THM processes.

From the aspect of predicting fault water inrush, many scholars have undertaken many studies. Some monitoring techniques have been applied to predict fault water inrush, such as the micro-seismic monitoring technique [11] and multi-field information monitoring technique [12]. Zhou et al. [13] predicted the fault water inrush under the impact of mining depth, fluid pressure, fault dip, and fault length. Xue et al. [14] predicted the water or mud inrush from a fault based on the cusp catastrophe model, and the method of risk prediction was successfully used in the Qingdao Kiaochow Bay subsea tunnel. However, the coupled THM processes has not been considered in the recent research on predicting fault water inrush.

Nonlinear flow in fractured rock has also been widely studied. Liu et al. [15] presented a fractal length distribution model to characterize hydraulic properties of rock fracture networks. Cherubini et al. [16] analyzed nonlinear flow in fractured media through hydraulic tests and numerical simulations. Liu et al. [17] developed a numerical approach to study the hydro-mechanical properties of rock fractures, and obtained the critical condition of quantifying the transition from a linear flow regime to a nonlinear flow regime in 2D fracture networks. According to Yang et al. [18], water inrush through fractured porous media experiences the Darcy flow in a confined aquifer, the non-Darcy flow in the fractured zone of a fault, and the Navier-Stokes' turbulent flow in coal seams. The non-Darcy flow in the fractured zone in a fault can be governed by the Brinkman equation or Forchheimer equation [19,20]. However, very little of the recent research on nonlinear flow in fractured rock has considered the impact of the coupled THM processes.

The processes of the fault water inrush include the nonlinear water flow process in the fractured zone of the fault and the coupled process of complex geological systems including high ground temperature, high ground water pressure, and high ground stress. In order to predict the fault water inrush, it is necessary to establish a coupled THM model that considers the effect of nonlinear flow in the fractured zone of the fault. In light of this, a coupled thermal-nonlinear hydraulic-mechanical (THM) model for fault water inrush was established in this paper. The governing equations of the THM coupling model couples the particle transport equation, the nonlinear flow equation, the mechanical equation, and the heat transfer equation. The nonlinear flow equation consists of the Darcy-Brinkman-NS equations. By setting different boundary conditions, the mechanical model, HM coupling model, and THM coupling model were established, respectively, to study the nonlinear water flow process in the fractured zone of the fault and the impact of water velocity on the temperature in the coupled THM process. The research results can be used to predict the fault water inrush. This coupled THM model can provide a numerical simulation method to describe the coupled process of the complex geological systems, which can be used to predict the fault water inrush induced by coal mining activities.

2. Governing Equations of Coupled THM Model

The governing equations of the coupled THM model couple the particle transport equation, nonlinear flow equation, mechanical equation, and heat transfer equation. The nonlinear flow equation consists of the Darcy-Brinkman-NS equations [21].

2.1. The Particle Transport Equation

The system of porous media of the fault includes fluid media, solid media, and the particles from solid According to Yao et al. [22], the convection-diffusion equation governs the particle transport. Taking the solid particles as the research object, the governing equation of particle transport was established based on the convection-diffusion equation. Figure 1 shows the micro-control volume for the particle transport equation in rectangular coordinates.

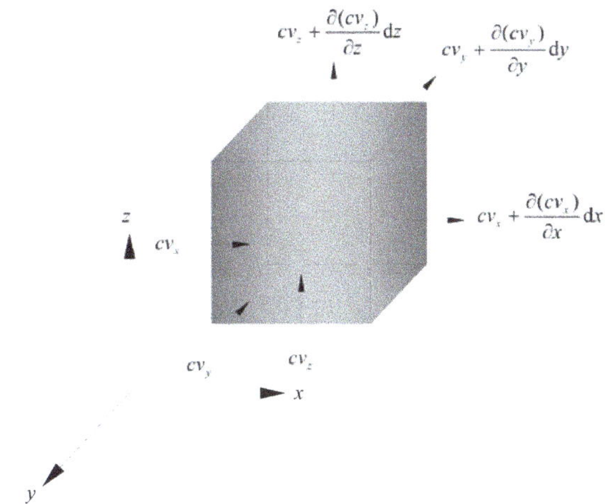

Figure 1. The micro-control volume for the particle transport equation.

The process of particle transport in the fractured zone of the fault can be treated as the transport of solute in the porous media. As the porous media is non-continuum and the convection diffusion equation is based on fluid continuous media, only the particles and fluid are treated as continuous media, so the convection diffusion equation can be used to establish the particle transport equation. Therefore, the solid media can be treated as a fluid. Based on this assumption, the porous media becomes continuous. Therefore, the equivalent mass concentration of particles is

$$c = \frac{dm_{pa}}{dV} = \frac{dm_{pa}}{dV_{po}/\varepsilon} = \varepsilon c_{pa} \tag{1}$$

where m_{pa} is the mass of particles, kg; dV is the volume of the micro-control volume, m^3; dV_{po} is the volume of pore, m^3; and c_{pa} is the actual mass concentration of particles, kg/m^3.

The net mass flux increase of particles in the x, y, and z directions is:

$$\begin{aligned}
&-\frac{\partial(cv_x)}{\partial x}dxdydz - \frac{\partial(cv_y)}{\partial y}dxdydz - \frac{\partial(cv_z)}{\partial z}dxdydz \\
&= -\frac{\partial(cv_x)}{\partial x}dV - \frac{\partial(cv_y)}{\partial y}dV - \frac{\partial(cv_z)}{\partial z}dV
\end{aligned} \tag{2}$$

The mass flux increase can be expressed by the mass concentration increase. Therefore, the net mass flux increase of the particles expressed by the equivalent mass concentration increase of particles in the process of particle transport in unit time is:

$$\left(c + \frac{\partial c}{\partial t}\right)dV - cdV = \frac{\partial c}{\partial t}dV \tag{3}$$

Solving and simplifying the compatibility of Equations (2) and (3) can obtain:

$$\frac{\partial c}{\partial t} + \nabla(cv_p) = 0 \tag{4}$$

$$\nabla = -\frac{\partial}{\partial x} - \frac{\partial}{\partial y} - \frac{\partial}{\partial z} \tag{5}$$

where v_p is the particle velocity, m/s.

The particle transport in the continuous fluid is driven by the effect of convection and diffusion. The effect of convection drives the particles to move at fluid velocity and the effect of diffusion drives the particles to move at diffusion velocity. The diffusion velocity is the difference between the particle velocity and fluid velocity. Therefore, the particle velocity is:

$$v_p = v + v_d \tag{6}$$

where v is the fluid velocity, m/s; and v_d is the diffusion velocity, m/s.

Substituting Equation (6) into Equation (4) can obtain:

$$\frac{\partial c}{\partial t} + \nabla(cv) + \nabla(cv_d) = 0 \tag{7}$$

According to Fick's law, the diffusion flux is:

$$cv_d = -D_d \nabla c \tag{8}$$

where D_d is diffusion coefficient, m^2/s.

Equation (7) is the convection diffusion equation. In fact, as mentioned previously, the process of particle transport in the porous media is non-continuum, which makes the diffusion coefficient of particles less than that in the continuous media. The effect of diffusion in the fractured zone of the fault can be replaced by the effect of hydrodynamic dispersion. Based on Fick's law, the hydrodynamic dispersion flux is:

$$J = -D \nabla c \tag{9}$$

where D is the hydrodynamic dispersion coefficient, m^2/s.

Substituting Equations (1) and (9) into Equation (7) can obtain:

$$\frac{\partial(\varepsilon c_{pa})}{\partial t} + \nabla(\varepsilon c_{pa}v) - \nabla(D\varepsilon \nabla c_{pa}) = 0 \tag{10}$$

Equation (10) is the particle transport equation for the coupled THM model.

2.2. The Nonlinear Flow Equation

The nonlinear flow equation consists of the Darcy-Brinkman equation-NS equations. According to Yang et al. [20], Darcy's law is:

$$\nabla\left[-\frac{k}{\mu}(\nabla p_d + \rho_l g \nabla z)\right] = Q_s \tag{11}$$

$$v_D = -\frac{k}{\mu}\nabla p_d \tag{12}$$

where k is the permeability, m^2; μ is the fluid dynamic viscosity, $N \cdot s/m^2$; p_d is the fluid pressure in the confined aquifer, Pa; v_D is the Darcy velocity in the confined aquifer, m/s; ρ_l is the fluid density, kg/m^3; z is a unit vector in the vertical direction; g is the gravitational acceleration, m/s^2; and Q_s is the volumetric flow rate, 1/s.

According to Yang et al. [19], the Brinkman equation is:

$$\left(-\nabla \cdot \frac{\mu}{\varepsilon}\left(\nabla \boldsymbol{v}_b + (\nabla \boldsymbol{v}_b)^T\right)\right) - \left(\frac{\mu}{k}\boldsymbol{v}_b + \nabla p_b\right) = 0 \tag{13}$$

$$\nabla \cdot \boldsymbol{v}_b = 0 \tag{14}$$

where ε is the porosity; $\boldsymbol{v}_b$ is the fluid velocity in the fault, m/s; and p_b is the fluid pressure in the fault, Pa.

According to Yang et al. [19], the Navier-Stokes equation is:

$$-\nabla \cdot \mu\left(\nabla \boldsymbol{v}_n + (\nabla \boldsymbol{v}_n)^T\right) + \rho \boldsymbol{v}_n \cdot \nabla \boldsymbol{v}_n - \nabla p_n = 0 \tag{15}$$

$$\nabla \cdot \boldsymbol{v}_n = 0 \tag{16}$$

where $\boldsymbol{v}_n$ is the fluid velocity in the coal seam, m/s; and p_n is the fluid pressure in the coal seam, Pa.

2.3. The Mechanical Equation

The mechanical equation is:

$$-\nabla \cdot \sigma = F \tag{17}$$

where σ is the total stress, Pa; and F is the volume forces, N/m^3.

2.4. The Heat Transfer Equation

The heat transfer equation is based on the heat transport theory in porous media of Bear et al. [23]. The heat transfer equation for the coupled THM model is established by rectangular coordinates. Figure 2 shows the micro-control volume for the heat transfer equation in the rectangular coordinates.

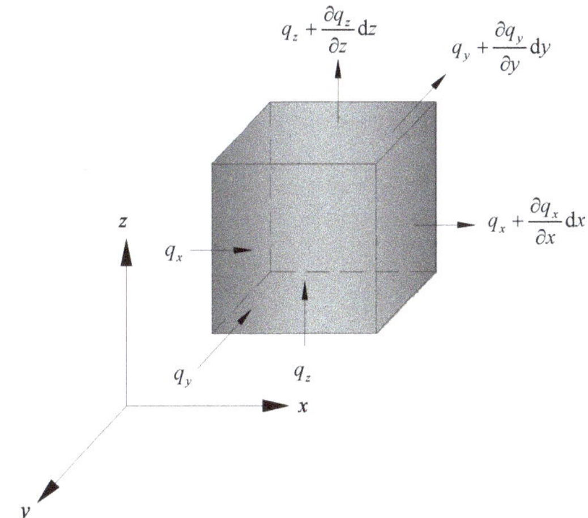

Figure 2. The micro-control volume for the heat transfer equation.

The net heat flux increase of micro-control volume in the x, y, and z directions is:

$$\begin{aligned}
&-\frac{\partial q_x}{\partial x}dxdydz - \frac{\partial q_y}{\partial y}dxdydz - \frac{\partial q_z}{\partial z}dxdydz \\
&= -\frac{\partial q_x}{\partial x}dV - \frac{\partial q_y}{\partial y}dV - \frac{\partial q_z}{\partial z}dV
\end{aligned} \tag{18}$$

The net heat flux increase of micro-control volume can be also expressed as:

$$(\rho_l c_l)_{eq} \frac{\partial T}{\partial t} dV \tag{19}$$

$$(\rho_l c_l)_{eq} = \theta_p k_p c_p + \varepsilon \rho c_p \tag{20}$$

$$\theta_p + \varepsilon = 1 \tag{21}$$

where c_l is the specific heat capacity of fluid at constant pressure, J/(kg·K); ρ_l is the fluid density, kg/m^3; $(\rho c_l)_{eq}$ is the equivalent volumetric heat capacity at constant pressure, J/(m^3·K); T is the temperature of porous media, K; θ_p is the volume fraction of solid material; and ε is the volume fraction of fluid or equivalently the porosity.

Solving and simplifying the compatibility of Equations (18) and (19) can obtain:

$$(\rho_l c_l)_{eq} \frac{\partial T}{\partial t} + \nabla q = 0 \tag{22}$$

$$\nabla = -\frac{\partial}{\partial x} - \frac{\partial}{\partial y} - \frac{\partial}{\partial z} \tag{23}$$

where q is the heat flux density, J/(s·m^2).

The process of heat transfer in porous media includes heat conduction, convection, and dispersion. Based on Fourier's law, the heat flux caused by heat conduction is:

$$q_c = -k_c \cdot \nabla T \tag{24}$$

$$k_c = \theta_p k_p + \varepsilon k_l \tag{25}$$

where k_c is the thermal conductivity of the solid-fluid system, W/(m·K); and k_p and k_l are the solid and fluid thermal conductivity respectively, W/(m·K).

The heat flux caused by heat dispersion can be expressed as:

$$q_d = -k_d \cdot \nabla T \tag{26}$$

where k_d is the thermal dispersion coefficient of the solid-fluid system, W/(m·K).

The sum of the thermal conductivity coefficient and the thermal dispersion coefficient of the solid-fluid system is expressed as:

$$k_{eq} = k_c + k_d \tag{27}$$

The heat flux caused by heat convection is proportional to the fluid velocity, which is expressed as:

$$q_a = \rho_l c_l v T \tag{28}$$

Therefore, the heat flux density is:

$$q = q_a + q_c + q_d = \rho_l c_l v_n T - k_{eq} \cdot \nabla T \tag{29}$$

Substituting Equation (29) into Equation (22) can obtain:

$$(\rho_l c_l)_{eq} \frac{\partial T}{\partial t} + \rho_l c_l v_n \cdot \nabla T - \nabla \cdot (k_{eq} \cdot \nabla T) = 0 \tag{30}$$

Equation (30) is the heat transfer equation for the coupled THM model.

3. Model Setup

Before the model setup, some simplifying assumptions for the models are listed as follows: the rock in the models was assumed to be a porous continuous medium; the heat transfer among the solid, liquid, and gas that actually exists in coal mining was assumed to be the heat transfer between the solid and liquid; and the load of the rock above the top boundary of the model on the model was assumed to be the uniform load.

In order to study the nonlinear water flow process in the fractured zone of the fault and the impact of water velocity on the temperature in the coupled THM process, three different kinds of coupling methods were established including the mechanical model (M), nonlinear hydraulic-mechanical (HM) coupling model, and thermal-nonlinear hydraulic-mechanical (THM) coupling model.

In order to study the impact of the working face advanced distance and the water pressure on the temperature, the working face proceeded as per the following steps. At the first step, the working face was excavated to 25 m. Then, each step excavated 10 m along the coal seam, for 10 steps, which moved the working face 125 m forward. The water pressure in the confined aquifer was applied in a monotonically increasing mode with an increment of 1 MPa per step, which made the water pressure increase from 0 MPa to 10 MPa.

3.1. The Engineering Background and the Geologic Model

The 2307 working face of coal seam 3 in the Anju coal mine of China is at an elevation of about −980 m. The distance between the limestone aquifer and the coal floor is about 89.5 m. The water pressure is about 4 MPa. Coal Seam 3 is attached to fault F23. Advancing the working face may cause the fault to activate, thus, fault water inrush may occur. Using the Anju coal mine as the background, a numerical simulating model for the coupled THM environments was established by using COMSOL Multiphysics 5.1 (COMSOL Co., Ltd., Shanghai, China). The geometric model is shown in Figure 3 [24]. In order to study the distribution of water velocity and temperature, two monitoring lines were given, as shown in Figure 3. Monitoring line 1 was used to study the temperature and monitoring line 2 was used to study water velocity. According to the hydrogeological report of the Anju coal mine, the mechanical parameters are listed in Table 1.

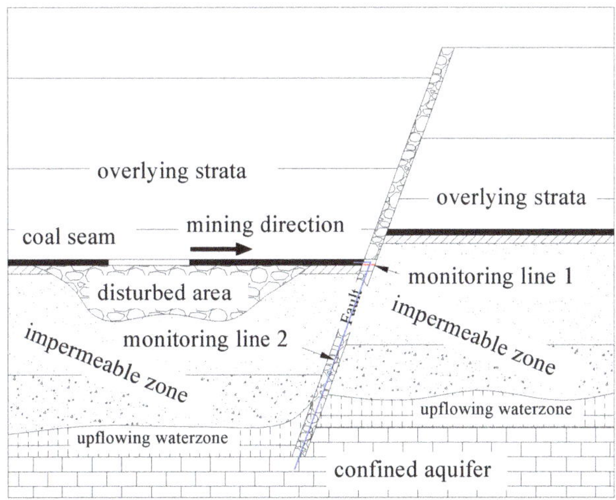

Figure 3. Geological profile along Coal Seam 3.

Table 1. Rock mechanical parameters.

Rock	Thickness (m)	Density (kg/m³)	Bulk Modulus (GPa)	Shear Modulus (GPa)	Permeability (m²)	Porosity
Sandstone 1	20	2660	9.896	8.051	1.1×10^{-14}	0.05
Sandstone 2	45	2650	9.756	7.257	1.2×10^{-14}	0.06
Main roof	30	2480	8.730	4.264	1.8×10^{-13}	0.12
Immediate roof	15	2502	8.872	6.031	1.7×10^{-14}	0.10
Coal seam	3	1400	5.455	1.295	6.1×10^{-13}	0.20
Immediate floor	4.2	2430	8.217	4.126	2×10^{-14}	0.13
Sandstone 3	50.3	2600	9.572	7.029	2.5×10^{-14}	0.12
Mudstone	35	2490	8.530	4.162	2.1×10^{-13}	0.21
confined aquifer	25	2620	10.417	5.952	5.1×10^{-11}	0.28
Fault	—	1500	2	1.5	1.2×10^{-11}	0.26

3.2. Boundary Condition Setting

3.2.1. Mechanical Model

Figure 4 shows the boundary condition setting for the mechanical model.

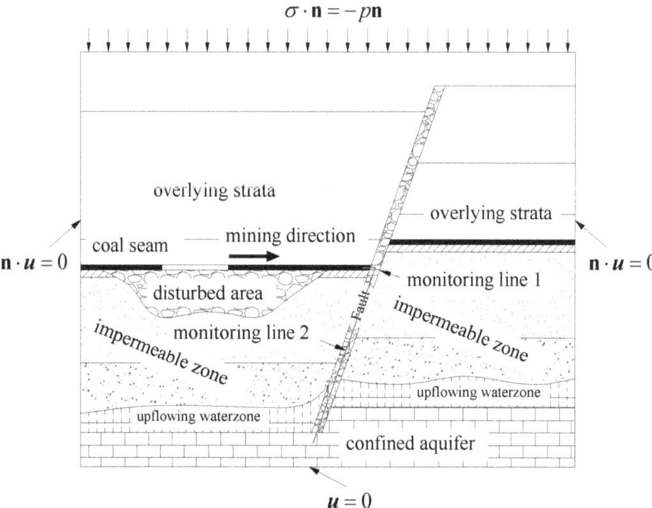

Figure 4. Boundary condition setting for mechanical model.

The boundary condition of the top boundary of the geologic model is:

$$\sigma \cdot \mathbf{n} = -p\mathbf{n} \tag{31}$$

where **n** is the unit vector normal to the boundary; and p is the pressure, Pa. The top boundary of the model was at an elevation of about -867 m, thus the thickness of the overlying strata was about 867 m, which made the boundary load of the top boundary of the model about 22 MPa.

The bottom boundary of the geologic model was set to be the fixed constraint condition:

$$\boldsymbol{u} = 0 \tag{32}$$

where $\boldsymbol{u}$ is the displacement vector, m.

The remaining boundaries of the geologic model were set as the slipping constraint condition:

$$\mathbf{n} \cdot \boldsymbol{u} = 0 \tag{33}$$

3.2.2. HM Coupling Model

The HM coupling model couples seepage field and stress field. Darcy's law, the Brinkman equation, and the Navier-Stokes equation are used to govern the water flow in the confined aquifer, the fault, and the coal seam, respectively. Figure 5 shows the boundary condition setting for the HM coupling model.

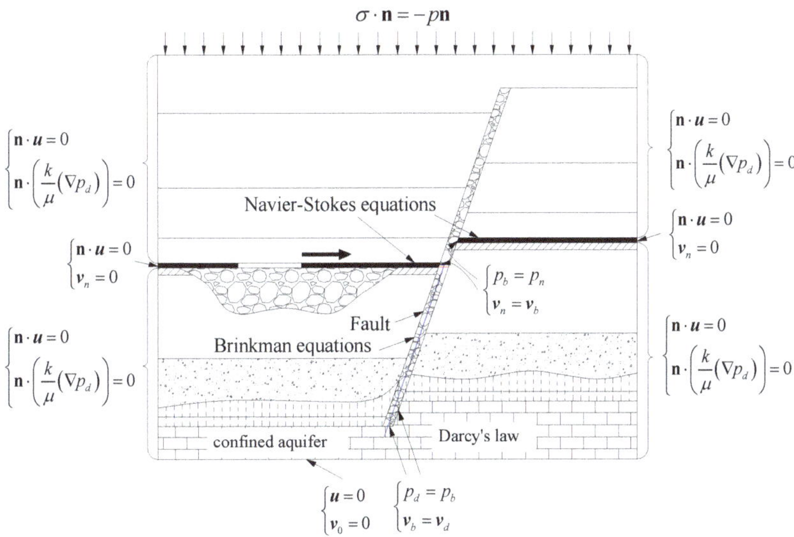

Figure 5. Boundary condition setting for the HM coupling model.

The bottom boundary of the confined aquifer is the inlet boundary, which was set as $v_0 = 1 \times 10^{-4}$ m/s. The common boundaries between the confined aquifer and the fault are the outlet boundary of the confined aquifer and the inlet boundary of the fault. In order to obtain a continuous solution at the common boundaries, the pressure and the velocity from the confined aquifer must equal the pressure and velocity from the fault. Therefore, the common boundaries were set as:

$$p_d = p_b \tag{34}$$

$$v_b = v_d \tag{35}$$

The right and left boundaries of the geologic model, except for the boundaries of the coal seam, are no flow boundaries, which were set as:

$$\mathbf{n} \cdot \left(\frac{k}{\mu}(\nabla p_d) \right) = 0 \tag{36}$$

The common boundaries between the fault and the coal seam are the outlet boundary of the fault and the inlet boundary of the coal seam. The pressure and the velocity from the fault must equal the pressure and the velocity from the coal seam. Therefore, the common boundaries between the fault and the coal seam were set as:

$$p_b = p_n \tag{37}$$

$$v_n = v_b \tag{38}$$

The right and left boundaries of the coal seam are the outlet boundary, which were set as:

$$v_n = 0 \tag{39}$$

3.2.3. THM Coupling Model

Figure 6 shows the boundary condition setting for the THM coupling model.

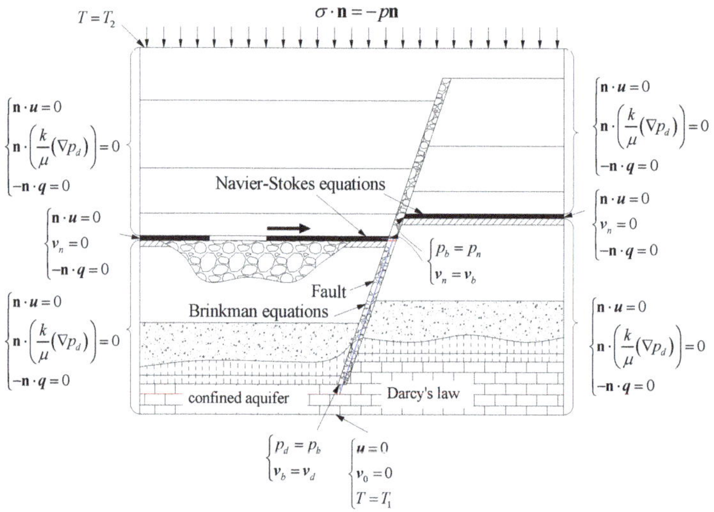

Figure 6. Boundary condition setting for the THM coupling model.

The boundary condition of the top and bottom boundaries of the model was set as:

$$T = T_1 \tag{40}$$

$$T = T_2 \tag{41}$$

where T_1 and T_2 are the initial temperatures of the top and bottom boundary of the model respectively, K, which were set as 39 °C and 43.9 °C, respectively. In addition, the boundaries of the working face was set as 37 °C.

The remaining boundaries of the model were set as the no heat flux boundary condition:

$$-\mathbf{n} \cdot \boldsymbol{q} = 0 \tag{42}$$

4. Results and Discussions

The change law of water velocity and the temperature were obtained. The temperature change was subjected to the interaction of the water pressure and the working face advanced distance. The change of the temperature of the water-rock environment near the fault can reflect the change of seepage field in the fault and confined aquifer, which was used to divide the warning levels of fault water inrush.

4.1. Simulation Results

4.1.1. The Change Law of Water Velocity

Figure 7 shows the change law of water velocity when the working face was excavated to 65 m. The water velocity increased as the water pressure increased. When the water pressure was 2 MPa, there was little water in the coal seam, which indicated that the water inrush channel had not been formed. When the water pressure was 4 MPa and 10 MPa, the water velocity had a rapid increase at the junction of the confined aquifer, fault, and coal seam, which indicated nonlinear flow processes in the fractured zone of the fault.

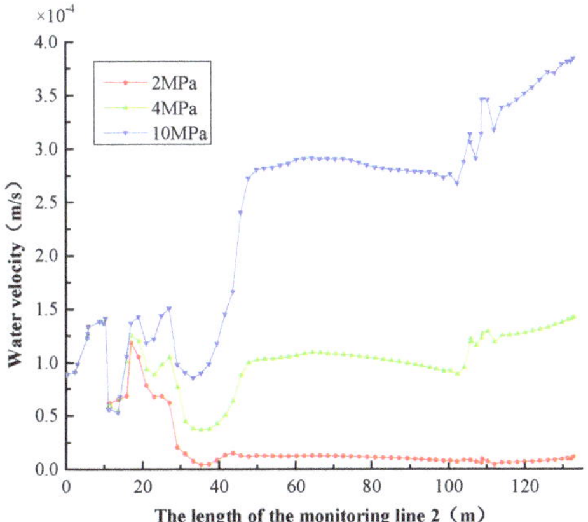

Figure 7. The water velocity when the working face was excavated to 65 m.

4.1.2. The Distribution of Temperature

1. The Impact of the Water Pressure on the Temperature

When the working face was excavated to 105 m, the distribution of seepage velocity and the temperature at different water pressure are shown in Figures 8 and 9, respectively.

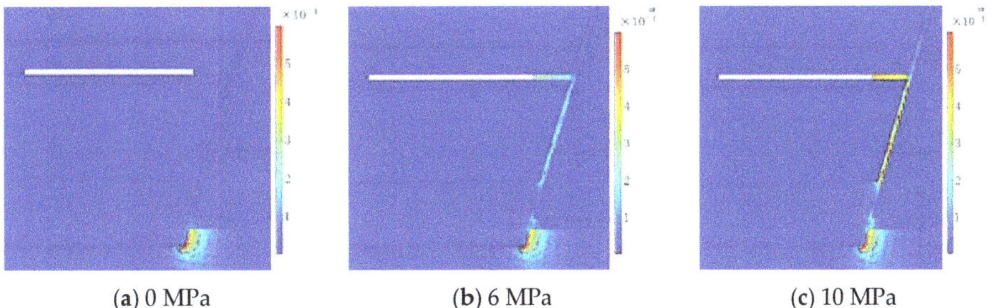

(a) 0 MPa   **(b)** 6 MPa   **(c)** 10 MPa

Figure 8. The seepage velocity distribution at different water pressure: (**a**) At 0 MPa, there is little water in the fault and the coal seam; (**b**) At 6 MPa, the water flows into the fault and the coal seam from the confined aquifer; and (**c**) At 10 MPa, the seepage velocity in the fault and the coal seam is higher than at 6 MPa.

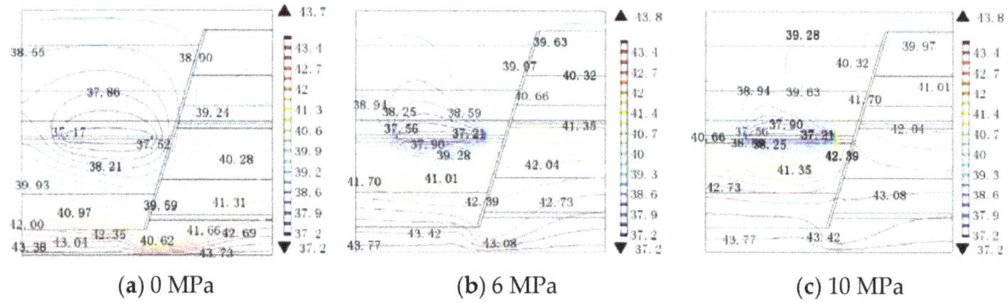

(a) 0 MPa   (b) 6 MPa   (c) 10 MPa

Figure 9. The temperature contours at different water pressure: (**a**) at 0 MPa, the temperature contours have a smooth shift through the fault plane and the coal seam; (**b**) at 6 MPa, the temperature contours is a convex curve through the fault plane and the coal seam; and (**c**) at 10 MPa, the convex degree of the temperature contours is higher than at 6 MPa.

When the water pressure is 0 MPa, there is little water in the fault and the coal seam. The temperature contours have a smooth shift through the fault plane. When the water pressure is 6 MPa and 10 MPa, the water flows into the fault and the coal seam from the confined aquifer and the seepage velocity increases with the increasing water pressure. The temperature contours are a convex curve through the fault plane and the coal seam, and the convex degree increases with the increasing water pressure. Comparing the results of Figures 8 and 9, the convex curve of the temperature contours and the seepage velocity had the same direction and change law. This is because, under the effect of the water pressure, the confined aquifer with higher temperature flows upwards along the fault plane, making the temperature of the water-rock environment near the fault plane increase. The higher the water pressure, the greater the seepage velocity will be, and there will be a shorter time of the heat transfer between the water and the rock in the fault, which makes the convex degree of the temperature contours increase as the water pressure increases. Therefore, the change of the temperature of the water-rock environment near the fault can reflect the change of seepage field in the fault and confined aquifer, which can be used as a new method of predicting fault water inrush.

The relationship between the maximum temperature on the monitoring line and the water pressure at different excavation steps is shown in Figure 10, from which it can be seen that the temperature increased with the increasing water pressure at every excavation step. The change range of the temperature increased as the distance between the working face and the fault plane decreased.

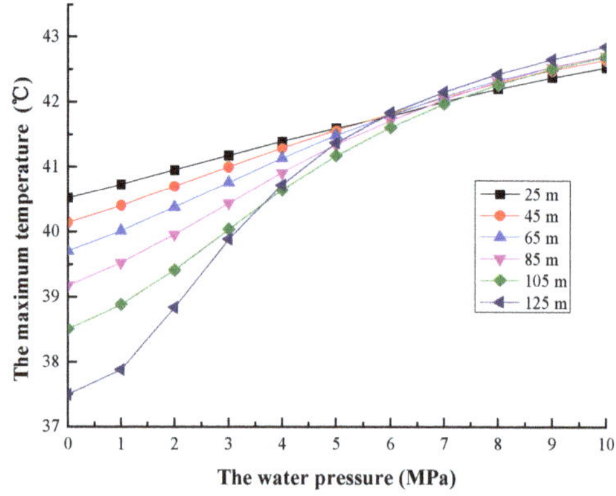

Figure 10. The maximum temperature on the monitoring line versus the water pressure when the working face advanced distance was 25 m, 45 m, 65 m, 85 m, 105 m, and 125 m, respectively.

2. The Impact of the Working Face Advanced Distance on the Temperature

When the water pressure was 6 MPa, the distribution of seepage velocity and temperature at the different excavation steps are shown in Figures 11 and 12, respectively. The seepage velocity increased as the working face advanced distance increased. The temperature contours were a convex curve through the fault plane, and the convex degree increased as the working face advanced distance increased. Comparing the results of Figures 11 and 12, the convex curve of the temperature contours and the seepage velocity had the same direction and change law. This is because the damage of the fault and the coal seam induced by coal mining activities increased with the increase of the working face advanced distance, which made the seepage velocity in the fault and the coal seam increase. Then, the increase of the seepage velocity made the temperature of the water-rock environment near the fault plane increase. Therefore, the convex degree of the temperature contours increased with the increase of the working face advanced distance.

The relationship between the maximum temperature on the monitoring line and the working face advanced distance at different water pressure is shown in Figure 13.

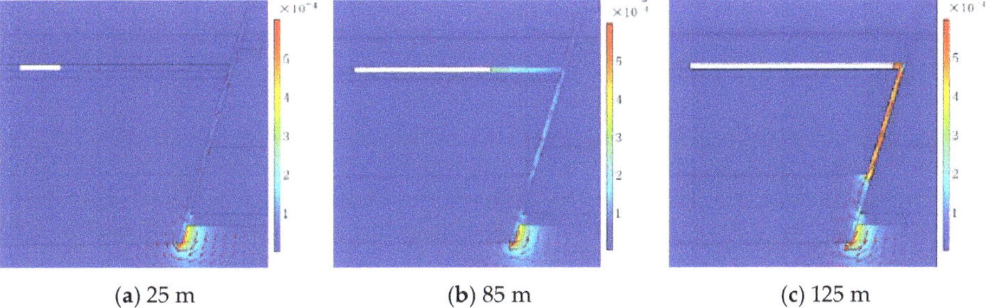

(a) 25 m   (b) 85 m   (c) 125 m

Figure 11. The seepage velocity distribution at different working face advanced distances: (**a**) At 25 m, the seepage velocity in the fault and the coal seam is very low; (**b**) at 85 m, the seepage velocity in the fault and the coal seam is higher than at 25 m; and (**c**) at 125 m, the seepage velocity in the fault and the coal seam is higher than at 85 m.

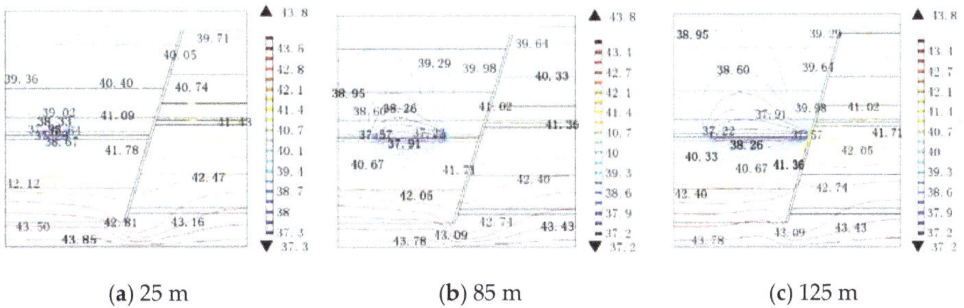

(a) 25 m   (b) 85 m   (c) 125 m

Figure 12. The temperature contours at different working face advanced distances: (**a**) At 25 m, the temperature contours is a convex curve through the fault plane and the coal seam; (**b**) at 85 m, the convex degree of the temperature contours is higher than at 25 m; and (**c**) at 125 m, the convex degree of the temperature contours is higher than at 85 m.

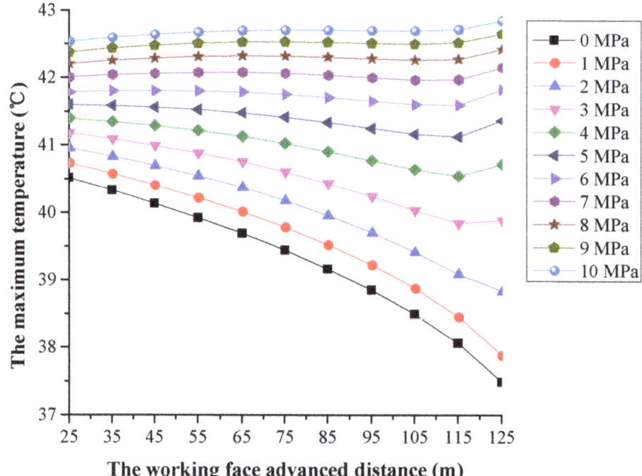

Figure 13. The maximum temperature on the monitoring line versus the working face advanced distance when the water pressure is from 0 MPa to 10 MPa, respectively.

Comparing the results of Figures 10 and 13, the maximum temperature on the monitoring line decreased as the working face advanced distance increased at lower water pressure, which indicated that the influence of the working face advanced distance on the temperature was greater than that of water pressure on the temperature. The decrease of the temperature decreased with the increasing water pressure. This is because at the working face of the coal mine, it is necessary to provide ventilation to guarantee enough oxygen and proper temperature for workers, which makes the temperature lower in the working face than the surrounding area of the working face. At lower water pressure, as the working face advanced distance increased, the lower seepage velocity made the temperature in the fault increase slightly. Therefore, the increase of the working face advanced distance reduced the temperature in the fault due to the lower temperature at the working face. At higher water pressure, the maximum temperature on the monitoring line increased as the working face advanced distance increased, which indicates that at higher water pressure, the influence of water pressure on the temperature is greater than that of the working face advanced distance on the temperature. This is because at higher water pressure, the higher seepage velocity makes the temperature in the fault have a higher increase. In addition, the decrease of the temperature gradually decreases with the water pressure increasing. This is because as the water pressure increases, the difference between the decrease of the temperature caused by the increase of the working face advanced distance and the increase of the temperature caused by the increase of the water pressure will decrease. In a word, the temperature change is subjected to the interaction of the water pressure and the working face advanced distance.

4.1.3. The Warning Level of Water Inrush

As mentioned previously, the change of the temperature of the water-rock environment near the fault can reflect the change of seepage field in the fault and confined aquifer, which can be used as a new method to predict fault water inrush. In order to use the temperature to predict fault water inrush, it is necessary to find a correspondence between the temperature change and danger degree of water inrush, which is used to divide the warning level of water inrush. Two methods were used as follows to find the correspondence.

1. The Method Based on the Water Inrush Coefficient

The conception of the water inrush coefficient was proposed based on the statistical analysis of long-term water inrush data and defined in the *Regulation for Coal Mine Water Prevention and Control* [25], where the water inrush coefficient is expressed as an empirical formula [26]:

$$K = \frac{P}{M} \tag{43}$$

where K is the water inrush coefficient (MPa/m); P is the water pressure of the confined aquifer (MPa); and M is the thickness of the aquifuge (m). The aquifuge means the zone between the coal seam and confined aquifer. If there is a fault in the aquifuge, the water inrush will occur when the water inrush coefficient is over 0.06 MPa/m, while the water inrush will not happen when the water inrush coefficient is less than or equal to 0.06 MPa/m.

In order to take more detailed corresponding measures to predict water inrush, the warning level of water inrush was divided into three grades: safe level with water inrush coefficient less than or equal to 0.06 MPa/m, dangerous level with water inrush coefficient greater than 0.06 MPa/m and less than or equal to 0.1 MPa/m, and more dangerous level with water inrush coefficient greater than 0.1 MPa/m.

According to the hydrogeological report of the Anju coal mine, the thickness of the aquifuge is from 71.85 m to 110.19 m. To obtain the minimum critical water pressure, the thickness of the aquifuge should be 71.85 m, which was substituted into Equation (43). Then, by substituting the water inrush coefficient of 0.06 MPa/m and 0.1 MPa/m into Equation (43), the water pressure was calculated to be 4.311 MPa and 7.815 MPa, respectively.

Using the quartic polynomial to fit the curves as shown in Figure 10, the fitting polynomials for the relation curves between maximum temperature and water pressure are listed in Table 3. By substituting the water pressure of 4.311 MPa and 7.815 MPa into the fitting polynomials, respectively, the calculated results of temperature are listed in Table 3.

The original temperature of the rock is 40.09 °C before the coal seam is excavated. The differences between the maximum temperature and the original temperature are listed in Table 2. At 4.311 MPa and 7.815 MPa, the difference between the maximum temperature and the original temperature is about 1 °C and about 2 °C, respectively, which can be used as the thresholds to divide the warning level of water inrush.

Table 2. The identification criteria for the water inrush warning level.

The Warning Level	The Water Inrush Coefficient (MPa/m)	The Water Pressure (MPa)	The Temperature Change (°C)
Safe	T ≤ 0.06	P ≤ 4.311	ΔT ≤ 1
Dangerous	0.06 < T ≤ 0.1	4.311 < P ≤ 7.185	1 < ΔT ≤ 2
More dangerous	T > 0.1	P > 7.185	ΔT > 2

2. The Method Based on the Water-Resisting Thickness of Floor

The theory of the water-resisting thickness of floor was established based on the "Down Three Zone" theory. The safe coal and rock pillars can be represented as:

$$h_a = h_1 + h_2 + h_3 \tag{44}$$

where h_1, h_2, and h_3 are the thickness of the disturbed area in the floor, impermeable zone, and confined upflowing waterzone, respectively, in meters. There are three kinds of empirical formulas to calculate h_1:

Table 3. The fitting polynomial, the maximum temperature, and the temperature difference.

The Working Face Advanced Distance	The Fitting Polynomial	4.311 MPa		7.185 MPa		8.997 MPa	
		The Maximum Temperature	The Temperature Difference	The Maximum Temperature	The Temperature Difference	The Maximum Temperature	The Temperature Difference
25 m	$y = -5 \times 10^{-5}P4 + 0.0007P3 - 0.0048P2 + 0.2279P + 40.514$	41.45	1.36	42.03	1.94	42.36	2.27
35 m	$y = -1 \times 10^{-5}P4 - 0.0003P3 + 0.0025P2 + 0.2475P + 40.329$	41.41	1.32	42.10	2.01	42.47	2.38
45 m	$y = 5 \times 10^{-5}P4 - 0.0018P3 + 0.0115P2 + 0.2653P + 40.132$	41.36	1.27	42.10	2.01	42.47	2.38
55 m	$y = 0.0001P4 - 0.0034P3 + 0.0223P2 + 0.2793P + 39.919$	41.30	1.21	42.08	1.99	42.42	2.33
65 m	$y = 0.0002P4 - 0.0052P3 + 0.0346P2 + 0.2904P + 39.692$	41.24	1.15	42.17	2.08	42.63	2.54
75 m	$y = 0.0003P4 - 0.0076P3 + 0.0506P2 + 0.295P + 39.441$	41.15	1.06	42.15	2.06	42.62	2.53
85 m	$y = 0.0004P4 - 0.0106P3 + 0.0701P2 + 0.295P + 39.164$	41.03	0.94	42.04	1.95	42.39	2.30
95 m	$y = 0.0006P4 - 0.0144P3 + 0.0955P2 + 0.2883P + 38.854$	40.93	0.84	42.11	2.02	42.62	2.53
105 m	$y = 0.0008P4 - 0.0198P3 + 0.1294P2 + 0.2769P + 38.495$	40.79	0.70	41.95	1.86	42.28	2.19
115 m	$y = 0.0012P4 - 0.0284P3 + 0.18P2 + 0.2697P + 38.051$	40.70	0.61	41.95	1.86	42.23	2.14
125 m	$y = 0.0022P4 - 0.0474P3 + 0.2724P2 + 0.3337P + 37.432$	40.90	0.81	42.17	2.08	42.38	2.29

$$h_1 = 0.7007 + 0.1079L \tag{45}$$

$$h_1 = 0.303L^{0.8} \tag{46}$$

$$h_1 = 0.00851H + 0.1665\alpha + 0.1079L - 4.3579 \tag{47}$$

where L is the length of the working face, m; H is the mining depth, m; and α is the angle of coal seam. Substituting 253 m, 980 m, and 0° into H, L, and α of the empirical formulas, respectively, can obtain values of h_1 at 27.9994 m, 25.3478 m, and 31.2806 m. In order to ensure safe coal seam mining, h_a should be as large as possible. Therefore, h_1 is 31.2806 m.

According to borehole No. AJGL3-1 of the Anju coal mine, there is no obviously confined upflowing waterzone. Therefore, h_3 is 0 m.

The h_2 can be calculated by the water-resisting coefficient:

$$Z_c = \frac{\sum Z_i h_i}{h_c} \tag{48}$$

$$h_2 = P/Z_c \tag{49}$$

where P is the critical water pressure, MPa; Z_i is the water-resisting coefficient of the ith rock, MPa/m; Z_c is the weighted average water-resisting coefficient, MPa/m; and h_i is the average thickness of the ith rock, m. According to the hydrogeological report of the Anju coal mine, the distance between coal seam No. 3 and the Shanxi Formation is from 46.30 m to 71.26 m, and the average distance is 54.5 m, and the rock is mainly medium sandstone and fine sandstone. The water-resisting coefficient of the medium sandstone and fine sandstone is 0.3 MPa/m. The distance between the Shanxi Formation and the confined aquifer is from 25.55 m to 38.93 m, and the average distance is 35 m, and the rock is mainly mudstone and siltstone. The water-resisting coefficient of the medium mudstone and siltstone is 0.1 MPa/m. Substituting the water-resisting coefficient of medium sandstone, fine sandstone, mudstone, and siltstone into Equation (48), Z_c is 0.2609 MPa/m. When the distance between the coal seam and confined aquifer is more than or equal to h_a, the coal seam mining is safe:

$$h_d \geq h_a = h_1 + h_2 + h_3 \tag{50}$$

where h_d is the distance between the coal seam and confined aquifer, m. According to the hydrogeological report of the Anju coal mine, h_d is from 71.85 m to 110.19 m. In order to obtain the minimum critical water pressure, h_2 must be the maximum, which requires h_d is equal to h_a and the minimum. Therefore, h_d is 71.85 m. Substituting h_1, h_3, and h_a into Equation (50), h_2 is 40.5694 m. By substituting h_2 into Equation (49), critical water pressure is 8.997 MPa.

By substituting the critical water pressure into the fitting polynomials respectively, the calculated results of the temperature are listed in Table 3. All of the differences between the maximum temperature and the original temperature were more than 2 °C. Therefore, the 2 °C of temperature rise can be used as the thresholds to divide the dangerous and more dangerous warning levels of fault water inrush. Bai et al. [12] measured the temperature of water in a fault from a confined aquifer in a coal seam roof and used the 1.5 °C and 2.4 °C of temperature change to divide the dangerous and more dangerous warning levels of fault water inrush. However, these results were not analyzed mathematically. This model can provide theoretical support for the work of Bai et al.

The temperature range calculated by this model was too narrow because the geothermal gradient of Anju coal mine was 21.54 °C/km and not high, and the temperature difference between the confined aquifer and coal seam was not large. One or two degrees Celsius is difficult to measure with such precision in practical engineering. Therefore, there is some uncertainty in using this model to deal with engineering problems with low geothermal gradients. However, in many deep engineering practices, the geothermal gradient is generally 30 °C/km to 50 °C/km. In some areas such as near a fault or with high thermal conductivity, the geothermal gradient is sometimes as high as 200 °C/km [27]. In engineering with high geothermal gradients, the temperature difference between the confined aquifer and coal seam will be large, which will make the temperature rise calculated by this model have a greater range. Using this model to deal with engineering problems with high geothermal gradients is more reliable. Whether in high or low geothermal gradient engineering, this model can be used as a single indicator for identifying the conditions for water inrush from a fault in coal mining, but this model should be combined with other approaches to predict fault water inrush.

Although this model can provide a proper numerical simulation method to study the processes of nonlinear water flow in the fractured zone of the fault and the multi-fields coupling of the complex geological systems of the fault water inrush, this model has some limitations:

1. There is some uncertainty in using this model to deal with engineering problems with low geothermal gradients.
2. The disturbed area may extend the fault zone and aggravate fault activation, which was not considered in this model.
3. The comparison between the measured results in reality and simulated results was not studied.
4. Although this model was established based on the engineering background of a coal mine, it can be applied to tunneling in the vicinity of a fault. This model can be used as a method to study the water-rock-temperature interactions in tunneling in the vicinity of a fault. However, as the water inrush coefficient and the water-resisting thickness are applicable only in coal mines, they cannot be used to find the thresholds to divide the warning levels of fault water inrush in tunneling. A new method to divide the warning levels of fault water inrush in tunneling should be found if this model is used to predict fault water inrush in tunneling.

Future research will be done to overcome the given limitations.

5. Conclusions

This paper proposed a coupled THM model to predict fault water inrush induced by coal mining activities. The main conclusions were obtained as follows:

1. The Darcy-Brinkman-NS equations can properly describe the nonlinear water flow process in the fractured zone of the fault. The water velocity increases with the increasing water pressure. The water velocity has a rapid increase at the junction of the confined aquifer, fault, and coal seam instead of a linear increase.
2. Temperature change of the fault zone is subjected to the interaction of the water pressure and the working face advanced distance. At a lower water pressure, the influence of the working face advanced distance on the temperature of the fault zone is greater than that of the water pressure on the temperature. At a higher water pressure, the influence of water pressure on the temperature of the fault zone is greater than that of the working face advanced distance on the temperature. When the working face advanced distance is constant, the temperature of the fault zone increases with the increasing water pressure. The range of the temperature of the fault zone increases with the distance between the working face and the fault plane decreasing.
3. The temperature change of the fault zone can reflect the change of the seepage field in the fault and confined aquifer. Monitoring the temperature rise of the fault zone, based on the conception of the water inrush coefficient and the water-resisting thickness of floor, the temperature increases of 1 and 2 °C were used as the thresholds to divide the warning levels of water inrush.

Author Contributions: W.L. planned the establishment step of the model. W.L. and J.Z. wrote the main manuscript. J.Z. established the mathematical model and prepared all figures and tables. R.N., Y.L., and Y.D. reviewed the manuscript. All authors contributed in the conclusion.

Funding: This research was funded by the National Natural science Foundation of China (grant 51274135), the National High Technology Research and Development Program (863 Program) of China (grant 2015AA016404-4), the State Key Research and Development Program of China (grant 2017YFC0804108), and the SDUST Research Fund (grant 2018TDJH102).

Conflicts of Interest: The authors declare no conflict of interest.

References

1. Andrés, C.; Ordóñez, A.; Álvarez, R. Hydraulic and Thermal Modelling of an Underground Mining Reservoir. *Mine Water Environ.* **2017**, *36*, 24–33. [CrossRef]
2. Uhlík, J.; Baier, J. Model Evaluation of Thermal Energy Potential of Hydrogeological Structures with Flooded Mines. *Mine Water Environ.* **2012**, *31*, 179–191. [CrossRef]
3. Xi, Y.; Li, J.; Liu, G.; Tao, Q.; Lian, W. A new numerical investigation of cement sheath integrity during multistage hydraulic fracturing shale gas wells. *J. Nat. Gas Sci. Eng.* **2018**, *49*, 331–341.
4. Rutqvist, J.; Wu, Y.S.; Tsang, C.F.; Bodvarsson, G. A modeling approach for analysis of coupled multiphase fluid flow, heat transfer, and deformation in fractured porous rock. *Int. J. Rock Mech. Min. Sci.* **2002**, *39*, 429–442. [CrossRef]
5. Zhang, Y.J.; Yang, C.S.; Xu, G. FEM Analyses for T-H-M-M Coupling Processes in Dual-Porosity Rock Mass under Stress Corrosion and Pressure Solution. *J. Appl. Math.* **2012**, *10*, 3129–3138. [CrossRef]
6. Tsang, C.F.; Jing, L.; Stephansson, O.; Kautsky, F. The DECOVALEX III project: A summary of activities and lessons learned. *Int. J. Rock Mech. Min. Sci.* **2005**, *42*, 593–610. [CrossRef]
7. Bond, A.E.; Bruský, I.; Cao, T.; Chittenden, N.; Fedors, R.; Feng, X.T.; Gwo, J.P.; Kolditz, O.; Lang, P.; McDermott, C.; et al. A synthesis of approaches for modelling coupled thermal–hydraulic–mechanical–chemical processes in a single novaculite fracture experiment. *Environ. Earth Sci.* **2017**, *76*, 1–19. [CrossRef]
8. Graupner, B.J.; Shao, H.; Wang, X.R.; Nguyen, T.S.; Li, Z.; Rutqvist, J.; Chen, F.; Birkholzer, J.; Wang, W.; Kolditz, O.; et al. Comparative modelling of the coupled thermal–hydraulic-mechanical (THM) processes in a heated bentonite pellet column with hydration. *Environ. Earth Sci.* **2018**, *77*, 1–16. [CrossRef]
9. Sheng, J. Fully coupled thermo-hydro-mechanical model of saturated porous media and numerical modeling. *Chin. J. Rock Mech. Eng.* **2006**, *25*, 3028–3033.
10. Zhu, W.C.; Wei, C.H.; Tian, J.; Yang, T.H.; Tang, C.A. Coupled thermal-hydraulic-mechanical model during rock damage and its preliminary application. *Rock Soil. Mech.* **2009**, *30*, 3851–3857.
11. Sun, J.; Wang, L.G.; Tang, F.R.; Shen, Y.F.; Gong, S.L. Micro seismic monitoring failure characteristics of inclined coal seam floor. *Rock Soil. Mech.* **2011**, *32*, 1589–1595.
12. Bai, J.; Li, S.; Liu, R.; Zhang, Q.; Zhang, H.; Sha, F. Multi-field information monitoring and warning of delayed water bursting in deep rock fault. *Chin. J. Rock Mech. Eng.* **2015**, *34*, 2327–2335.
13. Zhou, Q.; Herrera, J.; Hidalgo, A. The numerical analysis of fault-induced mine water inrush using the extended finite element method and fracture mechanics. *Mine Water Environ.* **2018**, *37*, 185–195. [CrossRef]
14. Xue, Y.; Wang, D.; Li, S.; Qiu, D.; Li, Z.; Zhu, J. A risk prediction method for water or mud inrush from water-bearing faults in subsea tunnel based on cusp catastrophe model. *KSCE J. Civ. Eng.* **2017**, *21*, 2607–2614. [CrossRef]
15. Liu, R.; Li, B.; Jiang, Y. A fractal model based on a new governing equation of fluid flow in fractures for characterizing hydraulic properties of rock fracture networks. *Comput. Geotech.* **2016**, *75*, 57–68. [CrossRef]
16. Cherubini, C.; Giasi, C.I.; Pastore, N. Bench scale laboratory tests to analyze non-linear flow in fractured media. *Hydrol. Earth Syst. Sci. Discuss.* **2012**, *9*, 2511–2522. [CrossRef]
17. Liu, R.; Li, B.; Jiang, Y.; Yu, L. A numerical approach for assessing effects of shear on equivalent permeability and nonlinear flow characteristics of 2-D fracture networks. *Adv. Water Resour.* **2018**, *111*, 289–300. [CrossRef]
18. Yang, T.; Shi, W.; Li, S.; Xin, Y.; Yang, B. State of the art and trends of water-inrush mechanism of nonlinear flow in fractured rock mass. *J. China Coal Soc.* **2016**, *41*, 1598–1609.
19. Yang, T.; Chen, S.; Zhu, W.; Meng, Z.; Gao, Y. Water inrush mechanism in mines and nonlinear flow model for fractured rocks. *Chin. J. Rock Mech. Eng.* **2008**, *27*, 1411–1416.
20. Yang, T.H.; Shi, W.H.; Liu, H.L.; Yang, B.; Xin, Y.; Liu, Z.B. A non-linear flow model based on flow translation and its application in the mechanism analysis of water inrush through collapse pillar. *J. China Coal Soc.* **2017**, *42*, 315–321.

21. Liu, W.; Zhao, J.; Li, Q. A MHC–NF Coupling Model for Water Inrush from Collapse Columns and Its Numerical Simulation. *Geotech. Geol. Eng.* **2018**, *36*, 2637–2647. [CrossRef]
22. Yao, B.; Mao, X.; Wei, J.; Wang, D. Study on coupled fluid-solid model for collapse columns considering the effect of particle transport. *J. China Univ. Min. Technol.* **2014**, *43*, 30–35.
23. Bear, J.; Bachmat, Y. Introduction to modeling of transport phenomena in porous media. *Theory Appl. Transp. Porous Med.* **1990**, *4*, 481–516.
24. Zhou, Q.; Herrera-Herbert, J.; Hidalgo, A. Predicting the risk of fault-induced water inrush using the adaptive neuro-fuzzy inference system. *Minerals* **2017**, *7*, 1–15. [CrossRef]
25. State Administration of Work Safety of China. *Coal Mine Water Prevention and Control Regulations*; China Coal Industry Publishing House: Beijing, China, 2009; pp. 84–85. (In Chinese)
26. Li, W.; Liu, Y.; Qiao, W.; Zhao, C.; Yang, D.; Guo, Q. An improved vulnerability assessment model for floor water bursting from a Confined Aquifer Based on the Water Inrush Coefficient Method. *Mine Water Environ.* **2017**, *37*, 11–19. [CrossRef]
27. He, M.; Lu, X.; Jing, H. Characters of surrounding rock mass in deep engineering and its nonlinear dynamic-mechanical design concept. *Chin. J. Rock Mech. Eng.* **2002**, *21*, 1215–1224.

Article

Numerical Simulation of Hydraulic Fracture Propagation in Coal Seams with Discontinuous Natural Fracture Networks

Shen Wang [1,2] , Huamin Li [1,3,]* and Dongyin Li [1,3,]*

[1] School of Energy Science and Engineering, Henan Polytechnic University, Jiaozuo 454000, China; shen.wang3@mail.mcgill.ca or wangwuhua@126.com

[2] Department of Mining and Materials Engineering, McGill University, Montreal, QC H3A 0E8, Canada

[3] State and Local Joint Engineering Laboratory for Gas Drainage & Ground Control of Deep Mines, Henan Polytechnic University, Jiaozuo 454000, China

[*] Correspondence: lihuamin2007@163.com (H.L.); lidongyin@126.com (D.L.); Tel.: +86-0391-3987921 (H.L.); +86-0391-3987908 (D.L.)

Received: 29 June 2018; Accepted: 30 July 2018; Published: 1 August 2018

Abstract: To investigate the mechanism of hydraulic fracture propagation in coal seams with discontinuous natural fractures, an innovative finite element meshing scheme for modeling hydraulic fracturing was proposed. Hydraulic fracture propagation and interaction with discontinuous natural fracture networks in coal seams were modeled based on the cohesive element method. The hydraulic fracture network characteristics, the growth process of the secondary hydraulic fractures, the pore pressure distribution and the variation of bottomhole pressure were analyzed. The improved cohesive element method, which considers the leak-off and seepage behaviors of fracturing liquid, is capable of modeling hydraulic fracturing in naturally fractured formations. The results indicate that under high stress difference conditions, the hydraulic fracture network is spindle-shaped, and shows a multi-level branch structure. The ratio of secondary fracture total length to main fracture total length was 2.11~3.62, suggesting that the secondary fractures are an important part of the hydraulic fracture network in coal seams. In deep coal seams, the break pressure of discontinuous natural fractures mainly depends on the in-situ stress field and the direction of natural fractures. The mechanism of hydraulic fracture propagation in deep coal seams is significantly different from that in hard and tight rock layers.

Keywords: hydraulic fracture network; cohesive element method; coal seams; fracture propagation; discontinuous natural fracture; secondary fracture

---

1. Introduction

Coal seam gas that mainly contains methane is an economical and promising solution for the world's energy crisis [1]. Global coalbed methane (CBM) resources are estimated to range from 84.38 Tm^3 to 262.21 Tm^3, of which approximately 14.44 Tm^3 are recoverable [2]. China has the world's third largest CBM reserve, totaling approximately 37 Tm^3 [3,4]; however, as they are influenced by coalification, geological structure and depth, most Chinese CBM reservoirs have low permeability [5–7], which makes gas extraction difficult, and even leads to coal and gas outbursts [8].

Hydraulic fracturing is an efficient stimulation method that is widely used to enhance CBM production [9–13] by creating a new fracture network to improve the reservoirs' permeability and by disturbing the gas in its absorbed state to promote gas desorption from coal [9,14]. Moreover, hydraulic fracturing has also been applied to destress coal seams to prevent coal and gas outbursts. Other significant applications of hydraulic fracturing include the measurement of geo-stress [15,16], stimulation in geothermal energy reservoirs [17,18], hard rock stratum breaking for ground control and mining [19,20], and underground radioactive waste disposal [21]. The application of hydraulic fracturing can be traced back to the 1860s, when liquid nitroglycerine was injected into boreholes to break rocks in the US [22]. In 1968, high-volume hydraulic fracturing was first used to extract crude oil [23]. To date, approximately 2.5 million fracturing treatments have been carried out [24]. Hydraulic fracturing for CBM extraction projects started in the US in the 1970s and in Australia in the 1990s [25].

Due to coalification and geological movement, coal seams in general include porous coal matrix, cleat systems and joint networks [26]. The cleat systems develop in each coalification stage, and are usually well-developed in low volatile bituminous coal, but poorly-developed in the lowest ranking coal or anthracite coal [27,28]. In most coal seams, discontinuous cleat systems cut across the coalbed at about 90° [28,29]. The cleat systems usually occur in two sets of roughly mutually-perpendicular cleats: face cleats and butt cleats [30,31]. Both cleats and joints constitute the main fracture network of coal seams. Natural fractures in coal seams are usually discontinuous and strength-heterogeneous due to differences in the roughness, opening, mineral fillings and extension length of fractures.

High-pressure fluid penetrates the porous coal matrix and natural fractures, resulting in a potential effect on pore pressure distribution and natural fracture propagation [32,33]. The hydraulic fracture may change its propagation direction in contact with natural fractures [34]. Therefore, hydraulic fracture propagation can be significantly affected by the pores and the discontinuous natural fracture network in coal seams [11,35,36]. Investigation of hydraulic fracture propagation and its interaction with discontinuous natural fracture (i.e., pre-existing fracture) networks is crucial to deepening the knowledge of hydraulic fracture propagation mechanisms in naturally fractured coal seams, leading to the better application of hydraulic fracturing in coal seams.

In the past decades, significant research has been focused on the mechanism of hydraulic fracture initiation and propagation [10,13,33,36–45]. Theoretically, in intact rocks or rocks with few natural macro-fractures, it is well accepted that the initial hydraulic fractures are roughly perpendicular to the direction of minimum principal stress. In this respect, geo-stress conditions have a significant effect on hydraulic fractures. The fracture initiation pressure depends on the geo-stress conditions, tensile strength, pore pressure, and porosity of coal or rock layers [26]. Several theoretical models have been put forward to predict the hydraulic fracture initiation pressure for different rock properties and pore pressure conditions [46–49].

To describe hydraulic fracture propagation in natural fracture networks, several theoretical models have been proposed based on laboratory experiments and field investigations, and can be classified into four main models [50]: pure opening mode (POM), pure shear stimulation (PSS), primary fracturing with shear stimulation leak-off (PFSSL), and mixed-mechanism stimulation (MMS). POM is a classic concept for the prediction of new and propagating hydraulic fractures that has been widely used in oil and gas extraction [10,51–53]. In the POM model, it is assumed that the rock crack is induced by pure tensile failure. POM is suitable for predicting hydraulic fracture propagation in reservoirs that contain only few natural fractures. PFSSL considers the leak-off behavior of fracturing liquid when the main continuous fractures propagate and meet with natural fractures [50]. Warpinski et al. [54], Palmer et al. [55], Rogers et al. [56] and Nagel et al. [57] used PFSSL to predict hydraulic stimulation in shale. MMS is based on the idea that continuous flow paths include both new and natural fractures, and that the propagating new fractures terminate in contact with natural fractures [35,58,59]. Therefore, in the MMS model, due to the natural fracture network, it is difficult to form large-scale continuous fractures, and the fracturing liquid is forced to flow through a more complex fracture network that involves both new and natural fractures. Affected by the local stress concentrations induced by the opening and slipping of surrounding fractures, hydraulic fractures not perpendicular to the minimum principal stress are forced to open at least partially [50]. Hydraulic fracture propagation and termination in MMS have been analyzed in laboratoy experiments [32,39,60,61], field investigations [62–64], and numerical simulations [61]. PSS is widely used in enhanced geothermal systems (EGS), and it is commonly assumed that the hydraulic liquid flow pathways mainly involve natural fractures, and that new fracture growth is usually not considered [50]. In PSS, the artificial hydraulic fracture network is formed by natural fractures connecting to each other.

Beginning with Khristianovic et al.'s work [65] in 1955, significant efforts have been applied to the development of numerical modeling techniques for hydraulic fracturing. The early studies, such as the axisymmetric penny-shaped model [66], the 2D plane strain PKN [51] and KGD [52] models, focused on the fluid–solid coupling analytical methods based on the assumption of a fracture profile. Those models solved the hydraulic fracture question by simplifying the fracture profile and the hydraulic pressure distribution. A well-known pseudo-3D model was proposed in the 1980s [67], which can model hydraulic fracture propagation for more realistic fracture profiles. After that, other analytical approaches for 2D and 3D hydraulic fractures have been proposed [10,68–73].

Furthermore, many numerical simulations using the finite element method (FEM), the boundary element method (BEM), the extended finite element method (XFEM), the cohesive element method and the discrete element method (DEM), have been performed to study hydraulic fracturing. To model hydraulic fracture propagation, the numerical governing equations should include: (1) the non-linear partial differential equations that describe the relationship between the fluid flow rate in fractures, the fracture opening width and the fluid pressure gradient; (2) the fracture initiation and propagation criteria; (3) the leak-off equations that define the diffusion process of the fracturing liquid from fractures in the coal or rock matrix [74]; and (4) seepage equations that describe the fluid seepage process in the coal or rock matrix and the variations of porosity and pore pressure. Wang et al. [75] proposed a numerical method combining FEM and a meshfree method to investigate crack propagation driven by hydraulic pressure. Yoon et al. [34,76–79] implemented and developed the fluid flow and seismicity algorithms using particle flow code 2D (PFC2D) software to model the hydraulic fracturing process in a natural pre-existing fracture network. Those algorithms allow higher permeability parameters to natural fractures that are represented by a smooth joint contact set in the PFC2D software. In addition, the PFC2D software was also adopted to simulate hydraulic fracturing by Al-Busaidi et al. [80], Shimizu [81] and Wang et al. [36,82]. Zhao et al. [83] investigated the effects of different types of natural fractures on hydraulic fractures in rock using the UDEC software. Based on the maximum tensile strain criterion, Huang [84] studied hydraulic fracture propagation and coalescence in coal seams using the RFPA software. Chen [74] modeled a single 2D hydraulic fracture using the cohesive element method and compared the numerical simulation results with analytical results, demonstrating that the cohesive element method is highly accurate when modeling hydraulic fracturing. Dahi Taleghani et al. [33] improved a traction–separation law for cohesive zone model to simulate hydraulic fracture propagation in naturally fractured formations. Zhao et al. [85] analyzed 3D hydraulic fracture propagation in coal, also using the cohesive element method.

Compared to other analytical methods and numerical simulation methods for modeling hydraulic fracturing, the cohesive element method has several advantages. (1) The cohesive element method adopts a non-linear damage zone to represent the local stress field around the crack tip, and avoids stress singularity in the crack tip, therefore providing good convergence for the fracture process numerical simulation; (2) The hydraulic fracture propagation process can be regarded as a dynamic moving process for fracture surfaces driven by fracture liquid as the propagation path is unknown in advance. Using the cohesive element method, the previously-unknown hydraulic fracture path can be dynamically calculated as a natural solution [74]; (3) The cohesive element method models microstructural damage mechanisms well (hydraulic fracture initiation, propagation, branching and coalescence with other fractures); (4) The cohesive element method can consider the leak-off behavior of fracturing liquid.

These above studies developed the simulation methods and algorithms of hydraulic fracture propagation. However, they mainly focused on hydraulic fracturing in intact rock or in continuous pre-existing joint networks. To date, hydraulic fracture propagation and interaction with the discontinuous fracture network in coal seams is still unclear.

This study proposes a hydraulic fracture propagation simulation method using the ABAQUS Software for discontinuous natural fracture networks and a preliminary analysis of hydraulic fracture morphology in coal seams. An innovative finite element meshing scheme was used to model hydraulic fracturing based on the cohesive element method, featuring global cohesive element embedding and global pore pressure node sharing techniques. A simulation methodology including leak-off and seepage effects was developed. The hydraulic fracture network characteristics, the growth mechanism of secondary hydraulic fractures, the pore pressure distribution in coal seams, and the variation of fracturing liquid pressure were investigated. Overall, this study aims to deepen the understanding of the mechanism of hydraulic fracture propagation in coal seams.

2. Simulation Methodology for Hydraulic Fracturing in Discontinuous Natural Fracture Networks

2.1. Concept and Methodology of the Hydraulic Fracturing Simulation Using the Cohesive Element Method

The cohesive element method is based on damage mechanics and the cohesive zone model (CZM). By embedding zero-thickness cohesive elements into a mesh, dynamic and unrestricted fracture propagation (especially crack branching, coalescence and swerve) can be modeled. Figure 1 illustrates the process of generating zero-thickness cohesive elements into a solid element mesh. This algorithm was coded with the Python programing language and implemented as an ABAQUS plugin to generate cohesive elements. By executing the plugin, cohesive elements are generated at every solid element interface. Therefore, the final model consists of solid elements and cohesive elements. The solid element and its neighboring cohesive element are connected by their sharing two common nodes, and the two adjacent cohesive elements connect to each other by sharing one node, as shown in Figure 1. Under internal/external traction loads, cohesive elements experience loading, damage, stiffness degradation and cracking. Crack initiation and propagation are represented by the damage and failure of cohesive elements. The coal/rock micro-elastic–plastic deformation and seepage behaviors are dominated by solid elements.

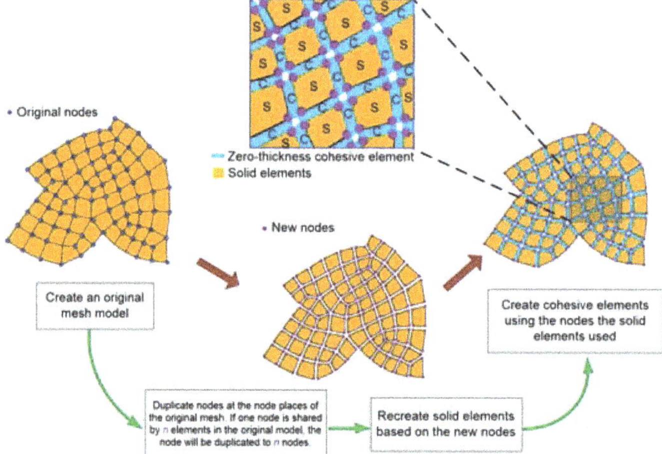

Figure 1. The process of generating zero-thickness cohesive elements at every solid element interface. Note that a solid element and its neighboring cohesive element are connected by sharing two common nodes, and two adjacent cohesive elements connect to each other by sharing one node.

In this study, the cohesive pore pressure element (CPPE), a special type of cohesive element, was used to simulate hydraulic fracture propagation and leak-off behavior from hydraulic fractures to the porous coal matrix. As illustrated in Figure 2a, a 2D zero-thickness CPPE has six nodes and two integration points. The connection of nodes 1 and 2 forms the bottom face of a cohesive element, and nodes 3 and 4 form the top face. The relative displacements between the top and bottom faces (both in 1-direction and 2-direction) represent crack opening and slipping. Therefore, Nodes 1 to 4 are used to calculate the relationship between tractions and separations.

Nodes 5 and 6 are the pore pressure nodes, to calculate the fracturing liquid flow from one CPPE to the next CPPE (tangential flow) and leak-off from cohesive elements to solid elements (normal flow). The CPPEs are zero-thick, so the pore pressure nodes of the discrete adjacent CPPEs occupy the same coordinates after generating the discrete CPPEs in a mesh, as shown in Figure 2a (pore pressure nodes $p1, p2, p3, p4$ have the same coordinates $P(x, y)$). The fracturing liquid flow and transmission need to go through the pore pressure nodes. All the adjacent CPPEs need to share one common pore pressure node to connect to each other and to allow the fracturing liquid to flow in the fracture from one CPPE to the next. Therefore, after creating the discrete CPPEs, the pore pressure nodes at the same coordinates were merged, as shown in Figure 2b.

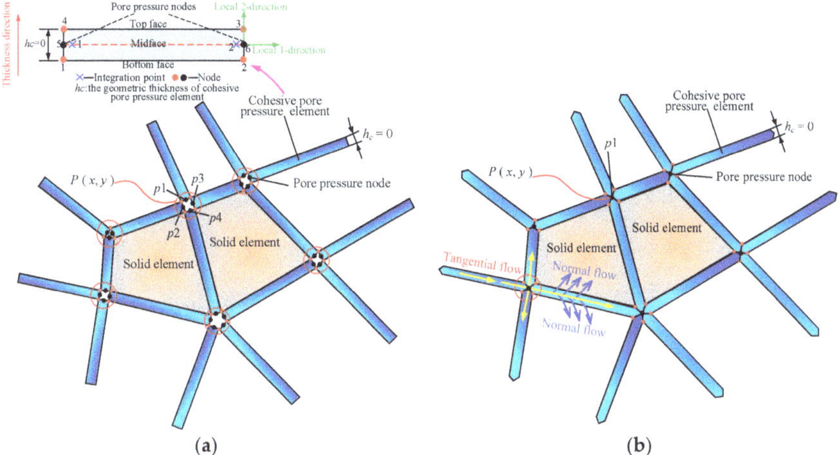

Figure 2. (**a**) Schematic of the CPPE and the mesh structure by embedding discrete CPPEs; (**b**) the mesh structure after merging the pore pressure nodes at the same coordinates and the concept of fracturing liquid flow in fractures. Note that the CPPEs are zero-thick, and the pore pressure nodes $p1, p2, p3, p4$ have the same coordinates $P(x, y)$. The fracturing liquid flow has two components: the tangential flow from one cohesive element to the next cohesive element, and the normal flow from cohesive elements to solid elements.

2.2. Discontinuous Natural Fracture Networks

As discussed in the introduction, natural fractures in coal seams usually have obvious directionality, thus coal seams are divided into regular hexahedral masses by the discontinuous natural fracture network. In fact, the fracture direction varies widely. This study assumed that the angle between the two sets of natural fractures was 90°, and its angle bisector was parallel to the maximum principal stress, as shown in Figure 3. The discontinuous fracture network was created as follows:

(1)   A 2D plane model was created, then this plane was partitioned by two sets of lines that are orthogonal to each other.
(2)   This plane was meshed to generate solid elements.
(3)   CPPEs were embedded in this mesh model and numbered. The serial numbers of the CPPEs that were on the two sets of partition lines into set A were picked.

(4)   A certain proportion of elements from set A was randomly selected. By assigning very low mechanical properties, these selected elements were used to represent the discontinuous fractures. The above process for the creation of a discontinuous fracture network was executed using a Python script program in ABAQUS.

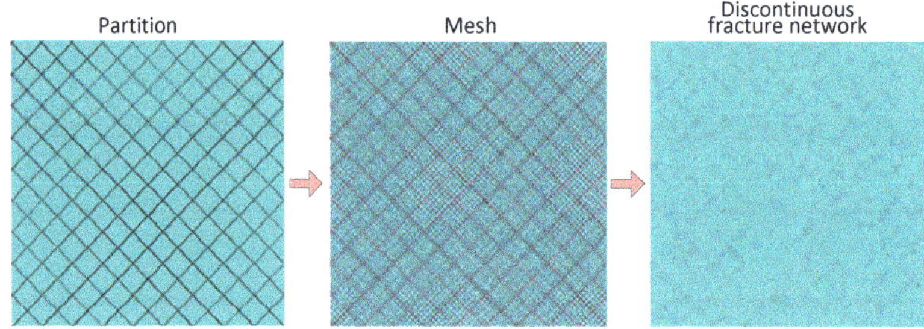

Figure 3. The process of creating the discontinuous fracture network.

3. Seepage and Hydraulic Fracture Equations

3.1. Seepage–Stress Coupling Equation for the Coal Matrix

Coal matrix is a typical porous medium that contains solid coal materials, gas-filled pores, and fracturing leak-off liquid that permeates the coal matrix during fracturing. In the seepage–stress coupling analysis, the effective stress principle was used, given by [86]:

$$\overline{\sigma} = \sigma - \chi u_w I \tag{1}$$

where $\overline{\sigma}$ is the effective stress matrix at a point, Pa; σ is the total stress matrix, Pa; χ is the Biot coefficient; u_w is the pore pressure, Pa; I is a unit matrix.

The porosity of the coal matrix, n, is defined as:

$$n = \frac{dV_v}{dV} \tag{2}$$

where dV_v is the void volume, m^3; and dV is the total volume in the current configuration, m^3.

The void ratio e is defined as:

$$e = \frac{dV_v}{dV - dV_v} = \frac{n}{1-n} \tag{3}$$

Saturation, s, is the ratio of the total free seepage volume of liquid and gas, dV_w, to the void volume, and is defined as:

$$s = \frac{dV_w}{dV_v} \tag{4}$$

3.1.1. Discretized Equilibrium Equation

Based on the virtual work principle, the stress equilibrium equation [87] in the current configuration can be expressed as:

$$\int_V \overline{\sigma} : \delta\varepsilon dV = \int_S t \cdot \delta v dS + \int_V f \cdot \delta v dV + \int_V s n \rho_w g \cdot \delta v dV \tag{5}$$

where δv is the virtual velocity field, m/s; $\delta\varepsilon$ is the virtual rate of deformation tensor, s^{-1}; t is the surface traction tensor, Pa; f is the body force except the seepage liquid weight, N/m^3; ρ_w is the density of the seepage liquid, kg/m^3; g is the gravity acceleration, m/s^2; V is the integration domains of volume, m^3; S is the integration domains of area, m^2. Note that the last term in Equation (5) is the weight of the seepage liquid.

In finite element analysis, the equilibrium equation is discretized into a set of equations with interpolation functions. The virtual velocity field and the virtual rate of deformation are interpolated as:

$$\begin{cases} \delta v = N^N \delta v^N \\ \delta \varepsilon = \text{sym}\left(\frac{\partial \delta N^N}{\partial x}\right)\delta v^N \end{cases} \tag{6}$$

where N^N is the matrix of interpolation functions defined with respect to the material coordinates; and x is the unit coordinate vector, m.

By substituting Equation (6) into Equation (5), the virtual work equation is discretized as

$$\delta v^N \int_V \beta^N : \overline{\sigma} dV = \delta v^N \left(\int_S N^N \cdot t dS + \int_V N^N \cdot f dV + \int_V sn\rho_w N^N \cdot g dV \right) \tag{7}$$

Note that the term $\int_V \beta^N : \overline{\sigma} dV$ is the internal force array, I^N, and $\int_S N^N \cdot t dS + \int_V N^N \cdot f dV + \int_V sn\rho_w N^N \cdot g dV$ is the external force array, P^N. Therefore, the equilibrium Equation (7) can be expressed as:

$$I^N - P^N = 0 \tag{8}$$

3.1.2. Continuity Equation of Seepage

The porous coal matrix can be modeled approximately by attaching the solid element mesh to the solid materials. Thus, the liquid flows through the mesh. The time rate of the change of liquid mass in a control volume is:

$$R = \frac{d}{dt}\left(\int_V \rho_w sn dV\right) = \int_V \frac{1}{J}\frac{d}{dt}(J\rho_w sn)dV \tag{9}$$

where R is the time rate of change of liquid mass in a control volume, kg/s; t is time, s; J is the rate of volume change of the porous coal matrix, defined as the ratio of volume in the current configuration, dV, to the volume in the reference configuration, dV_0:

$$J = \left|\frac{dV}{dV_0}\right| \tag{10}$$

The liquid mass which crosses a surface, S, and flows into the control volume per unit time is:

$$F = -\int_S \rho_w sn(n \cdot v_w)dS \tag{11}$$

where v_w is the liquid seepage velocity, m/s; n is the outward normal vector to S.

Obviously, the liquid mass change per unit time, R, is equal to the liquid mass flowing into the control volume per unit time, F, that is [85,88]:

$$\int_V \frac{1}{J}\frac{d}{dt}(J\rho_w sn)dV = -\int_S \rho_w sn(n \cdot v_w)dS \tag{12}$$

According to the divergence theorem, Equation (12) can be transformed to an equivalent weak form:

$$\int_V \delta u_w \frac{1}{J}\frac{d}{dt}(J\rho_w sn)dV + \int_V \delta u_w \frac{\partial}{\partial x}(\rho_w sn v_w)dV = 0 \tag{13}$$

Equation (13) is the continuity equation of the seepage. In this study, the seepage in the porous coal matrix obeys Darcy's law, which is given by [89]:

$$sn v_w = -\frac{1}{\rho_w g}k\left(\frac{\partial u_w I}{\partial x} - \rho_w g\right) \tag{14}$$

3.2. Flow Equation of the Fracturing Liquid Flow in CPPE

The constitutive response of the fracturing liquid flow in the CPPE combines tangential flow within the fracture and normal flow across the fracture, as illustrated in Figure 4. The fracturing liquid is assumed to be incompressible Newtonian. The tangential flow within the fracture is governed by [74,90]:

$$qw = -\frac{w^3}{12\mu}\nabla p_w \tag{15}$$

where q is the fracturing liquid flow rate per unit area, m/s; w is the opening width of the hydraulic fracture, m; μ is the viscosity of the fracturing liquid, Pa·s; ∇p_w is the pressure gradient along the CPPE, Pa/m. The term $\frac{w^3}{12\mu}$ is equivalent to the permeability or the resistance to fluid flow in the fracture.

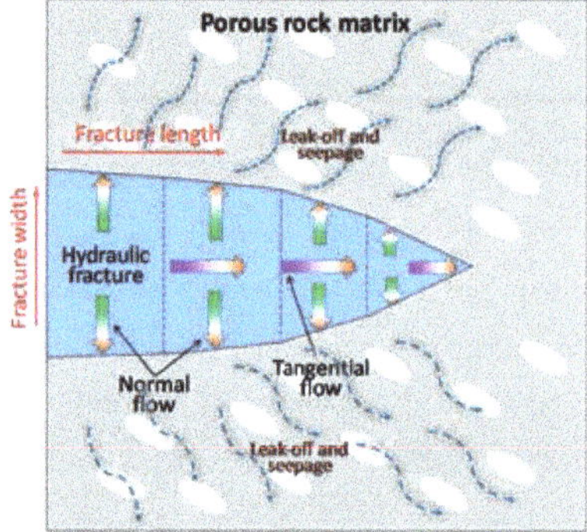

Figure 4. Schematic of fluid flow patterns, including tangential flow and normal flow.

Normal flow, representing the leak-off behavior from fractures into porous coal matrix, is defined as [74,85]:

$$\begin{cases} q_t = c_t(p_m - p_t) \\ q_b = c_b(p_m - p_b) \end{cases} \tag{16}$$

where q_t and q_b are the leak-off flow rates in the normal direction from the mid-face into the top face and bottom face of CPPE respectively, m/s; c_t and c_b are the corresponding leak-off coefficients, m/Pa·s; p_m is the mid-face pressure, Pa; p_t and p_b are the pore pressures on the top and bottom faces respectively, Pa. The leak-off coefficient can be interpreted as the permeability of the hydraulic fracture surfaces (i.e., the top and bottom faces of CPPE). According to Equation (16), the normal flow rate depends on the leak-off coefficient, and the pressure difference between the fracture (CPPE) and the porous coal matrix (solid element).

3.3. Damage Initiation and Evolution Law of Hydraulic Fractures

Crack initiation and propagation are governed by the traction–separation law. Figure 5 illustrates the fracture propagation driven by the fracturing liquid based on the cohesive element method. The traction–separation law defines the relationship between traction force, T, and the relative separation displacement between the top and bottom faces, δ. In the red cohesive zone in Figure 5, the cohesive elements are in an elastic state or in an incomplete damage state. δ_m^o is the critical separation displacement at the initial damage point, and δ_m^f is the critical displacement at the complete failure point. When $\delta \geq \delta_m^o$, the cohesive element starts to damage, and its stiffness decreases as T increases. When $\delta \geq \delta_m^f$, the stiffness of the cohesive element reduces to zero, which indicates that the fracture propagates forward. The cohesive elements in the blue fracture zone are in the complete damage state. Even if those CPPEs in the complete damage state close again from an open state, their tensile stiffness cannot recover. The concept of the crack tip was referenced from the literature [91]. The point of $\delta = \delta_m^f$ is the material crack tip, while the point of $\delta = \delta_m^o$ is the cohesive crack tip. The cohesive elements between the two crack tips are in the damaged state. The traction force, T, is the resultant force of hydraulic pressure in the cohesive element, pore pressure in the adjacent solid element, and geo-stress. Therefore, the hydraulic fracture is affected by fracturing pressure, the discontinuous natural fracture network, the seepage process in the coal matrix and geo-stress.

To model crack initiation and propagation in different materials using cohesive elements, various traction–separation models were developed. In this work, the irreversible bilinear model was used to simulate hydraulic fracture opening and closing behaviors. This model was also used by Chen [74] and Zhao et al. [85] to model hydraulic fracturing in rock. The model assumes that the mechanical response of cohesive elements satisfies a reversible linear elastic behavior prior to the initial damage point. The bearing capacity of cohesive elements decreases linearly in the damage evolution stage, and their stiffness cannot recover during unloading. The irreversible bilinear model was introduced as follows.

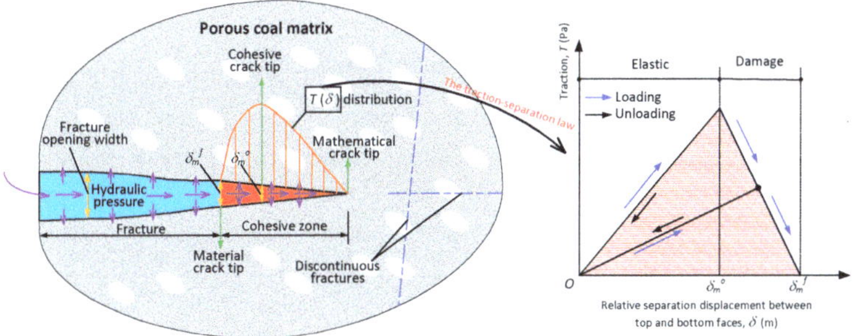

Figure 5. Hydraulic fracture propagation driven by fracturing liquid, and the traction–separation law for cohesive elements. In the blue zone, the cohesive elements are in a complete damage state, whereas those in red zone are in an elastic state or in an incomplete damage state.

3.3.1. Damage Initiation

The tension component of traction force is mainly induced by hydraulic pressure, while the shear component of traction force may be caused by geo-stress differences and the local shear stress concentration around the natural fracture. Therefore, it is significant and necessary to consider the combined effects of the tension force and the shear force when modeling hydraulic fracture initiation and propagation in coal seams. Thus, a quadratic stress criterion that incorporates the tension and shear mixed modes was used to predict damage initiation occurrence. The 2D crack damage initiates when the stress state of the cohesive elements satisfies [33,85]:

$$\left(\frac{t_n}{\sigma_t}\right)^2 + \left(\frac{t_s}{\sigma_s}\right)^2 = 1 \tag{17}$$

where σ_t and σ_s are tensile and shear strength, respectively, Pa; t_n and t_s represents normal traction (along the local 2-direction in Figure 2a) and shear traction (along the local 1-direction in Figure 2a), respectively, Pa.

3.3.2. Damage Evolution

A scalar damage variable, D, is defined to represent the overall damage degree in one cohesive element. D evolves from 0 to 1 monotonically upon further loading after damage initiation. Stress components are affected by the damage according to [92]:

$$
\begin{cases}
t_n = (1 - D)\bar{t}_n, \text{tensile state} \\
t_n = \bar{t}_n, \text{compressive state} \\
t_s = (1 - D)\bar{t}_s
\end{cases}
\tag{18}
$$

where $\bar{t}_n$ and $\bar{t}_s$ are stress components predicted by the linear elastic law at the current separation displacement, Pa, as shown in Figure 6a. Note that in Equation (18), the tensile stiffness of cohesive elements is degraded at the damage stage, whereas the compressive stiffness is unchanged.

For the linear damage softening evolution, D is expressed as [33,85,93]:

$$
D = \frac{\delta_m^f \left(\delta_m^{\max} - \delta_m^o \right)}{\delta_m^{\max} \left(\delta_m^f - \delta_m^o \right)}
\tag{19}
$$

where $\delta_m^{\max}$ is the maximum value of the relative separation displacement recorded during the loading history, m. As illustrated in Figure 6b, the loading history is the loading along the red path (path 1) or reloading via the green path (path 2). Both δ_m^o and δ_m^f consist of tensile and shear separation displacement components.

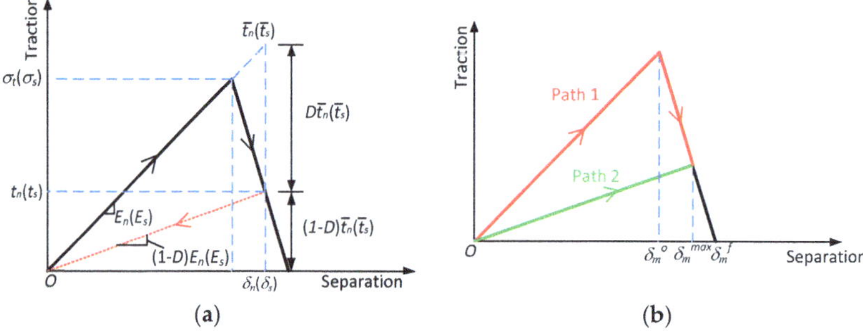

Figure 6. The relationship between traction and separation. (**a**) Tensile and shear components; (**b**) resultant traction. In (**a**), E_n and E_s are the initial tensile stiffness and shear stiffness of CPPE, Pa/m; δ_n and δ_s are tensile and shear separation components, respectively. (**a**) interprets the concept of D, and (**b**) illustrates the concept of loading history (path 1 or path 2).

The power law criterion of fracture energy is widely used to predict the complete damage state, given by [94]:

$$
\left(\frac{G_I}{G_{IC}} \right)^\alpha + \left(\frac{G_{II}}{G_{IIC}} \right)^\alpha = 1
\tag{20}
$$

where G_I and G_{II} are energy components of tension and shear respectively, Pa·m; G_{IC} and G_{IIC} are I-mode and II-mode fracture energies respectively, Pa·m. When G_I and G_{II} satisfy Equation (20), the damage variable D reaches 1.

Note that the stress state and the corresponding displacement at the initial damage point is determined by Equation (17) and the relationship between T and δ; therefore, the separation displacement components at the initial damage point of each cohesive element can be different. $\delta_m^{\max}$ is a function of G_I and G_{II}. δ_m^f is calculated using the fracture energies G_{IC} and G_{IIC}. If σ_t, σ_s, G_{IC} and G_{IIC} are given, for a displacement state $\delta_m^{\max}$, its corresponding damage variable D can be calculated using Equation (19), and then the stress components are updated by substituting D into Equation (18). Hydraulic fracture propagation is simulated using this damage law.

4. Numerical Simulation Procedure

4.1. The Mesh Model of Coal Seams with Discontinuous Natural Fracture Networks

By setting the out-of-plane dimension as 3 m, the 2D numerical coal seam model was equivalent to a 3D model with a thickness of 3 m. As shown in Figure 7, the model is a square with the dimensions 400 m². To avoid the boundary effect, the discontinuous natural fracture network zone is at the center of the model, with a size of 10 m × 10 m. The two fracture sets are orthotropic with each other, and their bisector is parallel to the direction of the principle stress. The fracture spacing was set to 0.5 m. The fracture network zone's mesh consists of 10,000 pore fluid/stress elements (the solid elements) of size 0.1 m and 20,200 CPPEs. The external zone's mesh consists of 5612 pore fluid/stress elements of size 0.1 m to 0.6 m. The pore fluid/stress elements used were four-node plane strain quadrilateral, bilinear displacement, bilinear pore pressure elements (called CPE4P in ABAQUS).

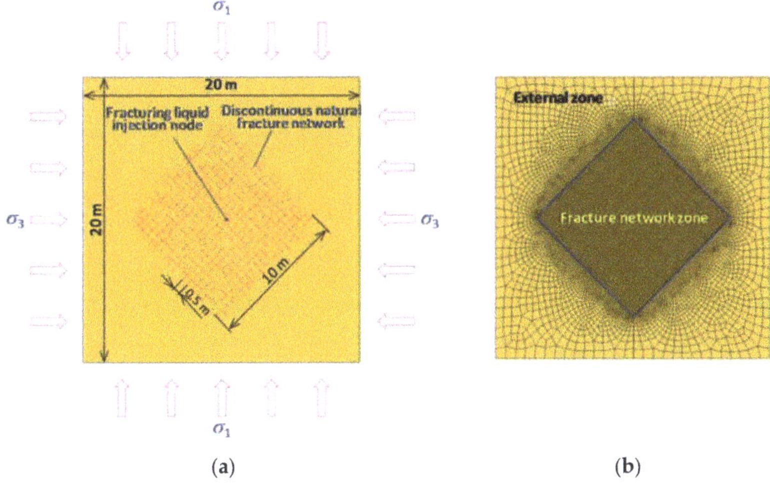

Figure 7. The mesh model of coal seams with discontinuous natural fracture networks. (**a**) shows the model size and the locations of the discontinuous natural fracture network and the injection node; (**b**) shows the mesh density. The fracture network zone is at the center of this model, with a size of 10 m × 10 m. The two fracture sets are orthotropic with each other, and their bisector is parallel to the principle stress direction. The fracture spacing is set to 0.5 m. The fracture network zone's mesh consists of 10,000 pore fluid/stress elements of size 0.1 m and 20,200 CPPEs. The external zone's mesh consists of 5612 pore fluid/stress elements of size 0.1 m to 0.6 m.

4.2. Mechanical Properties of the Coal Matrix, Discontinuous Natural Fracture and Fracturing Liquid

Many studies show that hydraulic fracturing has significant effects on the permeability, leak-off coefficient, and porosity of coal seams [85,95–97]. According to Xu et al. [38], the dynamic permeability k, the dynamic porosity n and the dynamic leak-off coefficient c can be calculated as follows:

$$k = k_0 \exp(3C_f \Delta p) \tag{21}$$

$$n = n_0(1 + C_f \Delta p) \tag{22}$$

$$c = c_0 \sqrt{1 + C_f \Delta p} \exp(\frac{3}{2} C_f \Delta p) \tag{23}$$

Here, k_0 is the initial permeability, m/s; Δp is the pressure difference between the fracture (CPPE) and the porous coal matrix (solid element), Pa; n_0 is the initial porosity; c_0 is the initial leak-off coefficient, m/Pa·s; C_f is the pore compressive coefficient; and $C_f = (n - n_0)/(n \cdot \Delta p)$, Pa^{-1}.

The user subroutine was used to model the dynamic variation of the coal seepage properties. In one numerical increment step, if Δp is determined, n can be calculated using Equation (22), and then C_f can be obtained. k and c are then calculated using Equations (21) and (23). Until this point, k, n and c were updated dynamically. In the next incremental step, these parameters were set as the updated initial values and substituted into the governing equations for the seepage to model the dynamic fracturing process.

The mechanical properties used in this numerical model are listed in Table 1. Note that the mechanical strengths of the discontinuous natural fractures are significantly weaker than those of the coal matrix.

Table 1. The mechanical properties of the discontinuous fractures, coal matrix and fracturing liquid.

Object	Mechanical Parameter	Value
	Elastic modulus (GPa)	5
	Poisson's ratio	0.28
	Tensile strength (MPa)	0.77
	Shear strength (cohesion) (MPa)	6.42
Coal matrix	Initial permeability (m/s)	5×10^{-8}
	Initial porosity	0.15
	I-mode fracture energy (N/m)	29.25
	II-mode fracture energy (N/m)	35.97
	Initial leak-off coefficient (m/Pa·s)	5×10^{-14}
	Tensile strength (MPa)	0.005
	Shear strength (MPa)	0.04
Discontinuous natural fractures	I-mode fracture energy (N/m)	0.19
	II-mode fracture energy (N/m)	0.23
	Initial leak-off coefficient (m/Pa·s)	5×10^{-14}
Fracturing liquid	Viscosity (Pa·s)	0.005
	Density (kg/m^3)	1000

4.3. Initial Conditions

In this study, the injection rate of the fracturing liquid via the injection node was 0.003 m^3/s; the initial gas pore pressure in the coal seams was 5 MPa. To analyze the hydraulic fracture propagation and its interaction with natural fractures under different ground stress conditions, four different stress conditions were applied, as shown in Table 2.

Table 2. Stress conditions.

Group	Minimum Principal Stress, σ_3 (MPa)	Maximum Principal Stress, σ_1 (MPa)	Stress Difference (MPa)
A	10	12.5	2.5
B	10	15	5
C	10	17.5	7.5
D	10	20	10

4.4. Solving Procedures, Convergence Criterion and Error Control

Figure 8 illustrates the pre-processing and the solving procedures of the ABAQUS/Standard solver. The model first calibrates geostatic stress to a balance state and then computes hydraulic fracturing. For the two steps above, the procedure types were 'geostatic' and 'soils', respectively. The module 'soils' in ABAQUS allows users to model hydraulic fracturing and fluid flow through porous media. In the 'soils' module, the transient analysis method and the direct backward difference operator were used to optimize the accuracy of the continuity equation.

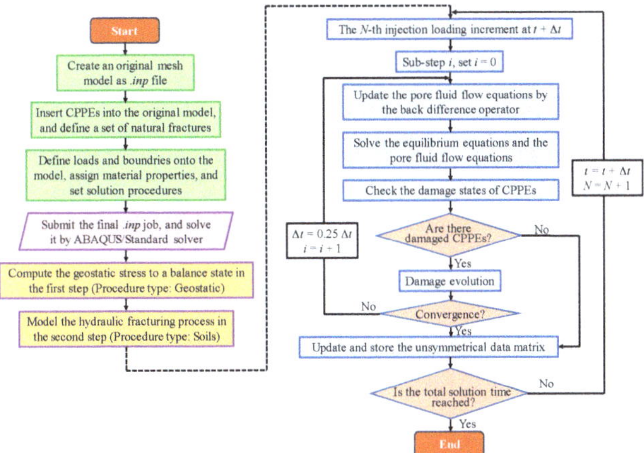

Figure 8. The preprocessing and the solving procedures of the ABAQUS/Standard solver. The module 'soils' was used to model hydraulic fracturing. The direct backward difference operator was used to optimize the accuracy of the continuity equation.

The solution technique used was the full Newton method for nonlinear equilibrium equations. In pore liquid flow analysis, pore pressure, u_w, was selected as a field index to evaluate the accuracy of the Jacobian matrix solution. According to the ABAQUS documentation [98], most nonlinear calculations are sufficiently accurate if the error of the largest field index residual is less than 0.005. In this study, the convergence criterion was also used:

$$r_{\max}^{u_w} \leq 0.005 \tilde{q}^{u_w} \tag{24}$$

where $r_{\max}^{u_w}$ is the largest residual in the pore pressure balance equation, m^3/s; $\tilde{q}^{u_w}$ is the overall time-averaged value of a typical flux for u_w until this point during this step, including the current increment, m^3/s.

To guarantee the accuracy of the fracture propagation when using CPPE, the cohesive element size should be smaller than the cohesive zone length d_c, which is determined by the material mechanical properties, and is calculated as [74]:

$$d_c = \frac{9\pi}{32} \frac{E}{(1-v^2)} \frac{G_{IC}}{\sigma_t^2} \tag{25}$$

where E is the Young's modulus, Pa; v is the Poisson's ratio. By substituting the parameter values from Table 1 into Equation (25), d_c = 0.182 m. The cohesive element size used in this study was 0.1 m, which is significantly less than d_c. This suggests good error control for modeling hydraulic fracture propagation.

5. Results and Discussion

5.1. Hydraulic Fracture Network Characteristics

The CPPEs with fracturing liquid flowing were selected and marked as hydraulic fractures during the fracturing process. Figure 9 shows the hydraulic fracture network characteristics under different stress conditions. According to this figure, hydraulic fractures in coal seams are not just a single crack, but a complex fracture network. Micro-seismic monitoring in the field also confirms that hydraulic fractures in coal seams usually form a network of fracture branches using the horizontal well technique [99].

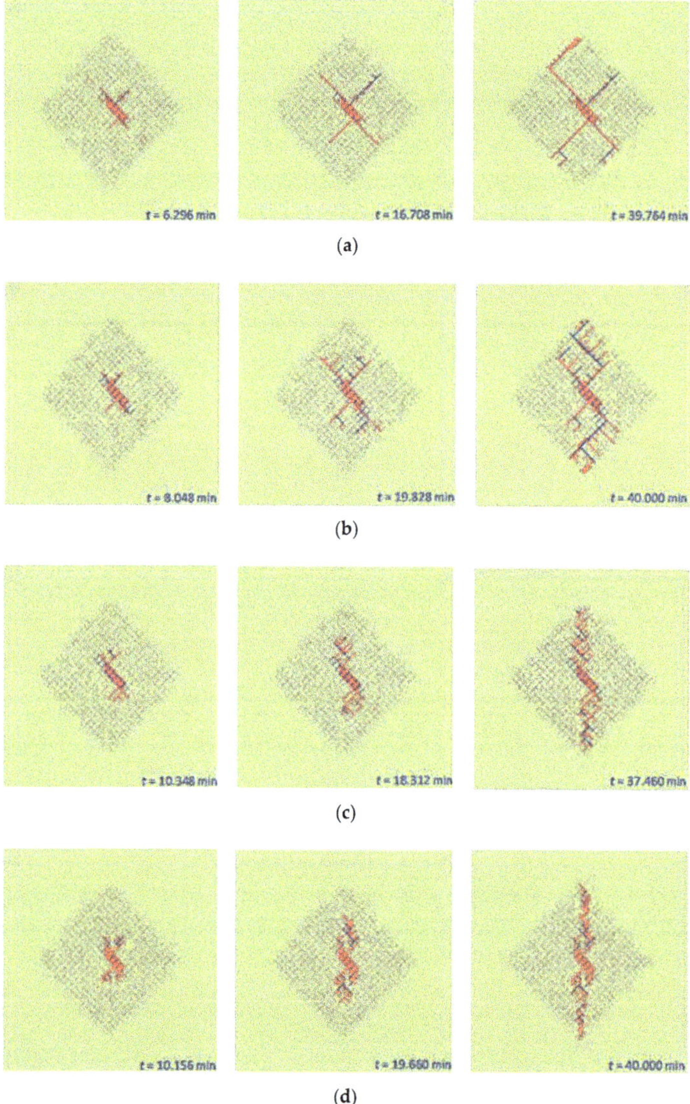

Figure 9. The hydraulic fracture propagation process in a discontinuous natural fracture network under different stress conditions. (a) $\sigma_1 - \sigma_3$ = 2.5 MPa; (b) $\sigma_1 - \sigma_3$ = 5 MPa; (c) $\sigma_1 - \sigma_3$ = 7.5 MPa; (d) $\sigma_1 - \sigma_3$ = 10 MPa. The red lines represent the new fractures induced by hydraulic fracturing, the blue lines represent the natural fractures with hydraulic liquid flowing, and the purple lines represent the discontinuous natural fracture network.

Figure 9 shows that the hydraulic fracture network has a main fracture trunk with many fracture branches. However, the main fractures are not strictly defined. Generally, this usually refers to the more prominent fractures with a relatively large opening. In hydraulic fracturing, the main fractures are the main liquid flow fracture channels. The fracture opening degree and the stiffness of the CPPEs are indicators of main hydraulic fractures. The main hydraulic fractures were identified by magnifying the hydraulic fracture opening 20 times, as shown in Figure 10.

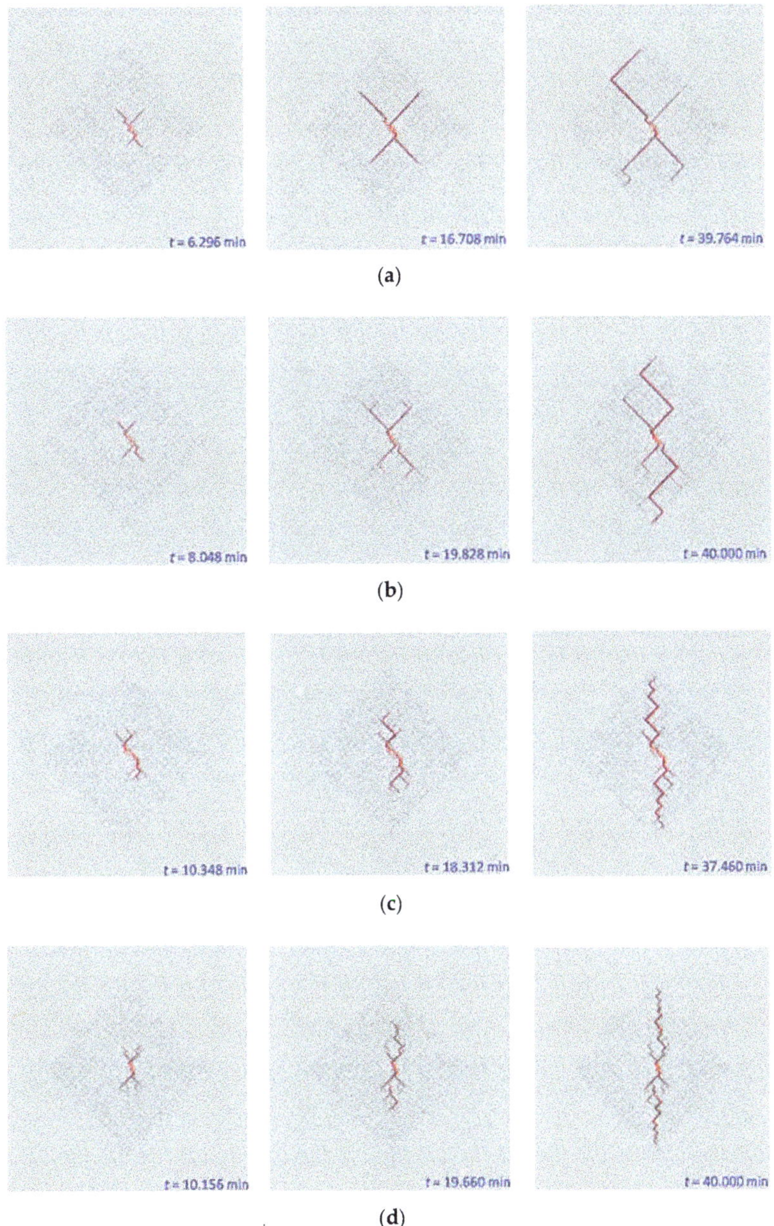

Figure 10. The propagation process of the main hydraulic fractures under different stress conditions. (**a**) $\sigma_1 - \sigma_3 = 2.5$ MPa; (**b**) $\sigma_1 - \sigma_3 = 5$ MPa; (**c**) $\sigma_1 - \sigma_3 = 7.5$ MPa; (**d**) $\sigma_1 - \sigma_3 = 10$ MPa.

During the initial hydraulic fracturing stage, the total fracture volume is small, thus a reticulated secondary fracture structure is created around the injection node (i.e., the injection borehole) when subject to the impact of high-pressure fracturing liquid, as shown in Figure 9. Then the total volume of fractures increases gradually, which in turn increases the fracturing liquid leak-off and weakens fluid impact. During the later stage, the secondary fractures grow from the main fracture trunk like branches.

In Figure 9, the maximum width of the hydraulic fracture network gradually narrows as the stress difference increases, indicating that the general direction of the fracture network follows σ_1 under high stress difference conditions. However, when $\sigma_1 - \sigma_3 = 2.5$ MPa, the main hydraulic fractures propagate straight along the natural fractures. These results suggest that the general direction of hydraulic fracture networks in coal seams is jointly influenced by geo-stress differences and pre-existing natural fracture networks. When the ground stress difference is insignificant, the natural fracture network dominates the direction of the main hydraulic fractures. On the contrary, the maximum principal stress shows strong control over the general direction of the hydraulic fracture network, but the hydraulic fracture network still has many secondary fractures.

When the stress difference is 7.5 MPa or 10 MPa, the hydraulic fracture network is spindle-shaped. As the fractures propagate, the general direction of the hydraulic fracture network gradually converges to the direction of σ_1, and the number of secondary fractures gradually decreases. To better understand how the stress field and natural fractures affect the hydraulic fracture morphology, tests that consider the direction and discontinuous degree of the natural fractures must be carried out in the future. The improved cohesive element method used in this study is leading to continuing research on these topics.

The total lengths of hydraulic fractures under different stress conditions were counted, as illustrated in Figure 11. According to the figure, the total lengths of the hydraulic fractures were between 82.4 m (when $\sigma_1 - \sigma_3 = 2.5$ MPa) and 100.7 m (when $\sigma_1 - \sigma_3 = 5$ MPa). There was no obvious correlation between the total length and stress difference. In each stress condition, the total length of the secondary fractures was significantly bigger than that of the main fracture. The ratio of secondary fracture total length to that of the main fracture was between 2.11 and 3.62, which shows no obvious correlation with stress difference. The results indicate that: (1) hydraulic fracturing can create a large number of secondary fractures in coal seams; (2) stress difference has no significant effect on the total length of the hydraulic fracture and the ratio of the secondary fracture length to the main fracture length.

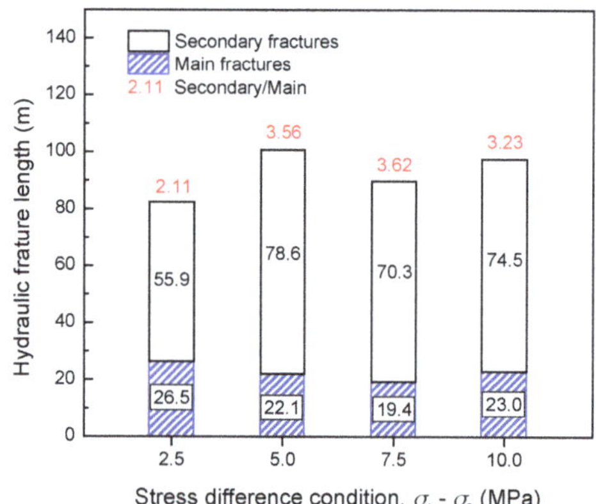

Figure 11. The hydraulic fracture lengths under different stress conditions.

5.2. Growth Process of the Secondary Hydraulic Fractures

According to Figures 9 and 10, the secondary fractures contain new fractures and natural fractures. There are two causes of secondary cracks: change to local stress fields caused by hydraulic fracture propagation and seepage in the porous coal matrix, and pre-existing low-intensity natural fractures. Figure 12 shows the growth process of secondary fractures during hydraulic fracturing when $\sigma_1 - \sigma_3 = 7.5$ MPa. As the main fracture propagates, the secondary fractures gradually branch out of the main fracture trunk. At $t = 8.380$ min, fracture branch 1 appears. At $t = 13.244$ min, fracture branches 2 to 6 appear, and branch 1 grows longer due to the fracturing liquid. At $t = 21.696$ min, new fractures 7 to 13 appear as the main fracture propagates. Note that fractures 7, 9 and 10 grow on the secondary fractures and are considered micro-secondary fractures. The hydraulic fracture network shows a multi-level branch structure.

The main fracture in Figure 12 propagates from natural fractures to the coal matrix, and then swerve back along the natural fracture. Affected by the stress field and discontinuous natural fracture network, some secondary fractures, such as fracture branches 5, 6, 8, 12, propagate along the discontinuous natural fractures, whereas the other secondary fractures start in the coal matrix. Although the tensile strength of the coal matrix is higher than that of the natural fracture, the difficulty of secondary fracture initiation in the coal matrix is almost equal to that in natural fractures due to the high geo-stress conditions. This result suggests that differences in geo-stress play an important role in inducing secondary fractures.

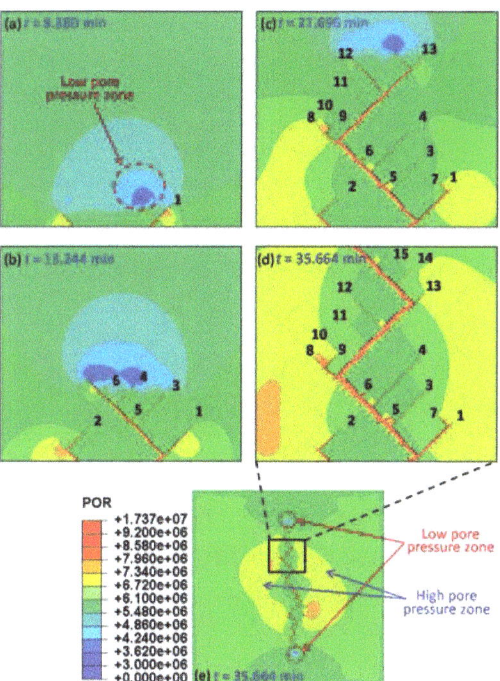

Figure 12. The numerical simulation results for the secondary hydraulic fracture growth process and pore pressure distribution. The serial numbers represent the order of secondary fracture initiation.

5.3. Pore Pressure Distribution Characteristics

Figure 12 also shows pore pressure distribution in the coal matrix and in hydraulic fractures. Note that there are two low pore pressure zones (the blue zones) in front of the two main fracture tips and two high pore pressure zones (the yellow zones) at both sides of the main fractures. Figure 13 shows pore pressure variation in the coal matrix on both sides of the hydraulic fractures. In Figure 13, the distance between the pore pressure peak and main fractures is approximately 1.7 m. The spindle-shaped zone, where the hydraulic fracture network is located, can be regarded as a coal damage zone that changes the transfer path of σ_1 from the far field to near field, as shown in Figure 14. The compressive stress transfer path bypasses the damage zone, resulting in a compressive stress concentration on both sides of the damage zone, which increases pore pressure. In front of the main hydraulic fractures, the coal matrix is subjected to tensile stress by hydraulic fracturing, therefore its volume tends to expand, lowering the pore pressure.

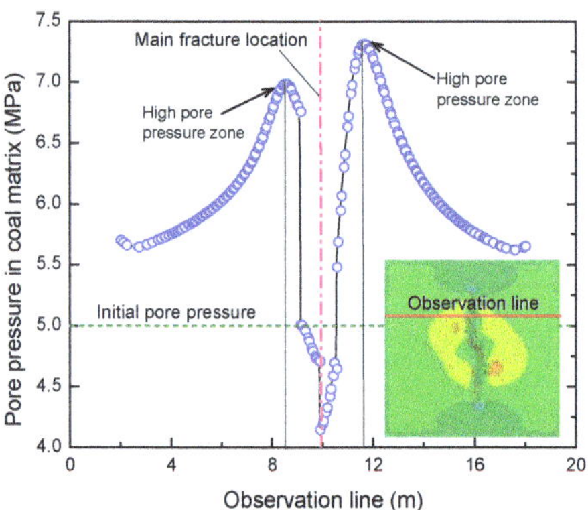

Figure 13. The pore pressure distribution along the observation line.

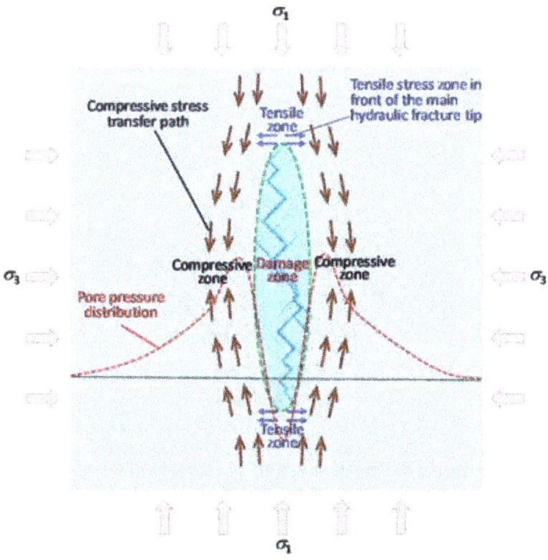

Figure 14. The schematic of pore pressure distribution around the hydraulic fractures.

5.4. Injection Fluid Pressure

5.4.1. Effect of the Injection Rate on Bottomhole Pressure

Due to the large scale of the model (400 m^2), it is difficult to consider the injection borehole profile, which is usually less than 0.04 m^2 in the field; thus an injection node was used instead.

The bottomhole pressure variation during hydraulic fracturing was captured by the injection node. Figure 15 illustrates the bottomhole pressure variations at different injection rates under the stress condition of σ_1 = 17.5 MPa and σ_3 = 10 MPa. According to Figure 15, within the first two minutes of hydraulic fracturing, the bottomhole pressure increased rapidly due to the injection fluid impact effect, and then decreased rapidly due to hydraulic fracture initiation. The peak pressure is called the break-down pressure. After three minutes, the bottomhole pressure stabilized. The stable bottomhole pressure and the break-down pressure increased as the injection rate increased, indicating that hydraulic fracture propagation is a dynamic process. Under high injection rates, the fracture volume rate increase is slower than the injection liquid volume rate increase. This explains why the bottomhole pressure increased as the injection rate increased.

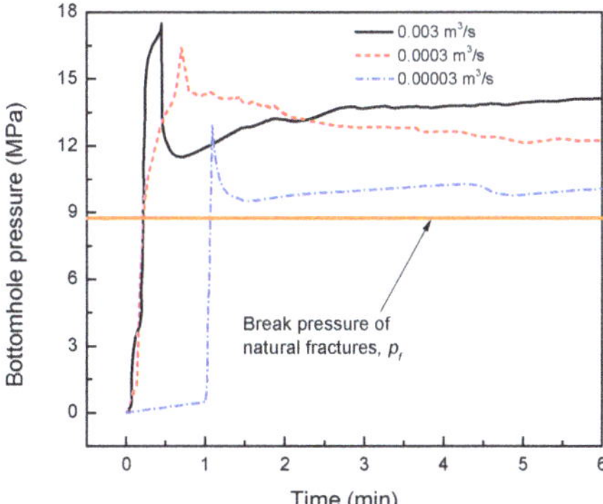

Figure 15. The bottomhole pressure variations at different injection rates. The figure shows that the stable bottomhole pressure and the break-down pressure increased as the injection rate increased.

5.4.2. Break Pressure of the Discontinuous Natural Fracture

As shown in Figure 16, when the hydraulic fracture propagates along discontinuous natural fractures, the break pressure of the fracture initiation p_f should overcome the sum of the normal compressive stress on discontinuous natural fractures p_n and tensile strength T_f. That is:

$$p_f = p_n + T_f \tag{26}$$

Based on the stress balance principle, p_n can be calculated as:

$$p_n = \frac{\overline{\sigma}_1 + \overline{\sigma}_3}{2} + \frac{\overline{\sigma}_1 - \overline{\sigma}_3}{2}\cos 2\theta \tag{27}$$

where θ is the angle between the normal direction of discontinuous natural fractures and the σ_1 direction. $\overline{\sigma}_1$ and $\overline{\sigma}_3$ are the effective stresses corresponding to σ_1 and σ_3, respectively. The effective stresses can be calculated using Equation (1).

By substituting Equation (27) into Equation (26), p_f can be calculated as:

$$p_f = \frac{\overline{\sigma}_1 + \overline{\sigma}_3}{2} + \frac{\overline{\sigma}_1 - \overline{\sigma}_3}{2}\cos 2\theta + T_f \tag{28}$$

Equation (28) clearly states that the break pressure, p_f, depends on the ground stress conditions, tensile strength and direction of discontinuous natural fractures.

When $\overline{\sigma}_1$ = 10 MPa, $\overline{\sigma}_3$ = 17.5 MPa, u_w (pore pressure) = 5 MPa, and θ = 45°, p_f was approximately 8.755 MPa for the natural fractures, and 9.52 MPa for other parts of the discontinuous natural fractures that have the same tensile strength as the coal matrix. Compared to the ground stress, the coal seam tensile strength was very low. According to Equation (28), in deep coal seams, σ_1 and σ_3 are much greater than T_f, so the break pressure of discontinuous natural fractures is mainly determined by the ground stress conditions and the direction of the discontinuous natural fractures.

Hard and tight rock layers, such as geothermal energy reservoirs, have an intact rock tensile strength significantly higher than that of the joints, so hydraulic fractures tend to propagate along natural fractures. When the fracturing liquid is full of natural fractures, the fracture shear strength is drastically reduced and eventually fails. This is why the PSS model is widely used in EGS to predict fracture propagation, but is not suitable for the prediction of hydraulic fractures in deep coal seams.

Note that the above analytical process is based on the assumption of static fracturing, which is different from true dynamic fracturing. Comparing the above analytical results with the numerical simulation results in Figure 15, it can be seen that the simulated bottomhole pressure was higher than the static analytical break pressure result.

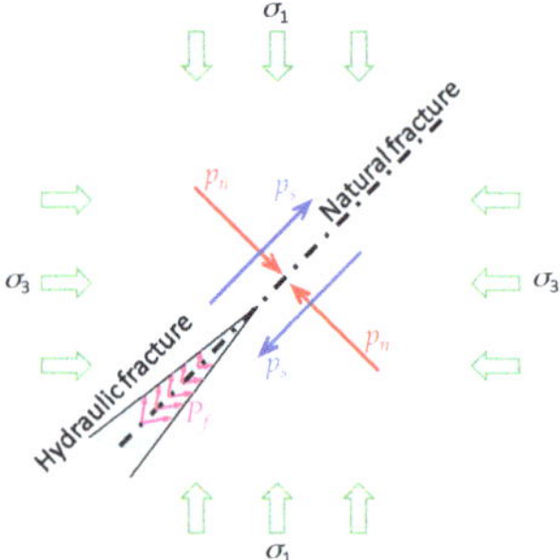

Figure 16. Schematic of the analytical model for calculating the break pressure of discontinuous natural fractures. p_f is the break pressure; p_n and p_s are the normal compressive stress and the tangential shear stress of discontinuous natural fractures.

6. Conclusions

In this study, an innovative finite element meshing scheme using cohesive elements (including global cohesive element embedding and global pore pressure node sharing techniques) was used to model hydraulic fracturing. Fracture propagation and interaction with discontinuous natural fracture networks in coal seams were simulated using the cohesive element method. The hydraulic fracture network characteristics under different stress conditions, the growth process of secondary hydraulic fractures, the pore pressure distribution in coal seams, and the variation of fracturing liquid pressure were investigated in detail. The numerical simulation results show that the improved cohesive element method in this study is capable of modeling hydraulic fracturing in naturally fractured formations.

In coal seams, hydraulic fracturing generally creates a complex hydraulic fracture network that is composed of a main fracture trunk and many secondary fracture branches. The morphology of the fracture network is determined by the geo-stress field and the discontinuous natural fracture network. Under high stress difference conditions, the hydraulic fracture network is spindle-shaped, and shows a multi-level branch structure. The higher the geo-stress difference, the narrower the width of the spindle-shaped hydraulic fracture network.

The ratio of secondary fracture total length to main fracture total length was 2.11~3.62, indicating that the secondary fractures are an important part of the hydraulic fracture network. The geo-stress difference had no significant effect on the ratio of secondary fracture length to main fracture length.

The stable bottomhole pressure and the break-down pressure increased with the injection rate, indicating that fracture propagation is a dynamic process in coal seams. In deep coal seams, the break pressure of discontinuous natural fractures mainly depends on the in-situ stress field and the direction of natural fractures. The mechanism of hydraulic fracture propagation in deep coal seams is significantly different from that in hard and tight rock layers. The PSS model, which is more frequently used in EGS to predict hydraulic fracture propagation, is not suitable for deep coal seams.

Author Contributions: S.W., H.L. and D.L. created the numerical simulations; S.W. performed the numerical simulation work; S.W. and H.L. discussed the numerical simulation results; S.W. and D.L. analyzed the data; S.W. wrote the manuscript; S.W., H.L. and D.L. reviewed the paper before submission.

Funding: This research was funded by the National Natural Science Foundation of China grant number [51604093 and 51474096], the Program for Innovative Research Team at the University of Ministry of Education of China (IRT_16R22), the Doctoral Research Fund Project of Henan Polytechnic University, China (B2017-42), and the Scientific and Technological Key Project of Henan province (172107000016).

Conflicts of Interest: The authors declare no conflict of interest.

References

1. Moore, T.A. Coalbed methane: A review. *Int. J. Coal Geol.* **2012**, *101*, 36–81. [CrossRef]
2. White, C.M.; Smith, D.H.; Jones, K.L.; Goodman, A.L.; Jikich, S.A.; LaCount, R.B.; DuBose, S.B.; Ozdemir, E.; Morsi, B.I.; Schroeder, K.T. Sequestration of carbon dioxide in coal with enhanced coalbed methane recovery a review. *Energy Fuels* **2005**, *19*, 659–724. [CrossRef]
3. Cai, Y.D.; Liu, D.M.; Pan, Z.J.; Yao, Y.B.; Li, J.Q.; Qiu, Y.K. Pore structure and its impact on ch4 adsorption capacity and flow capability of bituminous and subbituminous coals from northeast china. *Fuel* **2013**, *103*, 258–268. [CrossRef]
4. Cheng, W.M.; Liu, Z.; Yang, H.; Wang, W.Y. Non-linear seepage characteristics and influential factors of water injection in gassy seams. *Exp. Therm. Fluid Sci.* **2018**, *91*, 41–53. [CrossRef]
5. Zou, Q.L.; Lin, B.Q.; Liu, T.; Zhou, Y.; Zhang, Z.; Yan, F.Z. Variation of methane adsorption property of coal after the treatment of hydraulic slotting and methane pre-drainage: A case study. *J. Nat. Gas Sci. Eng.* **2014**, *20*, 396–406. [CrossRef]
6. Cheng, W.M.; Hu, X.M.; Xie, J.; Zhao, Y.Y. An intelligent gel designed to control the spontaneous combustion of coal: Fire prevention and extinguishing properties. *Fuel* **2017**, *210*, 826–835. [CrossRef]
7. Zhou, G.; Fan, T.; Ma, Y.L. Preparation and chemical characterization of an environmentally-friendly coal dust cementing agent. *J. Chem. Technol. Biotechnol.* **2017**, *92*, 2699–2708. [CrossRef]
8. Ni, G.H.; Cheng, W.M.; Lin, B.Q.; Zhai, C. Experimental study on removing water blocking effect (wbe) from two aspects of the pore negative pressure and surfactants. *J. Nat. Gas Sci. Eng.* **2016**, *31*, 596–602. [CrossRef]
9. Wright, C.A.; Tanigawa, J.J.; Mei, S.; Li, Z. Enhanced hydraulic fracture technology for a coal seam reservoir in Central China. In Proceedings of the International Meeting on Petroleum Engineering, Beijing, China, 14–17 November 1995; Society of Petroleum Engineers: Beijing, China, 1995. [CrossRef]
10. Adachi, A.; Siebrits, E.; Peirce, A.; Desroches, J. Computer simulation of hydraulic fractures. *Int. J. Rock Mech. Min. Sci.* **2007**, *44*, 739–757. [CrossRef]
11. Yuan, Z.G.; Wang, H.T.; Liu, N.P.; Liu, J.C. Simulation study of characteristics of hydraulic fracturing propagation of low permeability coal seam. *Disaster Adv.* **2012**, *5*, 717–720.
12. Hou, B.; Chen, M.; Wang, Z.; Yuan, J.B.; Liu, M. Hydraulic fracture initiation theory for a horizontal well in a coal seam. *Pet. Sci.* **2013**, *10*, 219–225. [CrossRef]
13. Li, D.Q.; Zhang, S.; Zhang, S.A. Experimental and numerical simulation study on fracturing through interlayer to coal seam. *J. Nat. Gas Sci. Eng.* **2014**, *21*, 386–396. [CrossRef]
14. Eshiet, K.I.I.; Sheng, Y. Carbon dioxide injection and associated hydraulic fracturing of reservoir formations. *Environ. Earth Sci.* **2014**, *72*, 1011–1024. [CrossRef]
15. Bredehoeft, J.; Wolff, R.; Keys, W.; Shuter, E. Hydraulic fracturing to determine the regional in situ stress field, piceance basin, colorado. *Geol. Soc. Am. Bull.* **1976**, *87*, 250–258. [CrossRef]
16. Abou-Sayed, A.; Brechtel, C.; Clifton, R. In situ stress determination by hydrofracturing: A fracture mechanics approach. *J. Geophys. Res. Solid Earth* **1978**, *83*, 2851–2862. [CrossRef]
17. Tenma, N.; Yamaguchi, T.; Zyvoloski, G. The hijiori hot dry rock test site, japan evaluation and optimization of heat extraction from a two-layered reservoir. *Geothermics* **2008**, *37*, 19–52. [CrossRef]
18. Parker, R. The rosemanowes hdr project 1983–1991. *Geothermics* **1999**, *28*, 603–615. [CrossRef]

19. Jeffrey, R.; Zhang, X.; Settari, A.; Mills, K.; Detournay, E. Hydraulic fracturing to induce caving: Fracture model development and comparison to field data. In *DC Rocks 2001, The 38th US Symposium on Rock Mechanics (USRMS)*; American Rock Mechanics Association: Alexandria, VA, USA, 2001.

20. Van As, A.; Jeffrey, R.G. Hydraulic fracturing as a cave inducement technique at northparkes mines. In Proceedings of the Massmin 2000, Brisbane, Austrilia, 29 October–2 November 2000; Australasian Institute of Mining and Metallurgy: Parkville Victoria, Australia, 2000; Volume 2000, pp. 165–172.

21. Sun, R.J. Theoretical size of hydraulically induced horizontal fractures and corresponding surface uplift in an idealized medium. *J. Geophys. Res.* **1969**, *74*, 5995–6011. [CrossRef]

22. Harper, J.A. The marcellus shale: An old "new" gas reservoir in Pennsylvania. *Pa. Geol.* **2008**, *38*, 2–13.

23. Gandossi, L. An overview of hydraulic fracturing and other formation stimulation technologies for shale gas production. *Eur. Comm. Jt. Res. Cent. Tech. Rep.* **2013**, *26347*, 4–29.

24. Montgomery, C.T.; Smith, M.B. Hydraulic fracturing: History of an enduring technology. *J. Petrol. Technol.* **2015**, *62*, 26–40. [CrossRef]

25. McDaniel, B.W. Hydraulic fracturing techniques used for stimulation of coalbed methane wells. In Proceedings of the SPE Eastern Regional Meeting, Columbus, OH, USA, 31 October–2 November 1990; Society of Petroleum Engineers: Columbus, OH, USA, 1990. [CrossRef]

26. Sampath, K.H.S.M.; Perera, M.S.A.; Ranjith, P.G.; Matthai, S.K.; Rathnaweera, T.; Zhang, G.; Tao, X. Ch4-CO$_2$ gas exchange and supercritical CO$_2$ based hydraulic fracturing as cbm production-accelerating techniques: A review. *J. CO$_2$ Util.* **2017**, *22*, 212–230. [CrossRef]

27. Ting, F.T.C. Origin and spacing of cleats in coal beds. *J. Press. Vessel Technol.* **1977**, *99*, 624–626. [CrossRef]

28. Su, X.B.; Feng, Y.L.; Chen, J.F.; Pan, J.N. The characteristics and origins of cleat in coal from western north china. *Int. J. Coal Geol.* **2001**, *47*, 51–62. [CrossRef]

29. Busse, J.; de Dreuzy, J.R.; Torres, S.G.; Bringemeier, D.; Scheuermann, A. Image processing based characterisation of coal cleat networks. *Int. J. Coal Geol.* **2017**, *169*, 1–21. [CrossRef]

30. Golab, A.; Ward, C.R.; Permana, A.; Lennox, P.; Botha, P. High-resolution three-dimensional imaging of coal using microfocus X-ray computed tomography, with special reference to modes of mineral occurrence. *Int. J. Coal Geol.* **2013**, *113*, 97–108. [CrossRef]

31. Rodrigues, C.F.; Laiginhas, C.; Fernandes, M.; de Sousa, M.J.L.; Dinis, M.A.P. The coal cleat system: A new approach to its study. *J. Rock Mech. Geotech. Eng.* **2014**, *6*, 208–218. [CrossRef]

32. Blanton, T.L. An experimental study of interaction between hydraulically induced and pre-existing fractures. In Proceedings of the SPE Unconventional Gas Recovery Symposium, Pittsburgh, PA, USA, 16–18 May 1982; Society of Petroleum Engineers: Pittsburgh, PA, USA, 1982. [CrossRef]

33. Dahi Taleghani, A.; Gonzalez-Chavez, M.; Yu, H.; Asala, H. Numerical simulation of hydraulic fracture propagation in naturally fractured formations using the cohesive zone model. *J. Petrol. Sci. Eng.* **2018**, *165*, 42–57. [CrossRef]

34. Yoon, J.S.; Zang, A.; Stephansson, O.; Hofmann, H.; Zimmermann, G. Discrete element modelling of hydraulic fracture propagation and dynamic interaction with natural fractures in hard rock. *Procedia Eng.* **2017**, *191*, 1023–1031. [CrossRef]

35. Damjanac, B.; Gil, I.; Pierce, M.; Sanchez, M.; Van As, A.; McLennan, J. A new approach to hydraulic fracturing modeling in naturally fractured reservoirs. In Proceedings of the 44th U.S. Rock Mechanics Symposium and 5th U.S.-Canada Rock Mechanics Symposium, Salt Lake City, UT, USA, 27–30 June 2010; American Rock Mechanics Association: Salt Lake City, UT, USA, 2010.

36. Wang, T.; Hu, W.R.; Elsworth, D.; Zhou, W.; Zhou, W.B.; Zhao, X.Y.; Zhao, L.Z. The effect of natural fractures on hydraulic fracturing propagation in coal seams. *J. Petrol. Sci. Eng.* **2017**, *150*, 180–190. [CrossRef]

37. Deng, J.Q.; Yang, Q.; Liu, Y.R.; Liu, Y.; Zhang, G.X. 3d finite element modeling of directional hydraulic fracturing based on deformation reinforcement theory. *Comput. Geotech.* **2018**, *94*, 118–133. [CrossRef]

38. Xu, L.L.; Cui, J.B.; Huang, S.P.; Tang, J.D.; Cai, L.; Yu, P. Analysis and application of fracture propagated model by hydraulic fracturing in coal-bed methane reservoir. *J. China Coal Soc.* **2014**, *39*, 2068–2074. [CrossRef]

39. Zhou, J.; Chen, M.; Jin, Y.; Zhang, G.Q. Analysis of fracture propagation behavior and fracture geometry using a tri-axial fracturing system in naturally fractured reservoirs. *Int. J. Rock Mech. Min. Sci.* **2008**, *45*, 1143–1152. [CrossRef]

40. Shilova, T.; Patutin, A.; Rybalkin, L.; Serdyukov, S.; Hutornoy, V. Development of the impermeable membranes using directional hydraulic fracturing. *Procedia Eng.* **2017**, *191*, 520–524. [CrossRef]

41. He, Q.; Suorineni, F.T.; Ma, T.; Oh, J. Effect of discontinuity stress shadows on hydraulic fracture re-orientation. *Int. J. Rock Mech. Min. Sci.* **2017**, *91*, 179–194. [CrossRef]

42. Liu, K.; Gao, D.L.; Wang, Y.B.; Yang, Y.C. Effect of local loads on shale gas well integrity during hydraulic fracturing process. *J. Nat. Gas Sci. Eng.* **2017**, *37*, 291–302. [CrossRef]

43. Yao, Y.; Wang, W.H.; Keer, L.M. An energy based analytical method to predict the influence of natural fractures on hydraulic fracture propagation. *Eng. Fract. Mech.* **2018**, *189*, 232–245. [CrossRef]

44. Tan, P.; Jin, Y.; Hou, B.; Han, K.; Zhou, Y.C.; Meng, S.Z. Experimental investigation on fracture initiation and non-planar propagation of hydraulic fractures in coal seams. *Pet. Explor. Dev.* **2017**, *44*, 470–476. [CrossRef]

45. Guo, T.K.; Rui, Z.H.; Qu, Z.Q.; Qi, N. Experimental study of directional propagation of hydraulic fracture guided by multi-radial slim holes. *J. Petrol. Sci. Eng.* **2018**, *166*, 592–601. [CrossRef]

46. Haimson, B.; Fairhurst, C. Initiation and extension of hydraulic fractures in rocks. *SPE J.* **1967**, *7*, 310–318. [CrossRef]

47. Schmitt, D.; Zoback, M. Poroelastic effects in the determination of the maximum horizontal principal stress in hydraulic fracturing tests—A proposed breakdown equation employing a modified effective stress relation for tensile failure. *Int. J. Rock Mech. Min. Sci. Geomech. Abstr.* **1989**, *26*, 499–506. [CrossRef]

48. Schmitt, D.R.; Currie, C.A.; Zhang, L. Crustal stress determination from boreholes and rock cores: Fundamental principles. *Tectonophysics* **2012**, *580*, 1–26. [CrossRef]

49. Atkinson, B.K. *Fracture Mechanics of Rock*; Academic Press Inc.: London, UK, 1987; Volume 2, pp. 33–45.

50. McClure, M.W.; Horne, R.N. An investigation of stimulation mechanisms in enhanced geothermal systems. *Int. J. Rock Mech. Min. Sci.* **2014**, *72*, 242–260. [CrossRef]

51. Perkins, T.; Kern, L. Widths of hydraulic fractures. *J. Petrol. Technol.* **1961**, *13*, 937–949. [CrossRef]

52. Geertsma, J.; Klerk, F.D. A rapid method of predicting width and extent of hydraulically induced fractures. *J. Petrol. Technol.* **1969**, *21*, 1571–1582. [CrossRef]

53. Nordgren, R.P. Propagation of a vertical hydraulic fracture. *SPE J.* **1972**, *12*, 306–314. [CrossRef]

54. Warpinski, N.; Wolhart, S.; Wright, C. Analysis and prediction of microseismicity induced by hydraulic fracturing. In Proceedings of the SPE Annual Technical Conference and Exhibition, San Antonio, TX, USA, 5–8 October 1997; Society of Petroleum Engineers: Richardson, TX, USA, 2001. [CrossRef]

55. Palmer, I.D.; Moschovidis, Z.A.; Cameron, J.R. Modeling shear failure and stimulation of the barnett shale after hydraulic fracturing. In Proceedings of the SPE Hydraulic Fracturing Technology Conference, The Woodlands, TX, USA, 23–25 January 2018; Society of Petroleum Engineers: College Station, TX, USA, 2007. [CrossRef]

56. Rogers, S.; Elmo, D.; Dunphy, R.; Bearinger, D. Understanding hydraulic fracture geometry and interactions in the horn river basin through dfn and numerical modeling. In Proceedings of the Canadian Unconventional Resources and International Petroleum Conference, Calgary, AB, Canada, 19–21 October 2010; Society of Petroleum Engineers: Richardson, TX, USA, 2010. [CrossRef]

57. Nagel, N.B.; Gil, I.; Sanchez-Nagel, M.; Damjanac, B. Simulating hydraulic fracturing in real fractured rocks-overcjavascript: Iterm()oming the limits of pseudo3D models. In Proceedings of the SPE Hydraulic Fracturing Technology Conference, The Woodlands, TX, USA, 23–25 January 2018; Society of Petroleum Engineers: Richardson, TX, USA, 2011. [CrossRef]

58. Weng, X.; Kresse, O.; Cohen, C.E.; Wu, R.; Gu, H. Modeling of hydraulic fracture network propagation in a naturally fractured formation. In Proceedings of the SPE Hydraulic Fracturing Technology Conference, The Woodlands, TX, USA, 23–25 January 2018; Society of Petroleum Engineers: Richardson, TX, USA, 2011. [CrossRef]

59. Wu, R.; Kresse, O.; Weng, X.; Cohen, C.E.; Gu, H. Modeling of interaction of hydraulic fractures in complex fracture networks. In Proceedings of the SPE Hydraulic Fracturing Technology Conference, The Woodlands, TX, USA, 23–25 January 2018; Society of Petroleum Engineers: The Woodlands, TX, USA, 2012. [CrossRef]

60. Renshaw, C.E.; Pollard, D.D. An experimentally verified criterion for propagation across unbounded frictional interfaces in brittle, linear elastic-materials. *Int. J. Rock Mech. Min. Sci. Geomech. Abstr.* **1995**, *32*, 237–249. [CrossRef]

61. Gu, H.; Weng, X. Criterion for fractures crossing frictional interfaces at non-orthogonal angles. In Proceedings of the 44th US Rock Mechanics Symposium and 5th US-Canada Rock Mechanics Symposium, Salt Lake City, UT, USA, 27–30 June 2010; American Rock Mechanics Association: Alexandria, VA, USA, 2010.

62. Warpinski, N.; Lorenz, J.; Branagan, P.; Myal, F.; Gall, B. Examination of a cored hydraulic fracture in a deep gas well. *SPE Prod. Facil.* **1993**, *8*, 150–158. [CrossRef]
63. Mahrer, K.D. A review and perspective on far-field hydraulic fracture geometry studies. *J. Petrol. Sci. Eng.* **1999**, *24*, 13–28. [CrossRef]
64. Jeffrey, R.G.; Zhang, X.; Thiercelin, M.J. Hydraulic fracture offsetting in naturally fractures reservoirs: Quantifying a long-recognized process. In Proceedings of the SPE Hydraulic Fracturing Technology Conference, The Woodlands, TX, USA, 5–7 February 2019; Society of Petroleum Engineers: Richardson, TX, USA, 2009. [CrossRef]
65. Khristianovic, S.; Zheltov, Y. Formation of vertical fractures by means of highly viscous fluids. In Proceedings of the 4th World Petroleum Congress, Rome, Italy, 6–15 June 1955; Volume 2, pp. 579–586.
66. Abé, H.; Mura, T.; Keer, L.M. Growth rate of a penny-shaped crack in hydraulic fracturing of rocks. *J. Geophys. Res.* **1976**, *81*, 5335–5340. [CrossRef]
67. Settari, A.; Cleary, M.P. Three-dimensional simulation of hydraulic fracturing. *J. Petrol. Technol.* **1984**, *36*, 1177–1190. [CrossRef]
68. Zhang, X.; Detournay, E.; Jeffrey, R. Propagation of a penny-shaped hydraulic fracture parallel to the free-surface of an elastic half-space. *Int. J. Fract.* **2002**, *115*, 125–158. [CrossRef]
69. Zhang, X.; Jeffrey, R.G. The role of friction and secondary flaws on deflection and re-initiation of hydraulic fractures at orthogonal pre-existing fractures. *Geophys. J. Int.* **2006**, *166*, 1454–1465. [CrossRef]
70. Lecampion, B.; Detournay, E. An implicit algorithm for the propagation of a hydraulic fracture with a fluid lag. *Comput. Methods Appl. Mech. Eng.* **2007**, *196*, 4863–4880. [CrossRef]
71. Dean, R.H.; Schmidt, J.H. Hydraulic-fracture predictions with a fully coupled geomechanical reservoir simulator. *SPE J.* **2009**, *14*, 707–714. [CrossRef]
72. Ji, L.J.; Settari, A.; Sullivan, R.B. A novel hydraulic fracturing model fully coupled with geomechanics and reservoir simulation. *SPE J.* **2009**, *14*, 423–430. [CrossRef]
73. Zhang, Z.; Ghassemi, A. Simulation of hydraulic fracture propagation near a natural fracture using virtual multidimensional internal bonds. *Int. J. Numer. Anal. Methods Geomech.* **2011**, *35*, 480–495. [CrossRef]
74. Chen, Z.R. Finite element modelling of viscosity-dominated hydraulic fractures. *J. Petrol. Sci. Eng.* **2012**, *88–89*, 136–144. [CrossRef]
75. Wang, J.G.; Zhang, Y.; Liu, J.S.; Zhang, B.Y. Numerical simulation of geofluid focusing and penetration due to hydraulic fracture. *J. Geochem. Explor.* **2010**, *106*, 211–218. [CrossRef]
76. Yoon, J.S.; Zimmermann, G.; Zang, A. Numerical investigation on stress shadowing in fluid injection-induced fracture propagation in naturally fractured geothermal reservoirs. *Rock Mech. Rock Eng.* **2015**, *48*, 1439–1454. [CrossRef]
77. Yoon, J.S.; Zimmermann, G.; Zang, A. Discrete element modeling of cyclic rate fluid injection at multiple locations in naturally fractured reservoirs. *Int. J. Rock Mech. Min. Sci.* **2015**, *74*, 15–23. [CrossRef]
78. Yoon, J.S.; Zimmermann, G.; Zang, A.; Stephansson, O. Discrete element modeling of fluid injection–induced seismicity and activation of nearby fault. *Can. Geotech. J.* **2015**, *52*, 1457–1465. [CrossRef]
79. Yoon, J.S.; Zang, A.N.; Stephansson, O. Numerical investigation on optimized stimulation of intact and naturally fractured deep geothermal reservoirs using hydro-mechanical coupled discrete particles joints model. *Geothermics* **2014**, *52*, 165–184. [CrossRef]
80. Al-Busaidi, A.; Hazzard, J.F.; Young, R.P. Distinct element modeling of hydraulically fractured lac du bonnet granite. *J. Geophys. Res. Solid Earth* **2005**, *110*. [CrossRef]
81. Shimizu, H. Distinct Element Modeling for Fundamental Rock Fracturing and Application to Hydraulic Fracturing. Ph.D. Thesis, Kyoto University, Kyoto, Japan, 2010.
82. Wang, T.; Zhou, W.B.; Chen, J.H.; Xiao, X.; Li, Y.; Zhao, X.Y. Simulation of hydraulic fracturing using particle flow method and application in a coal mine. *Int. J. Coal Geol.* **2014**, *121*, 1–13. [CrossRef]
83. Zhao, X.Y.; Wang, T.; Elsworth, D.; He, Y.L.; Zhou, W.; Zhuang, L.; Zeng, J.; Wang, S.F. Controls of natural fractures on the texture of hydraulic fractures in rock. *J. Petrol. Sci. Eng.* **2018**, *165*, 616–626. [CrossRef]
84. Huang, B. Research on Theory and Application of Hydraulic Fracture Weakening for Coal-Rock Mass. Ph.D. Thesis, China University of Mining Technology, Xuzhou, China, 2009.
85. Zhao, H.F.; Wang, X.H.; Liu, Z.Y.; Yan, Y.J.; Yang, H.X. Investigation on the hydraulic fracture propagation of multilayers-commingled fracturing in coal measures. *J. Petrol. Sci. Eng.* **2018**, *167*, 774–784. [CrossRef]

86. Nur, A.; Byerlee, J.D. An exact effective stress law for elastic deformation of rock with fluids. *J. Geophys. Res.* **1971**, *76*, 6414–6419. [CrossRef]

87. Zhang, G.M.; Liu, H.; Zhang, J.; Wu, H.A.; Wang, X.X. Three-dimensional finite element numerical simulation of horizontal well hydraulic fracturing. *Eng. Mech.* **2011**, *28*, 101–106.

88. Zhang, G.; Liu, H.; Zhang, J.; Biao, F.; Wu, H.; Wang, X. Simulation of hydraulic fracturing of oil well based on fluid-solid coupling equation and non-linear finite element. *Acta Pet. Sin.* **2009**, *30*, 113–116.

89. Zhang, G.M.; Liu, H.; Zhang, J.; Wu, H.A.; Wang, X.X. Mathematical model and nonlinear finite element equation for reservoir fluid-solid coupling. *Rock Soil Mech.* **2010**, *31*, 1657–1662.

90. Batchelor, G.K. *An Introduction to Fluid Dynamics*; Cambridge University Press: Cambridge, UK, 2000. [CrossRef]

91. Shet, C.; Chandra, N. Analysis of energy balance when using cohesive zone models to simulate fracture processes. *J. Eng. Mater. Technol.* **2002**, *124*, 440–450. [CrossRef]

92. Guo, T.K.; Qu, Z.Q.; Gong, D.Q.; Lei, X.; Liu, M. Numerical simulation of directional propagation of hydraulic fracture guided by vertical multi-radial boreholes. *J. Nat. Gas Sci. Eng.* **2016**, *35*, 175–188. [CrossRef]

93. Turon, A.; Camanho, P.P.; Costa, J.; Davila, C.G. A damage model for the simulation of delamination in advanced composites under variable-mode loading. *Mech. Mater.* **2006**, *38*, 1072–1089. [CrossRef]

94. Camanho, P.P. Failure Criteria for Fibre-Reinforced Polymer Composites. 2002. Available online: https://paginas.fe.up.pt/~stpinho/teaching/feup/y0506/fcriteria.pdf (accessed on 27 June 2018).

95. Geng, Y.G.; Tang, D.Z.; Xu, H.; Tao, S.; Tang, S.L.; Ma, L.; Zhu, X.G. Experimental study on permeability stress sensitivity of reconstituted granular coal with different lithotypes. *Fuel* **2017**, *202*, 12–22. [CrossRef]

96. Liu, R.; Li, B.; Jiang, Y.; Yu, L. A numerical approach for assessing effects of shear on equivalent permeability and nonlinear flow characteristics of 2-d fracture networks. *Adv. Water Resour.* **2018**, *111*, 289–300. [CrossRef]

97. Liu, R.; Li, B.; Jiang, Y. A fractal model based on a new governing equation of fluid flow in fractures for characterizing hydraulic properties of rock fracture networks. *Comput. Geotech.* **2016**, *75*, 57–68. [CrossRef]

98. Abaqus 2017 Documentation. Available online: https://www.3ds.com/products-services/simulia/support/documentation/ (accessed on 27 June 2018).

99. Liu, X. Mechanisms and Technical Features of Hydraulic Disturbance Method for Improvement of Coal and Rock Seams Permeability. Ph.D. Thesis, Henan Polytechnic University, Jiaozuo, China, 2015.

Article

A High-Order Numerical Manifold Method for Darcy Flow in Heterogeneous Porous Media

Lingfeng Zhou [1,2], Yuan Wang [1,*] and Di Feng [1,2]

[1] Key Laboratory of Ministry of Education for Geomechanics and Embankment Engineering, Hohai University, Nanjing 210098, China; zhoulf@hhu.edu.cn (L.Z.); fengdi@126.com (D.F.)

[2] College of Civil and Transportation Engineering, Hohai University, Nanjing 210098, China

[*] Correspondence: wangyuanhhu@163.com; Tel.: +86-139-5100-2717

Received: 19 June 2018; Accepted: 29 July 2018; Published: 1 August 2018

Abstract: One major challenge in modeling Darcy flow in heterogeneous porous media is simulating the material interfaces accurately. To overcome this defect, the refraction law is fully introduced into the numerical manifold method (NMM) as an a posteriori condition. To achieve a better accuracy of the Darcy velocity and continuous nodal velocity, a high-order weight function with a continuous nodal gradient is adopted. NMM is an advanced method with two independent cover systems, which can easily solve both continuous and discontinuous problems in a unified form. Moreover, a regular mathematical mesh, independent of the physical domain, is used in the NMM model. Compared to the conforming mesh of other numerical methods, it is more efficient and flexible. A number of numerical examples were simulated by the new NMM model, comparing the results with the original NMM model and the analytical solutions. Thereby, it is proven that the proposed method is accurate, efficient, and robust for modeling Darcy flow in heterogeneous porous media, while the refraction law is satisfied rigorously.

Keywords: Darcy flow; heterogeneity; numerical manifold method; high-order; refraction law

---

1. Introduction

Heterogeneity is the natural property of the geologic structure in groundwater systems. A real groundwater system can be regarded as a combination of several homogeneous subdomains. However, the contrast between the hydraulic conductivities of two adjacent subdomains presents a great challenge for numerical modeling. The two major limitations of modeling Darcy flow in heterogeneous porous media by conventional numerical methods are as follows. (1) The precision of the Darcy velocity (also known as specific discharge) is deficient, as the existing method usually solves the hydraulic head first, then computes the Darcy velocity by taking the derivation of the hydraulic head field. (2) The refraction law, which is expressed as the continuity of normal flux and discontinuity of tangential flux on the interface [1], can be hardly achieved. For the first limitation, the precision of the Darcy velocity decreases to a low order after derivation, and that leads to the discontinuity of the Darcy velocity around a node. For the second limitation, Cainelli et al. [2] summarized the accuracy of classic numerical schemes, such as the finite element method (FEM) [3,4] and finite volume method (FVM) [5–7], for modeling flow in heterogeneous formations. Results show that the above methods cannot satisfy the refraction law rigorously when the contrast of conductivities is large, hence the local computation error can be non-negligible.

In order to improve the first limitation (i.e., the lack of precision of the velocity solution), there are two different approaches. The first way is using some post-processing techniques to improve the velocity solution after the hydraulic head is solved [8–10]. These methods are efficient, as they can be easily modified from the original FEM approach. For example, in Yeh's model [9], the head solution is substituted into the variational formula of Darcy's law, which is then directly solved by the same FEM approach. Although normal flux across the interface is continuous, the tangential component still cannot satisfy refraction law when facing contrasting hydraulic conductivities. The other way is to solve the head and velocity simultaneously, which is known as the mixed finite element method (MFEM) [11,12]. However, iterations are still required for MFEM, making the calculation time-consuming [13].

Imposing the refraction law to the governing equation as an additional constraint condition is an efficient approach to dealing with the second limitation. For example, Zhou et al. [14] proposed a modified Yeh's method using a post-processing scheme to enforce the refraction law across material interfaces. Thus, iterations are required for the scheme to achieve convergence. Additionally, Xie et al. [15] developed a domain decomposed finite element method for solving the Darcy velocity in heterogeneous porous media without iterations. The basic idea is to separate the problem domain into several subdomains by material interfaces with coercive refraction law. Nevertheless, the above FEM-based methods are still mesh-dependent, which is inconvenient for pre-processing, especially when material interfaces are complex.

The numerical manifold method (NMM) first presented by Shi [16,17] can be regarded as a generalization to the conventional FEM. Due to the feature of solving both continuous and discontinuous problems in a unity frame, NMM is widely used in mechanical analysis, such as crack propagation [18–21], rock slopes failure [22–24], and dynamic problems [25–28]. NMM also played an important role in fluid flow modeling. Ohnishi et al. [29] presented a node shift method based on NMM for 2D saturated–unsaturated unsteady seepage simulation. Zhang et al. [30] solved Navier–Stokes equations directly by NMM, modeling unsteady incompressible viscous flow. Furthermore, NMM models for unconfined seepage flow in homogeneous media were reported [31,32]. However, it cannot exploit the properties of NMM to deal with homogeneous problems, compared with other continuous methods like FEM. Aiming to simulate Darcy flow in heterogeneous media, Hu et al. [33–35] developed a continuous approach with jump functions and a discontinuous approach with Lagrange multipliers, by enforcing the continuity of normal flux across material interfaces, while the refraction law for tangential flux is ignored. Zheng et al. [36] also proposed an NMM model for unconfined seepage flow by adopting the moving least squares (MLS) interpolation. Hence, NMM is an efficient approach for heterogeneous seepage problems. However, improving the accuracy of modeling, that is still a great challenge.

In this work, we developed a high-order NMM model for Darcy flow in heterogeneous porous media with a nodal-continuous weight function, as well as a local cover function derived by first order approximation of Taylor expansion. Then we introduced refraction law into the NMM model as an a posteriori constraint condition entirely to ensure that material interfaces are simulated correctly. The proposed NMM model has the following advantages. (1) The accuracy of the Darcy velocity solution is high and keeps continuous around a node, both of which are significant for modeling underground flow as well as solute transport. (2) The treatment of material interfaces in the NMM model is accurate and efficient, especially when facing various materials with interfaces intersecting each other, by taking full advantage of NMM's dual cover systems. More details of the proposed NMM model will be discussed in the following sections.

The framework of this paper is organized as follows. Section 2 briefly introduces some basic concepts of the original NMM theory. Then, the new NMM formation for Darcy flow in heterogeneous porous media is derived in Section 3. With the aim of testing the reliability of the proposed NMM model, numerical examples are simulated in Section 4. Finally, main conclusions and discussions are given in Section 5.

2. Basic Theory of NMM

Basic concepts of NMM are introduced in this section, including the dual cover systems and the manifold elements. The local approximation on physical covers and the global approximation on manifold elements are proposed next. This section shows that the NMM is suitable for modeling discontinuous problems because of the efficient meshing algorithm and the treatment of discontinuous boundaries.

2.1. Dual Cover Systems and Manifold Elements

The basic concepts of NMM include mathematical cover (MC), physical cover (PC), and manifold element (ME). The mathematical cover system is a set of patches which completely cover the whole physical domain. Thus, each patch is an MC. PCs are generated from MCs, cutting via boundaries or cracks and material interfaces inside the physical area. More remarkably, the shape of MC may be arbitrary, which is not necessary in coinciding with the boundaries as long as the whole domain is covered. Here, standard FEM mesh is used to generate MCs for convenience, and the density of mesh decides computational accuracy.

A simple example is given to illustrate the above concepts. As shown in Figure 1, a rectangular domain is divided into two different materials by a vertical interface. A regular triangular mesh is used to form MCs in this example. Each node in the mesh is defined as a star, and MCs are formed by triangles adjacent to a star, denoted as P_i, where the subscript i is the number of the star. Thus, the standard shape of MC in this example is a hexagon (i.e., P_5 and P_6) combined by six triangles. However, quadrangular or triangular MCs may be made using fewer triangles on the boundaries (i.e., P_9).

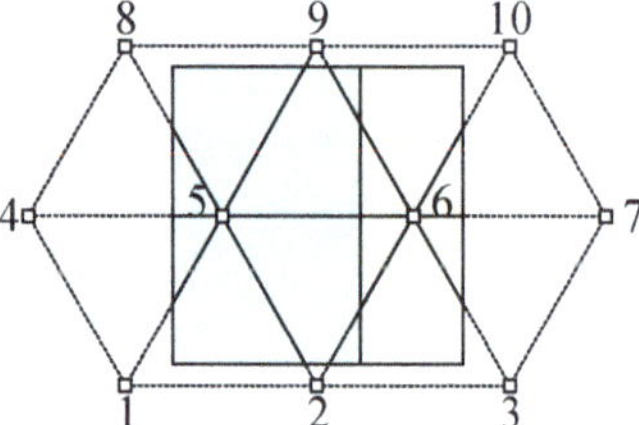

Figure 1. Finite element mesh and physical domain.

MCs are divided into independent parts by outside boundaries or material interfaces. Each part of MCs is regarded as a PC. This is denoted as P_i^j, where the superscript j is the number of the PC belonging to the i-th MC. As illustrated in Figure 2, P_5 is divided into P_5^1 and P_5^2 by boundaries and material interface. Similarly, P_6 is divided into P_6^1 and P_6^2, and P_9 is divided into P_9^1 and P_9^2.

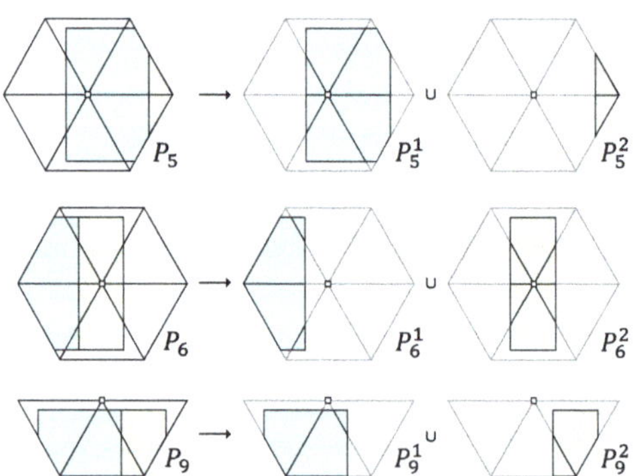

Figure 2. The dual cover systems: mathematical covers and physical covers.

After generating the cover systems, manifold elements—the basic integral region of the weak form of the problems—are defined as the overlapping areas of PCs. As exhibited in Figure 3, MEs are regarded as the common area of three layers of PCs when adopting triangular finite element mesh. Therefore, the star of the corresponding PCs is considered as a generalized node of the ME in NMM. This is different to the concept of nodes in FEM. For simplicity, the generalized node will be termed as "node" for the rest of this paper. In addition, the shape of MEs are generally triangular, except for the MEs adjacent to boundaries or interfaces, which may have an arbitrary shape. As long as the MCs are based on triangular mesh, the number of generalized nodes of an ME is three. As shown in Figure 3, ME is overlapped by P_5^1, P_6^1, and P_9^1. Thus, ME is termed as $P_5^1 P_6^1 P_9^1$, and ME is similarly termed as $P_5^2 P_6^2 P_9^2$. In this way, the discontinuity of the triangular element cut by boundaries or material interface is presented by NMM.

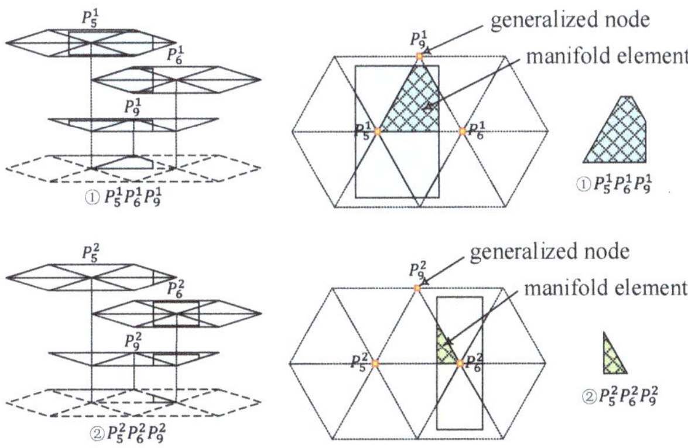

Figure 3. Manifold elements generated by physical covers.

Theoretically, NMM is a kind of partition of unity method (PUM). Its important feature is that the global approximation is established by combining local approximations using a PU function. According to the theory of NMM, local approximation is defined on each PC as a cover function, while global approximation on each ME is built through combining cover functions using weight functions, which are the same as PU functions. Details of NMM approximation is introduced in the following two subsections.

2.2. Local Approximation on Physical Covers

Local approximation is defined on each PC, representing the local field variables. That is, normally expressed as

$$\varphi_i(\mathbf{x}) = \mathbf{p}^T(\mathbf{x}) \cdot \mathbf{a}, \ i \in \left[1, n^{PC}\right] \tag{1}$$

where n^{PC} is the total number of PCs, $\mathbf{x}$ are coordinates of points within the i-th PC, $\mathbf{a}$ is a vector of unknown coefficients, and $\mathbf{p}^T(\mathbf{x})$ is the vector of polynomials bases. The latter can be expressed as

$$\begin{aligned}
\mathbf{p}^T(\mathbf{x}) &= [1, x, y] \\
\mathbf{p}^T(\mathbf{x}) &= \left[1, x, y, x^2, xy, y^2\right] \\
\mathbf{p}^T(\mathbf{x}) &= \left[1, x, y, x^2, xy, y^2, \cdots, x^n, \cdots, y^n\right]
\end{aligned} \tag{2}$$

for the first, second, and n-th order approximation, respectively, where n is a positive integer representing the approximation order.

However, in contrast with polynomial bases, specific bases can be selected as $\mathbf{p}^T(\mathbf{x})$ on each PC for achieving batter approximation. As a result, NMM has the ability to combine analytical solution with numerical method, that is to say, corresponding analytical solution can be chosen as the local approximation in the key position of problem domain, while polynomials bases are adopted on the rest of domain. Thus, a more accurate and efficient solution can be achieved. For example, Williams' displacement series are usually used as the local approximation near a crack tip for crack problems, making the stress solution around tip more accurate and reliable [18,19].

2.3. Global Approximation on Manifold Elements

Based on the NMM theory, global approximation is defined on each ME. The field variables φ on MEs are obtained by taking a weighted average of local approximations on corresponding PCs. Therefore, global approximation can be expressed as

$$\varphi(x,y) = \sum_{i=1}^{N_{PC}} w_i(x,y)\varphi_i(x,y) \tag{3}$$

where N_{PC} is the number of PCs on the ME and $w_i(x,y)$ is weight function of the i-th PC, which satisfies

$$\begin{cases} w_i(x,y) > 0 \ (x,y) \in U_i \\ w_i(x,y) = 0 \ (x,y) \notin U_i \end{cases} \tag{4}$$

where U_i is the geometric range of the i-th PC.

3. Steady Darcy Flow in Heterogeneous Porous Media by NMM

In this section, NMM formulation of steady Darcy flow in heterogeneous porous media is derived, including treatments for boundaries and material interfaces.

3.1. Governing Equations

Considering 2D steady Darcy flow in heterogeneous porous media, the continuity equation is derived using the conservation of mass, expressed as

$$\nabla^T v + Q = 0 \tag{5}$$

where ∇ is the gradient operator, v is the Darcy velocity or specific discharge, and Q is the source term. The constitutive equation for Darcy flow is Darcy's law, which means there is a linear relationship between the Darcy velocity and the hydraulic gradient. This is given as

$$v = -K\nabla h \tag{6}$$

where K is the permeability tensor and h is the hydraulic head. Substituting Equation (6) into Equation (5), the governing differential equation can be rewritten as

$$-\nabla^T(K\nabla h) + Q = 0. \tag{7}$$

As shown in Figure 4, the associated boundary conditions can be written as

$$h = \bar{h} \tag{8}$$

for the Dirichlet condition on Γ_h,

$$q_n = \bar{q} \tag{9}$$

for the Neumann condition on Γ_q, and

$$h^+ = h^-$$
$$v_n^+ = v_n^-$$
$$v_s^+ \cdot k^- = v_s^- \cdot k^+$$

(10)

for the refraction law across the material interface Γ_m, where $\bar{h}$ is the given hydraulic head on Γ_h, q_n is normal flux on Γ_q, $\bar{q}$ is the given flux, and v_n and v_s are normal and tangential components of the Darcy velocity, respectively. Superscript $+$ and $-$ represent the two sides of the material interface Γ_m. In addition, v_n and v_s can be expressed as

$$v_n = \boldsymbol{v} \cdot \boldsymbol{n}$$

(11a)

$$v_s = \boldsymbol{v} \cdot \boldsymbol{s}$$

(11b)

where $\boldsymbol{n}$ and $\boldsymbol{s}$ are the normal unit vector and tangential unit vector, respectively.

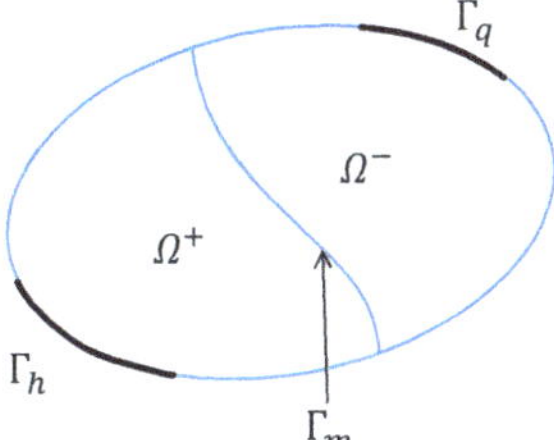

Figure 4. Physical domain and boundaries of a Darcy flow problem.

The first two conditions of refraction law can be easy deduced. They represent the continuity of the hydraulic head and normal flux. Here we briefly prove the third condition for tangential flux. As shown in Figure 5, a rectangular domain is divided into two different materials Ω_1 and Ω_2, both of which are homogeneous and isotropic media. The permeability coefficients are k^+ and k^-. Suppose fluid flows upside to downside, and θ_1 and θ_2 are the angles between the direction of the Darcy velocity and the normal direction of the material interface in Ω_1 and Ω_2, respectively. From Equation (10) it follows that

$$j_s^+ = j_s^-$$

(12)

across the material interface, where j is hydraulic gradient. Thus, the tangential component of the hydraulic gradient is continuous across the material interface. According to Darcy's law, the tangential component of the Darcy velocity can be expressed as

$$v_s^+ = k^+ \cdot j_s^+$$
$$v_s^- = k^- \cdot j_s^-$$

(13)

Substituting Equation (13) into Equation (12) and considering the geometric condition, the tangential velocity in two sides of the interface satisfies

$$\frac{v_s^+}{v_s^-} = \frac{k^+}{k^-} = \frac{\tan \theta_1}{\tan \theta_2}$$

(14)

which is equivalent to the third condition in Equation (10).

Through this subsection, NMM formation for Darcy flow in heterogeneous media is derived. The main purpose of this study is to solve the differential equation of Darcy flow, Equation (7) with constraint conditions of Equations (8)–(10) for the hydraulic head and the Darcy velocity using NMM.

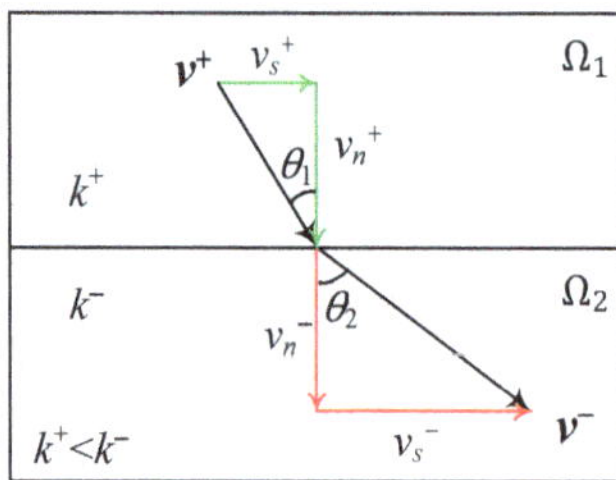

Figure 5. Refraction law across the material interface.

3.2. NMM Interpolations

In this paper, Taylor series expansion was applied to establish the local approximation on each PC. Thus, the base function on a PC can be adopted by first order approximation of the Taylor expansion of the corresponding node. According to Equation (1), the local approximation of the hydraulic head on each PC, $h_i(\mathbf{x})$, can be written as

$$h_i(\mathbf{x}) = \mathbf{p}^T(\mathbf{x}) \cdot \mathbf{a}, \ i \in \left[1, n^{PC}\right] \tag{15}$$

where a is a vector of unknown coefficients, and bases are expressed as

$$\mathbf{p}^T(\mathbf{x}) = [1, x - x_i, y - y_i] \tag{16}$$

where x_i and y_i are coordinates of node i.

Substituting Equation (16) into Equation (15), the local approximation on PCs can be rewritten as

$$h_i(\mathbf{x}) = h_i(x_i, y_i) + (x - x_i)\frac{\partial h_i(x_i, y_i)}{\partial x} + (y - y_i)\frac{\partial h_i(x_i, y_i)}{\partial y}. \tag{17}$$

Comparing with conventional polynomials bases, the exact physical meanings of the unknown coefficients, the corresponding hydraulic head, and the gradient on each node are achieved.

Then, global approximation on each ME can be derived as

$$h(\mathbf{x}) = \sum_{i=1}^{3} w_i(\mathbf{x}) h_i(\mathbf{x}) = \mathbf{N}\mathbf{a} \tag{18}$$

where N is the shape function, expressed as

$$\mathbf{N} = \sum_{i=1}^{3} w_i(\mathbf{x})\mathbf{p}^T(\mathbf{x}) \tag{19}$$

and $w_i(\mathbf{x})$ is the weight function on the i-th PC, originally used in mechanics problems [37,38]. This is expressed as

$$w_i(\mathbf{x}) = L_i + L_i^2(L_j + L_m) - L_i\left(L_j^2 + L_m^2\right) \tag{20}$$

where L_i ($i = 1, 2, 3$) are area coordinates of the triangular finite element. The above weight function has four important features:

(1)   the Kronecker delta property ($w_i(\mathbf{x}_j) = \delta_{ij}$ ($i, j = 1, 2, 3$)),

(2)   the non-negative property ($0 \leq w_i(\mathbf{x}) \leq 1$ ($\mathbf{x} \in U_i$)),

(3)   the partition of the unity condition ($\sum_{i=1}^{3} w_i(\mathbf{x}) = 1$),

(4)   and a nodal gradient of zero of the weight function.

Therefore, the gradient of the hydraulic head solution at a generalized node is continuous, because of the last property, which is of great importance in this study. A continuous Darcy velocity can be then derived by Darcy's law at nodes. However, the discontinuity of velocity still exists on element edges.

3.3. Discrete Equations

The weak form of governing Equation (7) is derived by the weighted residual method, and then the functional expression derived by the variational principle can be written as [39]

$$\Pi(h) = \int_{\Omega} \left[\frac{1}{2} (\nabla h)^T (K \nabla h) - hQ \right] d\Omega - \int_{\Gamma_q} h \bar{q} d\Gamma. \tag{21}$$

In each ME, the hydraulic head, h, is defined as

$$h = Na \tag{22}$$

where N is the shape function, and a is the unknowns.

The hydraulic gradient, j, is defined as

$$j = \nabla h = \nabla Na = Ba \tag{23}$$

where $B = \nabla N$.

The Darcy velocity, v, is expressed as

$$v = -Kj = -KBa \tag{24}$$

where K is the permeability tensor.

By substituting Equations (22)–(24) into Equation (21), the functional expression of each ME can be obtained.

$$\Pi^e = \int_{\Omega^e} \frac{1}{2} a^T B^T KBa d\Omega - \int_{\Omega^e} a^T N^T Qd\Omega - \int_{\Gamma_q} a^T N^T \bar{q} d\Gamma \tag{25}$$

where Ω^e is the domain of each ME. Therefore, the functional expression of the whole problem domain, Equation (21), can be rewritten in a matrix form

$$\Pi = \tfrac{1}{2} \begin{pmatrix} a_1 & \cdots & a_n \end{pmatrix} \begin{pmatrix} C_{11} & \cdots & C_{1n} \\ \vdots & \ddots & \vdots \\ C_{n1} & \cdots & C_{nn} \end{pmatrix} \begin{pmatrix} a_1 \\ \vdots \\ a_n \end{pmatrix} - \begin{pmatrix} a_1 & \cdots & a_n \end{pmatrix} \begin{pmatrix} F_1 \\ \vdots \\ F_n \end{pmatrix} = \tfrac{1}{2} a^T Ca - a^T F \tag{26}$$

where n is the number of unknowns in the problem, C is the conductivity matrix with $n \times n$ components, and F is the flux term with n components, expressed as

$$C = \int_{\Omega} B^T KB d\Omega \tag{27}$$

$$F = \int_{\Omega} N^T Qd\Omega + \int_{\Gamma_q} N^T \bar{q} d\Gamma \tag{28}$$

respectively. Their components can be derived by taking the derivatives with the unknowns, expressed as

$$C_{ij} = \frac{\partial^2 \Pi}{\partial a_i \partial a_j} \tag{29}$$

$$F_i = -\frac{\partial \Pi}{\partial a_i}. \tag{30}$$

According to the variational principle, the solution of the weak form of the differential equation is equivalent to a function a which makes Π stationary with respect to arbitrary changes of δa, that is

$$\delta \Pi = \frac{\partial \Pi}{\partial a_1} \delta a_1 + \frac{\partial \Pi}{\partial a_2} \delta a_2 + \ldots + \frac{\partial \Pi}{\partial a_n} \delta a_n = 0 \tag{31}$$

where $\delta a_1, \delta a_2, \ldots, \delta a_n$ are arbitrary changes. Therefore, Equation (7) reduces to a standard linear form as

$$\frac{\partial \Pi}{\partial a} = \left\{ \begin{array}{c} \frac{\partial \Pi}{\partial a_1} \\ \vdots \\ \frac{\partial \Pi}{\partial a_n} \end{array} \right\} = 0 \tag{32}$$

which can be rewritten as a matrix form

$$Ca = F \tag{33}$$

and

$$\begin{pmatrix} C_{11} & \cdots & C_{1n} \\ \vdots & \ddots & \vdots \\ C_{n1} & \cdots & C_{nn} \end{pmatrix} \left\{ \begin{array}{c} a_1 \\ \vdots \\ a_n \end{array} \right\} = \left\{ \begin{array}{c} F_1 \\ \vdots \\ F_n \end{array} \right\}. \tag{34}$$

Therefore, the governing differential equation, Equation (7) is transformed a system of linear equations (Equation (34)), which can be easily solved. In addition, the hydraulic head and Darcy velocity can be obtained using Equations (22) and (24), respectively.

3.4. Treatments for Boundaries and Material Interfaces

In Section 3.3, the Dirichlet boundary condition and refraction law across material interfaces are not involved. The treatments for boundaries and material interfaces are introduced in this subsection. Generally, two basic methods to impose additional constraints are the Lagrange multiplier method and the penalty method. Although the Lagrange multiplier method can achieve an accurate solution, additional degrees of freedom must be introduced into the governing function, which usually leads to computational difficulties. The penalty method is more convenient as it does not require a change of the size of the coefficient matrix and is free of the above drawbacks of the Lagrange multiplier method. Note that the penalty number should be neither too small nor too large to enforce the constraints and avoid computational difficulties. In this study, the penalty method is adopted, and penalty number α is determined as 10^6. The accuracy and robustness are verified by a series of numerical tests.

In addition, the treatments of material interfaces by existing methods are inadequate. In this paper, additional constraints of the Dirichlet boundary condition and refraction law across the material interface are introduced by forming a new functional

$$\Pi^* = \Pi + \Pi_D + \Pi_m \tag{35}$$

in which Π_D is the Dirichlet boundary condition term and Π_m is material interface term.

For the Dirichlet boundary condition, Π_D is expressed as

$$\Pi_D = \frac{\alpha}{2} \int_{\Gamma_D} \left(h - \bar{h} \right)^2 d\Gamma \tag{36}$$

where α is the penalty number. Thus, the components of the conductivity matrix, C_{ij}, and the flux term, F_i, for the equilibrium equation can be derived as

$$C_{ij} = \frac{\partial^2 \Pi_D}{\partial a_i \partial a_j} = \alpha \int_{\Gamma_D} N^T N d\Gamma \tag{37}$$

$$F_i = -\frac{\partial \Pi}{\partial a_i} = \alpha \int_{\Gamma_D} N^T \bar{h} d\Gamma. \tag{38}$$

For the material interface, Π_m can be written into three parts

$$\Pi_m = \Pi_m^a + \Pi_m^b + \Pi_m^c. \tag{39}$$

Each part of Π_m can be expressed as

$$\Pi_m^a = \frac{\alpha}{2} \int_{\Gamma_m} \left(h^+ - h^- \right)^2 d\Gamma \tag{40a}$$

$$\Pi_m^b = \frac{\alpha}{2} \int_{\Gamma_m} \left(v_n^+ - v_n^- \right)^2 d\Gamma \tag{40b}$$

$$\Pi_m^c = \frac{\alpha}{2} \int_{\Gamma_m} \left(\frac{v_s^+}{k^+} - \frac{v_s^-}{k^-} \right)^2 d\Gamma. \tag{40c}$$

Thus, the components of the conductivity matrix can be respectively derived as

$$C_{ij} = \frac{\partial^2 \Pi_m^a}{\partial a_i \partial a_j} = \alpha \int_{\Gamma_m} \left(\begin{array}{cc} \left(N^+\right)^T N^+ & -\left(N^+\right)^T N^- \\ -\left(N^-\right)^T N^+ & \left(N^-\right)^T N^- \end{array} \right) d\Gamma \tag{41a}$$

$$C_{ij} = \frac{\partial^2 \Pi_m^b}{\partial a_i \partial a_j} = \alpha \int_{\Gamma_m} \left(\begin{array}{cc} \left(K^+ B^+ n\right)^T \left(K^+ B^+ n\right) & -\left(K^+ B^+ n\right)^T \left(K^- B^- n\right) \\ -\left(K^- B^- n\right)^T \left(K^+ B^+ n\right) & \left(K^- B^- n\right)^T \left(K^- B^- n\right) \end{array} \right) d\Gamma \tag{41b}$$

$$C_{ij} = \frac{\partial^2 \Pi_m^c}{\partial a_i \partial a_j} = \alpha \int_{\Gamma_m} \left(\begin{array}{cc} \left(B^+ s\right)^T \left(B^+ s\right) & -\left(B^+ s\right)^T \left(B^- s\right) \\ -\left(B^- s\right)^T \left(B^+ s\right) & \left(B^- s\right)^T \left(B^- s\right) \end{array} \right) d\Gamma. \tag{41c}$$

Because the constraints of refraction law are fully introduced into the NMM model as a posteriori conditions, the material interface can be simulated accurately. The following section will present some numerical examples to verify the new NMM model for Darcy flow in heterogeneous porous media.

4. Numerical Examples

To verify the proposed NMM model, a set of numerical examples are carried out. First, two rectangular homogeneous models with different boundaries are studied. They are compared with analytical solutions in order to demonstrate the accuracy in modeling homogeneous Darcy flow. Then, a heterogeneous model, with hydraulic conductivity varying smoothly within the computational domain, is simulated to verify the capability and accuracy of the model for continuous-heterogeneous problems. Lastly, the new model is applied to a heterogeneous problem with material interfaces, showing the accuracy and efficiency for Darcy flow in heterogeneous porous media.

4.1. Example 1: Verification of Homogeneous Darcy Flow with an Analytical Solution

In order to verify the accuracy for a Darcy flow simulation before modeling heterogeneous problems, a homogeneous model is studied. As shown in Figure 6a, a 1×1 m rectangular domain with Dirichlet boundary conditions is as follows: $h(0, y) = 0$, $h(x, 0) = 0$, $h(1, y) = y$, and $h(x, 1) = x$. The model is homogeneous and isotropic with a permeability coefficient of $k(x, y) = 1$ m/s. The analytical solution of the hydraulic head is $h(x,y) = xy$. Thus, the Darcy velocity can be derived by Darcy's law or Equation (6) as follows: $v_x = -k_x \frac{\partial h}{\partial x} = -y$, $v_y = -k_y \frac{\partial h}{\partial y} = -x$.

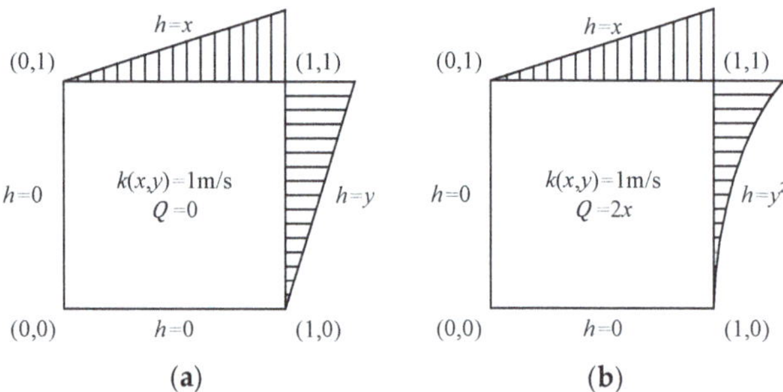

Figure 6. A 1×1 m rectangular model for homogeneous Darcy flow with different boundary conditions: (**a**) $h(0,y) = 0, h(x,0) = 0, h(1,y) = y, h(x,1) = x$; (**b**) $h(0,y) = 0, h(x,0) = 0, h(1,y) = y^2, h(x,1) = x$.

The numerical results solved by the proposed method with a 10-layers mesh are shown in Figure 7. Figure 7a shows the distribution of the hydraulic head in the problem domain, and Figure 7b,c are the distributions of the Darcy velocity component v_x and v_y, respectively. It is obvious that the hydraulic head and Darcy velocity are both continuous in the problem domain. To realize the advanced feature of the new model, a contrast simulation using the conventional NMM model is studied. Figure 8a is the result of the hydraulic head, which is similar to Figure 7a. Compared with the continuous Darcy velocity in Figure 7b,c, the velocity field obtained though the conventional model is discontinuous both around nodes and across element edges (Figure 8b,c). Therefore, it is proven that the proposed NMM model can achieve a continuous nodal velocity.

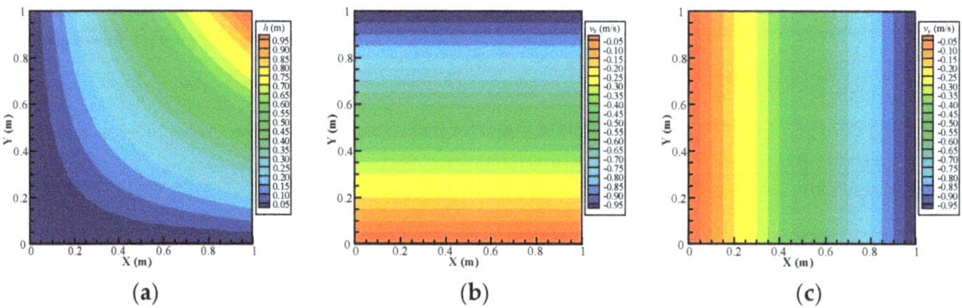

Figure 7. Computed contour map of the (**a**) hydraulic head, (**b**) Darcy velocity component in the x direction, and (**c**) Darcy velocity component in the y direction using the proposed NMM model with a 10-layer mesh.

Since the analytical solution is known in this problem, we can compare the numerical result with its exact result. Figure 9a shows the comparison of the hydraulic head solution along a profile located at $x = 0.3$ m, and Figure 9b,c are the comparison of the Darcy velocity component v_x and v_y, respectively. It is indicated that the accuracy of the new NMM model is close to the analytical solution without a refined mesh.

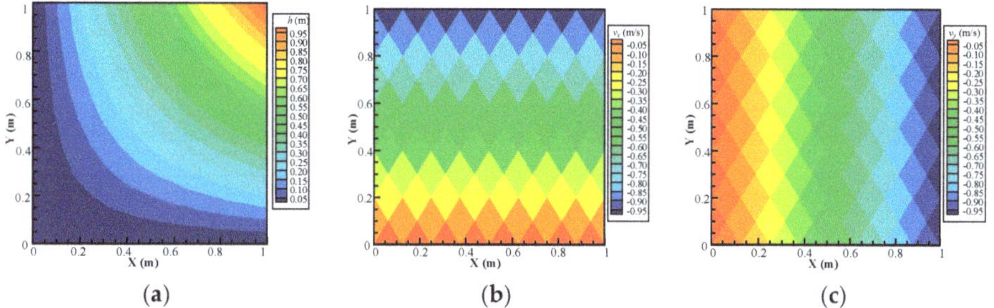

Figure 8. Computed contour map of the (**a**) hydraulic head, (**b**) Darcy velocity component in the x direction, (**c**) Darcy velocity component in the y direction using the conventional NMM model with a 10-layer mesh.

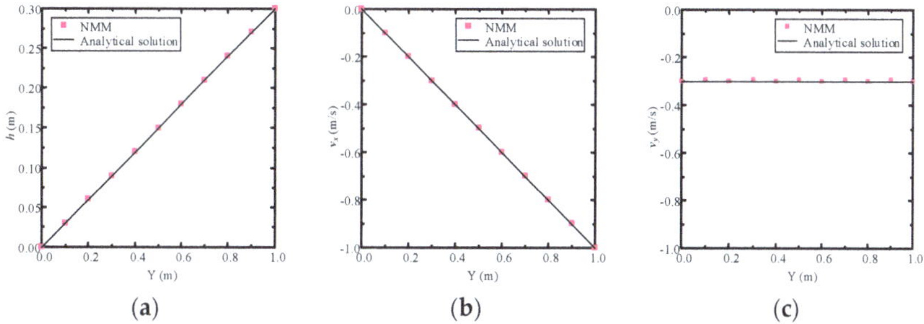

Figure 9. The comparison of solutions computed by the proposed numerical manifold method (NMM) model using a 10-layer mesh with an analytical solution along a profile located at $x = 0.3$ m: the (**a**) hydraulic head, (**b**) Darcy velocity in the x direction, and (**c**) Darcy velocity in the y direction.

Furthermore, in order to analyse the accuracy and convergence of the numerical solution, the relative error of the hydraulic head in the L^2 norm can be written as

$$\eta_h = \frac{\left[\int_\Omega \left(h - \hat{h}\right)^2 \mathrm{d}\Omega\right]^{\frac{1}{2}}}{\left[\int_\Omega \hat{h}^2 \mathrm{d}\Omega\right]^{\frac{1}{2}}} \times 100\% \tag{42}$$

where h is the numerical solution of the hydraulic head and $\hat{h}$ is the analytical solution. Similarly, the relative error of the Darcy velocity in the L^2 norm is defined as

$$\eta_v = \frac{\left[\int_\Omega (v - \hat{v})^{\mathrm{T}} \cdot (v - \hat{v}) \mathrm{d}\Omega\right]^{\frac{1}{2}}}{\left[\int_\Omega \hat{v}^{\mathrm{T}} \cdot \hat{v} \mathrm{d}\Omega\right]^{\frac{1}{2}}} \times 100\% \tag{43}$$

where v is the numerical solution of the Darcy velocity and $\hat{v}$ is the analytical solution.

After convergence analysis for both the new NMM model and the conventional NMM model, the convergence curves of the hydraulic head and Darcy velocity are plotted in Figure 10a,b, respectively. As shown in Figure 10a, the two curves are almost coincident, indicating that they have the same accuracy and convergence rate for the solution of the hydraulic head. However, Figure 10b shows that the velocity solution of the new NMM model is more accurate than the convention NMM model, and their convergence rates are still the same. The values of relative error and convergence rate of both hydraulic head and Darcy velocity by the two NMM models are shown in Tables 1 and 2, respectively.

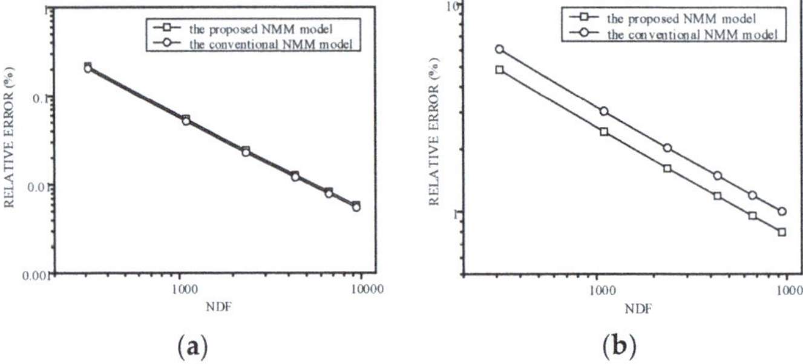

(a)   (b)

Figure 10. The comparison of convergence curves in the L^2 norm by the proposed NMM model and the conventional NMM model: (**a**) the relative error of the hydraulic head and (**b**) the relative error of the Darcy velocity. (NDF: number of degrees of freedom).

Table 1. Error in the L^2 norm of hydraulic head and Darcy velocity by the proposed NMM model.

kh [1]	L^2 Error of Hydraulic Head		L^2 Error of Darcy Velocity	
	Relative Error	Convergence Rate	Relative Error	Convergence Rate
5	2.18×10^{-3}	–	4.84×10^{-2}	–
10	5.47×10^{-4}	-1.10	2.42×10^{-2}	-0.55
15	2.44×10^{-4}	-1.06	1.61×10^{-2}	-0.53
20	1.29×10^{-4}	-1.04	1.19×10^{-2}	-0.50
25	8.36×10^{-5}	-1.03	9.55×10^{-3}	-0.52
30	5.86×10^{-5}	-1.02	7.98×10^{-3}	-0.52

[1] kh means the half number of mesh layers in the vertical direction.

Table 2. Error in the L^2 norm of hydraulic head and Darcy velocity by the convention NMM model.

kh	L^2 Error of Hydraulic Head		L^2 Error of Darcy Velocity	
	Relative Error	Convergence Rate	Relative Error	Convergence Rate
5	2.05×10^{-3}	–	6.07×10^{-2}	–
10	5.14×10^{-4}	-1.10	3.04×10^{-2}	-0.55
15	2.29×10^{-4}	-1.06	2.03×10^{-2}	-0.53
20	1.21×10^{-4}	-1.04	1.49×10^{-2}	-0.50
25	7.84×10^{-5}	-1.03	1.20×10^{-2}	-0.52
30	5.49×10^{-5}	-1.02	1.00×10^{-2}	-0.52

To demonstrate the capability of handling complex boundaries by the proposed method, the same model with a quadratic Dirichlet boundary condition is simulated, as shown in Figure 6b. The boundary conditions are $h(0,y) = 0$, $h(x,0) = 0$, $h(1,y) = y^2$, and $h(x,1) = x$, and the permeability coefficient remains unchanged. Note that the source term in this model is $Q = 2x$. The analytical solutions of the hydraulic head and Darcy velocity are $h(x,y) = xy^2$, $v_x = -y^2$, and $v_y = -2xy$, respectively.

First, the problem is simulated by the proposed NMM model with a 60-layers mesh. As shown in Figure 11, the distributions of the hydraulic head and Darcy velocity are presented. From Figure 11a, an obvious difference in hydraulic head distribution caused by the change of boundary conditions can be observed compared to Figure 8a. Furthermore, the velocity field is also continuous in the problem domain (Figure 11b,c).

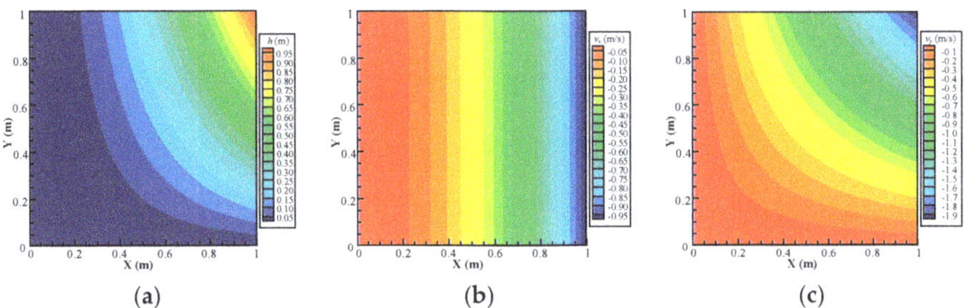

Figure 11. Computed contour map of the (**a**) hydraulic head, (**b**) Darcy velocity component in the x direction, and (**c**) Darcy velocity component in the y direction using the conventional NMM model with a 60-layers mesh.

By using different mesh densities, the accuracy of the proposed method for complex boundary conditions is verified. As shown in Figure 12a, the hydraulic head solution along a profile located at x = 0.3 m, in spite of different mesh densities, almost achieves its analytical solution. The Darcy velocity solution is also reliable compared with its analytical result (Figure 12b,c). In addition, when the mesh density reaches 60 layers, the Darcy velocity solution almost achieves its analytical solution.

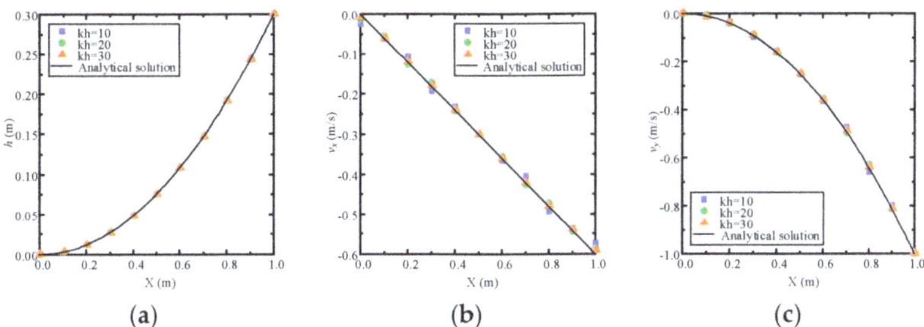

Figure 12. The comparison of solutions computed by the proposed NMM model using different mesh densities with an analytical solution along a profile located at x = 0.3 m: the (**a**) hydraulic head, (**b**) Darcy velocity in the x direction, and (**c**) Darcy velocity in the y direction.

Convergence analysis is performed both for hydraulic heads and velocities. Figure 13a shows the convergence curve of the hydraulic head, and Figure 13b shows the convergence curve of the Darcy velocity. However, there is no big difference with Figure 10, except that the relative errors of both the hydraulic head and Darcy velocity are larger than last examples, since the problem has more complex boundary conditions.

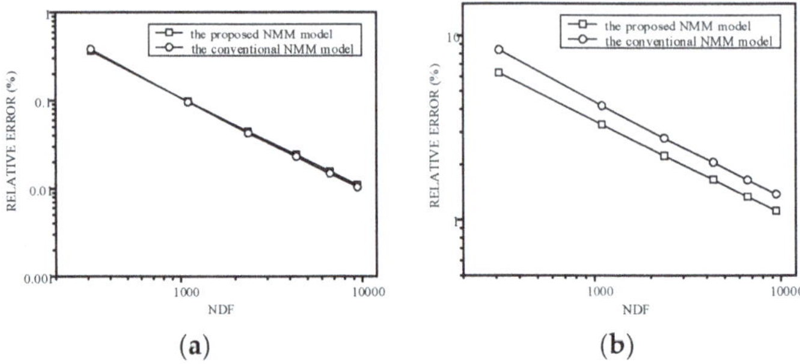

Figure 13. The comparison of convergence curves in the L^2 norm by the proposed NMM model and the conventional NMM model: (**a**) the relative error of the hydraulic head and (**b**) the relative error of the Darcy velocity.

4.2. Example 2: Accuracy Darcy Velocity Solution for Continuous-Heterogeneous Media

A continuous-heterogeneous model, as shown in Figure 14, is simulated in this example. The boundary conditions are as follows: $h(x, -5) = h(x, 5) = x^2 - 25$ and $h(-5, y) = h(5, y) = 25 - y^2$. The hydraulic conductivity varies smoothly within the problem domain, expressed as $k(x, y) = 1e^{-3} \times (x^2 + y^2)$. In addition, a source term $Q = 4e^{-3} \times (x^2 - y^2)$, is included in this model. The analytical solution of hydraulic head is $h(x, y) = x^2 - y^2$, and the solution of the velocity field can be derived as $v_x = -2e^{-3} \times (x^3 + xy^2)$ and $v_y = 2e^{-3} \times (x^2 y + y^3)$.

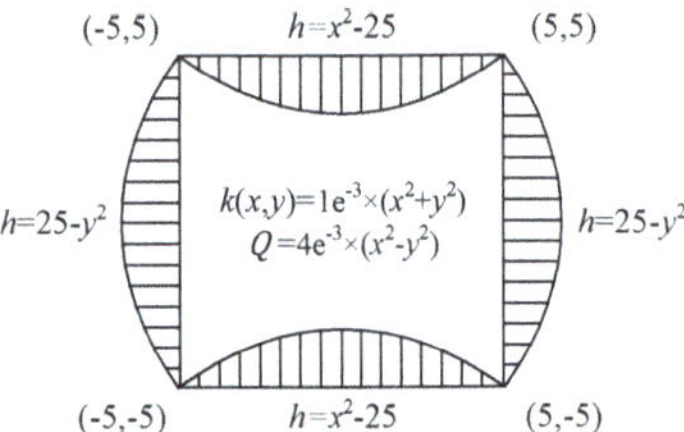

Figure 14. A continuous-heterogeneous model for Darcy flow: model geometry and boundary conditions.

Figure 15a shows the distribution of hydraulic head, whereas Figure 15b,c show the distribution of the Darcy velocity v_x and v_y, respectively. The contour plots show that the distribution of numerical results is symmetric and continuous by the proposed model. In order to study the robustness of the cover system, numerical results along a profile at $y = 0$ m with 20, 40, and 60 layers of mesh are compared with the analytical solution. Figure 16a shows the comparison of the hydraulic head solution between the proposed NMM model and the analytical solution along a profile located at $y = 0$ m, indicating high precision for the hydraulic head solution. Furthermore, the Darcy velocity solution also shows good agreement with the analytical solution in this continuous-heterogeneous media in Figure 16b,c.

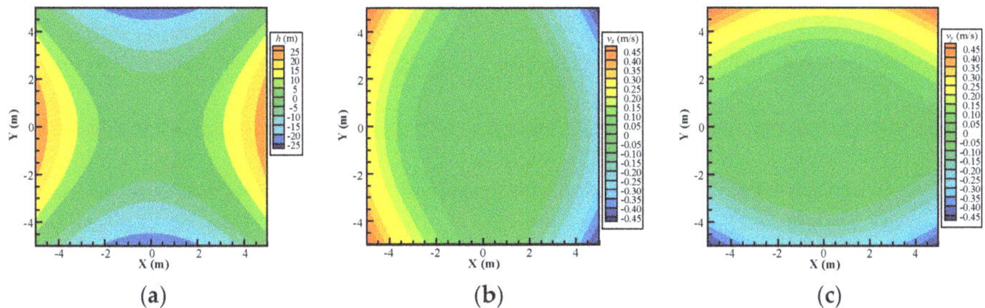

Figure 15. Computed contour map of the (**a**) hydraulic head, (**b**) Darcy velocity component in the *x* direction, and (**c**) Darcy velocity component in the *y* direction using the proposed NMM model with a 60-layer mesh.

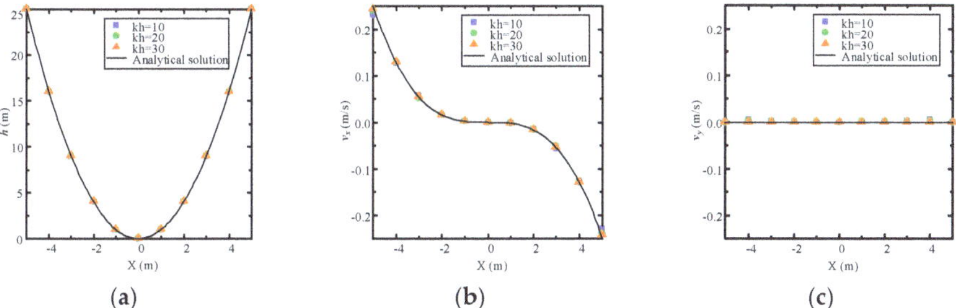

Figure 16. The comparison of solutions computed by the proposed NMM model using different mesh densities with an analytical solution along a profile located at $y = 0$ m: the (**a**) hydraulic head, (**b**) Darcy velocity component in the *x* direction, and (**c**) Darcy velocity component in the *y* direction.

For this continuous-heterogeneous problem, convergence curves of both the hydraulic head and velocity are shown in Figure 17. According to Figure 17a, the hydraulic head solution found using the proposed NMM model is more accurate than that of the conventional NMM model, and the convergence rate is still the same. Moreover, the proposed high-order NMM model also shows better accuracy of the Darcy velocity solution than the conventional model, while their convergence rates are almost the same.

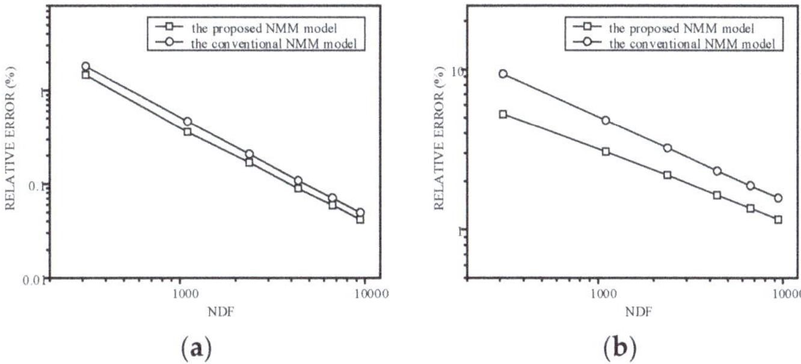

Figure 17. The comparison of convergence curves in the L^2 norm by the proposed NMM model and the conventional NMM model: the (**a**) the relative error of the hydraulic head and (**b**) the relative error of the Darcy velocity.

4.3. Example 3: Verification of Refraction Law for Discontinuous-Heterogeneous Media

The last example is to verify the refraction law across material interfaces [14], as shown in Figure 18a. Considering a rectangular domain of 120 × 120 m with material interfaces AB and CD, the permeability coefficients are $k_1 = 100$ m/d and $k_2 = 10$ m/d, respectively. The hydraulic heads are fixed with 1 m on the left side and 0 m on the right side, while the top boundary has a specified inflow flux of $q = 1$ m/d. The bottom of the model is impermeable. Due to the dual cover systems, the NMM mesh (Figure 18b) does not need to conform with the material interfaces, unlike the conventional FEM mesh (Figure 18c).

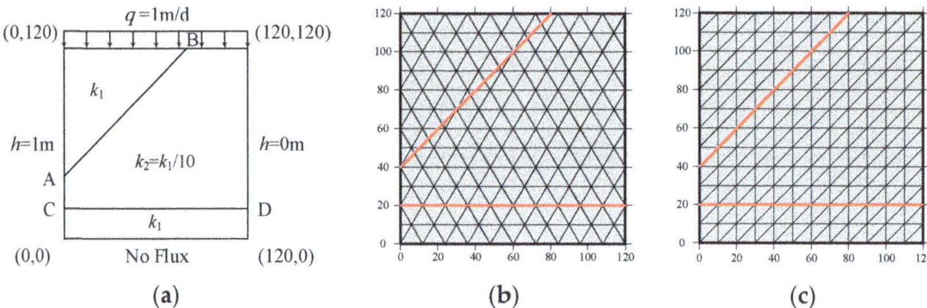

Figure 18. A discontinuous-heterogeneous model with material interfaces: (**a**) model geometry and boundary conditions, (**b**) non-conforming mesh based on the proposed NMM model, and (**c**) conforming mesh based on Zhou's FEM model [14].

The comparison of the hydraulic head distribution simulated by the proposed NMM model and Zhou's FEM model with same mesh size is shown in Figure 19. The results show that the hydraulic head is continuous across the material interfaces AB and CD.

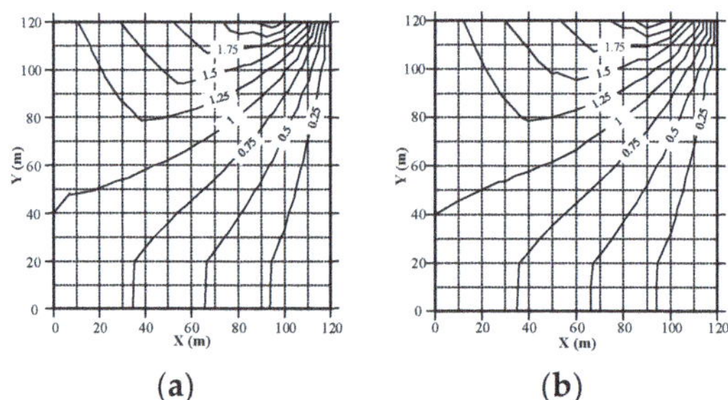

Figure 19. Numerical results of hydraulic head distribution simulated by (**a**) the proposed NMM model and (**b**) Zhou's FEM model [14].

In order to study the influence of the material interface, numerical results with two different mesh densities along profiles at $x = 100$ m and $y = 80$ m are plotted in Figures 20 and 21. As shown in Figure 20, the hydraulic head across the material interface is continuous, while the Darcy velocity is discontinuous. Similar results are shown by Figure 21, however, the Darcy velocity in the y direction is continuous as the material interface happens to be perpendicular to the y-axis.

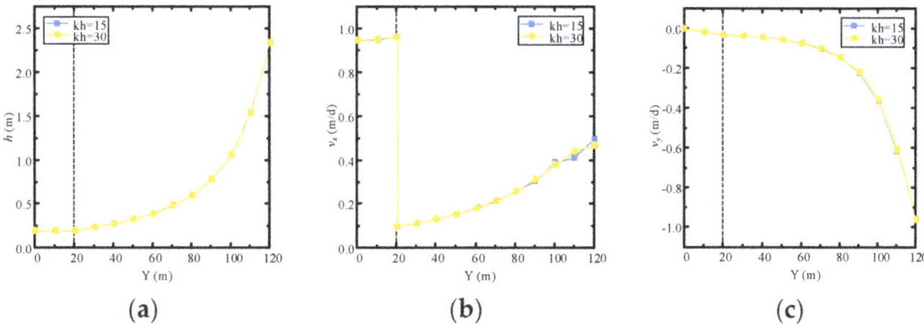

Figure 20. The comparison of solutions computed by the proposed NMM model using different mesh densities with an analytical solution along a profile located at $x = 100$ m: the (**a**) hydraulic head, (**b**) Darcy velocity component in the x direction, and (**c**) Darcy velocity component in the y direction.

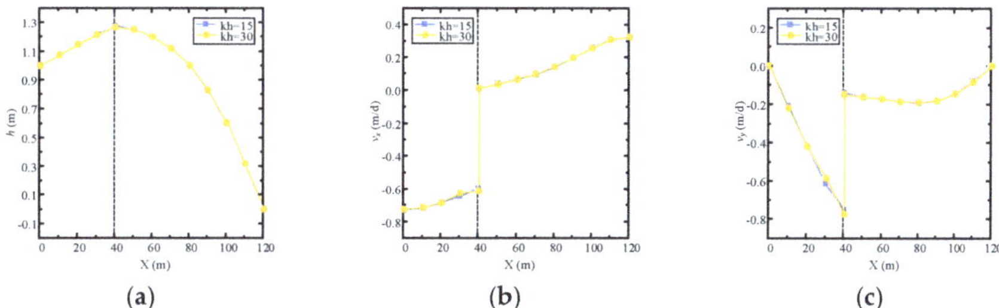

Figure 21. The comparison of solutions computed by the proposed NMM model using different mesh densities with an analytical solution along a profile located at $y = 80$ m: the (**a**) hydraulic head, (**b**) Darcy velocity component in the x direction, and (**c**) Darcy velocity component in the y direction.

For the purpose of verifying the accuracy of modeling refraction law across material interfaces, the velocity components of element nodes on the different sides of the material interface AB are plotted in Figures 22 and 23. The results show that the requirement of the refraction law cannot be strictly met using the conventional NMM model (Figure 22). Thus, the numerical results may be unreliable. As shown in Figure 23, the velocity components solved by the proposed model fully satisfy the refraction law, as well as Zhou's model [14].

As a result, the proposed NMM model can achieve the same precision of the Darcy velocity as Zhou's method without iterations, while the dual cover systems of NMM model are more flexible and convenient without the need of conforming to material interfaces. Moreover, the new model is proven to be insensitive to mesh size, so that reliable results can be achieved through a sparse mesh. Therefore, the proposed NMM model is very suitable for modeling the Darcy flow in heterogeneous porous media.

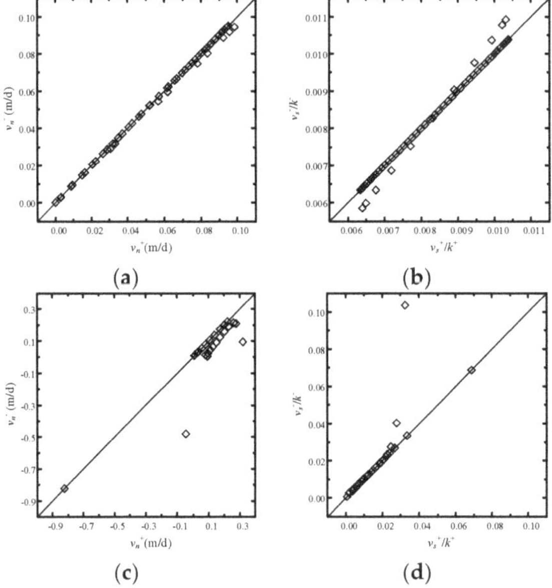

Figure 22. Verification of the refraction law on the material interface using the conventional NMM model: the (**a**) normal component on AB, (**b**) tangential component on AB, (**c**) normal component on CD, and (**d**) tangential component on CD.

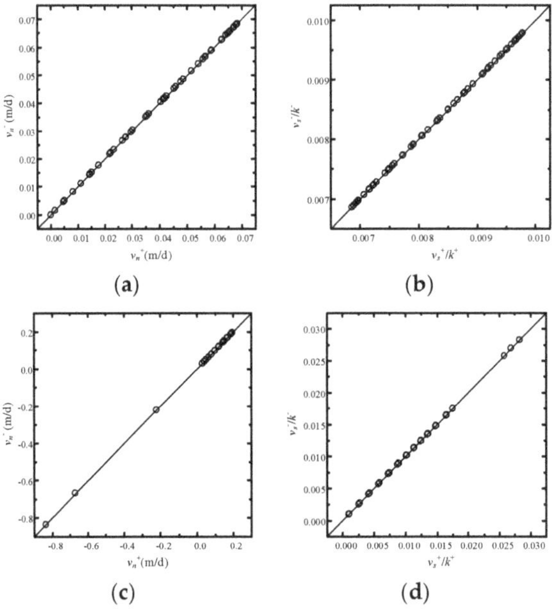

Figure 23. Verification of the refraction law on the material interface using the conventional NMM model: the (**a**) normal component on AB, (**b**) tangential component on AB, (**c**) normal component on CD, and (**d**) tangential component on CD.

5. Discussion and Conclusions

In this paper, a new high-order NMM model for the Darcy flow in heterogeneous porous media is presented. Due to the lack precision of measuring the velocity solution by ordinary methods, a new weight function with a continuous nodal gradient was adopted into the original NMM model. In order to achieve best precision when modeling material interfaces, the refraction law is fully introduced into the NMM model as an a posteriori requirement. Compared to other numerical schemes, the proposed NMM model has three major advantages as follows:

(1) the solution of Darcy velocity is accurate, and continuous at each node;

(2) the material interfaces exactly satisfy the refraction law, and thus the simulation results can be accurate and reliable; and

(3) the pre-processing work is efficient, due to the two independent cover systems of NMM, in contrast with other numerical schemes. Especially for complex heterogeneous problems even with intersecting material interfaces, the proposed model provides a significant advantage.

Additionally, a set of numerical examples are modeled to verify the accuracy, efficiency, and robustness of the proposed high-order numerical manifold method. First, Darcy flow in homogeneous porous media with different boundary conditions is simulated. The numerical results show good precision compared with analytical solutions, which is the foundation for modeling heterogeneous problems. Then, a continuous-heterogeneous example with hydraulic conductivity varying smoothly in the problem domain is demonstrated. Compared with the analytical solution, the developed method shows the ability to model heterogeneous problem without material interfaces. Moreover, convergence analysis shows that the new model has better accuracy for the Darcy velocity solution, especially in heterogeneous problems. Finally, the new high-order NMM model is applied into a heterogeneous problem with material interfaces and different boundary conditions. Results show that refraction law is exactly satisfied by the proposed model, while the computational cost, especially the mesh effort, is less than that of the existing models.

Furthermore, the proposed high-order NMM model can be extended to 3D cases and applied into fluid flow and contaminant transportation in heterogeneous porous media.

Author Contributions: Conceptualization, Y.W.; Methodology, L.Z.; Software, L.Z.; Validation, L.Z.; Writing—Original Draft Preparation, L.Z.; Writing—Review & Editing, Y.W. and D.F.; Visualization, L.Z.

Funding: This research was financially supported by the National Natural Science Foundation of China (U1765204, 41772340).

Conflicts of Interest: The authors declare no conflict of interest.

References

1. Bear, J. *Dynamics of Fluids in Porous Media*; American Elsevier Publishing Company: New York, NY, USA, 1972.
2. Cainelli, O.; Bellin, A.; Putti, M. On the accuracy of classic numerical schemes for modeling flow in saturated heterogeneous formations. *Adv. Water Resour.* **2012**, *47*, 43–55. [CrossRef]
3. Bellin, A.; Salandin, P.; Rinaldo, A. Simulation of dispersion in heterogeneous porous formations: Statistics, first-order theories, convergence of computations. *Water Resour. Res.* **1992**, *28*, 2211–2227. [CrossRef]
4. Neuman, S.; Orr, S.; Levin, O.; Paleologos, E. Theory and high-resolution finite element analysis of 2-d and 3-d effective permeabilities in strongly heterogeneous porous media. In Proceedings of the 9th International Conference on Computational Methods in Water Resources, Denver, CO, USA, 1 June 1992.
5. Ababou, R.; McLaughlin, D.; Gelhar, L.W.; Tompson, A.F. Numerical simulation of three-dimensional saturated flow in randomly heterogeneous porous media. *Transp. Porous Media* **1989**, *4*, 549–565. [CrossRef]
6. Chin, D.A.; Wang, T. An investigation of the validity of first-order stochastic dispersion theories in isotropie porous media. *Water Resour. Res.* **1992**, *28*, 1531–1542. [CrossRef]
7. Tompson, A.F.; Gelhar, L.W. Numerical simulation of solute transport in three-dimensional, randomly heterogeneous porous media. *Water Resour. Res.* **1990**, *26*, 2541–2562. [CrossRef]

8. Batu, V. A finite element dual mesh method to calculate nodal darcy velocities in nonhomogeneous and anisotropic aquifers. *Water Resour. Res.* **1984**, *20*, 1705–1717. [CrossRef]
9. Yeh, G.T. On the computation of Darcian velocity and mass balance in finite elements modeling of groundwater flow. *Water Resour. Res.* **1981**, *17*, 1529–1534. [CrossRef]
10. Zhang, Z.; Xue, Y.; Wu, J. A cubic-spline technique to calculate nodal Darcian velocities in aquifers. *Water Resour. Res.* **1994**, *30*, 975–981. [CrossRef]
11. D'Angelo, C.; Scotti, A. A mixed finite element method for darcy flow in fractured porous media with non-matching grids. *ESAIM-Math* **2012**, *46*, 465–489. [CrossRef]
12. Masud, A.; Hughes, T.J.R. A stabilized mixed finite element method for darcy flow. *Comput. Meth. Appl. Mech. Eng.* **2002**, *191*, 4341–4370. [CrossRef]
13. Mosé, R.; Siegel, P.; Ackerer, P.; Chavent, G. Application of the mixed hybrid finite element approximation in a groundwater flow model: Luxury or necessity? *Water Resour. Res.* **1994**, *30*, 3001–3012. [CrossRef]
14. Zhou, Q.; Bensabat, J.; Bear, J. Accurate calculation of specific discharge in heterogeneous porous media. *Water Resour. Res.* **2001**, *37*, 3057–3069. [CrossRef]
15. Xie, Y.F.; Wu, J.C.; Xue, Y.Q.; Xie, C.H.; Ji, H.F. A domain decomposed finite element method for solving darcian velocity in heterogeneous porous media. *J. Hydrol.* **2017**, *554*, 32–49. [CrossRef]
16. Shi, G.H. Manifold method of material analysis. In Proceedings of the Transactions of the 9th Army Conference on Applied Mathematics and Computing, Minneapolis, MN, USA, 18–21 June 1991; pp. 57–76.
17. Shi, G.H. Modeling rock joints and blocks by manifold method. In Proceedings of the 33th U.S. Symposium on Rock Mechanics, Santa Fe, NM, USA, 3–5 June 1992; pp. 639–648.
18. Ma, G.W.; An, X.M.; Zhang, H.H.; Li, L.X. Modeling complex crack problems using the numerical manifold method. *Int. J. Fract.* **2009**, *156*, 21–35. [CrossRef]
19. Zhang, H.H.; Li, L.X.; An, X.M.; Ma, G.W. Numerical analysis of 2-d crack propagation problems using the numerical manifold method. *Eng. Anal. Bound. Elem.* **2010**, *34*, 41–50. [CrossRef]
20. Wu, Z.J.; Wong, L.N.Y. Frictional crack initiation and propagation analysis using the numerical manifold method. *Comput. Geotech.* **2012**, *39*, 38–53. [CrossRef]
21. Zheng, H.; Xu, D.D. New strategies for some issues of numerical manifold method in simulation of crack propagation. *Int. J. Numer. Methods Eng.* **2014**, *97*, 986–1010. [CrossRef]
22. Ning, Y.J.; An, X.M.; Ma, G.W. Footwall slope stability analysis with the numerical manifold method. *Int. J. Rock Mech. Min. Sci.* **2011**, *48*, 964–975. [CrossRef]
23. He, L.; An, X.M.; Ma, G.W.; Zhao, Z.Y. Development of three-dimensional numerical manifold method for jointed rock slope stability analysis. *Int. J. Rock Mech. Min. Sci.* **2013**, *64*, 22–35. [CrossRef]
24. An, X.M.; Ning, Y.J.; Ma, G.W.; He, L. Modeling progressive failures in rock slopes with non-persistent joints using the numerical manifold method. *Int. J. Numer. Anal. Methods Geomech.* **2014**, *38*, 679–701. [CrossRef]
25. Fan, L.F.; Yi, X.W.; Ma, G.W. Numerical manifold method (NMM) simulation of stress wave propagation through fractured rock mass. *Int. J. Appl. Mech.* **2013**, *5*, 20. [CrossRef]
26. Wu, Z.J.; Wong, L.N.Y.; Fan, L.F. Dynamic study on fracture problems in viscoelastic sedimentary rocks using the numerical manifold method. *Rock Mech. Rock Eng.* **2013**, *46*, 1415–1427. [CrossRef]
27. Wong, L.N.Y.; Wu, Z.J. Application of the numerical manifold method to model progressive failure in rock slopes. *Eng. Fract. Mech.* **2014**, *119*, 1–20. [CrossRef]
28. Zhao, G.F.; Zhao, X.B.; Zhu, J.B. Application of the numerical manifold method for stress wave propagation across rock masses. *Int. J. Numer. Anal. Methods Geomech.* **2014**, *38*, 92–110. [CrossRef]
29. Ohnishi, Y.; Tanaka, M.; Koyama, T.; Mutoh, K. Manifold method in saturated-unsaturated unsteady groundwater flow analysis. In Proceedings of the Third International Conference on Analysis of Discontinuous Deformation (ICADD-3), Vail, CO, USA, 3–4 June 1999; pp. 221–230.
30. Zhang, Z.R.; Zhang, X.W.; Yan, J.H. Manifold method coupled velocity and pressure for navier-stokes equations and direct numerical solution of unsteady incompressible viscous flow. *Comput. Fluids* **2010**, *39*, 1353–1365. [CrossRef]
31. Jiang, Q.H.; Deng, S.S.; Zhou, C.B.; Lu, W.B. Modeling unconfined seepage flow using three-dimensional numerical manifold method. *J. Hydrodyn.* **2010**, *22*, 554–561. [CrossRef]
32. Wang, Y.; Hu, M.S.; Zhou, Q.L.; Rutqvist, J. Energy-work-based numerical manifold seepage analysis with an efficient scheme to locate the phreatic surface. *Int. J. Numer. Anal. Methods Geomech.* **2014**, *38*, 1633–1650. [CrossRef]

33. Hu, M.S.; Wang, Y.; Rutqvist, J. On continuous and discontinuous approaches for modeling groundwater flow in heterogeneous media using the numerical manifold method: Model development and comparison. *Adv. Water Resour.* **2015**, *80*, 17–29. [CrossRef]
34. Hu, M.S.; Wang, Y.; Rutqvist, J. An effective approach for modeling fluid flow in heterogeneous media using numerical manifold method. *Int. J. Numer. Methods Fluids* **2015**, *77*, 459–476. [CrossRef]
35. Hu, M.S.; Wang, Y.; Rutqvist, J. Development of a discontinuous approach for modeling fluid flow in heterogeneous media using the numerical manifold method. *Int. J. Numer. Anal. Methods Geomech.* **2015**, *39*, 1932–1952. [CrossRef]
36. Zheng, H.; Liu, F.; Li, C.G. Primal mixed solution to unconfined seepage flow in porous media with numerical manifold method. *Appl. Math. Model.* **2015**, *39*, 794–808. [CrossRef]
37. Yang, Y.T.; Zheng, H. A three-node triangular element fitted to numerical manifold method with continuous nodal stress for crack analysis. *Eng. Fract. Mech.* **2016**, *162*, 51–75. [CrossRef]
38. Yang, Y.T.; Xu, D.D.; Sun, G.H.; Zheng, H. Modeling complex crack problems using the three-node triangular element fitted to numerical manifold method with continuous nodal stress. *Sci. China Technol. Sci.* **2017**, *60*, 1537–1547. [CrossRef]
39. Zienkiewicz, O.C.; Taylor, R.L.; Zhu, J.Z. *The Finite Element Method: Its Basis and Fundamentals*, 7th ed.; Butterworth-Heinemann: Oxford, UK, 2013.

 *processes*

Article

Experimental Study on the Creep Characteristics of Coal Measures Sandstone under Seepage Action

Ziheng Sha [1], **Hai Pu [1,2,*]**, **Ming Li [1,3]**, **Lili Cao [1]**, **Ding Liu [1]**, **Hongyang Ni [1]** and **Jingfeng Lu [3]**

[1] State Key Laboratory for Geomechanics and Deep Underground Engineering,
    China University of Mining and Technology, Xuzhou 221116, China;
    zhsha@cumt.edu.cn (Z.S.); mingl@cumt.edu.cn (M.L.); caolili429@126.com (L.C.);
    liuding@cumt.edu.cn (D.L.); nhyang@cumt.edu.cn (H.N.)
[2] School of Mining Engineering, Xinjiang Institute of Engineering, Urumqi 830023, China
[3] School of Mechanics and Civil Engineering, China University of Mining and Technology,
    Xuzhou 221116, China; jingfeng_lu@cumt.edu.cn
* Correspondence: haipu@cumt.edu.cn; Tel: +86-0516-8399-5398

Received: 28 June 2018; Accepted: 24 July 2018; Published: 1 August 2018

Abstract: The seepage action of underground water accelerates the deformation of roadway surrounding rock in deep mines. Therefore, the study of creep characteristics of surrounding rock under seepage action is the basis for the stability control of roadway surrounding rock in deep water-rich areas. In this paper, a seepage-creep coupling test system for complete rock samples was established. Combined with a scanning electron microscopy (SEM) test system, the seepage-creep law of coal measures sandstone and the damage mechanism were revealed. The study results showed that the maximum creep deformation of sandstone under natural and saturation state decreased gradually with the increase of confining pressure, and the maximum creep deformation under saturation state was greater than the corresponding value under natural state when the confining pressure was same. When the confining pressure was constant, the creep deformation, the constant creep deformation rate and the accelerated creep deformation rate of sandstone increased rapidly with the increase of infiltration pressure. With the change of time, the change of permeability parameters went through three cycles; each cycle was divided into two stages, slow change stage and rapid change stage, and the rate of variation increased with the increase of the seepage pressure. Based on the macroscopic and microscopic characteristics of sandstone rupture, the connection between macroscopic and microscopic mechanism on sandstone rupture was established. The results in this paper can provide a theoretical basis for stability control of roadway surrounding rock in water-rich areas.

Keywords: coal measures sandstone; creep characteristics; seepage pressure; seepage-creep; microscopic morphology

1. Introduction

At present, many coal mines in China have entered the stage of deep mining. Compared with shallow coal mines, deep rock masses have significant rheological properties [1–7]. At the same time, the seepage action caused by underground water has also become one of the main factors which affect the stability control of roadway surrounding rock [8,9]. Especially in the process of deep mining, the creep characteristics of coal and rock medium under seepage are more pronounced. Therefore, to ensure the safety of deep mining, it is of great significance to understand the seepage-creep characteristics of rock thoroughly.

The study on rock creep properties originated from Griggs [10], who pointed out in 1939 that creep may occur when the load applied to the rock exceeds one-eighth of the damage load. Many studies [11,12] have shown that the rheological properties of rock are closely related to the complex stress conditions in which they are located. Maranini conducted uniaxial and triaxial compression creep tests on limestone and found that rock crack propagation under low confining pressure and pore collapse under high stress were the main causes of creep [13]. Fan carried out the triaxial compression creep tests of oil-bearing mudstone under low confining pressure; the results showed that there was an initial stress threshold of creep and failure stress threshold of creep [14]. Most of the laboratory tests on rock creep mainly focus on the study of compressive creep characteristics, but in fact, rock mass is often still in the complex stress conditions such as tension and shear. Therefore, some scholars [15–17] conducted researches on the creep characteristics of rock under different stress conditions. In recent years, many other scholars [18–24] have studied the rheological characteristics of rock from the microscopic level by using scanning electron microscopy (SEM) test system, CT (Computed Tomography) testing system and other equipment.

To study the seepage characteristics of fractured rock mass, many researchers have done a lot of work on seepage-stress coupling. Louis studied the seepage-stress relationship of a single fracture based on the results of fields water injection tests, and pointed out that the permeability coefficient of rock mass had a negative exponential relationship with normal stress [25]. Jones [26], Nelson [27], and Kranz [28] studied the permeability of different rocks respectively and established corresponding empirical formulas for permeability coefficients. Gale proposed an empirical formula which was similar to Cubic Law by conducting laboratory tests on three kinds of jointed rock masses [29]. Iwai found that the influence of joint roughness on the seepage law was mainly related to the contact rate of the fissure wall area [30]. Mordecai measured a 20% increase in permeability through a sandstone breaking test [31]. Li and Miao studied the seepage law and creep law of broken rock [32,33]. He conducted creep-seepage coupling tests on coal samples to reveal the consistency of permeability changes and the creep damage of coal samples [34]. Zhu studied the effect of seepage pressure on the damage and deterioration of brittle rock, and he also explored the change law of permeability characteristics during deformation process; the results showed that the seepage pressure had a certain degree of aggravating effect on the joint opening and expanded deformation of rock cracks [35]. Zhang conducted penetration tests on different rock samples and found that as the strain increases, the permeability increased more significantly [36]. Combined with damage mechanics model and numerical simulation tools, some researchers have also conducted some theoretical studies on seepage flow [37,38]. Based on the experimental results of rock stress-strain-permeability tests, Yang established a seepage-stress-damage coupling model with brittle characteristics from the microscopic view [39]. Pu studied the creep law of the overlying strata through a seepage-creep coupling model which was derived by himself [40]. Souley studied the relationship between excavation-disturbed zone of surrounding rock in an underground cavern and permeability through underground nuclear waste disposal tests in Canadian [41].

The creep test methods of rock have developed well. Relatively speaking, rock seepage-creep test technology is still in the development stage. Due to the low permeability of the complete rock samples, most of the studies used fractured rock or broken rock as the research object to carry out rock seepage tests or seepage-creep tests [42–44]. In recent years, some researchers [8,45] have used self-made test systems to conduct seepage-creep test studies on complete rock samples. However, systematic research results are rare, because of the superior difficulty of such experiments. Therefore, in this paper, a rock seepage-creep coupling test system is built for the test requirements of complete rock samples. The creep tests under natural state and saturation state and the seepage-creep tests under different seepage pressures are carried out for the complete coal measures sandstone samples. The effects of seepage pressure on the creep properties of the rock and the whole process are investigated. This study provides a theoretical basis for the control of roadway surrounding rock in deep coal mines.

2. Materials and Methods

2.1. Material Characterization

The rock samples were roof sandstone, taken from the overlying strata of a working face of a mine in Anhui. X-ray diffraction analysis was performed on the coal measures sandstone, and the mineral content of sandstones is shown in Table 1. It is observed that Kaolinite is a typical clay mineral and it is a binder between the internal material particles of the sandstone; when Kaolinite meets water, it will expand and soften. The feldspathic material is the main component of this type of sandstone, and its material particles will dilate and fracture under the long-term action of water, resulting in the macroscopic destruction of the rock [46].

According to the test plan, the sandstone was made into standard specimens ($\Phi \times h = 50 \times 100$ mm) in accordance with the 'Test Procedures for the Physical and Mechanical Properties of Rocks', and the machining accuracy of the specimens should meet the requirements of 'Standard for Test Method of Engineering Rock Mass' (GB50218-94) (Figure 1). The physical and mechanical properties of sandstone were tested, and its basic physical and mechanical parameters were obtained (Table 2). It is worth noting that all the parameters in Table 2 were obtained by taking the average of the results of multiple tests. Before the seepage-creep coupling test, the finished specimens were placed in water under a room temperature (20 °C) until the weight of the specimens no longer changed, then the specimens were considered saturated.

Table 1. Mineral content of sandstones.

Anorthite—Ca(Al$_2$Si$_2$O$_8$)	Microcline Maximum—K(AlSi$_3$) O$_8$	Kaolinite—Al$_2$Si$_2$O$_5$(OH)$_4$	Quartz—SiO$_2$
45.2%	31.4%	13.8%	9.6%

Table 2. Basic physical and mechanical properties of sandstones.

ρ/(kg·m^{-3})	v/(m·s^{-1})	E/GPa	μ	σ_u/MPa	σ_T/MPa	σ_t/MPa	φ/°
2540	2650	37.65	0.26	122.45	149.29	2.85	38

Remarks: ρ, density; v, longitudinal wave velocity; E, elasticity modulus; μ, Poisson's ratio; σ_u, uniaxial compression strength; σ_T, triaxial compression strength (confining pressure is 3 MPa); σ_t, tensile strength; φ, the angle of internal friction.

Figure 1. Part of finished samples.

2.2. Testing System

To meet the requirements of seepage-creep coupling tests for intact rock specimens, the existing equipment and devices were used to design and set up a rock seepage-creep coupling test system for standard rock specimens. The entire system consists of five parts: the axial pressure loading system, the confining pressure loading system, the seepage pressure loading system, the rock seepage system, and the data collecting and processing system, as shown in Figure 2.

1. The axial pressure loading system uses the Electro-hydraulic servo rock mechanics test system of MTS816 to provide a stable and continuous axial load for the specimen during the test, as shown in Figure 2a.
2. The confining pressure loading system applies a certain pressure to the specimen by using the pump to press the hydraulic oil into the cylinder.
3. The seepage pressure loading system provides a continuous and stable seepage pressure for the specimen based on the injector concept, and this is an important part to achieve the seepage-creep coupling effect, as shown in Figure 2b.
4. The rock seepage system is the core part of the rock seepage-creep coupling test system. It is the central system of the axial loading system, confining pressure loading system and seepage pressure loading system. Figure 2c shows the structural compositions of the rock seepage system. As can be seen from the picture, the entire testing system is roughly composed of 14 parts.
5. The data collecting and processing system includes a paperless recorder, a pressure transmitter, a flow sensor, and a computer terminal.

(**a**) Electro-hydraulic servo rock mechanics test system of MTS816.

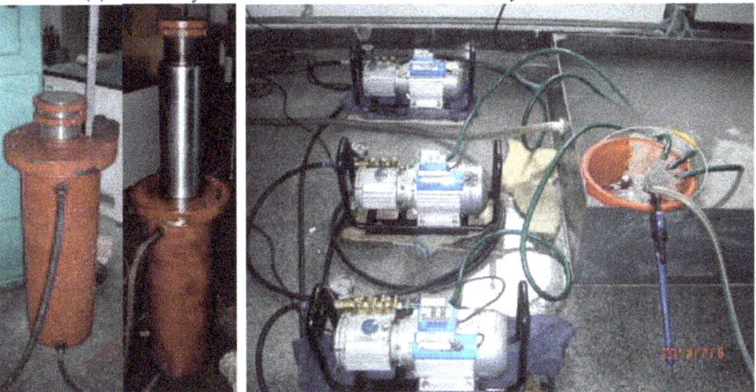

(**b**) Equipment of seepage pressure loading system.

Figure 2. *Cont.*

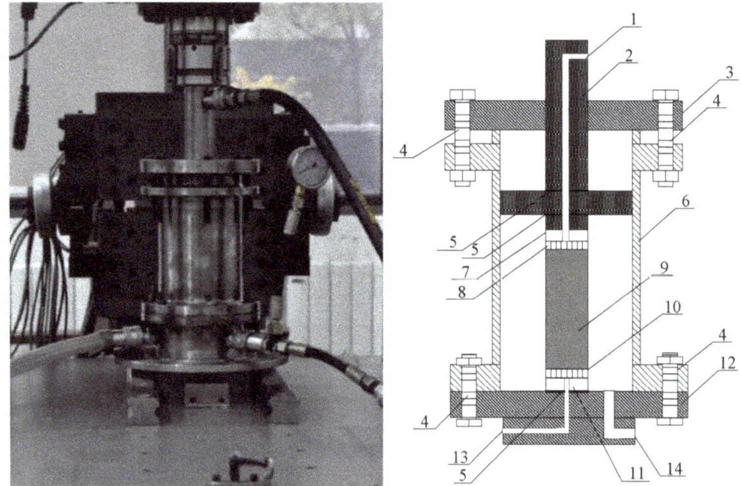

1—water injection hole; 2—piston; 3—head cover; 4—high Strength bolt; 5—O-type sealing ring; 6—cylinder block; 7—universal joint; 8—overlying porous disc; 9—rock sample; 10—lower porous disc; 11—pedestal; 12—baseplate with holes; 13—water outlet; 14—entrance of hydraulic oil

(**c**) Rock seepage system and its structure composition.

Figure 2. Rock seepage-creep coupling test system.

2.3. Test Scheme

The creep test adopts the method of gradation loading by applying different stress levels to the same rock specimens step by step. The final creep deformation of a specimen is the superposition of the creep deformation under each stage of load.

By using the rock seepage-creep coupling test system, the creep tests of coal measures sandstone under different seepage pressures will be tested. The specific experimental scheme is as follows (Figure 3):

1. Triaxial compression tests of sandstone. Triaxial tests of sandstone will be carried out under the natural state and the saturation state and the confining pressures are 1 MPa, 2 MPa, 3 MPa, and 4 MPa. Then the compressive strengths of sandstone under four confining pressures in two states are determined. According to the triaxial compressive strength, the load classification scheme of sandstone creep experiment can be obtained.

2. Triaxial creep tests of sandstone. According to the different loading levels, the creep tests with confining pressures of 1 MPa, 2 MPa, 3 MPa, and 4 MPa are carried out under natural and saturation conditions respectively, and the creep characteristics of sandstone under two conditions are obtained finally.

3. Seepage-creep coupling tests of sandstone. Water saturated rock specimens are used as the targets to perform the rock creep tests under different seepage pressures. The confining pressure is 4 MPa, and the seepage pressure p_0 is 0.5 MPa, 1.5 MPa, 2.5 MPa, and 3.5 MPa. The effect of seepage pressure on the creep deformation and permeability of sandstone can be obtained.

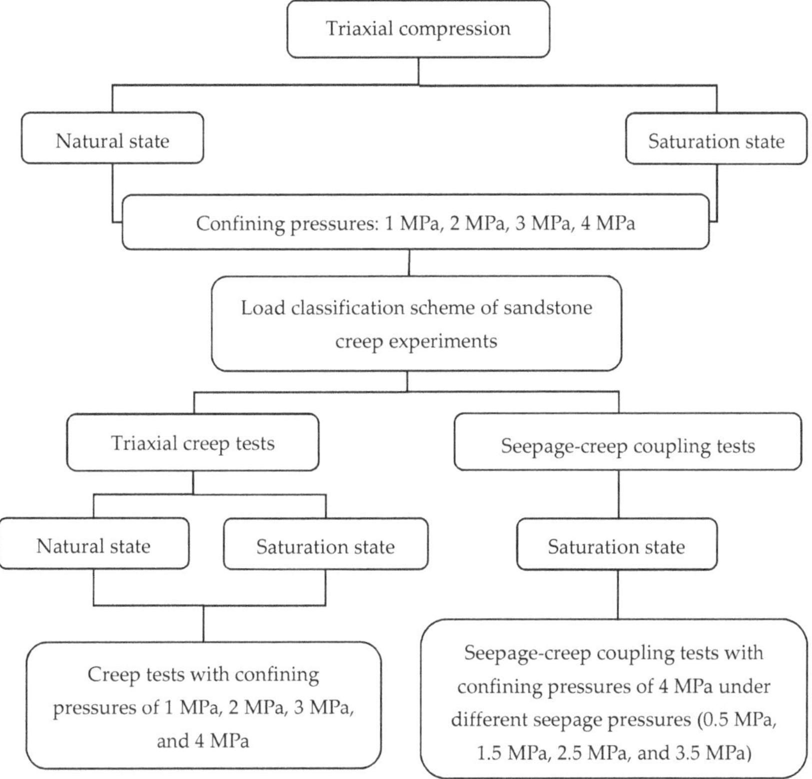

Figure 3. Test scheme.

2.4. Load Grading

Firstly, the triaxial compression tests of sandstones in natural and saturated conditions under different confining pressures (p = 1~4 MPa) were carried out. Figure 4 shows the variation of triaxial compression strength with confining pressure under two conditions. With the increase of confining pressure, the peak strength of sandstone under two states all increases linearly. When the confining pressure increases from 1 MPa to 4 MPa, the peak strength of sandstone in natural state increases from 128.69 MPa to 158.74 MPa, and the increase amplitude is 23.35%. Under the saturation condition, the peak strength of sandstone increases from 69.01 MPa to 115.26 MPa, and the increase amplitude is 67.02%. Comparing the two sets of curves, under the same confining pressure conditions, the peak strength of natural sandstone is greater than the peak strength of sandstone under the saturation condition, and the difference value between them is in the range of 42.54~59.68 MPa. The slope of the peak intensity change curve of sandstone under the saturation state is greater than that of the curve under the natural state, which indicates that the peak intensity of sandstone under the saturation condition is more sensitive to the confining pressure.

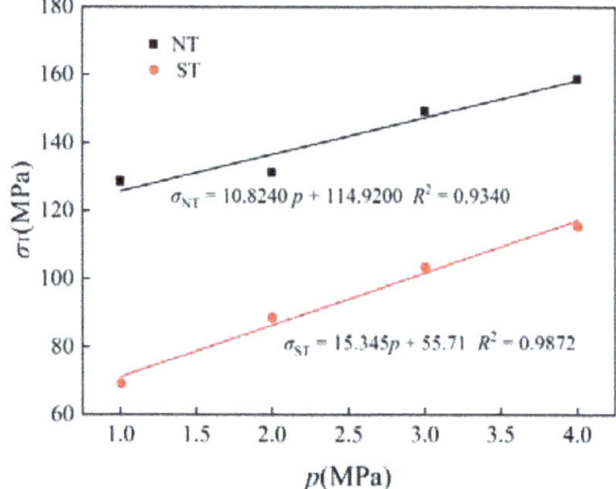

Figure 4. Change of maximum stress of sandstone with confining pressure.

According to the load classification principle of creep tests, combined with the results of the triaxial tests, the maximum stress is 90% of the peak strength, and then 6 equal parts are used to determine the graded loadings. The loading time t of each level is 4 h; if the rock specimen does not break down finally, the load will increase to the peak stress to continue the test. Table 3 shows the specific load classification scheme.

Table 3. Load classification scheme of sandstone creep experiment.

Hierarchical Load	Natural Condition				Saturation Condition			
	$p = 1$ MPa	$p = 2$ MPa	$p = 3$ MPa	$p = 4$ MPa	$p = 1$ MPa	$p = 2$ MPa	$p = 3$ MPa	$p = 4$ MPa
σ_1 (MPa)	20	20	23	25	10	14	15	18
σ_2 (MPa)	40	40	46	50	20	28	30	36
σ_3 (MPa)	60	60	69	75	30	42	45	54
σ_4 (MPa)	80	80	92	100	40	56	60	72
σ_5 (MPa)	100	100	115	125	50	70	75	90
σ_6 (MPa)	120	120	138	150	60	84	90	108

2.5. The Method of Microscopic Damage Test

(1) Test Equipment

The SEM technique was used to observe the fracture morphology of the damaged sandstone specimens. The equipment we used was a TESCAN VEGA3 (TESCAN, Brno, Czech Republic) type scanning electron microscope system, as shown in Figure 5. The principle of SEM test is to scan the surface of the specimens by using an electron beam emitted by the electron gun. Then specimens need to be excited to produce various physical signals. After detection, video amplification and signal processing, scanned images that can reflect various features of the specimen surface can be obtained on the viewing screen.

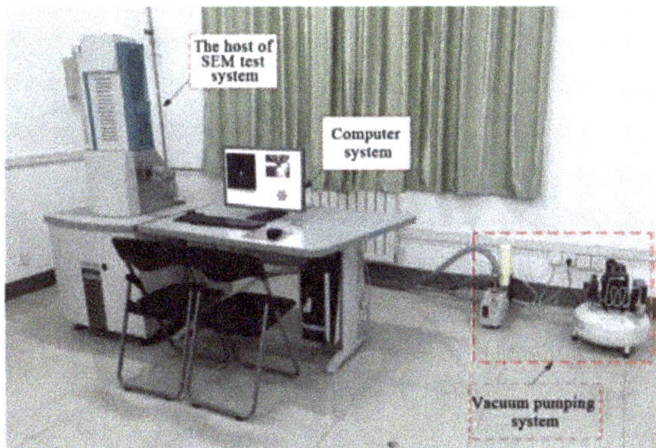

Figure 5. Scanning electron microscopy (SEM) test system.

(2)   Sample Preparation

First, we picked out the sample fragments that have flat surface after destruction, and then selected the appropriate size ($\Phi \times h = 10 \times 2$ mm) fragments from the middle of the sample rupture surfaces as the observation samples. The lower surfaces (non-observed surface) of the observation specimens need to be sanded with sandpaper to ensure that the specimens can be fixed on the sample platform steadily. Next, we needed to clean the observation surfaces by using a special equipment and then attached the observation samples to the sample platform. Finally, the surfaces of the observation samples were sputtered by an SBC-12 (KYKY Technology Co., Ltd., Beijing, China) ion sputtering equipment and then the sample preparation was completed (Figure 6).

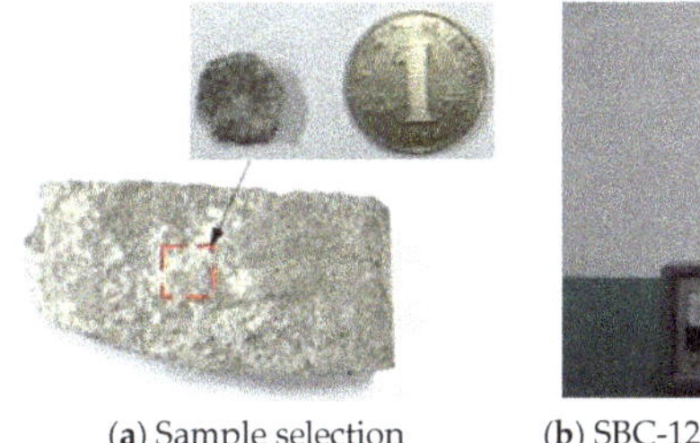

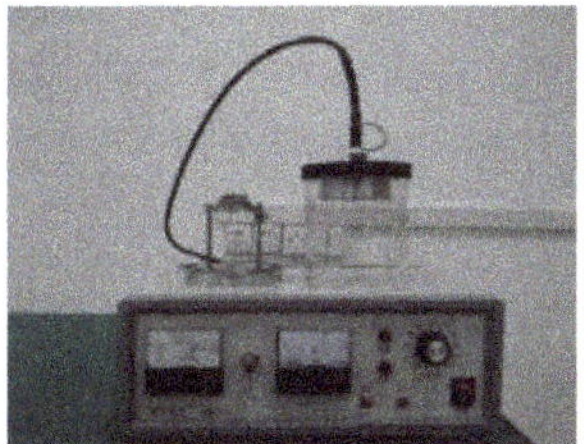

(**a**) Sample selection   (**b**) SBC-12 ion sputtering equipment   (**c**) Finished samples

Figure 6. Observe sample preparation

3. Results and Discussion

3.1. Creep Properties of Sandstone under Natural State and Saturation State

3.1.1. Creep Deformation Law of Sandstone under Two States

According to the test scheme, the creep tests of sandstone under natural and saturation conditions were carried out respectively, and the strain-time curves of sandstone under different confining pressures were obtained too. Finally, according to the principle of graded loading, the creep curves of sandstone were obtained, as shown in Figures 7 and 8.

As can be seen from Figures 7 and 8, the creep characteristic curves of sandstone all show a stepwise upward trend. Under each load, sandstone undergoes instantaneous deformation and stable deformation (or accelerated deformation under high loads), and more than 90% of the final deformation occurs instantaneously.

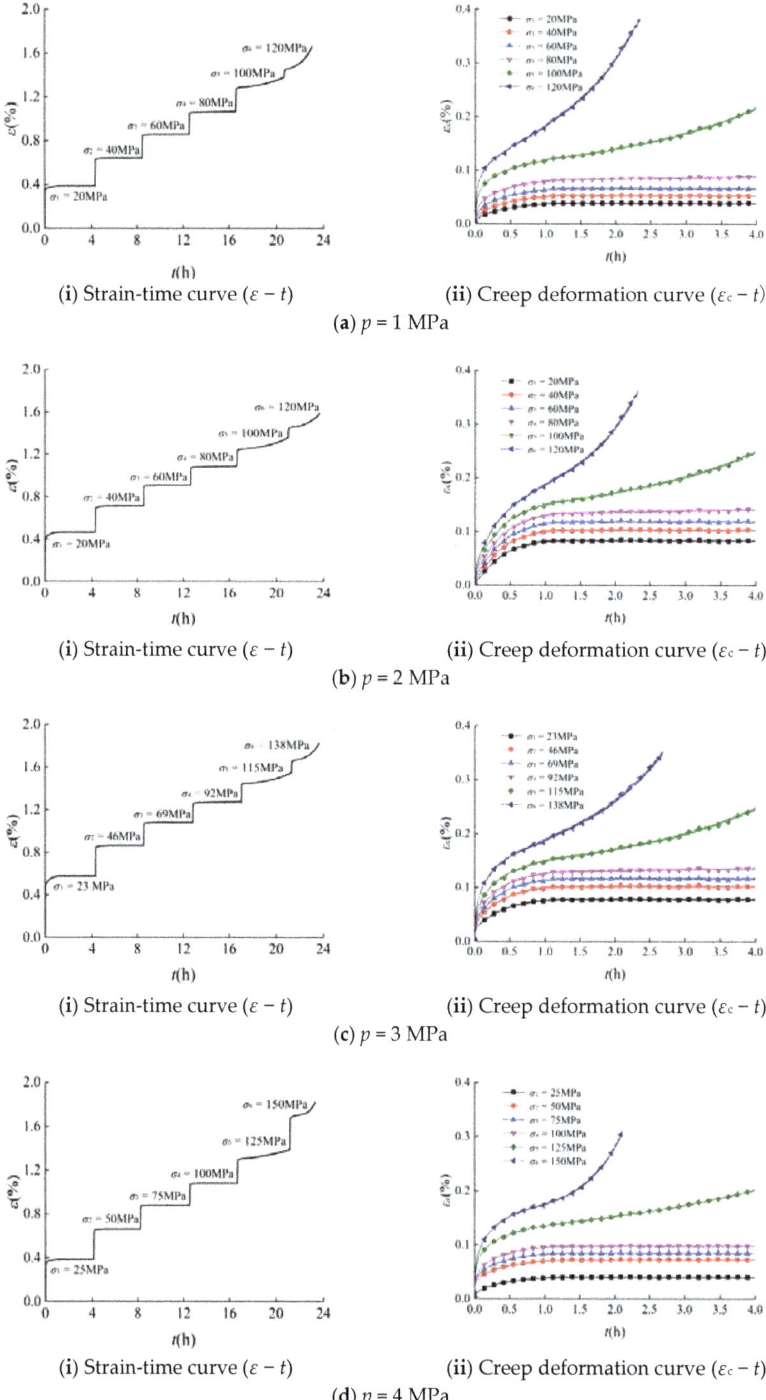

(**i**) Strain-time curve ($\varepsilon - t$)   (**ii**) Creep deformation curve ($\varepsilon_c - t$)

(**a**) $p = 1$ MPa

(**i**) Strain-time curve ($\varepsilon - t$)   (**ii**) Creep deformation curve ($\varepsilon_c - t$)

(**b**) $p = 2$ MPa

(**i**) Strain-time curve ($\varepsilon - t$)   (**ii**) Creep deformation curve ($\varepsilon_c - t$)

(**c**) $p = 3$ MPa

(**i**) Strain-time curve ($\varepsilon - t$)   (**ii**) Creep deformation curve ($\varepsilon_c - t$)

(**d**) $p = 4$ MPa

Figure 7. Creep characteristic curves of sandstone in natural condition.

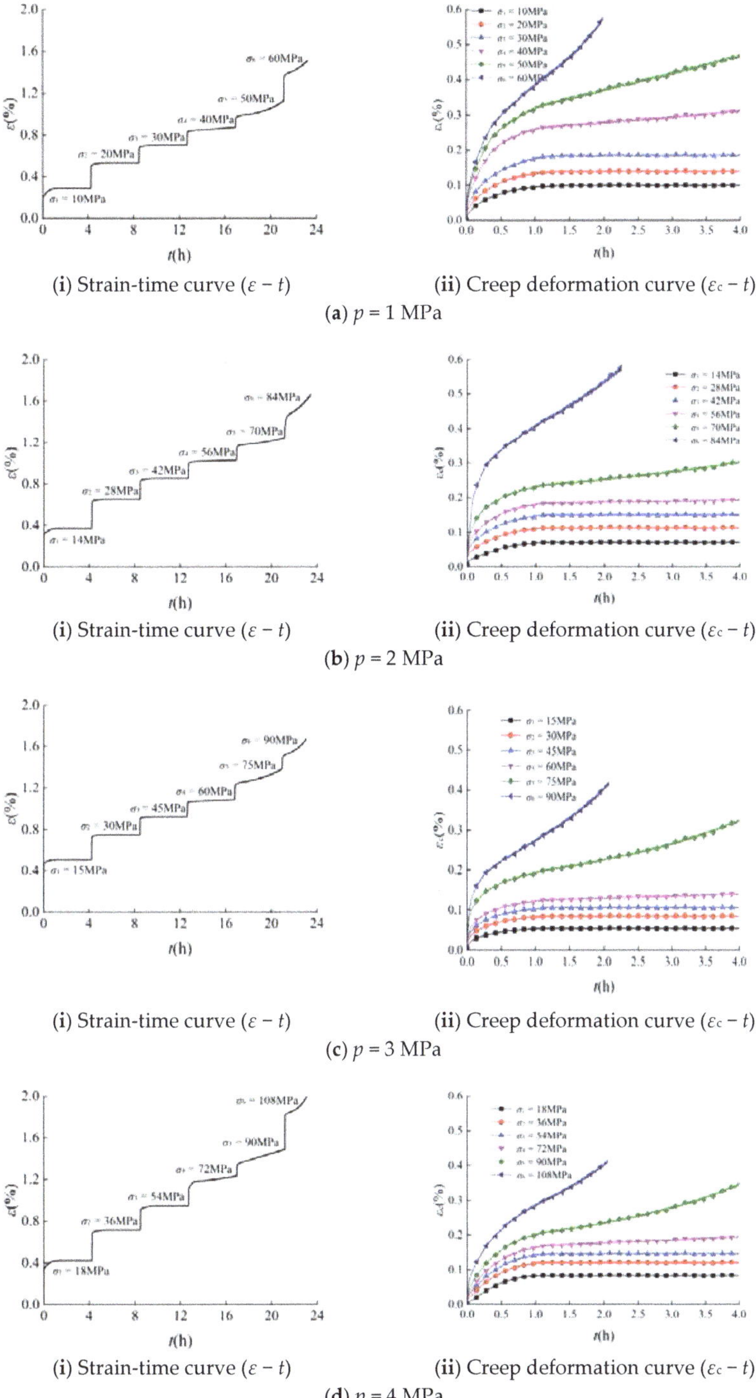

(**i**) Strain-time curve ($\varepsilon - t$) (**ii**) Creep deformation curve ($\varepsilon_c - t$)

(**a**) $p = 1$ MPa

(**i**) Strain-time curve ($\varepsilon - t$) (**ii**) Creep deformation curve ($\varepsilon_c - t$)

(**b**) $p = 2$ MPa

(**i**) Strain-time curve ($\varepsilon - t$) (**ii**) Creep deformation curve ($\varepsilon_c - t$)

(**c**) $p = 3$ MPa

(**i**) Strain-time curve ($\varepsilon - t$) (**ii**) Creep deformation curve ($\varepsilon_c - t$)

(**d**) $p = 4$ MPa

Figure 8. Creep characteristic curves of sandstone in saturation condition.

Figure 9 shows that when the confining pressure increases from 1 MPa to 4 MPa, the final strain of natural and saturated sandstone increases from 1.66% and 1.51% to 1.82% and 2.01%, respectively. With the increase of confining pressure, the final deformation of sandstone gradually increases, which is mainly determined by the graded loads, because the graded load develops with the confining pressure, resulting in greater deformation of sandstone. Under the same axial pressure conditions, the greater the confining pressure, the smaller the final deformation of the sandstone.

Comparing with the creep characteristic curves, the confining pressure increases from 1 MPa to 4 MPa, and the final creep strain of natural and saturated sandstones decreases from 0.38% and 0.58% to 0.30% and 0.41%, respectively. The final deformation of sandstone gradually decreases with the increase of confining pressure, as shown in Figure 10. Comparing the maximum creep deformation curves of sandstone under two conditions, we can see that under the same confining pressure condition, the final creep deformation of sandstone under saturation state is greater than the corresponding value under natural state, which reflects the physical erosion and softening effect of water on the sandstone.

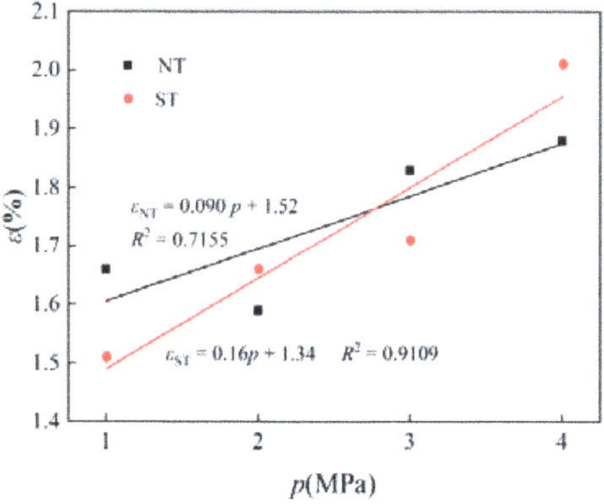

Figure 9. Change of the maximum deformation of sandstone.

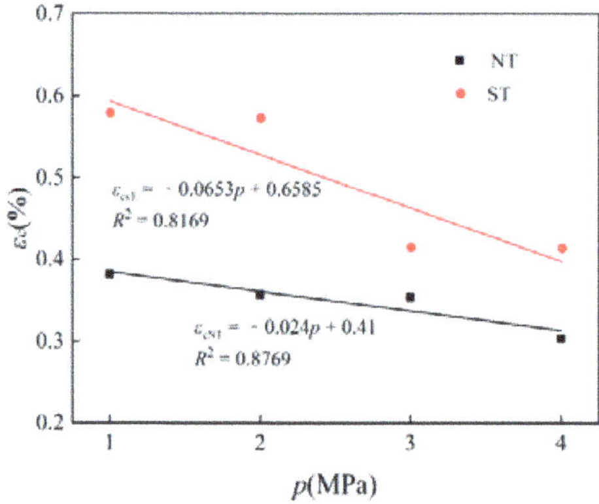

Figure 10. Change of the maximum creep deformation of sandstone.

3.1.2. Creep Deformation Rate of Sandstone under Two States

Deformation rate is an important indicator to reflect the creep deformation characteristics of sandstone. The deformation rate of stable creep and accelerated creep (which means the slopes of the creep characteristic curves) of sandstone under level 5 load and level 6 load are studied. Figure 11 shows the variation law of the slope of stable creep and accelerated creep deformation curves of sandstone under two states with confining pressure when the rock samples are under level 5 load and level 6 load. The effect of confining pressure on the creep deformation rate of sandstone is significant, but the characteristics of their deformation rates are different under different load levels. Under the level 5 load, the creep deformation rate of sandstone increases with the confining pressure, while the slope of the curve of the creep deformation rate of sandstone under the saturation condition is greater. Under the level 6 load, the stable creep and accelerated creep deformation rates of sandstone decrease with the increase of confining pressure.

The variation law of creep deformation rate of sandstone with confining pressure under different loading levels shows that: under the level 5 load, the creep deformation rate of sandstone is mainly affected by the load size, while the creep deformation rate is mainly affected by the confining pressure under the level 6 load. The cause of this phenomenon is that the confining pressure is an important factor to prevent the axial deformation of the rock. Under the same load, the larger the confining pressure is, the smaller the deformation rate and deformation of the rock are. In addition, the greater the axial pressure is, the greater the deformation rate is, indicating that the axial load plays an important role in the creep deformation of sandstone. Also, the larger the confining pressure is, the smaller the deformation rate is, which indicates that the effect of confining pressure on creep deformation is more significant.

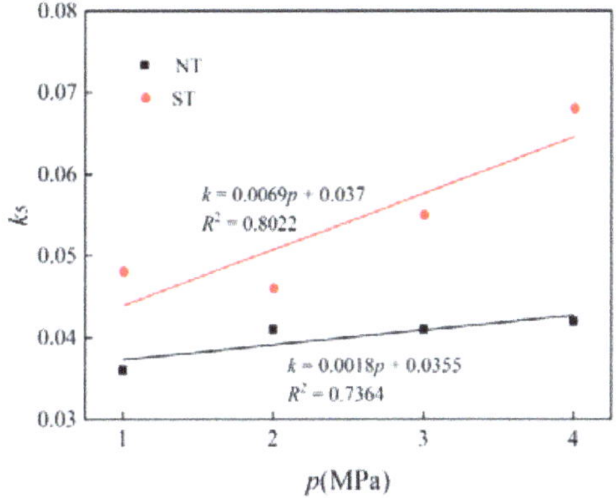

(**a**) Isokinetic creep deformation rate under level 5 load

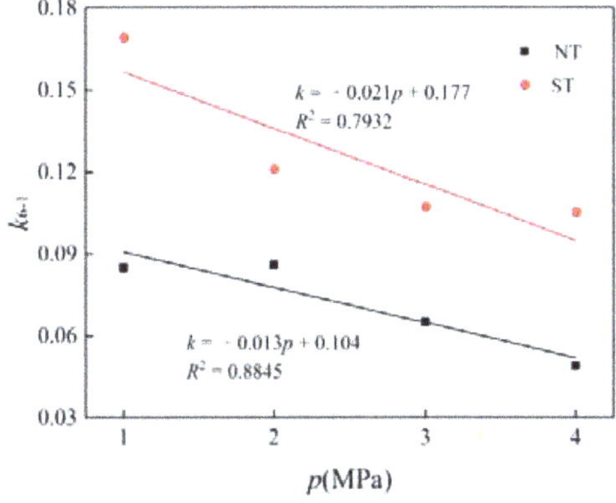

(**b**) Isokinetic creep deformation rate under level 6 load

Figure 11. *Cont.*

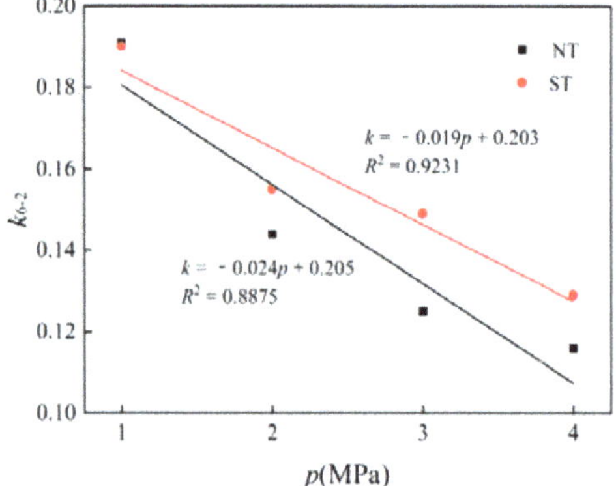

(**c**) Accelerated creep deformation rate under level 6 load

Figure 11. Change of creep deformation rate of sandstone.

3.2. Influence of Seepage Pressure on Seepage-Creep Properties of Sandstone

3.2.1. Variation of Creep Deformation of Sandstone with Seepage Pressure

Under the confining pressure of 4 MPa, creep tests of sandstone under the action of seepage pressures of 0.5 MPa, 1.5 MPa, 2.5 MPa and 3.5 MPa were carried out respectively, and the strain-time curves of sandstone were obtained, as shown in Figure 12. The curve with the seepage pressure of 0 MPa refers to the creep characteristic curve of the sandstone in the saturation state under no seepage pressure. The creep curves of sandstone show a stepwise upward trend, and the shape of each curve is basically the same.

Figure 13 shows the creep curves of sandstone under different seepage pressures. As can be seen from the whole curves, as the seepage pressure increases, the instantaneous deformation of sandstone under the same loading level increases gradually. Compared with the other four groups of curves, when the seepage pressure is 3.5 MPa, accelerated deformation and failure will occur under the level 5 load. With the increase of seepage pressure, the final deformation and maximum creep deformation of sandstone increase gradually, and the slopes of stable creep stage and accelerated creep stage change significantly.

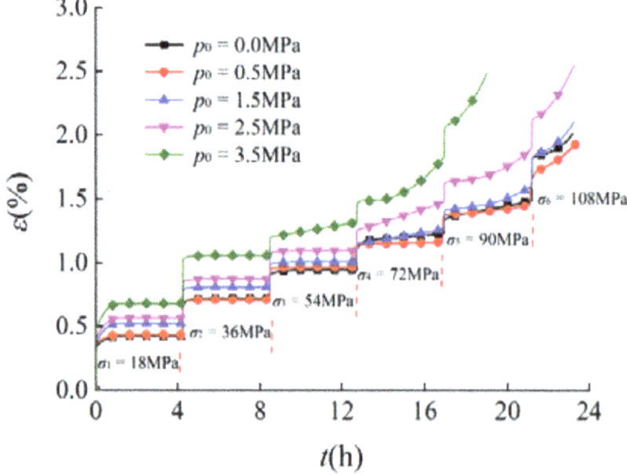

Figure 12. Seepage-creep strain-time curve of sandstone under difference seepage pressures.

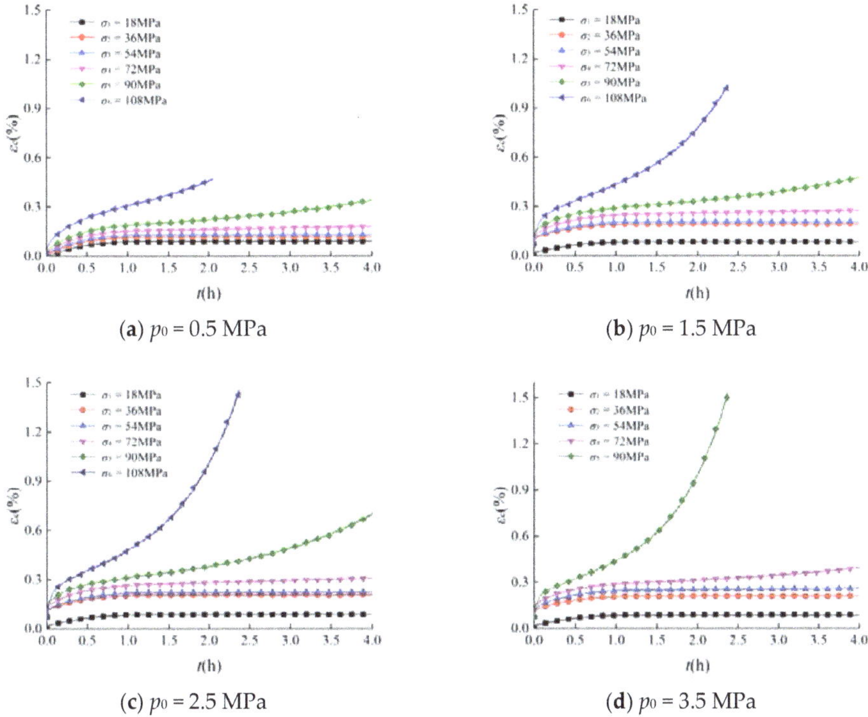

(**a**) p_0 = 0.5 MPa

(**b**) p_0 = 1.5 MPa

(**c**) p_0 = 2.5 MPa

(**d**) p_0 = 3.5 MPa

Figure 13. Seepage-creep deformation curve of sandstone under difference seepage pressures.

According to the creep characteristic curves, the curve of maximum deformation and maximum creep deformation of sandstone with seepage pressure was obtained, as shown in Figure 14. With the increase of the seepage pressure, the final deformation of the sandstone and the maximum creep deformation increase approximately linearly. It should be noted that the sandstone has been destroyed under the seepage pressure of 3.5 MPa at level 5 load; therefore, the maximum deformation is slightly reduced.

The slope of the creep deformation curve of the sandstone under the last two loading levels was studied. From Figure 15, it can be seen that, as the seepage pressure increases, the slope of each creep curve gradually increases, indicating that the creep deformation rate of sandstone increases steadily. The seepage pressure has three types of effects on the creep deformation of rock: First, the seepage pressure can directly act on the sandstone in the form of axial pressure. Under the same confining pressure, the greater the axial pressure, the greater the deformation and deformation rate of sandstone. Second, the seepage pressure can provide an extended pressure on the internal crack tip of the sandstone. The greater the seepage pressure is, the faster the crack tip develops, and the sandstone deformation and deformation rate increase correspondingly. In addition, the seepage pressure can produce a certain scouring effect on the crack surface. The greater the seepage pressure, the greater the corresponding scouring effect, resulting in a smaller surface roughness of the crack. As a result, it is easier to slide on both sides of the crack, and the macroscopic performance is that the deformation rate of sandstone gradually increases.

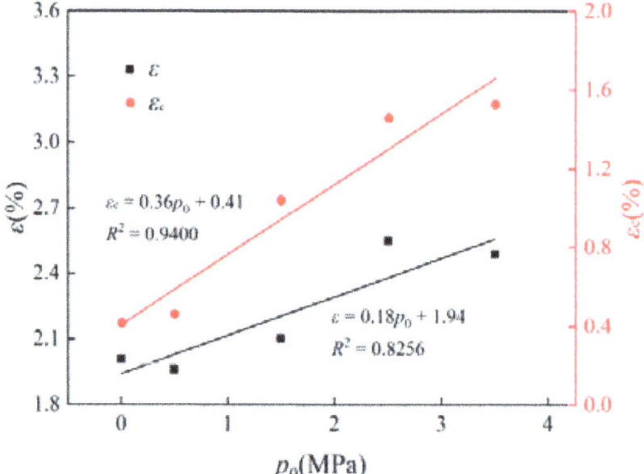

Figure 14. Change of maximum deformation of sandstone with seepage pressure.

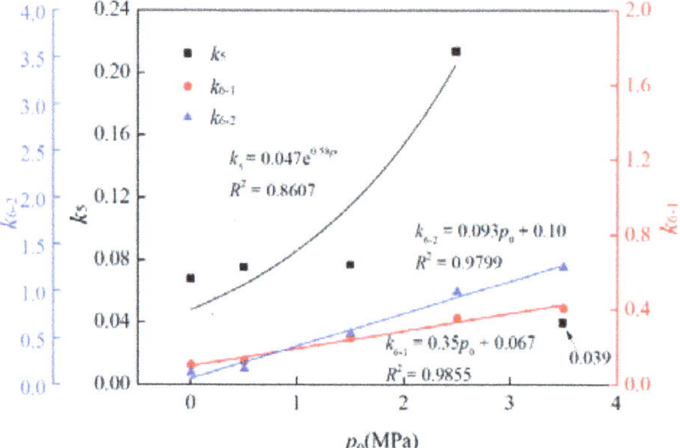

Figure 15. Change of creep slope of sandstone with seepage pressure.

3.2.2. Variation of Seepage Properties of Sandstone with Seepage Pressure

Figure 16 shows the curve of the permeability and the rate of permeability of sandstone with time under different seepage pressures during the tests; Figure 17 shows the curves of the permeability parameters of sandstone with time.

From Figure 17, it can be seen that the permeability parameters of sandstone undergo three cycles of change over time. In each cycle, the permeability parameters of sandstone gradually increase with time. At the junction of two cycles, these permeability parameters all drop rapidly. This change trend is the same as the change of permeability in Figure 16. On the other hand, the change of permeability of sandstone under different seepage pressures is particularly significant in the vertical dimension.

The time when the permeability parameters were measured for the first time during the tests shows a gradual advancement as the seepage pressure increases. In particular, under the seepage pressure of 3.5 MPa, the corresponding permeability parameters can be measured at the level 3 load (54 MPa), which the time is 8.42 h. In addition, under the other seepage pressures, the permeability parameters were measured at the level 4 load (72 MPa) for the first time through the tests. The specific moments are 12.08 h (p_0 = 2.5 MPa), 13.58 h (p_0 = 1.5 MPa), and 15.50 h (p_0 = 0.5 MPa), respectively. In the process of the tests, the change of the permeability characteristics is reflected by the change of the permeable capacity, and the time for the first measurement of the permeable capacity is different, which indicates that the time for the first appearance of the permeable channel in the sandstone is different. To be more specific, with the increase of the seepage pressure, the time for the appearance of permeable channels within the sandstone becomes faster. Comparing with the creep deformation curves, stable creep and decelerating creep all occur during the deformation stages that can measure the permeability of water. Therefore, sandstone had already experienced stable creep at the first measurement of permeable capacity, which can be reflected in the creep deformation curves.

Within each variation period, the maximum permeability parameters of sandstone increase with the increase of the seepage pressure. At the end of each loading cycle, the permeability of the permeable channels in the sandstone increases with the increase of the seepage pressure, and the internal cracks develop more abundantly.

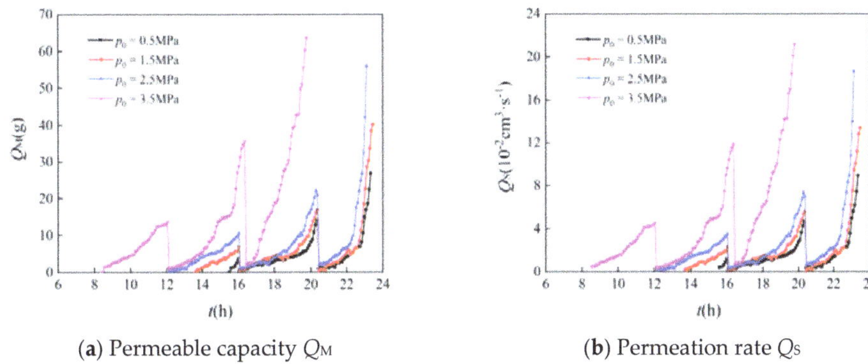

(**a**) Permeable capacity Q_M (**b**) Permeation rate Q_S

Figure 16. Change of permeability of sandstone with seepage pressure.

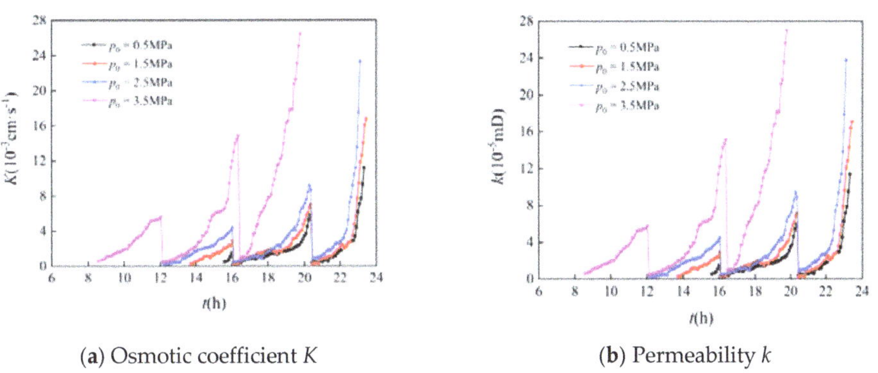

(**a**) Osmotic coefficient K (**b**) Permeability k

Figure 17. Change of permeability parameters of sandstone with seepage pressure.

During the last variation period of sandstone permeability parameters (Figure 17), the permeability parameters are generally divided into three changing processes. The first phase lasts from 20.38 h to around 20.70 h. The permeability parameters are at a smaller level (about 100 times of the permeability parameter of sandstone under the natural state), and this process lasts about 20 min. The second phase begins at 20.70 h and ends at approximately 23.31 h. The permeability parameters increase in a straight line with a fixed slope over time, and as the seepage pressure increases, the slope of the line increases gradually, and the change time in this stage accounts for about 80% of the entire change period. The last phase starts at 23.31 h and stops until the end of the test. The sandstone permeability parameters increase rapidly and reach a maximum in a very short time during this period. In fact, the change in the permeability characteristics of the sandstones in these three stages reflects the three processes of sandstone creep deformation under the final loading level. During the decelerating creep stage, internal cracks in the sandstone were closed; the internal structure was in a relatively stable state, and the permeability characteristics were stable, resulting in the permeation rate remaining equable and at a relatively small level. During the stable creep stage, the internal cracks in the sandstone expanded steadily and the permeation parameters continued to increase. During the accelerated creep stage, the internal cracks of the sandstone rapidly expanded and macroscopic permeable channels were formed, which leaded to a rapid increase in the permeability properties of the sandstone and a greater amount of permeable water.

3.3. Macroscopic and Microscopic Mechanism Study on Coal Measures Sandstone Rupture under Seepage Creep Coupling

3.3.1. Macroscopic Characteristics of Sandstone Rupture

Figure 18 shows the macroscopic morphology after the creep failure of sandstone under different seepage pressures. Generally speaking, under the same confining pressure, the number of through cracks on the surface of specimens after creep failure increases with the increase of seepage pressure, indicating that the degree of macroscopic damage gradually increases.

When the seepage pressure is 0.5 MPa, there is a main crack that runs through the surface of the specimen, and there are many small cracks that run parallel to the main crack. The failure mode is a typical shear failure. When the seepage pressure increases to 1.5 MPa, there are several macroscopic cracks penetrating the surface of the specimen after the creep failure, and the cracks are parallel to the axial loading direction of the specimen. It can be determined that the failure mode is pulling damage. When the seepage pressure rises to 2.5 MPa, multiple flaky pieces are peeled off from the surface of the specimen, and both shear stress and tensile stress appear on the failure surface. The creep failure mode of the specimen is the form of pulling damage and shear failure. When the seepage pressure is 3.5 MPa, there is a shear failure plane that penetrates the sample. After the shear failure, many fragments exist on the tensile rupture surface which is parallel to the axial loading direction. The shear failure and tensile failure constitute the failure mode of the specimen together.

According to the macroscopic damage characteristics of the specimens, we can see that the seepage pressure has a significant impact on the creep failure of sandstone. Especially when the seepage pressure reaches 2.5 MPa or more, the seepage pressure would act directly on the fracture plane after a macroscopic shear failure plane appears on the specimen under the action of axial load, resulting in many macroscopic fracture planes appearing in the block below the existing shear fracture plane.

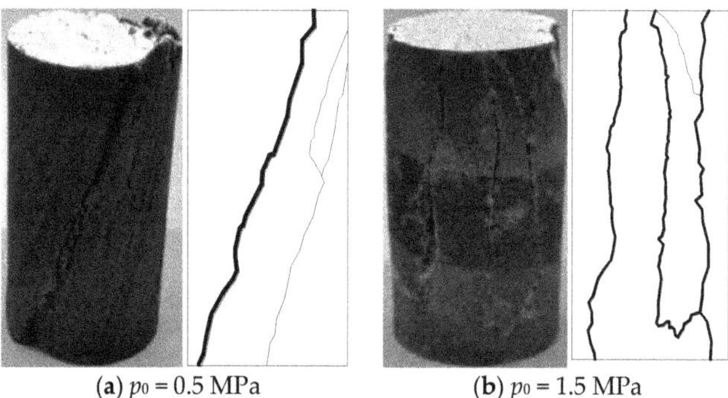

(**a**) p_0 = 0.5 MPa   (**b**) p_0 = 1.5 MPa

Figure 18. *Cont.*

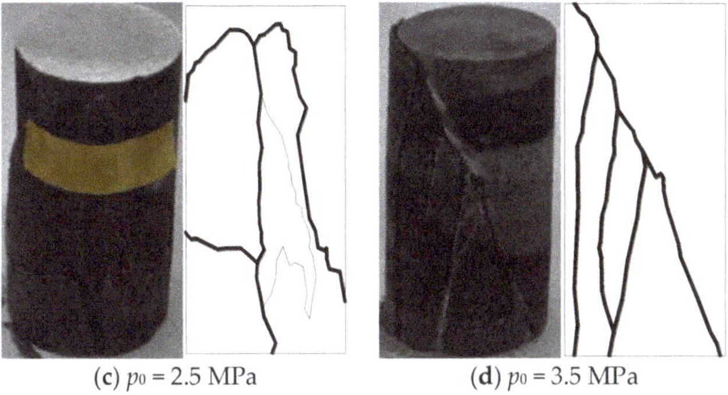

(**c**) p_0 = 2.5 MPa　　　　　　　(**d**) p_0 = 3.5 MPa

Figure 18. Creep failure characteristics of sandstone with different seepage pressures.

3.3.2. Microscopic Characteristics of Sandstone Rupture

Figures 19–22 show the observations of the fracture surface microscopic morphology of sandstones under different seepage pressures. It is obvious that the seepage pressure also has a significant effect on the micro-morphology of the sandstone creep failure.

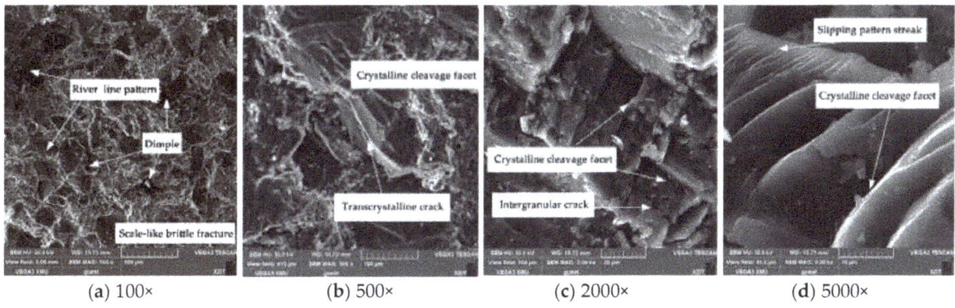

(**a**) 100×　　　　　(**b**) 500×　　　　　(**c**) 2000×　　　　　(**d**) 5000×

Figure 19. Micro-morphology of sandstone fracture in creep under p_0 = 0.5 MPa.

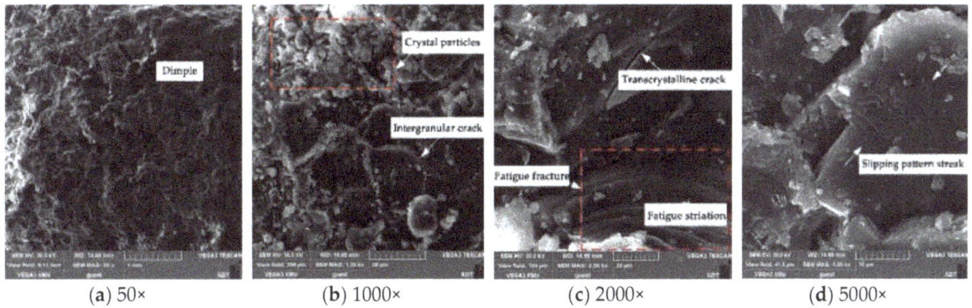

(**a**) 50×　　　　　(**b**) 1000×　　　　　(**c**) 2000×　　　　　(**d**) 5000×

Figure 20. Micro-morphology of sandstone fracture in creep under p_0 = 1.5 MPa.

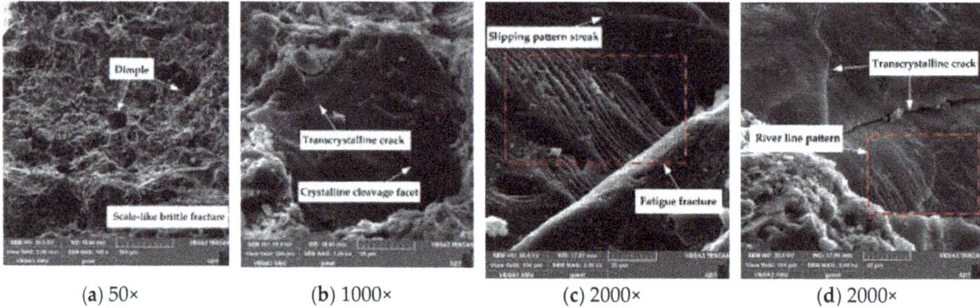

(a) 50×   (b) 1000×   (c) 2000×   (d) 2000×

Figure 21. Micro-morphology of sandstone fracture in creep under p_0 = 2.5 MPa.

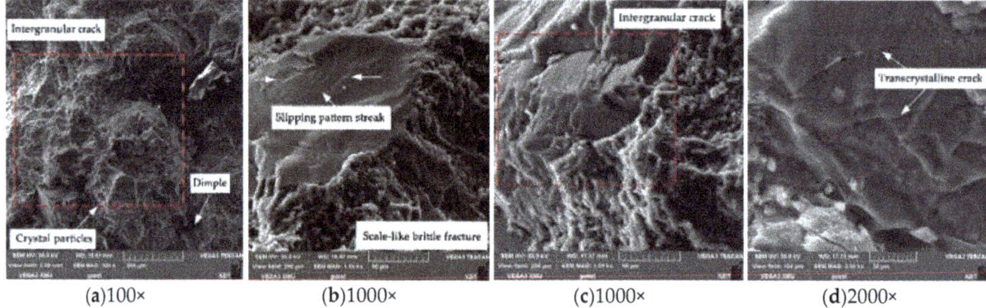

(a)100×   (b)1000×   (c)1000×   (d)2000×

Figure 22. Micro-morphology of sandstone fracture in creep under p_0 = 3.5 MPa.

First, sandstone samples were observed with the SEM at 50/100 times magnification. Under the seepage pressure of 0.5 MPa (Figure 19a), the fracture morphology of sandstone is dominated by river line pattern and scale-like brittle fracture, and there are sporadic dimples at the fracture. The fracture morphology of sandstone is mainly in the form of dimple under the seepage pressure of 1.5 MPa (Figure 10a). When the seepage pressure increases to 2.5 MPa (Figure 21a) and 3.5 MPa (Figure 22a), brittle fractures such as scale-like brittle fracture and intergranular crack are abundantly present in the fractures of sandstone after creep failure. However, there are sporadic dimples locally. With the increase of seepage pressure, the microscopic mode of sandstone creep failure gradually changes from brittle fracture mode (p_0 = 0.5 MPa) to ductile fracture mode (p_0 = 1.5 MPa), and then to brittle fracture mode (p_0 = 2.5, 3.5 MPa).

Next, the magnification of SEM was increased to 500/1000 times. Under the seepage pressure of 0.5 MPa (Figure 19b), there is a significant crystalline cleavage facet in the fracture of sandstone. Many scale-like brittle fractures are distributed around the fracture and small transcrystalline cracks are spread over the fracture surface. When the seepage pressure becomes 1.5 MPa (Figure 20b), several large-scale transcrystalline cracks separate the crystal particles from each other and the brittle fracture characteristics are remarkable. There are some small transcrystalline cracks around the crystalline cleavage facet under the seepage pressure of 2.5 MPa (Figure 21b). When the seepage pressure increases to 3.5 MPa (Figure 22b,c), many scale-like brittle fractures are covered around the section and a small number of slipping patterns exist in the middle region. Generally, the microscopic mode of sandstone creep failure gradually changes from the brittle-ductile coupling fracture mode (p_0 = 0.5 MPa) to the single brittle fracture mode (p_0 = 1.5 MPa), and finally transforms into brittle-ductile coupling fracture mode again (p_0 = 2.5, 3.5 MPa).

In the final step, sandstone samples were studied through the SEM at more than 2000 times magnification. When the seepage pressure is 0.5 MPa (Figure 19c,d), the crystalline cleavage facets cover the surface of the fracture, and there are a set number of slipping pattern streaks in the local area. When the seepage pressure increases to 1.5 MPa (Figure 20c,d), a new microscopic morphology appears on the fracture surface—fatigue fractures. A group of fatigue striations extend parallel to each other, and there are several significant transcrystalline cracks above the fatigue fracture surface. The fracture morphology of the sandstone under the seepage pressure of 2.5 MPa is similar to that under the seepage pressure of 1.5 MPa (Figure 21c,d). When the seepage pressure increases to 3.5 MPa (Figure 22d), the fracture morphology of sandstone is relatively simple. Intergranular cracks become the main morphology of sandstone due to brittle failure, while there are some small transcrystalline cracks in the local area. In summary, as the seepage pressure increases step by step, the evolution of secondary microscopic fracture mode of sandstone creep failure is from brittle fracture (p_0 = 0.5 MPa) to fatigue fracture (p_0 = 1.5, 2.5 MPa), and finally becomes brittle fracture (p_0 = 3.5 MPa). There are also several significant ductile fractures in some areas.

3.3.3. Connection between Macroscopic and Microscopic Mechanism on Sandstone Rupture

Under the influence of exterior load, the macroscopic failure of the rock is determined by the mesoscopic fracture of rock. The initiation, development, and connectivity of internal cracks in rock lead to local fractures. When the microscale damage and fracture accumulate to a certain extent, the macroscopic failure of the rock will occur.

With the increase of seepage pressure, the process of sandstone microscopic fracture mode evolves from low-energy brittle fracture (p_0 = 0.5 MPa) to moderate energy-consuming ductile fracture (p_0 = 1.5 MPa), and finally returns to low-energy brittle fracture (p_0 = 2.5, 3.5 MPa). During the process of macroscopic deformation, the input energy of the sandstone sample is the same. When the energy consumption of microscopic fracture is low, the macroscopic failure mode of sandstone will inevitably lead to shear failure with low energy consumption, and the number of macroscopic fracture surfaces and the fracture degree are higher. On the other hand, when the energy consumption of microscopic fracture is higher, the macroscopic failure mode of sandstone becomes tensile failure, and the fracture degree will be lower. Therefore, after the creep failure of sandstone with seepage pressure p_0 of 0.5, 2.5, and 3.5 MPa, the failure mode of the samples is dominated by shear failure, and the number of macroscopic cracks on the surface of the samples is very large and the scale of the cracks is also larger. When the seepage pressure p_0 is 1.5 MPa, the typical pulling damage dominates the creep failure of the sandstone, and the number of macroscopic cracks on the surface of the samples is relatively less and the crack scale is also smaller.

4. Conclusions

This paper took the complete coal measures sandstone samples as the research object, and designed a rock seepage-creep coupling test system which is suitable for standard rock specimens. The creep tests under natural state and saturation state and the seepage-creep tests under different seepage pressures of coal measures sandstone were carried out. The effects of seepage pressure on the creep properties and permeability characteristics of sandstone were studied. By establishing the relationship between meso-fracture and macroscopic damage of sandstone, the mechanism of sandstone creep failure under seepage pressure is revealed. The main conclusions are as follows:

1. Based on the triaxial compression tests of coal measures sandstone, the creep tests of sandstone under natural and saturation state were carried out through the grading loading tests. The results showed that: The creep characteristic curves of sandstone in two conditions were the same, both including decelerating creep and steady creep at lower loading levels. When the loading level was higher, the creep curve included three complete processes: decelerating creep, isokinetic creep and accelerated creep.

2. The seepage pressure had a significant effect on the creep deformation and permeability characteristics of coal measures sandstone. When the confining pressure was invariant, the constant creep deformation rate and accelerated creep deformation rate of sandstone increased rapidly with the increase of seepage pressure. With the change of time, the change of sandstone

permeability parameters underwent three cycles, and the characteristics of change in each cycle were also divided into two stages: slow change stage and rapid change stage. The change rate gradually increased with the increase of seepage pressure. As the permeation pressure increased, the permeation parameters measured for the first time gradually decreased accordingly.

3. When the confining pressure was invariant, with the increase of seepage pressure, the macroscopic damage of sandstone under natural and saturation conditions had a tendency of transition from tensile failure to shear failure. As the seepage pressure raised, the microscopic fracture mode of sandstone had changed from brittle fracture to ductile fracture and finally to brittle fracture. According to the macroscopic characteristics of sandstone rupture, microscopic morphology and fracture modes, the connection between macroscopic and microscopic mechanism on sandstone rupture was established.

In this article, we mainly used experimental methods to study the seepage-creep characteristics of coal measures sandstone, but the corresponding theoretical research still need to be continued. It is of urgent importance to establish a seepage-creep coupling model of coal measures sandstone based on the test results.

Author Contributions: Z.S., H.P., M.L. and L.C. conceived and designed the experiments; D.L., H.N. and J.L. performed the experiments; Z.S., M.L. and L.C. analyzed the data; H.P., D.L. and H.N. contributed analysis tools; Z.S., H.P., M.L. and L.C. wrote the paper.

Funding: This work was supported by the National Basic Research Program of China (2015 CB251601).

Acknowledgments: The authors are very grateful to the authors of all the references. The authors also express their gratitude to graduate students Xinghui Huo and Senhao Liu for their help in the revision of the paper.

Conflicts of Interest: The authors declare no conflict of interest.

References

1. He, M.; Jing, H.; Sun, X. Research progress of soft rock engineering geomechanics in china coal mine. *J. Eng. Geol.* **2000**, *8*, 46–62.
2. He, M. *Engineering Mechanics of Soft Rock*; Science Press: Beijing, China, 2002; pp. 52–109, ISBN 7-03-010340-8.
3. He, M.; Xie, H.; Peng, S.; Jiang, Y. Study on rock mechanics in deep mining engineering. *Chin. J. Rock. Mech. Eng.* **2005**, *24*, 2803–2813.
4. He, M. Conception system and evaluation indexes for deep engineering. *Chin. J. Rock. Mech. Eng.* **2005**, *24*, 2854–2858.
5. Zhou, H.; Xie, H.; Zuo, J. Research progress of rock mechanics behavior under deep high ground stress. *Adv. Mech.* **2005**, *35*, 91–99.
6. Xie, H.; Peng, S.; He, M. *Basic Theory and Engineering Practice in Deep Mining*; Science Press: Beijing, China, 2006; pp. 12–58, ISBN 7-03-016386-9.
7. Sun, J. Rock rheological mechanics and its advance in engineering applications. *Chin. J. Rock. Mech. Eng.* **2007**, *26*, 1081–1106.
8. He, F. The Research on Mechanism of Rock Creep-Seepage Coupling. Ph.D. Thesis, Liaoning Technical University, Fuxin, China, 2010.
9. Yang, H.; Xu, J.; Nie, W.; Peng, S. Experimental study on creep of rocks under step loading of seepage pressure. *Chin. J. Geotech. Eng.* **2015**, *37*, 1613–1619.
10. Griggs, D.T. Creep of Rocks. *J. Geol.* **1939**, *37*, 225–251. [CrossRef]
11. Wang, G. *Advanced Rock Mechanics Theory*; Metallurgical Industry Press: Beijing, China, 1996.
12. Zhou, W. *Advanced Rock Mechanics*; China Water & Power Press: Beijing, China, 1990.
13. Maranini, E.; Brignoli, M. Creep behavior of a weak rock: Experimental characterization. *Int. J. Rock. Mech. Min.* **1999**, *36*, 127–138. [CrossRef]
14. Fan, Q.; Li, S.; Gao, Y. Experimental study on creep properties of soft rock under triaxial compression. *Chin. J. Rock. Mech. Eng.* **2007**, *7*, 1381–1385.
15. Chen, Y.; Sun, J. Rheological fracture characteristics of rock. *Chin. J. Rock. Mech. Eng.* **1996**, *15*, 20–24. [CrossRef]

16. Li, J. Experimental study on the rheological shear properties of rock. *Chin. J. Geotech. Eng.* **2000**, *3*, 299–303.
17. Zhang, Z.; Xu, W.; Zhao, H.; Jian, B. Investigation on shear creep experiments of sandstone with weak plane in xiangjiaba hydropower station. *Chin. J. Rock. Mech. Eng.* **2010**, *S2*, 3693–3698.
18. Yang, G.; Xie, D.; Zhang, C.; Pu, Y. CT identification of rock damage properties. *Chin. J. Rock. Mech. Eng.* **1996**, *15*, 48–54.
19. Yang, G.; Lu, Z.; Pu, Y. The mechanic characteristics of damage propagation of rock under tri-axial stress condition. *J. Xi'an Univ. Sci. Technol.* **2000**, *20*, 101–104.
20. Ge, X.; Ren, J.; Pu, Y.; Ma, W.; Zhu, Y. A real-in-time CT triaxial testing study of meso-damage evolution law of coal. *Chin. J. Rock. Mech. Eng.* **1999**, *18*, 497–502.
21. Ge, X.; Ren, J. Study on the real intime CT test of the rock meso-damage. *Sci. China* **1999**, *2*, 104–111.
22. Cao, S.; Liu, Y.; Zhang, L.; Jiang, Y. Experimental on acoustic emission of outburst-hazardous coal under uniaxial compression and creep. *J. China Coal. Soc.* **2007**, *32*, 1264–1268.
23. Chen, W.; Zhao, F.; Gong, H. Study of triaxial creep mechanism of mica-quartz schist based on microscopic test. *Chin. J. Rock. Mech. Eng.* **2010**, *29*, 3578–3584.
24. Zhang, Q.; Yang, W.; Chen, F.; Li, W.; Wang, J. Long-term strength and microscopic failure mechanism of hard brittle rocks. *Chin. J. Geotech. Eng.* **2011**, *33*, 1910–1918.
25. Louis, C. *Rock Hydraulics in Rock Mechanics*; Springer Vienna: New York, NY, USA, 1974; pp. 36–108.
26. Jones, F.O. A laboratory study of the effects of confining pressure on fracture flow andstorage capacity in carbonate rocks. *J. Pet. Technol.* **1975**, *27*, 21–27. [CrossRef]
27. Nelson, R.A. Fracture Permeability in Porous Reservoirs: An Experimental and Field Approach. Ph.D. Thesis, Texas A&M University, College Station, TX, USA, 1975.
28. Kranz, R.L.; Frankel, A.D.; Engelder, T. The permeability of whole and jointed Barre granite. *Int. J. Rock. Mech. Min.* **1979**, *16*, 225–234. [CrossRef]
29. Gale, J.E. The effects of fracture type (induced versus natural) on the stress-fracture closure-fracture permeability relationships. In Proceedings of the 23rd U.S Symposium on Rock Mechanics (USRMS), Berkeley, CA, USA, 25–27 August 1982.
30. Iwai, K. Fundamental Studies of Fluid Flow through a Single Fracture. Ph.D. Thesis, University of California, Oakland, CA, USA, 1976.
31. Mordecai, M.; Morris, L.H. An investigation into the changes of permeability occurring in a sandstone when failed under triaxial stress conditions. In Proceedings of the 12th U.S. Symposium on Rock Mechanics (USRMS), Rolla, MI, USA, 16–18 November 1970.
32. Li, S.; Chen, Z.; Miao, X.; Liu, Y. Experimental study on the properties of time-dependent deformation-seepage in water-saturated broken sandstone. *J. Min. Saf. Eng.* **2011**, *28*, 542–547.
33. Miao, X.; Li, S.; Chen, Z.; Liu, W. Experimental study of seepage properties of broken sandstone under different porosities. *Transp. Porous Med.* **2011**, *86*, 805–814. [CrossRef]
34. He, F.; Wang, L.; Wang, Z.; Yao, Z. Experimental study on creep seepage coupling law of coal (rock). *J. China Coal. Soc.* **2011**, *36*, 930–933.
35. Zhu, Z.; Xu, W.; Zhang, A. Mechanism analysis and testing study on Damage and fracture of brittle rock. *Chin. J. Rock. Mech. Eng.* **2003**, *22*, 1411–1416.
36. Zhang, S.; Cox, S.F.; Paterson, M.S. The influence of room temperature deformation on porosity and permeability in calcite aggregates. *J. Geophys. Res.* **1994**, *99*, 15761–15775. [CrossRef]
37. Liu, R.; Li, B.; Jiang, Y. A fractal model based on a new governing equation of fluid flow in fractures for characterizing hydraulic properties of rock fracture networks. *Comput. Geotech.* **2016**, *75*, 57–68. [CrossRef]
38. Li, B.; Liu, R.; Jiang, Y. Influences of hydraulic gradient, surface roughness, intersecting angle, and scale effect on nonlinear flow behavior at single fracture intersections. *J. Hydrol.* **2016**, *538*, 440–453. [CrossRef]
39. Yang, T.; Tang, C.; Zhu, W.; Feng, Q. Coupling analysis of seepage and stresses in rock failure process. *Chin. J. Geotech. Eng.* **2001**, *23*, 489–493.
40. Pu, H.; Cao, L.; Qiu, Y.; Qiu, P. Study of overlying strata creep affected by seepage in backfilling mining process. *J. Min. Saf. Eng.* **2015**, *32*, 846–852.
41. Souley, M.; Homand, F.; Pepa, S.; Hoxha, D. Damage-induced permeability changes in granite: A case example at the URL in Canada. *Int. J. Rock. Mech. Min.* **2001**, *38*, 297–310. [CrossRef]
42. Ma, D. Water-Rock-Sand Mixture Flow Theory of Crushed Rocks and Temporal-Spatial Evolution. Ph.D. Thesis, China University of Mining and Technology, Xuzhou, China, 2017.

43. Yao, B. Research on Variable Mass Fluid-Solid Coupling Dynamic Theory of Broken Rock Mass and Application. Ph.D. Thesis, China University of Mining and Technology, Xuzhou, China, 2012.
44. Wang, L. Accelerated Experimental Study on Permeability for Broken Mudstone with Mass Loss. Ph.D. Thesis, China University of Mining and Technology, Xuzhou, China, 2014.
45. Yan, Y. Research on Rock Creep Tests under Seepage Flow and Variable Parameters Creep Equation. Ph.D. Thesis, Tsinghua University, Beijing, China, 2009.
46. Li, Y.; Ma, Z.; He, Y.; Wang, B.; Liang, X. Experimental Research on Permeability of Rocks of Coal-Bearing Strata. *J. Exp. Mech.* **2006**, *21*, 129–134. [CrossRef]

Article

Effect of Pore Fluid Pressure on the Normal Deformation of a Matched Granite Joint

Qiang Zhang [1,2] , **Xiaochun Li** [1,*], **Bing Bai** [1,*] , **Shaobin Hu** [3] **and Lu Shi** [1]

[1] State Key Laboratory of Geomechanics and Geotechnical Engineering, Institute of Rock and Soil Mechanics, Chinese Academy of Sciences, Wuhan 430071, China; zhangqiang02016@163.com (Q.Z.), 3660899@163.com (L.S.)

[2] University of Chinese Academy of Sciences, Beijing 100049, China

[3] College of Civil and Transportation Engineering, Hohai University, Nanjing 210098, China; hsbcumt@126.com

* Correspondence: xcli@whrsm.ac.cn (X.L.); bai_bing2@126.com (B.B.)

Received: 1 July 2018; Accepted: 24 July 2018; Published: 1 August 2018

Abstract: The influence of pore fluid pressure on the normal deformation behaviors of joints is vital for understanding the interaction between hydraulic and mechanical processes of joints. The effect of pore fluid pressure on the normal deformation of a granite matched joint was investigated by laboratory experiments. Experimental results indicate pore fluid pressure significantly affects the normal deformation of jointed sample, and the relative normal deformation of jointed sample during fluid injection consists of the opening of the joint and the dilation of host rock. The action of pore fluid pressure on the joint follows the Terzaghi's effective stress law. The normal deformation of the joint can be well quantitated by the generalized exponential model. The relative normal deformation of host rock during fluid injection would have a linear relationship with pore fluid pressure, and if affected by gas is more pronounced than water.

Keywords: laboratory experiment; pore pressure; hydro-mechanical coupling; fracture closure; constitutive model; effective stress

---

1. Introduction

The normal deformation of joints or fractures is one of the important mechanical behaviors for rock mass, which significantly affects the transport properties of rock mass [1–4]. The term "joint" is defined as a crack or fracture in rock when there has been little or no transverse displacement [5]. The aperture of joints or fractures depends on the normal stress and on the pore fluid pressure in some engineering [6–10], such as oil or gas exploitation, enhanced geothermal system, and CO_2 geological storage, underground disposal of nuclear waste, and so on. The normal deformation behaviors of joints play an important role between hydraulic and mechanical coupling processes, because the variation of pore fluid pressure or normal stress will change the aperture, and then the aperture of fracture dominates the hydraulic properties [11,12]. The normal deformation behaviors of a joint under normal stress and pore fluid pressure is a basic issue to understand the role of hydro-mechanical (HM) coupling of joints. Many works have been focused on the normal deformation behaviors under normal stress. However, the effect of pore fluid pressure on the normal deformation of a joint had never been systematically investigated in the literature, such as the role of pore pressure, validity of the effective stress and determination of effective stress coefficient. Therefore, it is vital for understanding the effect of pore fluid pressure on the normal deformation behaviors of a joint and developing the accurate HM model to descript this interaction between fluid pressure and deformation of a joint.

The normal deformation of a joint under normal stress has been extensively studied and many normal stress-deformation models have been proposed [2,13–18]. Under normal stress, the normal deformation behavior of a joint is non-linear and the stiffness of a joint increases with the increasing of closure. To describe these behaviors, some classical empirical models were developed by Shehata [19], Goodman [17], Swan [14], Barton and Bandis [13], and Sun [16,20]. These empirical models can be divided into three types: (1) logarithm model [19]; (2) exponential model [14,16,20]; and (3) hyperbolic model [13,17]. With further research on the normal deformation of a joint, Malama and Kulatilake [2] modified the exponential model, Yu et al. [21] and Rong et al. [18] developed new models based on the relation of compliance and deformation to better fit with experimental data in the medium and higher normal stress stage. Desai and Ma [22] proposed the disturbed-state concept (DSC) model based on the disturbed state concept. However, the action of pore pressure has not been considered in these models.

In addition, the HM behaviors of fractured rock has also received wide attention since the 1960s in the fields of geotechnics and geological engineering, due to the fact that joints or fractures in rock masses serve as flow channels, which govern their integrity and fluid transport efficiency [11]. The noted cube law of flow in a single idealized fracture was firstly developed to describe the relation of between flow flux and closure [23]. Thereafter, many researchers conducted many works to modify the cube law of flow in order to eliminate the deviation of between natural roughness fracture and idealized smooth surface fracture [24,25]. Thanks to the roughness of joints, there is a deviation between hydraulic aperture and physics aperture, the relation of both have widely been studied to improve the accuracy of this model.

However, most models tend only to quantify the variation of aperture as a function of normal stress, without considering the influence of pore fluid pressure. Despite the importance of pore fluid pressure, even fewer models considered the influence of pore fluid pressure on the normal deformation of a joint [6,8,10,26]. Some hydro-mechanical coupling models (H-M models) were based on the Terzaghi's effective stress [9,27–30] or the Biot's pore elasticity theory [8], and assuming that the normal deformation path of a joint during pore fluid injection is overlapping with the normal stress loading processes. Except for several field experiments, the validation of these models is rarely verified by experiments [10,30]. The Biot's coefficient α is an important coefficient introduced by Biot into the Biot's pore elasticity theory and is known as the effective stress coefficient [5,31]. The effective stress coefficient is related to the processes [32,33]. The effective stress coefficient for the normal deformation of a joint is unclear. Therefore, experimental study on the effect of pore fluid pressure on the normal deformation of a joint is essential.

In order to reveal the action of pore fluid pressure on the normal deformation of a joint, the normal deformation response of a granite matched joint under pore fluid pressure was investigated by the hydrostatic compression tests and fluid injection tests. The matched joint is when the contact surface fits perfectly and it differs from the unmatched joints. The action of pore fluid pressure on the normal deformation was observed in the fluid injection tests to verify the effective stress law. Furthermore, the present constitutive model was modified to quantify the action of pore pressure on the normal deformation of a joint.

2. Experimental Apparatus and Principle

In order to determine the role of pore fluid pressure on the normal deformation, two types of experiments were performed on granite sample, including (1) hydrostatic compression test of the intact and jointed sample, and (2) fluid injection tests of jointed sample. This study was focused on the deeper formation, which subjected to higher pore fluid pressure. The applied normal stress and pore pressure was up to 40 MPa and 35 MPa, respectively.

2.1. Material

For the sake of comparing with previous models and avoiding the effect of water on some mechanical sensitive minerals, granite was used in this study. The granite was quarried from Hubei province, China, then cut into cuboids with a height of 100 mm, a length of 50 mm and a width of 50 mm, and polished to achieve a flatness of 0.01 mm (Figure 1a). The composition of granite and the key mechanical parameters were compiled in Tables 1 and 2, respectively. After the hydrostatic compression test of intact rock, the intact sample was split into two halves perpendicular to its axial using Brazilian test to simulate a natural joint, as shown in Figure 1b.

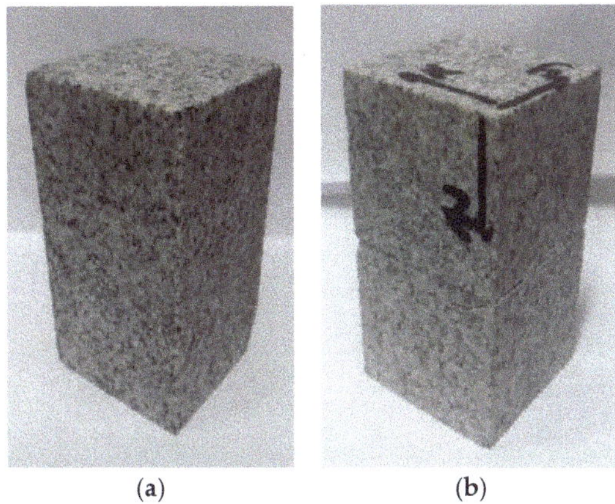

(a)　　　　　　**(b)**

Figure 1. The granite samples. (**a**) in intact state, (**b**) in split state.

Table 1. The composition of granite.

Minerals	Quarts	Albite	Microcline	Muscovite	Orthoclase
wt %	27.51	30.83	20.60	13.69	7.37

Table 2. Mechanical parameter of intact granite.

Size/mm	Uniaxial Strength/MPa	Tensile Strength/MPa	Elasticity Modulus/GPa	Poisson's Ratio	Lame Elastic Constant μ/GPa	Shear Modulus λ/GPa
$50 \times 50 \times 100$	172.5	4.59	57.29	0.226	23.36	19.27

2.2. Test Apparatus and Principle

All tests were conducted on the true tri-axial mechanical testing system of rocks, which could couple with high pore pressure [34,35]. This system was composed of a confining pressure cell, two loading frames and a pore fluid injection unit. The maximum confining pressure is 100 MPa. A syringe pump (Teledyne Technologies Incorporated, Lincoln, NE, USA) was connected with sample to provide pore fluid pressure as shown in Figure 2. The maximum of pore pressure is 68.9 MPa.

The sample was sandwiched by two pairs of steel plates and then sealed by polyurethane as shown in Figure 2b,c. Note that two copper sheets were pasted on the freed surface before being applied to polyurethane in order to avoid the polyurethane invading into the joint. Two linear variable differential transformers (LVDT) (Macro Sensors, Pennsauken, NJ, USA) were fixed on the axial steel plate to monitor the axial deformation of sample as shown in Figure 2a,b. The normal deformation of the joint was calculated based on the normal deformation of the sample in intact state and in split state under hydrostatic compression test.

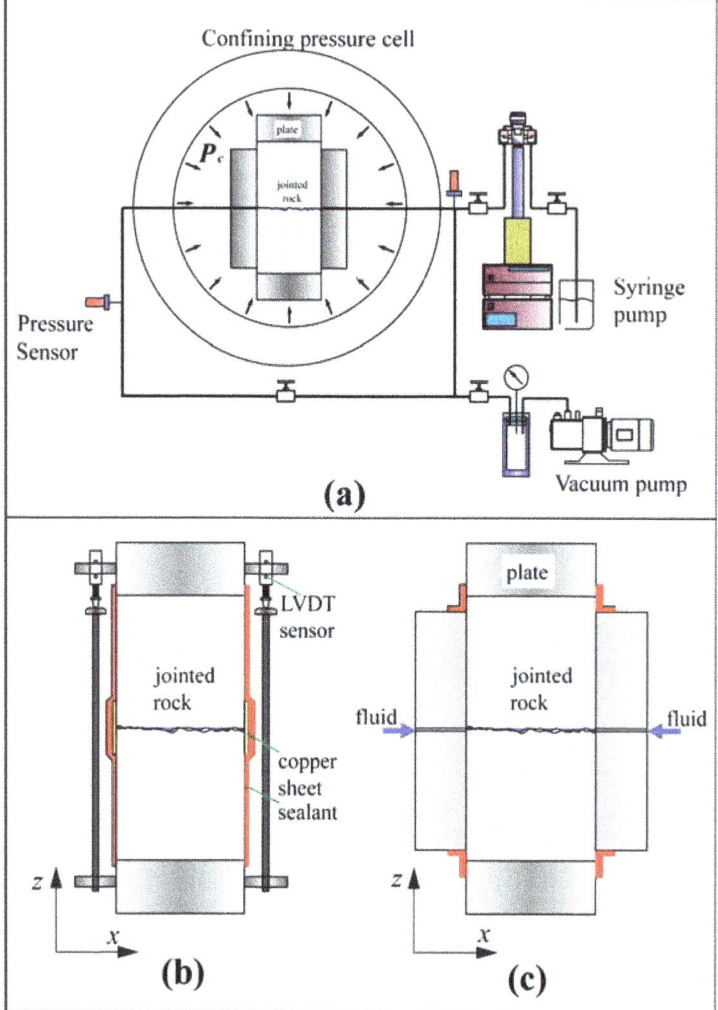

Figure 2. Schematic diagram of hydrostatic compression and injection test of joint. (**a**) Diagram of test system; (**b**) Left view of tested sample; (**c**) Front view of tested sample.

In this work, hydrostatic compression tests and fluid injection tests were performed on the same sample in intact state and in split sate. The actual axial deformation was taken as the average of two LVDTs. For the deformation of sample, compression is positive, dilation is negative. The normal deformation of intact sample (δ_i) is defined as the axial deformation of sample in intact state under hydrostatic compression tests. The total normal deformation of jointed sample (δ_j) is defined as the axial deformation of sample in split state under hydrostatic compression tests. The normal deformation of the joint (δ_n) is defined as the closure of the joint, which is calculated by the normal deformation of jointed sample minus that of intact sample:

$$\delta_n = \delta_j - \delta_i \tag{1}$$

The relative normal deformation of jointed sample ($\Delta\delta_j$) is defined as the relative axial deformation of jointed sample during fluid injection tests. Terzaghi's effective stress was first used to describe the influence of pore fluid pressure:

$$\sigma_n' = \sigma_n - P_p \tag{2}$$

where σ_n' is the effective normal stress, σ_n is the normal stress, P_p is the pore fluid pressure.

2.3. Test Procedure

Hydrostatic compression test was first performed on the sample in intact state and in split state, then fluid injection test was performed on the sample in split state, including the balanced pressure injection of N_2, the balanced pressure injection of water, and the constant pressure rate injection of water. The detailed procedure was as follows:

- Hydrostatic compression test

The intact sample was sealed by polyurethane and was curing for 48 h at room temperature. Then LVDTs were fixed on the axial plates as in Figure 2b. The prepared sample was installed into the confining pressure cell and the confining pressure was exerted on the sample at the rate of 0.01 MPa/s. When the confining pressure arrive the desired value, the confining pressure unloaded to 0 MPa at the same rate.

After that, the intact sample was split using Brazilian test to prefabricate the joint. The jointed sample was prepared and tested as the intact sample to gain the normal deformation of jointed sample under hydrostatic pressure.

- The balanced pressure injection of N_2

After the hydrostatic compression test of the jointed sample, the confining pressure was exerted on the jointed sample at the same rate until the desired value and remains at this pressure. N_2 was injected into sample from one port to anther port at a constant injection pressure. Until the differential pressure between two ports was eliminated, the variation of axial deformation was recorded and the injection pressure was raised to next level of pressure. When the maximum pore pressure arrived, the pore pressure was decreased to every level of pressure step by step in turn. The differential pressure between two ports should be eliminated at every level of pressure. In the end, the confining pressure was unloaded.

- The balanced pressure injection of water

After the balanced pressure injection of N_2, the jointed sample had a rest of 24 h. The procedures of the balanced pressure injection of water as the balanced pressure injection of N_2.

- The constant pressure rate injection of water

After the balanced pressure injection of water, the jointed sample also had a rest of 24 h. The confining pressure was exerted on the jointed sample at the same rate until the desired value and kept a constant pressure. Water was simultaneously injected into sample from two ports at the constant pressure rate of 0.01 MPa/s. During the injection, the normal deformation of sample was measured by axial LVDTs.

3. Test Results and Analysis

3.1. Hydrostatic Compression Test

Figure 3 shows the curves of normal stress-normal deformation of sample, in intact state and in split sate, and the joint. The matched joint exhibits the non-linear response under hydrostatic pressure condition. With increasing of hydrostatic pressure, the normal deformation of the joint gradually slows down and reaches maximum deformation at high hydrostatic pressure. These results are consistent with the normal deformation of a joint under uniaxial compression [13,20]. The lateral stress has no effect on the normal deformation.

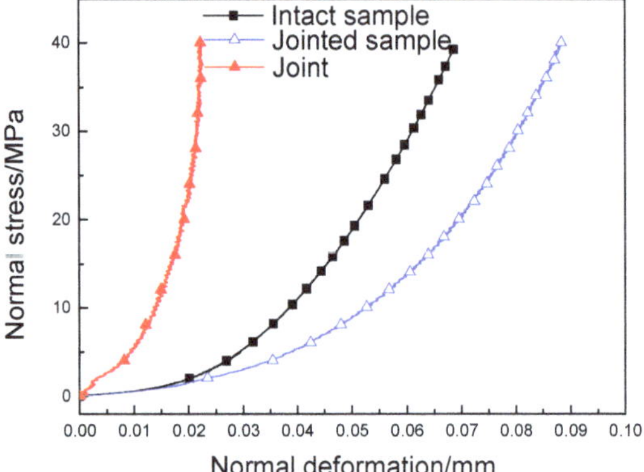

Figure 3. The curve of normal stress-normal deformation of sample in intact state and in split sate, and joint. The normal deformation of the joint δ_n was calculated by the normal deformation of jointed sample minus that of intact sample.

Figure 4 shows the curves of normal stress-normal deformation of intact and jointed sample under stress-loading and -unloading condition. Figure 5 shows the curves of normal stress-normal deformation of jointed sample under multicycle stress-loading and -unloading condition. In Figure 4, the intact and jointed sample exhibit a hysteresis loop between load- and unload-path. For the intact and jointed sample, the irreversible deformation is observed after unloading, and that of jointed sample significantly greater than the intact sample. However, after the first cycle of loading and unloading, the following cycle curves overlap as shown Figure 5. These indicate that the jointed sample exhibits non-linear elastic response of normal deformation under hydrostatic pressure as in the experimental result of Cook [36].

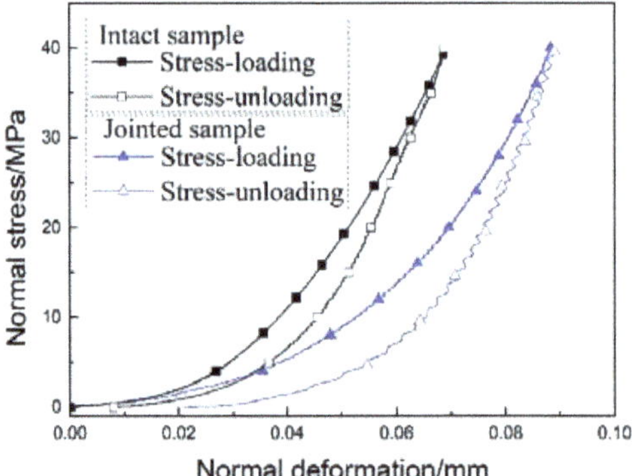

Figure 4. The curve of normal stress-normal deformation of intact and jointed sample under stress-loading and -unloading.

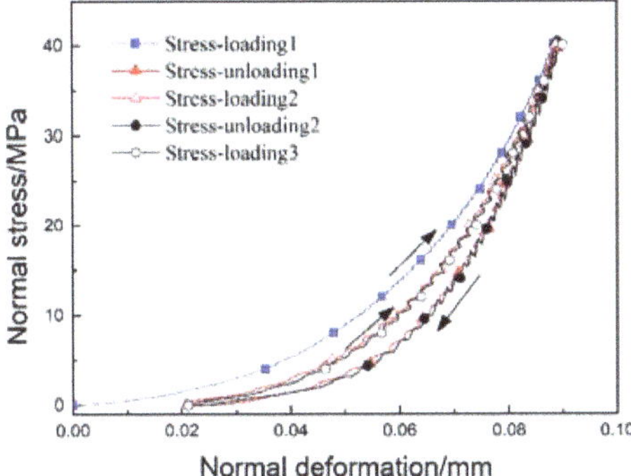

Figure 5. The curve of normal stress-normal deformation of jointed sample multiple under multicycle stress-loading and -unloading condition.

3.2. N_2 Injection Test of Jointed Sample

In the N_2 injection test, the pressure balanced injection was adopted and the hydrostatic pressure is constant at 40 MPa. Figure 6 shows the curve of the relative normal deformation of jointed sample ($\Delta \delta_j$) with effective normal stress/pore pressure. With decreasing of effective normal stress (i.e., increasing of pore N_2 pressure), the jointed sample continuously dilate and the normal deformation lags behind the curve of hydrostatic compression. To facilitate comparison between $\Delta \delta_j$ and the normal deformation of the joint and to find their differences in variation trend, the relative deformation of jointed sample was translated. It exceeds the normal deformation of the joint under hydrostatic compression. However, during N_2 releasing, the jointed sample continuing to compress with the decreasing of pore pressure and the variation trend is similar to that of the joint under hydrostatic compression. When the pressure of N_2 completely dissipates, a remarkable amount of dilation is observed. These indicate that pore N_2 pressure significantly affects the deformation of the joint and the host rock surrounding the joint.

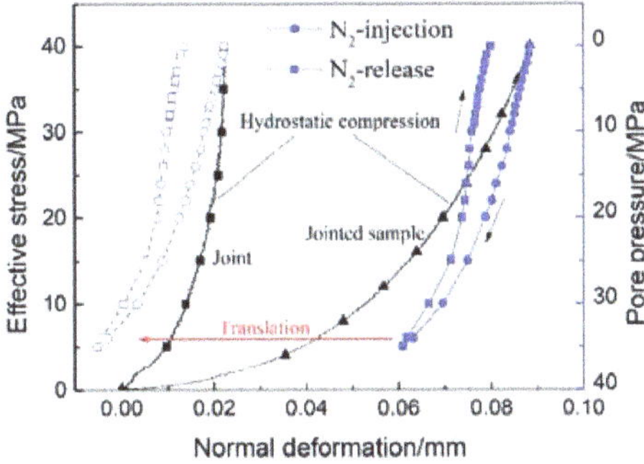

Figure 6. The relative normal deformation of jointed sample during N_2 injection and release.

3.3. Water Injection Test of Jointed Sample

In the water injection test, the pressure balanced injection and the constant pressure rate injection were all adopted and the hydrostatic pressure is constant at 40 MPa. Figures 7 and 8 show the results of the pressure balanced injection and the constant pressure rate injection, respectively.

For the pressure balanced injection (as shown in Figure 7), with decreasing of effective normal stress, the jointed sample also continuously dilates and the normal deformation lags behind the curve of hydrostatic compression. If the relative deformation of jointed sample was translated and compared with the joint deformation under hydrostatic compression, it also exceeds the normal deformation of the joint. At the early stage of the injection, the normal deformation of jointed sample basically along the curve of hydrostatic compression, gradually deviates from the curve with the increasing of pore water pressure. During pressure reduction, the relative normal deformation of jointed sample linear increase to the initial value. When the pressure of water completely dissipates, no remarkably residual dilation of jointed sample is observed. These indicate pore water pressure also significantly affects the deformation of the joint. The effect of pore water pressure on the host rock surrounding the joint is weaker than N_2.

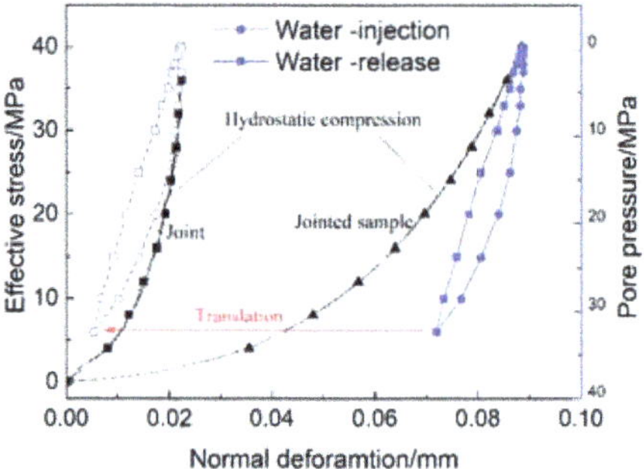

Figure 7. The relative normal deformation of jointed sample during the water injected in pressure balanced injection.

For the constant pressure rate injection (as shown in Figure 8), the pore water pressure did not significantly affect the relative normal deformation of jointed sample with pore water pressure less than 20 MPa. With increasing of pore water pressure, the relative normal deformation of jointed sample affected by pore water pressure was significantly enhanced. The relative normal deformation of jointed sample during injection is lags behind the normal deformation of jointed sample and the joint under hydrostatic compression. At high pore pressure, the relative normal deformation of jointed sample rapidly decreases with the pore pressure, and it coincides with the curves of the joint under hydrostatic compression when the pore pressure more than 35 MPa. This indicates that water has little effect on the host rock surrounding the joint, and the effect of pore water pressure on the joint is remarkable. Meanwhile, as the permeability of granite is lower than that with a fracture/joint [37], the host rock was viewed as impermeable within 1 h for the constant pressure rate injection. For the constant pressure rate injection, pore pressure mainly affects the joint.

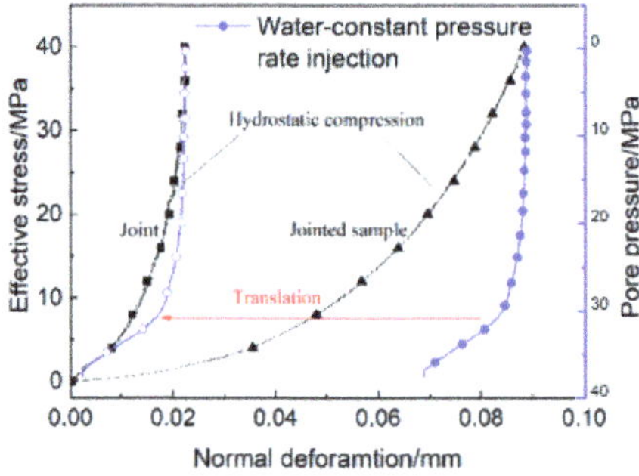

Figure 8. The relative normal deformation of jointed sample during the water injected in constant pressure rate injection.

The relative deformation of jointed sample at the initial stage of the pressure balanced injection and at the last stage of the constant pressure rate injection coincide with the normal deformation curve of the joint under hydrostatic compression. This indicates that the effect of pore water pressure on the joint follows the Terzaghi's effective stress and is consistent with the curve of the joint under hydrostatic compression. During fluid balanced injection, pore fluid pressure affects the deformation of host rock.

4. Discussion

The above experimental results indicate that the pore fluid pressure significantly affects the normal deformation of jointed sample and the joint. The influence of pore fluid pressure on the joint and host rock were analyzed and quantified in this section.

4.1. Impacts of Pore Fluid Pressure on Jointed Sample

When fluid was injected into the jointed sample, the normal deformation of the jointed sample was significantly influenced by pore pressure, as show in Figures 6–8. In the water constant pressure rate injection, the relative normal deformation of jointed sample exhibits a threshold pressure for opening of the joint as the result of in-situ experiment on fracture [26]. In this study, the threshold pressure results from the low permeability of the matched joint under high normal stress, which limit liquid water flow into the space between two joint planes in a short time. When the injection pressure exceeds the threshold pressure, water flows into the space between two joint planes. Pore pressure builds up rapidly, and the joint rapidly opens. Due to the limitation of injection time, the effect of pore pressure on host rock surrounding the joint is negligible.

The effective stress is a conceptual average stress and is defined as the combinations of total stress and pore pressure which produce effects [38]. The generalization effective stress, σ'_{ij} is expressed as:

$$\sigma'_{ij} = \sigma_{ij} - \alpha P_p \delta_{ij} \tag{3}$$

where, σ_{ij} is the total stress tensor, α is the effective stress coefficient, $0 \leq \alpha \leq 1$. The effective stress coefficient α can be considered the ratio of the area of applied pressure to the total area base on the definition of effective stress law. In this case, the joint is a throughout fracture, so fluid could exist in the total area of joint. Thus, it is reasonable that α takes 1. The generalization effective stress can be expressed as Terzaghi's effective stress:

$$\sigma'_{ij} = \sigma_{ij} - P_p \delta_{ij} \tag{4}$$

This deduction is in agreement with the experimental results that the action of pore fluid pressure on the joint follows the Terzaghi's effective stress law.

For the balanced pressure injection, the duration is longer than that of the constant pressure rate injection, the relative normal deformation of jointed sample being obviously larger than that of the joint under hydrostatic compression, as show in Figures 7 and 8. This indicates that pore fluid pressure influences the deformation of host rock, and this influence cannot be ignored. The influence of pore N_2 pressure on host rock is more significant than water as the lower viscosity coefficient of N_2. Therefore, the relative deformation of jointed sample ($\Delta\delta_j$) during fluid injection consists of the opening of joint ($\Delta\delta_n$) and the dilation of host rock ($\Delta\delta_r$) as shown in Figure 9.

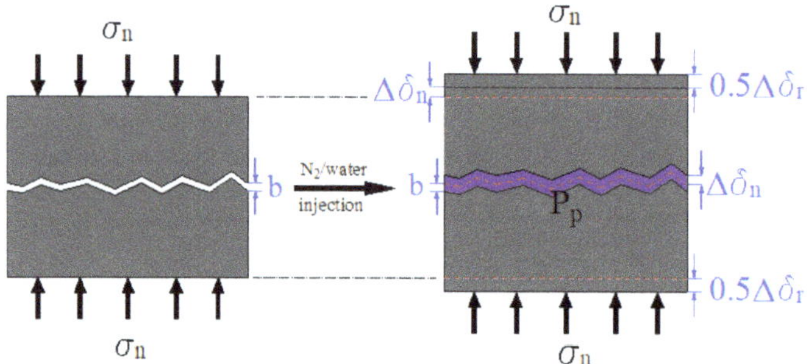

Figure 9. Schematic diagram of the normal deformation of jointed sample affected by pore pressure.

4.2. Quantification the Normal Deformation of the Joint Under Pore Pressure Condition

The normal deformation of the joint under normal stress can be quantified by several mathematical models, such as the hyperbolic model developed by Barton and Bandis [13] (hereinafter, BB model), the generalized exponential model [2] (hereinafter, Malama model), and so on. The BB model and the Malama model are characteristic of simplicity and clarity in parameters. Thus, these two models are used to compare to the experimental data of the joint under hydrostatic compression. According to Barton and Bandis et al. (1985), the BB model is given by Equation (5),

$$\delta_n = \frac{\sigma_n V_m}{\sigma_n + K_{ni} V_m} \tag{5}$$

where δ_n is the normal deformation, V_m is the maximum closure of the joint, K_{ni} is the initial stiffness, σ_n is the normal stress. According to Malama and Kulatilake (2003), the Malama model is given by Equation (6),

$$\delta_n = V_m [1 - \exp(-\ln 2 \left(\frac{\sigma_n}{\sigma_{1/2}}\right)^n)] \tag{6}$$

where δ_n is the normal deformation, V_m is the maximum closure of the joint, $\sigma_{1/2}$ is the half-closure stress (the normal stress when the normal deformation arrive half-maximum closure), n is the parameter, $0 < n \leq 1$. Based on our experimental data, the key parameters of the joint are compiled in Table 3. The curves of stress-deformation based on the BB model, Malama model and experimental data are presented in Figure 10. The error sum of squares in normal deformation between the BB model/Malama model and experimental result is 0.025/0.00042. The Malama model is better in describing the normal deformation characteristics of the joint than the BB model. Thus, the Malama model was adopted to analyze the normal deformation of the joint in the injection test. Section 4.1 indicates the normal deformation of the joint following Terzaghi's effective stress law at the action of pore pressure. Because the jointed sample exhibits non-linear elastic response under normal stress, it is reasonable to assume that the normal deformation path of jointed sample is the same when stress-loading and -unloading. According to the Terzaghi's effective stress, an improved Malama model can be modified:

$$\begin{cases} \sigma_n' = \sigma_n - P_p \\ \delta_n = V_m [1 - \exp(-\ln 2 \left(\frac{\sigma_n'}{\sigma_{1/2}}\right)^n)] \end{cases} \tag{7}$$

Because the relatively normal deformation of the jointed sample during fluid injection consists of the opening of joint plane and the dilation of host rock. The normal deformation of host rock under the action of pore pressure can be calculated by Equation (8):

$$\begin{cases} \sigma_n' = \sigma_n - P_p \\ \delta_n = V_m\left[1 - \exp\left(-\ln 2\left(\frac{\sigma_n'}{\sigma_{1/2}}\right)^n\right)\right] \\ \Delta\delta_r = \Delta\delta_j - \delta_n \end{cases} \tag{8}$$

where, $\Delta\delta_r$ is the relative normal deformation of host rock, $\Delta\delta_j$ is the relative normal of jointed sample and measured in the injection test. According to Equation (8), the normal deformation of host rock in N_2 and the water injection test were calculated and shown in Figures 11 and 12, respectively.

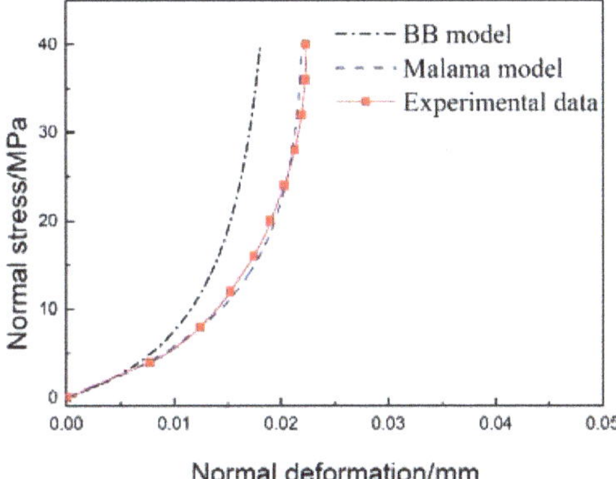

Figure 10. Comparison between experimental data of the joint and previous models for normal deformation versus normal stress.

Table 3. The parameters of the joint derived from experimental data.

Maximum Normal Stress σ_{max}/MPa	Half-Closure Stress $\sigma_{1/2}$/MPa	Maximum Closure δ_{max}/mm	Initial Normal Stiffness K_{ni}/MPa·mm^{-1}	Parameter n
40	6.76	0.0223	516.31	1

Figure 11. Decomposition of normal deformation of jointed sample induced by pore N_2 pressure: (**a**) component of host rock; (**b**) total normal deformation of jointed sample; and (**c**) component of the joint.

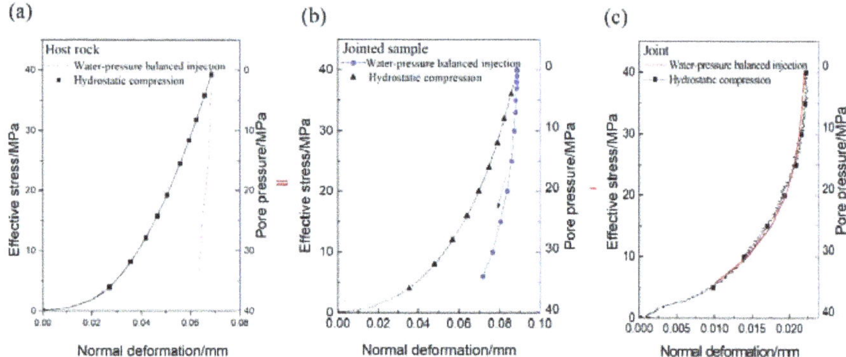

Figure 12. Decomposition of normal deformation of jointed sample induced by pore water pressure: (**a**) component of host rock; (**b**) total deformation of jointed sample; and (**c**) component of the joint.

As shown in Figures 11 and 12, the normal deformation of host rock has a linear relationship with pore pressure. When the confining pressure maintains at 40 MPa, host rock dilates with the increasing of pore pressure (decrease of effective stress). The effect of N_2 on host rock is obviously larger than water. However, the total amount of host rock deformation during N_2/water injection is far less than that during hydrostatic compression. The differences in the effect on host rock between N_2 and water may be a result of the difference in viscosity, where N_2 has a lower viscosity than water, it is easy to penetrate the host rock and build-up the pore pressure inside host rock. Host rock may be unsaturated when the pore pressure of the joint achieves a balance. The pore N_2 pressure is higher than the pore water pressure and it affects the deformation of the host rock. Thus, the dilation of host rock under N_2 injection is obviously larger than water.

4.3. Quantification of Host Rock Deformation under Pore Pressure Condition

Although granite has a low porosity, it is a porous material and its mechanical behavior is thus significantly affected by pore pressure [39]. The deformation behavior of rock under pore fluid pressure could described by the pore elastic theory [40,41]:

$$\sigma_{ij} = 2\mu\varepsilon_{ij} + \lambda\varepsilon_{ii}\delta_{ij} - \alpha P_p\delta_{ij} \tag{9}$$

where, μ is Lame elastic constant, λ is shear modulus, σ_{ij} is the stress tensor, ε_{ii} is the strain tensor, and α is the effective stress coefficient. α of host rock could derived by Equation (9):

$$\alpha = \frac{\sigma_{ij} - 2\mu\varepsilon_{ij} - \lambda\varepsilon_{ii}\delta_{ij}}{P_p\delta_{ij}} \tag{10}$$

In this test, the sample was subjected to a constant hydrostatic pressure and pore pressure, $\sigma_{ij} = 0, (i \neq j), \varepsilon_{ij} = 0, (i \neq j)$. Considering granite is homogeneous, $\varepsilon_{11} = \varepsilon_{22} = \varepsilon_{33}, \varepsilon_{ii} = 3\varepsilon_{11}$. The estimated value α of host rock could be expressed as:

$$\alpha = \frac{-(2\mu + 3\lambda)d\varepsilon_{11}}{dP_p} = \frac{-(2\mu + 3\lambda)d\delta_r}{ldP_p} \tag{11}$$

where, δ_r is the normal deformation of host rock during fluid injection, l is the height of intact sample, $l = 100$ mm. μ and λ are basic mechanical parameters and compiled in Table 2. When the estimated value α is constant, the normal deformation of host rock is proportional to the pore fluid pressure. Thus, the normal deformation of host rock has a linear relationship with pore pressure. According to the calculated normal deformation of host rock during N_2 and water pressure balanced injection, the estimated value α is 0.412, 0.168 respectively. From the range of α, these estimated values of α are reasonable for the host rock. Noting that whether the rock is saturated during reinjection is currently uncertain, the actual α of host rock needs further study. This indicates the improved Malama model is reasonable and suitable to describe the normal deformation behavior of the joint under pore fluid pressure condition. In addition, in the process of HM, the effect of pore fluid pressure on host rock is non-negligible, and the degree of its influence is related to types of fluid.

5. Conclusions

The effect of pore fluid pressure on normal deformation behavior of a granite matched joint was investigated by laboratory experiments. Several main conclusions can be drawn; they are as follows:

The matched granite joint has a characteristic of non-linear elastic response under hydrostatic compression. The Malama model is closer to the curve of normal deformation versus normal stress than the BB model. The error sum of squares in normal deformation between the BB model/Malama model and the experimental result is 0.025/0.00042.

In water constant pressure rate injection, experimental results indicate that the action of pore fluid pressure on normal deformation of the joint follows the Terzaghi's effective stress law. Due to low permeability of the joint under high normal stress, a threshold pore pressure is observed in the opening of the joint. When the duration of injection is enlarged, the influence of pore pressure on deformation of host rock is observed.

The effect of pore fluid pressure on the normal deformation of the jointed sample consists of two parts: the opening of the joint ($\Delta\delta_j$) and the dilation of host rock ($\Delta\delta_r$). The relative normal deformation of the joint could be quantitated by the Malama model and the Terzaghi's effective stress law, and the normal deformation of host rock can be quantified by pore elastic theory.

The relative normal deformation of host rock linearly increased with the pore fluid pressure and the effect of N_2 on host rock is obviously larger than water. In the process of hydro-mechanical coupling, the effect of pore fluid pressure on host rock is non-negligible, and the degree of its influence is related to types of fluid.

Author Contributions: Q.Z. and X.L. conceived and designed the experiments; Q.Z. performed the experiments; Q.Z. and B.B. analyzed the data; S.H., X.L. and L.S. contributed reagents/materials/analysis tools; Q.Z. wrote the paper.

Funding: This research is supported by the National Key Research and Development Program (No. 2016YFB0600805, 2018YFB0605601), the National Natural Science Foundation of China (No. 41672252), and the General Program of National Natural Science Foundation of China (No. 41472236).

Conflicts of Interest: The authors declare no conflict of interest.

References

1. Xia, C.C.; Yue, Z.Q.; Tham, L.G.; Lee, C.F.; Sun, Z.Q. Quantifying topography and closure deformation of rock joints. *Int. J. Rock Mech. Min. Sci.* **2003**, *40*, 197–220. [CrossRef]
2. Malama, B.; Kulatilake, P.H.S.W. Models for normal fracture deformation under compressive loading. *Int. J. Rock Mech. Min. Sci.* **2003**, *40*, 893–901. [CrossRef]
3. Pyrack-Nolte, L. *Hydraulic and Mechanical Properties of Natural Fractures in low-Permeability Rock*; International Society for Rock Mechanics and Rock Engineering: Montreal, QC, Canada, 1987.
4. Desai, C.S.; Rigby, D.B. Cyclic Interface and Joint Shear Device Including Pore Pressure Effects. *J. Geotech. Geoenviron. Eng.* **1997**, *123*, 568–579. [CrossRef]
5. Jaeger, J.C.; Cook, N.G.W.; Zimmerman, R. *Fundamentals of Rock Mechanics*, 4th ed.; John Wiley & Sons: Hoboken, NJ, USA, 2007.
6. Cappa, F.; Guglielmi, Y.; Rutqvist, J.; Tsang, C.-F.; Thoraval, A. Hydromechanical modelling of pulse tests that measure fluid pressure and fracture normal displacement at the Coaraze Laboratory site, France. *Int. J. Rock Mech. Min. Sci.* **2006**, *43*, 1062–1082. [CrossRef]

7. Rutqvist, J.; Stephansson, O. The role of hydromechanical coupling in fractured rock engineering. *Hydrogeol. J.* **2003**, *11*, 7–40. [CrossRef]

8. Bart, M.; Shao, J.F.; Lydzba, D.; Haji-Soutoudeh, M. Coupled hydromechanical modeling of rock fractures under normal stress. *Can. Geotech. J.* **2004**, *41*, 686–697. [CrossRef]

9. Yang, H.; Xie, S.Y.; Secq, J.; Shao, J.F. Experimental study and modeling of hydromechanical behavior of concrete fracture. *Water Sci. Eng.* **2017**, *10*, 97–106. [CrossRef]

10. Ji, S.H.; Koh, Y.K.; Kuhlman, K.L.; Lee, M.Y.; Choi, J.W. Influence of pressure change during hydraulic tests on fracture aperture. *Groundwater* **2013**, *51*, 298–304. [CrossRef] [PubMed]

11. Kamali-Asl, A.; Ghazanfari, E.; Perdrial, N.; Bredice, N. Experimental study of fracture response in granite specimens subjected to hydrothermal conditions relevant for enhanced geothermal systems. *Geothermics* **2018**, *72*, 205–224. [CrossRef]

12. Vogler, D.; Amann, F.; Elsworth, D. Permeability Evolution in Natural Fractures Subject to Cyclic Loading and Gouge Formation. *Rock Mech. Rock Eng.* **2016**, *49*, 3463–3479. [CrossRef]

13. Barton, N.; Bandis, S.; Bakhtar, K. Strength, deformation and conductivity coupling of rock joints. *Int. J. Rock Mech. Min. Sci. Geomech. Abstr.* **1985**, *22*, 121–140. [CrossRef]

14. Swan, G. Determination of stiffness and other joint properties from roughness measurements. *Rock Mech. Rock Eng.* **1983**, *16*, 19–38. [CrossRef]

15. Sun, Z. Study on the deformation vs stress properties of discontinuous. *Chin. J. Rock Mech. Eng.* **1987**, *4*, 287–300.

16. Sun, G.; Lin, W. The principle of closed deformation of structural plane and elastic deformation constitutive equation of rock mass. *Chin. J. Rock Mech. Eng.* **1982**, *2*, 177–180.

17. Goodman, R.E. *Methods of Geological Engineering in Discontinuous Rocks*; West Publishing Company: Eagan, MN, USA, 1976.

18. Rong, G.; Huang, K.; Zhou, C.B.; Wang, X.J.; Peng, J. A new constitutive law for the nonlinear normal deformation of rock joints under normal load. *Sci. Chin. Technol. Sci.* **2011**, *55*, 555–567. [CrossRef]

19. Sharp, J.C.; Maini, Y. Fundamental considerations on the hydraulic characteristics of joints in rock. In Proceedings of the Symposium of Percolation through Fissured Rock, Stuttgart, Germany, 18–19 September 1972.

20. Sun, Z.; Gerrard, C.; Stephansson, O. Rock joint compliance tests for compression and shear loads. *Int. J. Rock Mech. Min. Sci. Geomech. Abstr.* **1985**, *22*, 197–213. [CrossRef]

21. Yu, J.; Zhao, X.; Zhao, W.; Li, X.; Guan, Y. Improved nonlinear elastic constitutive model for normal deformation of rock fractures. *Chin. J. Geotech. Eng.* **2008**, *30*, 1316–1321.

22. Desai, C.S.; Ma, Y. Modelling of joints and interfaces using the disturbed-state concept. *Int. J. Numer. Anal. Method. Geomech.* **1992**, *16*, 623–653. [CrossRef]

23. Snow, D.T. Anisotropie Permeability of Fractured Media. *Water Resour. Res.* **1969**, *5*, 1273–1289. [CrossRef]

24. Witherspoon, P.A.; Wang, J.S.Y.; Iwai, K.; Gale, J.E. Validity of Cubic Law for fluid flow in a deformable rock fracture. *Water Resour. Res.* **1980**, *16*, 1016–1024. [CrossRef]

25. Amadei, B.; Illangasekare, T. A mathematical model for flow and solute transport in non-homogeneous rock fractures. *Int. J. Rock Mech. Min. Sci. Geomech. Abstr.* **1994**, *31*, 719–731. [CrossRef]

26. Cornet, F.H.; Li, L.; Hulin, J.-P.; Ippolito, I.; Kurowski, P. The hydromechanical behaviour of a fracture: An in situ experimental case study. *Int. J. Rock Mech. Min. Sci.* **2003**, *40*, 1257–1270. [CrossRef]

27. Duan, X.; Li, J.; Liu, J. Numerical simulation of flow motion of fractured rock mass under the interaction of stress field and seepage field. *J. Dalian Univ. Technol.* **1992**, *32*, 712–717.

28. Chen, P.; Zhang, Y. Seepage and stress coupling analysis of fractured rock mass. *Chi. J. Rock Mech. Eng.* **1994**, *13*, 299–308.

29. Runslätt, E.; Thörn, J. Fracture Deformation When Grouting in Hard Rock: In Situ Measurements in Tunnels under Gothenburg and Hallandsås. Available online: http://publications.lib.chalmers.se/records/fulltext/126954.pdf (accessed on 30 June 2018).

30. Schweisinger, T.; Svenson, E.J.; Murdoch, L.C. Hydromechanical behavior during constant-rate pumping tests in fractured gneiss. *Hydrol. J.* **2011**, *19*, 963–980. [CrossRef]

31. Nur, A.; Byerlee, J.D. An exact effective stress law for elastic deformation of rock with fluids. *J. Geophys. Res.* **1971**, *76*, 6414–6419. [CrossRef]

32. Brace, W.F.; Martin, R.J. A test of the law of effective stress for crystalline rocks of low porosity. *Int. J. Rock Mech. Min. Sci. Geomech. Abstr.* **1968**, *5*, 415–426. [CrossRef]

33. Berryman, J.G. Effective stress for transport properties of inhomogeneous porous rock. *J. Geophys. Res.* **1992**, *97*, 17409–17424. [CrossRef]

34. Shi, L.; Li, X.; Bing, B.; Wang, A.; Zeng, Z.; He, H. A Mogi-Type True Triaxial Testing Apparatus for Rocks with Two Moveable Frames in Horizontal Layout for Providing Orthogonal Loads. *Geotech. Test. J.* **2017**, *40*, 542–558. [CrossRef]

35. Hu, S.; Li, X.; Bai, B.; Shi, L.; Liu, M.; Wu, H. A modified true triaxial apparatus for measuring mechanical properties of sandstone coupled with CO_2-H_2O biphase fluid. *Greenh. Gas. Sci. Technol.* **2017**, *7*, 78–91. [CrossRef]

36. Cook, N.G. Natural joints in rock: Mechanical, hydraulic and seismic behaviour and properties under normal stress. *Int. J. Rock Mech. Min. Sci. Geomech. Abstr.* **1992**, *29*, 198–223. [CrossRef]

37. Zoback, M.D.; Byerlee, J.D. The effect of microcrack dilatancy on the permeability of westerly granite. *J. Geophys. Res.* **1975**, *80*, 752–755. [CrossRef]

38. Walsh, J.B. Effect of pore pressure and confining pressure on fracture permeability. *Int. J. Rock Mech. Min. Sci. Geomech. Abstr.* **1981**, *18*, 429–435. [CrossRef]

39. Cheng, C.; Li, X.; Li, S.; Zheng, B. Failure Behavior of Granite Affected by Confinement and Water Pressure and Its Influence on the Seepage Behavior by Laboratory Experiments. *Materials* **2017**, *10*, 798. [CrossRef] [PubMed]

40. Rice, J.R.; Cleary, M.P. Some basic stress diffusion solutions for fluid-saturated elastic porous media with compressible constituents. *Rev. Geophys.* **1976**, *14*, 227–241. [CrossRef]

41. Zhou, X.; Burbey, T.J. Fluid effect on hydraulic fracture propagation behavior: a comparison between water and supercritical CO_2-like fluid. *Geofluids* **2014**, *14*, 174–188. [CrossRef]

Article

Coal Anisotropic Sorption and Permeability: An Experimental Study

Yulong Chen [1], Xuelong Li [2,*] and Bo Li [3,*]

[1] State Key Laboratory of Hydroscience and Engineering, Tsinghua University, Beijing 100084, China; chen_yl@tsinghua.edu.cn

[2] State Key Laboratory of Coal Mine Disaster Dynamics and Control, Chongqing University, Chongqing 400030, China

[3] Key Laboratory of Karst Environment and Geohazard, Ministry of Land and Resources, Guizhou University, Guiyang 550000, China

* Correspondence: lilxcumt@126.com (X.L.); libo1512@163.com (B.L.)

Received: 27 June 2018; Accepted: 23 July 2018; Published: 30 July 2018

Abstract: Knowledge of the bedding plane properties of coal seams is essential for the coalbed gas production because of their great influence on the inner flow characteristics and sorption features of gas and water. In this study, an experimental study on the anisotropic gas adsorption–desorption and permeability of coal is presented. The results show that during the adsorption–desorption process, an increase in the bedding plane angle of the specimen expands the length and area of the contact surface, thereby increasing the speed and quantity of adsorption and desorption. With an increase in the bedding angle, the number of pores and cracks was found to increase together with the volumetric strain. The evolution of permeability of coal heavily depended on stress–strain stages. The permeability decreased with the increase of stress at the initial compaction and elastic deformation stages, while it increased with the increase of stress at the stages of strain-hardening, softening and residual strength. Initial permeability increased with increasing bedding angle.

Keywords: anisotropy; bedding plane orientation; coal; gas; adsorption–desorption

---

1. Introduction

Coalbed gas extraction has been considered as an attractive solution to control coal-mine disasters and provide new energy, and it has found great success in many countries (e.g., Australia, Canada). Coalbed gas extraction is a complicated process that is influenced by many factors, such as the mechanical properties of the formation (e.g., modulus, strength), sorptive capability, and transport properties (e.g., diffusivity, permeability) [1]. A better understanding of mechanical behavior and transport properties of coal is necessary to optimize field development. Among the transport properties, sorption and permeability are the most important parameters to influence gas production in coalbed gas formations [2–4].

A significant number of studies [5–10] have been conducted to investigate coal sorption and permeability under different conditions. It is found that coal sorption and permeability is impacted by many factors, such as fracture geometry [11–13], stress [14–16], water content [17,18], and temperature [19,20]. These experiments have focused on the isotropic characteristics of intact or powdered coals.

As a sedimentary rock with cleats, coal exhibits a notable anisotropic geometry and anisotropic mechanical properties [21,22]. The gas adsorption–desorption and permeability properties of coal are also found to be anisotropic [23], although many studies consider coal sorption and permeability as isotropic to simplify the analysis [13]. The anisotropic gas adsorption–desorption and permeability of coal remain unclear. To gain a better understanding of these two issues, an experimental study was carried out, and detailed results on the anisotropic gas adsorption–desorption and permeability of coal are presented in this study.

2. Experimental Method

In this study, a coal block was taken from the #7 coal seam in the Dabaoding coal mine in Sichuan Province, China. High-quality thermal coal was present with medium and ash, and had good thermal stability and high calorific value. The main characteristic parameters of coal are shown in Table 1.

Table 1. Status table of coal.

Index	Value
Moisture content (%)	0.71
Ash content (%)	10.11
Volatile matter content (%)	12.01
C (%)	90.91
H (%)	3.94
O (%)	2.54
N (%)	1.44

A series of cylindrical samples with a diameter of 50 mm and a length of approximately 100 mm were machined from this coal block. The coal samples were categorized based on the included angle β between the radial direction of the drill and the direction of the bedding surface, namely $\beta = 0°, 30°, 60°,$ and $90°$, as shown in Figure 1.

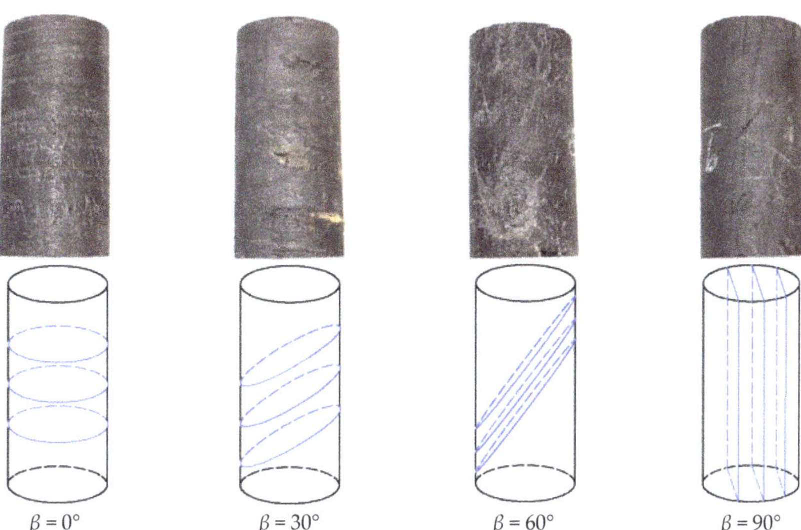

Figure 1. Coal samples and corresponding bedding angles.

Gas adsorption–desorption and permeability tests were performed on the prepared coal samples using a triaxial cell. Figure 2 shows the apparatus, where Figure 2a shows the main components: confining pressure and gas pressure control system, seepage pipe and holder system and automatic data recording system. Figure 2b is a photo of this apparatus. This cell is capable of accepting membrane-sheathed cylindrical samples (50 mm diameter). Axial and hydrostatic stresses up to 56 MPa were independently applied using two ISCO-260D pumps (Teledyne ISCO, Lincoln, NE, USA) with control up to 61 kPa. The gas pressure in the upstream was controlled by an ISCO-500D pump.

The flow rate was measured using a mass flowmeter with an accuracy of 4 mL/min. The flowmeter was connected with the outlet end of the fluid pipelines and the atmosphere. Temperature was controlled, and its variation was within ±0.1 °C during testing.

The cylindrical sample was inserted into a polyvinyl chloride (PVC) rubber jacket and then sandwiched in the holder by two cylindrical stainless-steel loading platens. The two platens with through-going flow connections and flow distributors were respectively connected with both ends of the triaxial cell.

Coal is sensitive to water [24,25]; it not only reduces the effective porosity but also induces swelling of the coal matrix. To minimize the effect of water on coal, the samples were dried in a vacuumed oven at 70 °C for two days before performing the gas adsorption–desorption and permeability tests. It is preferable to dry samples at lower temperatures for a longer time rather than at a high temperature for a shorter time because cracks or mineral alterations can occur at high temperatures [26]. To reduce the discreteness of the test data, three samples of coal, which have no macroscopic cracks, are of similar quality and have uniform features during CT scanning, were selected for the experiments. The details of the tests are described in the following section.

- Gas adsorption–desorption experiment: The samples were placed in the triaxial cell and subjected to negative pressure for 24 h. Next a gas injection pressure of 3 MPa and confining pressure of 3 MPa were applied for gas adsorption for 24 h. Subsequently, gas was desorbed until the desorption balance was achieved when the pressure was reduced to 0.5 MPa.

- Stress–strain–permeability experiment: Samples were placed under a confining pressure of 3 MPa at a speed of 0.1 MPa/s. Gas was allowed to enter the samples under a pressure of 2 MPa, which was maintained for 30 min with stable gas flux. Then the confining pressure and gas pressure were maintained, and the gas flux was monitored while the axial stress was continuously applied by stress control at a speed of 0.05 MPa/s until coal specimen failure. After failure, the stress control shifted to displacement control at a speed of 0.1 mm/min until the residual strength of the samples remained stable.

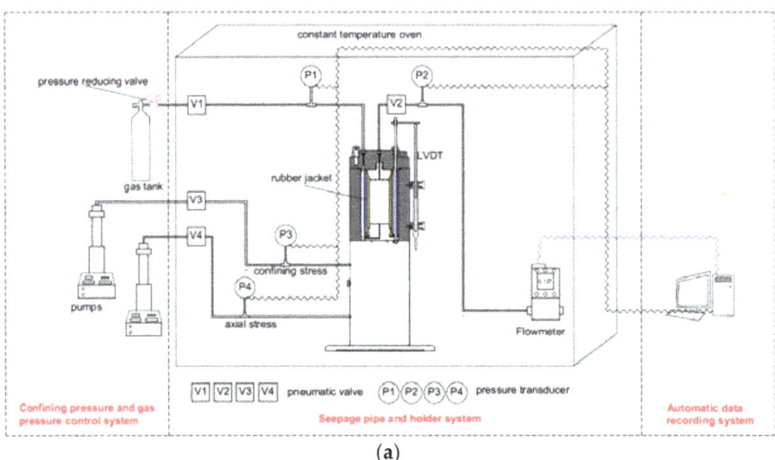

(a)

Figure 2. *Cont.*

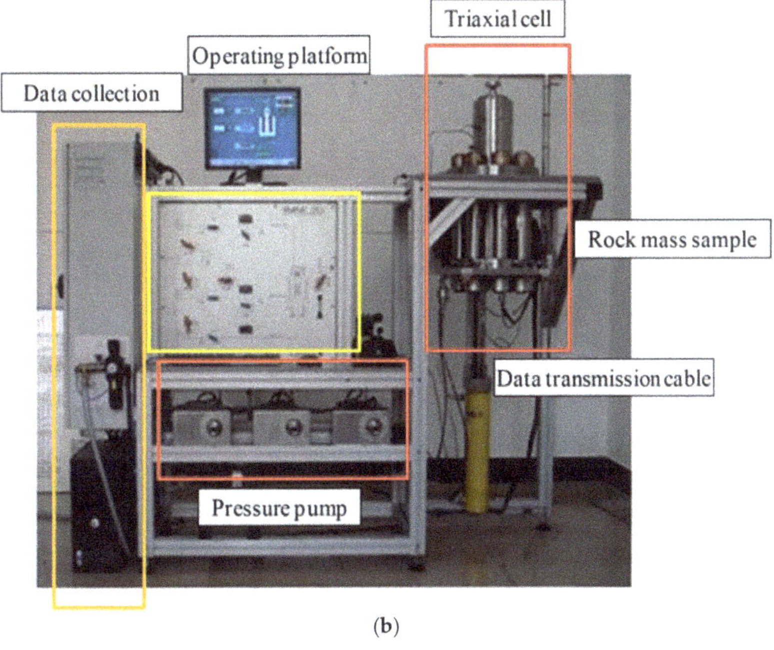

(**b**)

Figure 2. (**a**) Schematic diagram; (**b**) Experimental apparatus.

3. Results

3.1. Adsorption–Desorption Experiment

The final adsorption capacity, desorption capacity, and deformation of the samples are presented in Table 2. The raw coal samples exhibited strong heterogeneity during the adsorption–desorption process test. Even for those samples with the same bedding angle, the adsorption–desorption characteristics might still be different. Some differences emerged in the results among samples with the same bedding angles, but the differences of adsorption and desorption capacity and deformation were within 10%. This pattern indicates that the specific surface area of the matrix and the Van der Waals force for the samples were similar. Under the same adsorption–desorption pressure, pores in the matrix had the same ability to adsorb gas for the different samples. Therefore, the volumetric strain appeared to be similar for samples with different angles because all samples adsorbed equal amounts of gas.

A comparison of the standard adsorption–desorption curves for samples with different bedding angles is shown in Figures 3 and 4. After the gas entered the samples at 3 MPa of pressure, the interior of each sample was in a state of negative pressure. Due to the pressure difference and concentration difference, gas began to seep in and diffuse quickly; once the gas had seeped into a sample, the gas pressure difference and concentration difference gradually decreased, thus reducing the adsorption speed until a saturation state was reached. Gas desorption is the inverse process of gas adsorption; however, because microstructures existed within the coal, microcracks and pores of the coal matrix could close and affect the seepage and diffusion channels. Ultimately, the desorption quantity was reduced to less than 50% of the adsorption quantity, and the time to reach the desorption balance was twice that required for adsorption.

Table 2. Final adsorption capacity, desorption capacity, and deformation of the samples in the adsorption–desorption experiment.

Bedding Angle	Adsorption Capacity (mL)	Desorption Quantity (mL)	Volumetric Strain after Adsorption ($\mu\varepsilon$)	Volumetric Strain after Desorption ($\mu\varepsilon$)
0°	400	217	−1427	−337
	440	196	−1839	−279
	470	286	−2101	−310
30°	391	252	−1866	−286
	455	182	−1927	−310
	470	202	−1900	−347
60°	450	220	−1409	−382
	431	179	−1988	−315
	467	218	−1845	−336
90°	500	210	−1280	−299
	408	184	−1543	−340
	425	211	−1994	−326

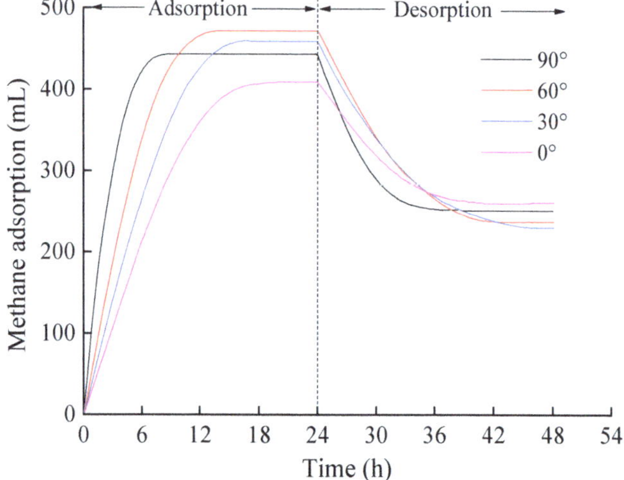

Figure 3. Gas adsorption–desorption over time.

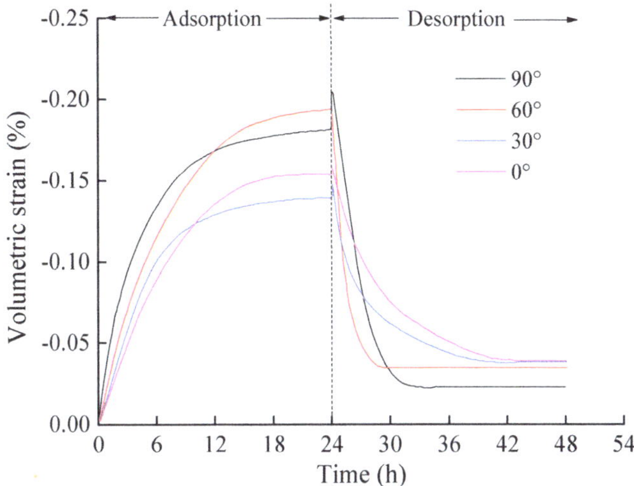

Figure 4. Volumetric strain during the gas adsorption–desorption process.

The adsorption–desorption process of samples at different bedding angles exhibited certain differences. As shown in Figure 3, for the same adsorption time, gas adsorption was fastest for the sample with a 90° angle. The time to reach adsorption stability for this sample was half that required for the sample with a 0° angle; the adsorption time was shorter for the sample with a 60° angle than for that with a 30° angle. The gas adsorption speed was the lowest at an angle of 0°. The gas adsorption quantities of the samples when $\beta = 0°$, 30°, 60°, and 90° were 440 mL, 470 mL, 450 mL, and 408 mL, respectively. Little difference therefore existed in the adsorption amounts, but the bedding angle was influential. Regardless of the adsorption or desorption process, the larger the bedding angle, the faster the adsorption or desorption.

Gas adsorption and desorption processes at a macroscopic level represent interactions occurring in the coal microstructure. The reason for such significant differences in the length and area of the bedding surface and the crack distribution could be the different bedding angles. The facet length and contact surface area of the bedding plane increased as the angle increased. Although the samples exhibited strong anisotropy with regard to volumetric strain during the adsorption–desorption process, volumetric strain increased in line with the bedding angle. These pores and cracks provided a seepage channel and a contact surface for adsorption in the inner matrix of the sample, which facilitated gas adsorption. The larger the bedding angle of the samples, the larger the pores and cracks, which explains why the gas adsorption–desorption speed increased as the bedding angle increased while the adsorption amount and volumetric strain remained the same.

3.2. Stress–Strain–Permeability Experiment

Many geotechnical engineering practices and experimental results have demonstrated that coal permeability is closely related to its stability and failure mode. During the process of becoming unstable (i.e., as cracks develop and failure occurs), the degree of extension of the original joint crack and its ongoing development determine the seepage features of gas in coalbed. The existence of bedding joints has been shown to greatly influence the mechanical properties of coal samples. Among multiple experiments, this study selected a representative sample, shown in Table 3.

Table 3. Photos of post-compression specimens and stress–strain–permeability curves.

Bedding Angle	Failure Mode	Failure Pattern	Stress–Strain–Permeability
0°	Splitting failure		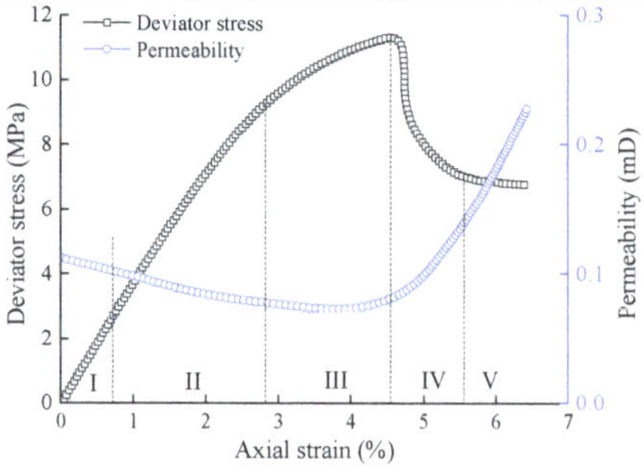
30°	Compound shear and splitting failure		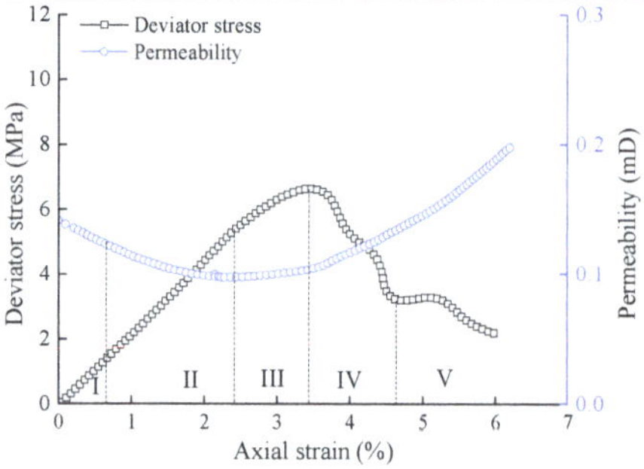
60°	Shear failure		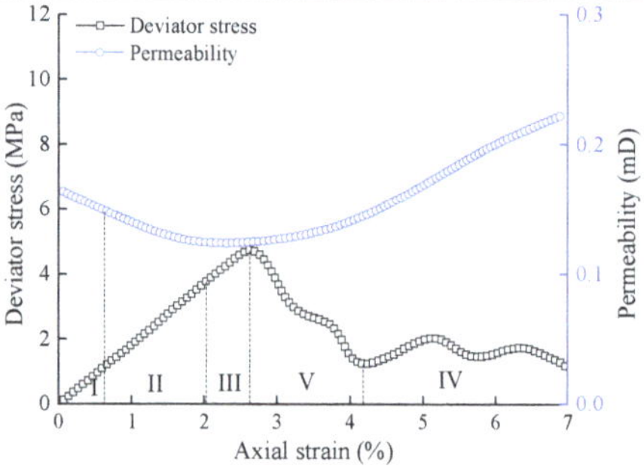
90°	Splitting failure		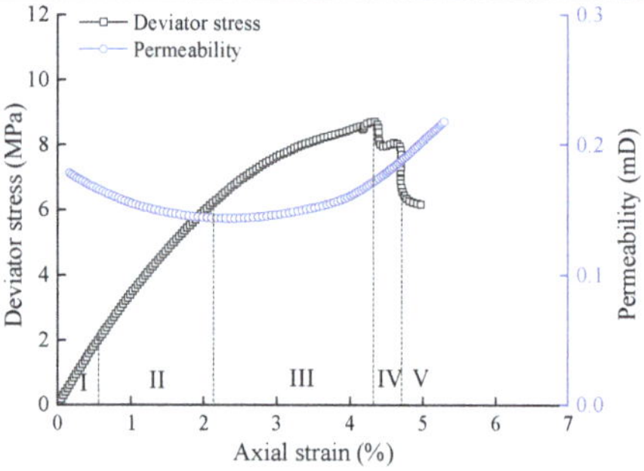

3.2.1. Axial Stress–Axial Strain–Permeability Curve of Coal

The compression experiment of samples with different angles demonstrated the stress–strain relationship and corresponding permeability–strain relationship, the forms of which were fundamentally consistent. After considering the relationship of these two curves, $\beta = 0°$ was used as an example; the stress–strain curve and permeability–strain can be divided into five stages and are denoted as I–V in Table 3.

(1) Initial compaction stage: The stress–strain curve moved upward slightly, and the primary pores and cracks in the coal sample gradually closed under pressure; this change decreased the gas seepage channels and resulted in reduced permeability.

(2) Elastic deformation stage: The stress–strain curve was nearly straight. Due to higher loading, the primary pores and cracks inside the coal sample continued to shrink. Shear and tensile stresses created new microcracks on the sample. This stage is the stable extension stage of cracks; as the strain increased, the effects on new cracks increased and permeability decreased as evidenced by the permeability–strain curve.

(3) Strain-hardening stage: In this stage, the stress–strain curve exhibited significant nonlinearity. As the normal loading continued to increase, shear and tension cracks developed rapidly, leading to irreversible deformation. By this time, the permeability curve had reached its minimum and had begun to increase slowly. During this stage, more microcracks began to extend and connect, and permeability increased gradually.

(4) Strain-softening stage: After the bearing capacity of the coal sample reached peak strength, the inner structure was damaged, but the sample retained its overall shape. In this stage, stress declined rapidly. Tension and shear failure developed quickly, macroscopic fractures appeared on the surface, the sample became unstable, and failure occurred, as reflected in the permeability–strain relationship. The gas permeability increased substantially due to the creation of new cracks that provided seepage channels for the gas.

(5) Residual strength stage: The deformation of the sample consisted of slippage along the mass of macroscopic fractures; stress decreased and stabilized as strain increased. This pattern indicates that the damage and cracks due to compression increased, additional gas channels were created, and gas permeability continued to rise.

3.2.2. Mechanical Properties and Gas Permeability of Coal

Compression strength of coal is closely related to the bedding angle. Under conditions of loading and gas seepage, an increase in the bedding plane angle resulted in an initial decrease in the compressive strength of the sample followed by an increase. Compressive strength was largest at 11.5 MPa when $\beta = 0°$. At $\beta = 30°$ and $\beta = 60°$, the compressive strength decreased to 5.5 MPa and 2.5 MPa, respectively. At $\beta = 90°$, the compressive strength was 8.8 MPa, larger than the value for $\beta = 60°$. This finding demonstrates that the mechanical property was anisotropic with regard to the bedding angle. The axial strain at peak strength, referred to as limit strain, was proportional to the peak strength; that is, the higher the peak strength, the higher the limit strain. The limit strain values of the four bedding angles were 4.6%, 3.4%, 2.5%, and 4.3%, respectively. The division between the brittleness and ductility of each sample was based on axial strain at peak strength; when the limit strain was lower than 1%, brittle failure occurred. Brittle ductile failure tended to occur in the range of 1–5%; all samples in this study exhibited brittle ductile failure.

As shown in Figures 5 and 6, the initial permeability varied widely among samples with different angles. Generally, initial permeability increased with an increase in the angle. The initial permeability values of the samples at angles of $30°$, $60°$, and $90°$ were respectively 1.27-, 1.46-, and 1.59-times that of the sample at an angle of $0°$. As the axial load increased, axial strain and elastic energy increased as well. When the axial strain of the samples at angles of $0°$, $30°$, $60°$, and $90°$ reached 0.9, 0.84, 0.7, and 0.5 of limit strain, respectively, gas permeability decreased to the respective minimum values of 0.0730 mD, 0.0973 mD, 0.124 mD, and 0.143 mD, representing 65%, 69%, 76%, and 80% of the initial permeability.

These results indicate that the original damage of the bedding plane of the coal sample affected the mechanical properties as well as permeability of the sample. The original damage to the coal sample involved the number of pores and cracks in the seam. If gas-containing coal samples possessed the same porosity, the degree to which fractures developed influenced the mechanical properties and seepage properties of the coal mass. As the gas permeability increased in line with the bedding angle, the number of seepage channels increased as well.

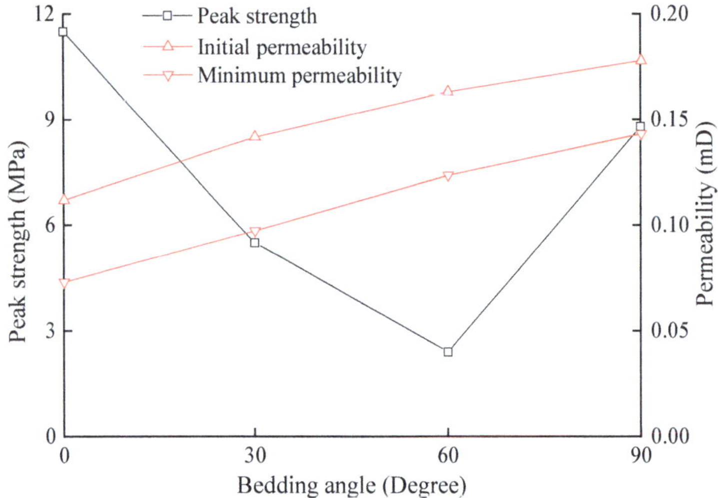

Figure 5. Peak strength and permeability at different bedding angles.

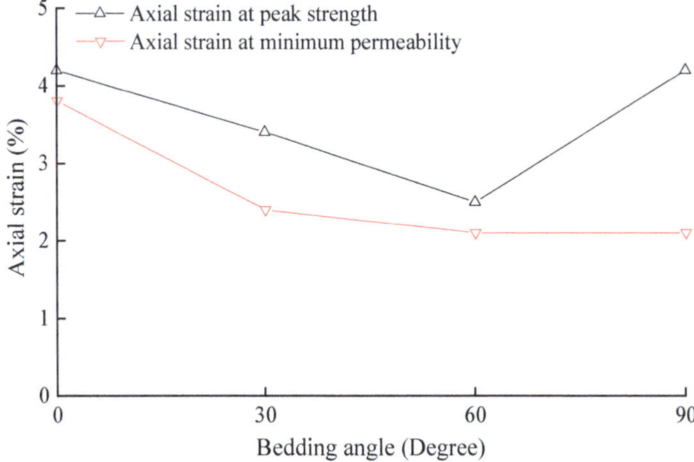

Figure 6. Axial strains at peak strength and minimum permeability at different bedding angles.

With an increase in axial strain, the pores in the matrix of the coal mass were compressed and gas seepage channels were affected, causing permeability to decline with an increase in stress. As a result, shear stress and tensile stress were apparently concentrated near the cracks in the interior of the sample, and the stress eventually extended to the surrounding area. The compression of pores in the matrix and crack extension occurred simultaneously. Once axial stress increased to a certain value, the influence of the gas seepage channel extension became greater than that of the compression of cracks on permeability. After gas permeability reached its minimum, it increased as stress increased. However, for samples with different angles, large differences appeared in shear stress and tensile stress during compression; this finding resulted in clear differences between the stress–strain curves and permeability evolution curves. The cracks represented the main seepage channel of gas distribution; therefore, the bedding contact surface was easily damaged. As more cracks developed and extended, the effective area of the bedding contact surface increased. As the angle of the coal sample increased and gas permeability decreased to the minimum, the strain capacity was reduced; however, the compressive capacity of the coal sample at a 90° angle was apparently larger than that of the sample at the 60° angle.

3.2.3. Failure Mode of Coal

The coal samples experienced the highest loading values under peak stress, and cracks developed continuously during the loading process; the critical point of failure was reached at peak stress. After that, localized microcracks coalesced and propagated to form a fracture. The fracture occurred along the main crack in the coal sample, after which point the stress declined sharply. Due to local through-going cracks and stress release following from pore compression in the matrix, effective stress was reduced and gas seepage channels were extended at the same time. Therefore, gas permeability demonstrated a dramatic increase.

The bedding angle affected the macroscopic failure of coal samples as well as crack distributions, resulting in large differences in the failure mode and permeability of different samples. The following responses were observed. (1) When $\beta = 0°$ and 90°, tensile splitting failure occurred, tensile stress affected the sample along the radial direction, and cracks appeared in the axial direction perpendicular and parallel to the bedding surface, respectively. The fracture was extensive, exhibiting numerous cracks. (2) When $\beta = 60°$, shear slippage failure and a single main fracture occurred along the bedding surface. (3) When $\beta = 30°$, the sample underwent compound shear and splitting failure. The crack began near the bedding plane and extended in the axial direction to the side end of the sample before ultimately splitting off the sample.

The bedding inclination caused some differences in the deformation and fracture properties, stimulating more complexity in the stress state along the bedding planes. The final failure mode of the sample reflected the influence of the bedding angle and was directly related to ultimate permeability. When $\beta = 0°$, the elastic energy released intensively, and the fracture was severe; therefore, the change rate of permeability, and the final permeability when residual stress stabilized, was the largest of all samples. When $\beta = 90°$, the sample was damaged by tension along the axial direction; dual splitting and cutting occurred in the sample, and the stress declined rapidly. Therefore, the final permeability was similar to the values of the sample with $\beta = 0°$. When $\beta = 60°$, due to the failure mode, the permeability did not exhibit a large jump, and the final permeability was smallest among all samples.

3.2.4. Relationship between Permeability, Radial Strain, and Volumetric Strain of Coal

Figures 7 and 8 illustrate the relationship between permeability, radial strain, and volumetric strain, respectively. For the same values of radial strain and volumetric strain, permeability increased with an increase in the angle.

Strong anisotropy was observed in the mechanical properties and seepage behaviors of samples with different bedding angles. The adsorption–desorption experiment excluded the effect of porosity on permeability under the loading condition. Moreover, the inclination of beddings induced shear stress near the plane; therefore, the bedding plane was the most probable location of crack initiation and propagation. Cracks occurred more easily along the surface of the bedding and then extended further before finally resulting in total failure of the entire specimen; as a result, the bedding plane could form the most likely percolation channels to influence permeability. As such, the bedding plane should be considered when determining permeability.

In this section, the difference in permeability induced by the bedding angle was analyzed based on the plate fluid theory.

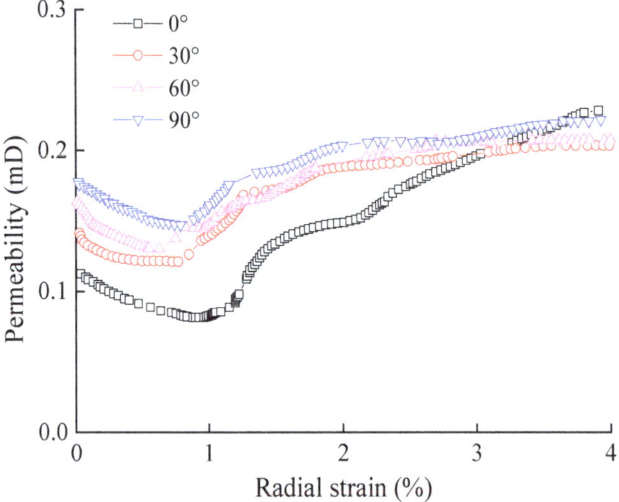

Figure 7. Relationship of permeability and radial strain.

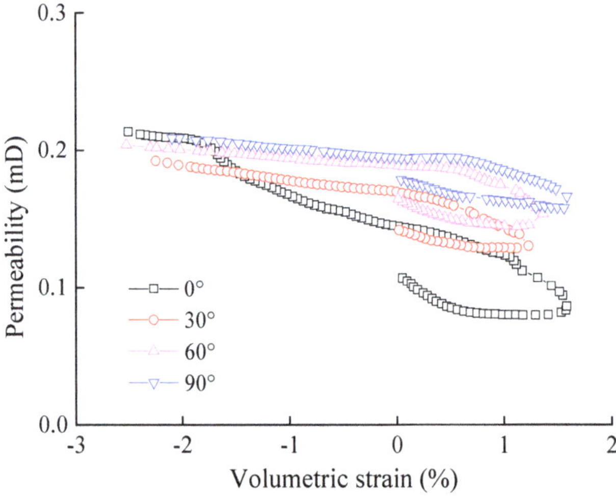

Figure 8. Relationship of permeability and volumetric strain.

The Hagen–Poiseuille flow function is written as follows:

$$Q = \frac{nd^3 l (P_1^2 - P_2^2)}{12\mu h P_2},$$

(1)

where n, d, l, and h denote the crack number (dimensionless), aperture (µm), length (m), and height (m), respectively. Q is the permeation rate (m^3/s), μ is the gas kinematic viscosity (Pa·s), P_1 is the gas pressure at the inlet of the specimen (Pa), and P_2 is the atmospheric pressure at the outlet of the specimen (Pa).

Darcy's law expresses the permeability k as

$$k = \frac{2\mu QLP_2}{A(P_1^2 - P_2^2)}.$$ (2)

Substituting Equation (1) into Equation (2) yields

$$k = \frac{nlLd^3}{6\mu Ah}.$$ (3)

When the sample deforms under stress, the radius R (m), height L (m), average aperture d_a (μm), length l_a (m), and height h_a (m) of the crack in the axial direction are respectively illustrated as=

$$\begin{cases} R = (1 - \varepsilon_r)R_0 \\ L = (1 - \varepsilon_a)L_0 \\ h_a = (1 - \varepsilon_a)h_{a0} \\ l_a = (1 - \varepsilon_r)l_{a0} \\ d_a = (1 - \varepsilon_r)d_{a0} \end{cases}$$ (4)

where R_0 and L_0 refer to the initial diameter (m) and height (m) of the specimen, respectively; and h_{a0}, l_{a0}, and d_{a0} refer to the initial average height (m), length (m), and aperture (μm) of the crack, respectively.

Substituting Equation (4) into Equation (3) yields the relationship between permeability and radial strain,

$$k = \frac{L_0}{6\mu A_0} \frac{n_a l_{a0} d_{a0}^3}{h_{a0}} (1 - \varepsilon_r)^2,$$ (5)

where $\frac{L_0}{6\mu A_0}$ is the formal kinematic viscous coefficient α, which is a constant. $\frac{n_a l_{a0} d_{a0}^3}{h_{a0}}$ is defined as the initial crack coefficient β_a, which is linked to the axial initial crack aperture.

Equation (5) can be written as

$$k = \alpha \beta_a (1 - \varepsilon_r)^2.$$ (6)

Permeability is parabolically related to radial strain. The permeability–radial strain data of coal samples with different bedding angles were fitted using the parabolic equation as shown in Figure 9, indicating that the parabolic equation clearly described the permeability–radial strain data of coal with different bedding angles.

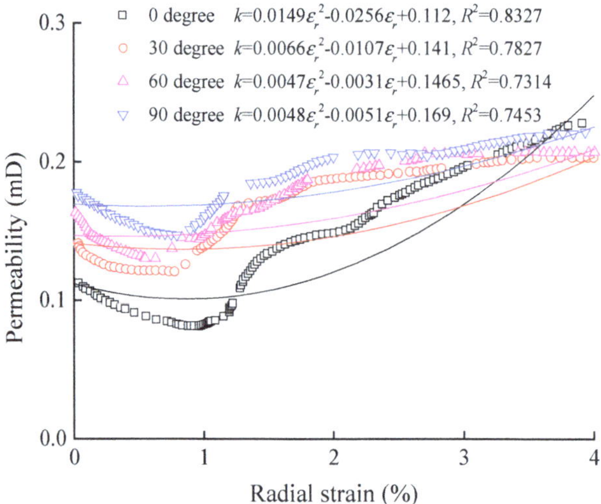

Figure 9. Fitting of permeability–radial strain data of coal samples with different bedding angles.

4. Conclusions

This study focused on the anisotropic adsorption–desorption and permeability of coal. A series of gas adsorption–desorption and permeability measurements were performed on specifically prepared coal samples. The following conclusions can be drawn from the experimental results:

(1) During the gas adsorption–desorption process, the length and area of the bedding contact surface of the sample increased with an increase in the bedding angle, resulting in faster adsorption/desorption.

(2) The larger the angle of the bedding plane in the sample, the larger the number of effective pores and cracks available for a seepage channel and a contact surface for adsorption in the inner matrix of the sample, which facilitated gas adsorption.

(3) The evolution of permeability varied across different stress–strain stages. Permeability declined with an increase in stress at the initial compaction and elastic deformation stages before increasing in line with stress during the strain-hardening, softening, and residual strength stages. Initial permeability increased with an increased bedding angle. Permeability was also found to be parabolically related to radial strain.

(4) The compression strength of the samples initially decreased and then rose with an increased bedding angle. Compression strength was largest at a bedding angle of 0°, followed by the compression strength when the bedding angle was 90°. Under these circumstances, splitting failure occurred in the samples. When the bedding angle was 60°, the compression strength of the samples was smallest, resulting in sliding failure along the bedding plane. The sample with a bedding angle of 30° exhibited compound splitting and shear failure.

Author Contributions: Y.C. designed and performed the tests and wrote the paper; X.L. and B.L. analyzed the data.

Funding: This work is supported by the China Postdoctoral Science Foundation under grant nos. 2017M620048 and 2018T110103.

Conflicts of Interest: The authors declare no conflict of interest.

References

1. Yang, D.; Qi, X.; Chen, W.; Wang, S.; Dai, F. Numerical investigation on the coupled gas-solid behavior of coal using an improved anisotropic permeability model. *J. Nat. Gas Sci. Eng.* **2016**, *34*, 226–235. [CrossRef]
2. Liu, R.; Li, B.; Jiang, Y. Critical hydraulic gradient for nonlinear flow through rock fracture networks: the roles of aperture, surface roughness, and number of intersections. *Adv. Water Resour.* **2016**, *88*, 53–65. [CrossRef]
3. Liu, R.; Li, B.; Jiang, Y. A fractal model based on a new governing equation of fluid flow in fractures for characterizing hydraulic properties of rock fracture networks. *Comput. Geotech.* **2016**, *75*, 57–68. [CrossRef]
4. Liu, R.; Jiang, Y.; Li, B.; Wang, X. A fractal model for characterizing fluid flow in fractured rock masses based on randomly distributed rock fracture networks. *Comput. Geotech.* **2015**, *65*, 45–55. [CrossRef]
5. Rodrigues, C.F.; Laiginhas, C.; Fernandes, M.; Sousa, M.J.L.D.; Dinis, M.A.P. The Coal Cleat System: A new approach to its study. *J. Rock Mech. Geotech. Eng.* **2014**, *6*, 208–218. [CrossRef]
6. Alexis, D.A.; Karpyn, Z.T.; Ertekin, T.; Crandall, D. Fracture permeability and relative permeability of coal and their dependence on stress conditions. *J. Unconv. Oil Gas Resour.* **2015**, *10*, 1–10. [CrossRef]
7. An, H.; Wei, X.R.; Wang, G.X.; Massarotto, P.; Wang, F.Y.; Rudolph, V.; Golding, S.D. Modeling anisotropic permeability of coal and its effects on CO_2 sequestration and enhanced coalbed methane recovery. *Int. J. Coal Geol.* **2015**, *152 Pt B*, 15–24. [CrossRef]
8. Beamish, B.B.; Crosdale, P.J. Instantaneous outbursts in underground coal mines: An overview and association with coal type. *Int. J. Coal Geol.* **1998**, *35*, 27–55. [CrossRef]
9. Chen, Y.; Wei, K.; Liu, W.; Hu, S.; Hu, R.; Zhou, C. Experimental characterization and micromechanical modelling of anisotropic slates. *Rock Mech. Rock Eng.* **2016**, *49*, 3541–3557. [CrossRef]
10. Dewhurst, D.N.; Siggins, A.F. Impact of fabric, microcracks and stress field on shale anisotropy. *Geophys. J. Int.* **2006**, *165*, 135–148. [CrossRef]
11. Korsnes, R.I.; Wersland, E.; Austad, T.; Madland, M.V. Anisotropy in chalk studied by rock mechanics. *J. Petrol. Sci. Eng.* **2008**, *62*, 28–35. [CrossRef]
12. Li, Y.; Tang, D.; Xu, H.; Meng, Y.; Li, J. Experimental research on coal permeability: The roles of effective stress and gas slippage. *J. Nat. Gas Sci. Eng.* **2014**, *21*, 481–488. [CrossRef]
13. Liu, J.; Chen, Z.; Elsworth, D.; Miao, X.; Mao, X. Linking gassorption induced changes in coal permeability to directional strains through a modulus reduction ratio. *Int. J. Coal Geol.* **2010**, *83*, 21–30. [CrossRef]
14. Liu, Q.; Liu, K.; Zhu, J.; Lu, X. Study of mechanical properties of raw coal under high stress with triaxial compression. *Chin. J. Rock Mech. Eng.* **2014**, *33*, 24–34.
15. Louis, L.; David, C.; Metz, V.; Robion, P.; Menendez, B.; Kissel, C. Microstructural control on the anisotropy of elastic and transport properties in undeformed sandstones. *Int. J. Rock. Mech. Min. Sci.* **2005**, *42*, 911–923. [CrossRef]
16. Ma, Y.; Pan, Z.; Zhong, N.; Connell, L.D.; Down, D.I.; Lin, W.; Zhang, Y. Experimental study of anisotropic gas permeability and its relationship with fracture structure of Longmaxi Shales, Sichuan Basin, China. *Fuel* **2016**, *180*, 106–115. [CrossRef]
17. Meng, Z.; Li, G. Experimental research on the permeability of high-rank coal under a varying stress and its influencing factors. *Eng. Geol.* **2013**, *162*, 108–117. [CrossRef]
18. Meng, Y.; Li, Z.; Lai, F. Experimental study on porosity and permeability of anthracite coal under different stresses. *J. Petrol. Sci. Eng.* **2015**, *133*, 810–817. [CrossRef]
19. Wang, S.; Elsworth, D.; Liu, J. Permeability evolution in fractured coal: The roles of fracture geometry and water-content. *Int. J. Coal Geol.* **2011**, *87*, 13–25. [CrossRef]
20. Wang, K.; Zang, J.; Wang, G.; Zhou, A. Anisotropic permeability evolution of coal with effective stress variation and gas sorption: Model development and analysis. *Int. J. Coal Geol.* **2014**, *130*, 53–65. [CrossRef]
21. Talesnick, M.L.; Hatzor, Y.H.; Tsesarsky, M. The elastic deformability and strength of a high porosity, anisotropic chalk. *Int. J. Rock Mech. Min. Sci.* **2001**, *38*, 543–555. [CrossRef]
22. Xu, X.; Sarmadivaleh, M.; Li, C.; Xie, B.; Iglauer, S. Experimental study on physical structure properties and anisotropic cleat permeability estimation on coal cores from China. *J. Nat. Gas Sci. Eng.* **2016**, *35*, 131–143. [CrossRef]
23. Liu, S.; Wang, Y.; Harpalani, S. Anisotropy characteristics of coal shrinkage/swelling and its impact on coal permeability evolution with CO_2 injection. *Greenh. Gases* **2016**, *6*, 615–632. [CrossRef]

24. Risnes, R.; Madland, M.V.; Hole, M.; Kwabiah, N.K. Water weakening of chalk—Mechanical effects of water–glycol mixtures. *J. Petrol. Sci. Eng.* **2005**, *48*, 21–36. [CrossRef]
25. Wang, S.; Elsworth, D.; Liu, J. Permeability evolution during progressive deformation of intact coal and implications for instability in underground coal seams. *Int. J. Rock Mech. Min. Sci.* **2013**, *58*, 34–45. [CrossRef]
26. Sirdesai, N.N.; Singh, T.N.; Gamage, R. Thermal alterations in the poro-mechanical characteristic of an Indian sandstone—A comparative study. *Eng. Geol.* **2017**, *226*, 208–220. [CrossRef]

Article

Experimental Study on the Damage of Granite by Acoustic Emission after Cyclic Heating and Cooling with Circulating Water

Dong Zhu [1,2], Hongwen Jing [1,*], Qian Yin [1] and Guansheng Han [1]

[1]   State Key Laboratory for Geomechanics and Deep Underground Engineering, China University of Mining and Technology, Xuzhou 221116, China; zhudong163@163.com (D.Z.); jeryin@foxmail.com (Q.Y.); Han_GS@cumt.edu.cn (G.H.)

[2]   College of Energy and Transportation Engineering, Jiangsu Vocational Institute of Architectural Technology, Xuzhou 221116, China

*   Correspondence: hwjingcumt@126.com

Received: 27 June 2018; Accepted: 19 July 2018; Published: 25 July 2018

Abstract: Hot dry rock is developed by injecting cold water into high-temperature rock mass. At the same time, cold water is heated in contact with the rock mass. With the continuous influx of cold water, the surrounding rock will undergo a rapid cooling process, which results in several cycles of heating and cooling. However, there is little research on the influence of cycles of heating and cooling with circulating water on the mechanical properties of rock, which is of great importance to the stability of rock mass engineering in the process of energy development. In this paper, the effects of cyclic heating and cooling with circulating water on the damage of granite are studied using uniaxial compressive, Brazilian and acoustic emission (AE) tests. The results show that heat treatment temperature and number of cycles have important effects on the mechanical properties of granite as follows: (1) at the same treatment temperature, an increase in the number of cycles means that the distribution of physical and mechanical parameters of the granite show an almost exponential downward trend. The uniaxial compression of granite results in its transformation from brittle to plastic, and the failure mode changes from slipping of the shear surface to plastic failure. With increased cycles of heating and cooling with circulating water, the tensile strength of granite also decreases; temperature has an obvious influence on physical and mechanical parameters, cracking of samples, and plays a controlling role in the failure mode of samples. In addition, (2) at the same temperature, the heating and cooling numbers N have a significant influence on the AE distribution characteristics of the sample under uniaxial compression and the number of AE collisions, and the cumulative number of AE decreases with the increase of N. (3) The concepts of mechanical damage and high-temperature and cold-water shock damage during uniaxial compression of samples were proposed based on AE, and the damage equations were established respectively. The curve equations of damage value (D) and cycle numbers N after thermal shock damage of high temperature and cold water were overlaid. The cracking mechanism of high-temperature and cold water impact on granite was analyzed, and the thermal shock stress equation of high temperature and water cooling was established.

Keywords: damage; cyclic heating and cooling; physical and mechanical parameters; failure mode; acoustic emission

---

1. Introduction

With the increased development of geothermal resources and nuclear waste storage technology, seepage mechanics of fractured rock mass and high-temperature rock damage has become a hot research topic [1–7].

In previous literature, researchers have carried out work that examines the mechanical properties of rock after exposure to high temperatures, including the deformation process, failure criterion of rock under pressure, and thermal cracking, and provided constitutive equations and carried out rock damage characterization in response [8–13]. Many scholars have paid attention to the study of high-temperature rock damage [14–17]. Yu et al. [18] built a mesostructure based on the numerical model for the analysis of rock thermal cracking based on elastic damage mechanics and thermal-elastic theory. In addition, the damage accumulation induced by thermal (T) and mechanical (M) loads is considered to modify the elastic modulus, strength and thermal properties of individual elements in line with the intensity of damage. Yang et al. [19] across an experimental investigation on thermal damage and failure of mechanical behavior of granite after exposure to different high-temperature treatments, found that the internal damage mechanism of rock is obviously affected by temperature.

Acoustic emission (AE) technology is widely used in material damage monitoring and evaluation [20–26]. The AE patterns of rocks at different temperatures were studied under uniaxial compression and the relationship between rock damage and AE was established. Wu et al. [27] studied the relationship between microstructure morphology and AE of granite found: with increase of temperature, there was more internal crack formation and internal damage for granite, and more frequent AE activity of granite under uniaxial compression. The mechanical properties and AE characteristics of granite with the formation of internal crack networks of granite has a corresponding relationship. The peak stress curve and ringing cumulative number curve of granite have a stable trend when crack expansion is slow. However, the peak stress and the ringing cumulative number curve of granite will have inflection point and lead to mutation when the crack network expands rapidly. Chen et al. [28] used AE technology to evaluate the fracture strength and energy value of hard rock and concluded that AE and index increased with the increase of rock temperature. A study on the mechanical characteristics of high-temperature rocks after cooling is in the center of scientific research interests [29–34]. They studied the effects of different cooling rates and methods on mechanical parameters, strength cracking and failure modes of high-temperature rocks. Shao et al. [35] studied the effects of the cooling rate and the constituent grain size on the mechanical behavior of heated rock. They found that larger-grain granite rocks are more affected by water quenching, rocks that have been slowly cooled have areas with greater crack growth, and the stress threshold of medium-grain rocks that have been slowly cooled is affected.

However, studies on the mechanical properties and damage to granite during water cycle cooling have rarely been conducted. A laboratory experiment was designed in view of the current engineering application in drilling and underground space engineering fire rescues [36], especially for the study of rock mechanics in the process of dry hot rock development. In this study, the effects of the cycles of heating and cooling with circulating water on the physical and mechanical parameters, damage of granite and failure mode are studied by carrying out uniaxial compression, and Brazilian and AE tests. In addition, we established the damage equation of rock samples based on the characteristics of AE data.

2. Materials and Methods

2.1. Sample Collection and Preparation

The granite samples used in the experiment originated from Zhangzhou City in Fujian Province, China. These rocks are naturally gray and white in color, and compact and uniform with no voids or cracks on their surface. The main minerals determined by X-ray diffraction are feldspar 35%, quartz 40%, amphibole 20% and mica 5%. The initial moisture content of the sample is 0.31%. The average density is 2.97 g/cm^3.

The samples were all taken from a cylindrical core drilled from a piece of granite stone which has uniform texture and were cut into normative Φ (50 $\pm$ 2) $\times$ (100 $\pm$ 2) mm cylinders.

Samples are as shown in Figure 1a. The dimensions of the sample for the Brazilian test are a standard (50 ± 2) mm $\times$ (25 ± 2) mm cylinders.

For this study, dry granite is exposed to temperatures of 250 °C, 350 °C, 450 °C, 550 °C, and 650 °C and the granite underwent many cycles of heating and then cooling with water, and then the mechanical properties of each sample were examined. Each sample underwent 0, 1, 5, 10, 15 and 20 cycles of heating and then cooling with water for each temperature respectively. Zero (0) denotes that the sample is naturally cooling at room temperature.

The samples were heated to the set temperature in a high-temperature furnace (type MXQ 1700) made in Shanghai (Micro-X Furnace Co. Ltd. (Shanghai, China)) at the rate of 10 °C/min and were maintained at the designated temperature for 2 h. The heating furnace has high precision of temperature control such that control accuracy error is ± 1 °C, which meets the requirements of these experiments, as shown in Figure 1b. Then, the samples were quickly taken out of the furnace by the crucible tongs and placed in a container filled with cold water and cooled to room temperature and dried. In this process, the contact temperature of the sample with cold water will be lower than the set temperature, which will influence the experimental results. This procedure represents one cycle of heating and then cooling with water. Figure 2 shows the scheme of circulating heating and water cooling of the sample.

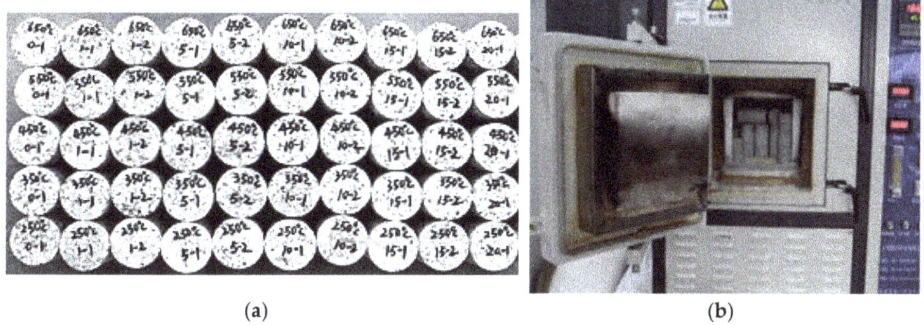

(a)   (b)

Figure 1. Granite sample and high-temperature furnace. (**a**) Granite samples; (**b**) High-temperature furnace.

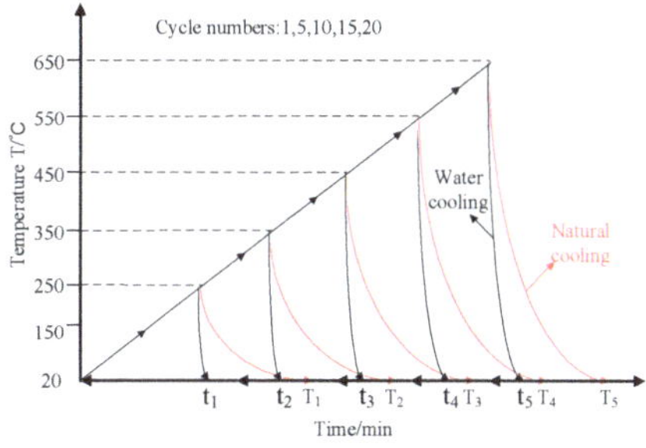

Figure 2. Test scheme of cyclic heating and cooling with circulating water.

2.2. Experimental Equipment

As shown in Figure 3, the experimental system comprises a loading system and a data acquisition system. The loading system is an MTS816 servo hydraulic testing machine that is located at the China University of Mining and Technology. The MTS816 testing machine has stable performance and a sensitive control system, and it is the most advanced laboratory testing equipment for rock mechanics in the world. The loading method is displacement control and the loading speed is 0.002 mm/min.

Prior to carrying out the uniaxial compression test, the sample was coated with a layer of lubricant to reduce the friction between the sample and the pressure head. The loading and data acquisition system mainly recorded the stress strain, AE signals and digital images. The latter were obtained with a high-speed camera which was used to capture the evolution of fractures on the samples during loading, while a DS2 acoustic emission signal analyzer recorded the AE signals of the samples.

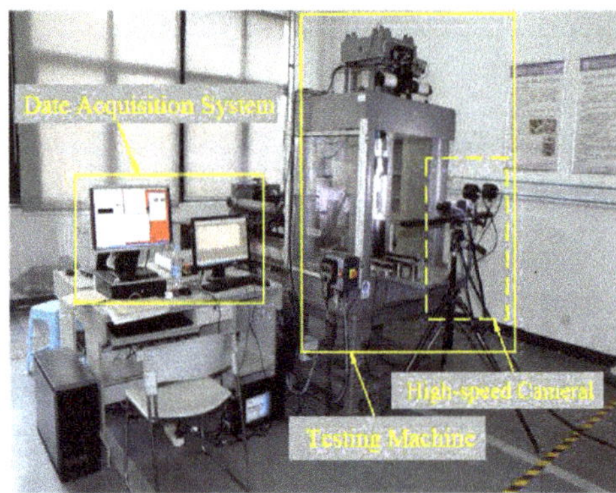

Figure 3. Experimental loading and data acquisition system.

3. Results and Discussion

3.1. Characteristics of Mechanical Strength of Granite

The stress-time curves of the two samples exposed to a temperature of 250 °C under uniaxial compression are shown in Figure 4a. The plot shows that the average peak strength, failure time, and average and secant modulus of the two samples are 123.09 MPa, 552.5 s, and 99.69 GPa and 69.25 GPa respectively, and the dispersion coefficients are 1.99×10^{-2}, 0.90×10^{-2}, 2.92×10^{-3} and 1.34×10^{-2} respectively. Figure 4b shows that the average tensile strength and failure time are 4.765 MPa and 119 s, respectively, and their dispersion coefficients are 0.11×10^{-2} and 3.4×10^{-2} respectively. The granite used in this study is quite homogeneous and the mechanical parameters show few differences, so the granite is suitable for quantitative analysis of its mechanical properties after cyclic heating and cooling. It can be seen from Figure 4 that there is no obvious yielding, which shows that the granite samples are highly brittle.

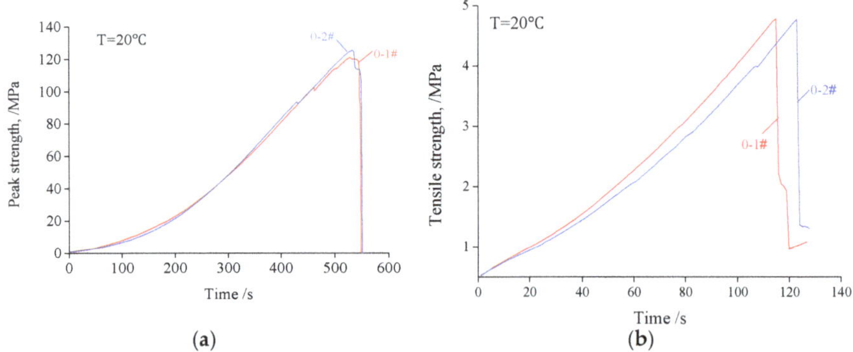

Figure 4. Peak and tensile intensity changes with time. (**a**) Peak intensity changes with time; (**b**) Tensile strength changes with time.

Figure 5a–e are plots of the strength of granite versus a different number of cycles of heating and cooling with circulating water at temperatures of 350 °C, 450 °C, 550 °C and 650 °C respectively. Figure 5f shows the peak strength based on the acoustic intensity test (peak intensity) at different temperatures and Figure 5g is the ratio of the peak strength after cooling with water to peak strength after cooling in air. Figure 5a–e show that the uniaxial compression of granite undergoes three stages of changes: pore compression, and elastic and strain softening deformations. Figure 5f shows that the cycles with water cooling have a significant effect on the peak strength of the granite and therefore cracking. With an increase in the number of cycles and increases in temperature, the elasticity and characteristics of the samples are gradually reduced and their strain softening behavior is increased. Therefore, with temperature increases, the granite is transformed from brittle to plastic. Figure 5a shows that at a temperature of 250 °C, the peak strength of the sample is 111.66 MPa while the peak intensity is 103.16 MPa after one cycle of heating and cooling with circulating water; the difference between the peak strength and peak intensity of the sample is 7.61%. The samples are very brittle. After 5 cycles of heating and cooling with circulating water, the peak strength before cracking of the sample of 90.51 MPa is equal to 81.06% of the peak strength of the sample after air cooling; the sample is obviously less brittle. After 5 cycles, the sample begins to transform from brittle to plastic. After 10 cycles, the peak strength of the sample is 71.69 MPa, which is equal to 64.20% of the peak strength of the sample after air cooling, and the material becomes obviously more plastic. After 15 and 20 cycles, the peak strength of the samples is 52.91 MPa and 50.52 MPa respectively, which is equal to 47.38% and 45.25% of the peak strength of the samples after air cooling. The samples obviously become more plastic, and the declining of the peak strength has stopped. The range of peak strengths that results in cracking is reduced after 15 cycles of heating at 250 °C and cooling with circulating water, and tends toward a constant.

The transformation from brittleness to ductility is not only affected by the number of cycles of heating and cooling with circulating water, but also the temperature. Figure 4a–e show that when the samples undergo 15 cycles at a temperature that is under 250 °C, there are no signs of brittle failure, but the material becomes more plastic. At temperatures of 350 °C, 450 °C, 550 °C, and 650 °C, the samples become more plastic after 5, 1, 1 and 0 cycles of heating and cooling with circulating water, respectively. Therefore, during the cycles, temperature plays a decisive role in the brittle to plastic transformation of granite. Figure 5f shows that with an increase in the number of cycles, the peak strength of granite at 250 °C decreases linearly. When the number of cycles is increased from 10 to 20, the peak strength shows a downward trend. With 20 cycles, the peak strength of the granite at temperatures of 350 °C to 650 °C undergoes three stages: Stage I is the rapid decline of the peak strength, Stage II is when the decline of the peak strength is reduced, and in Stage III, the peak strength is constant. In all three stages, there are further gradual decreases in the peak intensity of the samples with increases in temperature. The plot in Figure 5f is fitted with an exponential function in accordance with the parameters in Table 1.

Table 1. Fitting formulas of mechanical parameters after cyclic heating and cooling with circulating water.

Temperature	Mechanical Parameters	Fitting Formula (*N*:Cycle Numbers)	R^2
250 °C	Peak stress	$\sigma_{max} = 122.16 \exp(-N/25.55) - 11.69$	0.9921
350 °C	Peak stress	$\sigma_{max} = 60.65 \exp(-N/5.24) + 42.04$	0.9337
450 °C	Peak stress	$\sigma_{max} = 38.93 \exp(-N/11.34) + 34.45$	0.9056
550 °C	Peak stress	$\sigma_{max} = 19.31 \exp(-N/6.56) + 37.24$	0.8518
650 °C	Peak stress	$\sigma_{max} = 9.25 \exp(-N/0.31) - 6.79 \exp(N/6.17) + 6.79$	0.8674

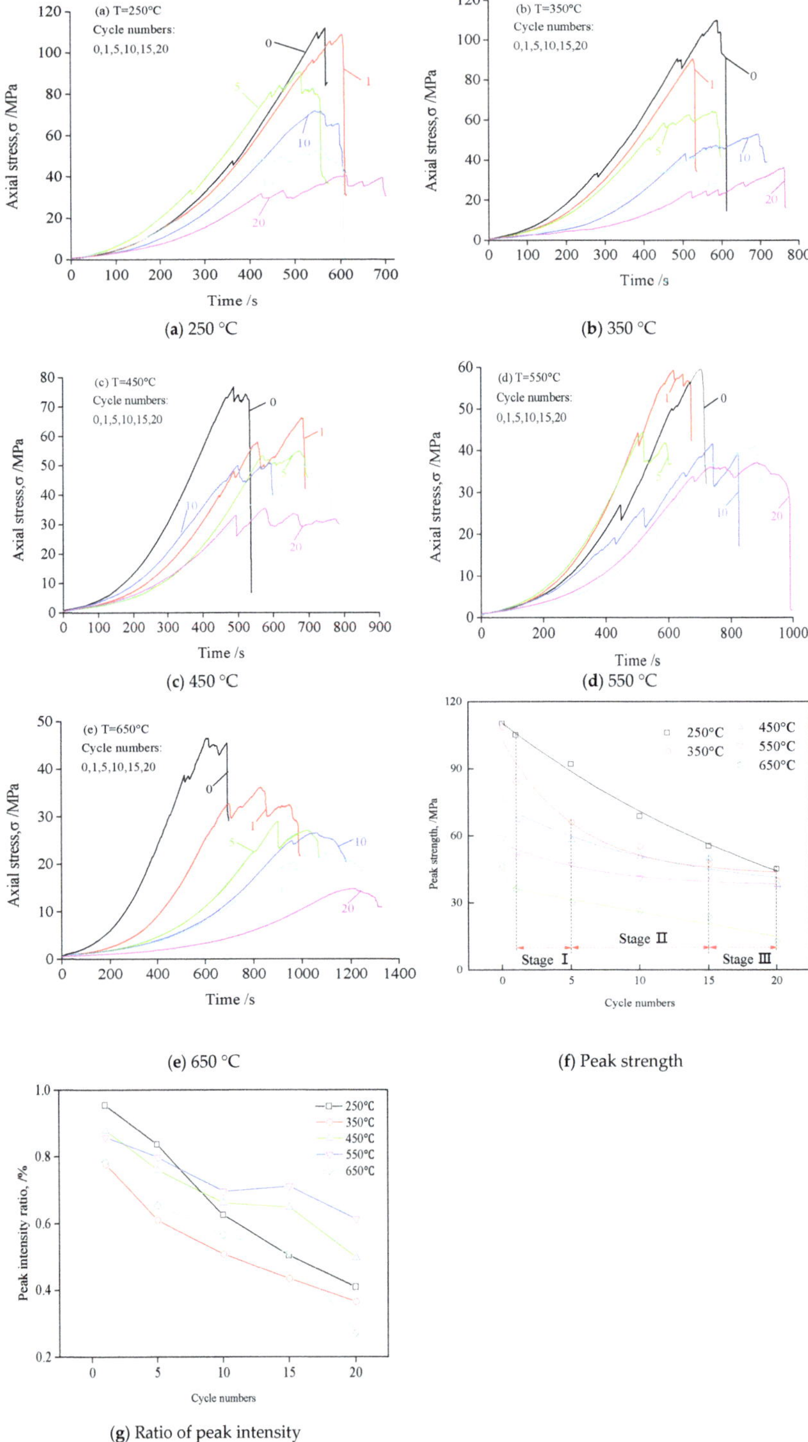

(**a**) 250 °C

(**b**) 350 °C

(**c**) 450 °C

(**d**) 550 °C

(**e**) 650 °C

(**f**) Peak strength

(**g**) Ratio of peak intensity

Figure 5. Strength curve and peak strength of granite.

3.2. Mechanical Tensile Strength of Granite after Cyclic Heating and Cooling with Circulating Water

Figure 6a shows that the tensile strength of granite is significantly reduced with increased number of cycles, and the effect of temperature on the tensile strength of granite is significant. After 10 cycles, cracking due to reduced tensile strength of the sample is minimal at a heating temperature of 250 °C, in which the tensile strength of the samples is reduced from 4.57 MPa to 4.39 MPa. The rate of cracking with lower tensile strength is higher at higher temperatures. This is especially true of the sample that was exposed to a temperature of 650 °C, in which the tensile strength is reduced from 2.22 MPa to 0.80 MPa, or a reduction of 64.19%. After 10 to 20 cycles, the rate of cracking with lower tensile strength is high at 250 °C; however, the tensile strength of the sample is slowly reduced with higher heating temperatures, in particular 650 °C, so that the rate of cracking is slowly increased. Thus, the cycles of heating and cooling with circulating water of granite carried out at high temperatures have a very significant impact on its tensile strength.

The tensile strength data from several experiments [37–39] are compared with the results of this study is shown in Figure 6b. It is observed that a faster rate of decline occurs at temperatures higher than 450 °C from Figure 6b. This is caused by changes in the internal structure of the granite minerals as a result of the heat. The volume expansion of quartz has increased dramatically between 450 and 550 °C. Due to the rapid expansion in the quartz volume, cracks occur between the quartz, feldspar and hornblende, causing the rapid decline in the strength at this stage.

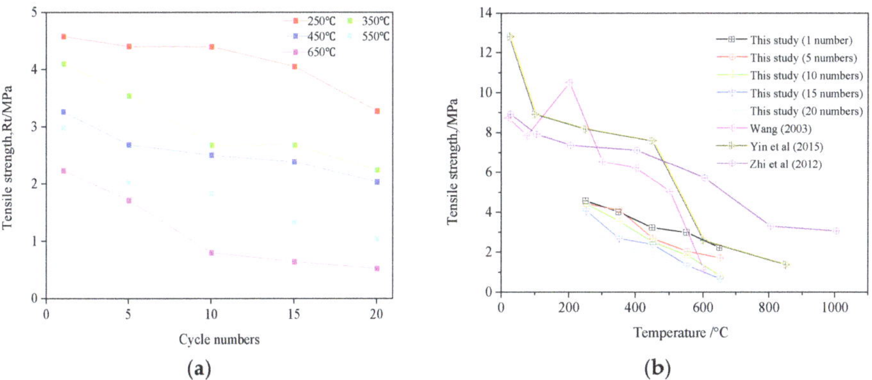

Figure 6. Tensile strength of granite after different temperature and cyclic treatment. (**a**) Tensile strength of granite; (**b**) Comparison of tensile strength of granite.

3.3. AE Characteristics of Rock Samples after Cyclic Heating and Water Cooling

AE impact is a signal whose amplitude exceeds the threshold value. The data obtained by the monitoring channel, the number of signals detected at a certain time, is the number of AE collisions, and the cumulative number of AE impacts is the cumulative number of impacts at each stage. The number of impacts which has a good correspondence with the development, expansion, confluence and transfixion of cracks in rocks reflects the activity of AE from rocks [19]. In this paper, we only analyze AE parameters of granite specimens under uniaxial compression after cyclic heating and cooling at 250 °C because of the length of the article. Figure 7 is the stress-time curve and AE characteristic curve of rock sample after cyclic heating and cooling at 250 °C.

Figure 7a is the rock sample in the natural state of cooling, i.e., the number of cycles is 0. In the initial compaction stage (oa), the number of AE collisions is small, which is generated by the closure of a small number of primary cracks, and the AE collisions are in a quiet period. Starting from the elastic stage (segment ab), the number of impacts began to increase, and a smaller number of AE events occur, and the number of AE stabilizes at a lower level. When entering the stable crack propagation stage (bc), the number of impacts increases rapidly, and a large AE event occurs, which indicates that the cracks in the rock sample are gradually expanding and converging, and the AE number enters the active period. When entering the unstable crack propagation stage (cd), the number of impacts suddenly rises to the peak, and the cumulative number of AE rises steeply. At point d of peak intensity, the maximum number of AE impacts is reached, and the rock sample corresponds to the penetration of macroscopic cracks after the peak (de), and the number of impacts decreased, but there was still a small amount of AE activity and maintained a high level. Figure 7a shows two stress drops in the stress curve. New macroscopic cracks can be observed during each stress drop during the test and meanwhile many AE events will appear on the AE characteristic curve showing that AE monitoring can reflect the rock damage process accurately.

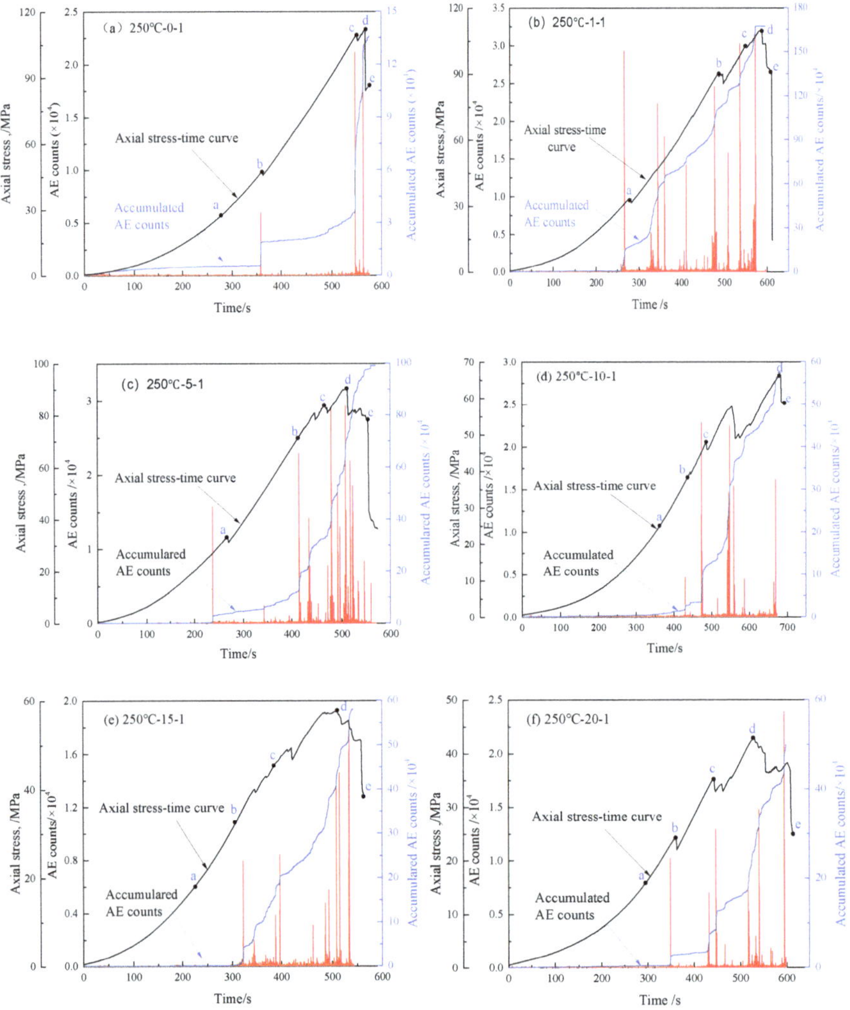

Figure 7. AE counts and accumulated of granite specimens with respect to cycle numbers at 250 °C.

As shown in Figure 7b–f, the AE characteristics of rock samples after circulating high temperature and water cooling, especially those after 10 cycles, are significantly different from those of cooling rock samples under natural conditions. Rock samples have very few AE signals at the initial compaction stage, indicating that there are fewer AE collisions, which indicates that the thermal damage caused by repeated high temperature and water cooling is greater and the internal pores and fractures of the rock sample are damaged. The number of impacts remained low when entering the elastic stage and AE numbers entered the quiet period. When the rock sample enters the stable crack propagation stage, the number of impacts increases, and the acoustic emission event is in the active period, which can be distinguished from the elastic phase; however, the AE activity of rock samples is not as strong as that of natural cooling. When the rock sample enters the unstable crack propagation stage, the number of AE collisions will suddenly rise to a small peak, but the increase degree of impact number is not as severe as that of natural cooling. The number of impacts does not reach the peak when the stress reaches the peak intensity, but lags behind the peak intensity. In general, after several cycles of high temperature and water cooling, the initial damage of rock sample becomes bigger and makes the AE signal inactive at the initial stage of loading. Because the thermal fracture increases the ductility of rock samples, starting with the elastic phase, AE impingement is less active than natural cooling, so the gradual failure process becomes slower than that of natural cooling as strain increases. With the increase of the number of cycles of high temperature and water cooling, the number of AE events accumulated in the sample under uniaxial compression decreased.

3.4. Rock Damage Analysis

3.4.1. Uniaxial Compression Mechanical Damage

Since the evolution of rock micro-defects is a random change, it can be considered that the strength distribution of microelements obeys Weibull distribution, and its distribution density function is:

$$\varnothing \varepsilon = m/\alpha \varepsilon^{m-1} \exp(-\varepsilon^m/\alpha), \tag{1}$$

$\varnothing \varepsilon$ is micro-element damage rate of sample under loading. Because the law of AE activity is basically consistent with the statistical distribution of internal defects of materials, when the whole cross section of the damage accumulation of AE is Ω_m, the accumulated AE of the sample compressed to destroy is:

$$\Omega = \Omega_m \int_0^\varepsilon \varnothing(x)dx, \tag{2}$$

Equation (1) is substituted into Equation (2) and then integrated to get Equation (3):

$$\Omega/\Omega_m = 1 - \exp(-\varepsilon^m/\alpha), \tag{3}$$

There is the following relationship between the damage parameter D and the probability density of micro-element failure:

$$D = \int_0^\varepsilon \varnothing(x)dx = 1 - \exp(-\varepsilon^m/\alpha), \tag{4}$$

Compared with Equation (3) and Equation (4), the uniaxial compression mechanical damage can be obtained as follows:

$$D = \Omega/\Omega_m, \tag{5}$$

Mechanical damage curves of rock samples at different temperatures are drawn based on AE experimental data, as shown in Figure 8, which shows that in the initial compaction stage, the primary pores and fractures in granite are compacted, and the damage variable is almost 0. Then, the damage variable increases slightly, and the deformation damage in this stage is mainly caused by elastic deformation of rock sample mineral particles. The mechanical damage of rock samples in these two stages is less affected by temperature and water-cooling times N. The rock is in the stage of unstable fracture when the stress reaches about 60% of the ultimate strength. The damage variable increases rapidly because of friction, dislocation occurs between granite particles, and new cracks occur. Therefore, $0.6\sigma_b$ can be used as the threshold stress value of granite damage [40]. When the ultimate strength is exceeded, the damage value gradually approaches 1 and the rock sample tends to be damaged. At the same time, it can be seen from Figure 8 that under the same temperature and stress, the mechanical damage becomes greater with the increase of the number N of high temperature and water cooling, indicating that high temperature and water cooling have caused thermal damage to the rock sample.

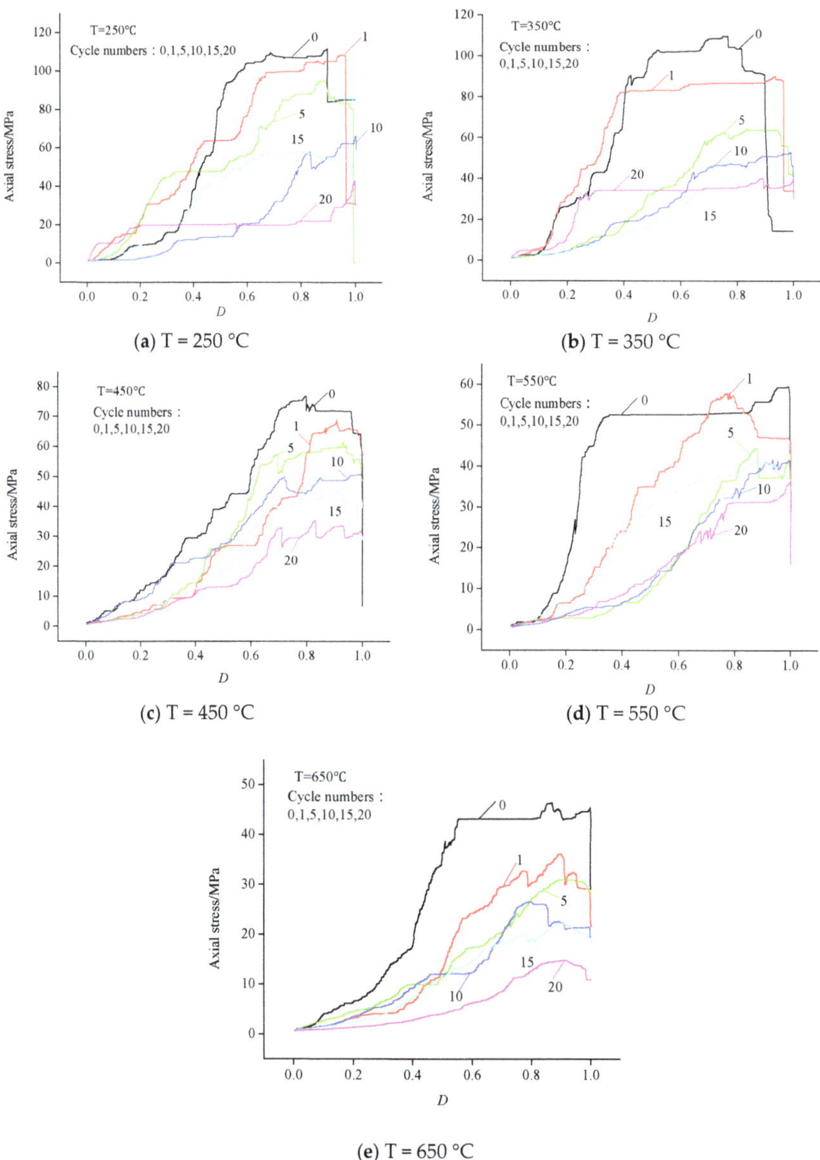

(**a**) T = 250 °C

(**b**) T = 350 °C

(**c**) T = 450 °C

(**d**) T = 550 °C

(**e**) T = 650 °C

Figure 8. Mechanical damage curve of samples after circulating high temperature and cooling water.

3.4.2. Heat Shock Damage of High Temperature and Cold Water

The elastic modulus is taken as the damage variable of rock sample to characterize the effect of the numbers of cyclic heating and water cooling on the mechanical properties of rock. Thermal shock damage of rock sample is 0 when the specimen is in the natural state at 25 °C. The elastic modulus of each sample at each test temperature was treated after 20 cycles of heating and water cooling; the thermal shock damage factor was defined as follows:

$$DT = 1 - E_T^{(N)} / E_0, \tag{6}$$

In the above formula, $E_T^{(N)}$ is the elastic modulus of granite after N times of circulating water cooling at temperature T, E_0 is the elastic modulus of granite in its natural state at 25 °C, shock damage D of heat−water at different temperatures is shown in Figure 8 and the damage fitting equation is shown in Table 2.

Table 2. Fitting curve equations of damage D value and cycle number N of samples.

Temperature	Thermal Shock	Fitting Formula (N:Cycle Numbers)	R^2
250 °C	D	$D = -0.0168\,N + 0.0124\,N^2 - 9.69 \times 10^{-2}\,N^3 + 2.261 \times 10^{-5}\,N^4$	0.9902
350 °C	D	$D = -0.0916\,N - 0.004\,N^2 + 8.47 \times 10^{-6}\,N^3 + 3013 \times 10^{-6}\,N^4$	0.9997
450 °C	D	$D = 0.0693\,N - 0.0065\,N^2 + 3.25 \times 10^{-4}\,N^3 - 5.868 \times 10^{-6}\,N^4$	0.9654
550 °C	D	$D = 0.0274\,N + 0.0015\,N^2 - 2.50 \times 10^{-4}\,N^3 + 6.94 \times 10^{-6}\,N^4$	0.9943
650 °C	D	$D = 0.08975\,N - 0.01308\,N^2 + 7.88 \times 10^{-4}\,N^3 - 1.614 \times 10^{-5}\,N^4$	0.9675

As can be seen from Figure 9, the D value of specimen damage increases with the increase of N at the same temperature. This is due to the appearance of tiny cracks in the surface of the granite specimen during the cold and thermal cycle, which is spread throughout the whole sample, cutting the rock mass and destroying the internal structure of the specimen leads to the reduction of the bearing capacity of the rock. The surface rock of the specimen is further softened by water action, which increases the accumulation of irreparable deformation, leads to the further expansion of the existing crack depth and the increases of the number of cracks, so the damage value increases gradually when the structural connection of rock is destroyed. At the same number of cycles of high temperature and cold water and as the temperature goes up, the damage value N gradually increases because the thermal tensile stress of granite exceeds the tensile strength of the material itself, the material will undergo thermal damage. The higher the temperature of the specimen itself, the greater the tensile stress on its surface. Tensile cracks will occur on the surface of the specimen when the tensile strength of the rock exceeds the tensile strength. The external temperature of the specimen decreases sharply but the internal temperature is relatively high during rapid cooling, the tensile crack on the surface of specimen and the excess part of material is caused by external shrinkage and internal expansion. In the heating process, more dorsal cracks are produced at the excess part of the material due to external expansion and internal contraction. Because the external temperature rises sharply, and the internal temperature decreases relatively, there will be more tiny cracks in the rock than expected.

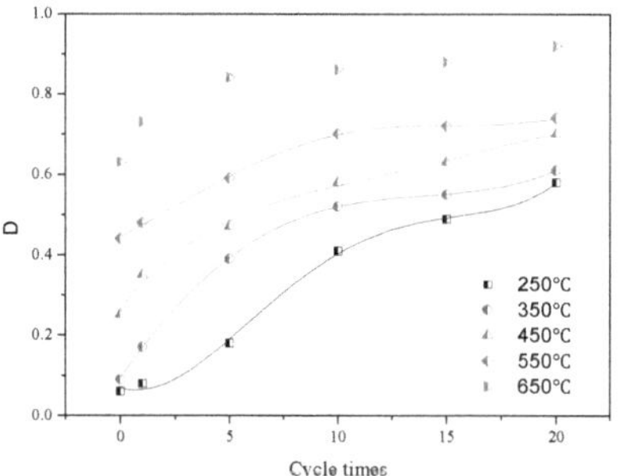

Figure 9. Samples damage fitting curve.

3.4.3. The Cracking Mechanism of Granite under High-Temperature and Cold-Water Impact

The total strain of the rock samples under different temperatures after cooling by cold water is ε. Suppose the rock sample contains n kinds of mineral particles, the thermal expansion coefficient of each mineral particle after N cycles of cold and heat is $\alpha_1^{(N)}$, $\alpha_2^{(N)}$ ... $\alpha_i^{(N)}$ ... $\alpha_n^{(N)}$, the elastic modulus are $E_1^{(N)}$, $E_2^{(N)}$ ... $E_i^{(N)}$ ... $E_n^{(N)}$ respectively. Suppose the granite sample is elastic material, the stress of each mineral particle in the sample after repeated heating and cooling can be obtained as follows:

$$\sigma_1^{(N)} = E_1^{(N)}(\alpha_1^{(N)}\Delta T - \varepsilon), \sigma_2^{(N)} = E_2^{(N)}(\alpha_2^{(N)}\Delta T - \varepsilon) \ldots \sigma_i^{(N)} = E_i^{(N)}(\alpha_i^{(N)}\Delta T - \varepsilon) \ldots \sigma_n^{(N)} = E_n^{(N)}(\alpha_n^{(N)}\Delta T - \varepsilon), \tag{7}$$

Due to in the process of rapid cooling there is only thermal shock and there is no other external force, the following Equation (8) can be established according to the mechanical equilibrium condition:

$$\sum_{i=1}^{n} E_i^{(N)} \left(\alpha_i^{(N)} \Delta T - \varepsilon \right) = 0 \tag{8}$$

The total strain ε can be solved by Equation (8):

$$\varepsilon = \sum_{i=1}^{n} E_i^{(N)} \alpha_i^{(N)} \Delta T / \sum_{i=1}^{n} E_i^{(N)}, \tag{9}$$

Substituting the strain ε into the stress expressions, the thermal shock stress of each mineral particle is:

$$\sigma_i^{(N)} = E_i^{(N)} \Delta T \left[\sum_{j=1}^{n} E_j^{(N)} \left(\alpha_i^{(N)} - \alpha_j^{(N)} \right) \right] / \sum_{i=1}^{n} E_i^{(N)} \tag{10}$$

In the process of high-temperature water cooling of rock sample, the compressive stress of highly deformed mineral particles is σ_{ic}, and the tensile stress of small deformed mineral particles is σ_{it}. Suppose σ_c and σ_t are the compressive strength and tensile strength of rock sample respectively. When $\sigma_{ic} \geq \sigma_c$ or $\sigma_{it} \geq \sigma_t$, thermal cracking occurs in rocks. Because granite is a relatively dense rock, when it is heated at high temperature and cooled by water, the deformation of mineral particles cannot be developed freely to a certain extent and the thermal stress of the structure is large. As a result, micro-fractures in rock samples increase. The deformation of mineral particles cannot be developed freely to a certain extent due to granite being a relatively dense rock during high-temperature heating and water cooling, therefore, the thermal stress of the structure produced by the particles is large, leading to the increase of micro-cracks in the rock sample. Even without external force, the thermal stress of the structure in rock sample increases with temperature is obvious. From what has been discussed above, after N cycles of high temperature and water cooling, the change from the original complete structure to the fractured rock structure caused by thermal shock stress accumulation is the main reason of rock strength damage.

3.5. Failure Mode of Granite at Different Temperatures

As shown in Figure 10a, brittle shear failure of the granite occurs after 1–10 cycles of heat treatment at 250 °C, such that the sample is broken into pieces. At the moment of failure, the sample releases large amounts of elastic energy, and breaks into large pieces, and the AE signals immediately increase. After 10 cycles, the brittleness of the samples is gradually reduced. During uniaxial loading, shear failure occurs in the granite, there is spalling on the surface, and no significant changes in the color of the sample.

As shown in Figure 10b, brittle failure occurs in the granite samples at a temperature of 350 °C after 1 to 5 cycles of heating and cooling with water. Slipping of the shear surface is mainly found in the samples, but the amount of damage is significantly reduced, and results in flaking and spalling in the middle of the sample. After 10–15 cycles, there is plastic failure of the samples, and the middle part flakes first during uniaxial loading. During the failure process, there is an increase in granite powder which shows that there is slipping of the shear surface. After 20 cycles, the color of the samples changes from a gray to pale yellow color.

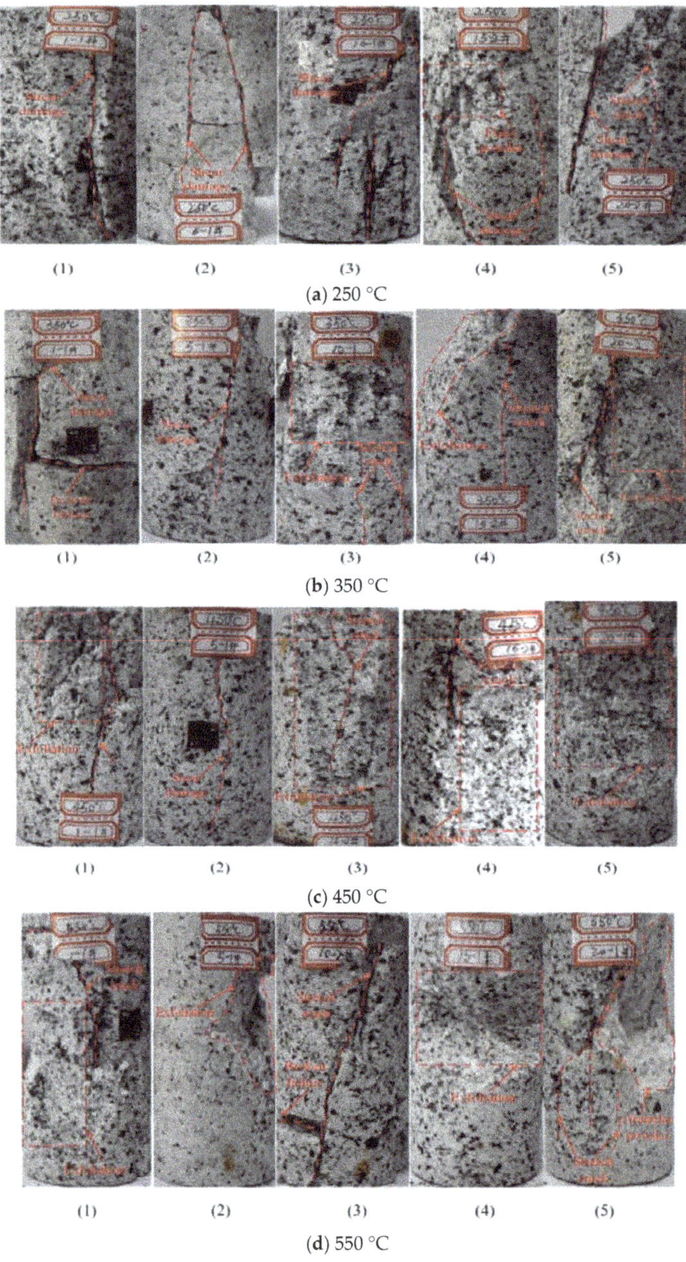

(a) 250 °C

(b) 350 °C

(c) 450 °C

(d) 550 °C

Figure 10. Failure mode of granite after cyclic heating and cooling with circulating water.

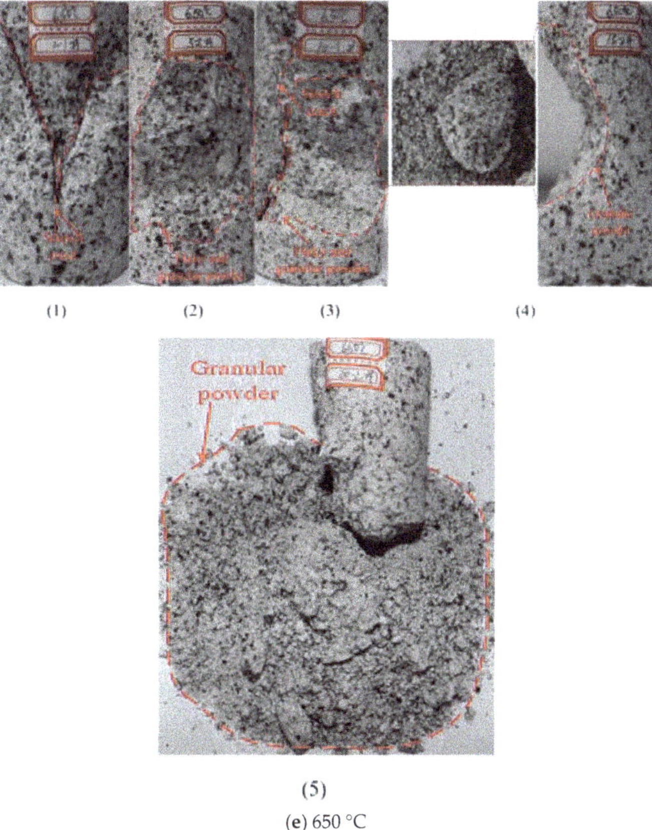

(5)

(**e**) 650 °C

Figure 10. *Cont.*

In Figure 10c,d, there is brittle failure of the granite samples at temperatures of 450 °C and 550 °C after 1 cycle and cracks are found on the samples. Shear failure is found in the samples under uniaxial compression, but there is release of the elastic energy and therefore the amount of damage is significantly less, with flaking of a small area of the surface material. After 5 cycles, plastic failure occurs in the samples, and the middle of the sample flakes first under uniaxial loading, and then there is spalling in the middle of the sample. In the process of failure, the size of the granite powder particles increases significantly with less shear slipping of the sample while incurring more shear failure. The color of the samples gradually changes from gray to yellow and then to pale yellow.

The mechanical parameters of granite at a temperature of 650 °C in Figure 10e are obviously changed after air cooling from brittle to plastic. After 1 cycle, a cross shear failure surface develops with a smaller number of layered blocks of granite. After 5 cycles, uniaxial loading causes the sample to change from flaking/spalling to particle detachment, and shows layers of a failed surface. With increases in the number of cycles, failure appears in the center of the sample. After the sample has undergone 15 cycles, the material has lost most of the cohesion and bonding between particles and the granite is changed to a granular material. After 5 cycles, the color of the sample changes from gray to light yellow and then to dark yellow.

In summary, the heat treatment temperature and cycles of heating and cooling with circulating water have important impacts on the failure mode of granite, with greater impacts on the internal structure of the granite at higher temperatures when rapid cooling with water takes place. The force of the thermal shock produces new cracks in the granite, and elongates the previous cracks. In addition, the granite particles are less compact when cooled rapidly with water, which is also the reason for the transformation of the granite from brittle to plastic. Thermal shock occurs during cycles of heating and cooling with circulating water at high temperatures. Higher temperatures result in more intense thermal shocks, as well as more obvious deterioration of the internal structure and changes in the chemical composition of the granite.

3.6. Analysis and Discussion

Figure 11 shows the normalized compressive strength data from several experiments [41–44] compared with the results of this study. Although the fluctuation of compressive strength of each type of granite is different, the normalized compressive strength can be observed decreasing below 200 °C in treatment temperature. When the treatment temperature exceeds 200 °C, the compressive strength quickly starts to decrease with increasing temperature. However, it is observed that a faster rate of decline occurs at temperatures higher than 400 °C (Figure 11). This is caused by changes in the internal structure of the granite minerals as a result of the heat. The sharp increase in the thermal expansion [45,46] of quartz at temperatures between 400 °C and 500 °C is shown in Figure 11. Due to the rapid expansion in the quartz volume, cracks occur between the quartz, feldspar and hornblende, causing the rapid decline in the strength at this stage.

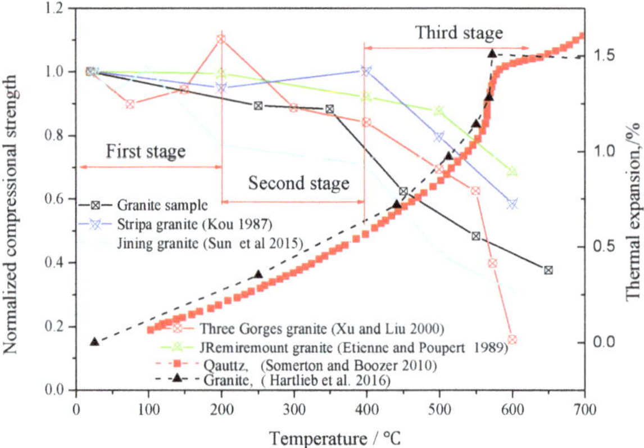

Figure 11. Differences in the normalized compressive strength of granite on different studies (source: three Gorges granite from Xu and Liu [41]; Remiremount granite from Etienne and Poupert [42]; Stripa granite from Kou [43]; Jining granite from Sun et al. [44]).

The decrease in some mechanical properties (compressive strength, and tensile strength) and the increase in the peak strain of heated granite are caused by thermally induced variations in internal structure. Because granite is composed of mineral particles with different thermal expansion coefficients and thermo-elastic characteristics, high temperature leads to inhomogeneous thermal expansion of mineral particles or the phase transition of some mineralogical components, generating internal stress and micro-cracks in granite [44]. Between 400 °C and 600 °C, and especially 500 and 600 °C, the minerals of granite have chemical changes [47]. At roughly 573 °C and under atmospheric conditions, quartz has a phase transformation from α phase to β phase, which can be used to explain the large variation of mechanical and physical properties.

The mechanical damage of samples after natural cooling from the literature [40] is compared with the results of this study as shown in Figure 12a. The mechanical damage curves of the samples studied in the literature have been obviously effected by temperature. In this paper, when the damage value of almost all samples was 0.7, the corresponding axial stress value reached the maximum. Figure 12b shows that the fluctuation of mechanical damage value of each sample is not obvious. When the stress value is 0–30% of the peak strength, the mechanical damage value of the unheated sample is the largest, and the D value is about 0.4; when the mechanical damage value of the 550 °C sample is the smallest, and the D value is about 0.2. When the stress value reached 30% of the peak strength, the mechanical damage value of the sample began to be significantly dispersed, and the mechanical damage value was significantly affected by temperature. Compared with the mechanical damage curves of the samples in the literature, we can find that the samples in this paper will suffer greater mechanical damage at a lower stress level which shows that the samples in the literature have higher strength and brittleness than the samples in this study. We found that the influence factors of mineral composition, processing precision, heating and cooling condition and stability of loading equipment are related to the difference tests of the mechanical damage curve.

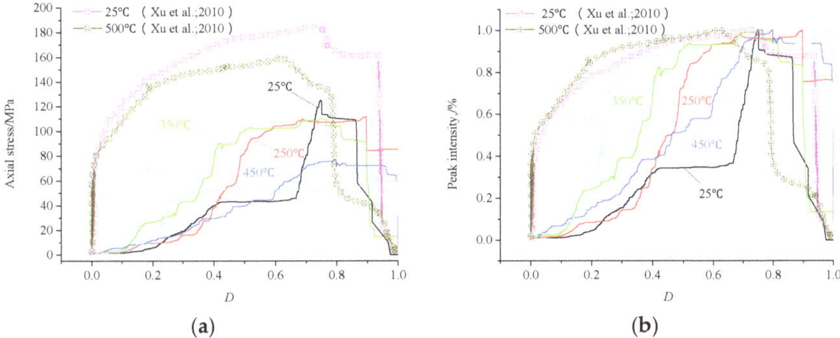

Figure 12. Mechanical damage curve of samples after natural cooling (source: Xu et al. [40]). (**a**) Mechanical damage curve under uniaxial compression; (**b**) Mechanical damage curve is represented by the ratio of axial stress to peak strength.

4. Conclusions

The following conclusions are made based on an experimental study of the mechanical properties of granite that have been exposed to temperatures from 250 °C to 650 °C under cyclic heating and cooling with circulating water, and cracking of the granite due to thermal shock with rapid water cooling.

(1) The uniaxial compression of granite undergoes three stages of changes after cyclic heating and cooling with circulating water at different heat treatment temperatures: pore compression, and elastic and strain softening deformations. The number of cycles has a significant influence on the peak stress, peak strain and uniaxial compressive strength. The transformation from brittleness to ductility is not only affected by the number of cycles, but also by the effects of temperature, which are more obvious. At a temperature of 250 °C, the granite becomes more plastic and is no longer brittle after 15 cycles. At heat treatment temperatures of 350 °C, 450 °C, 550 °C, and 650 °C, the samples become more plastic after 5, 1, 1 and 0 cycles, respectively. After heating and cooling with circulating water, the samples become even more plastic. During the cycles, temperature is seen to play a decisive role in the transformation of the granite samples from brittle to plastic.

(2) After cyclic heating and cooling with circulating water, the peak strength of the granite at different temperatures decreases, which can be described in three stages: Stage I is the rapid decline of the peak strength, Stage II is when the decline of the peak strength is reduced, and in Stage III, the strength of the sample is basically maintained at a constant level.

(3) The AE curve agrees well with the stress-time curve at each test temperature point, thus, mechanical damage can be defined by rock sample AE ringing counting rate and the threshold stress of rock damage is about 60% of the strength limit. The mechanical damage increases with the

increase of N water cooling at high-temperature cycles under the same stress. The thermal damage defined by the elastic modulus is generally increased, and can be synthesized by the four-order curve equation. The thermal stress of granite after cyclic high temperature and water cooling is caused by the thermal expansion effect of mineral particles in the structural composition and the thermal stress equation of particle structure was established, the main causes of damage were analyzed based on damage-cracking theory.

(4) With increases in the treatment temperature and more cycles of heating and water cooling, the failure mode of the granite transforms from brittle to plastic. The failure mode is transformed from slipping of the shear surface to flaking and particle spalling, and the color of the samples changes from gray to pale yellow. The temperature therefore affects the failure mode of the granite.

This paper attempts to study the mechanical parameter damage of granite samples after cyclic heating and cooling with cool water. However, the complexity of rock structures may lead to scattered data during experiments, so the interferences introduced by individual differences among the samples cannot be excluded. Therefore, these conclusions need further experiments for validation, and further investigation on their applicability under various conditions, so that more test methods can be developed for recognizing the damage process of rock failure.

Author Contributions: D.Z., H.J., Q.Y. and G.H. conceived and designed the experiments; D.Z. and Q.Y. performed the experiments; D.Z. and G.H. analyzed the data; H.J. contributed laboratory equipment; D.Z. and Q.Y. wrote the paper.

Funding: This research was funded by the National Natural Science Foundation of China (Grant Nos. 51734009, 51709260, 51704279), and the Natural Science Foundation of Jiangsu Province, China (No. BK20170276).

References

1. Nicholson, K. Environmental protection and the development of geothermal energy resources. *Environ. Geochem. Health.* **1994**, *16*, 86–87. [CrossRef] [PubMed]
2. Liu, R.; Li, B.; Jiang, Y. Critical hydraulic gradient for nonlinear flow through rock fracture networks: Theroles of aperture, surface roughness, and number of intersections. *Adv. Water Resour.* **2016**, *88*, 53–65. [CrossRef]
3. Liu, R.; Li, B.; Jiang, Y. A fractal model based on a new governing equation of fluid flow in fractures for characterizing hydraulic properties of rock fracture networks. *Comput. Geotech.* **2016**, *75*, 57–68. [CrossRef]
4. Liu, R.; Jiang, Y.; Li, B.; Wang, X. A fractal model for characterizing fluid flow in fractured rock masses based on randomly distributed rock fracture networks. *Comput. Geotech.* **2015**, *65*, 45–55. [CrossRef]
5. Li, B.; Liu, R.; Jiang, Y. Influences of hydraulic gradient, surface roughness, intersecting angle, and scale effect on nonlinear flow behavior at single fracture intersections. *J. Hydrol.* **2016**, *538*, 440–453. [CrossRef]
6. Tapponnier, P.; Brace, W.F. Development of stress-induced microcracks in Westerly Granite. *Int. J. Rock Mech. Min. Sci. Geomech. Abstr.* **1976**, *13*, 103–112. [CrossRef]
7. Wair, R.S.C.; Lo, k.Y.; Rowe, R.K. Thermal stress analysis in rock with nonlinear properties. *Int. J. Rock Mech. Min. Sci. Geomech. Abstr.* **1982**, *19*, 211–220.
8. Simpson, C. Deformation of granitic rocks across the brittle-ductile transition. *J. Struct. Geol.* **1985**, *7*, 503–511. [CrossRef]
9. Singh, B.; Ranjith, P.G.; Chandrasekharam, D.; Viete, D.; Singh, H.K.; Lashin, A.; Arifi, N.A. Thermo-mechanical properties of Bundelkhand granite near Jhansi, India. *J. Geomech. Geophys. Geo-Energy Geo-Resour.* **2015**, *1*, 35–53. [CrossRef]
10. Sygała, A.; Bukowska, M.; Janoszek, T. High Temperature Versus Geomechanical Parameters of Selected Rocks—The Present State of Research. *J. Sustain. Min.* **2013**, *12*, 45–51. [CrossRef]
11. Ranjith, P.G.; Viete, D.R.; Chen, B.J.; Perera, M.S. Transformation plasticity and the effect of temperature on the mechanical behaviour of Hawkesbury sandstone at atmospheric pressure. *Eng. Geol.* **2012**, *151*, 120–127.
12. Araújo, R.G.S.; Sousa, J.L.A.O.; Bloch, M. Experimental investigation on the influence of temperature on the mechanical properties of reservoir rocks. *Int. J. Rock Mech. Min. Sci.* **1997**, *34*, 298.e1–298.e16. [CrossRef]
13. Su, H.; Jing, H.; Du, M.; Wang, C. Experimental investigation on tensile strength and its loading rate effect of sandstone after high temperature treatment. *Arab. J. Geosci.* **2016**, *9*, 616–627. [CrossRef]

14. Meng, X.; Liu, W.; Meng, T. Experimental Investigation of Thermal Cracking and Permeability Evolution of Granite with Varying Initial Damage under High Temperature and Triaxial Compression. *Adv. Mater. Sci. Eng.* **2018**, *4*, 1–9. [CrossRef]

15. Chen, S.; Yang, C.; Wang, G. Evolution of thermal damage and permeability of Beishan granite. *Appl. Therm. Eng.* **2017**, *110*, 1533–1542. [CrossRef]

16. Guo, L.L.; Zhang, Y.B.; Zhang, Y.J.; Yu, Z.W.; Zhang, J.N. Experimental investigation of granite properties under different temperatures and pressures and numerical analysis of damage effect in enhanced geothermal system. Renew. *Energy* **2018**, *126*, 107–125. [CrossRef]

17. Rong, G.; Peng, J.; Yao, M.; Jiang, Q.H.; Wong, L.N.Y. Effects of specimen size and thermal-damage on physical and mechanical behavior of a fine-grained marble. *Eng. Geol.* **2018**, *232*, 46–55. [CrossRef]

18. Yu, Q.L.; Ranjith, P.G.; Liu, H.Y.; Yang, T.H.; Tang, S.B.; Tang, C.A.; Yang, S.Q. A Mesostructure-based Damage Model for Thermal Cracking Analysis and Application in Granite at Elevated Temperatures. *Rock Mech. Rock Eng.* **2015**, *48*, 2263–2282. [CrossRef]

19. Yang, S.Q.; Ranjith, P.G.; Jing, H.W.; Tain, W.L.; Ju, Y. An experimental investigation on thermal damage and failure mechanical behavior of granite after exposure to different high temperature treatments. *Geothermics* **2017**, *65*, 180–197. [CrossRef]

20. Uetsuji, Y.; Zako, M. On Evaluation Procedure to AE Test for Fiber Reinforced Composite Materials based on Damage Mechanics. *Trans. Jpn. Soc. Mech. Eng.* **1998**, *64*, 2938–2944. [CrossRef]

21. Venturini Autieri, M.R.; Dulieu-Barton, J.M. Initial Studies for AE Characterisation of Damage in Composite Materials. *Adv. Mater. Res.* **2010**, *13–14*, 273–280.

22. Ding, Q.L.; Ju, F.; Mao, X.B.; Ma, D.; Yu, B.Y.; Song, S.B. Experimental investigation of the mechanical behavior in unloading conditions of Sandstone after high-temperature treatment. *Rock Mech. Rock Eng.* **2016**, *49*, 2641–2653. [CrossRef]

23. Watanabe, H.; Murakami, Y.; Ohtsu, M. Quantitative Evaluation of Damage in Concrete Based on AE. *J. Soc. Mater. Sci. Jpn.* **2001**, *50*, 1370–1374. [CrossRef]

24. Suzuki, T.; Ohtsu, M. Quantitative damage evaluation of structural concrete by a compression test based on AE rate process analysis. *Constr. Build. Mater.* **2004**, *18*, 197–202. [CrossRef]

25. Sagar, R.V.; Prasad, B.K.R.; Kumar, S.S. An experimental study on cracking evolution in concrete and cement mortar by the b-value analysis of acoustic emission technique. *Cem. Concr. Res.* **2012**, *42*, 1094–1104. [CrossRef]

26. Vidya Sagar, R.; Raghu Prasad, B.K. A Review of recent development in parametric based acoustic emission techniques applied to concrete structures. *Nondestruct. Test. Eval.* **2012**, *27*, 47–68. [CrossRef]

27. Wu, G.; Zhai, S.T.; Wang, Y. Research on characteristics of mesostructure and acoustic emission of granite under high temperature. *J. Rock Soil Mech.* **2015**, *36*, 351–356.

28. Chen, G.Q.; Li, T.B.; Zhang, G.F.; Yin, H.Y.; Zhang, H. Temperature effect of rock burst for hard rock in deep-buried tunnel. *Nat. Hazard* **2014**, *72*, 915–926. [CrossRef]

29. Wang, J.S.Y.; Mangold, D.C.; Tsang, C.F. Thermal impact of waste emplacement and surface cooling associated with geologic disposal of high-level nuclear waste. *Environ. Geol. Water Sci.* **1988**, *11*, 183–239. [CrossRef]

30. Jiang, L.H.; Chen, Y.L.; Liu, M.L. Experimental study of mechanical properties of granite under high/low temperature freeze-thaw cycles. *Rock Soil Mech.* **2011**, *32*, 319–323.

31. Kim, K.; Kemeny, J.; Nickerson, M. Effect of Rapid Thermal Cooling on Mechanical Rock Properties. *Rock Mech. Rock Eng.* **2014**, *47*, 2005–2019. [CrossRef]

32. Kumari, W.G.P.; Ranjith, P.G.; Perera, M.S.A.; Chen, B.K.; Abdulagatov, I.M. Temperature-dependent mechanical behaviour of Australian Strathbogie granite with different cooling treatments. *Eng. Geol.* **2017**, *229*, 31–44. [CrossRef]

33. Isaka, B.L.A.; Gamage, R.P.; Rathnaweera, T.D.; Perera, M.S.A.; Chandrasekharam, D.; Kumari, W.G.P. An Influence of Thermally-Induced Micro-Cracking under Cooling Treatments: Mechanical Characteristics of Australian Granite. *Energies* **2018**, *11*, 1338. [CrossRef]

34. Xu, X.L.; Zhang, Z.Z. Acoustic Emission and Damage Characteristics of Granite Subjected to High Temperature. *Adv. Mater. Sci. Eng.* **2018**, *4*, 1–12. [CrossRef]

35. Shao, S.; Wasantha, P.L.P.; Ranjith, P.G.; Chen, B.K. Effect of cooling rate on the mechanical behavior of heated Strathbogie granite with different grain sizes. *Int. J. Rock Mech. Min. Sci.* **2014**, *70*, 381–387. [CrossRef]

36. Rickard, H. Fire behavior of mining vehicles in underground hard rock mines. *Int. J. Min. Sci. Technol.* **2017**, *27*, 627–634.

37. Wang, G. Experiment Research on the Effects of Temperature and Viscoelastoplastic Analysis of Beishan Granite. Ph.D. Thesis, Xi'an University of Science and Technology, Xi'an, China, 2003.

38. Zhi, L.P.; Xu, J.Y.; Liu, Z.Q.; Liu, S.; Chen, T.F. Research on ultrasonic characteristics and Brazilian splitting tensile test of granite under post-high temperature. *Rock Soil Mech.* **2012**, *33*, 61–66.

39. Yin, T.B.; Li, X.B.; Cao, W.Z.; Xia, K.W. Effects of Thermal Treatment on Tensile Strength of Laurentian Granite Using Brazilian Test. *Rock Mech. Rock Eng.* **2015**, *48*, 2213–2223. [CrossRef]

40. Xu, X.L.; Gao, F.; Ji, M. Damage Mechanical Analysis of Fracture Behavior of Granite Under Temperature. *J. Wuhan Univ. Technol.* **2010**, *32*, 143–147.

41. Xu, X. A preliminary study on basic mechanical properties for granite at high temperature. *Chin. J. Geotech. Eng.* **2000**, *22*, 332–335.

42. Homand-Etienne, F.; Houpert, R. Thermally induced microcracking in granites: characterization and analysis. *Int. J. Rock Mech. Min. Sci. Geomech. Abstr.* **1989**, *26*, 125–134. [CrossRef]

43. Kou, S.Q. Effect of thermal cracking damage on the deformation and failure of granite. *Acta Mech. Sin.* **1987**, *242*, 235–240.

44. Sun, Q.; Zhang, W.Q.; Xue, L.; Zhang, Z.Z.; Su, T.M. Thermal damage pattern and thresholds of granite. *Environ. Earth Sci.* **2015**, *74*, 2341–2349. [CrossRef]

45. Somerton, W.H.; Boozer, G.D. A method of measuring thermal diffusivities of rocks at elevated temperatures. *AIChE. J.* **2010**, *7*, 87–90. [CrossRef]

46. Hartlieb, P.; Toifl, M.; Kuchar, F.; Meisels, R.; Antretter, T. Thermo-physical properties of selected hard rocks and their relation to microwave-assisted comminution. *Miner. Eng.* **2016**, *91*, 34–41. [CrossRef]

47. Just, J.; Kontny, A. Thermally induced alterations of minerals during measurements of the temperature dependence of magnetic susceptibility: A case study from the hydrothermally altered Soultz-sous-Forêts granite, France. *Int. J. Earth Sci.* **2012**, *101*, 819–839. [CrossRef]

 *processes*

Article

Experimental Study of the Microstructural Evolution of Glauberite and Its Weakening Mechanism under the Effect of Thermal-Hydrological-Chemical Coupling

Shuzhao Chen [1],*, Donghua Zhang [2],*, Tao Shang [1] and Tao Meng [3]

[1]   College of Mining Engineering, China University of Mining and Technology, Xuzhou 221116, China; hedongsanjianke@sina.com
[2]   College of Mining Engineering, Taiyuan University of Technology, Taiyuan 030024, China
[3]   College of Chemical and Biological Engineering, Taiyuan University of Science and Technology, Taiyuan 030024, China; huyaoqing_tyut@163.com
*   Correspondence: hedongjianke@sina.com (S.C.); zhangdonghua@tyut.edu.cn (D.Z.)

Received: 19 June 2018; Accepted: 19 July 2018; Published: 24 July 2018

Abstract: The microstructures of rock gradually evolve with changes in the external environment. This study focused on the microstructure evolution of glauberite and its weakening mechanism under different leaching conditions. The porosity were used as a characteristic index to study the effect of brine temperature and concentration on crack initiation and propagation in glauberite. The research subjects were specimens of $\phi 3 \times 10$ mm cylindrical glauberite core, obtained from a bedded salt deposit buried more than 1000 m underground in the Yunying salt formation, China. The results showed that when the specimens were immersed in solution at low temperature, due to hydration impurities, cracks appeared spontaneously at the centre of the disc and the solution then penetrated the specimens via these cracks and dissolved the minerals around the crack lines. However, with an increase of temperature, the dissolution rate increased greatly, and crack nucleation and dissolved regions appeared simultaneously. When the specimens were immersed in a sodium chloride solution at the same concentration, the porosity s presented gradual upward trends with a rise in temperature, whereas, when the specimens were immersed in the sodium chloride solution at the same temperature, the porosity tended to decrease with the increase of sodium chloride concentration. In the process of leaching, the hydration of illite, montmorillonite, and the residual skeleton of glauberite led to the expansion of the specimen volume, thereby producing the cracks. The diameter expansion rate and the expansion velocity of the specimen increased with temperature increase, whereas, due to the common-ion effect, the porosity of the specimen decreases with the increase of sodium chloride solution concentration.

Keywords: glauberite cavern for storing oil & gas; thermal-hydrological-chemical interactions; temperature; brine concentration; microstructure; micro-CT

---

1. Introduction

Thermal-hydrological-chemical (THC) interactions are widely recognized and studied by researchers from different fields, including hydraulic engineering, civil engineering and geological environmental engineering [1–5]. The existing research on THC coupling has benefited these different engineering disciplines, with each of them having different key interacting factors influencing various projects. In the area of underground mineral resource development using the method of leach mining, pore structure of geo-materials, solution temperature, and chemical reaction kinetics are the key factors to solution migration, solute diffusion and recovery efficiency [6–11]. Leach mining is the extraction of useful minerals (elements or compounds) such as inorganic salt, copper, and uranium which are naturally dissolved in a leaching solution (water, chemical solvents, or microorganisms) [6–8]. The method was first used for sodium chloride and copper mining, and in the late 1970s, it was widely used for glauberite mining and refining. Glauberite ($Na_2Ca(SO_4)_2$) is an important sulphate deposit with potential industrial value because, not only can it be used as a substitute after the depletion of mirabilite deposits, but it is also recognized as the ideal medium for the storage of oil, natural gas, and carbon dioxide due to its low permeability, simple hydrogeology, and wide geographical distribution (see Figure 1) [12–14]. At present, the traditional mining method used in glauberite (blast-cave-leach mining) has many disadvantages, such as a high labour intensity, low recovery rate, and creation of environmental pollution. Therefore, some scholars proposed the adoption of in situ leaching to mine glauberite deposits [13]. In the glauberite mining (or leaching) process using the method of in situ leach mining, the glauberite ore is immersed in different concentrations of sodium chloride solution (or brine) at different temperatures for a few days. Under the effect of hot brine, the microstructure of glauberite minerals is changed due to the dissolution of sodium sulphate and the modification of calcium sulphate structure (i.e., the crystallization of calcium sulphate dihydrate) [13]. It is interesting to note that the leaching process is a typical THC multi-field coupling problem which involved the interaction and multi-influence of fluid migration, mineral dissolution, chemical reaction, solute diffusion, and heat exchange. In this complex process, these influence factors may greatly affect the pore microstructure, eventually resulting in a series of changes in macroscopic properties (i.e., mechanical strength and hydraulic conductivity), as happens with other rocky materials, such as cement-based ones [15–17]. Notably, the pore microstructure is closely related with the leaching efficiency and cavity stability of the glauberite deposit. Given all these, it raises questions about how the pore microstructure of glauberite changes under the effect of THC coupling. Will the solution temperature and concentration greatly affect the pore microstructure and eventually affect the mechanical strength and permeability? To address the questions, it is necessary to study the evolution of the pore microstructure of glauberite during the leaching process.

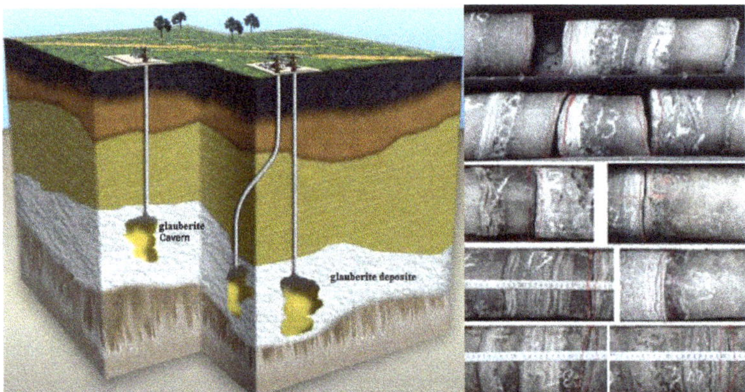

Figure 1. (**a**) In situ leach mining with a cavern for oil, natural gas and carbon dioxide storage; (**b**) Glauberite cores.

Several researchers previously concentrated on the effect of external environment on rock microstructures under the THC coupling effect. Meer et al. [18] have discussed the microstructure evolution of gypsum in the present of saline pore fluid and NaCl-free pore fluid. They concluded that pressure solution in gypsum is controlled by the precipitation reaction, and diffusion is unlikely to be the rate-controlling mechanism of pressure solution. Zhao et al. [19] investigated evolution characteristics of pores and the residual porous skeleton of glauberite, and concluded that three zones are present from dissolution interface to the outside in the process of glauberite dissolution. Zhang et al. [14] studied the cementation state and the grain arrangement mode of the halite and salt interlayers with scanning electron microscope analyses. They concluded that halite and argillaceous anhydrite are mixed together, and the pores of the halite are filled with argillaceous minerals. Liang et al. [20] have studied the mechanical properties of gypsum soaked in hot brine. They concluded that the weakening coefficient of gypsum increases with the increase of brine temperature, whereas it decreases with the increase of brine concentration. Yu et al. [21] experimentally presented the meso-structure of gypsum rock after treatment with pure water, half saturated brine, and saturated brine. They asserted that the weakening effect of half-saturated brine on gypsum is the most severe. Meng et al. [4] studied the weakening mechanism of gypsum under brine saturation. They concluded that the foremost reason for the weakening of gypsum is that the heat or water can easily separate the hydrogen bonds, which will lead to molecular chain and layer structure separation and dislocation.

In summary, the external environment may greatly affect the physical properties of rock, and the porosity is an important parameters to characterize the microstructure of rock [22–25]. However, all the research discussed above showed the microscopic damaging behaviours of rock salt, granite, and gypsum under different conditions and, to the authors' best knowledge, there is little research on microstructural evolution and the weakening mechanism of glauberite under the effect of THC coupling. It is well known that minerals variably respond in the presence of chemicals depending on their crystal structure and chemical composition [4]. Thus, in this study, the micro-computed tomography (MCT) scanning was conducted to obtain the porosity of specimens under different conditions and to discuss the corresponding weakening mechanisms. This study is of great significance to the development of micromechanics, in situ leach mining, and technological progress.

2. Methods and Materials

2.1. Test Pieces and Test Methods

Glauberite ores were selected with the main component (75%) being $Na_2Ca(SO_4)_2$, and the remaining components illite (10%), chlorite (5%), quartz (4%), mica (4%), and montmorillonite (2%). The core containing the glauberite was retrieved by the China National Petroleum Corporation in the Yingcheng Mine. We were not involved in the drilling work. However, we performed the work of core sampling and plug preparation. Note that we did not use the conventional method (a core-drilling machine) because it would cause unavoidable damage to the plug during the sample preparation. Therefore, we used a line cutting machine, which produces less damage, as shown in the figure below. The diameter of steel wire that was used in this paper is 0.3 mm. Glauberite cores were obtained from a bedded salt deposit buried more than 400 m underground in the Yunying salt formation, China, which belongs to a marine precipitate sequence. The Yunying Salt Mine is located in the middle of the Yunying Basin, in the northeast of the Jianghan Basin, Hubei Province. The Yunying Basin is an interior terrestrial faulted saline lake basin formed in Cretaceous and Tertiary systems. Well ZK-1053 was drilled by the China National Petroleum Corporation in the Yingcheng Mine to provide the formation data for the experimental analysis (see Figure 2).

In order to observe the evolution of the meso-structure clearly, the glauberite ores were processed into many small (ϕ3 × 10 mm) cylindrical specimens (see Figure 2b,c). The lithology of glauberite is chiefly middle-fine salt with good sorting and low texture maturity. The shape of the crystals is mainly subangular and sub-rounded, and the content of matrix is low; the particle is coarser, and the fractures are very clear; the phenocryst is mainly composed of sodium sulphate, calcium sulphate, and illite, and the matrix is quartz; and the rock is characterized by low porosity and low permeability.

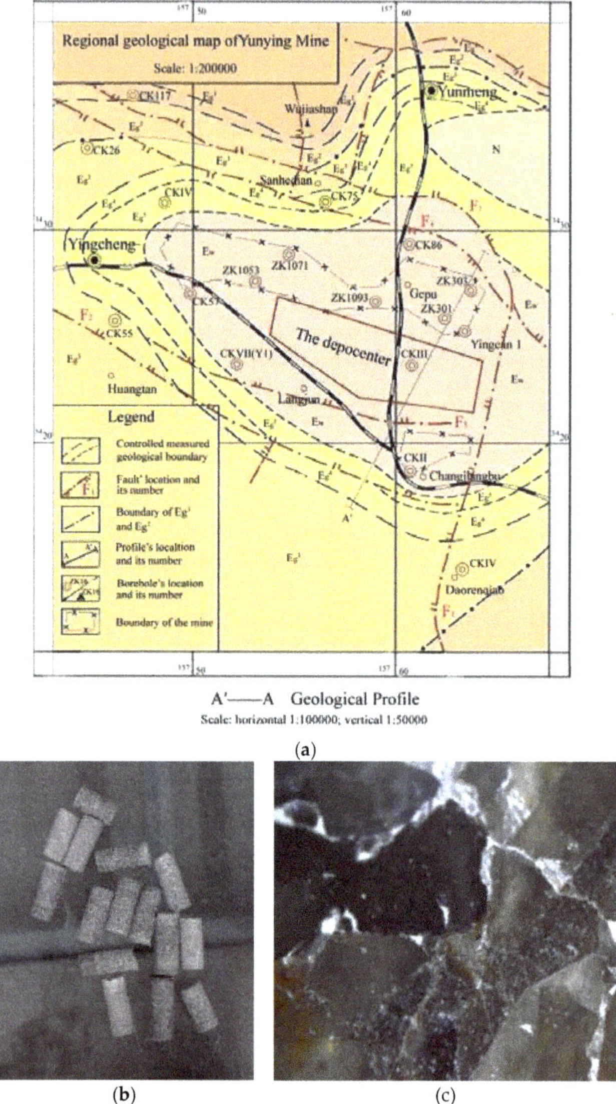

(a)

(b)   (c)

Figure 2. (**a**) Regional geological map of Yunying mine [14]; (**b**) glauberite ores ($\phi3 \times 10$ mm); (**c**) the thin section of glauberite (200×).

2.2. Test Equipment

The main equipment used in this study was the μCT225kVFCB high-accuracy (μm grade) CT test analysis system (the machine is made in Taiyuan University of Technology, Shanxi province, China) (see Figure 3a). It consists of a digital flat detector, a micro-focus X-ray machine, a high-precision work turntable and fixture, an acquisition and analysis system and other structural components, as shown in Figure 2. The micro-focus X-ray machine in the micro CT test system allowed weak currents (only 0.01–3.00 mA) which issued a small cone beam of X-ray scanning and then projected the scanned image to the digital flat-panel detector for display. The minimum focus size of the micro-focus X-ray machine was 3 μm, the focal length was 4.5 mm, and the maximum power was 320 W. In high-magnification tests for small specimens, a low power of 20–30 W ensured sharp focus. Due to all of these features, the system could realize three-dimensional CT scanning analysis of various metal and non-metallic materials. The magnification is 1–400 times, the rock sample size is φ1–50 mm, and the scanning unit resolution is 0.194 mm per magnification. For a specimen 400 times magnified, the size of the scan cell was 0.5 μm, which could distinguish 0.5 μm pores and cracks.

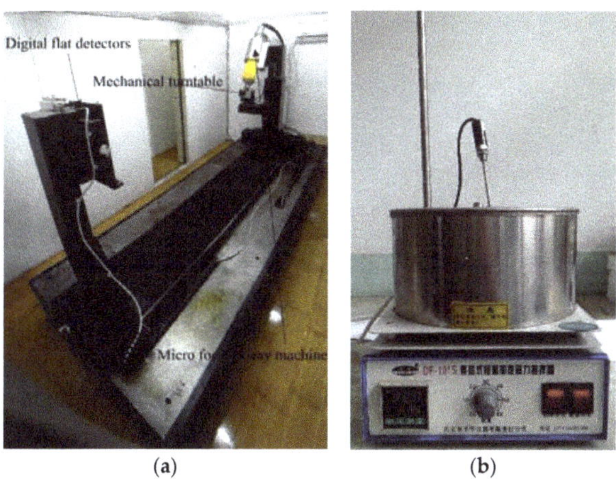

(a)   (b)

Figure 3. (**a**) μ-CT225kvFCB MCT test system. (**b**) Self-heated system with constant temperature (SHSWCT).

The basic principle of CT scanning relies on the penetrating power of the X-ray. In the scanned image, the dark areas relate to regions of low density, and the bright colours indicate the high-density regions—in grayscale, the whiter areas represent the higher density. Therefore, by scanning slices and unit analysis in all directions, the morphological distribution of the rock particles and the distribution and connectivity of the pore fissures would be clear.

2.3. Test Method

Then, the specimens were immersed in three different solutions and at three different temperatures; thus, there were nine different sets of experimental conditions. The solutions were pure water, half-saturated brine and saturated brine. The temperatures were 20, 50, and 80 °C. Six specimens were tested under each set of conditions; thus, 54 tests were performed in total.

Firstly in this experiment, MCT scanning was conducted for the specimens in a dry state, and then each specimen was immersed in pure water, semi-saturated, and saturated brine, respectively, at three different temperatures of 20, 50, and 80 °C. An intelligent water bath thermostatic device was used to accurately control the temperature throughout the process. Due to the changes of microstructure of specimens immersed in different conditions, it was important to conduct the MCT scanning of the specimens at regular intervals. The soaking duration for the specimens was 30 h, and the MCT scans of the column specimens were conducted while they were in the dry state and after they had been immersed for 10, 20, and 30 h. In order to ensure the comparability of the results before and after the scanning, the same part of the specimen was selected for comparison, and the corresponding microstructural evolutions were analysed. The detailed experimental procedures are shown below.

Experimental process: (1) Put 2000 mL of the pre-prepared liquid (pure water, half-saturated brine, or saturated brine) into dry and clean reagent bottles and mark them with numbers; (2) Heat each bottle using the SHSWCT (see Figure 3b) for approximately 20 min to ensure that the liquid temperature in each bottle rises to the predetermined temperature; (3) Gently place each specimen into the correct bottle filled with liquid and seal each bottle to ensure that the concentration of the solution in the bottles does not change via heating. Maintain a stable temperature for the predetermined duration; (4) Tweeze the specimens out of the bottles lightly to make sure that the specimens are not damaged by the tweezers, and dry the surface of each specimen with filter paper; (5) Dry the specimens using a drying oven for approximately 10 min at 30 °C to ensure that the specimens are desiccated; (6) Scan the specimens, including the original specimens using MCT; (7) After performing the MCT tests, put the same specimens into the bottle again for the next soaking saturation, and repeat the steps 4–7.

After the MCT tests, the MCT dataset needs to be constructed by binarization, in which a distinct set of classes is created and every single datapoint is assigned to a class. That is, the binarization of an image is to set the gray value of the pixel on the image to 0 (black) or 255 (white). However, the value of the final segmentation result is highly dependent on the thresholding techniques; hence, the porosity partially reflects the choices made by the operator. Note that conventional thresholding techniques (e.g., global thresholding) usually fail to extract narrow cracks and low porosity. In this paper we, therefore, chose to apply 3D voxel-based segmentation using the multiscale Hessian filter (MSHFF) of [26]. This approach, which involves low computational requirements, cannot only precisely extract the fractures and apertures from the MCT database but also better estimate the micro-porosity for rock types with smaller ranges of pore sizes. The MSHFF method has been implemented as macrocode in MATLAB. Second, the areas of the extracted fractures and apertures are calculated by using MATLAB. Then, the porosity may be obtained by dividing the fracture and aperture areas by the total area of the specimen.

3. Test Results

3.1. MCT Analysis of Microstructural Evolution

Figures 4–6 show the microstructure evolution of glauberite in pure water at different temperatures. Obviously, the microstructure evolution of glauberite varied greatly according to the pure water temperature. As can been seen in Figure 4, the structure of the specimen was compact with no obvious cracks after 10 h of immersion in pure water. However, when the specimens were immersed for 20 h, arbitrary branched and intersecting micro-cracks appeared inside the specimens. After 30 h, the aperture and length of the cracks increase considerably. In Figure 5, the water temperature was 50 °C. Micro-cracks emerged after 10 h of immersion, and the micro-crack density increased with the soaking time. In addition, after 30 h, banded dissolved areas were formed around the micro-cracks, and secondary branching cracks were nucleated in the main crack. Figure 6 indicates that the crack number greatly increased over time. After 30 h of immersion, the specimen broke up completely, and notches appeared at the edges. However, vastly distinct from Figures 4 and 5, large-scale dissolved areas (i.e., dark areas) were observed in Figure 5 when the specimens were only immersed for 10 h at 80 °C, whereas the dissolved areas in Figures 4 and 5 were negligible when immersed even for 30 h after 30 h of immersion. Overall, from the comparison of these MCT images, it could be determined that temperature may have accelerated the breakdown and dissolution rate of the glauberite.

Figures 6–8 display the microstructure evolution of the specimen in a 80 °C solution with various concentrations. It is clear in Figure 7 that there were a great number of internal cracks and point-like dissolved areas inside the specimen after immersion in half saturated brine for 10 h. Additionally, the crack density and dissolution areas increased over time until, finally, the specimen broke up thoroughly. However, as seen in Figure 7, the dissolution areas and the number of cracks were much smaller than those shown in Figure 6, and no notches on the edges occurred. Figure 8 shows that the dissolved areas were only concentrated around the crack lines. The location of dissolved areas was totally different from that of Figures 6 and 7. Furthermore, the dissolved areas appeared after immersion in saturated brine for 30 h. Overall, by comparing the variation trend of dissolved areas, it was concluded that the chloride ion inhibited the dissolution of glauberite.

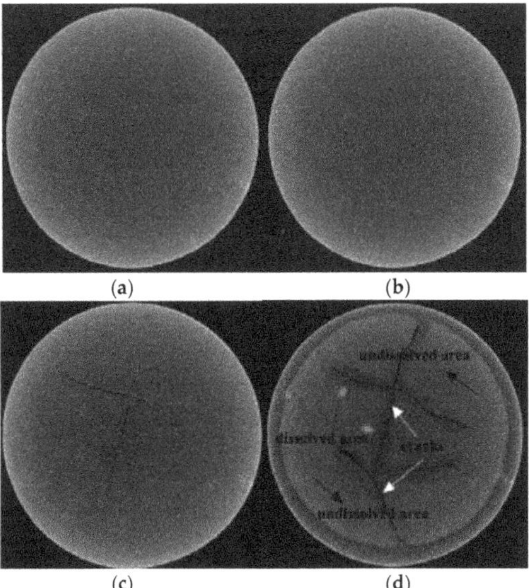

Figure 4. Microstructure evolutions of glauberite specimens immersed in pure water at 20 °C: (**a**) 20 °C pure water immersion, 0 h; (**b**) 20 °C pure water immersion, 10 h; (**c**) 20 °C pure water immersion, 20 h; and (**d**) 20 °C pure water immersion, 30 h.

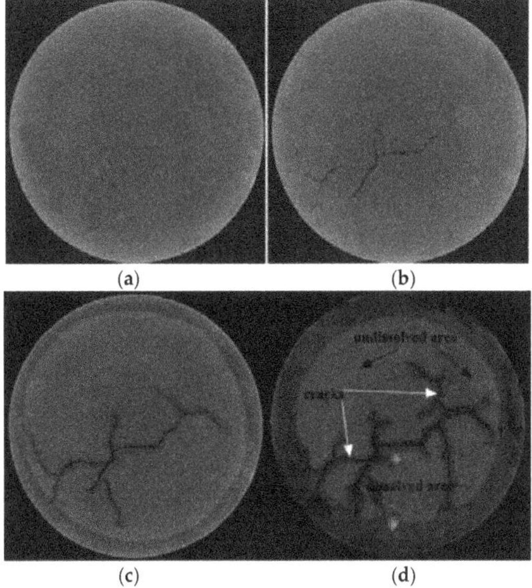

Figure 5. Microstructure evolutions of glauberite specimens immersed in pure water at 50 °C: (**a**) 50 °C pure water immersion, 0 h; (**b**) 50 °C pure water immersion, 10 h; (**c**) 50 °C pure water immersion, 20 h; and (**d**) 50 °C pure water immersion, 30 h.

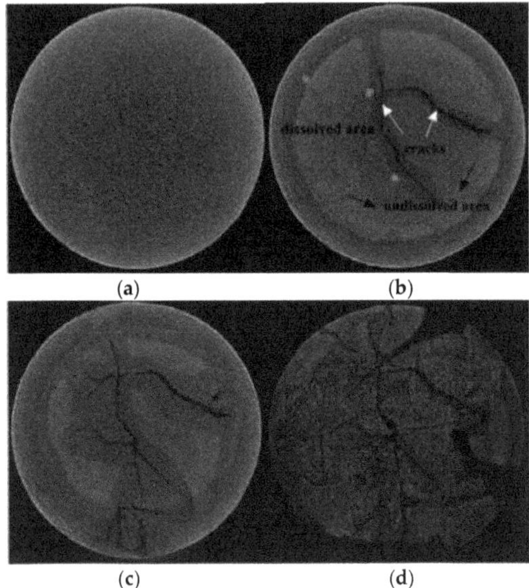

Figure 6. Microstructure evolutions of glauberite specimens immersed in pure water at 80 °C: (**a**) 80 °C pure water immersion, 0 h; (**b**) 80 °C pure water immersion, 10 h; (**c**) 80 °C pure water immersion, 20 h; and (**d**) 80 °C pure water immersion, 30 h.

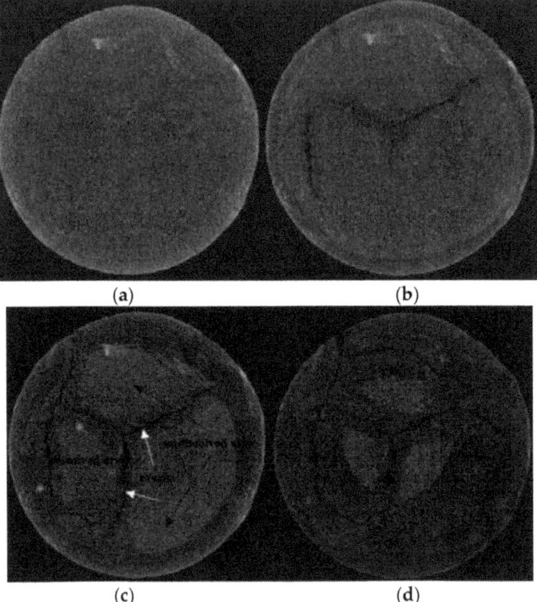

Figure 7. Microstructure evolutions of glauberite specimens immersed in half saturated brine at 80 °C: (**a**) 80 °C half saturated brine immersion, 0 h; (**b**) 80 °C half saturated brine immersion,10 h; (**c**) 80 °C half saturated brine immersion, 20 h; and (**d**) 80 °C half saturated brine immersion, 30 h.

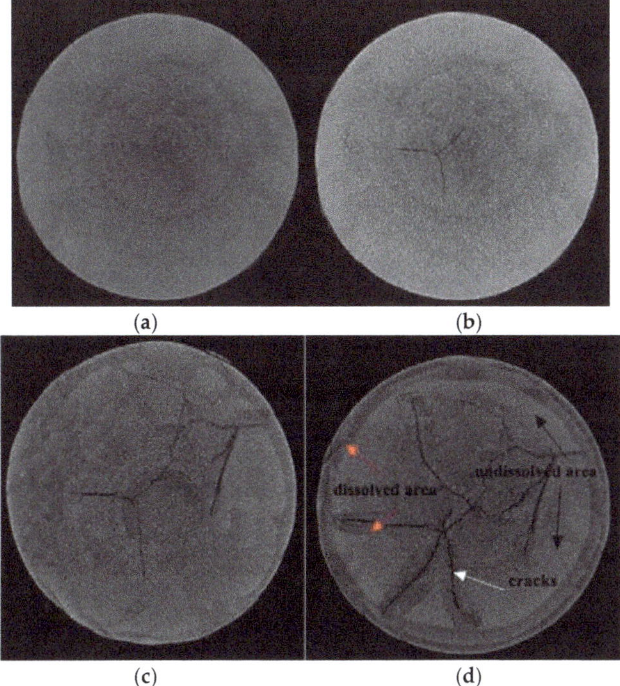

Figure 8. Microstructure evolutions of glauberite specimens immersed in saturated brine at 80 °C: (**a**) 80 °C saturated brine immersion, 0 h; (**b**) 80 °C saturated brine immersion, 10 h; (**c**) 80 °C saturated brine immersion, 20 h; and (**d**) 80 °C saturated brine immersion, 30 h.

3.2. The Porosity of Glauberite under Varying Conditions

Figure 9 and Table 1 shows the average porosity evolution of glauberite with different immersion times under different conditions. The porosity refers to the ratio of the total area of the minute voids in the cross section of the specimen to the total area of the cross section. In this paper, the self-developed MCT image analysis software was applied so as to statistically analyse the grey values of the glauberite MCT images, and determine the porosity under different conditions. As can be seen in Table 1, the porosity of glauberite specimens varied greatly depending on the solution temperature and concentration. When immersed in pure water at room temperature, the porosity of glauberite was 7.29% after immersing for 30 h. When the temperature was raised to 50 °C, the porosity of glauberite was 7.52% after immersing for 20 h and the porosity of glauberite after immersing for 10 h at 80 °C was 7.84%. Obviously, temperature played an important role in the dissolution of the glauberite and the evolution of pore microstructure. Furthermore, increasing the temperature shortened the dissolving time.

When comparing the porosity of specimens immersed in solution at 80 °C with different concentrations, it could be found that the porosity reached 19.48% after 30 h of immersion in pure water which was 16.23 times that of the initial porosity (dry condition), while the porosity was, respectively, 7.52% and 6.71% after immersion in half-saturated and saturated sodium chloride solution, which was 8.26 times and 3.92 times higher than the initial porosity. The differences between the porosity growth further proved that the concentration of a solution had an inhibiting effect on the evolution of glauberite pores. The higher the concentration the stronger the inhibition.

In summary, by analysing the micrographs and porosity of glauberite after immersion in pure water at different temperatures, it could be concluded that the dissolution was very slow at low temperature; the cracks appeared to initiate in the centre of the discs, and then acted as seepage channels for further dissolution. The dissolution rate increased greatly as the temperature rose, and the dissolved areas and kinked cracks were generated simultaneously in the specimens. Additionally, the crack density and the dissolution area increased with the temperature. By analysing the micrographs and porosity of glauberite after immersion in brine with varying concentrations, it could be concluded that the crack density and dissolution rate decreased with the increase in brine concentration.

Table 1. The average porosity of glauberite specimens in different immersion conditions.

Temperature	Immersing Condition	Porosity (%)				Regression Equations of Glauberite Porosity with Time
		0 h	10 h	20 h	30 h	
20 °C	Pure water	1.57	1.63	4.08	7.29	$y = 1.2898e^{0.0552x}$
	Half-saturated brine	1.88	1.91	1.97	2.12	$y = 1.8557e^{0.0039x}$
	Saturated brine	1.42	1.44	1.47	1.52	$y = 1.4135e^{0.0022x}$
50 °C	Pure water	1.73	3.29	7.52	10.97	$y = 1.7910e^{0.0637x}$
	Half-saturated brine	2.21	2.41	2.75	3.43	$y = 2.1417e^{0.0145x}$
	Saturated brine	1.86	1.91	1.99	2.14	$y = 1.8402e^{0.0046x}$
80 °C	Pure water	1.82	7.84	13.13	19.48	$y = 2.4756e^{0.0763x}$
	Half-saturated brine	0.91	2.39	4.86	7.52	$y = 1.0378e^{0.0705x}$
	Saturated brine	1.71	2.87	4.16	6.71	$y = 1.7491e^{0.0447x}$

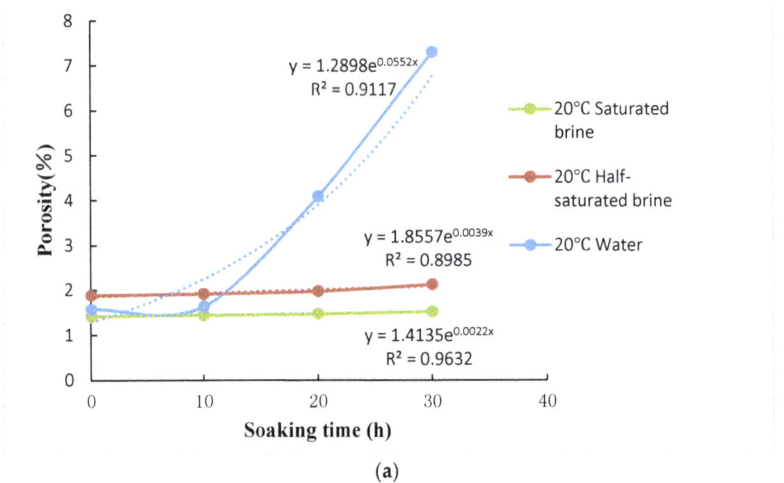

Figure 9. *Cont.*

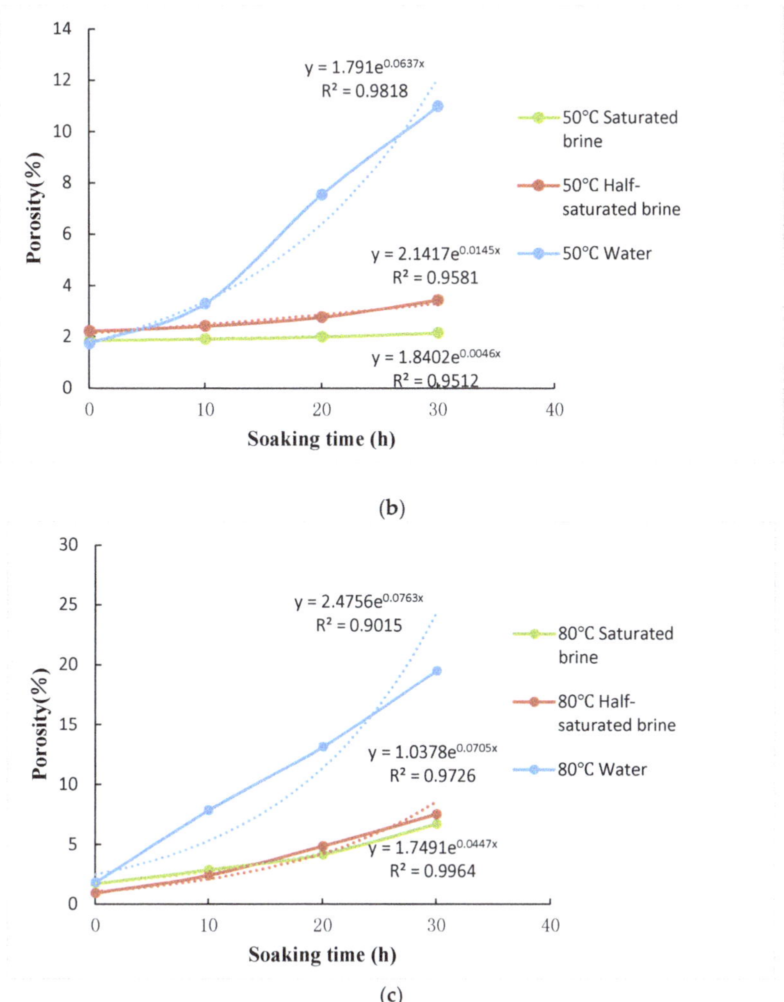

(b)

(c)

Figure 9. Variation in the average porosity of glauberite specimens with time under different immersion conditions. (**a**) 20 °C; (**b**) 50 °C; (**c**) 80 °C.

4. Discussion

4.1. Weakening Mechanism of Glauberite

Glauberite ($Na_2Ca(SO_4)_2$) is a special rock, with the main components of sodium sulphate (water-soluble) and calcium sulphate (slightly water-soluble) [12–14]. The microstructure evolution (or leaching) of glauberite in solution is a complicated process of dissolution-crystallization interaction [13]. Namely, the dissolution of sodium sulphate in glauberite is almost simultaneous with the crystallization of calcium sulphate dehydrate. When the specimen dissolution was totally complete, the surface of the specimen residuals were covered with a layer of white needle-like gypsum crystals. A detailed leaching process of glauberite is as follows:

There are two different kinds of evolution processes for the specimens soaked in liquid. The two evolutionary processes are shown below.

(1)   For the intact specimens with no cracks, the microstructural evolution process was very regular and orderly. As can be seen in Figures 10–12, the specimens were gradually dissolved from the outside to the inside, and the pores and three zones were formed during the leaching process. The zones were the dissolved area (or residual skeleton, porous structure), the undissolved area (i.e., dense and intact area), and the circular solid-liquid interface (i.e., circular line), respectively. As we all known, glauberite ($Na_2Ca(SO_4)_2$) is a special rock, with the main components of sodium sulphate ($NaSO_4$) (water-soluble) and calcium sulphate ($CaSO_4$) (slightly water-soluble). The sodium sulphate is water-soluble, whereas the calcium sulphate is poorly soluble in water. Hence, when the glauberite was in contact with water, the sodium ions and calcium ions (Na^+, SO_4^{2-}) in the crystal lattice were electrostatically attracted by the water molecules with an opposite charge, and if the attractive force of the water molecule to the ions was sufficient to overcome the attractive force between the ions in the crystal lattice, the glauberite crystal lattice was destroyed and the soluble minerals were dissolved by water. The chemical formula is as follows:

$$Na_2Ca(SO_4)_2 \rightarrow 2Na^+ + SO_4 + CaSO_4 \tag{1}$$

$$CaSO_4 \rightarrow Ca^{2+} + SO_4^{2-} \tag{2}$$

After dissolution, the remaining insoluble matter (i.e., $CaSO_4$ or $CaSO_4 \cdot 2H_2O$) in glauberite became the residual skeleton. On one hand, this increased the porosity of the specimen and, on the other hand, the penetration resistance of the solution was greatly reduced due to the seepage paths (i.e., residual skeleton) (see Figures 10–12). Then, the solution easily flowed via the preferential seepage paths through the dissolved zones and came into contact with the newly-exposed minerals to form a new solid-liquid interface. Eventually, the sodium sulphate would be completely dissolved and removed, whereas the calcium sulphate would completely become the residual skeleton.

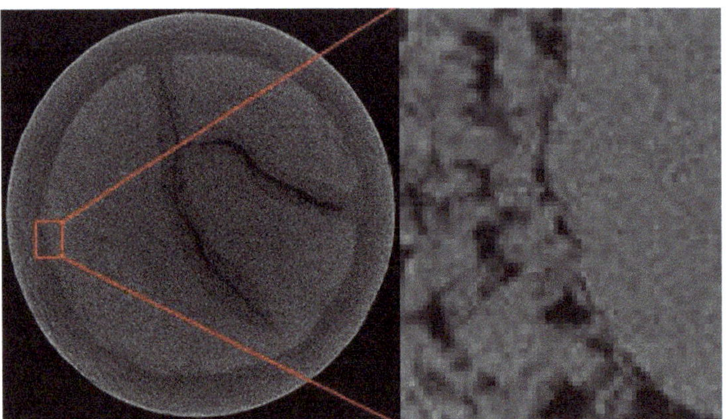

Figure 10. The dissolved areas (residual porous skeleton area) and undissolved areas.

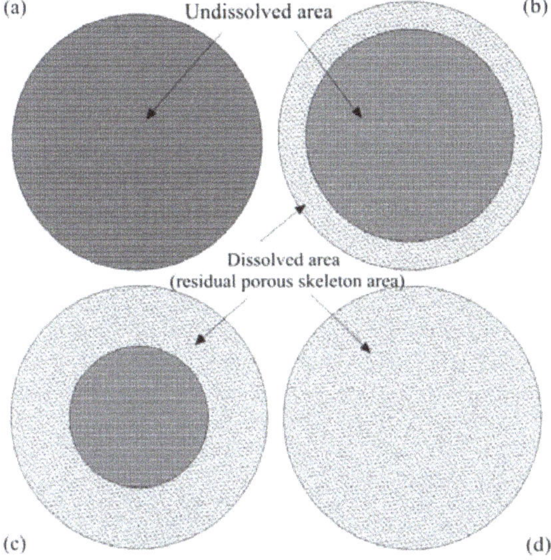

Figure 11. The spherical ore specimen (i.e., intact specimen with no cracks) dissolving process and ideal model diagram. (**a**) liquid immersion, 0 h; (**b**) liquid immersion, 10 h; (**c**) liquid immersion, 20 h; and (**d**) liquid immersion, 30 h.

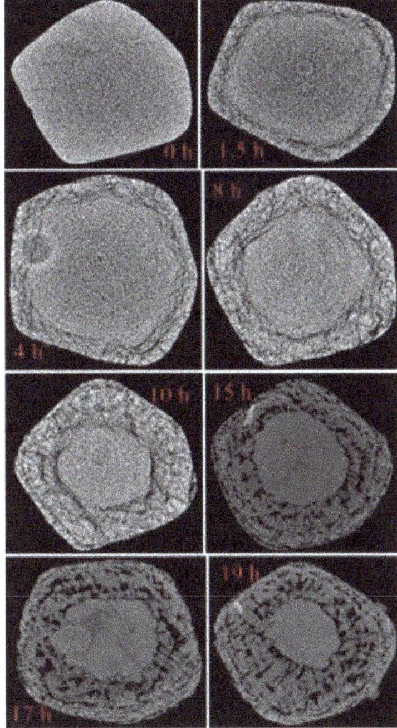

Figure 12. Dissolution of a small salt-gypsum specimen by pure water after different immersion times [19]. Images were made using MCT (250×).

(2)   For the specimen with cracks, the microstructural evolution process was very irregular and complex. By comparison with the microstructures evolution process at different soaking duration, there were four different zones formed in the MCT images. The zones were dissolved area (or residual skeleton), cracks area, undissolved area, and circular solid-liquid interface, respectively. The reason why the evolution processes is complex is that the cracks formed during the leaching process will greatly affect the flow field and path line of liquid. That is, the liquid will concurrently flow via the preferential seepage paths through the dissolved zones (or residual skeleton) and cracks zones, and come into contact with the newly-exposed minerals to form several new solid-liquid interfaces (i.e., the cracks will become the new solid-liquid interfaces). Then, the dissolution trajectories initiated at the previous circular interface and the crack lines, and spread to the surrounding undissolved areas (see in Figures 6–8 and 13). By comparison, there was only one interface during the whole leaching process for the specimens with no cracks. Hence, the evolution process was very irregular and complex. It is worth to note that, primarily due to the clay (i.e., mainly illite and montmorillonite) hydration, swelling and the thermal cracking (more details see in the Section 4.1), the cracking can easily occur to the specimens when soaked in liquid.

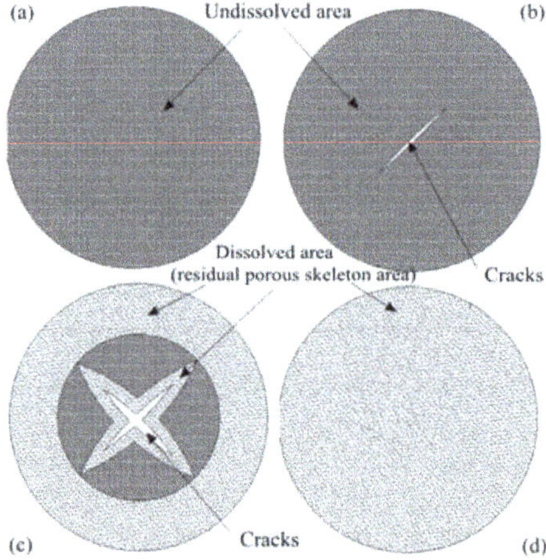

Figure 13. The spherical ore specimen (i.e., fractured specimen with cracks) dissolving process and ideal model diagram. (**a**) liquid immersion, 0 h; (**b**) liquid immersion, 10 h; (**c**) liquid immersion, 20 h; and (**d**) liquid immersion, 30 h.

As shown in Figure 14, due to the action of dissolution–penetration, sodium sulphate in the specimen was dissolved into the solution or water, and the insoluble calcium sulphate in the specimen formed the residual skeleton (the black parts represent the seepage pathways which resulted from the dissolution of sodium sulphate, whereas the white parts are the calcium sulphate or calcium sulphate dihydrate). Finally, the impermeable glauberite salt rock gradually became a permeable porous medium with calcium sulphate (or calcium sulphate dihydrate) as the solid skeleton, mechanically supporting and sustaining the network of seepage pathways and eventually changing and affecting the physical and mechanical properties of the glauberite. Additionally, the processes of crack initiation, propagation, and coalescence can also be observed on the petrographic thin section. For example, the cracks is not obvious for the dry specimen slice. However, after 5 min of dissolution, several cracks initiated at the centre of the petrographic thin section due to the impurity hydration and swelling. After 10 min of dissolution, the width of the cracks became obviously larger. This is because the pore liquid which infiltrated into the crack zones takes the crack lines as the new solid-liquid interfaces, and then preferentially dissolved the undissolved areas around the cracks. After 15 min of dissolution, the morphology of specimen underwent a considerable change; cracks completely disappeared; a variety of punctate pores were randomly distributed at the centre of microscope images; and a visible deformation occurred at the residual skeleton. This is because the calcium sulphate may have reacted with water to form calcium sulphate dihydrate in the process of leaching (i.e., crystallization, see in the Equation (3)), which will create the swelling force. Furthermore, the specimens contained a large percentage of impurities (mainly illite and montmorillonite) in addition to the glauberite. When the impurities were immersed in water, they absorbed moisture, leading to an increase in volume. Dékány et al. [27] asserted that the free swell ratio of illite and montmorillonite could reach 40% and 56%, respectively. Hence, the porous structures underwent an obvious deformation, and the cracks disappeared.

$$Ca_2SO_4 + 2H_2O \underset{\text{Dissolution}}{\overset{\text{Crystallization}}{\rightleftharpoons}} Ca_2SO_4 \cdot 2H_2O \tag{3}$$

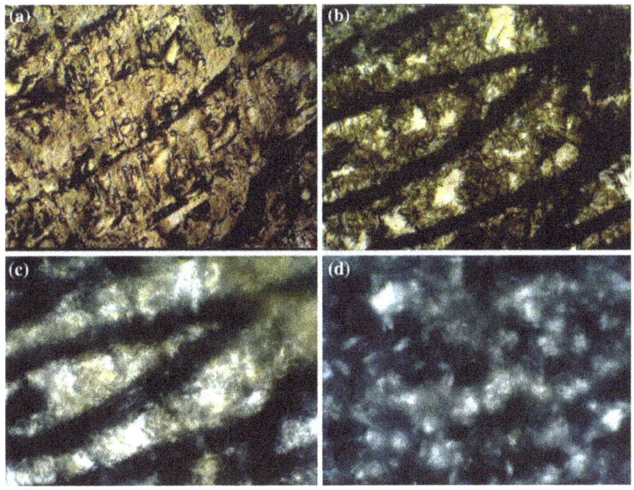

Figure 14. Microscope photos of glauberite sections dissolved by pure water at different times (250×) [13]. (**a**) Original dry slice; (**b**) after 5 min of dissolution; (**c**) after 10 min of dissolution; (**d**) after 15 min of dissolution.

Undoubtedly, all of these behaviours (i.e., chemical reactions and physical reactions) would greatly improve the porosity of the specimen and further accelerated the dissolution rate. For example, as shown in Table 1, compared with the porosity of the dry specimen, the porosity of the specimen soaked in pure water at 20 °C increased by 1.03 times after 10 h of immersion, 2.59 times after 20 h of immersion, and 4.64 times after 30 h of immersion. Obviously, this may have fully reflected the special permeability characteristics of glauberite during the leaching process.

4.2. The Effect of Temperature on the Microstructure of Glauberite

The MCT scan images of glauberite after immersing in pure water at 20 °C showed that the dissolution rate of glauberite was very slow. After 10 h of immersion, the porosity increased by only 3.8%; after 20 h of immersion, the porosity increased by 159.8%, and the cracks appeared to initiate at the centre of the discs, crisscrossed together. The reason for this is that the specimen contained a large percentage of impurities (mainly illite and montmorillonite, 12%) in addition to glauberite [13]. Dékány et al. [27] asserted that part of the impurities could be characterized by the super-hydrophilic performance and large expansibility factor. Based on their experimental results, the free swell ratio of a specimen could reach 40%. Therefore, when immersed in pure water, the specimen volume expanded, thus, generating cracks inside the specimen. Subsequently, the solution or water accumulated in the crack regions, and as a result, the crack faces (or crack lines) became the new solid-liquid contact surfaces. Thus, after being immersed for 30 h, new dissolving regions appeared around the crack lines (see Figure 4). When the specimens were immersed in pure water at 50 °C and 80 °C for only 10 h, complex cracks appeared at the centre of the discs, and the porosity growth rate of the pore number increased with the temperature. For example, the exponent (i.e., regression equations of glauberite porosity with time) in 80 °C pure water (immersing for 30 h) were 1.19 times and 1.20 times that of 50 °C, and 1.38 times and 1.43 times that of 20 °C. The main reasons are as follows:

(1) The dissolution rate of the soluble minerals in the specimen increased with the temperature, which eventually resulted in the increase of porosity. As shown in Figure 6, when immersed in pure water at 80 °C for 10 h, it dissolved along the edges from the surface inwardly on a large scale, instead of forming dissolving areas around the crack lines in the specimen; while in 20 °C pure water, the evolution process began with cracks, and then dissolved with the new solid-liquid contact surfaces created by the cracks, inwardly. By comparing the dissolution regions and areas, it could be concluded that the dissolution rate of glauberite increased with the temperature.

(2) As the temperature increased, the activity of the water and the impurities (illite and montmorillonite) increased, which accelerated the impurity/hydration reaction, and thereby exacerbated the expansion of the specimen [28].

(3) After the temperature rose, the chemical reaction rate of calcium sulphate and water accelerated. Lewis et al. [29] maintained that the specimen volume could increase by 30% when the calcium sulphate completely turned into dihydrate calcium sulphate, and the swelling force could reach 584–840 kPa. It was the volume expansion that enlarged the lattice spacing of the specimen, and thereby producing cracks.

For quantitative explanation on the expansion characteristics of the specimen at different temperatures, the diameter of the specimens after soaking in pure water at different temperatures were calculated. As shown in Figure 15, the diameter grew gradually with the increasing temperature, which was consistent with the variational trend of the porosity in the paper. Additionally, this strongly supports the analysis mentioned above.

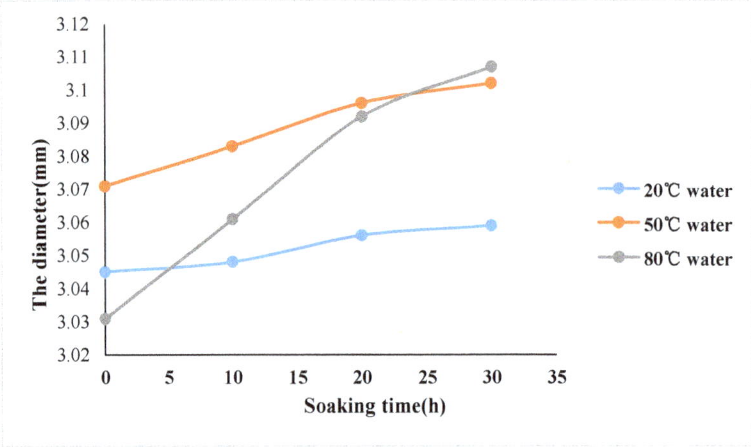

Figure 15. The relationship between the diameter of the glauberite specimen and the immersing time.

4.3. The Effect of Brine Concentration on Microstructure of Glauberite

By comparing the MCT scanning images of glauberite immersed in 80 °C solution with varying concentrations, it can be determined that, after 10 h of immersion in pure water, a wide range of dissolving areas appeared; after 30 h of immersion, the specimen broke completely, and the dissolution notch was observed. However, this phenomenon was not observed when immersed in semi-saturated brine and saturated brine. By analysing Figure 9 and the data of Table 1, it is revealed that the growth rate of glauberite pore varied greatly—the difference between them is large, even by an order of magnitude. For example, when immersed in pure water, the exponent in the regression equations of glauberite porosity with time was 0.0552; in semi-saturated solution, the exponent decreased to 0.0039; in saturated salt solution, it was as low as 0.0022. From Table 1, the porosity decreased gradually with the increasing brine concentration. For example, when immersed for 30 h, the porosity in pure water at 80 °C were 2.59 times higher than that of half-saturated brine, and were 2.90 times higher than that of saturated brine. The comparison strongly confirmed that the chloride ion had an inhibitory effect on sodium sulphate dissolution, and the phenomenon could be explained by the Debye-Huckel theory, namely, sodium chloride was a symmetrical electrolyte [30–33]. When the concentration of sodium chloride solution was increased, the activity coefficient of calcium sulphate decreased and calcium sulphate formed ion pairs. Thus, the short-range electrostatic interaction was enhanced, lowering the solubility of the calcium sulphate, thereby accelerating the dissolution rate of sodium sulphate. In summary, the solubility and dissolution rate of sodium sulphate decreased with the increase of sodium chloride concentration (see Figure 16), as did the corresponding porosity of glauberite.

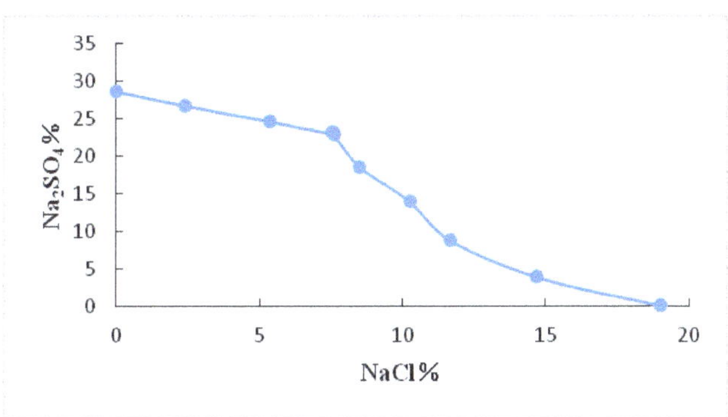

Figure 16. $Na_2SO_4-NaCl-H_2O$ phase diagram of three element system (20 °C).

5. Conclusions

Glauberite is a special kind of salt rock. Its microstructure evolution in the process of leaching is not only related to mining efficiency, but also to the deformation and stability of its mineral layer. This paper focused on the microstructural evolution of glauberite under different leaching conditions. The main conclusions are as follows:

(1)   When the specimens were immersed in a low-temperature solution, the dissolution rate was very slow. With the increase of temperature, the dissolution rate greatly improved, and the initiation of micro-cracks and the dissolution regions were produced spontaneously.

(2)   In the process of leaching, the effects of temperature and concentration of solution on the porosity of glauberite was significant. When immersed in brine with the same concentration, the porosity of the specimens increased considerably as the temperature increased; when at the same temperature, the porosity of the specimen decreased with the increase in brine concentration.

(3) When the specimens were immersed in pure water or brine, the sodium sulphate in glauberite dissolved and the remaining solid insoluble substances became the residual skeleton. The specimens are gradually dissolved from the outside to the inside, and the pores and three zones are formed during the leaching process. The zones are the dissolved area (or residual skeleton, porous structure), the undissolved area (i.e., dense and intact area), and the circular solid-liquid interface (i.e., circular line), respectively.

(4) With the increase of temperature, the hydration of illite and montmorillonite in glauberite and the hydration of calcium sulphate were improved, resulting in a gradual increase of the expansion rate of glauberite and fracture density increased as well. With the increase of sodium chloride concentration, the porosity of the specimen decreased.

Author Contributions: S.Z.C. contributed to the experimental design and sourcing of the laboratory equipment and raw materials; D.H.Z. contributed to the experimental operation, data collection, and writing of the manuscript; T.S. carried out the CT analysis and the correction and editing of the manuscript; T.M. also contributed in the experimental design and correction of the manuscript.

Funding: This paper was support by the Natural Science Funds for Young Scholar (2016YFC0501103), Major Research and Development Plans of Shanxi Province (grant no. 201603D121031), the School Fund of Tai Yuan University of Technology and Science (no. 20172018, 20182008), Science and technology innovation project of colleges and universities in Shanxi (Grant No. 2017158), Shanxi Scholarship Council of China (Grant No. 2017-086), State Key Laboratory Breeding Base of Coal Science and Technology Co-founded by Shanxi Province and the Ministry of Science and Technology, Taiyuan University of Technology(Grant No. mkx201702).

Acknowledgments: The authors gratefully acknowledge Tao Meng, who has given us many useful ideas and suggestions.

Conflicts of Interest: The authors declare no conflict of interest.

References

1. MacQuarrie, K.T.B.; Mayer, K.U. Reactive transport modeling in fractured rock: A state-of-the-science review. *Earth-Sci. Rev.* **2005**, *72*, 189–227. [CrossRef]
2. Yin, S.; Dusseault, M.B.; Rothenburg, L. Coupled THMC modeling of CO_2 injection by finite element methods. *J. Pet. Sci. Eng.* **2011**, *80*, 53–60. [CrossRef]
3. Meng, T.; Hu, Y.; Fang, R. Study of fracture toughness and weakening mechanisms in gypsum interlayers in corrosive environments. *J. Nat. Gas Sci. Eng.* **2015**, *26*, 356–366. [CrossRef]
4. Meng, T.; Hu, Y.; Fang, R. Weakening mechanisms of gypsum interlayers from Yunying salt cavern subjected to a coupled thermo-hydro-chemical environment. *J. Nat. Gas Sci. Eng.* **2016**, *30*, 77–89. [CrossRef]
5. Zhang, R.; Winterfeld, P.H.; Yin, X. Sequentially coupled THMC model for CO2 geological sequestration into a 2D heterogeneous saline aquifer. *J. Nat. Gas Sci. Eng.* **2015**, *27*, 579–615. [CrossRef]
6. Russo, A.J. *Solution Mining Code for Studying Axisymmetric Salt Cavern Formation*; Sandia National Labs: Albuquerque, NM, USA, 1981.
7. Ripley, E.A.; Redmann, R.E. *Environmental Effects of Mining*; CRC Press: Boca Raton, FL, USA, 1995.
8. Mudd, G.M. Critical review of acid in situ leach uranium mining: 2. Soviet Block and Asia. *Environ. Geol.* **2001**, *41*, 404–416. [CrossRef]
9. Guo, J.Y.; Lu, W.X.; Jiang, X. A quantitative model to evaluate mine geological environment and a new information system for the mining area in Jilin province, mid-northeastern China. *Arab. J. Geosci.* **2017**, *10*. [CrossRef]
10. Liu, R.; Li, B.; Jiang, Y. Critical hydraulic gradient for nonlinear flow through rock fracture networks: The roles of aperture, surface roughness, and number of intersections. *Adv. Water. Resour.* **2016**, *88*, 53–65. [CrossRef]
11. Liu, R.; Li, B.; Jiang, Y. A fractal model based on a new governing equation of fluid flow in fractures for characterizing hydraulic properties of rock fracture networks. *Comput. Geotech.* **2016**, *75*, 57–68. [CrossRef]
12. Salvany, J.M.; García-Veigas, J.; Orti, F. Glauberite-halite association of the Zaragoza Gypsum Formation (Lower Miocene, Ebro Basin, NE Spain). *Sedimentology* **2007**, *54*, 443–467. [CrossRef]
13. Liang, W.; Zhao, Y.; Xu, S. Dissolution and seepage coupling effect on transport and mechanical properties of glauberite salt rock. *Transp. Porous Media* **2008**, *74*, 185–199. [CrossRef]

14. Zhang, G.; Li, Y.; Yang, C. Stability and tightness evaluation of bedded rock salt formations for underground gas/oil storage. *Acta Geotech.* **2014**, *9*, 161–179. [CrossRef]
15. Ramezanianpour, A.A.; Malhotra, V.M. Effect of curing on the compressive strength, resistance to chloride-ion penetration and porosity of concretes incorporating slag, fly ash or silica fume. *Cem. Concr. Compos.* **1995**, *17*, 125–133. [CrossRef]
16. Ortega, J.M.; Sánchez, I.; Antón, C.; de Vera, G.; Climent, M.A. Influence of Environment on Durability of Fly Ash Cement Mortars. *ACI Mater. J.* **2012**, *109*, 647–656.
17. Joshaghani, A.; Balapour, M.; Ramezanianpour, A.A. Effect of controlled environmental conditions on mechanical, microstructural and durability properties of cement mortar. *Constr. Build. Mater.* **2018**, *164*, 134–149. [CrossRef]
18. Meer, S.; Spiers, C.J. Influence of pore-fluid salinity on pressure solution creep in gypsum. *Tectonophysics* **1999**, *308*, 311–330. [CrossRef]
19. Zhao, Y.; Yang, D.; Liu, Z. Problems of evolving porous media and dissolved glauberite micro-scopic analysis by micro-computed tomography: Evolving porous media (1). *Transp. Porous Media* **2015**, *107*, 365–385. [CrossRef]
20. Liang, W.; Yang, X.; Gao, H. Experimental study of mechanical properties of gypsum soaked in brine. *Int. J. Rock Mech. Min. Sci.* **2012**, *53*, 142–150. [CrossRef]
21. Yu, W.D.; Liang, W.G.; Li, Y.R.; Yu, Y.M. The meso-mechanism study of gypsum rock weakening in brine solutions. *Bull. Eng. Geol. Environ.* **2016**, *75*, 359–367. [CrossRef]
22. Wang, M.; Xue, H.; Tian, S.; Wilkins, R.W. Fractal characteristics of Upper Cretaceous lacustrine shale from the Songliao Basin, NE China. *Mar. Pet. Geol.* **2015**, *67*, 144–153. [CrossRef]
23. Ji, W.; Song, Y.; Jiang, Z. Fractal characteristics of nano-pores in the Lower Silurian Longmaxi shales from the Upper Yangtze Platform, south China. *Mar. Pet. Geol.* **2016**, *78*, 88–98. [CrossRef]
24. Yang, R.; He, S.; Yi, J.; Hu, Q. Nano-scale pore structure and fractal dimension of organic-rich Wufeng-Longmaxi shale from Jiaoshiba area, Sichuan Basin: Investigations using FE-SEM, gas adsorption and helium pycnometry. *Mar. Pet. Geol.* **2016**, *70*, 27–45. [CrossRef]
25. Shao, X.; Pang, X.; Li, Q.; Wang, P.; Chen, D. Pore structure and fractal characteristics of organic-rich shales: A case study of the lower Silurian Longmaxi shales in the Sichuan Basin, SW China. *Mar. Pet. Geol.* **2017**, *80*, 192–202. [CrossRef]
26. Voorn, M.; Exner, U.; Barnhoorn, A.; Baud, P.; Reuschlé, T. Porosity, permeability and 3D fracture network characterisation of dolomite reservoir rock samples. *J. Pet. Sci. Eng.* **2015**, *127*, 270–285. [CrossRef] [PubMed]
27. Dékány, I.; Szántó, F.; Nagy, L.G. Wetting and adsorption on organophilic illites and swelling montmorillonites in methanol-benzene mixtures. *Colloid Polym. Sci.* **1988**, *266*, 82–96. [CrossRef]
28. Buck, B.J.; Brock, A.L.; Johnson, W.H.; Ulery, A.L. Corrosion of depleted uranium in an arid environment: Soil-geomorphology, SEM/EDS, XRD, and electron microprobe analyses. *Soil Sediment Contam.* **2004**, *13*, 545–561. [CrossRef]
29. Lewis, K.N.; Thomas, M.V.; Puleo, D.A. Mechanical and degradation behavior of polymer-calcium sulfate composites. *J. Mater. Sci. Mater. Med.* **2006**, *17*, 531–537. [CrossRef] [PubMed]
30. Helgeson, H.C.; Kirkham, D.H. Theoretical prediction of the thermodynamic behavior of aqueous electrolytes at high pressures and temperatures; II, Debye-Huckel parameters for activity coefficients and relative partial molal properties. *Am. J. Sci.* **1974**, *274*, 1199–1261. [CrossRef]
31. Meng, T.; Bao, X.; Zhao, J. Study of mixed mode fracture toughness and fracture characteristic in gypsum rock under brine saturation. *Environ. Earth. Sc.* **2018**, *77*, 364. [CrossRef]
32. Bao, X.; Tao, M.; Zhao, J. Study of mixed mode fracture toughness and fracture trajectories in gypsum interlayers in corrosive environment. *Roy. Soc. Open. Sci.* **2018**, *5*, 171374. [CrossRef] [PubMed]
33. Tao, M.; Yechao, Y.; Jie, C. Investigation on the permeability evolution of gypsum interlayer under high temperature and triaxial pressure. *Rock Mech. Rock Eng.* **2017**, *50*, 2059–2069. [CrossRef]

Article

The Fracturing Behavior of Tight Glutenites Subjected to Hydraulic Pressure

Zhichao Li [1], Lianchong Li [2,]*, Zilin Zhang [3], Ming Li [3], Liaoyuan Zhang [3], Bo Huang [3] and Chun'an Tang [1]

[1] State Key Laboratory of Coastal and Offshore Engineering, Dalian University of Technology, Dalian 116024, China; lizhichaohn@163.com (Z.L.); tca@mail.neu.edu.cn (C.T.)
[2] School of Resources and Civil Engineering, Northeastern University, Shenyang 110819, China
[3] Shengli Oilfield Branch Company, SINOPEC, Dongying 257000, China; cyyylzzl@163.com (Z.Z.); li_ming2016@163.com (M.L.); zhang_liaoyuan@126.com (L.Z.); huang_bo2015@126.com (B.H.)
* Correspondence: li_lianchong@163.com; Tel.: +86-24-8368-7705

Received: 5 July 2018; Accepted: 19 July 2018; Published: 20 July 2018

Abstract: Tight glutenites are typically composed of heterogeneous sandstone and gravel. Due to low or ultra-low permeability, it is difficult to achieve commercial production in tight glutenites without hydraulic fracturing. Efficient exploitation requires an in-depth understanding of the fracturing behavior of these reservoirs. This paper provides a numerical method that integrates the digital image processing (DIP) technique into a numerical code rock failure process analysis (RFPA). This method could consider the glutenite heterogeneities, including intrarock and interrock heterogeneities, and the practicability is verified through two numerical tests. Two-dimensional (2D) simulations show hydraulic fractures (HFs) can penetrate or deflect to propagate along the gravels, depending on the magnitude of stress anisotropy and gravel strength. Three-dimensional (3D) simulations with the consideration of gravel distribution orientation, gravel size and axial ratio show HFs could propagate past the gravel with no deflection, forming a bypass fracture that is not easy to observe in common laboratory experiments. HFs could also deflect to propagate along the gravels. The impacts of the gravel distribution orientation, gravel size and axial ratio are discussed in detail. The main propagation modes of HFs intersecting the gravels are summarized as: (1) penetrating directly; (2) deflecting to propagate along the gravels to form distorted HFs; (3) propagating to bypass the gravels; (4) a combination of (1) and (2), or (2) and (3).

Keywords: glutenite; gravel; hydraulic fracture; numerical simulation; propagation

1. Introduction

Hydraulic fracturing is one of the primary methods of stimulation in unconventional hydrocarbon reservoirs. When high-pressure fluid is injected into the reservoirs, the rock can be fractured to form high-conductivity pathways for hydrocarbon migration and thus enhance production. Therefore, hydraulic fracturing has become a common practice for stimulation in hydrocarbon reservoirs since the 1940s. For researchers and engineers, it is important to understand and control effectively the hydraulic fracture (HF) characteristics such as fracture number, spacing and geometry (length, height, aperture and propagation mode) that are closely related to the reservoir fracturing behavior, which can contribute significantly to long-term hydrocarbon production. The reservoir fracturing behavior is influenced by numerous factors such as reservoir geostress, rock properties (strength, permeability, brittleness, etc.), reservoir heterogeneity (pores, natural fractures, other heterogeneous structures, etc.), fracturing fluid property (viscosity, leak-off, etc.) and pumping operation (pumping rate, time, etc.) [1–7]. In-depth understanding the reservoir fracturing behavior subjected to hydraulic pressure is crucial to the optimization of fracture design parameters.

Glutenite reservoirs are typically composed of sandstone and conglomerate gravels which are formed under rapid deposition in nearby provenances [8]. Tight glutenites are among those with low or ultra-low permeability that can range from 0.1 mD to 0.0001 mD [9]. The gravels in the tight glutenites display complex spatial distributions as indicated in Figure 1a which shows the Formation MicroScanner Image (FMI) result in the Ken761 block in Dongying, Shandong province, China [10]. Glutenite cores are drilled for further research and some of them are shown in Figure 1b. From closer observation of the drill cores such as those in Figure 2a,b, glutenites are found to generally develop gravels of irregular shapes, various contents, and different sizes. Since this type of tight reservoir generally develops with low or ultra-low permeability, low porosity and poor connectivity, it is hard to extract the hydrocarbon at a commercial rate without a fracturing operation. The presence of gravels raises the reservoir heterogeneity, complicates the propagation behavior of HFs to exhibit very different characteristics from those in other reservoirs, and makes it difficult to predict the fracture path [11,12].

Due to the random gravel distribution, it is almost impossible to make clear the fracturing behavior of a glutenite reservoir merely through theoretical analysis. On site, HF geometry is inspected indirectly through post-fracturing data acquisition methods such as microseismic monitoring, which is rather rough due to the inability to identify open-mode HFs [13,14], easy contamination by a variety of noises, signal attenuation through thick, faulty or soft formations [15], and common drawbacks in the interpretation of microseismic mapping [16]. Therefore, in view of the monitoring uncertainty and underground invisibility, it is utterly infeasible to obtain uncontroversial answers to this issue from field-monitoring applications.

Figure 1. (a) The Formation MicroScanner Image (FMI) result in the Ken761 block in Dongying, Shandong province, China; (b) the drill cores obtained [10].

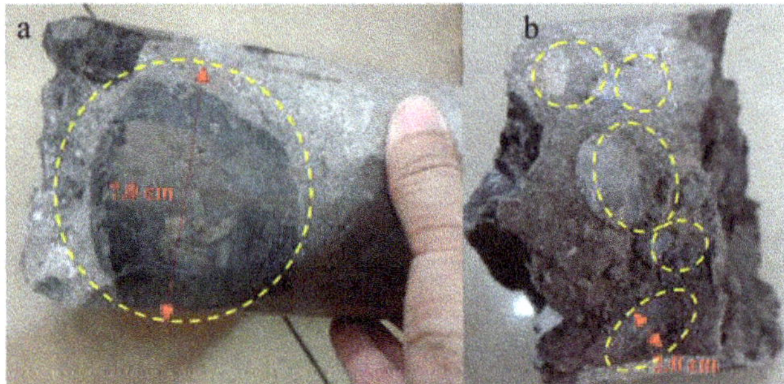

Figure 2. (**a**,**b**) drill cores from Figure 1b showing closer morphology of tight glutenite. The conglomerate gravels are shown inside the yellow circles [10].

To deal with this issue, laboratory experiments and numerical simulations are suitable approaches. Previous studies on this issue have been conducted through these two approaches. Laboratory experiment has always been a direct and compelling method to conduct such investigations and valuable related findings have been achieved [8,11,12,17–19]. However, similarly, limited by the inner invisibility of a rock specimen and lack of suitable observation methods, in most cases fractured samples after wellbore shut-in are cut open to observe the fracture geometry, which may partly destroy the HF geometry due to the tough manual separation, leave out small HFs or branches, and lead to loss of the emerge of the fracturing process that is quite essential to illuminate problems in some cases. All these technical restrictions and limitations urge the development of numerical tools for careful study. To conduct studies on this issue, numerical modeling tends to be a good choice that is a low-cost tool that can describe in considerable detail the phenomena observed in laboratory experiments and explore the answer in complicated situations that are difficult to carry out in those experiments. However, to date, few numerical models have considered the heterogeneity of rock reservoirs during hydraulic fracturing. In fact, rock is a heterogeneous geological material which contains a different scale of natural weaknesses, such as pores, grain boundaries and pre-existing cracks [20]. Increasing evidence from practical stimulations in the field shows that HFs can initiate and propagate in intricate ways that are highly influenced by the heterogeneity of rock reservoirs [21–24]. In tight glutenite, extensive material heterogeneity exists not only in the single rock, but in between the two types of rock, i.e., the sandstone and gravel create interrock heterogeneity. How to investigate the reservoir fracturing behavior with the consideration of the intrarock and interrock heterogeneities remains an enormous challenge.

A numerical code, known as rock failure process analysis (RFPA), is applied in this study. Based on a statistical model, RFPA is considered appropriate for modeling heterogeneous material such as rock. When integrated with the digital image processing (DIP) technique, RFPA is able to distinguish sandstone and gravel that show different colors in digital images, and then be applied to establish numerical models for analysis. Numerical investigations in this paper begin with 2D modeling with the consideration of the geostress anisotropy and gravel strength. Then, 3D models are established to further investigate the fracturing behavior in tight glutenite in detail, especially in situations beyond the 2D model's ability, with the consideration of gravel distribution, gravel size, and axial ratio. The propagation modes are summarized and the corresponding conditions for each mode to occur are then provided.

2. The Digital Image Processing (DIP) Technique in Rock Failure Process Analysis (RFPA) and Rock Heterogeneities

Since a detailed description of the coupled flow-stress-damage (FSD) model of RFPA has been previously presented [25], this section will briefly introduce the DIP technique in RFPA and then illuminate the issue of intrarock and interrock heterogeneities.

2.1. DIP Technique and Integration in RFPA

DIP is a technique used to capture a scene electronically by transforming it into a two-dimensional pixel image and then processing it so as to make the information available for mathematical algorithms [26]. Digital images can be numerically represented in many color systems such as the gray color space, true color space (i.e., the red-green-blue (RGB) space) or the hue, saturation, intensity (HSI) space. It is well known that in the true color space, three independent integer values (R, G and B, from 0 to 255, respectively) are required to describe the red, green and blue level at each pixel. The HSI space is also extensively used due to its close correlation with how humans perceive colors. The hue component H (from 0 to 360) represents the dominant wavelength of the color, which is the domain color feature perceived by humans. The saturation component S (value from 0 to 1) represents the purity of color, defined as how strongly it is polluted with white. The intensity component I (value from 0 to 1) stands for the color brightness or lightness that is unrelated to colors [27]. For convenience to display in the software, the three component variables H, S and I used in this paper are normalized to be from 0 to 255. Digital images can be identified and edited by some image processing tools. With the integration of the DIP technique, RFPA is able to identify the images in BMP (Bitmap) format and then can establish numerical models.

Digital images of rock can be obtained by photographing rock surfaces or the fresh cross-sections of rock samples, or by recovering maps from X-ray Computed Tomography (CT) scanning or other methods [27]. When such a BMP-format image of rock is imported into RFPA, it will be discretized into many square elements of identical size. Each element maps directly onto one finite element that is used for subsequent analysis. It is strongly recommended that the elements number should not be lower than the total image pixels so as to ensure that the element size is not larger than that a pixel represents, thus avoiding a loss of information the pixels contain. Since rock contains a variety of micro-structures (such as minerals, pores, pre-existing cracks, etc.) that may exhibit distinct perceived colors, elements mapping onto these micro-structures can be divided into several groups according to the color data (R, G, B and I) provided in RFPA. The groups number is suggested in agreement with the micro-structures number in cautious research. Once the grouping has been completed, the unique relation between micro-structures and relevant material properties can be established. It is important to note that, in this study, making clear which type each micro-structure in the digital image really is should be done upfront because it is impossible for a software like RFPA to figure out what a micro-structure really is just by their color data. Similarly, the mechanical parameters of micro-structures should be obtained before numerical simulation.

A digital image of a glutenite sample from the Ken761 block in Dongying, Shandong province, China, is shown in Figure 3a, in which a circular cross-section (Diameter: 110 mm) contains two micro-structures: the dark color represents conglomerate gravels and the bright color represents sandstone. A block is cut from the cross-section for analysis, as shown in Figure 3b. It consists of 500 × 500 pixels and the side length is 58 mm.

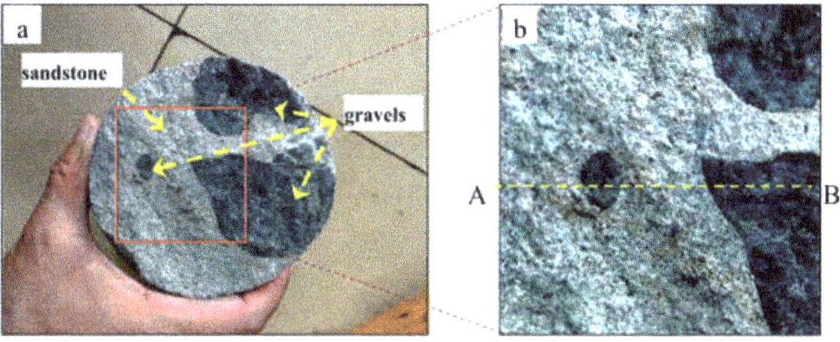

Figure 3. (**a**) shows a tight glutenite core from Ken761 block in Dongying, Shandong province, China [10]; (**b**) is a block cut from (**a**).

When this block is imported into RFPA, it will be discretized into 500 × 500 finite elements by default, with each pixel transformed into an element. RFPA can capture the color information in the image and then provide to the operator. Since the two micro-structures in the block exhibit distinct difference in brightness, it is quite convenient to divide element groups by the values of I. Figure 4a is a statistical graph provided by RFPA that describes the counts of elements with different I values. Color information of each element can be obtained from RFPA. For example, when an arbitrary line is drawn by the mouse in the RFPA interface, like A–B in Figure 3b, the color information of the elements the line passes will be presented to the operator. Figure 4b describes the I values of the 500 elements along line A–B, from which those in gravel sections are observed generally lower than those in sandstone sections as expected. After several trials, a threshold, 85, at the level that the blue line in Figure 4b denotes, is adopted to divide all the elements into two groups. Elements with I values lesser than 85 are classified into the gravel group and those with I values equal to or larger than 85 are classified into the sandstone group. After assigning reasonable mechanical parameters to the two element groups, the rough sketch of the numerical model will form as shown in Figure 5.

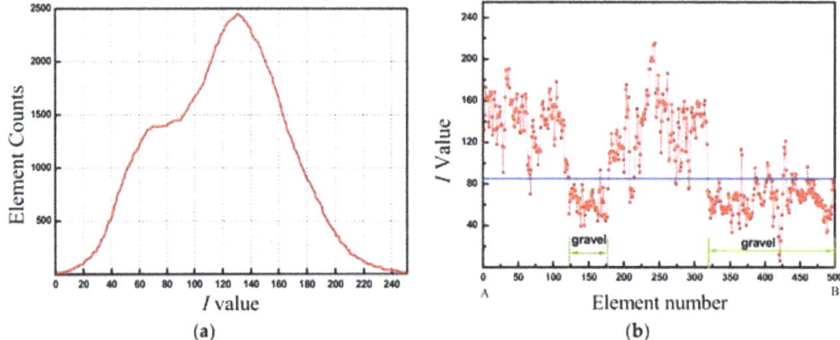

Figure 4. Elements color information provided by rock failure process analysis (RFPA). (**a**) Statistical element counts with different I values; (**b**) I values of elements along A–B.

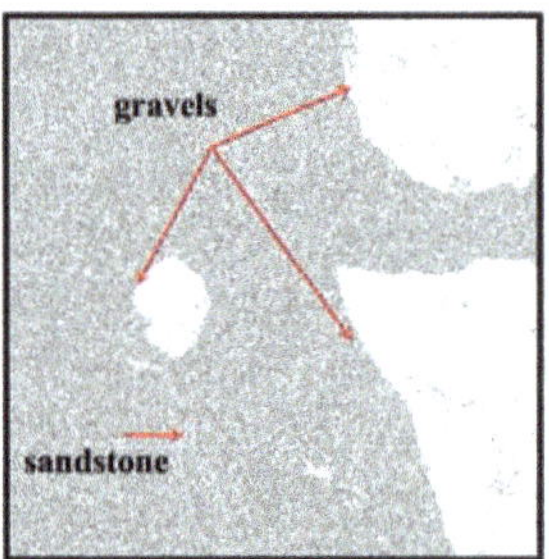

Figure 5. Sketch of the numerical model.

2.2. Intrarock and Interrock Heterogeneities

As one of the most important rock characteristics, heterogeneity significantly influences geostress distribution and rock failure by the existence of different minerals, pores and micro-cracks [28,29]. To numerically represent the influence, a statistical model based on Weibull distribution is introduced into RFPA to take into account this intrarock heterogeneity [30]. A simple numerical model is established in RFPA as shown in Figure 6 to reveal this influence. The rock specimen is 400 mm × 400 mm in size and it is discretized into 400 × 400 finite elements. Constant confining stress, σ_x (3.0 MPa) and σ_y (2.0 MPa), is applied to the model boundary. A hole with radius of 15 mm is preset in the center, into which an initial hydraulic pressure of 1.0 Mpa with an increment of 0.05 Mpa per loading step is imposed until the specimen breaks downs. Two cases are set for comparison. Case I: the homogeneity index m is set at 2.0, which represents a large dispersion of material parameters. Case II: the homogeneity index m is set at 20.0, which represents a concentration of material parameters. The rock physico-mechanical parameters are listed in Table 1.

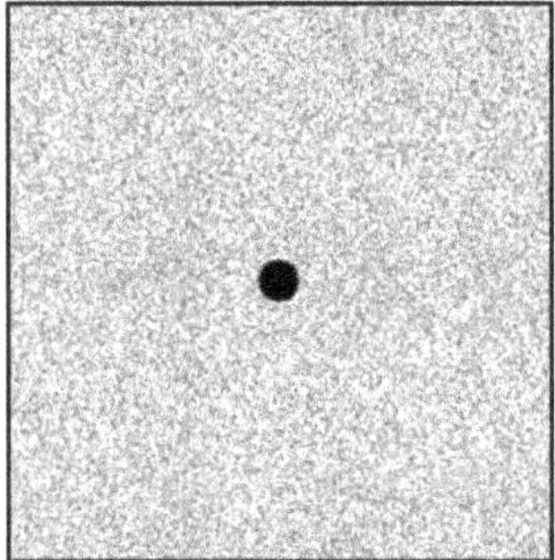

Figure 6. The numerical model.

Table 1. Physico-mechanical parameters.

Physico-Mechanical Parameters	Value
Young's modulus (E_0), GPa	30.0
uniaxial compressive strength (f_{c0}), MPa	45.0
Ratio of compressive and tensile strength (f_{c0}/f_{t0})	10
Poisson's ratio (ν)	0.22
Internal friction angle (φ),	31
Coefficient of the pore water pressure (α)	0.7
Permeability coefficient (k_0), m/s	1×10^{-10}

When the initial pressure 1.0 Mpa is imposed on the hole, the radial stress component σ_r of elements along the dashed radius shown in Figure 6 in the two cases are drawn in Figure 7. The theoretical solution [31] of the same model on homogeneous rock is also drawn for comparison. As shown in the figure, there is a good agreement between the numerical solution in Case II and the theoretical solution except some small fluctuations. However, in Case I, obvious stress fluctuations can be observed, although the average level floats close to the theoretical solution. This phenomenon indicates that rock heterogeneity significantly influences the stress field and that a lower homogeneity index (m) contributes to the larger stress dispersion and vice versa.

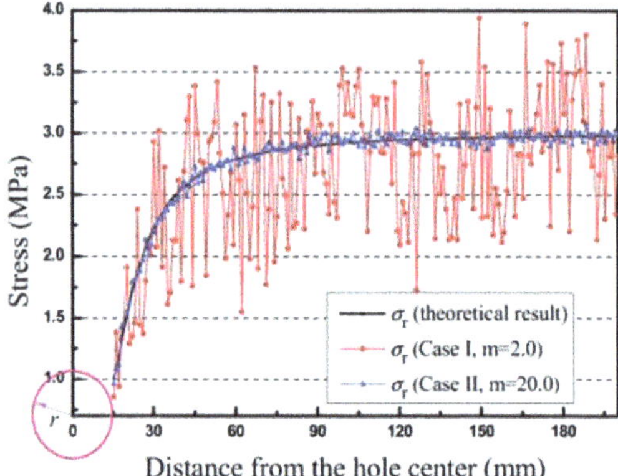

Figure 7. Stress distribution along the horizontal radius.

HFs generated in Case I and Case II are shown in Figure 8 (in which pure black elements represent those preset as cavity or have failed in the numerical calculation). Figure 8a reveals HFs in Case I do not rigorously initiate at the hole wall in the maximum principle stress direction, i.e., x axis direction. The right branch deviates slightly from the x axis direction and makes the HFs looks asymmetric. In Case II, as shown in Figure 8b, HFs initiate exactly at the hole wall along the x axis direction and propagate straight to form a typical symmetrical fracture. In Case I, some scattered failed elements are observed to appear on both sides of the HFs while few can be found in Case II. The comparison of the two cases demonstrates rock heterogeneity really influences the rock failure.

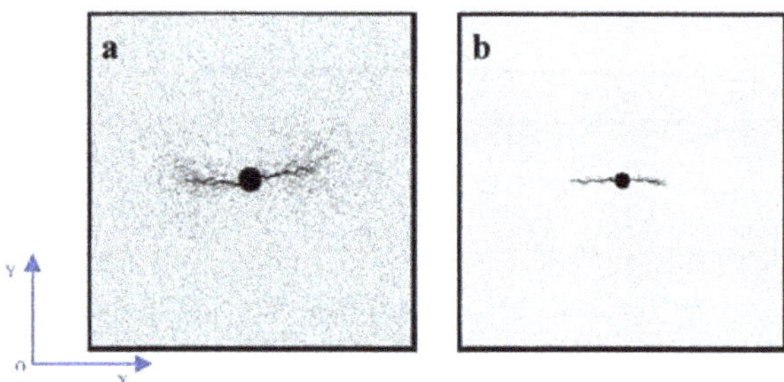

Figure 8. Hydraulic fracture (HF) geometries. (**a**) Corresponds to Case I (m = 2.0); (**b**) corresponds to Case II (m = 20.0).

In addition to intrarock heterogeneity, problems still exist in formations with variational lithologies such as the glutenite reservoirs. In these formations, multi-types of rock compose the formation and this interrock heterogeneity could also affect significantly rock failure. This is not difficult to understand but how to consider this heterogeneity appropriately is not easy to answer. Fortunately, the DIP technique could help to deal with this task as long as suitable images are available.

For example, uniaxial tensile numerical tests are conducted in RFPA on a pure sandstone specimen and a glutenite specimen and they are named as Case III and Case IV, respectively. The geometric size and boundary conditions are shown in Figure 9a,b. Each specimen is discretized into 250×500 finite elements. The glutenite specimen in Case IV is established in RFPA with the DIP technique from a sub-image from Figure 3a, in which bright regions represent gravel and dark regions represent sandstone. To exclude the influence of intrarock heterogeneity, the homogeneity indices m of both the sandstone and gravel in the two tests are set at 100.0 that is large enough to create a homogeneous material. The rock physico-mechanical parameters are listed in Table 2. Side notches are precut in the specimens as shown in Figure 9a,b so as to preferably catch the gravel effect. An external displacement Δu, at a constant rate of 0.0002 mm/step, is applied on the specimens in the axial direction.

The test results are shown in Figure 9c,d. A common tensile fracture forms in the homogeneous sandstone specimen along the preset notches (Figure 9c). In the glutenite specimen, however, when intersecting the gravel, the tensile fracture tends to deflect and propagate along the sandstone–gravel interface, forming a bending fracture. It can be seen that the glutenite specimen, although composed of homogeneous rocks, exhibits peculiar behavior under tension due to composition of two types of rock. Therefore, the interrock heterogeneity in formations with variational lithologies should be seriously considered and the DIP technique in RFPA can be useful for dealing with similar problems.

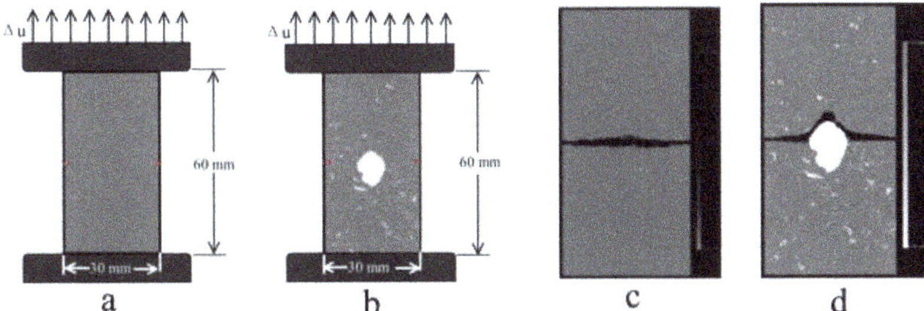

Figure 9. (**a**) The uniaxial tensile test on sandstone specimen in Case III; (**b**) the uniaxial tensile test on glutenite specimen in Case IV; (**c,d**) tests results of Case III, IV, respectively.

Table 2. Physico-mechanical parameters of sandstone and gravel.

Physico-Mechanical Parameters	Sandstone	Gravel
Young's modulus (E_0), GPa	30.0	55.0
uniaxial compressive strength (f_{c0}), MPa	45.0	130.0
Ratio of compressive and tensile strength (f_{c0}/f_{t0})	10	10
Poisson's ratio (ν)	0.22	0.25
Internal friction angle (ϕ),	31	33
Coefficient of the pore water pressure (α)	0.7	0.7
Permeability coefficient (k_0), m/s	1×10^{-10}	1×10^{-11}

In this study, both the glutenite intrarock and interrock heterogeneities are considered in the numerical simulations. The Weibull statistical model is applied to consider the intrarock heterogeneity and each homogeneous index represents a heterogeneity level. The interrock heterogeneity is considered through the application of the DIP technique, which relies mainly on the micro-structures in the digital images. After assigning physico-mechanical parameters and boundary conditions, the heterogeneous glutenite model can be set for calculation.

3. 2D Numerical Investigation

Based on the 2D sketch shown in Figure 5, numerical simulations are conducted in this section to investigate the fracturing behavior in tight glutenites with consideration of the stress anisotropy and rock strength.

The numerical model is shown in Figure 10, in which a wellbore is preset at the model center with a horizontal perforation. Constant confining stresses, σ_x and σ_y, are applied to the model boundaries. The homogeneity indices m of sandstone and gravel are set at 3.0 and 6.0, respectively. Other physico-mechanical parameters are also shown in Table 2. A hydraulic pressure with an increment of 0.1 MPa per loading step is imposed into the wellbore until the specimen breaks down. The model is simplified as a plane strain case during simulation.

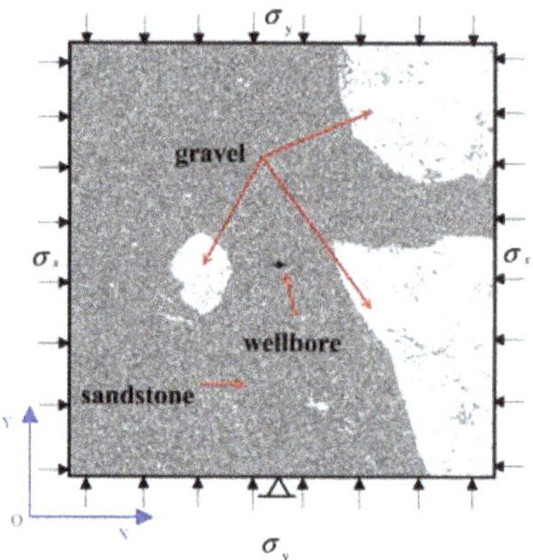

Figure 10. The 2D numerical model.

3.1. Impact of Stress Anisotropy

Geostress has been confirmed by numerous studies to be a key factor dominating HF propagation [32–36]. Stress anisotropy is crucial for fracturing treatment optimization because HFs always prefer to propagate in the maximum principal stress direction. Therefore, the stress anisotropy should be seriously considered during hydraulic fracturing. In this study, the horizontal stress difference $\Delta\sigma$, defined as $\Delta\sigma = \sigma_x - \sigma_y$, is applied to quantify the magnitude of stress anisotropy in the numerical model. Base on the stress difference, three numerical cases (A, B and C) as shown in Table 3 are set for simulation.

Table 3. Stress magnitude in cases A, B and C.

Case	σ_x (MPa)	σ_y (MPa)	$\Delta\sigma$ (MPa)
A	30.0	15.0	15.0
B	30.0	20.0	10.0
C	30.0	28.0	2.0

To display the fracturing process and HF geometries in detail, three pictures in each case are selected and shown in sequence in Figure 11a–c corresponding to those in Case A. Figure 11 d–f correspond to those in Case B. Figure 11g–i correspond to those in Case C.

In Case A, HFs initiate from the perforations in both sides and then propagate in the maximum horizontal stress direction, i.e., the x axis direction (Figure 11a). When intersecting the gravels, the HFs exhibit no obvious change in the path and continue to propagate into the gravels or penetrate the gravels (Figure 11b,c). During the entire process, the HFs propagate with no deflection and the gravels exert little influence on the propagation path.

Compared with that in Case A, the stress difference $\Delta\sigma$ decreases in Case B, and its control on the HFs propagation attenuates. On this occasion, the effect of rock heterogeneity is enhanced. That is why the initial HFs propagate in not as straight a way as that in Case A (Figure 11d). When intersecting the gravels, the left HF is observed to be deflected from its previous direction and propagates along the gravel until reorientating into its previous direction once again for propagation (Figure 11e,f). In this situation, the gravels act as obstacles for HF propagation. The right HF propagates into the gravel after a small deflection along the gravel (Figure 11f) and then bifurcates into multi-fractures. It should be specially explained that the gravels in the model are not totally monolithologic due to existence of interior defects that exhibit darker color than surrounding gravel elements (Figure 10). In reality, conglomerate gravels generally contain geological defects such as pores, fissures and fillings [37,38]. These defects could complicate the HFs path in gravels.

In Case C, the horizontal stress difference $\Delta\sigma$ is the lowest and its control is the weakest. The gravel still acts as obstacle for the left HF (Figure 11g,h). The difference is that, after deflection along the gravel, the left HF reorientates more slowly to the previous direction (Figure 11i), leading to a larger reorientation distance than that in Case B. In addition, the right HF deflects to propagate along the gravel, with new branches extending into the gravel and propagating along the preexisting defects, forming complex multi-fractures.

Therefore, the fracturing operation in tight glutenite could generate HFs with different degrees of complexity, from a single HF as in case A to complex multi-fracture as in Case C, relying on the stress anisotropy. A higher stress difference $\Delta\sigma$ facilitates the HFs to penetrate through the gravels and induces simplex HFs. A lower $\Delta\sigma$ usually leads to fracture deflections and bifurcations, resulting in multi-fractures or fracture networks that create higher connectivity among pores and offer easier pathways for the hydrocarbon to move towards the wellbore.

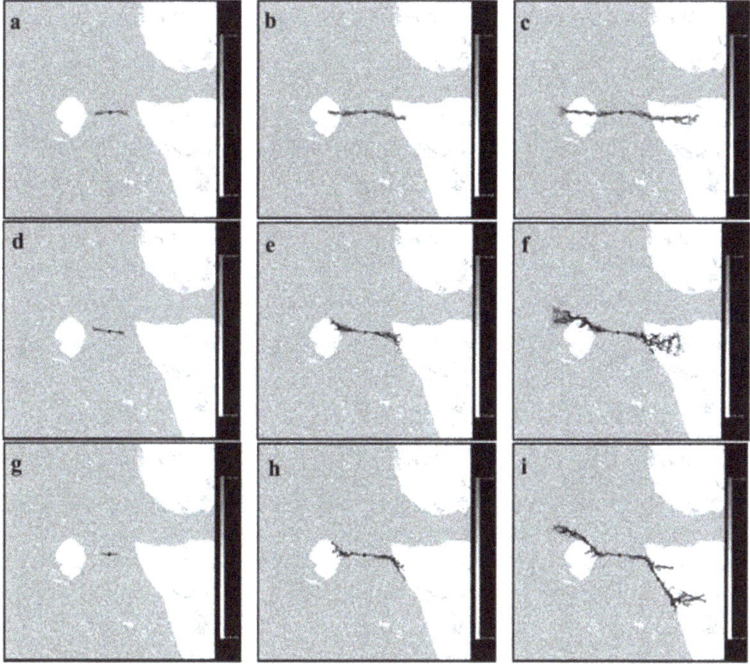

Figure 11. Numerically obtained fracturing process in Case A, B and C. (**a–c**) correspond to those in Case A, (**d–f**) in Case B, and (**g–i**) in Case C.

3.2. Impact of Gravel Strength

Due to differences in mineral composition, gravels even in the same glutenite reservoir can exhibit distinct strengths. For example, Ma et al. [8] observed two types of gravels in the glutenite specimens marked by gravel (A) and gravel (B). The brown-red gravel (B) containing a high amount of quartz has more than three times the tensile strength of the caesious gravel (A) containing a high amount of feldspar. To consider this gravel strength in this study, Case D and Case E are set for simulation by merely changing the gravel strength to compare with Case B. The gravel uniaxial compressive strength (f_{c0}) decreases to 60.0 MPa in Case D and increases to 200.0 Mpa in Case E compared with the 130.0 MPa in Case B, while keeping the ratio of compressive and tensile strength constant. The obtained HFs geometries in the three cases are shown in Figure 12. Obviously, the HFs are observed to penetrate the gravels with low strength (Figure 12b, Case D) but entirely deflect to propagate along the gravels with high strength (Figure 12c, Case E). The gravel strength actually plays a significant role in the fracturing behavior of the tight glutenite reservoirs.

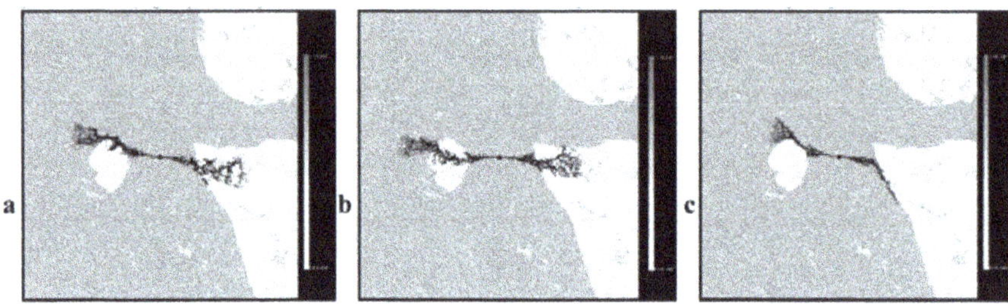

Figure 12. HF geometries in the numerical simulations. (**a**) Case B, f_{c0} = 130.0 MPa; (**b**) Case D, f_{c0} = 60.0 MPa; (**c**) Case E, f_{c0} = 200.0 MPa.

4. 3D Numerical Investigation

The propagation of HFs in tight glutenites can be simplified as a 2D model in some cases, but in some other cases a 3D model is essential because of the unique gravel characteristic such as gravel shape. This section focuses on the fracturing behavior in tight glutenites through 3D simulation. Numerous studies indicate conglomerate gravels are of various shapes and sizes. The gravels generally develop in quasi-spheroidal, ellipsoidal or irregular shapes. The gravel size, taking the Daxing conglomerate, for example [37], can range from 2 cm to 20 cm, and some big gravels can reach several dozen centimeters or even a few meters. Both the gravel shape and size can affect the HF propagation and geometry. In addition, the gravel distribution, which determines the positional relation of the gravel and the HF, is also considered in the numerical model.

4.1. The 3D Numerical Model

A 3D numerical model is established in RFPA with the sketch illustrated in Figure 13a. A cubic glutenite specimen, whose side length is 0.50 m, is discretized into 1,953,125 (125 × 125 × 125) finite elements. A wellbore, 0.024 m in diameter and 0.10 m in depth, into which an initial hydraulic pressure (20.0 MPa) with an increment (0.1 MPa per loading step) is imposed, is preset vertically (the wellbore axis is parallel to the z axis in the Cartesian coordinate system) at the cube center. Notches are precut on both sides of the borehole along the x axis as perforations. σ_x (30.0 MPa), σ_y (20.0 MPa) and σ_z (32.0 MPa), which represent the three principal stresses, are applied to the model boundaries. An ellipsoidal gravel locates in the cubic and its axial size is expressed by a, b and c (Figure 13b). The gravel shape and size change as the three variables change. For simplification, a and b are set equal in the model. Therefore, a (or b) and c are independent variables to control the gravel shape and size. Four numerical cases (Case F, G, H and I) are established according to the gravel characteristic with detailed description illustrated in Figure 13c–f and Table 4. The homogeneity indices m of sandstone and gravel are still set at 3.0 and 6.0, respectively. The gravels are all assigned with high strength (f_{c0} = 200.0 MPa, the same as in Case E) that are not easily destroyed under hydraulic pressure compared with sandstone. All the other physico-mechanical parameters are also shown in Table 2.

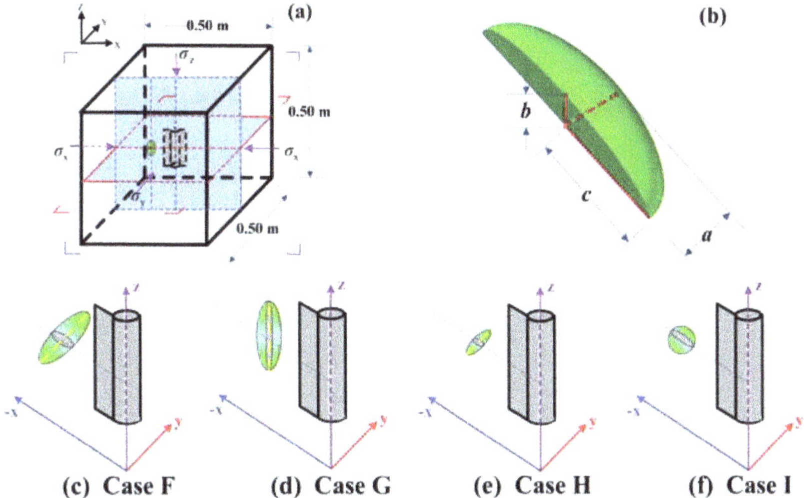

Figure 13. 3D numerical model and the four cases. (**a**) is the 3D model sketch. (**b**) is a quarter of the ellipsoidal gravel showing axial sizes. (**c–f**) are diagrams corresponding to Cases F, G, H and I, showing gravels in different sizes, axial ratios and distributions. The yellow dashed line shows the gravel's long axis. The right endpoint of the blue dashed line overlaps the cube center.

Table 4. Gravel size and distribution parameters in each case.

Gravel Parameters	Long Axis Direction	a (mm)	c/a Ratio
Case F	parallel to y	20	3
Case G	parallel to z	20	3
Case H	parallel to y	12	3
Case I	—	20	1

4.2. Results and Discussion

As the initial hydraulic pressure 20.0 MPa is imposed on the wellbore, the maximum principal stress σ_1 and the minimum principal stress σ_3 of model elements in Case F, G, H and I along the blue dashed lines in Figure 13c–f, respectively, are shown in Figure 14 in which the model boundary element is set as the abscissa origin and the perforation locates in the positive horizontal direction. This demonstrates that the principal stresses of sandstone elements in the four cases are nearly equal to each other and they exhibit some fluctuations. However, both the two principal stresses of gravel elements are found to be much higher than those of sandstone elements and they exhibit smaller fluctuations. The gravel differences in the four cases lead to stress differences in gravel elements and the cases (G and I) with higher values of σ_1 are observed to have lower values of σ_3.

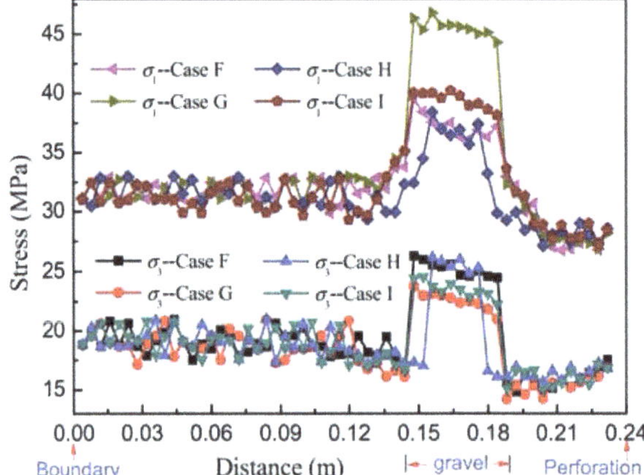

Figure 14. Principal stresses of elements along the blue dashed lines in Figure 13c–f. The boundary element is set as ordinate origin and the models are under initial hydraulic pressure 20.0 MPa.

To inspect the fracturing behavior carefully, a cross section and a longitudinal section both passing the cube center are selected for description (Figure 13a). Figure 15a,b, respectively, shows the cross section and the longitudinal section in Case F before fracturing. Elements in the figures exhibiting cool colors like blue or green represent materials with high Young's modulus while those exhibiting warm colors like red or orange represent materials with low Young's modulus. The gravel exhibits the coolest colors and can be easily distinguished from the sandstone. The HF propagation process in Case F is shown in Figure 15c–h, in which c, d and e correspond to the cross sections and f, g and h correspond to the longitudinal sections in the same loading steps as in the cross sections.

The HFs initiate from the perforations and extend in both length and height. Once intersecting the gravel, the fracture section right ahead the gravel stops but the sections above and below the gravel continue to propagate, as revealed in Figure 15c,f. Afterwards, the upper section radiates downwards and the lower section radiates upwards, leading to the coalescence behind the gravel (Figure 15d,g). In this manner, a bypass HF forms with the final geometry shown in Figure 15e,h.

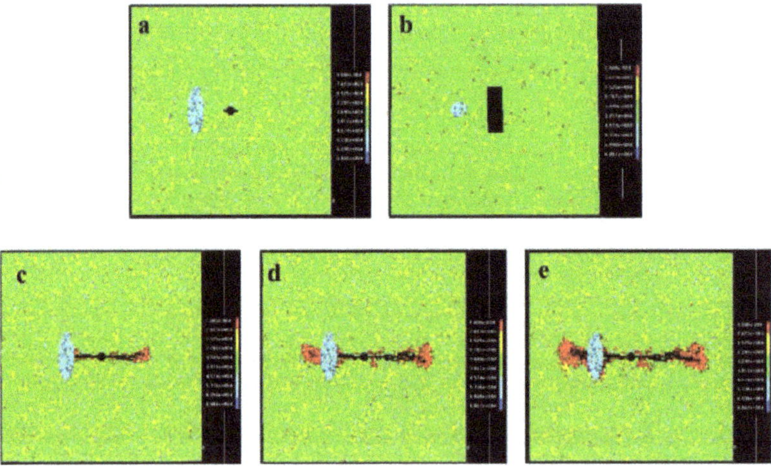

Figure 15. *Cont.*

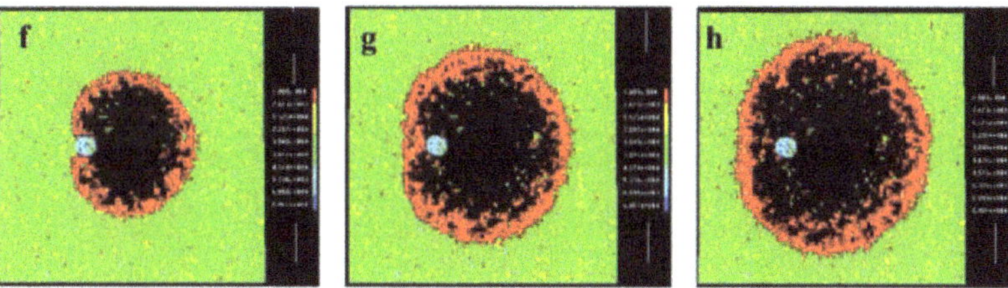

Figure 15. HF geometries during the fracturing process in Case F. (**a**,**b**) respectively show the cross section and longitudinal before fracturing; (**c**–**e**) correspond to those in the cross section; (**f**–**h**) correspond to those in the longitudinal section.

The diagrams in Figure 16a,b illustrate the formation of the bypass HF in Case F. Light and dark blue regions represent the fracture plane in front of the gravel and its extended plane behind the gravel, respectively. The white region shows the plane orthogonal to the blue region. Their intersecting line is highlighted in red. The black and blue arrows show the directions of the HF extending out from the wellbore and near the gravel, respectively. The 3D view and front view describe the HF path when intersecting the gravel. The gravel prevents the propagation of the HF right ahead but the upper and lower sections continue to propagate past the gravel. Then the upper and the lower sections radiate to coalesce and attain a full fracture height behind the gravel. This process can also be reflected through the stress evolution in Figure 17, in which the minimum principal stresses σ_3 of elements along the blue dashed line in Figure 13c in four loading steps are recorded. In step 260, the HF propagates to the location 0.212 m away from the boundary, before intersecting the gravel. Elements stress σ_3 before the HF are large enough to reach the tensile failure threshold (sandstone: $-f_{t0} = -4.5$ MPa; gravel: $-f_{t0} = -20.0$ MPa). In step 268, the HF propagates to exactly intersect the gravel. The stress σ_3 of elements just before the HF, i.e., the gravel elements stress σ_3, decrease dramatically from that in step 260. Afterwards, in step 277, both sandstone and gravel elements stress σ_3 before the HF continues to decrease but there are still no more failure elements in the figure because no more elements' stress σ_3 reach the threshold. In reality, this step is the time when the upper and lower HFs propagate past the gravel and radiate before coalescence. In step 278, the HF has bypassed the gravel and propagated to the location 0.104 m away from the boundary, with the stress σ_3 of sandstone elements before the HF building up, reaching the failure threshold and transferring to surrounding elements. This cyclic process of stress buildup and transfer signifies the continuous HF propagation. It is important to note that the gravel elements stress σ_3 keep above the gravel tensile failure threshold during the simulation. That is why the gravel in Case F remains unbroken in the whole fracturing process, as shown in Figure 15.

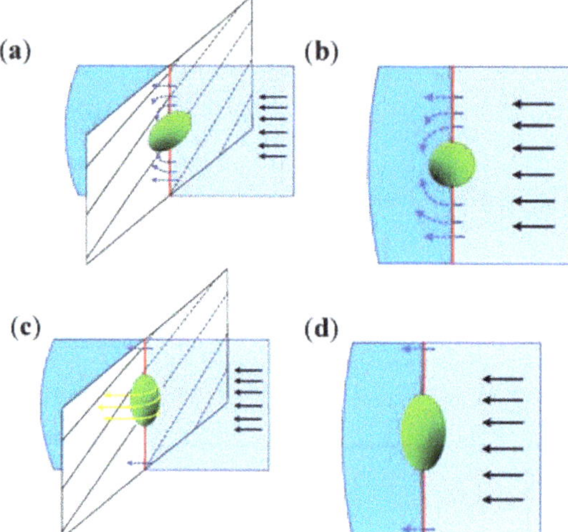

Figure 16. 3D view and front view of HF intersecting the gravel. (**a**,**b**) correspond to Case F; (**c**,**d**) correspond to Case G; (**a**,**c**) show the 3D view; (**b**,**d**) show the front view.

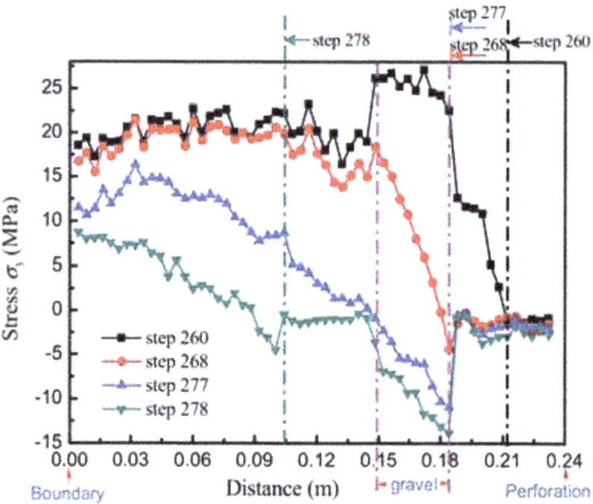

Figure 17. The minimum principal stress of elements in Case F along the blue dashed line in Figure 13c in four loading steps.

This bypass HF exhibits unique characteristic and its formation is quite different from those in conventional reservoirs. Firstly, this fracturing process contains not only the common process of outward fracture propagation in fracture length and height but also the regional process of fracture stop, inward radiation and coalescence. This behavior must be interpreted with caution through 3D simulations since this bypass facture is formed through complex 3D growth and cannot be captured in a 2D simulation. Secondly, the final HF is observed to be cut off at the gravel and it is discontinuous (Figure 15e). From the 2D view, the HF can actually propagate regularly with no deflection after bypassing the gravel in the 3D model. After the bypass, the gravel appears be embedded into the sandstone at two sides of the HF, with only the circle edge the HF bypasses are exposing in the fracturing fluid. On this occasion, even though at the time when the HF has completely penetrated the cube boundaries, it is still difficult to separate the cube specimen along the HF because the gravel embedded into the sandstone has not broken. However, in most laboratory experiments, the fractured specimens are often separated along the main HF by manual operation to observe the inner HF geometry. This operation, unfortunately, may break the gravel along the HF section or strip the gravel at the sandstone-gravel interface, which may be mistaken as evidences for the HF directly penetrating the gravel or deflecting to propagate along the interface, respectively. In reality, this operation has disturbed the observer's decision on the mode the HFs form in when they intersect gravels. This might partly be the reason why this mode of HF propagation is rarely stated during previous experiments.

The gravel in Case G changes only the distribution orientation with the long axis parallel to the z axis, compared with that in Case F. However, just such a small change leads to totally different simulation results as shown in Figure 18. When intersecting the gravel, the HF section right ahead of the gravel deflects to propagate (Figure 18a–c) while the upper and lower sections maintain their previous direction to propagate over the gravel with no deflection (Figure 18d–f). This process is illustrated in the diagrams in Figure 16c,d where the yellow arrows show the directions of HF deflections. The gravel merely deflects the HF section right ahead, ultimately resulting in a distorted HF.

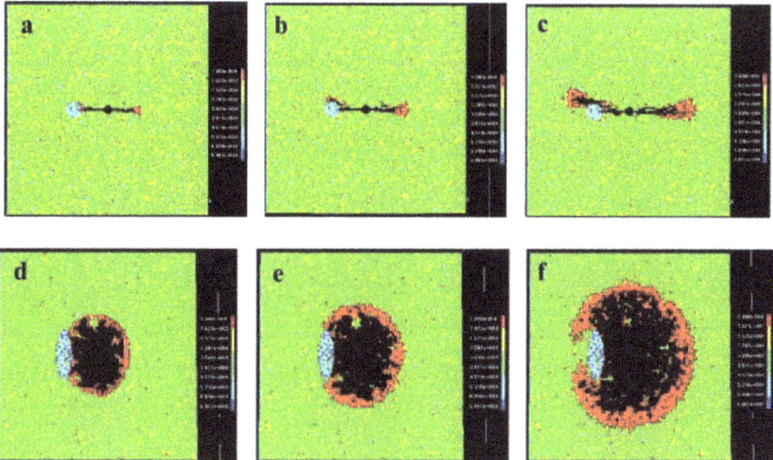

Figure 18. HF geometries during the fracturing process in Case G. (**a–c**) show the HF geometries during fracturing in the horizontal cross-section; (**d–f**) show the corresponding HF geometries in the vertical cross-section.

Compared with Case F, the gravel in Case H simply develops with a smaller size. The final HF geometries in Case H are shown in Figure 20a,b. It is easy to see that the HF propagates past the gravel with no deflection and it is a bypass fracture which is similar to that in Case F.

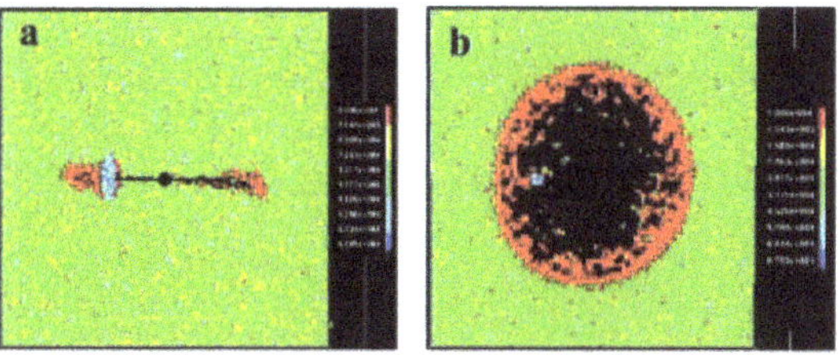

Figure 19. *Cont.*

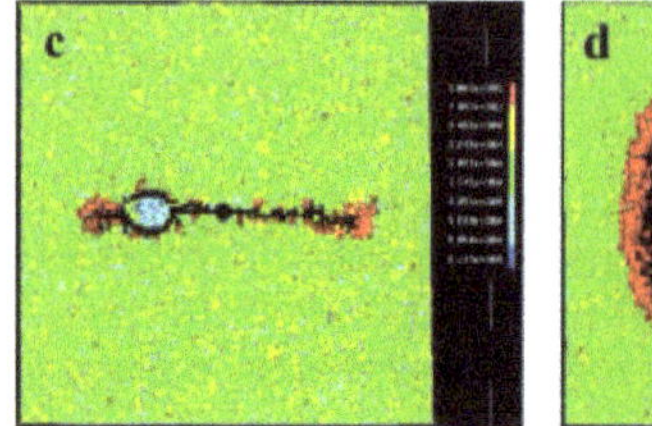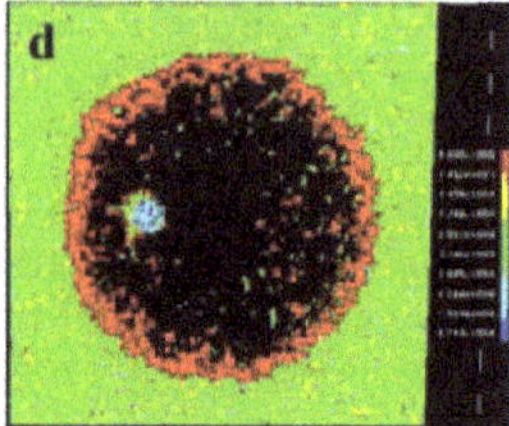

Figure 20. The final HF geometries in Case H and Case I. (**a**,**b**) correspond to Case H that show the cross section and longitudinal section respectively; (**c**,**d**) correspond to those of Case I.

In Case I, the axial ratio decreases to 1 and the gravel is actually a sphere. The final HF geometries in Case I are shown in Figure 20c,d. When intersecting the gravel, the HF right ahead the gravel bifurcates into two branches which deflect to propagate along the gravel. In this manner a distorted HF forms. The propagation mode is similar to that in Case G.

Some findings can be obtained from the comparison of the four numerical cases above. By comparing Cases F and H, in which the gravels share the same distribution orientation and axial ratio but different sizes, it can be seen that pure gravel size within certain limits has little impact on the propagation mode of HFs when intersecting gravels, since the HFs in the two cases show similar behavior under considered conditions. Similarly, by comparing Cases F and G, and Cases F and I, where HFs exhibit different behavior, the impacts of gravel distribution orientation and axial ratio are verified. When the gravel's long axis locates in agreement with the HF width, HF is preferential to bypass the gravel with no deflection, or penetrate the gravel directly in situations such as in Case A where the gravel strength is not as high as in the 3D simulations. When the gravel axial ratio decreases, for example, from 3 in Case F to 1 in Case I, the HFs deflect to propagate when intersecting the gravel. This phenomenon indicates a rounder gravel, i.e., a lower axial ratio, creating better opportunities for HFs deflection.

Through further analysis, the authors attribute the impacts of factors such as gravel distribution orientation and axial ratio to the competition of the gravel size in the HF height direction with that in the width direction. When the gravel size in the HF width direction prevails, it is easier for HFs to penetrate or bypass the gravel than to deflect to propagate for a long distance, because the vast majority of the fracturing fluid will still flow into the main fracture after passing the gravel, rather than flow into the tortuous path of the deflected fracture. Otherwise, HFs will deflect for some distance to propagate along the gravels, especially in situations when the gravels develop weak interfaces.

In summary, the propagation modes of HFs intersecting gravels in glutenite reservoirs are as follows:

(1)   Penetrating directly.
(2)   Deflecting to propagate along the gravels to form distorted HFs.
(3)   Propagating to bypass the gravels.
(4)   Combination of (1) and (2), or (2) and (3).

in which mode (1) prefers to occur in situations when the horizontal stress difference is high, the gravel strength is not large, and the gravel size along the HF width direction is not small; mode (2) prefers to occur when the horizontal stress difference is low, the gravel strength is large, and the gravel size along the HF width direction is not large; mode (3) prefers to occur when the horizontal stress difference is high, the gravel strength is large, and the gravel size along the HF width direction is not small; mode (4) occurs in situations at the intermediate levels correspond to (1) and (2), or (2) and (3), or when the gravels develop interior weaknesses.

5. Conclusions

This paper provides a numerical method to investigate fracturing behavior in tight glutenites. The DIP technique and its integration into RFPA are briefly introduced. Glutenite heterogeneities, including intrarock heterogeneity and interrock heterogeneity, are considered to affect HF propagation. RFPA is capable of considering the intrarock heterogeneity by applying the Weibull model and the interrock heterogeneity with the import of suitable images when integrated with the DIP technique. Its practicability is verified through numerical simulations of hole hydraulic fracturing tests and uniaxial tension tests that have simultaneously confirmed the significant influence of intrarock and interrock heterogeneities on rock stress and failure.

A section image of a glutenite sample was imported into RFPA for a 2D fracturing simulation. Results show that when intersecting the gravel, the HFs can penetrate or deflect, depending on the magnitude of stress anisotropy and gravel strength. A higher horizontal stress difference drives the HFs to directly penetrate the gravel, forming simplex fractures. While a lower horizontal stress difference encourages the HF to deflect along gravels, forming complex fractures. Gravels with interior defects provide opportunities to form complex multi-fractures. Gravels with high strength will behave as obstacles for HFs to penetrate and vice versa.

3D simulations are conducted with the consideration of gravel distribution, gravel size and axial ratio. HFs could propagate past the gravel with no deflection, forming a bypass fracture, which is not easy to observe in common laboratory experiments. They can also deflect to propagate along the gravels like in the 2D simulations. The impact of the gravel distribution orientation and the axial ratio are attributed to the competion of gravel size in the HF height direction with that in the width direction, through which the propagation behaviors of HFs intersecting the gravels can be clarified.

The propagation modes of HFs intersecting gravels in tight glutenite are summarized as: (1) penetrating directly; (2) deflecting to propagate along the gravels to form distorted HFs; (3) propagating to bypass the gravels; (4) a combination of (1) and (2), or (2) and (3). The corresponding conditions for each mode to occur are provided. This study is expected to shed light on the fracturing behavior of tight glutenites during hydraulic stimulations.

Author Contributions: Conceptualization, Z.L.; Methodology, L.L. and Z.Z.; Investigation, M.L.; Formal analysis, L.Z.; Validation, B.H.; Software, C.T.; Writing—Original Draft, Z.L.; Writing—Review and Editing, L.L.

Funding: The authors acknowledge support from the grants from National Science and Technology Major Project (Grant No. 2017ZX05072), National Natural Science Foundation of China (Grant Nos. 51761135102 and 51479024) and Anhui Province Science and Technology Project of China (Grant 17030901023).

References

1. Weng, X.; Kresse, O.; Chuprakov, D.; Cohen, C.-E.; Prioul, R.; Ganguly, U. Applying complex fracture model and integrated workflow in unconventional reservoirs. *J. Petroleum Sci. Eng.* **2014**, *124*, 468–483. [CrossRef]
2. Zou, Y.; Zhang, S.; Zhou, T.; Zhou, X.; Guo, T. Experimental Investigation into Hydraulic Fracture Network Propagation in Gas Shales Using CT Scanning Technology. *Rock Mech. Rock Eng.* **2015**, *49*, 1–13.
3. Dehghan, A.N.; Goshtasbi, K.; Ahangari, K.; Jin, Y.; Bahmani, A. 3D Numerical Modeling of the Propagation of Hydraulic Fracture at Its Intersection with Natural (Pre-existing) Fracture. *Rock Mech. Rock Eng.* **2016**, *50*, 1–20. [CrossRef]
4. Fu, W.; Ames, B.C.; Bunger, A.P.; Savitski, A.A. Impact of Partially Cemented and Non-persistent Natural Fractures on Hydraulic Fracture Propagation. *Rock Mech. Rock Eng.* **2016**, *49*, 4519–4526. [CrossRef]
5. Huang, B.; Liu, J. Experimental Investigation of the Effect of Bedding Planes on Hydraulic Fracturing under True Triaxial Stress. *Rock Mech. Rock Eng.* **2017**, *50*, 2627–2643. [CrossRef]
6. Liu, Z.; Wang, S.; Zhao, H.; Wang, L.; Li, W.; Geng, Y.; Tao, S.; Zhang, G.; Chen, M. Effect of Random Natural Fractures on Hydraulic Fracture Propagation Geometry in Fractured Carbonate Rocks. *Rock Mech. Rock Eng.* **2017**, *51*, 491–511. [CrossRef]

7. Llanos, E.M.; Jeffrey, R.G.; Hillis, R.; Zhang, X. Hydraulic Fracture Propagation through an Orthogonal Discontinuity: A Laboratory, Analytical and Numerical Study. *Rock Mech. Rock Eng.* **2017**, *50*, 2101–2118. [CrossRef]

8. Ma, X.; Zou, Y.; Li, N.; Chen, M.; Zhang, Y.; Liu, Z. Experimental study on the mechanism of hydraulic fracture growth in a glutenite reservoir. *J. Struct. Geol.* **2017**, *97*, 37–47. [CrossRef]

9. Zhu, W.; Wang, Z.; Li, A.; Gao, Y.; Yue, M. *The Seepage Theory and Exploitation Technology of Hydraulic Fracturing in Thin Interbed Tight Reservoirs*; Science Press: Beijing, China, 2016.

10. Li, L. *Size Effect Tests and Mechanical Properties of Low-Permeability Sandstone under Static and Dynamic Loadings*; Shengli Oilfield Branch Company: Dongying, China, 2017.

11. Meng, Q.M.; Zhang, S.C.; Guo, X.M.; Chen, X.H.; Zhang, Y. Aprimary investigation on propagation mechanism for hydraulic fractures in Glutenite formation. *J. Oil Gas Technol.* **2010**, *32*, 119–123.

12. Li, L.; Meng, Q.; Wang, S.; Li, G.; Tang, C. A numerical investigation of the hydraulic fracturing behaviour of conglomerate in Glutenite formation. *Acta Geotech.* **2013**, *8*, 597–618. [CrossRef]

13. Warpinski, N.R.; Mayerhofer, M.J.; Agarwal, K.; Du, J. Hydraulic fracture geomechanics and microseismic source mechanisms. In Proceedings of the SPE Annual Technical Conference and Exhibition (ATCE2012), San Antonio, TX, USA, 8–10 October 2012.

14. Haddad, M.; Sepehrnoori, K. XFEM-Based CZM for the Simulation of 3D Multiple-Cluster Hydraulic Fracturing in Quasi-Brittle Shale Formations. *Rock Mech. Rock Eng.* **2016**, *49*, 4731–4748. [CrossRef]

15. Ziarani, A.S.; Chen, C.; Cui, A.; Quirk, D.J.; Roney, D. Fracture and wellbore spacing optimization in multistage fractured horizontal wellbores: Learnings from our experience on Canadian unconventional resources. In Proceedings of the International Petroleum Technology Conference (IPTC2014), Kuala Lumpur, Malaysia, 10–12 December 2014.

16. Cipolla, C.; Weng, X.; Mack, M.; Ganguly, U.; Gu, H.; Kresse, O.; Cohen, C.E. Integrating microseismic mapping and complex fracture modeling to characterize fracture complexity. In Proceedings of the SPE Hydraulic Fracturing Technology Conference and Exhibition (HFTC2011), The Woodlands, TX, USA, 24–26 January 2011.

17. Zhao, Z.; Guo, J.; Ma, S. The Experimental Investigation of Hydraulic Fracture Propagation Characteristics in Glutenite Formation. *Adv. Mater. Sci. Eng.* **2015**, *2015*, 1–5. [CrossRef]

18. Ju, Y.; Liu, P.; Chen, J.; Yang, Y.; Ranjithd, P.G. CDEM-based analysis of the 3D initiation and propagation of hydrofracturing cracks in heterogeneous glutenites. *J. Nat. Gas Sci. Eng.* **2016**, *35*, 614–623. [CrossRef]

19. Li, N.; Zhang, S.; Ma, X.; Zou, Y.; Chen, M.; Li, S.; Zhang, Y. Experimental study on the propagation mechanism of hydraulic fracture in glutenite formations. *Chin. J. Rock Mech. Eng.* **2017**, *36*, 2383–2392. (In Chinese)

20. Wang, R.Q.; Kemeny, J.M. A study of the coupling between mechanical loading and flow properties in tuffaceous rock. In Proceedings of the 1st North American Rock Mechanics Symposium, Austin, TX, USA, 1–3 June 1994.

21. Mahrer, K.D. A review and perspective on far-field hydraulic fracture geometry studies. *J. Pet. Sci. Eng.* **1999**, *24*, 13–28. [CrossRef]

22. Beugelsdijk, L.J.L.; De Pater, C.J.; Sato, K. Experimental hydraulic fracture propagation in a multi-fractured medium. In *SPE Asia Pacific Conference on Integrated Modelling for Asset Management*; SPE 59419; Society of Petroleum Engineers: Richardson, TX, USA, 2000.

23. Jeffrey, R.G.; Bunger, A.; LeCampion, B.; Zhang, X.; Chen, Z.; van As, A.; Allison, D.P.; de Beer, W.; Dudley, J.W.; Thiercelin, E.S.M.J.; et al. Measuring hydraulic fracture growth in naturally fractured rock. In *SPE Annual Technical Conference and Exhibition*; SPE124919; Society of Petroleum Engineers: Richardson, TX, USA, 2009.

24. Liu, P.; Ju, Y.; Ranjith, P.G.; Zheng, Z.; Chen, J. Experimental investigation of the effects of heterogeneity and geostress difference on the 3D growth and distribution of hydrofracturing cracks in unconventional reservoir rocks. *J. Nat. Gas Sci. Eng.* **2016**, *35*, 541–554. [CrossRef]

25. Tang, C.A.; Tham, L.G.; Lee, P.K.K.; Yang, T.H.; Li, L.C. Coupled analysis of flow, stress and damage (FSD) in rock failure. *Int. J. Rock Mech. Min. Sci.* **2002**, *39*, 477–489. [CrossRef]

26. Tan, X.; Konietzky, H.; Chen, W. Numerical Simulation of Heterogeneous Rock Using Discrete Element Model Based on Digital Image Processing. *Rock Mech. Rock Eng.* **2016**, *49*, 1–8. [CrossRef]

27. Zhu, W.C.; Liu, J.; Yang, T.H.; Sheng, J.C.; Elsworth, D. Effects of local rock heterogeneities on the hydromechanics of fractured rocks using a digital-image-based technique. *Int. J. Rock Mech. Min. Sci.* **2006**, *43*, 1182–1199. [CrossRef]

28. Tang, C.A.; Liu, H.; Lee, P.K.K.; Tsui, Y.; Tham, L.G. Numerical studies of the influence of microstructure on rock failure in uniaxial compression—Part I: Effect of heterogeneity. *Int. J. Rock Mech. Min. Sci.* **2000**, *37*, 555–569. [CrossRef]

29. Yuan, S.C.; Harrison, J.P. Development of a hydro-mechanical local degradation approach and its application to modelling fluid flow during progressive fracturing of heterogeneous rocks. *Int. J. Rock Mech. Min. Sci.* **2005**, *42*, 961–984. [CrossRef]

30. Tang, C.A. Numerical simulation of progressive rock failure and associated seismicity. *Int. J. Rock Mech. Min. Sci.* **1997**, *34*, 249–261. [CrossRef]

31. Charlez, P.A. *Rock Mechanics (II:Petroleum Applications)*; Technical Publisher: Paris, France, 1991.

32. Blanton, T.L. *An Experimental Study of Interaction between Hydraulically Induced and Pre-Existing Fractures*; SPE10847; Society of Petroleum Engineers: Richardson, TX, USA, 1982.

33. Blanton, T.L. *Propagation of Hydraulically and Dynamically Induced Fractures in Naturally Fractured Reservoirs*; SPE 15261; Society of Petroleum Engineers: Richardson, TX, USA, 1986.

34. Abass, H.H. *Experimental Observations of Hydraulic Fracture Propagation through Coal Blocks*; SPE 21289; Society of Petroleum Engineers: Richardson, TX, USA, 1990.

35. Chuprakov, D.A.; Akulich, A.V.; Siebrits, E.; Thiercelin, M.J. *Hydraulic-Fracture Propagation in a Naturally Fractured Reservoir*; SPE 128715; Society of Petroleum Engineers: Richardson, TX, USA, 2010.

36. Lin, C.; He, J.; Li, X.; Wan, X.; Zheng, B. An Experimental Investigation into the Effects of the Anisotropy of Shale on Hydraulic Fracture Propagation. *Rock Mech. Rock Eng.* **2017**, *50*, 543–554. [CrossRef]

37. Liu, H.; Jiang, Z.; Zhang, R.; Zhou, H. Gravels in the Daxing conglomerate and their effect on reservoirs in the Oligocene Langgu Depression of the Bohai Bay Basin, North China. *Mar. Petroleum Geol.* **2012**, *29*, 192–203. [CrossRef]

38. Zhao, Z.; Xu, S.; Jiang, X.; Lin, C.; Cheng, H.; Cui, J.; Jia, L. Deep strata geologic structure and tight sandy conglomerate gas exploration in Songliao Basin, East China. *Petroleum Exp. Dev.* **2016**, *43*, 13–25. [CrossRef]

 *processes*

Article

A New Pseudo Steady-State Constant for a Vertical Well with Finite-Conductivity Fracture

Yudong Cui [1], Bin Lu [1], Mingtao Wu [2] and Wanjing Luo [1,*]

[1] School of Energy Resources, China University of Geosciences, Beijing 100083, China;
    CuiYudong1995@126.com (Y.C.); cadbin@163.com (B.L.)
[2] Beijing Key Laboratory of Unconventional Natural Gas Geological Evaluation and Development
    Engineering, China University of Geosciences, Beijing 100083, China; wumt@cugb.edu.cn
* Correspondence: luowanjing@cugb.edu.cn; Tel.: +86-188-1002-8882

Received: 19 June 2018; Accepted: 9 July 2018; Published: 19 July 2018

Abstract: The Pseudo Steady-State (PSS) constant b_{Dpss} is defined as the difference between the dimensionless wellbore pressure and dimensionless average pressure of a reservoir with a PSS flow regime. As an important parameter, b_{Dpss} has been widely used for decline curve analysis with Type Curves. For a well with a finite-conductivity fracture, b_{Dpss} is independent of time and is a function of the penetration ratio of facture and fracture conductivity. In this study, we develop a new semi-analytical solution for b_{Dpss} calculations using the PSS function of a circular reservoir. Based on the semi-analytical solution, a new conductivity-influence function (CIF) representing the additional pressure drop caused by the effect of fracture conductivity is presented. A normalized conductivity-influence function (NCIF) is also developed to calculate the CIF. Finally, a new approximate solution is proposed to obtain the b_{Dpss} value. This approximate solution is a fast, accurate, and time-saving calculation.

Keywords: Pseudo Steady-State (PPS) constant; finite-conductivity fracture; conductivity-influence function; normalized conductivity-influence function; circular closed reservoir

---

1. Introduction

Hydraulic fracturing has been widely used to enhance oil and gas recovery [1–9]. Some models have been introduced to describe fluid flow in hydraulic fractures, such as the uniform-flux fracture model [10–12], infinite-conductivity fracture model [10,13,14], and the finite-conductivity fracture model [15–19].

Pseudo steady-state (PSS) is a boundary-dominant flow regime created when pressure waves spread to the boundary in a closed drainage area. In this flow regime, the relationship of dimensionless wellbore pressure and dimensionless average pressure can be expressed as [20–24]:

$$p_{wD} - p_{avgD} = b_{Dpss} \tag{1}$$

where b_{Dpss} is the Pseudo Steady-State (PSS) constant [24]. This PSS constant b_{Dpss} has been widely used to define the appropriate dimensionless decline rate in many currently used production-decline rate analysis models [20,25–28].

When the pressure disturbance reaches to the boundary, the PPS flow regime occurs and the PSS constant b_{Dpss} can be obtained by running a long-term numerical simulation [20,21,23]. Pratikno et al. [20] presented a numerical solution for the PSS constant for a well with a finite-conductivity fracture in a circular closed reservoir (Figure 1).

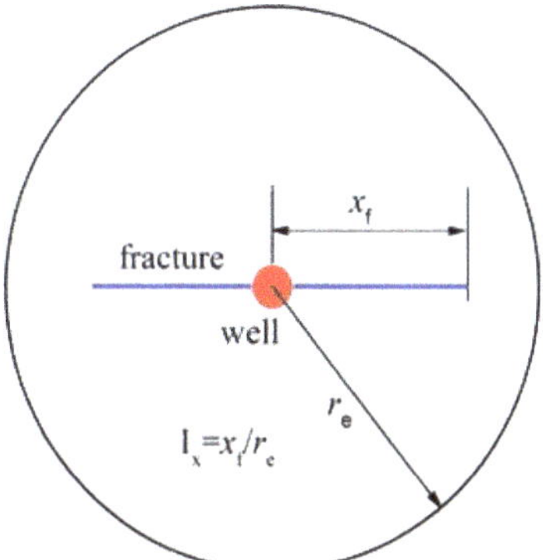

Figure 1. Schematic of a vertical well with a fracture in a circular reservoir.

For convenience, b_{Dpss} is usually approximated as an analytical expression [20,21,23]. Pratikno et al. [20] proposed the following approximate expression:

$$b_{Dpss} = f(r_{eD}) + f\left(C_{fD}\right) \tag{2}$$

where:

$$f(r_{eD}) = \ln(r_{eD}) - 0.049298 + 0.43464/r_{eD}^2 \tag{3}$$

$$f\left(C_{fD}\right) = \frac{0.936268 - 1.00489\psi + 0.319733\psi^2 - 0.0423532\psi^3 + 0.00221799\psi^4}{1 - 0.385539\psi - 0.0698865\psi^2 - 0.0484653\psi^3 - 0.00813558\psi^4} \tag{4}$$

and:

$$\psi = \ln C_{fD} \tag{5}$$

Equation (3) is the pressure drop of a well with an infinite-conductivity fracture in the PSS flow regime [11]. The second term in Equation (2), $f(C_{fD})$, is the additional pressure drop caused by the effect of fracture conductivity. We define $f(C_{fD})$ as the conductivity-influence function (CIF). Thus, the PSS constant b_{Dpss} can be expressed as the sum of the pressure drop of a well with an infinite-conductivity fracture and the conductivity-influence function (CIF). Note that the CIF in Equation (4) is only a function of the fracture conductivity.

Wang et al. [23] also introduced an approximate expression for $f(C_{fD})$ using regression for a circular reservoir.

$$f\left(C_{fD}\right) = \frac{0.95 - 0.56\psi + 0.16\psi^2 + 0.028\psi^3 + 0.0028\psi^4 - 0.00011\psi^5}{1 + 0.094\psi + 0.093\psi^2 + 0.0084\psi^3 + 0.001\psi^4 + 0.00036\psi^5} \tag{6}$$

For the low permeability and ultra-low permeability reservoirs, the elliptical boundary has been used to approximately represent circular a reservoir for calculation of b_{Dpss}. The corresponding conductivity-influence function for an elliptical reservoir was presented by Amini et al. [21]:

$$f\left(C_{fD}\right) = \frac{-4.7468 + 36.2492\psi + 55.0998\psi^2 - 3.98311\psi^3 + 6.07102\psi^4}{-2.4941 + 21.6755\psi + 41.0303\psi^2 - 10.4793\psi^3 + 5.6108\psi^4} \tag{7}$$

In addition, assuming an elliptical boundary, two analytical solutions for a well with an infinite-conductivity fracture and a finite-conductivity fracture were developed by Prats et al. [29] and Lu et al. [24], respectively.

As analyzed from a previous statement, three problems occur with the b_{Dpss} calculation.

(1) The assumption of elliptical flow is an approximate model of the circular boundary [29]. For a fractured well, the real pressure front should be a circle instead of an ellipse during the late-time flow regime.

(2) When C_{fD} trends to infinity, i.e., the infinite-conductivity fracture, Equations (4), (6), and (7) cannot meet the following condition, meaning the limit of conductivity-influence function $f(C_{fD})$ should be zero.

$$\lim_{C_{fD} \to \infty} f\left(C_{fD}\right) = 0 \tag{8}$$

(3) For the existing approximate models [20,21,23], the conductivity-influence function $f(C_{fD})$ is only relative to the fracture conductivity. For different penetration fracture ratios I_x, the distributions in the flow field around the fracture are different, which affects the value of conductivity-influence function (Figure 2). Thus, CIF is not only a function of conductivity, but also related to penetration ratio.

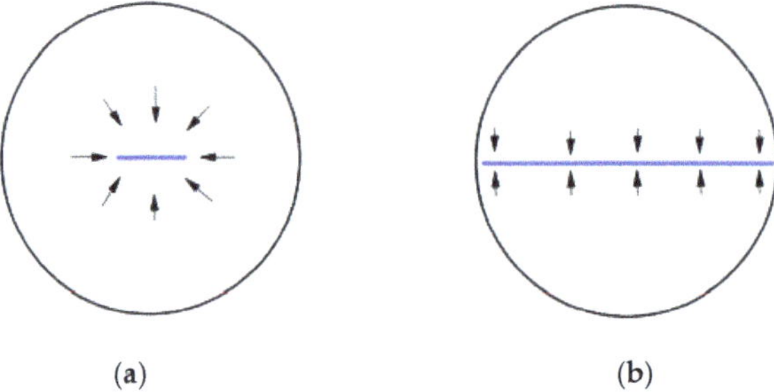

(a)   (b)

Figure 2. Schematic of the flow field around a fracture in a circular reservoir at late-time regime: (**a**) a short fracture with a low penetration ratio and (**b**) a long fracture with high penetration ratio.

In this paper, based on the assumption of a circular closed reservoir, we extended and corrected the work of Pratikno et al. [20]. The contributions of our work include: (1) a semi-analytical method is developed to calculate the b_{Dpss} by use of the PSS function instead of the transient-pressure function [20]; (2) based on the results from the semi-analytical method, a new conductivity-influence function (CIF) is introduced considering the effect of penetration ratio and fracture conductivity has been established; and (3) a new normalized conductivity-influence function (NCIF) is introduced to calculate the value of b_{Dpss}.

2. Mathematical Model

2.1. Basic Assumptions

As shown in Figure 1, a vertical fractured well is located in the center of the circular closed isotropic formation with radius r_e. The flow in the reservoir and fracture is assumed to be single phase and isothermal with a slightly compressible Newtonian fluid. The penetrate ratio I_x is defined as:

$$I_x = \frac{x_f}{r_e} = \frac{1}{r_{eD}} \tag{9}$$

2.2. Semi-Analytical Model for b_{Dpss} Calculation

Different from the transient-pressure-function method presented by Pratikno et al. [20], we derive the semi-analytical model with the PSS function. The advantages of our method include: (1) we can obtain the PSS pressure directly instead of using the long-term approximation of transient pressure, and (2) without the numeric inversion, our method is more accurate and is less time consuming.

2.2.1. Flow Model of the Reservoir

The fracture is equally divided into $2N$ segments. Each segment can be considered as a uniform-flux fracture [30]. Therefore, the uniform-flux solution of a fracture located in a closed circular reservoir can be used to calculate the pressure [11]. To consider the fracture symmetry, we focused on half of the fracture. According to the superposition principle, the dimensionless pressure of the ith segment in the reservoir can be written as:

$$p_{Di} = p_{avgD} + \sum_{j=1}^{2N} q_{Dj} \cdot F_{ij}(x_{Di}, y_{Di}, x_{wDi}, y_{wDi}, r_{eD}), i = 1, 2, 3 \cdots N \tag{10}$$

where F is the function presented by Ozkan [11] for a circular closed reservoir in the PSS flow regime.

Equation (10) can be written in matrix form as:

$$\vec{p}_D - \begin{pmatrix} p_{avgD} \\ p_{avgD} \\ \vdots \\ p_{avgD} \end{pmatrix} = F \cdot \vec{q}_D \tag{11}$$

2.2.2. Flow Model of the Fracture

Luo and Tang [30] derived a wing solution in the discretized form and the pressure of the ith segment, which can be expressed as:

$$P_{wD} - P_{fDi} = \left(\frac{2\pi}{C_{fD}}\right) \cdot \left[\begin{array}{c} r_{Di} \cdot \sum\limits_{k=1}^{N} q_{fDk} - \left(\frac{\Delta r_{Di}}{8}\right) \cdot q_{fDi} \\ - \sum\limits_{k=1}^{i-1} q_{fDk} \cdot \left[\frac{\Delta r_{Di}}{2} + (r_{Di} - k \cdot \Delta r_{Di})\right] \end{array} \right], i = 1, 2, \ldots N \tag{12}$$

with

$$r_{Di} = \sum_{k=1}^{i-1} L_{fDk} + L_{fDi}/2, \ \Delta r_{Di} = L_{fDi} \tag{13}$$

Equation (13) can be written in matrix form as:

$$\begin{pmatrix} p_{wD} \\ p_{wD} \\ \vdots \\ p_{wD} \end{pmatrix} - \vec{p}_{fD} = C \cdot \vec{q}_{fD} \tag{14}$$

2.2.3. Semi-Analytical Solution for b_{Dpss}

According to the continuity condition, which states that the pressure and flux must be continuous along the fracture surface, the following conditions must hold along the wing plane:

$$\vec{p}_{fD} = \vec{p}_D, \ \vec{q}_{fD} = \vec{q}_D \tag{15}$$

Substituting Equation (15) into Equations (11) and (14) yields:

$$\begin{pmatrix} p_{wD} \\ p_{wD} \\ \vdots \\ p_{wD} \end{pmatrix} - \begin{pmatrix} p_{avgD} \\ p_{avgD} \\ \vdots \\ p_{avgD} \end{pmatrix} = (C + F) \cdot \vec{q}_{fD} \tag{16}$$

Substituting Equation (1) into Equation (16), we obtain:

$$(C + F) \cdot \vec{q}_{fD} - b_{Dpss} = 0 \tag{17}$$

In addition, the total flow rate satisfies the following:

$$\sum_{j=1}^{N} q_{fDi} = \frac{1}{2} \tag{18}$$

b_{Dpss} can be obtained by solving Equations (17) and (18) with the Gauss elimination method.

2.3. Conductivity-Influence Function (CIF)

The procedure to calculate the conductivity-influence function are as follows: (1) Calculate the PSS constant b_{Dpss} for different I_x and C_{fD} using the semi-analytical model in Section 2.2. (2) Calculate the pressure drop of the infinite-conductivity fracture $f(r_{eD})$ for different I_x with Equations (3) and (9). (3) Calculate the difference of b_{Dpss} and $f(r_{eD})$ with Equation (2) and the value of conductivity-influence function (CIF) can be obtained.

Figure 3 presents the CIF for different penetration ratios and fracture conductivities. Different than the solutions obtained by Pratikno et al. [20], the CIF is not only dependent on fracture conductivity but also has a strong relationship with penetration ratio, especially when C_{fD} is less than 10. Additionally, fracture conductivity function tends to be zero when C_{fD} is greater than 300. Table 1 presents the value of the conductivity-influence function for I_x = 0.001–1 and C_{fD} = 0.1–1000.

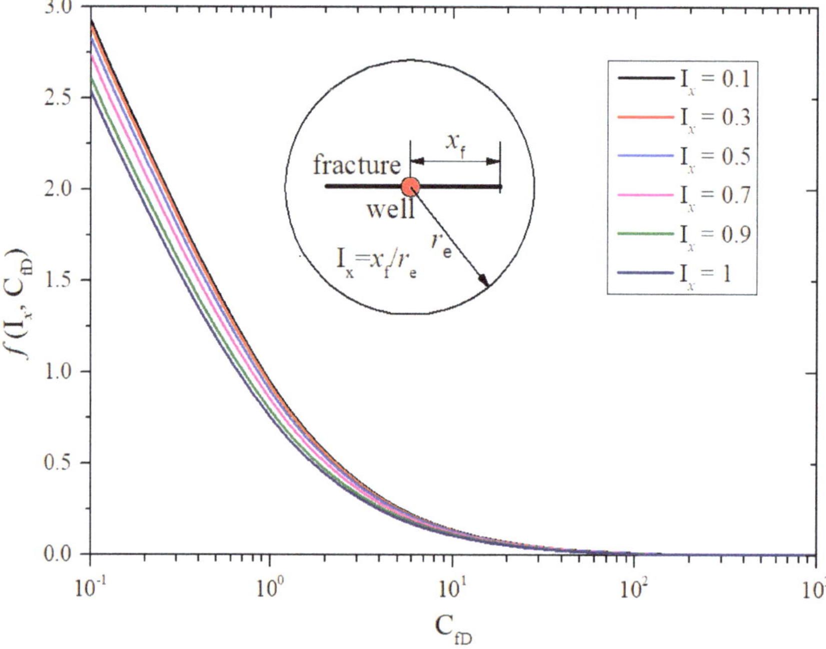

Figure 3. Conductivity-influence function with fracture conductivity (C_{fD}) and penetration ratio (I_x).

Table 1. Conductivity-influence function (CIF) of different penetration ratio (I_x) and fracture conductivity (C_{fD}).

C_{fD}	I_x								
	0.001	0.05	0.01	0.05	0.1	0.3	0.5	0.8	1
0.1	2.924955	2.923983	2.924917	2.923983	2.921063	2.889918	2.82763	2.675801	2.535652
0.125893	2.694126	2.693183	2.694089	2.693183	2.690353	2.660159	2.599771	2.452575	2.316702
0.158489	2.473614	2.472703	2.473578	2.472703	2.46997	2.440814	2.382503	2.240371	2.109172
0.199526	2.259775	2.258901	2.25974	2.258901	2.256277	2.228293	2.172325	2.035904	1.909976
0.251189	2.050804	2.049971	2.050771	2.049971	2.047474	2.02083	1.967543	1.837656	1.71776
0.316228	1.846441	1.845657	1.84641	1.845657	1.843302	1.818188	1.767959	1.645526	1.532512
0.398107	1.647651	1.64692	1.647622	1.64692	1.644726	1.621328	1.574532	1.460467	1.355176
0.501187	1.456232	1.45556	1.456206	1.45556	1.453543	1.432024	1.388986	1.284082	1.187247
0.630957	1.274395	1.273785	1.274371	1.273785	1.271955	1.252433	1.213389	1.118219	1.03037
0.794328	1.104337	1.103791	1.104315	1.103791	1.102154	1.084688	1.049755	0.964606	0.886007
1	0.9479	0.947418	0.947881	0.947418	0.945973	0.930557	0.899725	0.824572	0.7552
1.258925	0.806339	0.805919	0.806322	0.805919	0.80466	0.791229	0.764366	0.698889	0.638448
1.584893	0.680231	0.67987	0.680217	0.67987	0.678786	0.667223	0.644099	0.587732	0.535701
1.995262	0.569498	0.56919	0.569486	0.56919	0.568267	0.558422	0.538731	0.490735	0.446431
2.511886	0.473515	0.473256	0.473505	0.473256	0.472478	0.464176	0.447573	0.407101	0.369743
3.162278	0.391259	0.391042	0.39125	0.391042	0.390391	0.383452	0.369573	0.335742	0.304514
3.981072	0.321455	0.321276	0.321448	0.321276	0.320736	0.314979	0.303467	0.275404	0.2495
5.011872	0.262718	0.26257	0.262712	0.26257	0.262125	0.257382	0.247898	0.224778	0.203437
6.309573	0.213648	0.213526	0.213643	0.213526	0.213162	0.209279	0.201513	0.182584	0.16511
7.943282	0.17291	0.172811	0.172906	0.172811	0.172514	0.169353	0.16303	0.147618	0.133391
10	0.139273	0.139193	0.13927	0.139193	0.138953	0.136392	0.131272	0.118789	0.107268
12.58925	0.111637	0.111572	0.111634	0.111572	0.111379	0.109316	0.105189	0.095131	0.085847
15.84893	0.089036	0.088984	0.089034	0.088984	0.088829	0.087175	0.083866	0.075802	0.068357
19.95262	0.070637	0.070596	0.070636	0.070596	0.070472	0.069152	0.066513	0.060078	0.054139
25.11886	0.055731	0.055698	0.05573	0.055698	0.0556	0.054552	0.052457	0.047349	0.042633
31.62278	0.043716	0.043691	0.043715	0.043691	0.043613	0.042785	0.04113	0.037096	0.033371
39.81072	0.034091	0.03407	0.03409	0.03407	0.034009	0.033359	0.032058	0.028887	0.02596
50.11872	0.026434	0.026418	0.026433	0.026418	0.02637	0.025861	0.024844	0.022363	0.020074
63.09573	0.020396	0.020383	0.020395	0.020383	0.020346	0.01995	0.019157	0.017225	0.015441
79.43282	0.015687	0.015677	0.015687	0.015677	0.015648	0.015341	0.014725	0.013223	0.011838
100	0.012068	0.012061	0.012068	0.012061	0.012038	0.011799	0.011321	0.010155	0.009078
125.8925	0.009341	0.009335	0.009341	0.009335	0.009318	0.009131	0.008758	0.007849	0.007009
158.4893	0.007342	0.007337	0.007342	0.007337	0.007324	0.007177	0.006883	0.006166	0.005504
199.5262	0.005936	0.005933	0.005936	0.005933	0.005922	0.005803	0.005567	0.004991	0.004459
251.1886	0.005012	0.005009	0.005012	0.005009	0.005	0.004902	0.004706	0.004228	0.003786
316.2278	0.004479	0.004476	0.004478	0.004476	0.004468	0.004383	0.004213	0.003799	0.003417
398.1072	0.004259	0.004257	0.004259	0.004257	0.00425	0.004172	0.004017	0.003639	0.00329
501.1872	0.004292	0.004289	0.004292	0.004289	0.004283	0.004208	0.004059	0.003695	0.003359
630.9573	0.004525	0.004523	0.004525	0.004523	0.004516	0.00444	0.00429	0.003923	0.003584
794.3282	0.004917	0.004914	0.004916	0.004914	0.004907	0.004828	0.00467	0.004285	0.00393
1000	0.005431	0.005429	0.005431	0.005429	0.005421	0.005336	0.005167	0.004753	0.004372

2.4. New Approximate Solution of Pseudo Steady-State Constant b_{Dpss}

Based on the above work, we redefined the PSS constant as the sum of two parts: PSS constant for infinite fracture conductivity [11] and the conductivity-influence function. The PSS constant for infinite fracture conductivity $b_{Dpss,FC}$ is the function of the penetration ratio, and the conductivity-influence function is the function of the penetration ratio and conductivity.

$$b_{Dpss,FC}(I_x, C_{fD}) = b_{Dpss,IC}(I_x) + f(I_x, C_{fD}) \tag{19}$$

where

$$b_{Dpss,IC}(I_x) = -\ln(I_x) - 0.049298 + 0.43464(I_x)^2 \tag{20}$$

We firstly calculate a specific case ($I_x = 0$). As shown in Equation (9):

$$I_x = \lim_{r_e \to \infty} \frac{x_f}{r_e} = \lim_{r_{eD} \to \infty} \frac{1}{r_{eD}} = 0 \tag{21}$$

If the circular closed reservoir is replaced by an infinite-acting reservoir, b_{Dpss} can be approximately replaced by the dimensionless-pressure difference between the finite-conductivity fracture and infinite-conductivity fracture at the radial flow regime for the infinite reservoir [15].

Figure 4 illustrates the CIF of different conductivity at $I_x = 0$ (blue) and $I_x = 1$ (red). Two regression equations for CIF are obtained as follows.

For the case of $I_x = 0$:

$$f_0(C_{fD}) = f(I_x = 0, C_{fD}) = \frac{a_1 u^2 + a_2 u + a_3}{b_1 u^3 + b_2 u^2 + b_3 u + 1} \tag{22}$$

where:

$$u = \ln(C_{fD}) \tag{23}$$

and:

$$\begin{aligned} a_1 = 0.02705; \; a_2 = -0.3123; \; a_3 = 0.9479 \\ b_1 = 0.01736; \; b_2 = 0.1218; \; b_3 = 0.3539 \end{aligned} \tag{24}$$

For the case of $I_x = 1$, the hydraulic fracture fully penetrates the reservoir.

$$f_1(C_{fD}) = f(I_x = 1, C_{fD}) = \frac{a'_1 u^2 + a'_2 u + a'_3}{b'_1 u^3 + b'_2 u^2 + b'_3 u + 1} \tag{25}$$

where:

$$\begin{aligned} a'_1 = 0.02188; \; a'_2 = -0.2509; \; a'_3 = 0.7552 \\ b'_1 = 0.01702; \; b'_2 = 0.1233; \; b'_3 = 0.3798 \end{aligned} \tag{26}$$

We notice that in our model, Equations (22) and (25) meet the following condition:

$$\lim_{C_{fD} \to \infty} f_0(C_{fD}) = \lim_{C_{fD} \to \infty} f_1(C_{fD}) = 0 \tag{27}$$

Moreover, we define a normalized fracture conductivity function (NCIF) as follows:

$$\hat{f}(I_x, C_{fD}) = \frac{f(I_x, C_{fD}) - f_0(C_{fD})}{f_1(C_{fD}) - f_0(C_{fD})} \tag{28}$$

We calculate the CIF for $I_x = 0.001$–1 and $C_{fD} = 0.1$–1000 with the method in Sections 2.2 and 2.3. Ten values were calculated for each logarithmic period. Thus, 1200 values of CIF were obtained. The values of NCIF can be calculated using Equation (28).

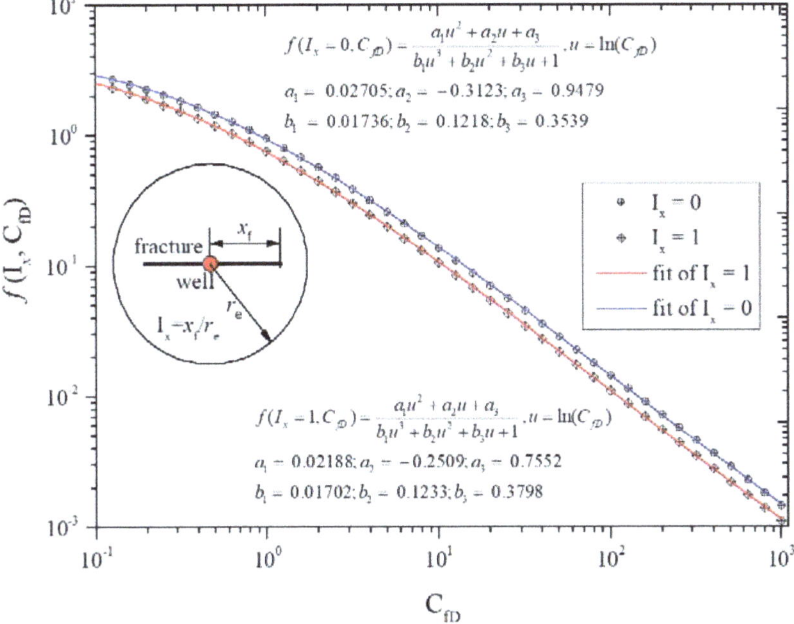

Figure 4. Regression of conductivity-influence function when the penetration ratio (I_x) equals 0 and 1.

Figure 5 shows the relationship between NCIF and I_x. Notably, all data fall in the same straight line in logarithmic coordinates. This means that the NCIF is solely dependent on penetration ratio I_x. A regression equation can be obtained for NCIF.

$$\hat{f}(I_x, C_{fD}) = \frac{f(I_x, C_{fD}) - f_0(C_{fD})}{f_1(C_{fD}) - f_0(C_{fD})} = I_x^{2.00592} \approx I_x^2 \tag{29}$$

Note that if $I_x = 0$, the NCIF is equal to 0 in Equation (29).
Recasting Equation (29) yields:

$$f(I_x, C_{fD}) = I_x^2[f_1(C_{fD}) - f_0(C_{fD})] + f_0(C_{fD}) \tag{30}$$

The limit of Equation (30) is equal to zero.

$$\lim_{C_{fD} \to \infty} f(I_x, C_{fD}) = 0 \tag{31}$$

Substituting Equations (20) and (30) into Equation (19), we obtain:

$$b_{Dpss,FC}(I_x, C_{fD}) = \underbrace{-\ln(I_x) - 0.049298 + 0.43464(I_x)^2}_{b_{Dpss,IC}(I_x)} + \underbrace{I_x^2[f_1(C_{fD}) - f_0(C_{fD})] + f_0(C_{fD})}_{f(I_x, C_{fD})} \tag{32}$$

Based on our work, a new b_{Dpss} is presented in Equation (32).

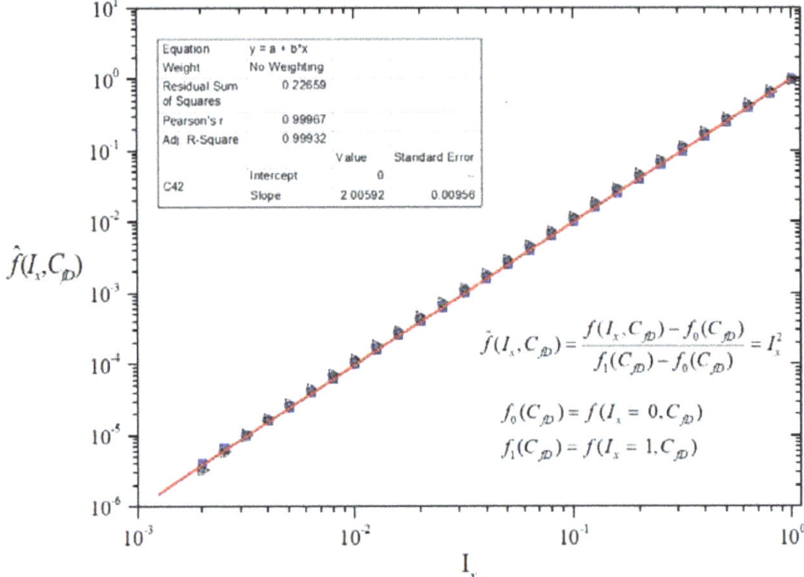

Figure 5. Regression of normalized conductivity-influence function (1200 data points).

3. Results

In this section, we will compare our approximate model with our semi-analytical model and Pratikno et al.'s approximate solutions [20] for different penetrate ratio and fracture conductivity. Table 2 presents the values of b_{Dpss} obtained by our semi-analytical method, Pratikno et al.'s method and our approximate model. As shown in Table 2, the maximum relative error is 0.4% at $I_x = 0.1$ and 1.93% at $I_x = 1$ between our approximate model and semi-analytical model. However, huge differences between Pratikno et al.'s method and our semi-analytical method can be observed at large penetration ratio, for example, relative error 16.57% at $C_{fD} = 0.631$ and $I_x = 1$.

We further show the comparisons in Figure 6. The circles with crosses, red lines, and blue lines correspond to the semi-analytical solutions (accurate solutions), Pratikno et al.'s solutions, and approximate solutions, respectively. As shown in Figure 6, an excellent agreement among the three methods is visible when the dimensionless fracture conductivity C_{fD} is greater than 10 for all penetration ratios. The lines cross the centers of circles at low penetration ratios, such as I_x values less than 0.3. For these cases, the differences among the three models can be ignored. Our approximate solutions (blue lines) match very well with semi-analytical solutions (circles with cross) for all penetration ratios and fracture conductivities. The red lines deviate from the circles and blue lines and huge differences between Pratikno et al.'s solutions and our solutions are noticeable at low conductivity ($C_{fD} < 10$) with a high penetration ratio ($I_x > 0.5$).

Table 2. Comparisons of b_{Dpss} for different calculation methods.

C_{fD}	Semi-Analytical Solutions (This Study)			H.Pratikno et al. Solutions (2003) SPE 84287			Approximate Solutions (This Study)			Relative Error between H.Pratikno et al. Solutions and Semi-Analytical Solutions, %			Relative Error between Semi-Analytical Solutions and Approximate Solutions, %		
	Ix = 0.1	Ix = 0.5	Ix = 1	Ix = 0.1	Ix = 0.5	Ix = 1	Ix = 0.1	Ix = 0.5	Ix = 1	Ix = 0.1	Ix = 0.5	Ix = 1	Ix = 0.1	Ix = 0.5	Ix = 1
0.1000	5.1692	3.5727	2.9135	5.1960	3.6909	3.3237	5.1787	3.5801	2.9210	0.52	3.31	14.08	0.18	0.21	0.26
0.1259	4.9454	3.3517	2.7003	4.9595	3.4544	3.0872	4.9480	3.3523	2.7020	0.29	3.06	14.33	0.05	0.02	0.06
0.1585	4.7252	3.1350	2.4928	4.7319	3.2268	2.8596	4.7276	3.1350	2.4945	0.14	2.93	14.72	0.05	0.00	0.07
0.1995	4.5091	2.9230	2.2915	4.5115	3.0064	2.6392	4.5139	2.9248	2.2953	0.05	2.85	15.18	0.11	0.06	0.17
0.2512	4.2979	2.7163	2.0972	4.2976	2.7924	2.4253	4.3051	2.7201	2.1031	0.01	2.80	15.64	0.17	0.14	0.28
0.3162	4.0924	2.5160	1.9111	4.0902	2.5850	2.2179	4.1009	2.5205	1.9179	0.05	2.74	16.05	0.21	0.18	0.35
0.3981	3.8940	2.3234	1.7340	3.8902	2.3851	2.0179	3.9024	2.3270	1.7405	0.10	2.66	16.37	0.22	0.16	0.37
0.5012	3.7040	2.1396	1.5672	3.6989	2.1938	1.8267	3.7112	2.1415	1.5726	0.14	2.53	16.55	0.19	0.09	0.34
0.6310	3.5242	1.9663	1.4118	3.5181	2.0130	1.6458	3.5296	1.9659	1.4157	0.17	2.37	16.57	0.15	0.02	0.27
0.7943	3.3560	1.8048	1.2687	3.3493	1.8442	1.4770	3.3598	1.8023	1.2713	0.20	2.18	16.42	0.11	0.14	0.21
1.0000	3.2008	1.6563	1.1386	3.1939	1.6888	1.3216	3.2036	1.6522	1.1405	0.21	1.96	16.08	0.09	0.25	0.17
1.2589	3.0597	1.5217	1.0218	3.0530	1.5479	1.1807	3.0623	1.5169	1.0238	0.22	1.72	15.55	0.09	0.32	0.19
1.5849	2.9333	1.4015	0.9186	2.9271	1.4220	1.0548	2.9364	1.3966	0.9210	0.21	1.46	14.83	0.10	0.35	0.27
1.9953	2.8219	1.2957	0.8285	2.8162	1.3111	0.9439	2.8259	1.2912	0.8318	0.20	1.18	13.94	0.14	0.35	0.40
2.5119	2.7249	1.2039	0.7508	2.7198	1.2147	0.8475	2.7301	1.2001	0.7551	0.19	0.89	12.89	0.19	0.32	0.57
3.1623	2.6418	1.1253	0.6847	2.6371	1.1320	0.7648	2.6480	1.1221	0.6899	0.18	0.60	11.71	0.24	0.29	0.75
3.9811	2.5713	1.0587	0.6290	2.5671	1.0620	0.6948	2.5784	1.0560	0.6348	0.16	0.31	10.47	0.28	0.26	0.93
5.0119	2.5121	1.0030	0.5825	2.5068	1.0017	0.6345	2.5198	1.0004	0.5888	0.21	0.12	8.93	0.30	0.26	1.06
6.3096	2.4630	0.9567	0.5441	2.4585	0.9534	0.5862	2.4708	0.9540	0.5505	0.18	0.35	7.73	0.31	0.28	1.15
7.9433	2.4225	0.9186	0.5126	2.4180	0.9129	0.5457	2.4301	0.9155	0.5187	0.19	0.62	6.46	0.31	0.33	1.19
10.0000	2.3894	0.8874	0.4868	2.3847	0.8796	0.5124	2.3966	0.8838	0.4926	0.20	0.88	5.26	0.30	0.41	1.18
12.5893	2.3625	0.8620	0.4659	2.3576	0.8525	0.4853	2.3690	0.8577	0.4712	0.20	1.11	4.17	0.28	0.51	1.12
15.8489	2.3406	0.8415	0.4490	2.3356	0.8305	0.4633	2.3465	0.8364	0.4537	0.21	1.31	3.19	0.25	0.61	1.04
19.9526	2.3230	0.8249	0.4354	2.3179	0.8128	0.4456	2.3281	0.8190	0.4395	0.22	1.47	2.35	0.22	0.72	0.93
25.1189	2.3088	0.8116	0.4244	2.3037	0.7985	0.4314	2.3132	0.8050	0.4280	0.22	1.61	1.64	0.19	0.83	0.83
31.6228	2.2974	0.8009	0.4157	2.2923	0.7872	0.4200	2.3012	0.7936	0.4187	0.22	1.72	1.04	0.17	0.92	0.73
39.8107	2.2883	0.7924	0.4086	2.2832	0.7781	0.4109	2.2916	0.7846	0.4113	0.22	1.80	0.56	0.15	0.99	0.65
50.1187	2.2810	0.7855	0.4030	2.2760	0.7709	0.4037	2.2840	0.7774	0.4054	0.22	1.86	0.18	0.13	1.05	0.59
63.0957	2.2752	0.7801	0.3986	2.2703	0.7652	0.3980	2.2780	0.7717	0.4008	0.21	1.90	0.13	0.12	1.09	0.56
79.4328	2.2705	0.7757	0.3950	2.2658	0.7607	0.3935	2.2733	0.7672	0.3972	0.21	1.93	0.36	0.12	1.10	0.55
100.0000	2.2668	0.7722	0.3921	2.2623	0.7572	0.3900	2.2697	0.7638	0.3944	0.20	1.95	0.55	0.12	1.10	0.58
125.8925	2.2639	0.7695	0.3899	2.2595	0.7543	0.3872	2.2670	0.7613	0.3924	0.20	1.96	0.69	0.13	1.08	0.63
158.4893	2.2616	0.7673	0.3881	2.2573	0.7521	0.3850	2.2650	0.7594	0.3908	0.19	1.97	0.80	0.15	1.04	0.71
199.5262	2.2597	0.7655	0.3866	2.2555	0.7504	0.3832	2.2636	0.7581	0.3898	0.19	1.98	0.89	0.17	0.98	0.81

Table 2. *Cont.*

	Semi-Analytical Solutions (This Study)			H.Pratikno et al. Solutions (2003) SPE 84287			Approximate Solutions (This Study)			Relative Error between H.Pratikno et al. Solutions and Semi-Analytical Solutions, %			Relative Error between Semi-Analytical Solutions and Approximate Solutions, %		
251.1886	2.2582	0.7641	0.3855	2.2541	0.7489	0.3818	2.2626	0.7572	0.3891	0.18	1.99	0.97	0.20	0.91	0.93
316.2278	2.2570	0.7630	0.3846	2.2529	0.7478	0.3806	2.2621	0.7567	0.3888	0.18	2.00	1.04	0.22	0.83	1.07
398.1072	2.2561	0.7621	0.3839	2.2519	0.7468	0.3796	2.2619	0.7565	0.3886	0.19	2.01	1.11	0.26	0.74	1.23
501.1872	2.2553	0.7614	0.3833	2.2511	0.7459	0.3788	2.2619	0.7566	0.3887	0.19	2.03	1.18	0.29	0.64	1.39
630.9573	2.2547	0.7609	0.3828	2.2503	0.7452	0.3780	2.2621	0.7568	0.3889	0.20	2.06	1.26	0.33	0.54	1.56
794.3282	2.2543	0.7604	0.3825	2.2496	0.7445	0.3774	2.2625	0.7572	0.3893	0.21	2.09	1.34	0.37	0.43	1.74
1000	2.2539	0.7601	0.3822	2.2490	0.7439	0.3767	2.2631	0.7577	0.3897	0.22	2.13	1.44	0.40	0.32	1.93

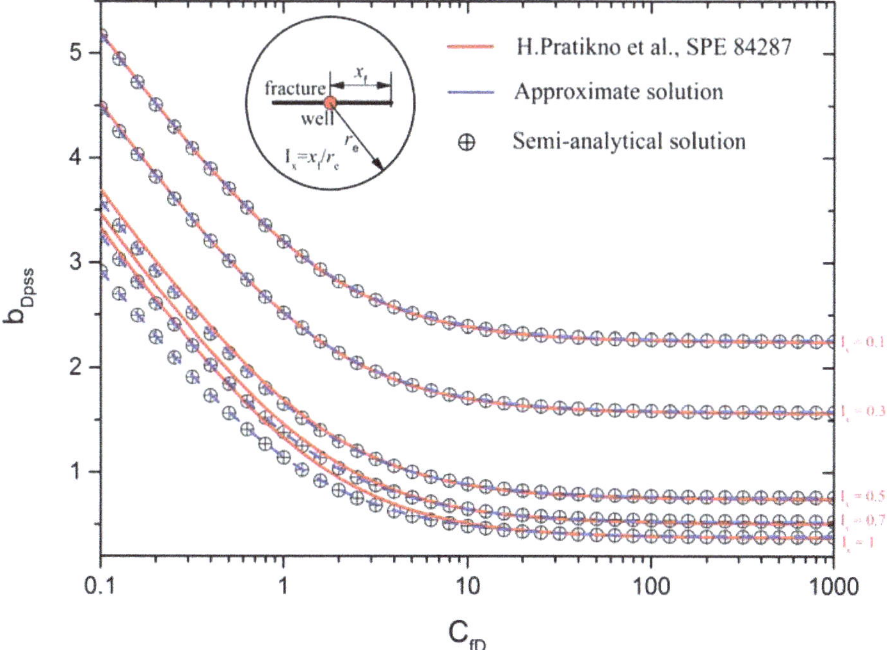

Figure 6. Comparisons of b_{Dpss} with our solutions and Pratikno et al.'s solutions.

4. Conclusions

The following conclusions can be drawn from this study: (1) Pratikno et al. [20] stated that only the fracture conductivity affected the conductivity-influence function (CIF) described in Equation (4). Based on our work, we found that both the fracture conductivity and penetration ratio exerted significant influences on CIF. As Figure 3 shows, CIF decreases with increasing fracture conductivity, and CIF tends to be zero when C_{fD} is greater than 300. Additionally, CIF has obviously differences, especially when C_{fD} is less than 10; the CIF decreases with increasing penetration ratio. (2) Based on the PSS function [11], a new semi-analytical model was proposed to directly calculate the b_{Dpss}, which consumes less time. Besides, our method is more accurate due to simultaneously considering both fracture conductivity and penetration ratio. (3) A new conductivity-influence function (CIF), considering the effect of penetration ratio and fracture conductivity, was developed. A normalized conductivity-influence function (NCIF) was also developed to calculate the value of CIF. Our work provides a fast, accurate, and time-saving method to evaluate b_{Dpss} for a well with a finite-conductivity fracture in a circular closed reservoir.

Author Contributions: Conceptualization, W.L.; Data curation, Y.C., B.L. and M.W.; Formal analysis, B.L.; Investigation, Y.C.; Methodology, W.L.; Project administration, W.L.; Software, Y.C. and M.W.; Supervision, W.L.; Validation, M.W.; Writing—original draft, Y.C.

Funding: This research was funded by [National Natural Science Foundation of China] grant number [51674227].

Acknowledgments: This work was supported by the National Natural Science Foundation of China (Grant No. 51674227) and the Fundamental Research Funds for the Central Universities.

Conflicts of Interest: The authors declare no conflict of interest.

References

1. Wasantha, P.L.P.; Konietzky, H. Fault reactivation and reservoir modification during hydraulic stimulation of naturally-fractured reservoirs. *J. Nat. Gas Sci. Eng.* **2016**, *34*, 908–916. [CrossRef]
2. Ghaderi, S.M.; Clarkson, C.R. Estimation of fracture height growth in layered tight/shale gas reservoirs using flowback gas rates and compositions—Part I: Model development. *J. Nat. Gas Sci. Eng.* **2016**, *36*, 1018–1030. [CrossRef]
3. Taheri-Shakib, J.; Ghaderi, A.; Hosseini, S.; Hashemi, A. Debonding and coalescence in the interaction between hydraulic and natural fracture: Accounting for the effect of leak-off. *J. Nat. Gas Sci. Eng.* **2016**, *36*, 454–462. [CrossRef]
4. Li, B.; Liu, R.; Jiang, Y. Influences of hydraulic gradient, surface roughness, intersecting angle, and scale effect on nonlinear flow behavior at single fracture intersections. *J. Hydrol.* **2016**, *538*, 440–453. [CrossRef]
5. Liu, R.; Li, B.; Jiang, Y. A fractal model based on a new governing equation of fluid flow in fractures for characterizing hydraulic properties of rock fracture networks. *Comput. Geotech.* **2016**, *75*, 57–68. [CrossRef]
6. Liu, R.; Li, B.; Jiang, Y. Critical hydraulic gradient for nonlinear flow through rock fracture networks: The roles of aperture, surface roughness, and number of intersections. *Adv. Water Res.* **2016**, *88*, 53–65. [CrossRef]
7. Haeri, F.; Izadi, M.; Zeidouni, M. Unconventional multi-fractured analytical solution using dual porosity model. *J. Nat. Gas Sci. Eng.* **2017**, *45*, 230–242. [CrossRef]
8. Al-Rbeawi, S. Analysis of pressure behaviors and flow regimes of naturally and hydraulically fractured unconventional gas reservoirs using multi-linear flow regimes approach. *J. Nat. Gas Sci. Eng.* **2017**, *45*, 637–658. [CrossRef]
9. Oyedokun, O.; Schubert, J. A quick and energy consistent analytical method for predicting hydraulic fracture propagation through heterogeneous layered media and formations with natural fractures: The use of an effective fracture toughness. *J. Nat. Gas Sci. Eng.* **2017**, *44*, 351–364. [CrossRef]
10. Gringarten, A.C.; Ramsey, H.J., Jr.; Raghavan, R. Unsteady Pressure Distribution Created by a Well with a Single Infinite-Conductivity Vertical Fracture. *SPE J.* **1974**, *14*, 347–360. [CrossRef]
11. Ozkan, E. Performance of Horizontal Wells. Ph.D. Thesis, University of Tulsa, Tulsa, OK, USA, 1988.
12. Ozkan, E.; Raghavan, R. Some new solutions to solve problems in well test analysis: I. Computational considerations and applications. *SPE Form. Eval.* **1991**, *6*, 359–368. [CrossRef]
13. Kuchuk, F.J.; Goode, P.A.; Wilkinson, D.J.; Thambynayagam, R.K.M. Pressure-transient behavior of horizontal wells with and without gas cap or aquifer. *SPE Form. Eval.* **1991**, *6*, 86–94. [CrossRef]
14. Hagoort, J. The productivity of a well with a vertical infinite-conductivity fracture in a rectangular closed reservoir. *SPE J.* **2009**, *14*, 715–720. [CrossRef]
15. Cinco-Ley, H.; Samaniego, V.F.; Dominguez, A.N. Transient Pressure Behavior for a Well with a Finite-Conductivity Vertical Fracture. *SPE J.* **1978**, *18*, 253–264. [CrossRef]
16. Cinco-Ley, H.; Meng, H.Z. Pressure transient of wells with finite conductivity vertical fractures in double porosity reservoirs. Presented at the SPE Annual Technical Conference and Exhibition, Houston, TX, USA, 2–5 October 1988.
17. Lee, S.T.; Brockenbrouth, J.R. A new approximate analytic solution for finite conductivity vertical fractures. Presented at the SPE annual technical conference and exhibition, San Francisco, CA, USA, 5–8 October 1986.
18. Blasingame, T.A.; Poe, B.D. Semianalytic solutions for a well with a single finite-conductivity vertical fracture. Presesnted at the SPE Annual Technical Conference and Exhibition, Houston, TX, USA, 3–6 October 1993.
19. Luo, W.; Tang, C. A semi-analytical solution of a vertical fractured well with varying conductivity under non-darcy-flow condition. *SPE J.* **2015**, *20*, 1028–1040. [CrossRef]
20. Pratikno, H.; Rushing, J.A.; Blasingame, T.A. Decline Curve Analysis Using Type Curves—Fractured Wells. In Proceedings of the SPE Annual Technical Conference and Exhibition. Society of Petroleum Engineers, Denver, Colorado, 5–8 October 2003.
21. Amini, S.; Liik, D.; Blasingame, T.A. Evaluation of the Elliptical Flow Period for Hydraulically-Fracture Wells in Tight Gas Sands-Theoretical Aspects and Practical Considerations. Presented at the SPE Hydraulic Fracturing Technology Conference, College Station, TX, USA, 29–31 January 2007.
22. Hagoort, J. Semisteady-State Productivity of a Well in a Rectangular Reservoir Producing at Constant Rate or Constant Pressure. *SPE Res. Eval. Eng.* **2011**, *14*, 677–686. [CrossRef]

23. Wang, L.; Wang, X.D.; Ding, X.M.; Chen, L. Rate Decline Curves Analysis of a Vertical Fractured Well with Fracture Face Damage. *J. Energy Resour. Technol.* **2012**, *134*, 032803. [CrossRef]
24. Lu, Y.; Chen, K.P. Productivity-Index Optimization for Hydraulically Fractured Vertical Wells in a Circular Reservoir: A Comparative Study with Analytical Solutions. *SPE J.* **2016**, *21*, 2208–2219. [CrossRef]
25. Fetkovich, M.J. Decline Curve Analysis Using Type Curves. *J. Pet. Technol.* **1980**, *32*, 1065–1077. [CrossRef]
26. Fetkovich, M.J.; Vienot, M.E.; Bradley, M.D.; Kiesow, U.G. Decline Curve Analysis Using Type Curves—Case Histories. *SPE Form. Eval.* **1987**, *2*, 637–656. [CrossRef]
27. Doublet, L.E.; Blasingame, T.A. Decline Curve Analysis Using Type Curves: Water Influx/Waterflood Cases. Presented at the 1995 Annual SPE Technical Conference and Exhibition, Dallas, TX, USA, 22–25 October 1995.
28. Agarwal, R.G.; Gardner, D.C.; Kleinsteiber, S.W.; Fussell, D.D. Analyzing Well Production Data Using Combined-Type-Curve and Decline-Curve Analysis Concepts. In Proceedings of the SPE Annual Technical Conference and Exhibition, Society of Petroleum Engineers, Houston, TX, USA, 3–6 October 1999; pp. 478–486.
29. Prats, M.; Hazebroek, P.; Strickler, W.R. Effect of vertical fractures on reservoir behavior–compressible-fluid case. *SPE J.* **1962**, *2*, 87–94. [CrossRef]
30. Luo, W.; Wang, X.; Tang, C.; Feng, Y.; Shi, E. Productivity of multiple fractures in a closed rectangular reservoir. *J. Pet. Sci. Eng.* **2017**, *157*, 232–247. [CrossRef]

Article

Critical Hydraulic Gradient of Internal Erosion at the Soil–Structure Interface

Quanyi Xie [ID], Jian Liu *, Bo Han *, Hongtao Li, Yuying Li and Xuanzheng Li

School of Civil Engineering, Shandong University, 17922, Jingshi Road, Jinan 250061, China; quanyixiesdu@163.com (Q.X.); 17865131577@163.com or lihongtaosdu@163.com (H.L.); liyuyingsdu@163.com (Y.L.); lixuanzheng1992@163.com (X.L.)
* Correspondence: liujianshanda@163.com (J.L.); bo.han@sdu.edu.cn (B.H.); Tel.: +86-136-1641-9012 (J.L.)

Received: 25 June 2018; Accepted: 16 July 2018; Published: 18 July 2018

Abstract: Internal erosion at soil–structure interfaces is a dangerous failure pattern in earth-fill water-retaining structures. However, existing studies concentrate on the investigations of internal erosion by assuming homogeneous materials, while ignoring the vulnerable soil–structure-interface internal erosion in realistic cases. Therefore, orthogonal and single-factor tests are carried out with a newly designed apparatus to investigate the critical hydraulic gradient of internal erosion on soil–structure interfaces. The main conclusions can be draw as follows: (1) the impact order of the three factors is: degree of compaction > roughness > clay content; (2) the critical hydraulic gradient increases as the degree of compaction and clay content increases. This effect is found to be more obvious in the higher range of the degree of soil compaction and clay content. However, there exists an optimum interface roughness making the antiseepage strength at the interface reach a maximum; (3) the evolution of the interface internal erosion develops from inside to outside along the interface, and the soil particles at the interface flow as a whole; and (4) the critical hydraulic gradient of interface internal erosion is related to the shear strength at the interface and the severity and porosity of the soil.

Keywords: soil–structure interface; internal erosion; critical hydraulic gradient; orthogonal tests

1. Introduction

Internal erosion is the transportation of soil particles induced by internal seepage [1,2]. The current studies broadly categorize internal erosion into four groups: (a) concentrated leak erosion; (b) backward erosion; (c) contact erosion; (d) suffusion. Concentrated leak erosion is the process of sweeping particles away from the side of the crack due to the effect of the seepage [3–5]. Backward erosion refers to the process of generating permeating channels from downstream to upstream due to the action of water flow in strong permeable layers [6–8]. Contact erosion occurs in the interface between particles with different diameters, and the small particles erode into the framework of large particles [9]. Suffusion refers to the phenomenon that small particles in the soil are flowed away from the pore between large particles [10,11]. However, internal erosion between soil and structure is not included in the four types of internal erosion discussed above.

Soil–structure interfaces widely exist in hydraulic structures and the associated interface internal erosion failures significantly threaten engineering safety. In particular, seepage channels can be easily developed through the weak interfaces due to the differential mechanical properties between soil and the structure [12]. This can eventually lead to the formation of pipes/conduits, cavities and unstable zones in earth-fill structures [13]. For instance, the Teton dam in US, with a height of 91.5 m, collapsed in June, 1976. After the accident investigation, the main reason of the dam failure was attributed to the internal erosion at the interface between the clay core wall and rock [14,15].

Although the phenomenon of soil–structure-interface internal erosion has been noticed, existing studies have concentrated on the investigations of internal erosion by assuming homogeneous materials, while ignoring the more vulnerable soil–structure-interface internal erosion in realistic cases. [16–19]. The failure mechanism of interface internal erosion can be more complex and dangerous due to the interaction with the internal affiliated structures [20,21]. The associated interface internal erosions have been frequently observed, such as at the interfaces between cut-off walls and earth-fill materials in dams [22–26], between cut-off walls and earth-fill materials in levees [27–29], and between retaining walls and backfill materials [30–32]. In these works, some empirical criteria are proposed and developed for evaluating the internal stability potential at the soil–structure interface. However, there are few studies on the effects and mechanism of internal erosion at the soil–structure interface.

Therefore, in this paper, a newly designed seepage apparatus is employed to investigate the failure mechanism of internal erosion at soil–structure interfaces. Both orthogonal tests and single-factor tests are designed to investigate the sensitivity of the critical hydraulic gradient of internal erosion subjected to three critical soil properties, that is, degree of compaction, clay content and roughness. The failure mechanism of interface internal erosion is studied by analyzing the observed failure phenomena and the variation of seepage behavior. Furthermore, the relationship between interface shear strength and critical hydraulic gradient is obtained by analyzing the forces on the soil–structure interface for the investigated cases.

2. Soil–Structure-Interface Internal Erosion Tests

2.1. Soil–Structure Interface Seepage Failure Apparatus

Figure 1 shows the designed soil–structure interface seepage failure apparatus. The dimensions of the apparatus are 600 × 300 × 1000 mm (length × width × height). The dimensions of the sample container are 500 × 300 × 800 mm. The sample container is made of acrylic plates, and it consists of two parts, the upper sample chamber and the lower seepage transition chamber. The soil and concrete blocks are placed in the upper chamber with the concrete blocks at the two sides and a soil specimen in the middle. The porous boards are divided into two parts, that is, an inclined porous board and a horizontal porous board. The slope of the inclined porous board is 1:1 and it can effectively filter any gas bubbles in the filled water. The inlet and outlet are set into the lower and upper parts of the sample container, respectively (as shown in Figure 1). When the tests are conducted, water flows upward. The testing apparatus is equipped with a constant-head water supply system and a data acquisition system (seepage discharge and hydraulic head are recorded).

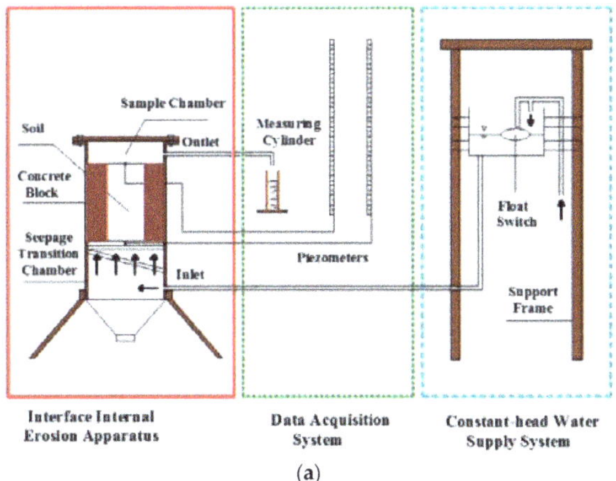

(**a**)

Figure 1. *Cont.*

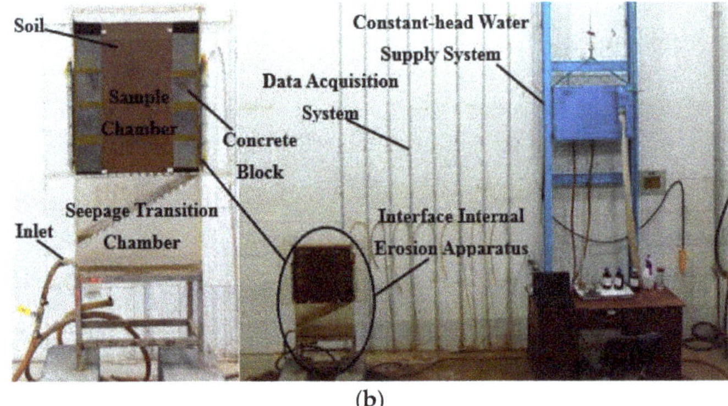

(b)

Figure 1. Testing apparatus. (**a**) Schematic graph of interface internal erosion testing system; (**b**) physical testing apparatus.

2.2. Testing Materials

In reality, when filling soil around concrete buildings, soil of high liquid-plastic limits and high clay particle content is usually used. Therefore, the selection of soil samples in the present work is based on these two factors. The samples for this study are obtained by mixing two soils: silt and clay from the Yellow River alluvial plain. The grain-size distribution curves of the four tested soils are shown in Figure 2. The ranges of clay content and liquid limit of the soil samples are 21.8–29.8% and 31.84–33.78%, respectively, and these cover the concerned soil property ranges for hydraulic engineering structures defined in the Chinese Embankment Dam Constructions Code (DL/T 5395 2007). The material properties and particle compositions of the four tested soils are listed in Table 1.

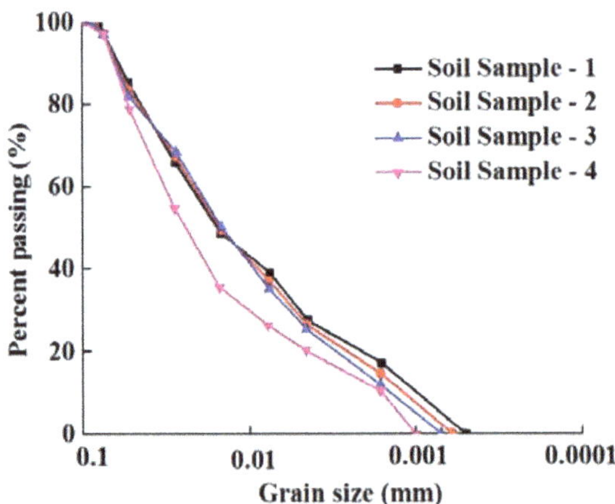

Figure 2. Grain-size distribution curves.

Table 1. Material properties for the tested soils.

Material Properties	Soil Sample-1	Soil Sample-2	Soil Sample-3	Soil Sample-4
Clay content (<0.005 mm) (%)	29.8	27.9	26.8	21.8
Liquid limit (%)	33.78	33.51	33.07	31.84
Plastic limit (%)	18.90	18.42	18.36	17.14
Specific gravity	2.74	2.74	2.74	2.74
Optimal water content (%)	20.3	18.7	18.3	18.1
Maximum dry density (g/cm^3)	1.660	1.672	1.678	1.683

2.3. Soil Sample Preparation

Silt and clay are first crushed and mixed in a certain proportion. The mass ratio of silt and clay of samples 1, 2, 3 and 4 are 3:7, 3.5:6.5, 4:6 and 5:5, respectively. In order to ensure full water absorption, optimum water content is maintained when producing the soil samples for 24 h. At the same time, in order to reduce the influence of moisture content in the test, the difference of moisture content between the two sets of soil samples for repeated tests is controlled within 1%.

In order to ensure uniformity of the tested soil samples, the obtained samples are subsequently filled into the upper chamber by controlling a certain degree of compaction. The thickness of each layer is 4 cm. The degree of compaction of soil samples is controlled by controlling the compaction quality of each layer.

The roughness of the soil–structure interface is defined by the height of the salient of the structure. The height of the salient varies by changing the diameter of sand particles that are attached to the structure. For instance, a roughness of 0.3 cm means that the diameter of the sand particles attached to the structure is 0.3 cm.

The hydraulic head in the constant-head water supply system is raised to saturate the soil samples before the tests. The hydraulic head is raised by 1 cm every 1 h. This relatively slow saturation process can reduce the seepage scouring effect on soil samples.

2.4. Testing Program

When the tests start, the inlet hydraulic head is gradually raised to the designed values. The piezometric levels of the outflow are recorded, and the outflow seepage discharge is measured every 5 min. When the outflow hydraulic heads and seepage discharge subjected to two sequent hydraulic head raises are sufficiently close, that is, the differences of two results are within 5%, it is considered to reach the steady state. Furthermore, typical failure phenomena, such as water turbidity and slight bulging, are also monitored as the failure criteria during the experiments. When any of the mentioned failure phenomena appear at the soil–structure interface, the tests are continuously observed for 1 to 2 h. Internal erosion is recognized when the hydraulic head cannot be further raised.

2.5. Testing Schemes

Orthogonal tests are carried out to analyze the sensitivity of failure mechanisms of interface internal erosion subjected to three critical soil properties (degree of compaction, clay content and roughness). Regardless of the interplay of the factors, the orthogonal table L9 (3^4) is used. The designed tables of the influence factor level and the orthogonal tests are listed in Tables 2 and 3, respectively.

On the basis of the orthogonal tests, the single-factor tests are also carried out to investigate in detail the influence of the three factors on interface internal erosion. The designed testing scheme of the single-factor tests is shown in Table 4. In order to ensure the reproducibility and accuracy of the test data, three independent tests are carried out for each test condition. The mean values of the data from the three tests are firstly calculated (that is, the hydraulic gradient-flow velocity curve). If all the data lie in the ±10% deviation range of the mean curve, they are considered to be reliable and the mean curve is used to represent the soil behavior under this specific test condition. Furthermore, a series of measures are also employed to enhance the reproducibility of the data. In particular, firstly, the deviations of water content and clay content are controlled within 0.5%; secondly, quality control is adopted when reconstituting the sample and the deviation of compactness is controlled within 1%; thirdly, the accuracy of the water supply system is 1 mm and the error is only 2%; last but not least, the accuracy of the seepage discharge measure unit is 0.01 cm^3 and the error is 1%.

Table 2. Level of influence factors.

Factor Level	A Degree of Compaction (%)	B Clay Content (%)	C Roughness (cm)
1	80	21.8	0
2	85	26.8	0.3
3	90	29.8	0.6

Table 3. Testing scheme of orthogonal tests.

Factor Test Number	A Degree of Compaction (%)	B Clay Content (%)	C Roughness (cm)
I-1	80	21.8	0.6
I-2	80	26.8	0.3
I-3	80	29.8	0
I-4	85	21.8	0.3
I-5	85	26.8	0
I-6	85	29.8	0.6
I-7	90	21.8	0
I-8	90	26.8	0.6
I-9	90	29.8	0.3

Table 4. Testing scheme of single-factor tests.

Test Number	Degree of Compaction (%)	Clay Content (%)	Roughness (cm)
II-1	90	29.8	0
II-2	87	29.8	0
II-3	85	29.8	0
II-4	80	29.8	0
II-5	85	27.9	0
II-6	85	26.8	0
II-7	85	21.8	0
II-8	85	29.8	0.6
II-9	85	29.8	0.4
II-10	85	29.8	0.3

3. Testing Results

3.1. Observed Test Phenomena

The observed phenomena of the soil–structure-interface internal erosion are shown in Figure 3 (soil sample on left, concrete block on right). The presented results show a three-stage failure evolution of interface internal erosion, that is, the stable, transition and failure stages. At the stable and transition stages, no particles flow from the surface of the soil–concrete interface, but the soil at the interface shows a slight bulging (uplift of soil) in the transition stage. At the failure stage, a large number of fine particles are transported along the interface. After the interface internal erosion, a crack is formed at the soil–structure interface (see Figure 3d).

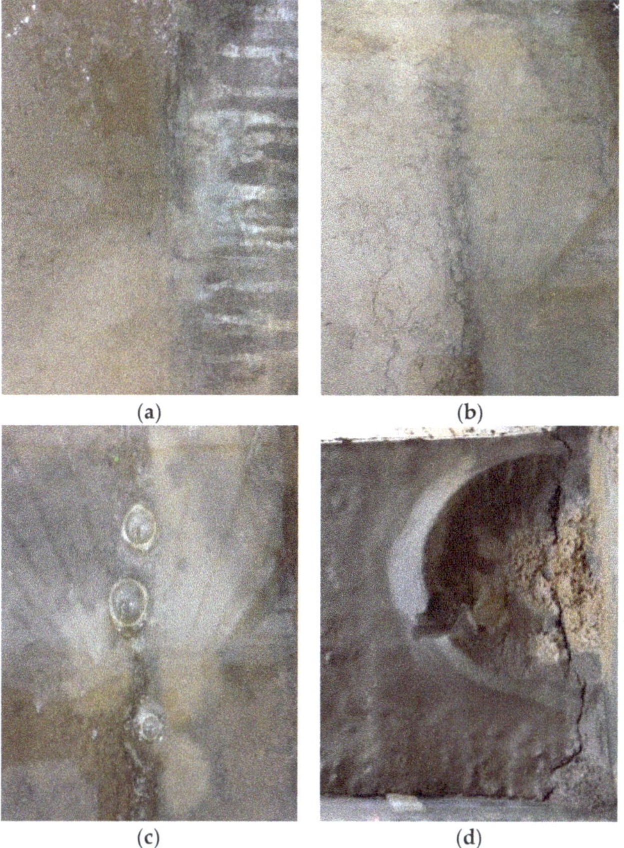

Figure 3. Phenomena of soil–structure-interface internal erosion. (**a**) Stable stage; (**b**) transition stage; (**c**) failure stage; (**d**) interface internal erosion.

3.2. Mechanism of Interface Internal Erosion

The relationships of the seepage velocity, hydraulic gradient, permeability coefficient, and the eroded soil mass against time of II-1 (degree of compaction = 90%, clay content = 29.8%, interface roughness = 0) are shown in Figures 4 and 5. In the experiment, the hydraulic gradient is applied by a certain increment. With the increase of hydraulic gradient, it can be seen that the seepage velocity starts to increase, but the average hydraulic coefficient does not change, and no soil particles are flowing from the interface. When the hydraulic gradient increases to 1.63, the average hydraulic coefficient starts to increase, but the soil particles have not been eroded from the interface. This indicates that the soil particles at the interface start to move due to seepage but have not been totally eroded from the interface. When the hydraulic gradient increases to 1.99, a large number of soil particles are rushed out from the interface. At this stage, the interface has been penetrated and destroyed. From Figure 3d, it can be seen that the interface internal erosion is observed as the overall flow of soil particles at the interface. In particular, the infiltration at the interface is developed from the inside to the outside, and a crack is formed at the soil–structure interface.

The critical hydraulic gradient is used to characterize the seepage stability of soil samples, related to soil porosity and density. Critical hydraulic gradient is defined by the hydraulic gradient where particles start to outflow from soil samples. The determination of the critical hydraulic gradient of internal erosion is based on the occurrence of "sand boil" or other indicating phenomena of seepage failure. Therefore, the threshold hydraulic gradient, when soil particles outflow from the interface, is defined as the critical hydraulic gradient for interface internal erosion.

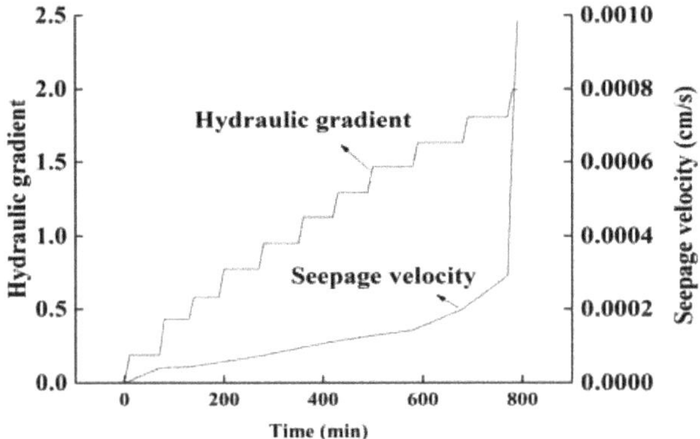

Figure 4. Hydraulic gradient and seepage velocity-t curve of II-1.

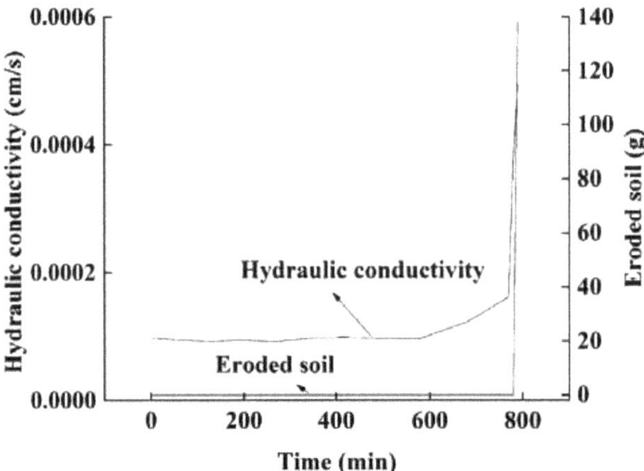

Figure 5. Hydraulic conductivity and eroded soil-t curve of II-1.

3.3. Orthogonal Test Results

The results from the orthogonal tests are analyzed and discussed in this section. It is noted that the error analysis, range analysis and variance analysis are carried out to analyze the effect of error and three critical soil property factors.

3.3.1. Error Analysis

The critical hydraulic gradients from orthogonal test results are listed in Table 5. SSj is the sum of squares of each factor and SSE is the error sum of squares. In the orthogonal tests, the average critical hydraulic gradient is 1.48. The sum square of degree of compaction is 1.83. The sum square of clay content is 0.23. The sum square of roughness is 0.45. The error sum square of the orthogonal test is 0.04. The error sum square of the orthogonal experiment is much smaller than the sum of squares of the factors. Therefore, the degree of the influence of error in the orthogonal experiment can be neglected.

Table 5. Critical hydraulic gradient of orthogonal test results.

	A Degree of Compaction (%)	B Clay Content (%)	C Roughness (cm)	E Empty Column	Critical Hydraulic Gradient
I-1	1(80)	1(21.8)	3(0.6)	1	0.98
I-2	1(80)	2(26.8)	2(0.3)	2	1.24
I-3	1(80)	3(29.8)	1(0)	3	1.06
I-4	2(85)	1(21.8)	2(0.3)	3	1.40
I-5	2(85)	2(26.8)	1(0)	1	1.00
I-6	2(85)	3(29.8)	3(0.6)	2	1.31
I-7	3(90)	1(21.8)	1(0)	2	1.81
I-8	3(90)	2(26.8)	3(0.6)	3	1.79
I-9	3(90)	3(29.8)	2(0.3)	1	2.74
SSj	1.83	0.23	0.45	0.04	

3.3.2. Range Analysis

The range analysis results are shown in Table 6. K_1, K_2 and K_3 are the average values of critical hydraulic gradient under the same test condition of different factors. The results from the range analysis show that $R_A = 1.02 > R_C = 0.50 > R_B = 0.36$, where RA, RB and RC are the ranges of degree of compaction, clay content and roughness, respectively. This demonstrates that the investigated interface internal erosion is most significantly affected by degree of compaction. The influence of the roughness is relatively less profound, while the clay content shows the least impact.

Table 6. Range analysis results of interface internal erosion.

	A Degree of Compaction (%)	B Clay Content (%)	C Roughness (cm)
K_1	1.09	1.40	1.36
K_2	1.24	1.34	1.79
K_3	2.11	1.70	1.29
Range R_i	1.02	0.36	0.5

3.3.3. Variance Analysis

The data of the variance analysis of orthogonal test results is shown in Table 7. The variance analysis results of orthogonal tests show that the F of degree of compaction is 42, which is greater than $F_{0.025}(2,2)$, and its effect on the critical hydraulic gradient is significant. The F of clay content is 6, which is greater than $F_{0.25}(2,2)$, and its effect on the critical hydraulic gradient is relevant. The F of roughness is greater than $F_{0.10}(2,2)$, and its effect on the critical hydraulic gradient is significant. This is also in agreement with the results from the variance analysis. In particular, the results from the variance analysis show that the significance level of degree of compaction is the highest, followed by roughness and clay content. Therefore, the impact order of the three factors is: degree of compaction > roughness > clay content.

Table 7. Variance analysis results of interface internal erosion.

	A Degree of Compaction (%)	B Fine Content (%)	C Roughness (cm)	E Empty Column
$K_{1j}{}^2$	3.28	4.19	4.08	4.72
$K_{2j}{}^2$	3.71	4.03	5.38	4.36
$K_{3j}{}^2$	6.34	5.11	3.87	4.25
Free degree	2	2	2	2
SS	1.83	0.23	0.45	0.04
MS	0.92	0.12	0.23	0.02
F	42	6	11.50	
$F_{0.01}(2,2)$	99	99	99	
$F_{0.025}(2,2)$	39	39	39	
$F_{0.05}(2,2)$	19	19	19	
$F_{0.10}(2,2)$	9	9	9	
$F_{0.25}(2,2)$	3	3	3	
Significance level	** (Greatly significant)	- (Relevant)	* (Significant)	

3.4. Effect of Degree of Compaction

Figure 6 shows the hydraulic gradient–seepage velocity curves under different degrees of compaction. The degrees of compaction of the tests II-1, II-2, II-3 and II-4 are 80%, 85%, 88% and 90%, respectively. The relationship between hydraulic gradient and seepage velocity of three degrees of compaction is characterized by three stages. Figure 7 plots the variations of the critical hydraulic gradient against degree of compaction. When the soil degree of compaction increases from 80% to 85%, the hydraulic gradient increases by 10.47%. However, when the degree of compaction increases from 85% to 90%, the hydraulic gradient increases by 70.86%. It indicates the strengthening effects of soil compaction against seepage-induced interface deformation. This effect is found to be more obvious in the higher range of the degree of soil compaction, that is, a higher increase rate.

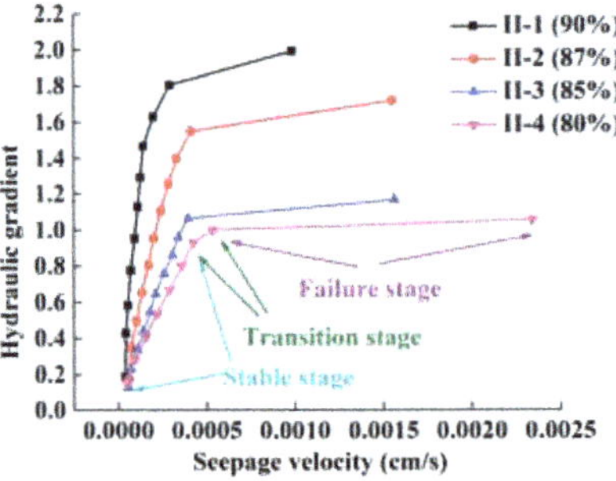

Figure 6. Hydraulic gradient–seepage velocity curves under different degrees of compaction.

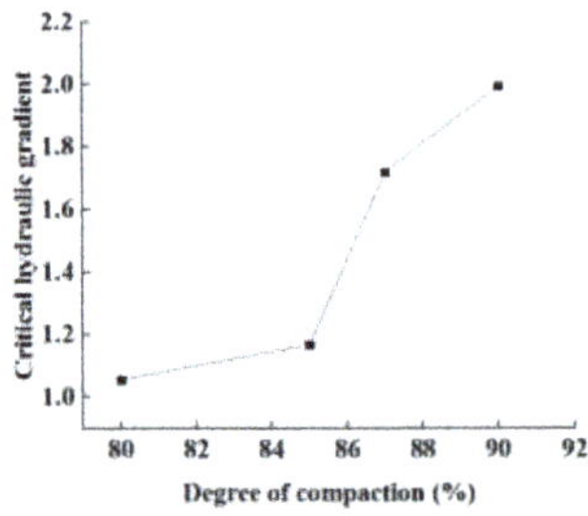

Figure 7. Critical hydraulic gradients for different degrees of compaction.

3.5. Effect of Interface Roughness

The hydraulic gradient–seepage velocity behavior under different roughness conditions is shown in Figure 8. The results reflect the impact of interface roughness (bonding between the soil and structure) on interface internal erosion. The interface roughnesses of the tests II-3, II-8, II-9 and II-10 are 0, 0.3, 0.4 and 0.6 mm, respectively. It can be seen that as interface roughness increases, the critical hydraulic gradient first increases, reaches the peak at 0.3 mm roughness and then decreases. This is also reflected by the critical hydraulic gradient–interface roughness relation in Figure 9. The presented results indicate an optimum interface roughness where the highest antiseepage strength can be obtained against interface internal erosion. The reason for the optimum interface roughness is that when the soil–structure interface is relatively smooth, soil particles can be easily transported by seepage water and therefore the critical hydraulic gradient is low. When interface roughness is higher, the antiseepage strength and the critical hydraulic gradient are larger as a consequence of a bigger friction at the interface. However, after reaching a threshold value, the voids between soil and structure are so large that a more significant water flow generates and therefore leads to a lower critical hydraulic gradient, as illustrated in Figure 10. The optimum roughness is found to be approximately 0.3 mm for the investigated cases.

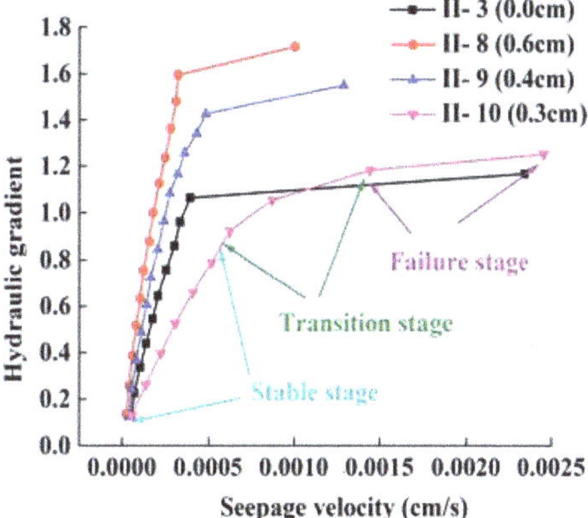

Figure 8. Hydraulic gradient–seepage velocity curves under different roughnesses.

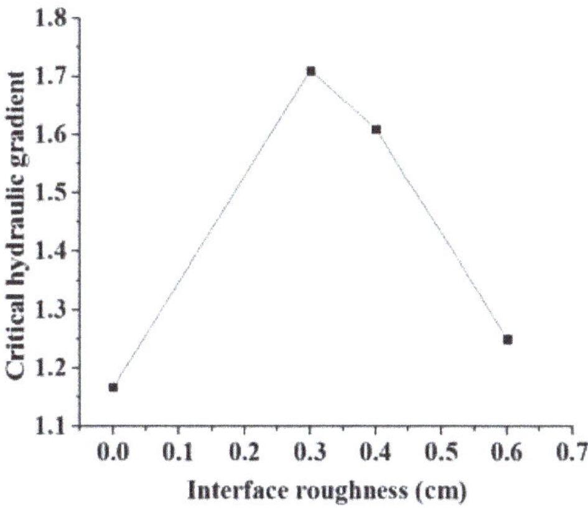

Figure 9. Critical hydraulic gradients for different roughnesses.

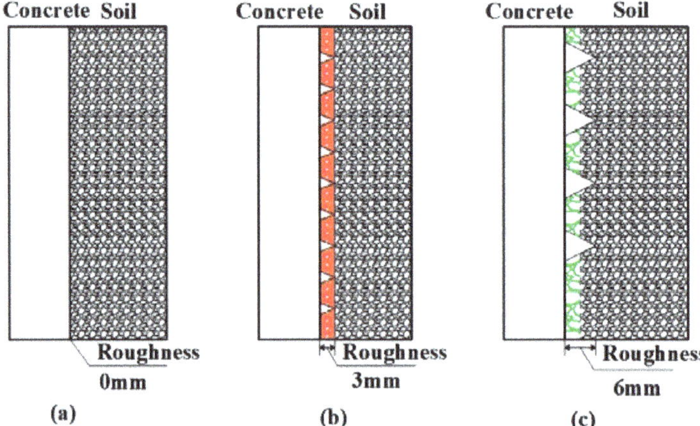

Figure 10. Schematic graph for different interface roughnesses. (**a**) Schematic graph for the interface roughness of 0 mm; (**b**) schematic graph for interface roughness of 3 mm; (**c**) schematic graph for interface roughness of 6 mm.

3.6. Effect of Clay Content

The hydraulic gradient–seepage velocity behavior plots under different clay contents are shown in Figure 11. It can be seen that the difference between the results from II-6 and II-7 tests is negligible, indicating the insignificant effect of clay content in its low range. However, by comparing the results from II-3, II-5 and II-6, it shows that in the higher range of clay content, the critical hydraulic gradient increases more significantly as the clay content increases, and the stable seepage stage is also obviously prolonged. Figure 12 further plots the variations of critical hydraulic gradients against clay contents. When the clay content of soil increases from 21.8% to 26.8%, the critical hydraulic gradient increment is negligible. However, when the clay content of soil increases from 26.8% to 29.8%, the critical hydraulic gradient increases by 18%. It is obvious that the critical hydraulic gradient of interface internal erosion presents a piecewise functional relationship with the increase of clay content. When the clay content increases to 26.8%, the increase of clay content of soil can significantly improve the critical hydraulic gradient under the experimental conditions.

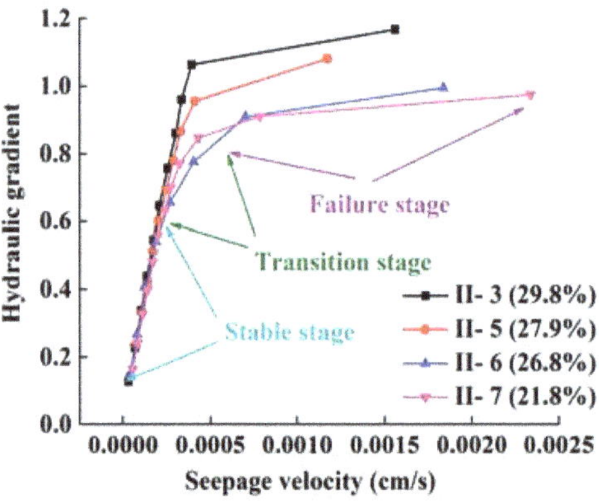

Figure 11. Hydraulic gradient–seepage velocity curves under different clay contents.

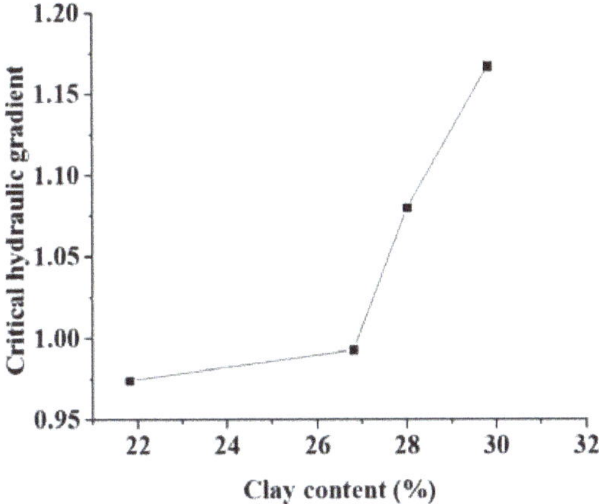

Figure 12. Critical hydraulic gradients for different clay contents.

4. Discussion of Critical Hydraulic Gradient

The critical hydraulic gradient of internal erosion is mostly calculated by limit balance equilibrium of forces in soil units. In this section, the critical hydraulic gradient of interface internal erosion is studied by analyzing the forces imposed on soil particles. Furthermore, a section of soil and concrete is selected as the control body (height: dz, thickness: da). The soil–structure shear strength is expressed by the maximum shear stress. The forces acting on soil particles in the control body are shown in Figure 13. They are discussed as follows:

The volume force acting on the soil particles can be expressed as:

$$f_z = r_\omega \frac{dh}{dz}.$$ (1)

The shear force between soil and concrete in the control body is defined as:

$$2\tau dadz.$$ (2)

The submerged unit weight of the soil particles at the interface is expressed as:

$$\gamma' = \gamma_\omega (Gs - 1)(1 - n).$$ (3)

When interface internal erosion occurs, the soil particles on the interface are in the limit equilibrium state, where the submerged unit weight of soil particles plus the shear force between soil and concrete is equal to the volume force loaded on the soil particles by water.

After substituting Equations (2) and (3) into Equation (1), the critical hydraulic gradient of interface internal erosion can be written as Equation (4):

$$i_{cr} = \frac{dh}{dz} = (Gs - 1)(1 - n) - \frac{2\tau da}{A\gamma_\omega}$$ (4)

The definition of symbols in the Equation (4) are shown in Table 8. According to Equation (4), critical hydraulic gradient is related to the shear strength of the interface and the severity and porosity of soil. The degree of compaction and clay content of soil affect the impermeability of soil–structure-interface internal erosion through changing the porosity and severity of soil. The interface roughness mainly affects the shear strength of the soil–structure interface. In order to improve the critical hydraulic gradient of soil–structure internal erosion, measures aiming to enhance the soil–structure shear stress or the impermeability of soil should be adopted.

Table 8. Definition of symbols.

Symbol	Definition
dh	the hydraulic head differentials between the two ends of the control body
dp	the water pressure differentials between the two ends of the control body
n	the porosity of soil
A	cross-sectional area of soil
dz	the height of control body
γ_w	unit weight of water
f_z	the volume force acting on the soil particles' unit volume
τ	the shear stress between soil and concrete
da	the thickness of control body
Gs	specific gravity of soil particles

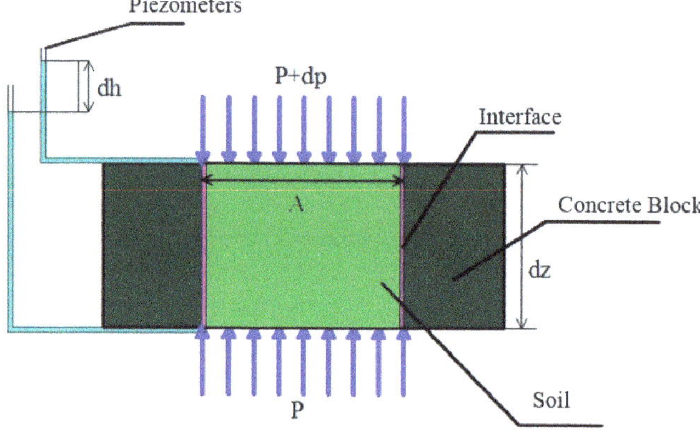

Figure 13. Forces acting on soil particles in the control body.

5. Conclusions

This paper employed a newly designed seepage apparatus to investigate the failure mechanism of internal erosion at the soil–structure interface. Orthogonal and single-factor tests were designed to investigate the sensitivity of the critical hydraulic gradient of internal erosion subjected to three critical soil properties, that is, degree of compaction, clay content and roughness. Furthermore, the limit equilibrium state method was used to analyze the critical hydraulic gradient of interface internal erosion. Based on the experimental results, the following conclusions can be drawn:

(1) The impact order of the three factors on the critical hydraulic gradient of interface internal erosion is: degree of compaction > roughness > clay content.

(2) The critical hydraulic gradient increases as the levels of degree of compaction and clay content increase. This effect is found to be more obvious in the higher range of the degree of soil compaction and clay content. However, there exists an optimum interface roughness where the highest anti seepage strength can be obtained against interface internal erosion. This optimum roughness is found to be approximately 0.3 mm for the investigated cases.

(3) The evolution of the interface internal erosion develops from inside to outside along the interface, and the soil particles on the interface flow as a whole.

(4) The critical hydraulic gradient of interface internal erosion is related to the shear strength of the interface and the severity and porosity of the soil. The degree of compaction and clay content of soil affect the impermeability of the soil–structure-interface internal erosion through changing the porosity and severity of soil. The interface roughness mainly affects the shear strength of the soil–structure interface.

Author Contributions: Q.X., J.L. and B.H. conceived of and designed the study. Q.X., Y.L., X.L. and H.L. performed the experiments. Q.X., J.L., B.H., Y.L., X.L. and H.L. wrote and modified the paper.

Funding: This research was funded by the National Science and Technology Support Program of China, grand number is 2015BAB07B05; National Natural Science Foundation of China, grand number is 41172267; and National Natural Science Foundation of China, grand number is 51508310.

Conflicts of Interest: The authors declare no conflicts of interest.

References

1. Sato, M.; Kuwano, R. Suffusion and clogging by one-dimensional seepage tests on cohesive soil. *Soils Found.* **2015**, *55*, 1427–1440. [CrossRef]

2. Horikoshi, K.; Takahashi, A. Suffusion-induced change in spatial distribution of fine fractions in embankment subjected to seepage flow. *Soils Found.* **2015**, *55*, 1293–1304. [CrossRef]

3. Poesen, J.; Luna, E.D.; Franca, A.; Nachtergaele, J.; Govers, G. Concentrated flow erosion rates as affected by rock fragment cover and initial soil moisture content. *Catena* **1999**, *36*, 315–329. [CrossRef]

4. Marot, D.; Bendahmane, F.; Rosquoet, F.; Alexis, A. Internal flow effects on isotropic confined sand-clay mixtures. *J. Soil Contam.* **2009**, *18*, 294–306. [CrossRef]

5. Mercier, F.; Bonelli, S.; Golay, F.; Anselmet, F.; Philippe, P.; Borghi, R. Numerical modelling of concentrated leak erosion during hole erosion tests. *Acta Geotech.* **2015**, *10*, 1–14. [CrossRef]

6. Beek, V.M.V.; Sellmeijer, J.B.; Barends, F.B.J.; Bezuijen, A. Initiation of backward erosion piping in uniform sands. *Géotechnique* **2014**, *64*, 927–941. [CrossRef]

7. Bendahmane, F.; Marot, D.; Alexis, A. Experimental parametric study of suffusion and backward erosion. *J. Geotech. Geoenviron. Eng.* **2008**, *134*, 57–67. [CrossRef]

8. Richards, K.S.; Reddy, K.R. Experimental investigation of initiation of backward erosion piping in soils. *Géotechnique* **2012**, *62*, 933–942. [CrossRef]

9. Germer, L.H. Physical processes in contact erosion. *J. Appl. Phys.* **1958**, *29*, 1067–1082. [CrossRef]

10. Ke, L.; Takahashi, A. Experimental investigations on suffusion characteristics and its mechanical consequences on saturated cohesionless soil. *Soils Found.* **2014**, *54*, 713–730. [CrossRef]

11. Yacine, S.; Didier, M.; Luc, S.; Alain, A. Suffusion tests on cohesionless granular matter. *Eur. J. Environ. Civ. Eng.* **2011**, *15*, 799–817. [CrossRef]

12. Luo, Y.L.; Zhan, M.L.; Sheng, J.C.; Qiang, W. Hydro-mechanical coupling mechanism on joint of clay core-wall and concrete cut-off wall. *J. Cent. South. Univ.* **2013**, *20*, 2578–2585. [CrossRef]

13. Kaoser, S.; Barrington, S.; Elektorowicz, M.; Ayadat, T. The influence of hydraulic gradient and rate of erosion on hydraulic conductivity of sand-bentonite mixtures. *Soil Sediment Contam. An Int. J.* **2006**, *15*, 481–496. [CrossRef]

14. Armando, B.; Scheffler, M.L. Numerical analysis of the teton dam failure flood. *J. Hydraul. Res.* **1982**, *20*, 317–328. [CrossRef]

15. Muhunthan, B.; Pillai, S. Teton dam, USA: Uncovering the crucial aspect of its failure. *Civ. Eng.* **2008**, *161*, 35–40. [CrossRef]

16. Boulon, M.; Nova, R. Modelling of soil–structure interface behaviour a comparison between elastoplastic and rate type laws. *Comput. Geotech.* **1990**, *9*, 21–46. [CrossRef]

17. Shahrour, I.; Rezaie, F. An elastoplastic constitutive relation for the soil–structure interface under cyclic loading. *Comput. Geotech.* **1997**, *21*, 21–39. [CrossRef]

18. Hu, L.; Pu, J. Testing and modeling of soil–structure interface. *J. Geotech. Geoenviron. Eng.* **2004**, *130*, 851–860. [CrossRef]

19. Zhang, G.; Zhang, J.M. Large-scale apparatus for monotonic and cyclic soil–structure interface test. *ASTM Geotech. Test. J.* **2006**, *29*, 401–408. [CrossRef]

20. Ferdos, F.; Wörman, A.; Ekström, I. Hydraulic conductivity of coarse rockfill used in hydraulic structures. *Transp. Porous Media* **2015**, *108*, 1–25. [CrossRef]

21. Kim, H.; Park, J.; Shin, J. Flow behaviour and piping potential at the soil–structure interface. *Géotechnique* **2018**, 1–6. [CrossRef]

22. Luo, Y.L.; Wu, Q.; Zhan, M.L.; Sheng, J.C. Study of critical piping hydraulic gradient of suspended cut-off wall and sand gravel foundation under different stress states. *Rock Soil Mech.* **2012**, *36*, 73–78. [CrossRef]

23. Maeda, K.; Sakai, H.; Sakai, M. Development of seepage failure analysis method of ground with smoothed particle hydrodynamics. *Struct. Eng.* **2006**, *23*, 307–319. [CrossRef]

24. Maeda, K.; Sakai, H. Seepage failure and erosion of ground with air bubble dynamics. *Geotech. Spec. Publ.* **2010**, *204*, 261–266. [CrossRef]

25. Liu, R.; Li, B.; Jiang, Y. Critical hydraulic gradient for nonlinear flow through rock fracture networks: The roles of aperture, surface roughness, and number of intersections. *Adv. Water Resour.* **2016**, *88*, 53–65. [CrossRef]

26. Liu, R.; Jiang, Y.; Li, B.; Wang, X. A fractal model for characterizing fluid flow in fractured rock masses based on randomly distributed rock fracture networks. *Comput. Geotech.* **2015**, *65*, 45–55. [CrossRef]

27. Ervin, M.C.; Benson, N.D.; Morgan, J.R.; Pavlovic, N. Melbourne's southbank interchange: A permanent excavation in compressible clay. *Can. Geotech. J.* **2004**, *41*, 861–876. [CrossRef]

28. Wang, B.T.; Chen, X.A. Research on effect of suspended cut-off wall with simulation test. *Chin. J. Rock Mech. Eng.* **2008**, *27*, 2766–2771.

29. Shao, S.J.; Yang, C.M. Research on the impermeability design method of the slurry protection diaphragm wall in the coarse-grained soil foundation. *J. Hydraul. Eng.* **2015**, *46*, 46–53.

30. Barrospérsio, L.A.; Santospetrucio, J. Coefficients of active earth pressure with seepage effect. *Can. Geotech. J.* **2012**, *49*, 651–658. [CrossRef]

31. Wang, S.; Chen, J.S.; Luo, Y.L.; Sheng, J.C. Experiments on internal erosion in sandy gravel foundations containing a suspended cut-off wall under complex stress states. *Nat. Hazards* **2014**, *74*, 1163–1178. [CrossRef]

32. Wang, J.; Zhang, H.; Liu, M.; Chen, Y. Seismic passive earth pressure with seepage for cohesionless soil. *Mar. Georesour. Geotechnol.* **2012**, *30*, 86–101. [CrossRef]

Article

Investigation of the Porosity Distribution, Permeability, and Mechanical Performance of Pervious Concretes

Rentai Liu *, Haojie Liu, Fei Sha, Honglu Yang, Qingsong Zhang, Shaoshuai Shi and Zhuo Zheng

Geotechnical and Structural Engineering Research Center, Shandong University, Jinan 250061, China; rlhaojie@163.com (H.L.); shafei_97@163.com (F.S.); yanghonglu321@163.com (H.Y.); zhangqingsong@sdu.edu.cn (Q.Z.); shishaoshuai@sdu.edu.cn (S.S.); 413708318@qq.com (Z.Z.)
* Correspondence: rentailiu@163.com

Received: 18 May 2018; Accepted: 20 June 2018; Published: 26 June 2018

Abstract: Pervious concrete is a kind of porous and permeable material for pavements and slope protection projects, etc. In this paper, a kind of pervious concrete was prepared with Portland cement, silica fume (SF), polycarboxylate superplasticizer (SP), and limestone aggregates. The performance of concrete, such as its porosity, pore distribution, permeability coefficients, and mechanical properties, were studied through laboratory tests. The volumetric porosity was measured by the water displacement method, and the planar porosity and pore size distribution were determined using image processing technology. A permeameter with a transparent sidewall and an exact sidewall sealing method were used to measure the permeability coefficients accurately. The results show that the segregation index and flow values of pastes increased with the increase of SP and water cement ratio (W/C). The measured porosity (volumetric porosity and planar porosity) of pervious concrete with a single-size aggregate was closer to the design porosity than that with the blended aggregate. Compared with the design porosity selected in this study, aggregate size was the main factor influencing the pore distribution of pervious concrete. The standard deviation of the permeability coefficient was less than 0.03 under different hydraulic gradients. It was found that the relationships between the permeability coefficient and volumetric porosity (or effective pore size d_{50}) approximately obey polynomial function. Based on the test results, the optimized parameters were suggested for practical engineering: W/C of 0.26–0.30; 0.5% SP content; 5% SF content; 15–21% design porosity; and aggregate sizes of 4.75–9.5 mm and 9.5–16 mm.

Keywords: pervious concrete; permeability coefficient; porosity; pore distribution characteristics; strength

---

1. Introduction

With the rapid development of economy and the increment of urban populations, many large-scale infrastructures and impermeable roads have been built, which has caused more and more serious environmental problems, such as urban hot island phenomena and the disappearance of groundwater. Pervious concrete, also referred to as porous ecological concrete, is a porous media material which consists of cement, water, aggregates, and admixtures, etc. Pervious concrete has a large porosity and its permeability can reach up to about 2–6 mm/s [1]. Rainwater can penetrate into pervious concrete quickly, so groundwater can be replenished and stored effectively. It can also reduce city temperatures and alleviate urban hot island phenomena [2].

Pervious concrete was first invented in Britain in 1824. It was first introduced into the United States in the mid-1970s and has since been rapidly developed [3]. The Sponge City was introduced in China in approximately 2012. Because of its huge environmental benefits, it was quickly and widely emphasized and generalized in the construction of China's Sponge City. To date, the properties of pervious concretes have been studied in many laboratories, including mixing design, conservation method, porosity characteristics, strength and durability, etc. [4–6]. Compared with traditional or common concrete, pervious concrete has an amount of porosity and pore structure features, which have been proven to have a great influence on the properties of pervious concretes [7,8]. Montes et al. [9] recommended using the water displacement method, based on Archimedes buoyancy principle, to obtain porosity. Neithalath et al. [8] used image analysis to analyze the pore structure features of pervious concrete. The permeability of pervious concrete is high due to the presence of high porosity, and the test methods of the permeability coefficient are especially important [10,11]. Kayhanian et al. [12] used a National Center for Asphalt Technology's (NCAT) falling head permeameter to evaluate the permeability coefficient of pervious concrete. Montes et al. [13] evaluated permeability by using a falling head permeameter system. The system could register the total volume flowing out of the permeable material per second. At the same time, Jiang Zhengwu and Yang et al. [13,14] showed that porosity, W/C, aggregate cement ratio, and aggregate size were the main factors influencing the permeability and strength of pervious concrete. William et al. [15] studied the effect of vertical porosity distributions on the permeability coefficient. Because of the low strength of the pervious concrete prepared by common materials and methods, various admixtures and fine aggregates were introduced to expand the application scope of pervious concrete. Yang and Jiang [16] used fine aggregates, organic intensifiers, and optimized mix proportions to produce a high strength pervious concrete. Huang [17] improved the compressive strength of pervious concrete by using polymer modification and sand.

Although many researches on pervious concretes have been conducted, studies concerning the properties of paste, as well as the accurate determination of the permeability coefficient and pore structure, are still limited. The rheological property of cement paste is one of the key factors responsible for the preparation of good pervious concrete [18,19]; however, it has not been studied sufficiently. Besides, when the permeability coefficient of pervious concrete is measured, the fluid leakage through the specimen-permeameter interface is not given significant attention. In this paper, the influences of polycarboxylate superplasticizer (SP), silica fume (SF), and W/C on the fluidity of cement pastes and the segregation index of pervious concretes were discussed in detail. The effect of design porosity and aggregate size on the porosity distribution of pervious concrete was studied. A visual permeameter and an exact sidewall sealing method were established to measure the permeability coefficients accurately. The mechanical behavior, such as the compressive strength, was also measured.

2. Materials and Test Methods

2.1. Materials

P.O 42.5 ordinary Portland cement was provided by Sunnsy Group, Jinan, China, and its chemical composition ingredients are shown in Table 1. Silica fume (SF) was provided by Elkem Company, Oslo, Norway, and its chemical composition is presented in Table 1. The polycarboxylate superplasticizer (SP) was provided by Shandong Academy of Building Research, Jinan, China, and the main properties of NC-J are shown in Table 2. According to the standard of CJJ/T135-200, the literature [13,20] and requirements of construction projects, limestone gravel was used as an aggregate, and aggregate sizes followed those listed in Table 3.

Table 1. Chemical composition of materials.

Name	Chemical Composition: %										
	SiO_2	CaO	Al_2O_3	Fe_2O_3	MgO	K_2O	Na_2O	SO_3	TiO_3	LOI	Others
Cement	23.52	61.47	7.59	2.42	1.70	0.57	0.27	0.65	0.11	0.84	0.86
Silica fume (SF)	97.2	0.42	0.27	0.08	0.55	0.48	0.19	0.51	-	-	0.30

* Chemical compositions of cement and silica fume were tested in the laboratory.

Table 2. Properties of Polycarboxylate Superplasticizer (SP).

Solid Content: % Al_2O_3	pH	Density: g/mL Na_2O	Na_2SO_4: % TiO_3	Cl^-: % Others	Water Reducing Rate: %
30.25 0.42 0.27	5.01	1.095	0.67 -	0.03 0.30	32.1

* Properties of NC-J came from the manufacturer.

Table 3. Gravel sizes.

	A	B	C	D	E	F	G
Gravel sizes: mm	4.75–9.5	9.5–16	16–19	75% 4.75–9.5 25% 9.5–16	50% 4.75–9.5 50% 9.5–16	50% 4.75–9.5 50% 16–19	50% 9.5–16 50% 16–19

2.2. Test Methods

2.2.1. Fluidity

The fluidity of cement paste was tested according to GB/T8077-2012. The truncated cone mold had a top diameter of 36 mm, a bottom diameter of 60 mm, and a height of 60 mm. Different SP contents (0.3%, 0.5%, and 0.7%), SF contents (5% and 10%), and W/C (0.22–0.32) were used to prepare the cement pastes. Then, the paste was poured into truncated cone mold which was placed on a glass pane. Next, the truncated cone mold was removed from the glass pane. The diffusion diameter of the paste was measured with a ruler in an orthogonal direction after 30 s. The average flow value was the initial diffusion diameter of the paste.

2.2.2. Mix Proportions

The mix proportion of pervious concrete was designed by the absolute volume method according to CJJ/T135-2009. The cement (m_c), water (m_w), and admixture (m_j) were calculated as follows:

$$\frac{m_g}{\rho_g} + \frac{m_c}{\rho_c} + \frac{m_w}{\rho_w} + \frac{m_j}{\rho_j} + P = 1 \tag{1}$$

where m_g is the mass of the aggregate; ρ_g, ρ_c, ρ_w, ρ_j are the apparent densities of aggregate, cement, water, and admixture, respectively; and P is the design porosity.

In the segregation tests of pervious concrete, the aggregate size was 4.75–9.5 mm and the design porosity was 15%. The W/C, SP, and SF were shown in the fluidity test.

In the porosity distribution tests, permeability tests, and mechanical performance tests of pervious concrete, the aggregate sizes were as shown in Table 3. The design porosities were 12%, 15%, 18%, 21%, 24%, and 27%; W/Cs were 0.24–0.32. The SP content was 0.5%. The SF content was 5%. As an example, the parameters of pervious concrete with a W/C of 0.28 are shown in Table 4.

Table 4. Mix proportions of pervious concrete.

Aggregate Size mm	Design Porosity %	Aggregate kg/m^3	Cement kg/m^3	SF (silica fume) kg/m^3	Water kg/m^3	SP(polycarboxylate superplasticizer) kg/m^3
	12	1489.0	546.1	28.7	161.0	2.87
	15	1489.0	496.3	26.1	146.3	2.61
4.75–9.5	18	1489.0	446.5	23.5	131.6	2.35
	21	1489.0	396.7	20.9	116.9	2.08
	24	1489.0	346.9	18.3	102.2	1.82
	27	1489.0	297.1	15.6	87.6	1.56
	12	1518.0	527.8	27.8	155.6	2.78
	15	1518.0	478.0	25.2	140.9	2.51
9.5–16	18	1518.0	428.2	22.5	126.2	2.25
	21	1518.0	378.4	19.9	111.5	1.99
	24	1518.0	328.7	17.3	96.9	1.73
	27	1518.0	289.4	15.2	85.3	1.52
	12	1529.0	519.5	27.3	153.1	2.73
	15	1529.0	469.7	24.7	138.4	2.47
16–19	18	1529.0	419.9	22.1	123.8	2.21
	21	1529.0	370.1	19.5	109.1	1.95
	24	1529.0	320.4	16.9	94.4	1.69
	27	1529.0	270.6	14.2	79.7	1.42

2.2.3. Molding and Curing

Concrete specimen preparation procedures were as follows. Firstly, the aggregate and 70% water were mixed for 60 s. Then, the 50% cement and admixture were mixed for another 60 s. The remaining cement and water were added and mixed for 2 min. In order for the material to uniformly fill in the mold, fresh concrete was divided into two layers into the mold. Each layer of concrete was vibrated for 20 s on the vibration table, and then the concrete was compacted with 0.08 MPa compressive stress. After molding, the specimen was placed in a curing box at 20 °C with a relative humidity of 97%. The specimen was removed from the mold after 24 h, and then placed in a curing box for 3 days, 7 days, or 28 days.

2.2.4. Segregation Index

The segregation index of the concrete was evaluated according to ASTMC1610; the test device is shown in Figure 1. The specimen was produced in a polyvinyl chloride (PVC) tube. The mass of the specimen in the top section of the column (m_t) and the mass (m_b) in the bottom section were measured after vibrating for 2 min on a vibration table. The segregation index (S) was calculated as follows:

$$S = 2 \times \left[\frac{m_b - m_t}{m_b + m_t} \right] \times 100\% \tag{2}$$

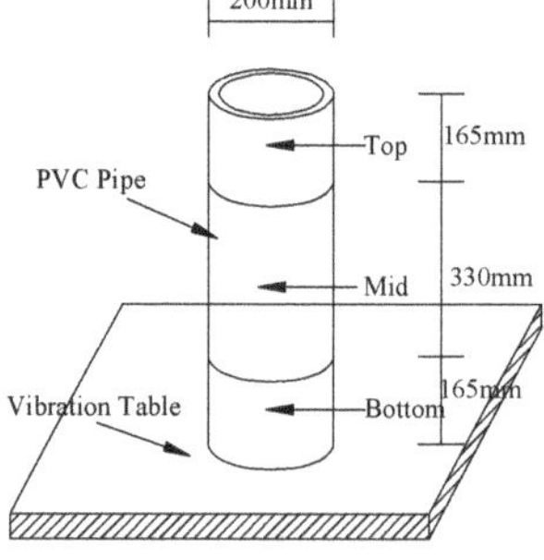

Figure 1. The segregation test device of pervious concrete.

2.2.5. Porosity

The volumetric porosity was measured by the water displacement method. The specimen was dried in a drying box at 110 °C for 24 h, and then put into water for 24 h, so the volume of the water (V_d) repelled by the specimen was acquired. The volume of open pores was calculated by subtracting V_d from the sample bulk volume (V_b) [21]. The volumetric porosity of the specimen was calculated as follows:

$$P = \left(\frac{V_b - V_d}{V_b}\right) \times 100\% \tag{3}$$

2.2.6. Pore Structure Features

The pore structure of pervious concrete was analyzed by Image-Pro Plus 6.0. For a concrete specimen, three to five cylinders were used, and the diameter and height of the specimen were 110 mm and 160 mm, respectively. In the process of the pore structure test, first of all, the specimen was sectioned into 20 mm thick slices. The surface of the specimen was grinded to obtain a flattened surface. An image of surface was captured by a camera, and the outer region of this image was cut to avoid edge effects. Finally, the image was processed by Image-Pro Plus 6.0 software, and an appropriate threshold was chosen to obtain the pore size distribution and planar porosity, according to Reference [8].

2.2.7. Permeability Coefficient

The permeability coefficient was measured by the permeameter that was designed with transparent plexiglass by our group. As shown in Figure 2, three outlets at different heights of the permeameter were installed to create different hydraulic gradients. The heights of the three outlets were 10 cm, 15 cm, and 20 cm. For a concrete specimen with same mix proportions, three cylinders were prepared, and the diameter and height of the specimen were 110 mm and 120 mm, respectively. In the process of the permeability test, the specimen was put into the permeameter, and the sidewall space between the specimen and the permeameter was sealed with Vaseline and transparent wrapping film, so it was easy to know whether there was sidewall leakage. Each specimen was tested three times under different hydraulic gradients to ensure the accuracy of the results. The permeability coefficient of pervious concrete K was calculated by Darcy's law as follows:

$$K = \frac{Qh}{HAt} \tag{4}$$

where h is the specimen height (cm); H is the hydraulic gradient (cm); and A is the specimen area (cm^2).

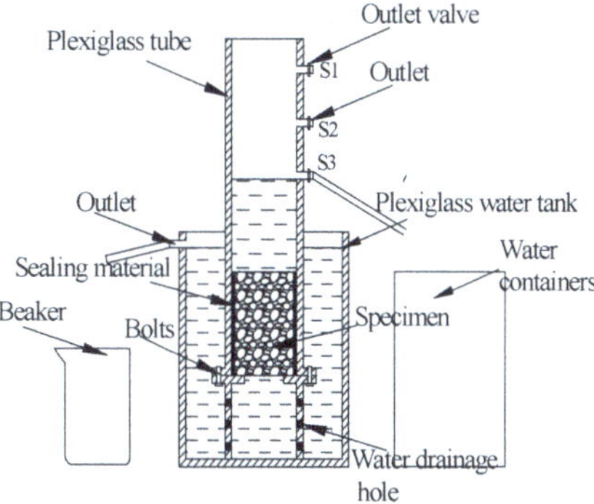

Figure 2. The permeability test device of pervious concrete.

2.2.8. Mechanical Strength

The flexural strength and compressive strength of the pervious concrete were tested according to GB/T 50081-2002. The size of the specimen used in the compressive strength test was 100 mm × 100 mm × 100 mm. The size of the specimen used in the flexural strength test was 100 mm × 100 mm × 400 mm. The mechanical strength of the specimen reached 70% of its peak strength during the test, after which the test was stopped. The compressive strength and flexural strength were tested after 3 days, 7 days, and 28 days.

3. Results, Analysis, and Discussion

3.1. Fluidity Performance

Figure 3 shows the measured flow values of pastes with different dosages of SP, SF, and W/C.

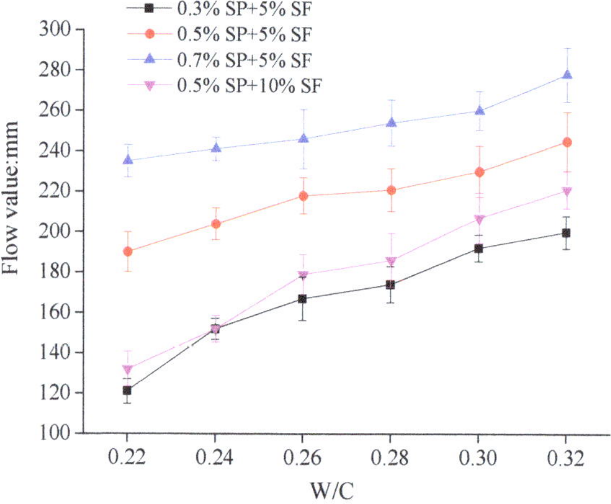

Figure 3. Flow values of pastes with different W/C (water cement ratio). Two to three pastes were prepared for each mix design and the figure shows an average of these together with the standard deviation.

It can be seen in Figure 3 that the flow values linearly increased with increase of the W/C. SP was usually added to improve the fluidity of the paste at a low W/C. The use of SP significantly affected the flow values of the pastes; a small increase of SP could make the flow value of paste reach the desired value under a lower W/C. SF was introduced to improve strength of pervious concrete. However, because of the smaller particle size and large specific surface area of SF, the flow values of pastes significantly decreased when the content of SF increased.

Figure 4 shows the effect of SP, SF, and W/C on the segregation index of pervious concretes, for which a 4.75–9.5 mm aggregate size was used.

In Figure 4, it was obvious that the segregation index increased as the W/C increased. When the W/C was constant, an increase of SP and a decrease of SF also increased the segregation index. The segregation index of pervious concrete with 0.7% SP and 5% SF was the largest. This was mainly related to the flow value of the paste; the increase of the flow value of the paste can increase the segregation index. Moreover, it also can be seen from Figures 3 and 4 that the effects of the W/C on the flow value of the paste and segregation index of pervious concretes were found to be not so obvious compared with those of SP.

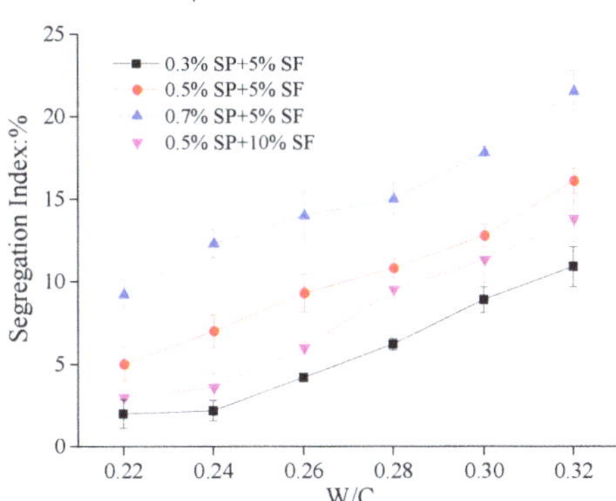

Figure 4. Segregation index of pervious concretes. (Three concrete specimens were prepared for each mix design and the figure shows an average of these together with the standard deviation).

Combined with the above test and strength test, it was found that pervious concrete with 0.3% SP was difficult to mold. The paste easily flowed down and gathered in the bottom of the pervious concrete with 0.7% SP. When the SF content increased from 5% to 10%, the flow values of the pastes drastically decreased, so there were some difficulties to molding pervious concrete when the W/C was less than 0.28. At same time, the strength of the pervious concrete also decreased. For the pervious concrete with 0.5% SP and 5% SF, when the W/C was less than 0.26, the flow values of the pastes were relatively small, the cohesive force of the pervious concrete was small, and the compressive strength was low; however, when the W/C was greater than 0.30, the pastes easily flowed down and gathered in the bottom of the pervious concrete, which was unfavorable for water permeability. The flow value of the cement paste and segregation index of pervious concretes were satisfactory with a W/C of 0.26–0.30, 0.5% SP, and 5% SF. The W/C of 0.28 was emphatically selected for the following study.

3.2. Porosity and Pore Distribution of Pervious Concrete

3.2.1. Porosity of Pervious Concrete

As shown by the results given in Figure 5a, three different sizes of the aggregate were used to study the effect of the specimen surface at different locations on the planar porosity. Figure 5b shows the volumetric porosity and average planar porosity of pervious concrete with different aggregate sizes.

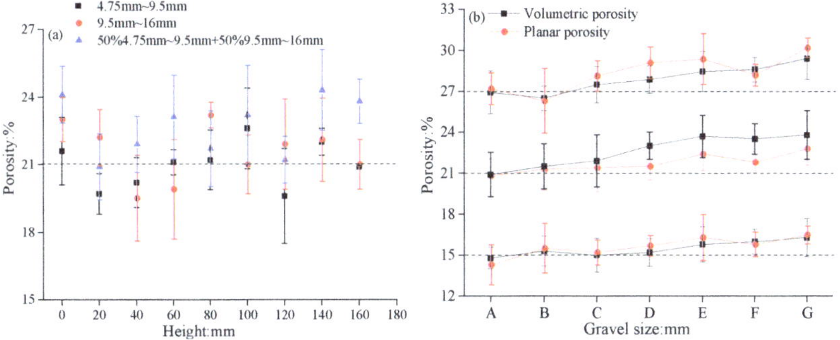

Figure 5. Porosity of pervious concrete. (**a**) Planar porosity, the design porosity was 21%; (**b**) volumetric porosity and average planar porosity, the design porosity was 15%, 21%, and 27%. The figure shows an average value and standard deviation of three to five specimens.

As shown in Figure 5a, the difference value between the planar porosity and design porosity was within ±3%, so the planar porosity of the pervious concrete had less fluctuation at each cross-section. This indicated that the quality of the specimen using both the vibration and compaction method was more satisfactory. It can be seen from Figure 5b that the volumetric porosity and planar porosity increased when the design porosity increased. The difference value between the volumetric porosity and planar porosity was within ±3%. Additionally, the volumetric porosity and planar porosity of concrete with single-size aggregates were closer to the designed porosity. The pervious concretes with blended aggregates had higher porosity than those with single sizes of aggregates. The porosity of the pervious concretes with 50% 9.5–16 mm and 50% 16–19 mm aggregate sizes was the largest. The reason for this may be that a loosening effect may have occurred when big aggregates were pushed by small aggregates [18], and this increased the porosity of the pervious concretes.

3.2.2. Pore Distribution of Pervious Concrete

Figure 6 shows the effect of a single-size aggregate on the pore distribution of pervious concrete. Figure 6a shows the effect of a single-size aggregate on the pore size histogram. Figure 6b shows the effect of a single-size aggregate on the cumulative frequency distribution.

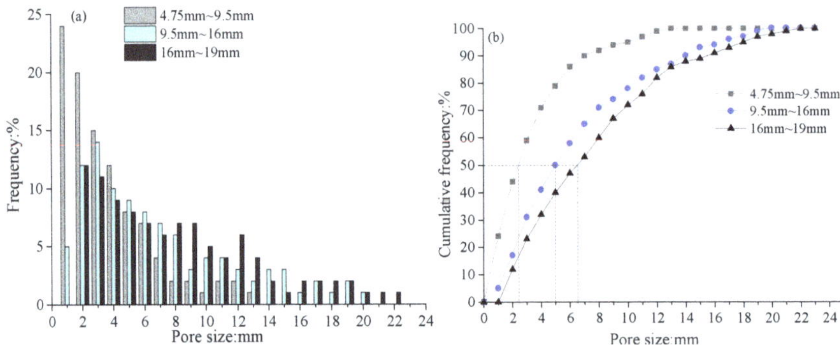

Figure 6. Effect of a single-size aggregate on the pore distribution of pervious concrete. (**a**) Frequency distribution of pore sizes; (**b**) cumulative frequency distribution of pore sizes. The design porosity was 15%. The figure shows the average value of three to five specimens.

In Figure 6a, the pervious concrete with smaller aggregates had a higher proportion of smaller pores, whereas the pervious concrete with larger aggregates had a relatively large proportion of large pores. In Figure 6b, the pervious concrete with larger aggregates had a larger effective pore size (the pore size corresponding to 50% of the cumulative frequency distribution [8]), and its cumulative frequency curve increased more slowly. For example, when the aggregate size of concrete was 4.75–9.5 mm, the pore sizes of pervious concrete were mainly concentrated within 7 mm; 71% of the pores were smaller than 5 mm and the effective pore size was about 2.4 mm. When the aggregate size of concrete was 9.5–16 mm, the pore sizes of pervious concrete were mainly concentrated within 10 mm and the effective pore size was about 5 mm. However, 40% of the pores in the specimen with 9.5–16 mm aggregates were smaller than 5 mm, 28% of the pores were larger than 10 mm, and the effective pore size was about 6.5 mm. This was because the smaller aggregate resulted in a larger surface area and a more regular shape of the aggregate. Therefore, as the number of contact points between aggregates increased, the internal pore size decreased and the pore number increased.

Figure 7 shows the effect of a blended aggregate on the pore distribution of pervious concrete. Figure 7a shows the effect of a blended aggregate on the pore size histogram. Figure 7b shows the effect of a blended aggregate on the cumulative frequency distribution.

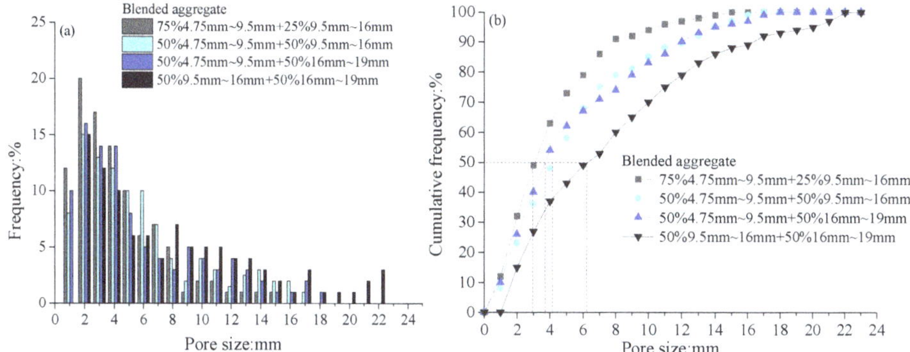

Figure 7. Effect of a blended aggregate on the pore distribution of pervious concrete. (**a**) Frequency distribution of pore sizes; (**b**) cumulative frequency distribution of pore sizes. The design porosity was 15%. The figure shows the average value of three to five specimens.

The pore size distribution of pervious concrete with the blended aggregate size was close to that of concrete with the single-size aggregate. The addition of smaller aggregates in pervious concrete could improve the internal pore size distribution. With the increase of smaller aggregates in concrete, the proportion of larger pores decreased; at the same time, the effective pore size also decreased.

Figure 8 shows the effect of the design porosity on the pore distribution of pervious concrete. Figure 8a shows the effect of the design porosity on the pore size histogram. Figure 8b shows the effect of the design porosity on the cumulative frequency distribution.

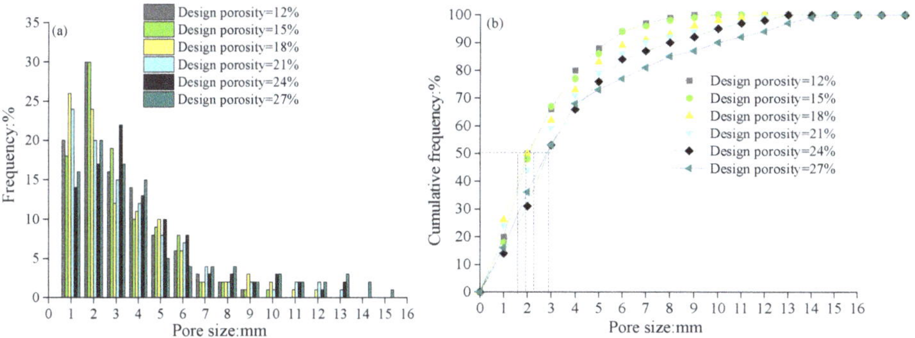

Figure 8. Effect of the design porosity on the pore distribution of pervious concrete. (**a**) Frequency distribution of pore sizes; (**b**) cumulative frequency distribution of pore sizes. The aggregate size was 4.75–9.5 mm. The figure shows the average value of three to five specimens.

In Figure 8b, the difference between all of the cumulative frequencies of pores was relatively low for different design porosities. The effective pore sizes were within 2–3 mm. In Figure 8a, there was only slight increase of large pores in the concrete with the larger design porosity. This might be attributed to the fact that the pervious concrete with a larger design porosity only resulted in a larger number of open and connected pores, but had little effect on the pore size of the concrete.

3.3. Permeability of Pervious Concrete

3.3.1. Accurate Determination of Permeability Coefficient

As shown by the results given in Figure 9, different design porosities and different aggregate sizes were used to study the effect of hydraulic gradients on the permeability coefficient, and the standard deviations of the permeability coefficient under three hydraulic gradients were calculated.

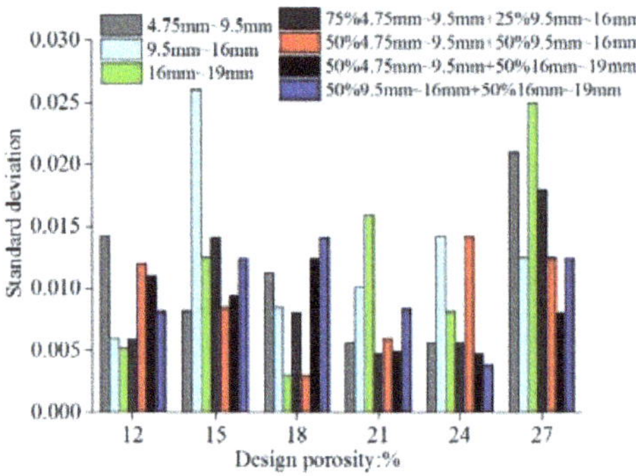

Figure 9. Standard deviation of the permeability coefficient of pervious concrete with different hydraulic gradients. The figure shows the average standard deviation values of three specimens.

It was obvious that the standard deviation values of pervious concrete with different mixture ratios were no more than 0.03, and the variation of the standard deviation values was not regular. This reflected that the dispersion degree of the permeability coefficients under different hydraulic gradients was very small. The use of a visual permeameter and the sealed sidewall method with Vaseline and transparent wrapping film could effectively solve the problem of the permeability coefficient changing, as the hydraulic gradient changed due to side wall leakage.

Figure 10 shows the permeability coefficient of pervious concrete measured by the new method and traditional method (no blocking on the sidewall of the specimen). Figure 10a shows the permeability coefficient of pervious concrete with different design porosities. Figure 10b shows the permeability coefficient of pervious concrete with different aggregate sizes.

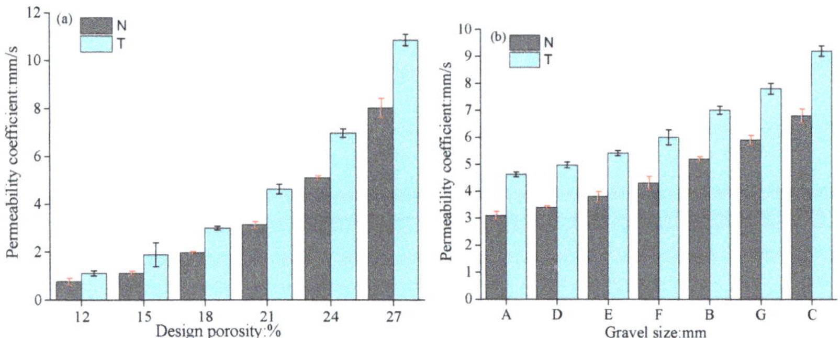

Figure 10. The permeability coefficient of permeable concrete under different test methods. (**a**) The design porosity of permeable concrete with 4.75–9.5 mm aggregate sizes; (**b**) the aggregate size of permeable concrete with 21% design porosity. The figure shows the average value and standard deviation of three specimens.

In Figure 10, the permeability coefficient tested by the traditional method was larger than that tested by the new method. Moreover, the larger design porosity and aggregate size resulted in a larger difference value of the permeability coefficient measured by the two methods. For example, in Figure 10a, the difference value of the permeability coefficient in pervious concrete with 12% design porosity was 0.35 mm/s, whereas the difference value increased to 2.8 mm/s for pervious concrete with 27% design porosity. In Figure 10b, the difference value of the permeability coefficient in pervious concrete with 4.75–9.5 mm aggregate size was 1.52 mm/s, whereas the difference value increased to 2.4 mm/s for pervious concrete with 16–19 mm aggregate size. This was because the larger design porosity and aggregate size resulted in a larger pore size and greater porosity of the sidewall surfaces. Thus, the leakage and permeability coefficient of pervious concrete increased in the traditional method.

3.3.2. Influential Factors of the Permeability Coefficient

As be seen from Figure 10a, the permeability coefficient increased as the design porosity increased. In particular, when the porosity was greater than 15%, a small change of porosity significantly changed the permeability coefficient. This was because the pervious concrete with a larger design porosity resulted in a larger measured porosity. When the measured porosity reached a critical value, the resistance of the internal structure of the concrete to water was greatly reduced, so the permeability coefficient increased rapidly. Figure 11a shows the relationship between the permeability coefficient and volumetric porosity of pervious concrete. It was found that the relationship between the permeability coefficient and volumetric porosity was right for a quadratic function, and the correlation coefficient was 0.97.

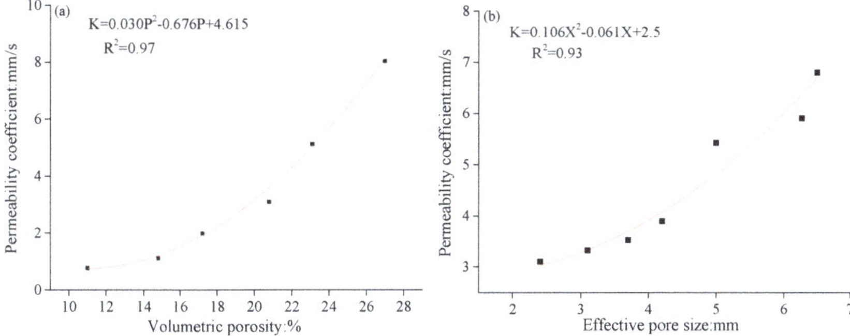

Figure 11. Permeability coefficient of pervious concrete. (**a**) The relationship between the permeability coefficient and volumetric porosity of pervious concrete with aggregate sizes of 4.75–9.5 mm; (**b**) the relationship between the permeability coefficient and effective pore size of pervious concrete with 21% design porosity.

As can be observed in Figure 10b, the effect of the aggregate size on the permeability coefficient was significant. The permeability coefficient increased with the increase of single-size aggregates, and the permeable coefficient of the pervious concrete with the blended aggregate was slightly lower than that of concrete with the single-size aggregate. As shown in Figure 5b, there was little difference in the porosity variation of pervious concrete with different mix ratios. Therefore, the permeability coefficient was not sensitive to the porosity. The reason may be related to the internal pore sizes of pervious concrete; the larger aggregate size in the pervious concrete result in a larger internal pore size. Figure 11b shows the relationship between the effective pore size and the permeability coefficient. It was found that the relationship between the permeability coefficient and the effective pore size approximately obey the polynomial function, and the correlation coefficient was 0.93.

3.4. Strength of Pervious Concrete

As shown in Figure 12, two different aggregate sizes and three different design porosities were used in the pervious concrete samples. The compressive strength and flexural strength of the pervious concrete were measured.

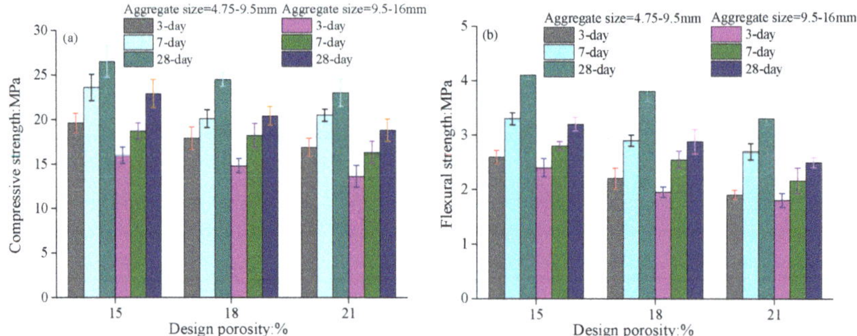

Figure 12. Effect of the design porosity on the strength of pervious concrete. (**a**) Compressive strength; (**b**) Flexural strength. The figure shows an average value and standard deviation of three specimens.

In Figure 12, it was obvious that the strength of the pervious concrete increased with the decrease of the design porosity. It was also found that the strength of the pervious concrete with 4.75–9.5 mm aggregate size was larger than that with 9.5–16 mm aggregate size. For the pervious concrete with 4.75–9.5 mm aggregate sizes, the compressive and flexural strengths of the concrete with 15% design porosity were about 27.2 MPa and 4.1 MPa, respectively. Meanwhile, they were about 23.5 MPa and 3.3 MPa for the pervious concrete with 21% design porosity, respectively. For the pervious concrete with 9.5–16 mm aggregate sizes, they were about 19 MPa and 2.4 MPa for pervious concrete with 21% design porosity, respectively. At the same time, the strengths of the pervious concrete were satisfactory at 0.26–0.30 W/C ratios. For example, the compressive strengths of pervious concretes with 4.75–9.5 mm aggregate sizes were about 25.2–27.2 MPa, and the related flexural strengths were about 3.7–4.1 MPa. The strength of pervious concrete gradually increased with the increase of the curing time. It could be deduced that the strength was mainly dependent on the bonding strength, as well as the number of bonding points and bonding areas between cementitious materials and aggregates. The number of bonding points, contact areas, and effective bonding points increased with the decrease of design porosity and aggregate size.

4. Conclusions

1. The segregation index of pervious concrete increased with the increase of SP and the W/C ratio, and the decrease of SF. The use of SP could significantly increase the flow value, while the use of SF signally reduced the flow value. The flow values of pastes and the segregation index for pervious concretes were obtained and found to be very suitable when the W/C of 0.26–0.30, 0.5% SP, and 5% SF were used.

2. In the paper, volume porosity, planar porosity, and pore sizes were obtained by the water displacement method and image processing technology. It was found that the planar porosity of pervious concrete was close to the volumetric porosity. There was little difference between the measured porosity (volumetric porosity and planar porosity) and the design porosity when a single-size aggregate was used. However, the difference was large when a blended aggregate was used. This may be due to the possible loosening effect of the blended aggregates. The aggregate size was the main influencing factor of the pore distribution of the pervious concrete, and the effective pore size increased with the increase of the aggregate size. The design porosity had little

effect on the pore distribution. The dosage of cementitious material decreased when the design porosity increased; correspondingly, the number of open and connected pores increased and the connections became difficult.

3. Based on Darcy's law, the permeameter and the sealed sidewall method can effectively solve the problem of sidewall leakage and the permeability coefficient can be accurately determined. The standard deviation of the permeability coefficient was less than 0.03 under different hydraulic gradients. The permeability coefficient of the pervious concrete was determined by the porosity and aggregate size jointly, and the permeability coefficient increased with increase of the aggregate size and design porosity. It was found that the relationships between the permeability coefficient and the volumetric porosity (or effective pore size d_{50}) approximately obey the polynomial function.

4. In the pervious concrete material system, the use of SF and SP improved the performance of pervious concrete, but the dosage of SF and SP should not be excessive. The strength of the concrete significantly increased with the increase of cementitious material and the decrease of aggregate size. When 15% design porosity and 4.75–9.5 mm aggregate size were used in concrete, the 28-day compressive strength was 27.2 MPa and the 28-day flexural strength was 4.1 MPa. Based on the performance of pervious concretes, the W/C of 0.26–0.30; 0.5% SP content; 5% SF content; 15–21% design porosity; and aggregate sizes of 4.75–9.5 mm and 9.5–16 mm are suggested.

Author Contributions: R.L., H.L. and F.S. conceived of and designed the study. H.L., F.S., and H.Y. performed the experiments. R.L., F.S. and Q.Z. developed the permeameter. R.L., H.L., F.S., H.Y., Q.Z., S.S. and Z.Z. wrote and modified the paper.

Funding: This research was funded by the General Program of the National Natural Science Foundation of China, grand number is 51779133; National Key R&D Plan of China, grand number is 2016YFC0801604; and Joint Funds of National Natural Science Foundation of China, grand number is U1706223.

Conflicts of Interest: The authors declare no conflict of interest.

References

1. Tennis, P.D.; Leming, M.L.; Akers, D.J. *Pervious Concrete Pavements*; EB302 Portland Cement Association: Skokie, IL, USA, 2004.

2. Schaefer, V.R.; Wang, K.; Suleiman, M.T.; Kevern, J.T. *Mix Design Development for Pervious Concrete in Cold Weather Climates, Final Report*; National Concrete Pavement Technology Center, Iowa State University: Ames, IA, USA, 2006.

3. Malhotra, V.M. No-fines concrete-its properties and applications. *ACI Mater. J.* **1976**, *73*, 628–644.

4. Liang, L. *Preparation, Porous Structure and Camouflage Performance of Porous Ecological Concrete*; Nanjing University of Aeronautics and Astronautics: Nanjing, China, 2010.

5. Kevern, J.T. Advancement of Pervious Concrete Durability. Ph.D. Thesis, Iowa State University, Ames, IA, USA, 2008.

6. Kevern, J.T. Mix Design Determination for Freeze–Thaw Resistant Portland Cement Pervious Concrete. Master's Thesis, Iowa State University, Ames, IA, USA, 2006.

7. Hu, J.; Stroeven, P. Application of image analysis to assessing critical pore size for permeability prediction of cement paste. *Image Anal. Stereol.* **2003**, *22*, 97–103. [CrossRef]

8. Deo, O.; Neithalath, N. Compressive response of pervious concrete proportioned for desired porosities. *Constr. Build. Mater.* **2011**, *25*, 4181–4189. [CrossRef]

9. Montes, F.; Valavala, S.; Haselbach, L. A new test method for porosity measurements of Portland cement pervious concrete. *J. ASTM Int.* **2005**, *2*, 73–82.

10. Liu, R.; Li, B.; Jiang, Y. A fractal model based on a new governing equation of fluid flow in fractures for characterizing hydraulic properties of rock fracture networks. *Comput. Geotech.* **2016**, *75*, 57–68. [CrossRef]

11. Liu, R.; Jiang, Y.; Li, B.; Yu, L. Estimating permeability of porous media based on modified Hagen–Poiseuille flow in tortuous capillaries with variable lengths. *Microfluid. Nanofluid.* **2016**, *20*, 120. [CrossRef]

12. Kayhanian, M.; Anderson, D.; Harvey, J.T. Permeability measurement and scan imaging to assess clogging of pervious concrete pavements in parking lots. *J. Environ. Manag.* **2012**, *95*, 114–123. [CrossRef] [PubMed]
13. Jiang, Z.; Sun, Z. Effects of Some Factors on Properties of Porous Pervious Concrete. *J. Build. Mater.* **2005**, *8*, 513–519.
14. Yang, Z.F. *Study on Material Design and Road Performances of Porous Concrete*; Wuhan University of Technology: Wuhan, China, 2008.
15. Martin, W.D., III; Kaye, N.B.; Putman, J.B. Impact of vertical porosity distribution on the permeability of pervious concrete. *Constr. Build. Mater.* **2014**, *59*, 78–84. [CrossRef]
16. Yang, J.; Jiang, G. Experimental study on properties of pervious concrete pavement materials. *Cem. Concr. Res.* **2003**, *33*, 381–386. [CrossRef]
17. Huang, B.; Wu, H.; Shu, X.; Burdette, E.G. Laboratory evaluation of permeability and strength of polymer-modified pervious concrete. *Constr. Build. Mater.* **2010**, *24*, 818–823. [CrossRef]
18. Tamai, M. Water permeability of hardened materials with continuous voids. *Cem. Assoc. Jpn. Rev.* **1990**, 446–449.
19. Mishima, N.; Tanigawa, Y.; Mori, H.; Kurokawa, Y.; Terada, K.; Hattori, T. Experimental study on rheological properties of dense suspension by shear box test. *J. Struct. Constr. Eng.* **2000**, *65*, 13–19. [CrossRef]
20. Lian, C.; Zhuge, Y.; Beecham, S. The relationship between porosity and strength for porous concrete. *Constr. Build. Mater.* **2011**, *25*, 4294–4298. [CrossRef]
21. Larrard, D.F. *Concrete Mixture Proportioning: A Scientific Approach*; E&FN Spon: London, UK, 1999; p. 421.

Article

Deformation and Control Countermeasure of Surrounding Rocks for Water-Dripping Roadway Below a Contiguous Seam Goaf

Changqing Ma [1,2,*], Pu Wang [3,4,*], Lishuai Jiang [4] and Changsheng Wang [4,5]

[1] Department of Mechanical and Electronic Engineering, Shandong University of Science and Technology, Tai'an 271019, China

[2] Shanxi Coking Coal Refco Group Ltd., Huozhou 031400, China

[3] Department of Resources and Civil Engineering, Shandong University of Science and Technology, Tai'an 271019, China

[4] State Key Laboratory of Mining Disaster Prevention and Control, Shandong University of Science and Technology, Qingdao 266590, China; skwp0701@hotmail.com(L.J.); cswang0635@163.com (C.W.)

[5] Graduate School of Engineering, Nagasaki University, Nagasaki 852-8521, Japan

[*] Correspondence: changqing.ma@uconn.edu (C.M.); tianwailaike911@163.com (P.W.); Tel.: +86-158-548-48872 (P.W.)

Received: 10 May 2018; Accepted: 21 June 2018; Published: 25 June 2018

Abstract: To solve the technical problem of supporting a water-dripping roadway below contiguous seams at the Tuanbai coal mine, the deformation law of surrounding rocks for the roadway was studied using Fast Lagrangian Analysis of Continua in Three Dimensions (FLAC3D) numerical simulation. Then, a mechanical model of water-dripping rock using a bolt support was established, and further, technical countermeasures to control the deformation of the roadway with a bolt and cable support are proposed. The results show that the erosion of the water dripping on the roadway was substantial and showed notable changes over time during roadway excavation and mining work. These effects caused the road to heave slightly, but it tended to be stable during roadway excavation. Moreover, the erosion of the roof and two ribs increased exponentially, and the floor heave increased with significant displacement oscillation during mining. The anchoring length of bolts and the rock weakening from water dripping had noticeable effects on the surrounding rocks of the roadway. The logical parameters of the bolt spacing and tightening force (the bolt line spacing was 0.7–0.9 m and the tightening force exceeded 40 kN) of the bolt supports were studied and optimized. Finally, a support scheme for water dripping on the roadway at the Tuanbai coal mine is proposed. The observation data regarding the deformation of the surrounding rocks, monitoring of bolt and cable stress, endoscopy results of roof failure, and roof bed separation monitoring were used to verify the reasonableness of the scheme and ensure the requirements for support were met. The study results can serve as a reference regarding the support for water dripping on a roadway under similar conditions.

Keywords: contiguous seams; water-dripping roadway; roadway deformation; bolt support

---

1. Introduction

Mining activities can change the structure and stress-state of coal and rocks, resulting in their deformation and destruction. Internal fractures of the rock masses develop and expand, eventually leading to the deformation of the roadway [1,2]. It is easy, then, to cause roof caving, rib spalling, and floor heaving or induce gas outburst and rock burst, possibly resulting in fatalities [3–7]. Several on-site photos of roadway instability are shown in Figure 1 [7].

Figure 1. Field photos after severe roadway damage.

Contiguous seams usually refer to coal seams having relatively small vertical distance and great shared effect when the coal seams are mined. The roof structure and the stress distribution of the lower coal seam in contiguous seams are affected significantly by upper coal mining, leading to the development of fractures that worsen gradually [8]. The main effect is that the strength of the surrounding rocks reduces, and the deformation of the roadway increases significantly [9]. When an aquifer occurs near the upper coal area, the water of the upper goaf affected by the aquifer failure can permeate into rock masses along the roof fractures of the lower coal seam. It weakens the strength of the surrounding rocks and aggravates the deformation failure of the roadway, thereby noticeably reducing the roadway stability further [10,11]. Hence, water dripping below a contiguous seam onto the roadway below results in the need to strengthen the roadway.

Bolt support, which has been widely used in roadway support, can form a stable bearing structure with the interaction between the bolt and the rock in the corresponding zone. It can change the stress state of surrounding rocks and increase confining pressure, thereby improving the bearing capacity of the surrounding rocks and restraining the development of the fractured zone [12]. Many studies show that bolt support can markedly enhance the compressive strength of the surrounding rocks of a roadway [13].

Many studies have been conducted regarding the evolution of mining-induced stress and deformation failure of the surrounding rocks for a roadway below a contiguous seam. Zhao [14] and Yang et al. [15] studied the evolution laws of deformation failure and mining-induced stress of the surrounding rocks for roadways, with or without water immersion during roadway excavation, by numerical simulation. Yang [16] and Wu [17] studied roadway stability below a goaf of a contiguous seam and found that the roadway was supported easily with a large coal pillar from the upper coal seam and a large distance away from the upper coal seam. Dunning et al. [18] showed water could change the structure and composition of rocks, resulting in the reduction of strength, cohesion, and the friction coefficient. Hawkins and McConnell [19] carried out uniaxial compression strength tests of 35 kinds of sandstone under dry and saturated conditions and found that water pressure had little effect on the rock strength. Chugh and Missavage [20] showed the uniaxial compression strength of rock specimens reduced significantly under the conditions of water and 100% humidity. Ren [21] studied the displacement field, stress field, and seepage field affected by water, utilizing several methods, such as numerical simulation, theoretical analysis, and field observation; moreover, Ren showed that the support of a composite structure with pre-stressed steel bolts and rocks can achieve a better active securing effect. Zhang et al. [22] suggested some support schemes based on the differences in the distance between roof rock and coal seam to be conscious of safe roadway excavation and mining work.

Currently, regarding the theories for and experiences in the control of roof strata with single coal seam mining or long-distance coal seam mining, considerable studies have been conducted; however, they cannot explain adequately the strata pressure appearance and mechanisms underlying the conditions of contiguous seam mining. Contiguous seam mining, stress, and deformation of the surrounding rocks for a roadway are often studied, but the stability of the surrounding rocks and the corresponding support scheme for the roadway affected by dripping water are rarely discussed. Hence, in view of the geological and mining conditions (dripping water and contiguous seams), such as the deformation of the surrounding rocks by the water dripping on a roadway below a contiguous seam goaf, the mechanical model of the surrounding rocks with bolt support is established here. In this paper, the effect of bolt anchorage length and water-dripping erosion on rock strength is studied. The combined support scheme of bolt and cable is suggested, and two important technical parameters of bolt support (bolt line spacing and tightening force) are examined in detail. Furthermore, feasible measures for support to control this geological condition are suggested. The study results not only can provide a theoretical basis and technical guidance for solving the issues of support and control of the surrounding rocks for water dripping on a roadway below a contiguous seam goaf, but also can serve as a reference for the application of roadway support under similar geological conditions; thus, they are worthy of further research.

2. Numerical Analysis of Deformation for a Water-Dripping Roadway

2.1. Engineering Overviews of Tuanbai Coal Mine

Coal seam no. 11 of the Tuanbai coal mine in the Shanxi province is below coal seam no. 10. The vertical distance between the two seams is 40 m, and they are contiguous seams. Due to the mining activities and an aquifer fracture, a lot of water accumulates in the goaf after the no. 10 coal seam mining exit, leading to floor strata fracture and conspicuously affecting the roof stability of the working area in the no. 11 coal seam.

Panel 11-101, which is the testing area for the no. 11 coal seam, is below the goaf of the no. 10 coal seam and is 5.4 m away from the upper goaf. The mining depth and height of Panel 11-101 are 300–357 m and 3.1–3.3 m, respectively. Regarding the geologic column shown in Figure 2, hard siltstone is between the two coal seams and is not deformed, expanded or affected by water.

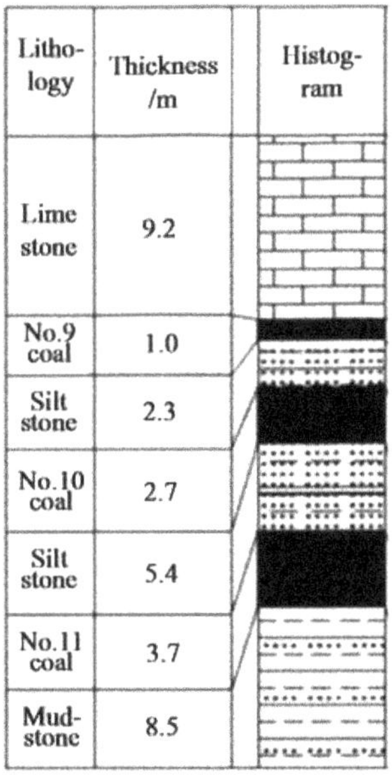

Litho-logy	Thickness /m	Histog-ram
Lime stone	9.2	
No.9 coal	1.0	
Silt stone	2.3	
No.10 coal	2.7	
Silt stone	5.4	
No.11 coal	3.7	
Mud-stone	8.5	

Figure 2. Geologic column of Panel 11-101.

2.2. Numerical Model Establishment

The numerical simulation methods, which are widely used in mining engineering, can be used to effectively solve complex engineering problems [23]. The Fast Lagrangian Analysis of Continua in Three Dimensions (FLAC3D) software can simulate the stress field and deformation field of a roadway with complex conditions during the process of excavation and has achieved many research results. FLAC3D was developed by the ITASCA consulting group of the United States of America (U.S.A.) and has become one of the most important numerical methods for rock and soil mechanics calculation. It can simulate the mechanical properties of three-dimensional soil, rock mass, and other geological materials, especially the plastic rheological properties when they reach their yield limit. It is also suitable for the analysis of progressive failure and large deformation. Moreover, it consists of 11 constitutive models of elastoplastic materials, including static, dynamic, creep, seepage, and temperature calculation models. Hence, it is widely used in many fields like slope stability evaluation, support design and evaluation, tunnel engineering, and mine engineering [24]. Therefore, it can meet the needs of this study. This software provides an idea and method for evaluating the stability and the optimization of a support system for the roadway [25]. Based on the geological conditions of Panel 11-101 of the Tuanbai coal mine, a FLAC3D numerical model was established concerning the goaf from the upper coal seam and a roadway for the lower coal seam experiencing a water-dripping condition.

To meet the requirements of the study in this paper, the thicknesses of the roof–floor strata of Panel 11-101 were adjusted. Considering the boundary effect, a simulation model with dimensions of 200 m (length) × 50 m (width) × 100 m (height) was established, as shown in Figure 3. Herein, following the upper coal seam of the no. 10 mines, the size of the goaf is 150 m (dip length) × 50 m (strike width) × 100 m (height); the size of the roadway in the lower coal seam, with a mining depth of 350 m and a distance from the upper coal seam of 40 m, is 4.3 m (width) × 2.7 m (height).

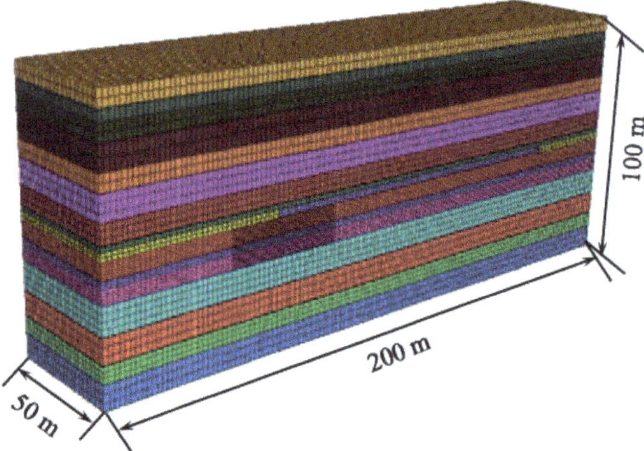

Figure 3. Numerical calculation model.

The mechanical properties assigned to the model were conventionally derived from laboratory testing programs [26]. Table 1 lists the parameters of the main rocks derived from the geological report for Panel 11-101. The Mohr–Coulomb model was used to simulate the failure of coal and rock masses. Considering the water accumulation in the goaf of the upper coal seam and its weakening effect on the rock strength, the double yield model was used to simulate the goaf of the upper coal seam.

Table 1. Mechanical properties of rock masses.

Lithology	Bulk/GP	Shear/GPa	Cohesive/MPa	Internal Angle/°	Tensile Strength/MPa
Coal seam	0.92	0.83	1.68	20	1.340
Limestone	22.6	11.1	4.72	32	7
Siltstone	6.11	4.84	2.30	30	4.47
Fine sandstone	5.87	3.38	3.26	28	3.19
Aluminous mudstone	3.06	1.89	1.80	26	3.89
Mudstone	2.17	1.21	1.30	25	2.78

The full displacement constraint boundary, in this numerical model, was used for the bottom, while the horizontal displacement constraint and uncontrolled vertical displacement were used at the left and right sides; moreover, the top boundary was unstructured. According to the calculation method of reference written by Wang et al. [7], the failed simulation of the overlying strata in this model was presumed to be the vertical stress of 7.50 MPa acting on the top boundary. Based on the geological report on Panel 11-101, the value of the horizontal stress applied to the two sides of the model with the trapezoidal distribution was half of the value of the vertical stress.

2.3. Numerical Scheme

(1) Stage of Roadway Excavation

The roadway in the no. 11 coal seam was excavated after the upper working phase of the no. 10 mine exit, and the total excavation length was set to be 200 m. A section of roadway at the starting position of the excavation and an arrangement of four monitoring points were selected—located in the roof, two ribs, and the floor—to monitor the deformation of the roadway. Hence, during the roadway excavation, the distance between the heading face and the surveyed section rose gradually.

(2) Stage of Working Face Mining

Panel 11-101 was mined after the completion of the roadway excavation. A section of roadway 50 m from the opening was selected, and four monitoring points similar to those in the previous step were arranged around the stage of the roadway excavation to monitor the deformation of the roadway affected by the mining activities. Thus, during mining, the distance between the work area and the monitoring section in the survey range decreased gradually.

2.4. Simulation Results of Roadway Deformation

Due to the roadway excavation or the mining phase, the roadway became deformed. Hence, the roadway deformation showed a noteworthy effect during the roadway excavation and the mining phases. Moreover, it presented significant stage characteristics, as shown in Figure 4.

Figure 4a,b show that the deformation variations of the roof subsidence and the two ribs were similar. The notably affected zone of the roadway mainly occurred at the initial stage of the roadway excavation and the stage during mining. The deformation variations rose sharply at the initial stage of roadway excavation, with the maximum values at 0.21 m and 0.11 m, represented by Stage I. Continuing with the roadway excavation, the subsidence values had little change and the subsidence rate was smaller (Stage II). Following the commencement of mining during the working phase, the deformations increase notably (Stage III); however, the subsidence rose with ladder-shape when the monitoring point was more distant from the work area, and then it was affected sharply by significant abutment stress.

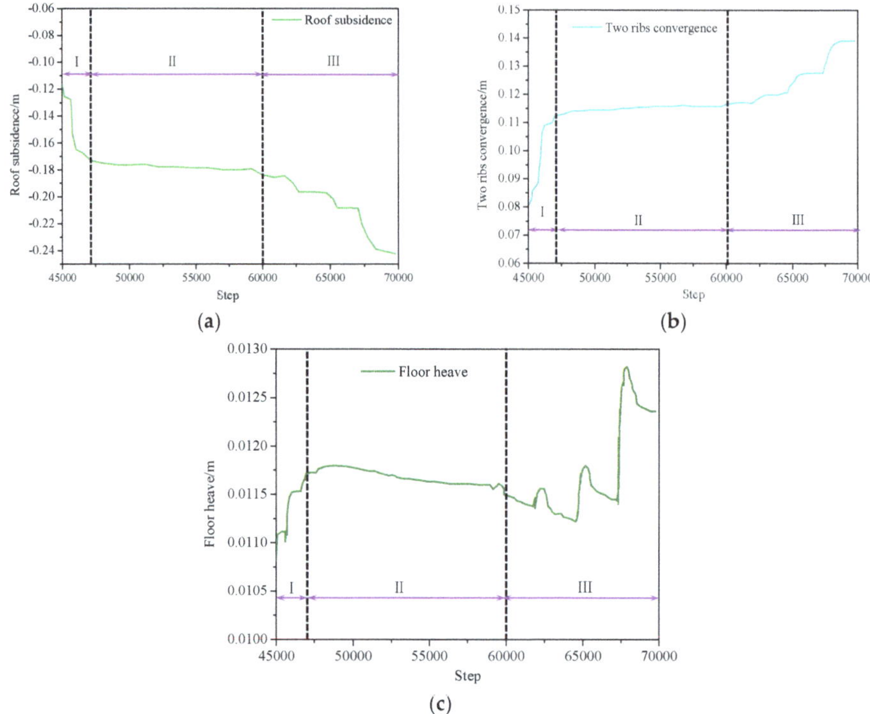

Figure 4. Variation curve of roadway deformation. (**a**) Roof subsidence; (**b**) Two ribs' convergence; (**c**) Floor heave deformation.

Figure 4c displays that the deformation variation of the floor heave found in Stage I was similar to that of the roof and two ribs' distortion. However, the variations of Stage II and Stage III were different, when comparing the warp of the floor heave, roof, and two ribs. During roadway excavation within Stage II (see Figure 4c), the deformation lessened slightly due to abutment stress transfers and reduction. During the mining phase of Stage III (see Figure 4c), the floor heave increased with significant displacement oscillation, and then dropped sharply when the surveyed section was already into the goaf of Panel 11-101.

Thus, it can be concluded that in the initial stage of the roadway excavation (corresponding to Stage I in Figure 4), the deformation of the roadway (roof subsidence, two ribs converging, and floor heave, for example) rose sharply and the increase rate was large. During the middle and late stages of the roadway excavation (Stage II in Figure 4), the deformation varied slightly; herein, the deformation of the roof and two ribs increased, while that of floor heave decreased. Stage III in Figure 4, shows that the deformation rose again as a whole when affected by mining activities; however, the deformation of the roof and two ribs showed ladder-shaped growth, while that of the floor heave presented substantial displacement oscillation.

3. Mechanical Analysis of Water-Dripping Rock with Bolt Support

Garnered from the simulation analysis in Section 2, the roadway of Panel 11-101 had large deformations and could not be used normally; hence, bolt support was used for roadway maintenance. Bolt support can control the deformation and the failure of the surrounding rocks and maintain the integrity to the maximum in the anchor zone, thereby enhancing the strength and stability of the surrounding rocks [27,28].

3.1. Mechanical Model of Water-Dripping Rock with Bolt Support

Figure 5 shows a mechanical model of fractured rock with a bolt support. The maximum principal stress σ_1 and minimum principal stress σ_3 wrought on the rock, and the normal stress and shear stress that take effect on the fracture plane (Plane AB), are set to σ and τ in this model. When a bolt support goes through Plane AB, the relationship between the shear stress and shear displacement can be expressed by Equation (1) as follows [29]:

$$\tau = Ku \tag{1}$$

where τ is the shear stress, MPa; u is the shear displacement, m; and K is the interface stiffness factor of the fracture Plane AB without water dripping obtained by a rock test.

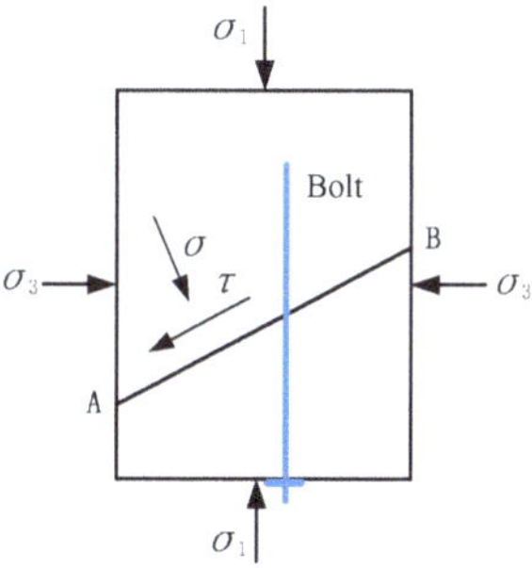

Figure 5. Mechanical model of fracture rock with bolt support.

When the rocks encounter water, their strength weakens. Hence, erosion factors of a rock fracture are marked as D^m (with water dripping) are defined by Equation (2) as follows:

$$D^m = \frac{K - K'}{K} \tag{2}$$

where K' is the interface stiffness factor of the fracture plane with water dripping. Herein, D^m is in the range of 0–1, and $D^m = 0$ means that the rock is affected by water, while $D^m = 1$ indicates that the rock fractures completely. Thus, the relationship between shear stress and shear displacement with water-dripping can be expressed by Equation (3).

$$\tau = (1 - D^m)Ku \tag{3}$$

When the shear stress of the interface between the bolt—coated with epoxy resin to prevent corrosion—and the surrounding rock is less than its shear strength, the interface is in a completely elastic state without relative displacement. Therefore, the bolt and the surrounding rock need to meet the deformation coordination conditions. Figure 6 shows the stress analysis of the bolt.

According to Hooke's law, the relationship between the displacement of the bolt $u(z)$ and the axial stress of the bolt $P(z)$ at the depth of z can be expressed by Equation (4) as follows:

$$\frac{du(z)}{dz} = -\frac{4P(z)}{\pi D^2 E_a} \tag{4}$$

where E_a is the equivalent elastic modulus of the bolt-coating, $E_a = \frac{E_m(D^2 - d^2) + E_b d^2}{D^2}$, MPa; E_m and E_b are elastic modulus of the bolt coating and bolt, MPa; D and d are the diameters of the bolt coating and bolt, respectively, m; and L_b is the anchorage length, m.

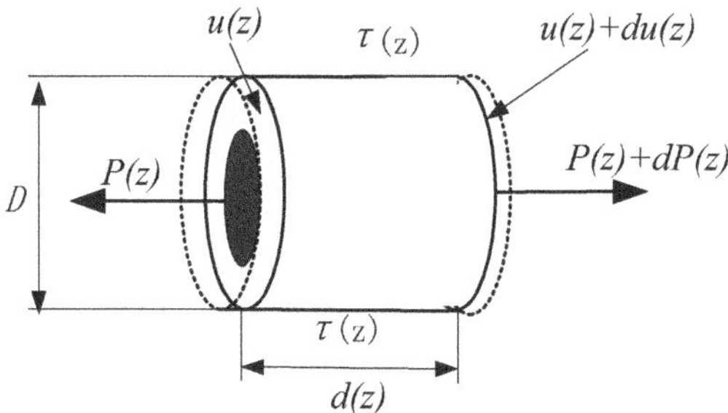

Figure 6. Stress analysis of bolt-coating body.

Based on the static equilibrium of a unit on the bolt coating, the shear stress of the bolt $\tau(z)$ with the depth of z can be expressed by Equation (5).

$$\frac{dP(z)}{dz} = -\pi D \tau(z) \tag{5}$$

According to the boundary conditions of the bolt coating in Equation (6), the shear stress of the bolt-$\tau(z)$ with the depth of z is obtained via Equation (7) as follows:

$$\begin{cases} \left. \dfrac{E_a D^2 \pi du(z)}{4dz} \right|_{z=0} = -P \\[3em] \left. \dfrac{E_a D^2 \pi du(z)}{4dz} \right|_{z=L_b} = 0 \end{cases} \tag{6}$$

$$\tau(z) = \frac{\lambda \cosh[\lambda(L_b - z)]}{\pi D \sinh(\lambda L_b)} P \tag{7}$$

where λ is a constant, $\lambda = \sqrt{\dfrac{4(1-D^m)K_1}{DE_a}}$

When the interface between the bolt linkage is in elastoplasticity, the shear stress of the bolt truss at $z = 0$ reaches its ultimate shear strength. Thus, the ultimate pulling force of the bolt fixture P_u is obtained as shown in Equation (8).

$$P_u = \frac{\pi D(1 - D^m)\tau \tanh(\lambda L_b)}{\lambda} \tag{8}$$

Thus, from Equation (8), the ultimate pulling force of the bolt P_u had a close relation to the anchoring length L_b, water-dripping erosion factors D^m, and other factors.

When the fractured rock supported by a bolt slipped along the fracture Plane AB, the upper part of the rock above the fracture plane was selected and analyzed. Figure 7 shows the mechanical analysis for the slipping of the fractured rock supported by a bolt.

Regarding the bolt, the shear stress along the fracture plane affected by the bolt was broken down into the shear stress τ_N and the tensile stress τ_P, as expressed in Equation (9).

$$\begin{cases} \tau_N = (\tau - c - \sigma \tan \varphi) \cos \alpha \\ \tau_P = (\tau - c - \sigma \tan \varphi) \sin \alpha \end{cases} \tag{9}$$

where c is the cohesive of the fracture Plane AB, measured in MPa; and φ is the internal angle of the fracture plane, measured in degrees.

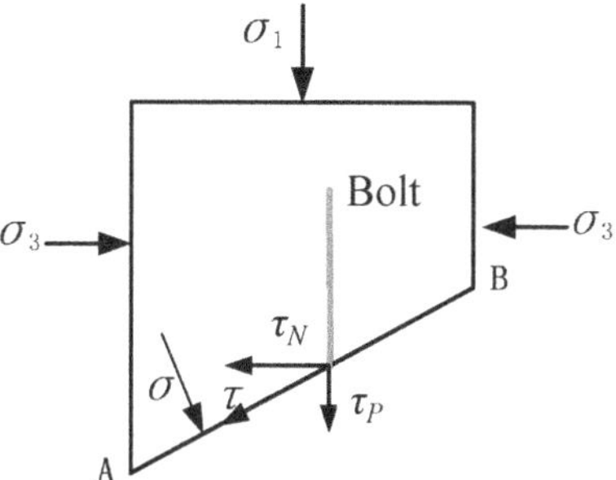

Figure 7. Mechanical model with the slipping of fracture rock supported by a bolt.

The stress of the fracture Plane AB, if Plane AB bears the osmotic stress p' affected by dripping water, can be obtained as shown in Equation (10).

$$\begin{cases} \sigma = \frac{1}{2}(\sigma_1 + \sigma_3) + \frac{1}{2}(\sigma_1 - \sigma_3)\cos 2\alpha - p\prime \\ \tau = \frac{1}{2}(\sigma_1 - \sigma_3)\sin 2\alpha \end{cases} \tag{10}$$

Concerning the fracture rock supported by the bolt, there are two failure modes when the bolt fails, for the most part. One is shear failure along the fracture Plane AB, and the other is the slipping failure between the bolt and the rock.

Assuming the shear strength of the bolt body is τ_b, the bolt will be in the limited equilibrium state of shear failure with $\tau_N = \tau_b$. Hence, Equation (11) can be obtained as follows:

$$\sigma_1 = \sigma_3 + \frac{2(c + \sigma_3 \tan\varphi - p'\tan\varphi + \tau_b/\cos\alpha)}{\sin 2\alpha - \cos 2\alpha \tan\varphi - \tan\varphi} \tag{11}$$

The bolt fixture and rock mass will be in the limited equilibrium state of slipping failure with $\tau_P \frac{\pi D^2}{4} = P_u$ if the anchorage interface is in a complete elastic state. Hence, Equation (12) can be obtained as follows:

$$\sigma_1 = \sigma_3 + \frac{2[c + \sigma_3 \tan\varphi - p'\tan\varphi + 4(1 - D^m)\tau_b \tanh(\lambda L_b)/D\lambda \sin\alpha]}{\sin 2\alpha - \cos 2\alpha \tan\varphi - \tan\varphi} \tag{12}$$

Thus, from the previous analysis, the strength of the fracture rock supported by the bolt drops due to deterioration by dripping water and the reduction of the anchoring bolt length.

3.2. Effect of Anchorage Length and Water-Dripping Erosion Factor on Rock Strength

3.2.1. Effect of Anchorage Length on Rock Strength

To study the effect of anchorage length on rock strength, different anchorage lengths of a bolt, such as 0.1 m, 0.2 m, 0.5 m, 1.0 m, and 2.0 m, were examined. Equations (11) and (12) show that the main parameters included the following: $\sigma_3 = 2$ MPa, $c = 0.5$ MPa, $\varphi = 22°$, $D = 0.03$ m, $p' = 0.5$ MPa, $\tau_b = 0.6$ MPa, $D^m = 0$, and $\lambda = 2.83$. Hence, the relationship between the maximum principal stress σ_1 and the dip angle of the fracture Plane AB with the different anchorage lengths is shown in Figure 8.

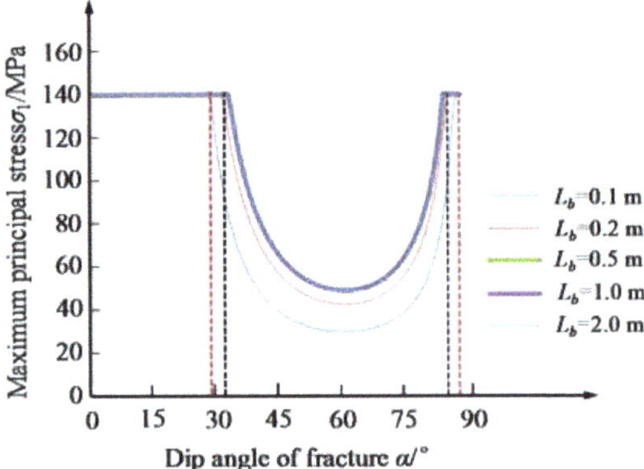

Figure 8. Effect of different anchorage lengths on rock strength.

In Figure 8, the variation curves of rock strength with different anchorage lengths are similar. However, the dip angle of the rock slipping along the failure face AB with different anchorage lengths is different, which indicates that the effects of anchorage lengths on rock strength are different.

The maximum principal stress increased significantly with $L_b \leq 0.5$ m, while having little change within 0.5 m $< L_b \leq 2.0$ m. Moreover, the affected range of the dip angle for the rock slipping along the failure Plane AB decreased with $L_b \leq 0.5$ m, while it had little change within 0.5 m $< L_b \leq 2.0$ m, as shown in Table 2. When $L_b = 0.1$ m, the rock failed along the fracture Plane AB within $29° \leq \alpha \leq 86°$, and the minimum value was 29.7 MPa, for example; when $L_b = 0.5$ m, the rock failed along the fracture plane within $34° \leq \alpha \leq 83°$, and the minimum value increased to 48.9 MPa.

Table 2. Dip angle of rock slipping along the failure face AB with different anchorage lengths.

Anchorage Length L_b/m	Dip Angle of Rock Slipping along the Failure Face α/°
0.1	$29° \leq \alpha \leq 86°$
0.2	$32° \leq \alpha \leq 84°$
0.5	$34° \leq \alpha \leq 83°$
1	$34° \leq \alpha \leq 83°$
2	$34° \leq \alpha \leq 83°$

Therefore, it can be concluded that there is an optimal value for the anchor length's capacity to support the roadway. When the anchor length is less than its optimal value, the support effect rises noticeably as the anchor length increases, while it is not obvious once the anchor length exceeds the optimal value.

3.2.2. Effect of Water-Dripping Erosion Factor on Rock Strength

To study the effect of the water-dripping erosion factor on rock strength, different water-dripping erosion factors, such as 0, 0.3, 0.6, and 0.9, were studied. Similar to the parameters in Section 3.2.1, some boundaries were added and adjusted, such as $K = 0.6$ GPa/m and $E_a = 10$ GPa. The relationship between the maximum principal stress σ_1 and the water-dripping erosion factor is shown in Figure 9.

Figure 9 shows the variation curves of rock strength with different water-dripping erosion factors are similar; the weakening effect on the rock strength and the affected range of the dip angle for rock slipping along the failure face AB increased markedly as the erosion factors rose. When $D^m = 0$, the rock failed along the fracture Plane AB within $34° \leq \alpha \leq 84°$ (as shown in Table 3) and the minimum value was 49 MPa, for example; when $D^m = 0.3$, the rock failed along the fracture Plane AB within $32° \leq \alpha \leq 84°$, and the minimum value fell to 41.9 MPa; when $D^m = 0.6$, the rock failed along the fracture Plane AB within $30° \leq \alpha \leq 86°$, and the minimum value was 32.9 MPa; when $D^m = 0.9$, the rock failed along the fracture Plane AB within $27° \leq \alpha \leq 87°$, and the minimum value fell to 19 MPa. The rock strength decreased from 41.9 MPa to 19 MPa as the erosion factor rose, making the weakening effect obvious.

Hence, it can be concluded that the rock strength can be enhanced by increasing the anchor length or by decreasing the water-dripping erosion factor, thereby enhancing the support for the roadway.

Table 3. Dip angle of rock slipping along the failure face with different water-dripping erosion factors.

Water-Dripping Erosion Factors D^m	Dip Angle of Rock Slipping along the Failure Face $\alpha/°$
0	$34° \leq \alpha \leq 84°$
0.3	$32° \leq \alpha \leq 84°$
0.6	$30° \leq \alpha \leq 86°$
0.9	$27° \leq \alpha \leq 87°$

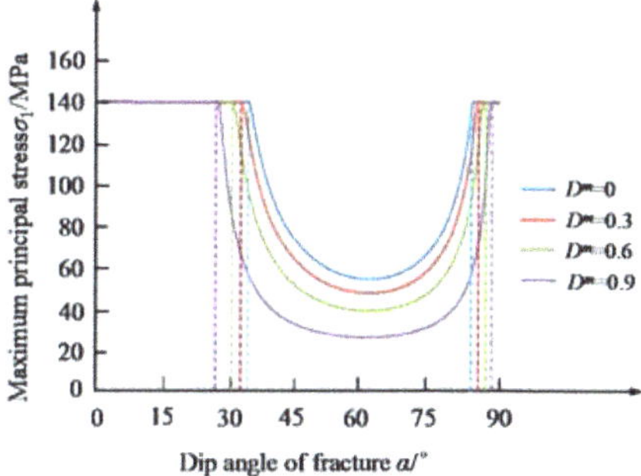

Figure 9. Effect of different water-dripping erosion factors on rock strength.

4. Numerical Analysis of Deformation of a Water-Dripping Roadway with Bolt Support

4.1. Results of Numerical Simulation

Deformation of a roadway with water dripping onto it is sizeable (see Section 2); however, bolt support can effectively restrain deformation (see Section 3). Hence, based on the numerical simulation used in Section 2, an alternate model of a roadway with bolt support is established and used to compare with the model without bolt support. Regarding the model with bolt support, the support parameters mainly included a bolt diameter of 18 mm, bolt length of 2200 mm, and bolt spacing of 900 × 900 mm.

To study the effect of the bolt support on the roadway, the index of deformation and failure of roadway, such as roof subsidence, maximum principal stress, and plasticity zone, were studied and discussed. Figures 10–12 depict the nephogram regarding these three indices of roadway, with or without bolt support.

Figures 10–12 display that the peak values of the three indices—roof subsidence, maximum principal stress, and plasticity zone distribution—without bolt support were larger than those with bolt support. The peak values of roof subsidence with or without bolt support were 48.37 mm and 60.63 mm, respectively, which reduced the amount of subsidence by 20% with bolt support, for example; the maximum principal stress without bolt support was 0.062 MPa of tensile stress, while that with bolt support was −0.0063 MPa of compressive stress. Moreover, the range of the plasticity zone without bolt support was substantial, mainly having tensile and shear failures, whereas that with bolt support decreased noticeably with mainly shear failure. The results indicate that bolt support can significantly reduce the roof subsidence and plasticity range of a roadway; meanwhile, it greatly drops the maximum principal stress and changes the stress state from tensile stress to compressive stress, which can enhance the rock strength markedly. Thus, bolt support has a good effect on controlling the deformation and failure of a roadway, thereby maintaining its stability.

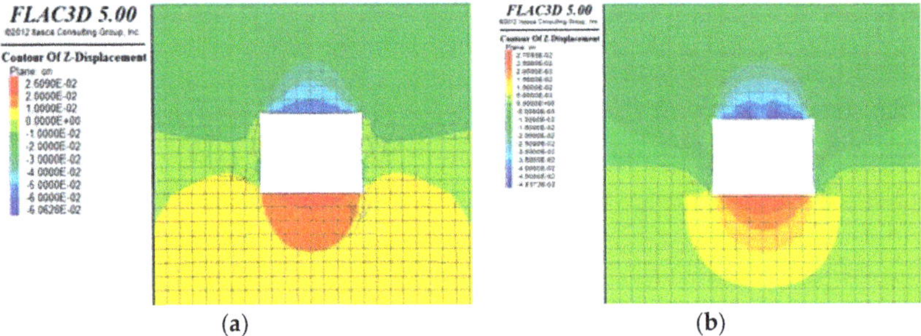

Figure 10. Nephogram of the roof subsidence with or without bolt support. (**a**) Without bolt support; (**b**) With bolt support.

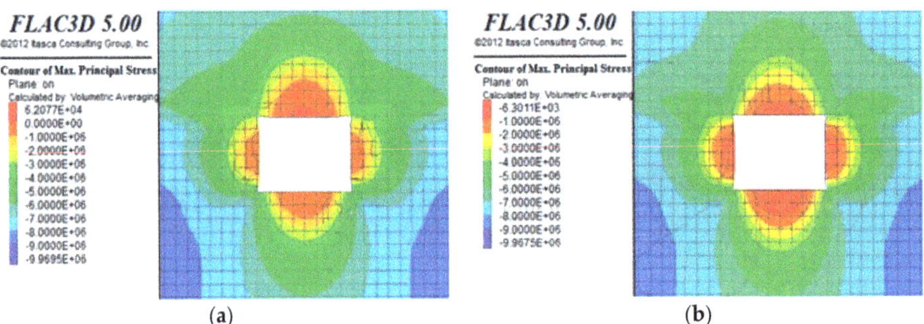

Figure 11. Nephogram of the maximum principal stress with or without bolt support. (**a**) Without bolt support; (**b**) With bolt support.

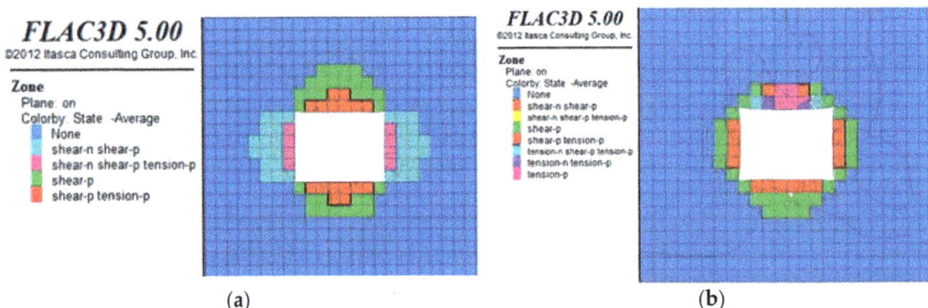

Figure 12. Nephogram of the plastic zone with or without bolt support. (**a**) Without bolt support; (**b**) With bolt support.

4.2. Effect of Bolt Parameters on Roadway Stability

According to the reference written by Kang et al. (2008), the line space and tightening force of bolts are the important technical parameters of bolt support for the roadway. The line spacing of bolts directly determines the range of the compressive stress of the surrounding rocks. The effect of different line spacing and different tightening force on the stability of a roadway was studied using the control variable method, which only changed line spacing or tightening force.

4.2.1. Effect of Line Space of Bolt on Roadway Stability

To study the effect of the spacing of the bolts on the roadway stability, the roof subsidence and the plasticity zone are displayed in Figures 13 and 14. Herein, the tightening force of a bolt was set to 40 kN and remained unchanged, and the different spacing of the bolts, such as 0.6 m, 0.7 m, 0.8 m, 0.9 m and 1.0 m, were the variable.

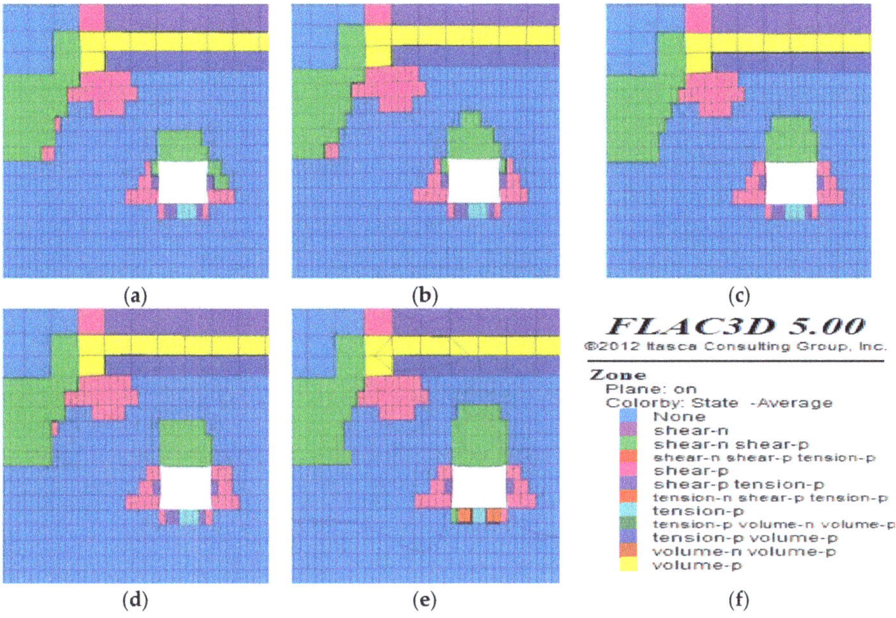

Figure 13. Range of the plasticity zone of the roadway with different line spacing of the bolts. (**a**) 0.6 m; (**b**) 0.7 m; (**c**) 0.8 m; (**d**) 0.9 m; (**e**) 1.0 m; (**f**) Rock failure state.

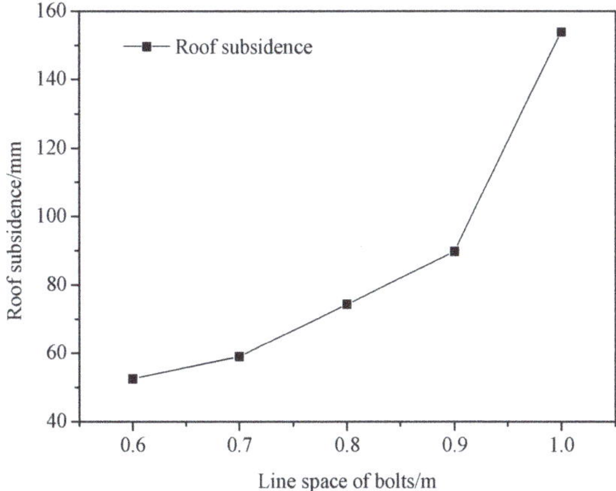

Figure 14. Variation of the roof subsidence for the roadway with different line spacing of bolts.

Figures 13 and 14, with the spacing of the bolts increasing from 0.6 m to 1.0 m, show the range of the plasticity zone, while the roof subsidence of the roof rock increased, indicating that the support effect of the bolt weakened. However, the two parameters increased gradually as the spacing escalated from 0.6 m to 0.9 m, with a significant rise in the spacing of 1.0 m. Herein, the roof subsidence rose slightly from 52.56 mm to 89.74 mm as the line spacing increased from 0.6 m to 0.9 m, and it markedly rose to 153.85 mm with the spacing of 1.0 m, with an increase of 63% compared to the spacing at 0.9 m. Thus, the closer spacing/greater frequency of the bolts has better support for the roadway; however, too little space between the bolts will significantly raise the cost.

Thus, from the analysis of the roof subsidence and plasticity range, the spacing of the bolts should be selected between 0.7 m and 0.9 m, according to this simulation.

4.2.2. Effect of Tightening Force of Bolt on Roadway Stability

To study the effect of the tightening force of the bolts on roadway stability, the roof subsidence and the plasticity zone are displayed in Figures 15 and 16. Herein, the spacing of bolts was set to 0.9 m and remained unchanged, and the different tightening force of the bolts, such as 20 kN, 40 kN, 60 kN and 80 kN, were the variable.

Figures 15 and 16, with the tightening force of the bolts increasing from 20 kN to 80 kN, show the range of the plasticity zone and the subsidence of the roof rock, which indicate that the bolt enhances the support effect. However, the two parameters dropped gradually with the tightening force of 20 kN and 40 kN, while they decreased noticeably with the tightening force of 60 kN and 80 kN. Herein, the roof subsidence dropped slightly from 95.15 mm (assuming the tightening force was 20 kN) to 88.98 mm (if the tightening force was 40 kN), and it markedly decreased to 55.28 mm with the tightening force of 80 kN. Hence, the stronger tightening force, exceeding 40 kN, had better support for the roadway.

Thus, the line spacing and the tightening force of the bolts have significant effect on the roadway support. Moreover, it can be concluded that the reasonable parameters of bolts with the spacing of 0.7–0.9 m and a tightening force exceeding 40 kN for roadway support are more effective.

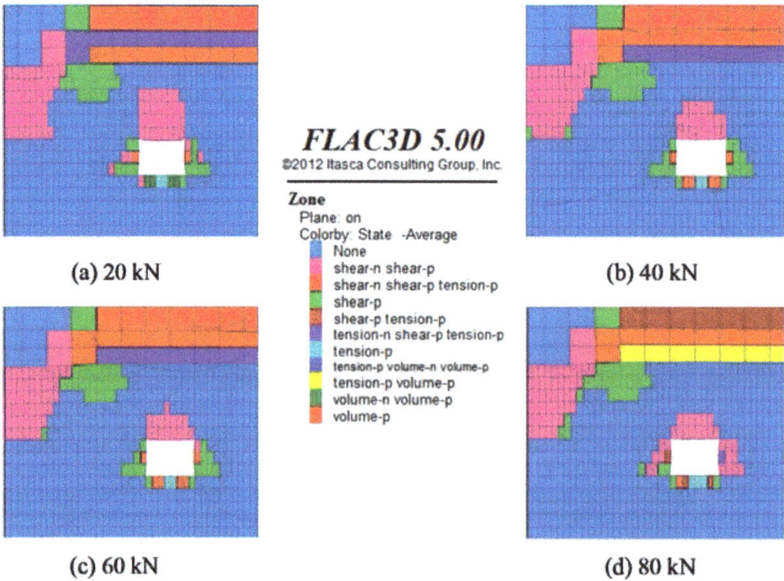

Figure 15. Range of the plasticity zone of the roadway with different tightening forces.

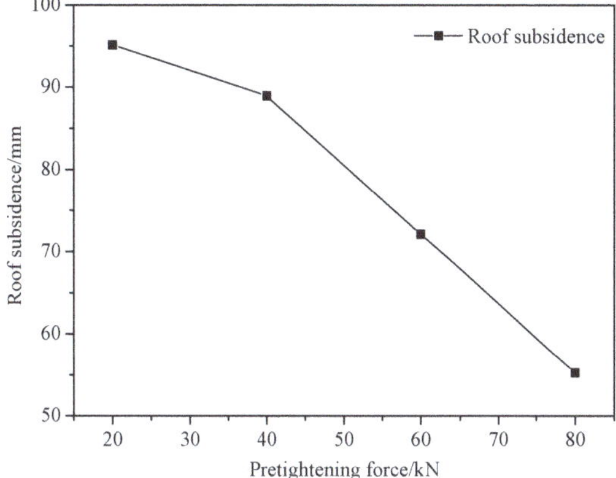

Figure 16. Variation of the roof subsidence for the roadway with different tightening forces.

5. Field Observations About Support Effect of Roadway with Dripping Water

5.1. Support Scheme

According to the geological report for Panel 11-101, a scheme of bolt and cable support for the roadway was carried out, as shown in Figure 17. Based on the previous studies, the mechanism of a cable support was similar to that of a bolt support; however, the support effect of the cable was more obvious than that of the bolt support. Hence, the bolt and cable support for the roadway could better control the deformation of the surrounding rocks [6,13].

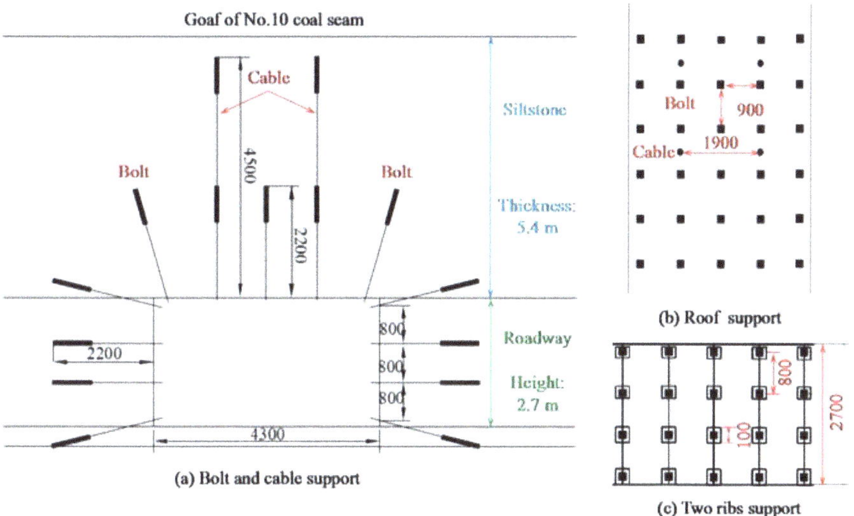

Figure 17. Scheme of bolt and cable support.

It can be divided into bolt support and cable support in this analysis, and the main parameters are as follows:

(1) Bolt Support System. The system used a high strength bolt made of screw steel, measuring $\Phi 18 \times 2200$ mm, and used a tightening force exceeding 40 kN. The anchoring agent marked as FSCK2340 had waterproof resin properties. The line and row spacing of the bolts were 800 mm and 900 mm, respectively.

(2) Cable Support System. The system used a cage-type cable, measuring $\Phi\ 17.78 \times 4500$ mm, with a tray size of $100 \times 100 \times 10$ mm. The anchoring agent marked as FSCK2340 $\times$ 4 was a waterproof resin.

5.2. Analysis of Field Observation

A section of the roadway was selected as the starting position for the excavation and the assembly of four monitoring points located in the roof, two ribs, and floor to monitor the deformation of the roadway. Moreover, the roof development and bolt and cable forces are discussed. The observation data are shown in Figures 18–20.

5.2.1. Observation with Roadway Excavating

(1) Deformation Monitoring of Roadway

Table 4 and Figure 18 show that, with roadway excavation, the variation of the roadway deformation, such as roof subsidence, two ribs' convergence, and floor heave, had similar growth tendencies. When the excavating distance exceeded 100 m, the roof subsidence and two ribs' convergence tended to be stable, while the floor heave also tended to be stable when the mining distance exceeded 200 m. Moreover, the roof subsidence and two ribs' convergence had small peak values, which indicated that the bolt and cable support effectively restrained the deformation of the roadway with good support; however, the floor heave was relatively large due to the significant effect of the dripping water.

Table 4. Deformation variation of the roadway with roadway excavating.

Excavating Distance/m	Roof Subsidence/mm	Floor Heave/mm	Two Ribs' Convergence/mm
1.5	0		0
9	0		3
40.5	1		4
49.5	1		4
57	1		4
72.5	2	0	6
86	2	1	16
101.5	4	1	16
105.3	4	28	16
143	4	47	16
176	4	52	16
209	5	54	16
240	5	57	16
269	5	57	16
342	5	57	16

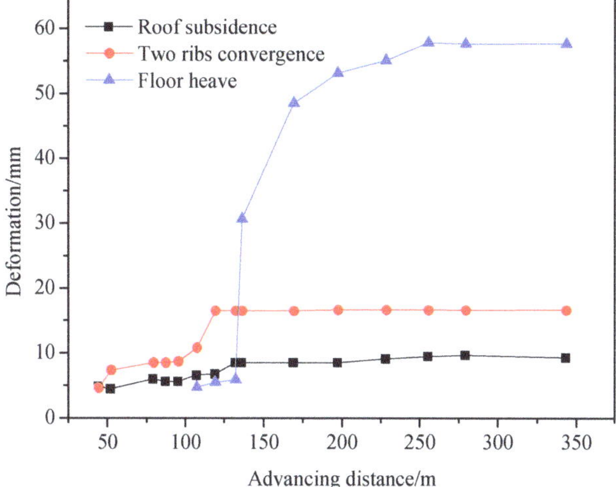

Figure 18. Deformation variation of the roadway with roadway excavating.

(2) Stress Monitoring of Bolt and Cable

Table 5 and Figure 19 allow for the conclusion that the stresses on the bolts and cables were small and showed little change during the roadway excavation. The stresses on the roof bolt, right rib bolt, and roof cable remained unchanged with tightening forces of 48 kN, 39 kN, and 68 kN, respectively. Moreover, the stress on the left rib bolt increased by 8 kN, but the value was still small. This indicates that the roadway was in a low stress state and shows the support of the bolt plus the cable is more effective.

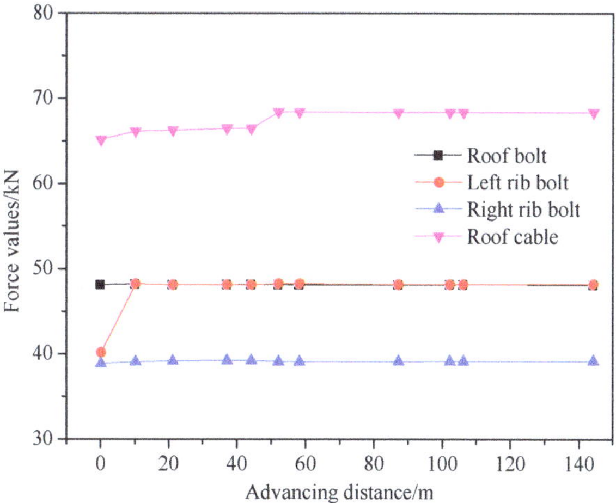

Figure 19. Stress variation of the bolt and cable with the roadway excavation.

Table 5. Stress variation of the bolt and cable with the roadway excavation.

Excavating Distance/m	Roof Bolt/kN	Right Rib Bolt/kN	Left Rib Bolt/kN	Roof Cable/kN
1.5	48	39	40	65
9	48	39	48	66
40.5	48	39	48	66
49.5	48	39	48	66
57	48	39	48	66
72.5	48	39	48	68
86	48	39	48	68
101.5	48	39	48	68
105.3	48	39	48	68
143	48	39	48	68
176	48	39	48	68
209	48	39	48	68
240	48	39	48	68
269	48	39	48	68
342	48	39	48	68

(3)   Endoscopy Results of Roof Failure

A drill hole was set up on the roadway roof and the rock failure with different depths depicted by an electronic drilling peep instrument, as shown in Figure 20.

Figure 20 shows that there was no severe damage to the roof except that a vertical crack at the depth of 2 m occurred. The drill hole was basically intact, and there was no bed separation observed in the roof rocks. This shows that the surrounding rocks of the roadway formed an effective load-bearing structure with the support of the bolt and cable, and the support was better.

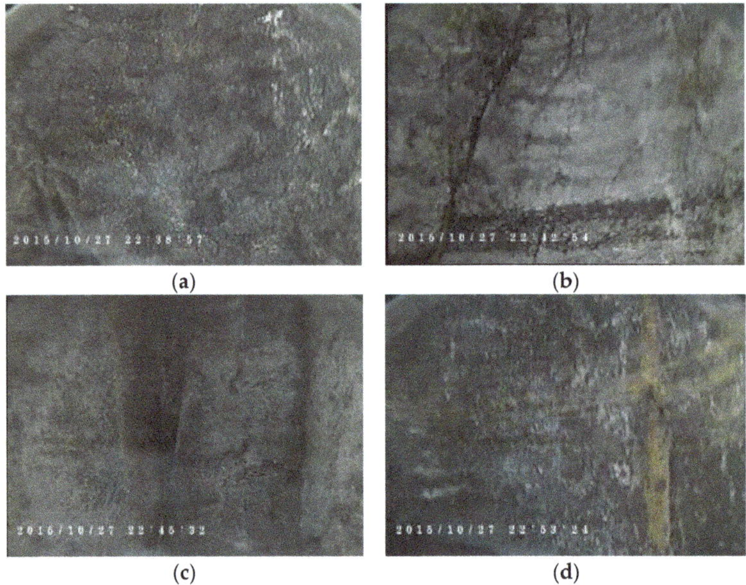

(a)   (b)   (c)   (d)

Figure 20. Rock failure with different depths depicted by electronic drilling peep instrument. (**a**) Drill hole depth of 1 m; (**b**) Drill hole depth of 2 m; (**c**) Drill hole depth of 3.5 m; and (**d**) Drill hole depth of 4.5 m.

(4)   Results of Roof Bed Separation

Two monitoring points were set up at the roadway roof with different heights—2.4 m and 6.0 m—to study the bed separation of the roof, as shown in Table 6 and Figure 21.

Table 6. Variation of the roof bed separation with the roadway excavation.

Excavation Distance/m	Roof Bed Separation Values/mm	
	Monitoring Point of Height 2.4 m	Monitoring Point of Height 6.0 m
20.7	0	0
37	0	2
40.5	0	2
49.5	0	2
57	0	2
66	0	2
86	0	2
101.5	0	2
105.3	0	4
143	0	4
209	0	4
240	0	4
269	0	4
342	0	4

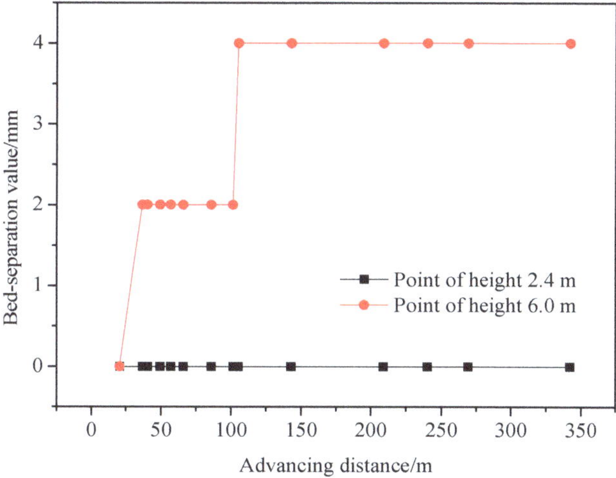

Figure 21. Rock failure with different depths depicted by electronic drilling peep instrument.

Regarding the roadway excavation, the maximum bed separation values of the two monitoring points on the roadway using the bolt and cable support were insignificant at only 0 mm and 4 mm, respectively. This indicates that bed separation basically did not occur in the roof using this support scheme and further shows that a bolt and cable fixture had high efficiency for initial support, laying a good foundation for maintaining the long-term stability of the roadway.

5.2.2. Observation with Working Face Mining

During mining, a section of the roadway with the distance from the open-off cut of 50 m was selected as a monitoring station to survey the roof subsidence and two ribs' convergence by the method of cross-cloth point.

Figure 22, with the mining area approaching the surveyed section, shows the two ribs' convergence rose exponentially and tended to be stable. However, the value with the distance of 10 m was only 5 mm, and it was in a low level. This indicates that the deformation of the two ribs was affected by the mining activities and shows certain periodicity; moreover, it also shows that the support effect of the bolt and cable applied to the roadway was noteworthy.

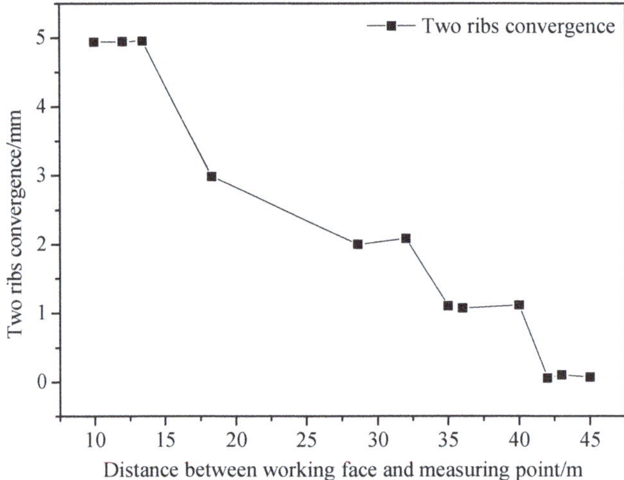

Figure 22. Variation curve of the two ribs' convergence for the roadway section.

Figure 23 shows that the roof subsidence rose slowly as the mining area approached the surveyed section, while the distance of 10 m had a value of 14 mm. This indicates that the abutment stress had no abrupt change, and the roof moved smoothly; moreover, it also shows that the support of the bolt and cable applied to the roadway was better and significant.

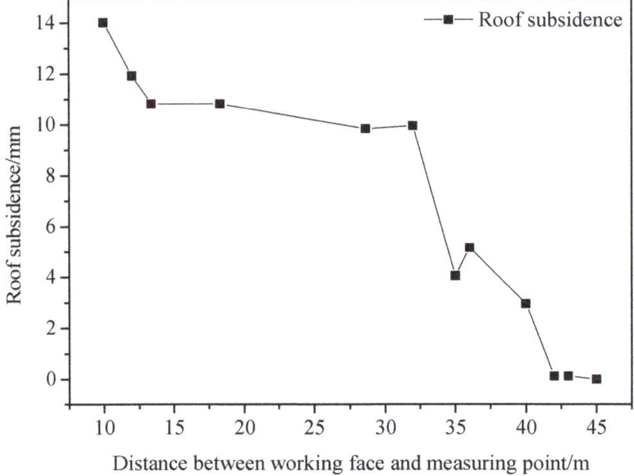

Figure 23. Variation curve of the roof subsidence for the roadway section.

Comparing the prior numerical simulation results to the support and field observation results after installing anchors, the field observation results of the roof subsidence and two rib convergence during the roadway excavation process (9 mm and 16 mm shown in Figure 18 and Table 4) had more significance than that of just the numerical simulation results (180 mm and 115 mm shown in Figure 4). The stresses of the bolts and cables were small with little variation, as shown in Figure 19; it was very close to the installation stress at the initial stage of excavation. The stress of the roof cable rose to 68 kN, which was close to the installation stress of 80–100 kN, for example. Moreover, Figure 21 and Table 6 show the maximum roof bed separation value was only 4 mm, which indicates that bed separation basically does not occur in the roof affected by this support scheme. Thus, when a scheme of bolt and cable support was applied to the roadway of Panel 11-101 in the Tuanbai coal mine, the roadway deformation was small, the bolt/cable stress was low, and the roof fracture was rarely observed. Hence, it can be concluded that the support scheme is reliable and effective to control the deformation of a roadway, thereby ensuring the safety of the roadway support. Moreover, the results of the field observation verify the aforementioned numerical simulation results and deserve further promotion and application.

6. Conclusions

Based on the technical problem of support for a water-dripping roadway below a contiguous seam goaf in the Tuanbai coal mine, this study on the deformation and control countermeasures of the surrounding rocks for water dripping on a roadway was carried out in detail by means of numerical simulation, theoretical analysis, and field observation. The main study contents and the conclusions are obtained as follows:

- The deformation of the water-dripping roadway was noticeable and showed marked effect over time, particularly during the roadway excavation and mining. It presented significant stage characteristics as follows: during the initial stage of roadway excavation, the roadway deformation rose sharply, and then it varied slightly in the middle and late stages of the roadway excavation, at which point the deformation of the roof and two ribs increased, while the floor heave decreased. During the stage of mining, the deformation rose again while affected by mining activities; however, the deformation of the roof and two ribs showed ladder-shaped growth, while that of the floor heave presented a significant displacement oscillation characteristic.

- Based on the large deformation of the surrounding rocks by the dripping water onto the roadway, the mechanical models of surrounding rocks with bolt supports was established; the effect of the bolt anchorage length and water-dripping erosion on the strength of rocks was studied. The authors found that there was an optimal value for the anchor length support for the roadway, and the support effect increased noticeably with an increase in length following the point where the length was less than the value. It was not obvious once the length exceeded the value; moreover, the effect weakened significantly as the water-dripping erosion factor on the bolt rose. Hence, a bolt with a reasonable anchor length and waterproof fixative agent can help to exert the needed support for the surrounding rocks of the water-dripping roadway.

- According to the above conclusions, a combined support scheme of bolt and cable was recommended, and two important technical parameters of bolt support (line spacing and tightening force) were studied in detail. Finally, the reasonable and optimized parameters of spacing and the tightening force of bolts were found to be 0.7–0.9 m and exceeding 40 kN, respectively.

- Combined with the above study results and the actual conditions at the Tuanbai coal mine, feasible measures for bolt and cable support were proposed. A field case regarding the deformation of the surrounding rocks, monitoring of anchor stress, endoscopy results of roof failure, and roof bed-separation monitoring were used to verify the logic of the scheme.

The study results can serve as a reference regarding supports for a roadway with water dripping onto it under similar conditions.

Author Contributions: C.M. and P.W. contributed to publishing this paper. L.J. and C.W. contributed to the revised paper.

Funding: The study was funded by the Key Research and Development Plan of Shandong Province (No. 2017GGX90102), the Science and Technology Plan of the Institution of Higher Learning in Shandong Province (No. J17KA217), and Funding for the Growth Plan of Young Teachers in Shandong.

Conflicts of Interest: The authors declare no conflicts of interest.

References

1. Wang, J.C.; Liu, F.; Wang, L. Sustainable coal mining and mining sciences. *J. China Coal Soc.* **2016**, *41*, 2651–2660.

2. Sui, W.H.; Hang, Y.; Ma, L.X.; Wu, Z.Y.; Zhou, Y.J.; Long, G.Q.; Wei, L.B. Interactions of overburden failure zones due to multiple-seam mining using longwall caving. *Bull. Eng. Geol. Environ.* **2015**, *74*, 1019–1035. [CrossRef]

3. Gu, R.; Ozbay, U. Numerical investigation of unstable rock failure in underground mining condition. *Comput. Geotech.* **2015**, *63*, 171–182. [CrossRef]

4. Coggan, J.; Gao, F.Q.; Stead, D.; Eimo, D. Numerical modelling of the effects of weak immediate roof lithology on coal mine roadway stability. *Int. J. Coal Geol.* **2012**, *90–91*, 100–109. [CrossRef]

5. Liang, Y. Scientific conception of precision coal mining. *J. China Coal Soc.* **2017**, *42*, 1–7.

6. Li, H.; Lin, B.Q.; Hong, Y.D.; Gao, Y.B.; Yang, W.; Liu, T.; Wang, R.; Huang, Z.B. Effects of in-situ stress on the stability of a roadway excavated through a coal seam. *Int. J. Min. Sci. Technol.* **2017**, *27*, 917–927. [CrossRef]

7. Wang, P.; Jiang, L.S.; Jiang, J.Q.; Zheng, P.Q.; Li, W. Strata Behaviors and Rock Burst–Inducing Mechanism under the Coupling Effect of a Hard, Thick Stratum and a Normal Fault. *Int. J. Geomech.* **2018**, *18*, 04017135. [CrossRef]

8. Zhang, N.C. *Stress Redistribution Law of Floor Strata under Chain Pillar and Its Application in Multi-Seam Mining*; China University of Mining and Technology: Xuzhou, China, 2016.

9. Yan, H.; Weng, M.Y.; Feng, R.M.; Li, W.K. Layout and support design of a coal roadway in ultra-close multiple-seams. *J. Cent. South Univ.* **2015**, *22*, 4385–4395. [CrossRef]

10. Liu, R.C.; Li, B.; Jiang, Y.J. A fractal model based on a new governing equation of fluid flow in fractures for characterizing hydraulic properties of rock fracture networks. *Comput. Geotech.* **2016**, *75*, 57–68. [CrossRef]

11. Liu, R.C.; Li, B.; Jiang, Y.J. Critical hydraulic gradient for nonlinear flow through rock fracture networks: The roles of aperture, surface roughness, and number of intersections. *Adv. Water Resour.* **2016**, *88*, 53–65. [CrossRef]

12. Hao, S.Z. *The Bolt Support Mechanism Study of Wake Rock Burst Roadway*; Chongqing University: Chongqing, China, 2016.

13. Li, B. *Research on the Layout Optimization and the Support System in Close Coal Seams*; China University of Mining and Technology: Beijing, China, 2012.

14. Zhao, Y.F. Numerical analysis on roof raining affected to surrounding rock deformation of mine gateway. *Coal Sci. Technol.* **2012**, *40*, 27–34.

15. Yang, R.S.; Li, Y.L.; Guo, D.M.; Yao, L.; Yang, T.M.; Li, T.T. Failure mechanism and control technology of water-immersed roadway in high-stress and soft rock in a deep mine. *Int. J. Min. Sci. Technol.* **2017**, *27*, 245–252. [CrossRef]

16. Yang, Z.W. On influence factors of stability of roadway under many gobs in ultra-close coal seams and supporting countermeasures. *China Coal* **2014**, *40*, 60–64.

17. Wu, A.M.; Zuo, J.P. Reasearch on the effects of multi-mining on activity law of rock strata in close coal seams. *J. Hunan Univ. Sci. Technol. (Nat. Sci. Ed.)* **2009**, *24*, 1–6.

18. Dunning, J.; Douglas, B.; Miller, M.; McDonald, S. The role of the chemical environment in frictional Deformation: Stress corrosion cracking and comminution. *Pure Appl. Geophys.* **1994**, *43*, 151–178. [CrossRef]

19. Hawkins, A.B.; Mcconnell, B.J. Sensitivity of sandstone strength and deformability to changes in moisture content. *Q. J. Eng. Geol. Hydrogeol.* **1992**, *25*, 115–130. [CrossRef]

20. Chugh, Y.P.; Missavage, R.A. Effects of moisture on strata control in coal mines. *Eng. Geol.* **1981**, *17*, 241–255. [CrossRef]

21. Ren, W.F. *Theory Research of Stress Field Displacement Filed and Seepage Filed and Study in Grouting Waterproofing of High Water Pressure Tunnel*; Central South University: Changsha, China, 2013.

22. Zhang, Z.W.; Wu, J.N.; Fan, M.J.; Gao, F.Q.; Zhang, L.H. Research on Roadway Support Technology Under Goaf of Close Coal Seam. *Coal Eng.* **2015**, *47*, 37–40.

23. Jiang, L.S.; Wang, P.; Zhang, P.P.; Zheng, P.P.; Xu, B. Numerical analysis of the effects induced by normal faults and dip angles on rock bursts. *C. R. Mecanique* **2017**, *345*, 690–705. [CrossRef]

24. Itasca Consulting Group. *FLAC3D: Fast Lagrangian Analysis of Continua 3D User's Guide*; Itasca Consulting Group: Minneapolis, MI, USA, 2005.

25. Gao, F.Q.; Stead, D.; Kang, H.P.; Wu, Y.Z. Discrete element modelling of deformation and damage of a roadway driven along an unstable goaf—A case study. *Int. J. Coal Geol.* **2014**, *127*, 100–110. [CrossRef]

26. Wang, W.; Cheng, Y.P.; Wang, H.F.; Liu, H.Y.; Wang, L.; Li, W.; Jiang, J.Y. Fracture failure analysis of hard-thick sandstone roof and its controlling effect on gas emission in underground ultra-thick coal extraction. *Eng. Fail. Anal.* **2015**, *54*, 150–162. [CrossRef]

27. Li, S.C.; Wang, H.T.; Wang, Q.; Jiang, B.; Wang, F.Q.; Guo, N.B.; Liu, W.J.; Ren, Y.X. Failure mechanism of bolting support and high-strength bolt-grouting technology for deep and soft surrounding rock with high stress. *J. Cent. South Univ.* **2016**, *23*, 440–448. [CrossRef]

28. Cheng, L.; Zhang, Y.D.; Ji, M.; Zhang, K.; Zhang, M.L. Experimental studies on the effects of bolt parameters on the bearing characteristics of reinforced rock. *SpringerPlus* **2016**, *5*, 866. [CrossRef] [PubMed]

29. Lister, J.R.; Kerr, R.C. Fluid-mechanical models of crack propagation and their application to magma transport in Dykes. *J. Geophys. Res.* **1991**, *96*, 10049–10077. [CrossRef]

 *processes*

Article

Investigation on Reinforcement and Lapping Effect of Fracture Grouting in Yellow River Embankment

Jian Liu, Zhi Wan, Quanyi Xie, Cong Li, Rui Liu, Mengying Cheng and Bo Han *

School of Civil Engineering, Shandong University, 17922, Jingshi Road, Jinan 250061, China;
lj75@sdu.edu.cn (J.L.); mscj521@163.com (Z.W.); quanyixiesdu@163.com (Q.X.); licon9@126.com (C.L.);
sdhzjclr@163.com (R.L.); cicmying@yahoo.com (M.C.)
* Correspondence: bo.han@sdu.edu.cn; Tel.: +86-531-8839-6007

Received: 23 May 2018; Accepted: 19 June 2018; Published: 22 June 2018

Abstract: Fracture grouting has been a mitigation measure widely used against seepage in the Yellow River Embankment. However, there is currently a lack of systematic investigations studying the anti-seepage effect of the fracture grouting employed in this longest river embankment in China. Therefore, in this work, laboratory and in situ experiments are carried out to investigate the reinforcement effect of fracture grouting in the Jinan section of the Yellow River Embankment. In particular, laboratory tests concentrate on studying the optimum strength improvement for cement–silicate grout by varying the content of backfilled fly ash and bentonite as admixtures. Mechanical strength and Scanning Electron Microscope photographs are investigated for assessing the strength and compactness improvement. Subsequently, based on the obtained optimum admixtures content, in situ grouting tests are carried out in the Jinan section of the Yellow River Embankment to evaluate the reinforcement and lapping effect of fracture grouting veins, where geophysical prospecting and pit prospecting methods are employed. Laboratory results show that, compared with pure cement–silicate grouts, the gelation time of the improved slurry is longer and gelation time increases as fly ash content increases. The optimum mixing proportion of the compound cement–silicate grout is 70% cement, 25% fly ash, and 5% bentonite, and the best volume ratio is 2 for the investigated cases. Geophysical prospecting including the ground penetrating radar and high-density resistivity method can reflect the lapping effect of fracture grouting veins on site. It shows that the grouting material mainly flows along the axial direction of the embankment. The treatment used to generate directional fracture is proved to be effective. The injection hole interval distance is suggested to be 1.2 m, where the lapping effect of the grouting veins is relatively significant. For the investigated cases, the average thickness of the grouting veins is approximately 6.0 cm and the corresponding permeability coefficient is averagely 1.6×10^{-6} cm/s, which meets the anti-seepage criterion in practice.

Keywords: fracture grouting; cement–silicate grout; geophysical prospecting; seepage; Yellow River Embankment

---

1. Introduction

The Yellow River Embankment is the longest river embankment in China, and is mainly constructed of Yellow River alluvial soils, i.e., silt with low surface strength, high porosity, and drastic capillarity. Many scholars [1–3] have investigated the Yellow River Embankment from different points of view. Seepage-induced instabilities such as infiltration, piping, and leakage frequently occurred in the Yellow River Embankment during its long-term operation period, mainly due to its unfavorable soil properties, animal interference, and water level change [4]. Among the potential seepage control methods, fracture grouting has been developing rapidly recently due to its low cost and high efficiency [5]. According to the statistics, more than 3000 dangerous reservoirs and 2000 km of embankments have been reinforced with fracture grouting in the past 20 years [6].

Researchers have carried out a series of studies on grouting materials [7–9], grouting mechanism, and fluid flow [10,11] against seepage through indoor experiments and theoretical analysis. In particular, Warner et al. [12] used compaction grouting to mitigate the sinkholes at WAC Bennett dam, showing satisfactory reinforcement effect. Grotenhuis [13] predicted the fracture length and thickness of grouting in sandy soil through analytical solutions. Tunçdemir et al. [14] investigated fracture grouting in fissured Ankara Clay with low-viscosity cement grouts, where the effect of water/solid ratio and the applied vertical stress on fracturing pressure was studied. Bezuijen [15] studied fracture grouting in sand by developing a conceptual and analytical model and obtained the relationship between crack dimensions and grouting material properties. Yoneyama et al. [16] discussed the mechanism by which cement grout controlled water permeation through fractures in rocks, focusing on the effectiveness of the clogging of cement particles for closing water paths. Yun et al. [17] studied the injection process of fracture grouting and proposed an analytical solution to determine fracture pressure, length, and thickness. Wang et al. [18] performed laboratory tests on loose sand under confined boundary conditions to explore the grouting evolution and diffusion process, with different grout water cement ratios and degrees of saturation of soil. They highlighted the influence of these two factors on the injected grout volume, grout density, and the characteristics of the grouted bulbs. Sun [19] put forward the important factors affecting diffusion radius and derived the calculation formula of fracture grouting diffusion radius based on the assumption that the soil is isotropic and the calculated model is an ideal parallel plane model. However, those introduced works mainly concentrate on investigating grouting in sand or some specific soil whose engineering properties are very different from the Yellow River silt. Therefore, the grouting parameters in existing studies cannot be directly used to investigate the anti-seepage grouting in the Yellow River Embankment. Furthermore, due to the complexity and randomness of grouting mechanism and material properties, there is a need for a systematic assessment for the grouting effectiveness in the Yellow River Embankment.

In existing studies, experimental or theoretical works have been conducted to investigate the properties of the Yellow River silt. In particular, Liu et al. [20] established the stress–strain behavior of the Yellow River silt under dynamic loading through dynamic triaxial and resonance column tests. Xiao et al. [21] investigated the basic physical–mechanical properties of the Yellow River silt by a series of laboratory experiments to treat the distress of silt rainfall of the Beijing–Jiulong railway in the Yellow River alluvial plain area. Song et al. [22] simulated the rainfall infiltration and capillary rising through laboratory tests and analyzed the hydrophilic characteristics of silty roadbed in the Yellow River alluvial plain. Zhu et al. [23] studied the attenuation law of strength and stiffness of silty embankment under capillary water and evaluate the reinforcement effect of the embankment by using the Soletanche method based on large-scale indoor model tests. Chen et al. [24] carried out a number of triaxial tests with stress-controlled monotonic loading and cyclic loading for the Yellow River silt under the CU condition to investigate factors that influence the mechanical behavior of the Yellow River silt. Shi [25] studied the bond performance between polymer anchorage body and silt through ultimate pullout tests on vertical polymer anchors. It is noted that the existing studies concentrate on investigating the mechanical properties of the Yellow River silt through indoor tests. However, laboratory tests cannot comprehensively reflect the in situ state of soil or reveal realistically the grouting mechanism of fracture grouting in the Yellow River Embankment.

Therefore, in this work, laboratory and in situ experiments are carried out to investigate the reinforcement effect of fracture grouting in the Jinan section of the Yellow River Embankment. In particular, firstly, the laboratory tests concentrate on studying the optimum strength and compactness improvement for cement–silicate grout by varying the content of backfilled fly ash and bentonite as admixtures. Flexural strength and Scanning Electron Microscope photographs are investigated for assessing the strength and compactness improvement. Subsequently, based on the obtained optimum admixtures content, in situ grouting tests are carried out in the Jinan section of the

Yellow River Embankment to investigate the anti-seepage effect of fracture grouting, where geophysical prospecting and pit prospecting methods are employed. The work provides references for determining the technical construction parameters of practical fracture grouting in river embankment, i.e., distance of grouting holes, the optimum mixing proportion of grouting material and termination condition.

2. Experimental Scheme and Testing Procedure

2.1. Grouting Material Laboratory Experiment

Grouting effectiveness significantly depends on the grouting materials used in practice. Cement–silicate grout has been widely used due to its short gelation time and high stone rate. However, it has some limitations, such as low-level stability and fluidity, which has restricted its further development. Therefore, experiments are designed to investigate the improvements on the reinforcement effect of the existing cement–silicate grout aiming for the optimum content of backfilled fly ash and bentonite as water reducing agent and expansive agent respectively.

2.1.1. Raw Material

The cement used in the experiment is a Portland cement (PC) graded 42.5 and produced by Sunnsy Group in Jinan. The fly ash (FA) can be classified as Class F according to ASTM C618-05 [26]. The compositions of PC and FA are shown in Table 1. The main characteristics and properties of bentonite (B) are shown in Table 2. The Baume degree and modulus of the silicate are 40 and 3.3 respectively.

Table 1. Main chemical composition of PC and FA.

Material	CaO (%)	SiO_2 (%)	Al_2O_3 (%)	Fe_2O_3 (%)
PC	62.60	22.61	4.35	2.46
FA	3.75	54.64	28.09	6.20

Table 2. Main characteristics and properties of bentonite and silicate.

Water Absorption	Swell Volume	Colloid Valence	Particle Size (75 µm)	Water Content
420% (2 h)	49 mL/g	630 mL/15 g	95%	9%

2.1.2. Experimental Scheme

To improve the grouting effects of cement–silicate grout, fly ash and bentonite are added as admixtures. In particular, first, construction cost can decrease when partially replacing cement with fly ash. Second, the fluidity of grout increases as fly ash particle is spherical and finer than the cement ones. Last but not the least, secondary reaction can occur between fly ash and the reactants of hydration action of cement, and low calcium hydrates like $CaSiO_3$ can be generated, which improves the anti-aqueous solubility of the concretion and promotes the durability of the anti-seepage curtain. According to Sha [27], the optimum percentage of fly ash ranges from 20% to 30%.

Bentonite can improve the stability and infiltration capacity of grout. However, bentonite affects the hydration of cement and lowers concrete strength [28]. Therefore, the mixing content of bentonite should be controlled in a certain range. Based on the research of Liu [29], the percentage of bentonite is controlled at 5% in this paper to achieve an optimum improvement.

In addition, the water over solid (W/S) ratio is controlled at 1.0 according to practical engineering. The volume ratios (VR) between cementitious suspensions and silicate are parametrically varied with 1:1, 2:1, 3:1, and 4:1. The specific proportion of mixture of experimental grout is shown in Table 3.

2.1.3. Testing Procedure

Inverted cup tests are firstly carried out to measure the gelation time of the slurry under different fly ash content (FA) and volume ratio between cementitious suspensions and silicate (VR). Secondly, the specimens are produced in laboratory and the dimensions of the specimen are $40 \times 40 \times 160$ mm according to GB/T 17671-1999 [30]. The specimens are subsequently maintained in a standard curing room (23 ± 3 °C and 100% R.H.) for seven and 28 days, two maintenance conditions. Thirdly, a WDW-100E mechanical press is used to measure the compressive and flexural strength of the specimen with a speed of 2 kN/min and 2 mm/min respectively. Finally, Scanning Electron Microscope (SEM) photographs are investigated for assessing the anti-seepage properties in a micro perspective.

Table 3. Proportion of mixture of experimental grout. VR: Volume ratio; FA: Fly ash; PC: Portland cement; B: Bentonite.

Number	VR	FA/%	PC/%	B/%
A1	1:1	20	75	5
A2	1:1	25	70	5
A3	1:1	30	65	5
B1	2:1	20	75	5
B2	2:1	25	70	5
B3	2:1	30	65	5
C1	3:1	20	75	5
C2	3:1	25	70	5
C3	3:1	30	65	5
D1	4:1	20	75	5
D2	4:1	25	70	5
D3	4:1	30	65	5

2.2. In Situ Grouting Test

2.2.1. Yellow River Silt

According to Cui [31], Yellow River silt consists of over 80% silt and a small amount of clay, with low surface strength, high porosity, and drastic capillarity.

Prior to the in situ tests, particle gradation tests, direct shear tests, and permeability tests are carried out. The result of the particle gradation testis shown in Figure 1. The non-uniform coefficient (C_u) and curvature coefficient (C_c) of Yellow River Silt are 3.6 and 1.51 respectively. It can be seen that the particle gradation of the Yellow River silt is relatively uniform and the particles in the size range 0.002–0.074 mm account for more than 80%. In addition, the gradation of the soil is poor as C_u is smaller than 5. Based on the results from the direct shear tests, the cohesion (c_q) and angle of shearing resistance (φ_q) are 10.78 kPa and 23.67°, respectively. The permeability of the Yellow River silt is obtained by the variable water head permeability test and the tested permeability coefficient is 5.913×10^{-5} cm/s.

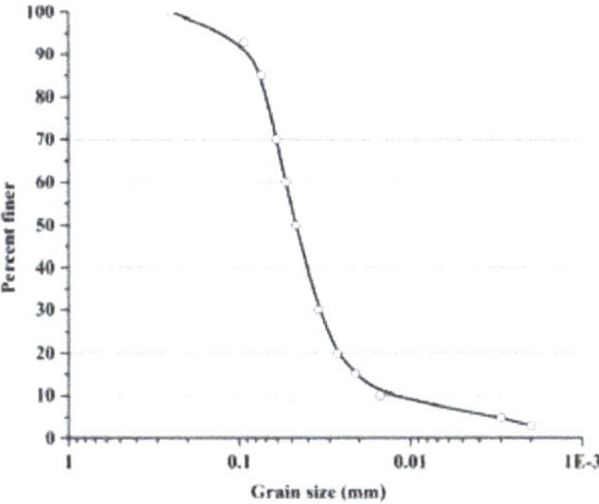

Figure 1. Particle size distribution curve.

2.2.2. Design of In Situ Grouting Tests

(1) Layout of grouting holes

The interval distance between grouting holes is determined by diffusion radius and grouting range, while diffusion radius is decided by rheological property of grouting material, grouting pressure and grouting time. However, there is a lack of theoretical relation for confirming the diffusion radius at present. As a consequence, the determination of the distance will be determined empirically based on practical applications.

When determining the interval distance, the lapping effect between two holes should be taken into consideration. Based on the fact that the permeability of Yellow River silt is relatively low, the fracture grouting would be the dominant type when conducting grouting experiment. According to the existing applications [32], the distance between two grouting holes can be determined as 1.0–2.0 m. In order to further decide the optimum distance, the interval distance is parametrically varied from 0.8 m to 1.5 m in the tests. The layout of the grouting holes and the schematic cross section of the grouting field are shown in Figure 2.

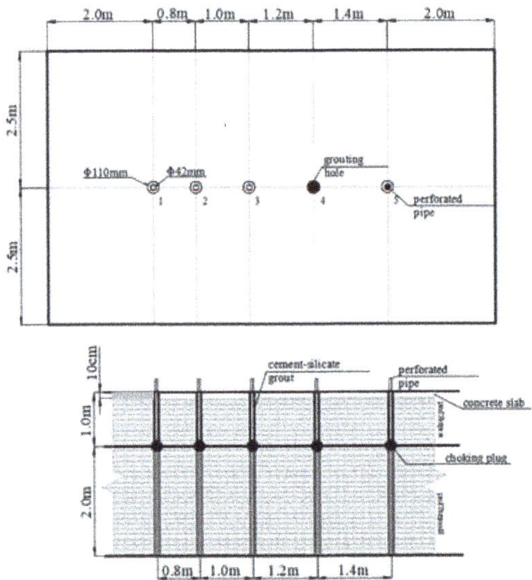

Figure 2. Layout of field test.

(2) Grouting pressure

It is crucial to control the grouting pressure during the grouting process, which affects the compactness of the soil. However, deformation or even destruction of earth structures can occur when grouting pressure exceeds a critical value. Therefore, the grouting pressure should be controlled in a certain range. The maximum allowed value of fracture grouting pressure can be calculated by Equation (1) [33]:

$$P_{max} = \gamma h + \sigma_t,\tag{1}$$

where γ, h and σ_t represent the specific weight of soil, the depth of grouting pipe and soil tensile strength respectively.

After substituting $\gamma = 15.5$ kN/m^3, $\sigma_t = 10$ kPa(empirical value) and h = 3 m into Equation (1), the maximum allowable pressure is determined to be 56.5 kPa.

Considering that emitting slurry phenomenon may happen when grouting in shallow strata, the ultimate grouting pressure can be calculated by Equation (2) [32]:

$$P_u \leq \gamma h \tan^2(45° + \varphi/2) + 2c \tan(45° + \varphi/2). \tag{2}$$

Based on the results in Section 2.2.1, the ultimate grouting pressure is calculated as 141.9 kPa.

Combining the calculated results, initiation pressure and the ultimate pressure are determined as 60 kPa and 150 kPa in the field grouting experiment.

2.2.3. Grouting Procedure

Perforated pipe grouting is used in in situ tests. The grouting procedure involves grouting pad construction, positioning and drilling of grouting holes, preparation of the grout, and, finally, injection of the grout. A flow chart of grouting is shown in Figure 3a.

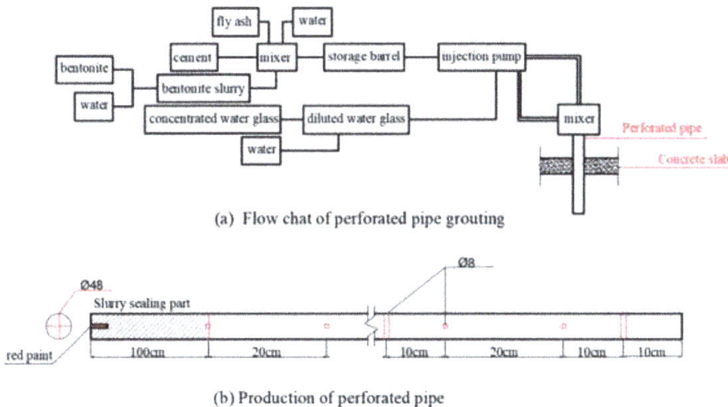

(a) Flow chat of perforated pipe grouting

(b) Production of perforated pipe

Figure 3. Flow chart of perforated pipe grouting and production of perforated pipe.

A concrete slab with dimensions of 8.4 m × 5 m × 10 cm (length × width × depth) is produced as the grouting pad a week before the grouting injection. After fabricating the grouting pad for three days, the grouting holes are excavated according to Figure 2. At the same time, the perforated pipes are produced with seamless steel tube with diameter of 48 mm and wall thickness of 3.5 mm. In order to generate directional grouting fractures along the axis, the hole distances on the pipe are designed to be different in two direction where one direction is 20 cm and the other is 40 cm. The direction of the dense holes on the pipes is marked with red oil paint. The detailed process is shown in Figure 3b.

The optimum admixture content is obtained from the laboratory tests and is adopted in the in situ tests. The cementitious suspensions and the water glass with Baume degree of 40 are placed in different agitated tanks. The prepared grout is agitated in case of sedimentation or setting. Meanwhile, drilling rig is used to bore the grouting holes according to the grouting holes positions.

After boring the grouting holes, perforated pipes are placed into the holes. It should be noted that the painted direction is parallel to the axial direction of the layout of the in situ grouting test. Before injection, cement–silicate grout with volume ratio of 2:1 is used to seal the holes to avoid slurry inflow.

After preparation, the grout is pumped into the grouting holes and the sequence of grouting holes number is 1, 3, 5, 2 and 4. Thin slurry is used at the beginning and the grouting pressure increases progressively until it reaches the fracture pressure (about 0.06 MPa). After spitting the soil, the grouting pressure is decreased and adjusted to the defined level. Meanwhile, the slurry is gradually densified. The injection process does not complete until one of the following two criteria is satisfied: either the volume of the injected grout is equal to the predefined volume (150–200 kg/m) or the inflow injection pressure reaches the ultimate pressure (about 0.15 MPa). When the injection completes, the circuits of the injection pump should be washed with clean water to avoid blocking. Besides, concentrated grout is injected to the grouting holes to seal the holes.

2.2.4. Geophysical Prospecting Method

In order to comprehensively evaluate grouting effectiveness, both the ground-penetrating radar (GPR) and high-density resistivity method (HDRM) are used to avoid the limitation and multiplicity induced by a single method, as shown in Figures 4 and 5. Pit prospecting is additionally carried out to investigate the reinforcement mechanism and the created grouting veins. The results of pit prospecting can verify the validity of the geophysical prospecting method and can further help to better understand the grouting mechanism in the soil.

The in situ test procedure is described as follows. First, the reinforcement condition is detected using GPR and HDRM two weeks after grouting, where the layout of geophysical prospecting is shown in Figure 6. After geophysical prospecting, the grouted field is excavated to observe the grouting veins. Finally, part of the grouting veins is sampled, and the thickness of the grouting veins is measured.

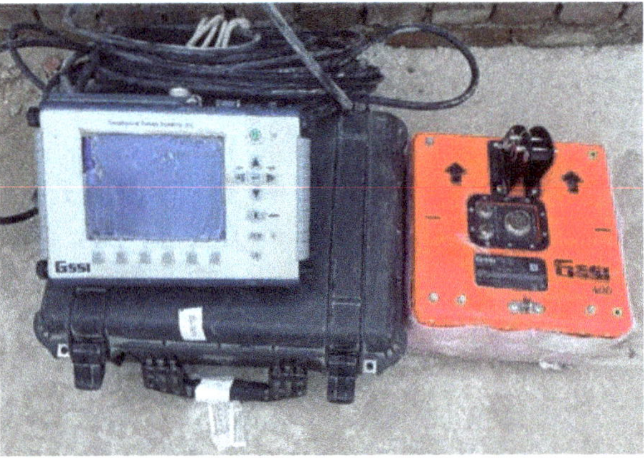

Figure 4. Ground-penetrating radar.

Figure 5. High-density resistivity method.

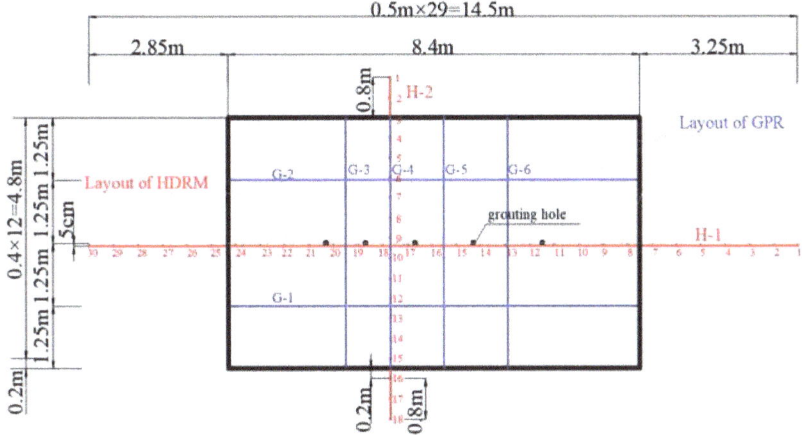

Figure 6. The layout of GPR and HDRM.

3. Discussion of Laboratory Results

3.1. Gelation Time of the Slurry

The test results in terms of gelation time of the slurry are shown in Figure 7. The gelation time prolongs as fly ash content increases. When partly replacing cement with fly ash and bentonite, the water over cement ratio (W/C), that is a key factor in determining the setting time, decreases. In particular, the backfilled bentonite influences the hydration of cement and prolongs the gelation time. In practice, a slurry with appropriate gelation time is of great significance to avoid small range diffusion or water erosion. Therefore, the gelation time of slurry needs to be adjusted without greatly changing the slurry property by adding specific amount of fly ash.

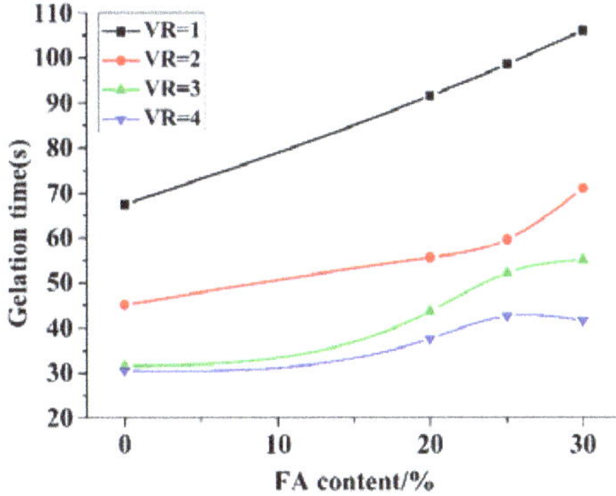

Figure 7. Test result of gelation time of the slurry (VR: the volume ratio).

3.2. Strength of the Specimen

The tested flexural strength and compressive strength of the specimens are shown in Figures 8 and 9 respectively. According to Figure 8, except for the case of 28-day maintenance and under the volume ratio of 1, all the other seven tests results show that the flexural strength firstly increases and then decreases as the fly ash content increases. The flexural strength reaches the peak when the fly ash content is in the range of approximately 25%. As to the result of compressive strength, it is noted that the compressive strength decreases as the fly ash content increases from 20 to 30%.

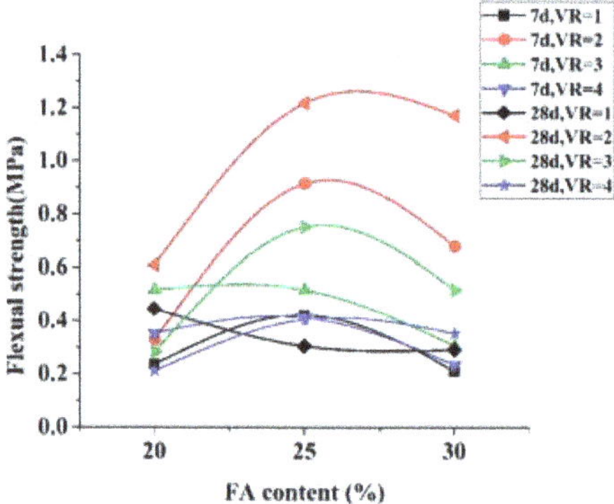

Figure 8. Test result of flexural strength of the specimen (VR: the volume ratio).

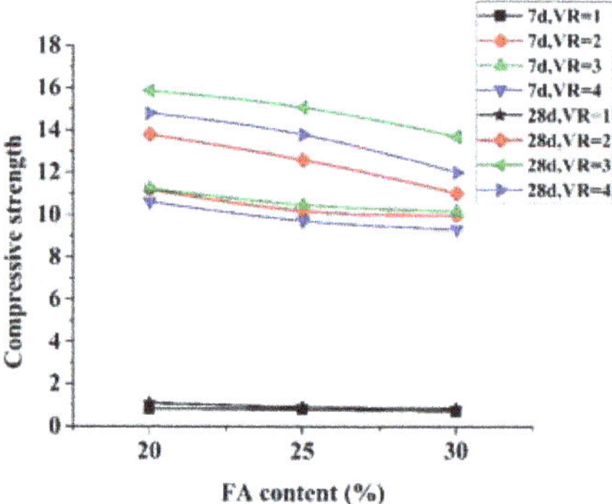

Figure 9. Test result of compressive strength of the specimen (VR: the volume ratio).

W/C influences concrete strength as a result of the weakening effects from cement hydration when incorporating fly ash and bentonite. While backfilling 20% fly ash, pozzolanic reaction between fly ash and the hydrate of cement (such as calcium hydroxide) can supplement concrete strength. In this case, fly ash is almost reacted and therefore its effect on improving the compactness of grout material is insignificant. When the percentage of fly ash increases to 25%, the hydration of fly ash further completes and the residual fly ash fills the pore in the concrete. However, the residual fly ash is redundant and plays a negative role to the compactness when the content of fly ash reaches to 30%. In seepage control engineering of embankment, the veins generated by fracture grouting tend to be relatively thin, which makes it more susceptible to bending [34]. Therefore, the optimum range of fly ash is in the range of approximately 25% where the flexural strength is relatively large under the test condition.

Regarding the influence of volume ratio, when it is in the range of 2, the hydration action is relatively complete and the strength maximizes. After a threshold value of approximately 2, the volume ratio is so large that the silicate is not sufficient to react with hydration products (like calcium hydroxide), which eventually results in slightly decrease of the mechanical strength of the specimens.

3.3. SEM Analysis of Specimen

According to the test result in 3.2, the specimen B1, B2 and B3 are selected to investigate the influence of fly ash on the microstructure for the volume ratio of 2 using the scanning electron microscope (SEM). The photographs of SEM of the specimens under different fly ash are shown in Figure 10.

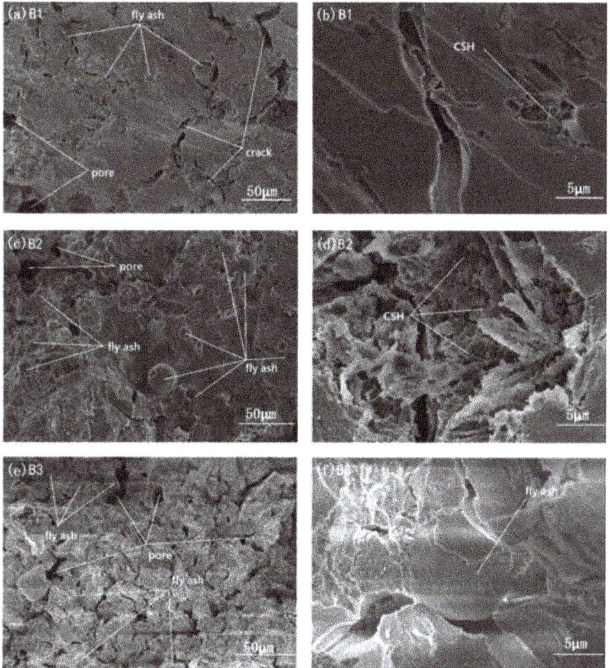

Figure 10. Photographs of SEM of the specimens under different fly ash content.

According to Figure 10e, it is noted that the pores on the concrete of B3 (30%FA + 5%B + 65%PC, VR = 2) are large and numerous. It is presented in Figure 10f that there is a large amount of fly ash particles in the surface of the microstructure of B3. According to Figure 10c, the compactness of B2 (25%FA + 5%B + 70%PC, VR = 2) significantly increases compared with B3, and the micro-aggregate effect of fly ash is obvious with different sizes of fly ash particles embedded in the hydrates and those fly ash particles fill some void of the concrete. Furthermore, it is shown in Figure 10d that a large amount of calcium silicate hydrates (CSH) form, which increases concrete strength of B2. In Figure 10a, there are more cracks and less fly ash particles on the surface of the concrete of B1 (20%FA + 5%B + 75%PC) compared with B2. The unreacted fly ash is insufficient and therefore the filling effect is not obvious as the microstructure of B2 in Figure 10c.

Considering the results of mechanical strength and SEM, the macroscopic analysis and microscopic analysis show good agreements. In particular, the optimum mixing proportion of the grout is 70% cement, 25% fly ash and 5% bentonite, and the best volume ratio is 2 for the investigated cases. The slurry under this mix proportion is selected for the subsequent in situ grouting experiment.

4. Discussion of In Situ Test Results

Based on the obtained optimum admixture content, in situ grouting tests are carried out in the Jinan section of the Yellow River Embankment to investigate the anti-seepage effect of fracture grouting, where geophysical prospecting and pit prospecting methods are employed. In order to comprehensively discuss the grouting effect, the geophysical prospecting results are divided into two parts: results in the axial direction and vertical direction.

4.1. Geophysical Prospecting for Grouting Effectiveness

4.1.1. Results in Axial Direction

The results from the GPR and HDRM tests in the axial direction are shown in Figures 11 and 12 respectively. According to Figure 11, the soil in the grouting area can be divided into three layers, with the grouting slab and miscellaneous filling in the first layer, the unreinforced Yellow River silt in the second layer and the strengthened silt in the third layer. The results show that there is continuous inhomogeneous medium close to the grouting pipe, which is defined as the grouting vein. It is noted that there is obvious inhomogeneous medium near the pipes, numbered 3 and 4 in the upper of the excavation line.

Based on the prospecting results in Figure 12, the distribution of resistivity includes three main horizontal layers in the grouting area. The first layer is a high-resistance region with an average thickness of 0.5 m, which consists of low water content soil in the upper layer. The second layer is low-resistance area with the depth of 1.5 m, without reinforcement. The third layer is a part with high resistivity compared with the second layer, where the overall water content is relatively low in this layer.

In summary, the grouting area can be divided into three layers and it shows good agreements with the actual grouting process. Regarding the grouting effectiveness, the mainly reinforcement area is the third layer where the water content is relatively low and there are grouting veins in the axial direction.

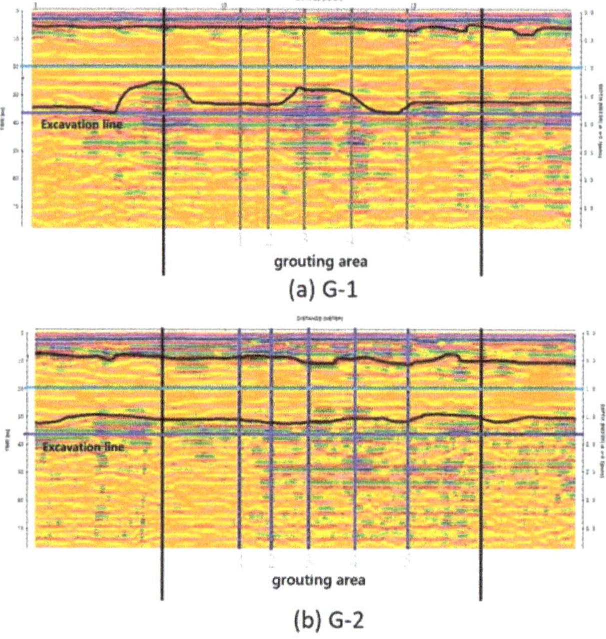

Figure 11. Photographs of GPR in the axial direction.

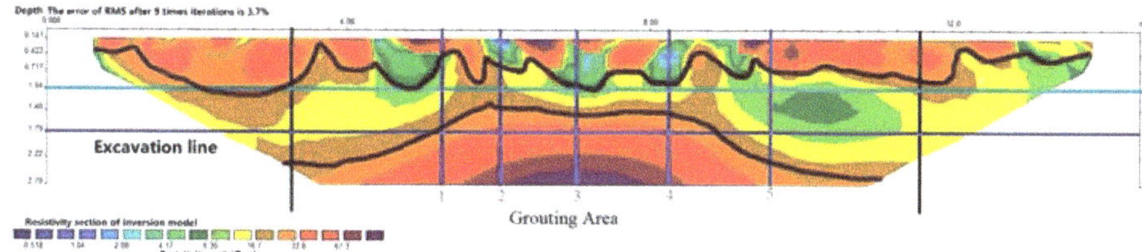

Figure 12. Photographs of HDRM in the axial direction.

4.1.2. Vertical Direction

The results from the GPR and HDRM tests in the vertical direction are presented in Figures 13 and 14 respectively. Similarly, the soil can be divided into three layers according to Figure 13. The depths of the first two layers are 0.30 m and 1.65 m respectively. The grouting materials are mainly found in the third layer. However, the grouting effectiveness is not as obviously observed as with the results in the axial direction, indicating that the grout mainly flows along the axial direction. Therefore, the treatment to the perforated pipe to induce the fracture direction is proved to be effective.

It is noted that the HDRM prospected depth in the vertical direction is only 1.91 m. Figure 14 shows that there are two layers in this range. The upper layer is a high-resistance area which is the grouting slab and miscellaneous filling. The lower part is a low-resistance region where there is no significant boundary in this layer in the depth of 1.5 m, which indicates that the water content of the soil keeps unchanged. In other words, the grouting effectiveness is low.

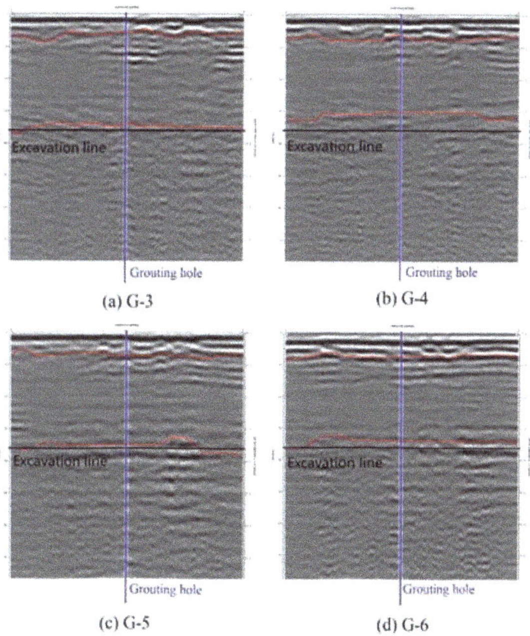

Figure 13. Photographs of GPR in the vertical direction.

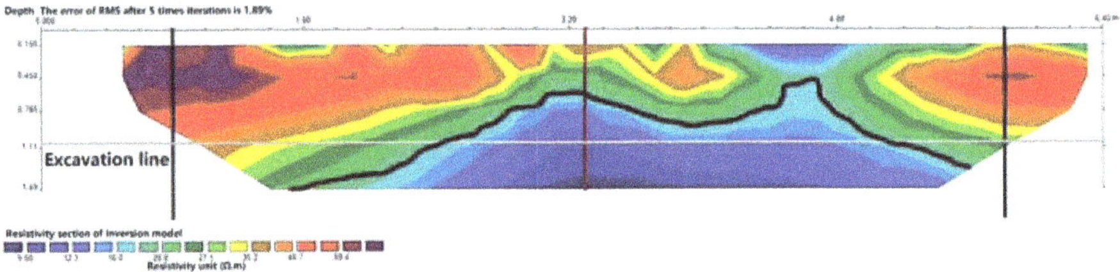

Figure 14. Photographs of HDRM in the vertical direction.

4.2. Pit Prospecting for Grouting Effectiveness

The pit prospecting result is presented in Figure 15. Considering that the groundwater level is 1.27 m and the mechanical strength of Yellow River silt is relatively low, the excavation depth is determined as 1.8 m. In Figure 15, the four color lines represent the grouting veins generated by grouting holes 2, 3, 4, and 5, respectively.

Based on Figure 15, it can be seen that the grouting type in the Yellow River Embankment is fracture grouting. The grouting veins between grouting holes 2 and 3 do not exactly follow the axial direction. The adjacent grouting overlaps at the vertical cross section of the grouting axis. When the interval distance of grouting holes is 1.2 m, the grouting veins induced by holes 3 and 4 laps and the thickness is relatively large, which indicates that the lapping effect is relatively satisfactory. When the grouting veins formed by grouting hole 5 laps with those of 4, it is evident that there is another grouting vein that is almost vertical to the axis. This is unfavorable for the reinforcement effect in generating anti-seepage curtain.

In order to quantitatively investigate the grouting effectiveness, the grouting veins are sampled and the thickness of the veins is measured in the laboratory. The test results are shown in Figure 16. It can be seen that the four grouting veins overlap and form an anti-seepage curtain in the ground, showing satisfactory anti-seepage effects. The thickness of the grouting veins decreases when they are further from the grouting hole. It is noted that the average thickness of the grouting vein induced by grouting holes 3 and 4 is relatively large, due to the appropriate grouting interval distance. After conducting permeability tests for the grouting vein samples from grouting veins 3 and 4, the permeability coefficient is found to be 1.6×10^{-6} cm/s on average. According to DL/T 5129-2013, the required permeability coefficient for an impervious body should be less than 1.0×10^{-5} cm/s. The modified cement silicate grouting material used in the in situ tests can meet the anti-seepage criterion in practical engineering.

Figure 15. Pit prospecting result.

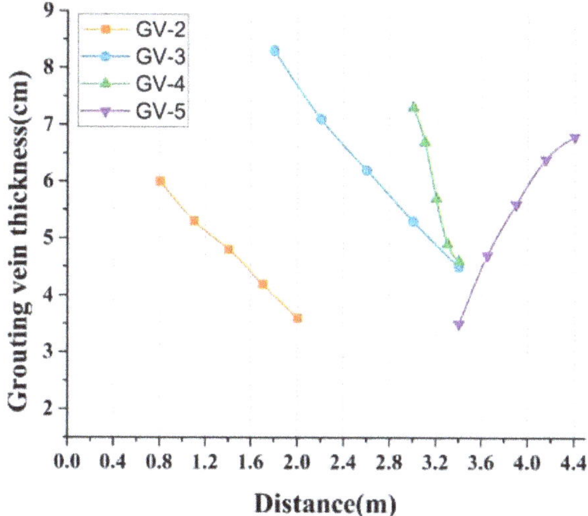

Figure 16. Measured value of the grouting veins. GV: Grouting vein.

According to the pit prospecting result, the pit test results show good agreement with the geophysical prospecting in the excavation range. The soil conditions within the excavation depth are well reflected by both the GPR and HDRM results. Therefore, the result of geophysical prospecting is reliable and the grouting effectiveness of the unexcavated part can be evaluated by geophysical prospecting. Moreover, when the interval hole distance is between 1.0 m and 1.2 m, the lap joint is relatively close to the grouting hole and the anti-seepage curtain can be easily controlled. When the interval distance is over 1.2 m, some grouting veins are almost vertical to the reinforcement direction and they are unfavorable to the reinforcement effect in generating anti-seepage curtain. Therefore, the optimum interval hole distance should be in the range of 1.2 m under the test conditions. After conducting a permeability test for the grouting vein samples from grouting veins 3 and 4, the permeability coefficient is found to be 1.6×10^{-6} cm/s on average, which meets the anti-seepage criterion in practical engineering. In summary, the grouting type in Yellow River silt is mainly fracture grouting. The grouting effectiveness of the field test is satisfactory based on the results of geophysical prospecting and pit prospecting tests.

5. Conclusions

In order to investigate grouting material and the anti-seepage effect of fracture grouting employed in the Yellow River Embankment, laboratory and in situ experiments were carried out to research the reinforcement effect of fracture grouting in the Jinan section of the Yellow River Embankment. In particular, laboratory tests concentrated on studying the optimum strength improvement for cement–silicate grout by varying the content of backfilled fly ash and bentonite as admixtures. Subsequently, in situ grouting tests were carried out in the Jinan section of the Yellow River Embankment to evaluate the anti-seepage effectiveness of fracture grouting, where geophysical prospecting and pit prospecting methods were employed. Based on the experimental results, the following conclusions can be drawn:

(1) Compared with pure cement–silicate grouts, the gelation time of the improved slurry is longer and gelation time increases as fly ash content increases. The optimum mixing proportion of the compound cement–silicate grout is 70% cement, 25% fly ash, and 5% bentonite, and the best volume ratio is 2 for the investigated cases.

(2) Good agreement is found between the ground-penetrating radar and high-density resistivity methods and the two geophysical prospecting methods can both reflect the anti-seepage effectiveness of fracture grouting on site.

(3) The pit prospecting result shows that grouting material mainly flows along the axial direction of the embankment, which means that the treatment used to generate directional fracture is proven to be effective. The injection hole interval distance is suggested to be 1.2 m, where the lapping effect of the grouting veins is relatively significant.

(4) For the investigated cases, the average thickness of the grouting veins is approximately 6.0 cm and the corresponding permeability coefficient is averagely 1.6×10^{-6} cm/s, which meets the anti-seepage criterion in practice.

Author Contributions: Conceptualization, Z.W. and J.L.; Methodology, J.L., Q.X. and B.H.; Software, C.L.; Validation, Z.W., J.L. and B.H.; Formal Analysis, Z.W.; Investigation, C.L. and R.L.; Resources, R.L.; Data Curation, M.C.; Writing—Original Draft Preparation, Z.W.; Writing—Review & Editing, B.H. and Q.X.; Visualization, C.L.; M.C.; Supervision, J.L.; Project Administration, J.L.; Funding Acquisition, J.L.; B.H.

Funding: This research was funded by the National Natural Science Foundation of China (Project No. 41172267), the National Science and Technology Support Program of China (Grant No. 2015BAB07B05), and the Natural Science Foundation of Shandong Province (Grant No. ZR2018QEE008).

Acknowledgments: We thank Hongtao Li, Yuying Li and Xuanzheng Li for their critical comments and suggestions.

Conflicts of Interest: The authors declare no conflicts of interest. The funding sponsors had no role in the design of the study; in the collection, analyses or interpretation of the data; in the writing of the manuscript; nor in the decision to publish the results.

References

1. Liu, H.N.; Wang, J.M.; Wang, S.J.; Liu, H.D.; Hu, B. Seepage analysis of unsaturated soil levee slopes in lower reaches of Yellow River. *Rock Soil Mech.* **2006**, *27*, 1835–1840. [CrossRef]

2. Zhao, Y.K.; Liu, H.D.; Li, Q.A. Slope stability analysis of lower reaches' dikes of Yellow River under flood immersion and water level rapid drawdown. *Rock Soil Mech.* **2011**, *32*, 1495–1499. [CrossRef]

3. Zhang, H.M.; Li, Z.B.; Yao, W.Y. Design method of drain system for dike in lower reaches of Yellow River. *J. Hydraul. Eng.* **2006**, *37*, 865–869. [CrossRef]

4. Zhang, X.Y. Study on Failure Mechanism and Methodology of Safety Assessment of the Lower Yellow River Dike. Ph.D. Thesis, Hohai University, Nanjing, China, 2005.

5. Bai, Y.N.; Liu, X.K. *Grouting of the Earth Dam Body and Embankment*; China Water Conservancy Press: Beijing, China, 1982; pp. 1–94.

6. Luo, C.Q. Discussion on the construction technique of spit grouting in body of earth dam. *Chin. J. Rock Mech. Eng.* **2005**, *24*, 1605–1611. [CrossRef]

7. Jorne, F.; Henriques, F.M.A. Evaluation of the grout injectability and types of resistance to grout flow. *Constr. Build. Mater.* **2016**, *122*, 171–183. [CrossRef]

8. Azadi, M.R.; Taghichian, A.; Taheri, A. Optimization of cement-based grouts using chemical additives. *J. Rock Mech. Geotech. Eng.* **2017**, *9*, 623–637. [CrossRef]

9. Pasian, L.; Secco, M.; Piqué, F.; Artioli, G.; Rickerby, S.; Cather, S. Lime-based injection grouts with reduced water content: An assessment of the effects of the water-reducing agents ovalbumin and ethanol on the mineralogical evolution and properties of grouts. *J. Cult. Herit.* **2018**, *30*, 70–80. [CrossRef]

10. Liu, R.C.; Li, B.; Jiang, Y.J. Critical hydraulic gradient for nonlinear through rock fracture networks: The roles of aperture, surface roughness, and number of intersections. *Adv. Water Resour.* **2016**, *88*, 53–65. [CrossRef]

11. Liu, R.C.; Jiang, Y.J.; Li, B.; Wang, X. A fracture model for characterizing fluid flow in fractured rock masses based on randomly distributed rock fracture networks. *Comput. Geotech.* **2015**, *65*, 45–55. [CrossRef]

12. Warner, J.; Jefferies, M.; Garner, S. Compaction Grouting for Sinkhole Repair at WAC Bennett Dam. In Proceedings of the International Conference on Grouting and Ground Treatment, New Orleans, LA, USA, 10–12 February 2003; pp. 869–880.

13. Grotenhuis, R. Fracture Grouting in Theory. Master's Thesis, Delft University of Technology, Delft, The Netherlands, 2004.

14. Tunçdemir, F.; Ergun, M.U. A Laboratory Study into Fracture Grouting of Fissured Ankara Clay. In Proceedings of the Geo-Frontiers Congress, Austin, TX, USA, 24–26 January 2005; pp. 1–12.

15. Bezuijen, A.; Grotenhuis, R.T.; Tol, A.F.V.; Bosch, J.W.; Haasnoot, J.K. Analytical model for fracture grouting in sand. *J. Geotech. Geoenviron. Eng.* **2011**, *137*, 611–620. [CrossRef]

16. Yoneyama, K.; Okuno, T.; Nakaya, A.; Tosaka, H. Experimental and numerical study on fracture grouting by fine particle cement. *Chem. Lett.* **2012**, *1994*, 9–12.

17. Yun, J.W.; Park, J.J.; Kwon, Y.S.; Kim, B.K.; Lee, I.M. Cement-based fracture grouting phenomenon of weathered granite soil. *KSCE J. Civil Eng.* **2016**, *21*, 232–242. [CrossRef]

18. Wang, Q.; Wang, S.; Sloan, S.W.; Sheng, D.; Pakzad, R. Experimental investigation of pressure grouting in sand. *Soils Found.* **2016**, *56*, 161–173. [CrossRef]

19. Sun, F.; Zhang, D.L.; Chen, T.L.; Zhang, X.P. Meso-mechanical simulation of fracture grouting in soil. *Chin. J. Geotech. Eng.* **2010**, *32*, 474–480.

20. Liu, Q.; Zheng, X.L.; Liu, H.J.; Wang, J.G.; Dong, S.Y. Experimental studies on liquefaction behavior of silt in the huanghe river delta. *World Earthq. Eng.* **2007**, *23*, 161–166. [CrossRef]

21. Xiao, J.; Liu, J.; Peng, L.; Chen, L. Effects of compactness and water yellow-river alluvial silt content on its mechanical behaviors. *Rock Soil Mech.* **2008**, *29*, 409–414. [CrossRef]

22. Song, X.G.; Zhang, H.B.; Wang, S.G.; Jia, Z.X.; Guan, Y.H. Hydrophilic characteristics and strength decay of silt roadbed in yellow river alluvial plain. *Chin. J. Geotech. Eng.* **2010**, *32*, 1594–1602.

23. Zhu, D.Y.; Guan, Y.H.; Liu, H.Z.; Wang, Q.; Zhang, Q.T. Model tests on fracture grouting reinforcement of silt embankment by using soletanche method. *Chin. J. Geotech. Eng.* **2012**, *34*, 1425–1431.

24. Chen, C.L.; Ma, S.X.; Li, L.L.; Cai, Z.P. Study on post-cyclic undrained deformation and strength characteristics of saturated silt in the floodplain of the Yellow river. *J. Hydraul. Eng.* **2014**, *45*, 801–808. [CrossRef]

25. Shi, M.S.; Xia, W.Y.; W, F.M.; L, H.; Pan, Y.H. Experimental study on bond performance between polymer anchorage body and silt. *Chin. J. Geotech. Eng.* **2014**, *36*, 724–730.

26. ASTM C618-05. *Standard Specification for Coal Fly Ash and Raw or Calcined Natural Pozzolan for Use in Concrete*; ASTM: West Conshohocken, PA, USA, 2005.

27. Sha, F.; Li, S.; Liu, R.; Li, Z.; Zhang, Q. Experimental study on performance of cement-based grouts admixed with fly ash, bentonite, superplasticizer and water glass. *Constr. Build. Mater.* **2018**, *161*, 282–291. [CrossRef]

28. Hwang, H.; Yoon, J.; Rugg, D.; Mohtar, C.S.E. Hydraulic Conductivity of Bentonite Grouted Sand. In Proceedings of the Geo-Frontiers Congress, Dallas, TX, USA, 13–16 March 2011; pp. 1372–1381.

29. Liu, J.; Hu, L.Q.; Xu, B.J.; Yue, X.L.; Qi, B.L.; Zhong, Q. Study on cement-based seepage grouting materials for earth-rock dam. *J. Shandong Univ. Eng. Sci.* **2016**, *47*, 9–15.

30. Chinese National Quality and Technical Supervision & Chinese Ministry of Construction. *Method of Testing Cements-Determination of Strength*; Standards Press of China: Beijing, China, 1999; pp. 1–12.

31. Cui, X. Traffic-induced settlement of subgrade of low liquid limit silt in yellow river delta. *China Civ. Eng. J.* **2012**, *45*, 154–162.

32. Collaboration of Rock and Soil Grouting Theory and Engineering Examples. *Rock and Soil Grouting Theory and Engineering Examples*; Science Press: Beijing, China, 2001; pp. 71–94.

33. National Development and Reform Commission of the People's Republic of China. *Technical Specification of Grouting for Earth Dam*; China Water & Power Press: Beijing, China, 2010; pp. 1–18.

34. Shi, M.S.; Yu, D.M.; Wang, F.M. Bending Properties of a polymer grout. *J. Mater. Sci. Eng.* **2010**, *28*, 514–517.

Article

Effects of Water Soaked Height on the Deformation and Crushing Characteristics of Loose Gangue Backfill Material in Solid Backfill Coal Mining

Junmeng Li [1,2], Yanli Huang [1,*], Ming Qiao [1], Zhongwei Chen [2], Tianqi Song [1], Guoqiang Kong [1], Huadong Gao [1] and Lei Guo [3]

[1] State Key Laboratory of Coal Resources and Safe Mining, School of Mines, China University of Mining & Technology, Xuzhou 221116, China; lijunmeng1201@163.com (J.L.); tb16020013b1@cumt.edu.cn (M.Q.); stq344359844@163.com (T.S.); kgq9569913@163.com (G.K.); gaohuadong1314@163.com (H.G.)

[2] School of Mechanical and Mining Engineering, the University of Queensland, St. Lucia, QLD 4072, Australia; uqjli30@uq.edu.au

[3] College of Mining and Safety Engineering, Shandong University of Science and Technology, Qingdao 266590, China; sdustgl@sdust.edu.cn

* Correspondence: 5306@cumt.edu.cn or huangyanli6567@163.com; Tel.: +86-139-0520-7498

Received: 23 April 2018; Accepted: 28 May 2018; Published: 30 May 2018

Abstract: In solid backfill coal mining (SBCM), loose gangue backfill material (LGBM) is used to backfill the goaf after coal resources are exploited from the underground mines. Under certain geological conditions, LGBM with a certain height may be soaked in the water, and then becomes saturated, significantly altering its mechanical properties. The confined compression experiments were used in this paper to analyze the deformation and the crushing characteristics of LGBM with varying water soaked heights in coal mines. The results showed that a large number of small holes that were distributed in the gangue blocks were the main reason why the material absorbed water and was softened. The crushing ratio and the maximum axial strain of LGBM samples gradually increased with the water soaked heights of the samples. In addition, there was a strong linear correlation between the crushing ratio and the maximum axial strain. When LGBM was used as a solid backfill material in SBCM, its deformation resistance would significantly decrease after it was soaked in the water. Higher water soaked height of LGBM led to lower deformation resistance and greater influence on the quality of backfilling. This research has great significance in getting a deep and better understanding of the mechanical properties of LGBM, as well as guiding engineering practice.

Keywords: solid backfill coal mining; goaf; water soaked height; loose gangue backfill material; deformation; crushing ratio

---

1. Introduction

Solid backfill coal mining (SBCM) is a green mining technology that can meet the challenges of gangue dumps on the surface and the coal mining under buildings, railways, and water bodies. This technology is aimed at using the solid waste material, such as loose gangue and fly ash, to backfill the goaf and to support the overburden strata after the coal resources are exploited from the underground. The main purpose is to stabilize the overlying strata and reduce surface subsidence in the process of recovering coal reserves that are located under buildings, railways, and water bodies, thus reducing piled gangue on the surface and the environmental pollution. Currently, this technology has become a preferable technological measure that is widely used in the modern green exploitation of coal resources [1,2]. In the process of SBCM, loose gangue backfill material (LGBM) and other solid backfill materials are transported by material-feeding shaft to the surge bin under the mine. When needed, these solid backfill materials are transported to the scraper conveyor suspended at the rear of the SBCM hydraulic support from the surge bin through an underground conveyor system. Then, these materials are offloaded from the discharge port with the flashboard at its bottom and filled into the goaf. After that, tamp mechanism, as a key structure in the backfilling process, pushes and compacts these materials to fully connect the immediate roof with required density and fully fill the goaf. Finally, these solid backfill materials become the main support bearing the overburden strata [3,4]. The basic principle of SBCM is shown in Figure 1.

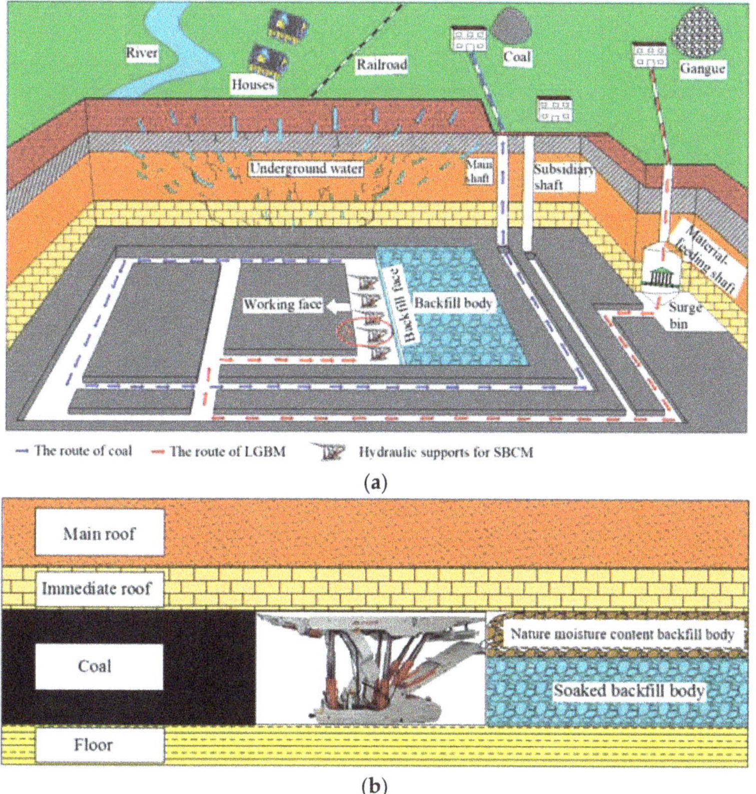

Figure 1. Basic principle of solid backfill coal mining (SBCM). (**a**) Overview; and, (**b**) Detailed view of hydraulic roof support at longwall/backfill face.

After the coal resources are exploited, water from the overburden layer seeps into the goaf through cracks and pores that exist in the roof. Wastewater that is produced in the process of coal production can also be stored in the goaf and soak in a portion of LGBM in the goaf. The amount of water in the goaf determines the soaked height of the LGBM, as is shown in Figure 1. The portion of the gangue that is soaked in water for a long time eventually becomes saturated, and its mechanical properties change dramatically. Thus, the mechanical strength of LGBM in bearing the overburden load varies with the water soaked height.

According to the research conducted by Yilmaz., and Fall., in terms of paste backfill, many factors affect its short-term and long-term strength and stability. They can be mainly classified into two categories: intrinsic factors and extrinsic factors. Intrinsic factors included all of the parameters that are related to the physical, chemical, and mineralogical properties of the three main components of paste backfill (the tailings, binder, and mixing water), as well as their mixing properties (e.g., water-to-cement ratio). Extrinsic factors are related to all phenomena that occurred in a stope filled with paste backfill and show interaction between the stope and the adjacent rock. Another extrinsic factor affecting the properties of cemented paste backfill is the in situ curing temperature [5]. In the process of SBCM, there were also many factors that would have influence on the filling quality, including the types and the mechanical properties of solid backfill materials, mining geological conditions, backfilling process design, etc. The mechanical properties of solid backfill materials are one of the most important factors determining the filling quality [3,4]. Water invariably presented in the goaf of coal mines. After the solid backfill, materials were soaked in the mine water and their mechanical properties would change significantly, which would affect the filling quality.

At present, a large number of researches on the mechanical properties of non-viscous bulk materials have been done. Marsal et al. [6–8] and Kong et al. [9] researched the crushing characteristic of rockfill materials. Their results showed that the crushing characteristic of particle is related to particle shape, strength, size, density, gradient, stress conditions, and other factors. Jaroslav [10] and Wang et al. [11] studied the stress and the strain of coarse grain materials, such as sand pebbles and rockfill, by using compression tests. They found that these materials clearly demonstrated their dilatancy effect. Bagherzadeh-Khalkhali [12] studied the effects of the maximum particle diameter on the shear strength of coarse soils. Kjaernsli et al. [13] and Hall et al. [14] carried out tri-axial compression tests on sand specimens and discovered that under fixed stress, irregular particles with rough surfaces are more easily to be crushed. In addition, well-graded sand specimens are less crushed than the poorly-graded sand specimens. Zhang et al. conducted the confined compaction tests on LGBM in the steel cylinder and obtained the relationships among strain, compaction, stress, as well as the influence of compaction time [15]. Huang et al. [16] used numerical simulations to study the macro-scale mechanical response and micro-scale movement characteristics of LGBM during tri-axial compression testing.

Clearly, there is an abundance of researches on non-viscous bulk materials with natural moisture content. However, very little researches have focused on the mechanical effects that are caused by soaking these materials in the water. After the solid backfill, materials are soaked in the mine water, and their mechanical properties will change dramatically, thus affecting the filling quality. As a result, the influence of water on the mechanical properties (the deformation and crushing characteristics) of LGBM will be studied in this paper. It is of great significance to guide the engineering practice of SBCM and to ensure the filling quality.

2. Materials and Test Procedure

2.1. Materials and Sample Preparation

The experimental study took SBCM face of Pingmei No. 12 mine (Henan province, China) as the project background. The goaf was backfilled with "excavation gangue" (gangue produced in the excavation process of rock roadway and coal roadway) piled on the surface, whose lithology is sandstone. The gangue was first crushed by a crusher and then filtered by a classifying screen to obtain five different distribution of particle sizes, including 10~20 mm, 20~30 mm, 30~40 mm, 40~50 mm, and 50~60 mm. When the samples were well prepared, the weight of gangue in each particle size group was 3.2 kg, and the mass ratio was 1:1:1:1:1. Then, the LGBM samples were mixed evenly, and their distribution of particle size is listed in Table 1. After that, they were divided into two parts. The first part was kept at natural condition and the other was immerged in water for 48 days, until the gangues were fully saturated.

Table 1. The particle size distributions of loose gangue backfill material (LGBM) samples.

Particle size/mm	10–20	20–30	30–40	40–50	50–60
Content/%	20	20	20	20	20

2.2. Equipment

A scanning electron microscope (SEM) was first used to analyse the density of the samples. In this experiment, FEI Quanta TM 250 SEM was adopted. Then, an X-ray diffractometer (D/Max-3B, Rigaku Company, Tokyo, Japan) was used to identify and to quantify the mineral and chemical contents of the samples. The equipment parameters were as follows: Incoming radiation mainly include copper target and $K\alpha$ radiation (λ = 1.5406 Å), the X-ray tube operating voltage was 35 kV, the tube current was 30 mA, the slit width was $1°$, and the scan rate was $8°$/min. Lastly, confined compression experiments were carried out to study the effects of water soaked height on the deformation and the crushing characteristics of LGBM. In the process of SBCM, when LGBM is filled into the goaf, it is surrounded by coal pillars, which limits the lateral displacement of LGBM, thus leading to LGBM only having vertical displacement, as is shown in Figure 1. As a consequence, the confined compression tests (the tri-axial compression was under uniaxial strain condition) are usually used to research the deformation characteristics of LGBM [17–20]. Before the experiment, LGBM was directly placed into the steel cylinder, as shown in Figure 2a. During the compression experiment, the lateral deformation of the sample was limited by the wall of the cylinder and the lateral stress was passively generated produced during the compression process due to Poisson's ratio effect. The actual lateral stress constantly changes with the change in vertical compression load.

(a)

(b)

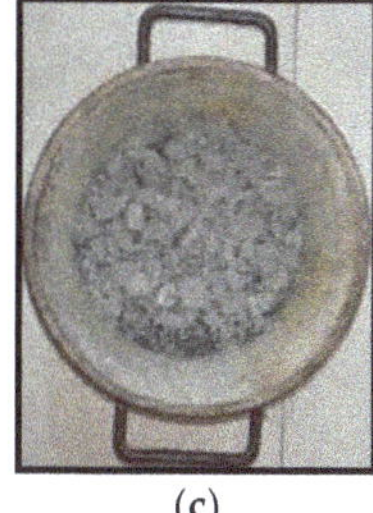
(c)

Figure 2. The process of confined compression tests. (**a**) Before experiment; (**b**) During confined compression experiment; and, (**c**) After experiment.

The loading control system that was used in the experiments was an MTS815 rock mechanics electrohydraulic servo test system (MTS company, Eden Prairie, MN, USA), as shown in Figure 2b. The container that was used in the confined compression tests of LGBM was a Q235 seamless steel drum that was specifically designed to test solid backfill materials that were used in coal mines. It had the following dimensions: the internal diameter is 350 mm, the depth is 378 mm, the loading platen thickness is 40 mm, and the maximum loading height capacity is 338 mm. The height of the material samples that were used in the following experiments was 338 mm.

2.3. Test Procedure

Firstly, the mineral composition and mesostructure of the sandstone-type "excavation gangue" samples were tested by using the scanning electron microscope and X-ray diffractometer. Then, the MTS815 rock mechanics electrohydraulic servo system was applied to measure the basic mechanical properties of the rock samples. Finally, the confined compression tests were employed to examine the effects of water soaked height on the deformation and the crushing characteristics of LGBM. When designing the loading conditions, two key factors were taken into account: the mining depth and the interaction of LGBM with surrounding coal pillars. The maximum loading stress was used to represent the mining depth of 600 m, and the uniaxial strain condition instead of conventional tri-axial condition, was applied for all of the tests in this work due to the restraint of its lateral movement from coal pillars. The same constant loading rate was applied for all of the tests and the loading rate was 1 kN/s.

In the process of testing, the solid backfill materials with different water soaked heights in the mine were simulated by varying the water saturated height of the gangue. When the water soaked heights of LGBM are small, the mechanical properties of LGBM are less affected. Therefore, this study only considered the conditions that the water soaked height was higher than 50% of the total samples. At the same time, the LGBM with a water soaked height of 0 served as a control group. The saturated gangue was loaded into the bottom of the steel drum and the gangue with natural moisture content was placed on the top of the saturated. The test schemes included four-group experiments, as are listed in Table 2 and shown in Figure 3

Table 2. Test design.

Test Case	Saturated Gangue Content (Height Ratio)/h = 338 mm
#1	0 h
#2	0.5 h
#3	0.66 h
#4	1 h

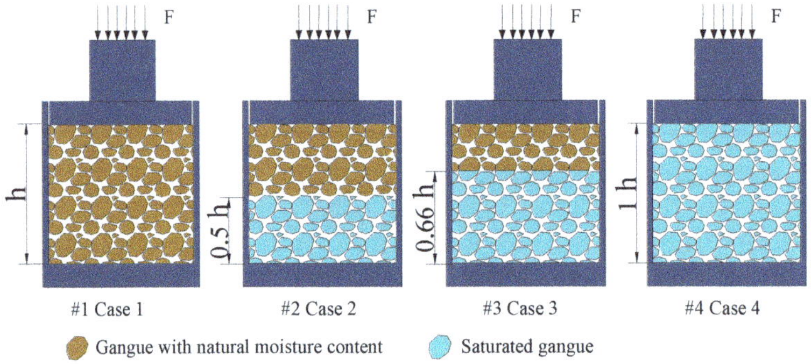

Figure 3. The schematic diagrams of test schemes.

3. Results and Analysis

3.1. Mineral Content and Mesostructure

The mineral and chemical composition of the sandstone type "excavation gangue" samples were tested by using the X-ray diffracrometer. The X-ray diffraction spectrum of the samples is shown in Figure 4. The mineral and chemical composition of the samples are shown in Figures 5 and 6, respectively.

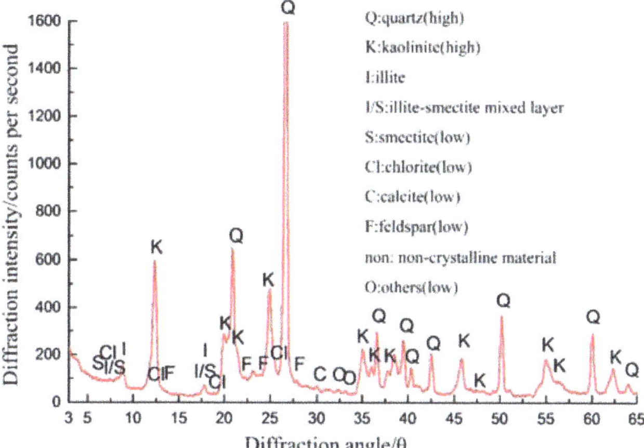

Figure 4. X-ray diffraction spectrum.

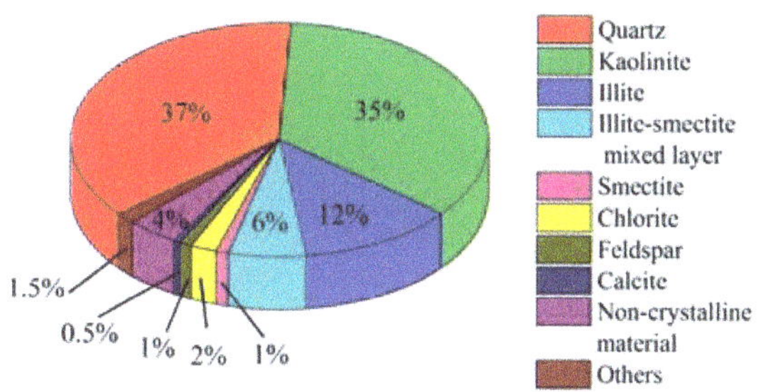

Figure 5. The mineral composition of the sandstone-type "excavation gangue" samples.

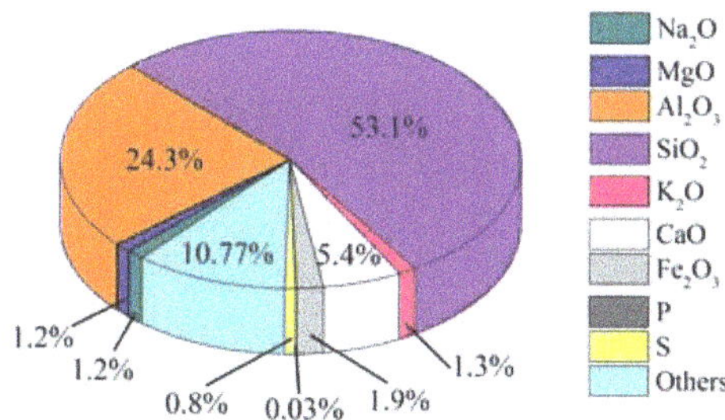

Figure 6. The chemical composition of the sandstone-type "excavation gangue" samples.

The study indicated that the mineral composition and the category of rocks are two main factors influencing the water absorption characteristics of rocks. The content of clay minerals, such as I/S (illite-smectite mixed layer), kaolinite, illite, and chlorite, is a key factor affecting water absorption characteristics of rocks. In particular, the montmorillonite has a stronger ability to absorb water [21]. Figures 4–6 show that the sandstone-type "excavation gangue" is composed mainly of 37% quartz, some illite, mixture of illite and smectite, as well as small amounts of chlorite. Therefore, the sandstone-type "excavation gangue" has a certain ability to absorb water. As for the chemical composition of the sandstone-type "excavation gangue" samples, the content of SiO_2 is 53.1%, which is relatively high, contributing to these samples having a rather high hardness and deformation resistance. In addition, the content of Al_2O_3 is 24.3%, which makes these samples prone to hydrolysis when the samples are immersed in water.

SEM images of LGBM under different magnifications are shown in Figure 7.

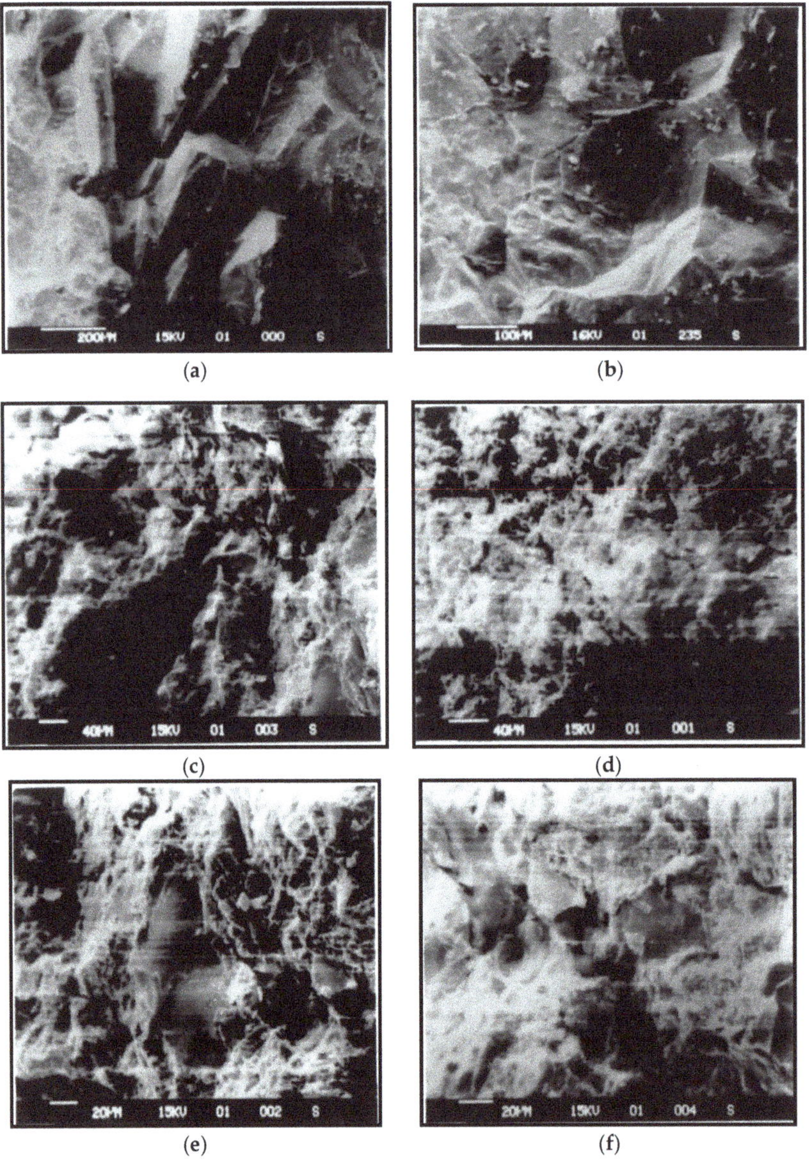

Figure 7. Scanning electron microscope (SEM) images of LGBM under various magnifications. (**a**) 80×; (**b**) 160×; (**c**) 200×; (**d**) 250×; (**e**) 350×; and, (**f**) 450×.

It is shown in Figure 7 that the coarse particles form a certain skeletal framework by being filled with smaller particles. The surface of particles is rough and irregular with a coarse texture. Many pores are visible. However, there are not many large holes or fissures. The existence of many small holes and crevices explains the reason why the material absorbs water and softens.

3.2. Mechanical Properties of Gangue Block

Basic mechanical parameters of the rocks, such as elastic modulus E (secant modulus), Poisson's ratio μ, uniaxial compressive strength σ_c, uniaxial tensile strength σ_t, cohesion c, and internal friction angle φ, were measured by uniaxial compression testing, point load splitting testing, and variable-angle shear testing under both saturated and natural-moisture-content conditions. In the experiments, a total of 42 samples were used and the details are listed in Table 3. Each experiment was repeated three times, and the average of the value of each mechanical parameter was taken as the ultimate value of each mechanical parameter. The ultimate results are listed in Table 4.

Table 3. The number of the samples.

Experiment Name	Uniaxial Compression Testing		Point Load Splitting Testing		Variable-Angle Shear Testing	
Sample type	Gangue with natural moisture content	Saturated gangue	Gangue with natural moisture content	Saturated gangue	Gangue with natural moisture content	Saturated gangue
Number	3	3	3	3	15	15

Table 4. Basic mechanical properties.

Sample Type	E/GPa	μ	σ_c/MPa	σ_t/MPa	c/MPa	φ/°
Gangue with natural moisture content	74.58	0.18	134.33	18.85	12.91	37.65
Saturated gangue	65.21	0.16	121.14	16.56	10.23	35.12

(1) Uniaxial compression testing

The sample was placed on MTS815 rock mechanics electrohydraulic servo test system and was compressed at the rate of 500 N/s, and the pressure was recorded. In this experiment, four axial strain gauges were employed to measure the axial strain and the lateral strain of the samples. Specifically, two strain gauges were used to monitor the axial strain, and the other two strain gauges were used to monitor the lateral strain. The layout of the strain gauges is shown in Figure 8. The tests were repeated three times, and the average of the axial strain values and lateral strain values that were obtained through three tests were taken as the ultimate result.

Figure 8. Uniaxial compression testing of gangue block.

(2) Point load splitting testing

The sample was placed on the point splitting clamp and split at the rate of 500 N/s, during which time the peak pressure was recorded. The tests were repeated three times, and the average of peak pressure obtained through three tests was taken as the ultimate result. The experimental process is shown in Figure 9.

Figure 9. Point load splitting testing of gangue block.

(3) Variable-angle shear testing

In the experiment, the shear angles were respectively taken $45°$, $50°$, $55°$, $60°$, and $65°$. The samples were loaded to damage at the rate of 500 N/s, and the peak pressure was recorded. The experimental process is shown in Figure 10.

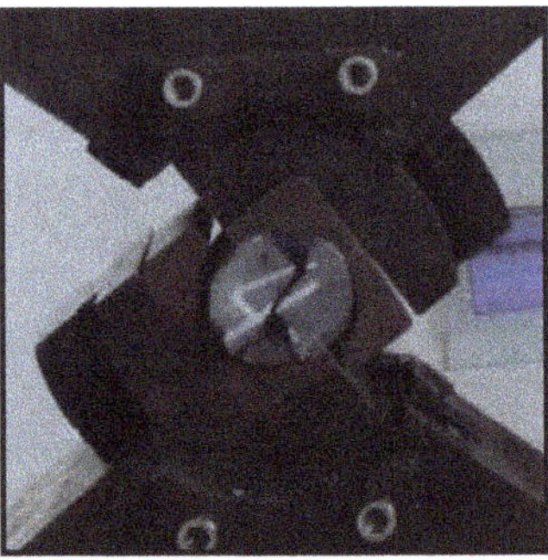

Figure 10. Variable-angle shear testing of gangue block.

Table 4 shows that the basic mechanical properties of the sandstone-type "excavation gangue" decreased after the gangue was soaked in the water and became saturated. To be specific, gangue's elastic modulus E, uniaxial compressive strength σ_c, and uniaxial tensile strength σ_t, respectively, decreased from 74.58 GPa, 134.33 MPa, and 18.85 MPa under a natural-moisture-content condition to 65.21 GPa, 121.14 MPa, and 16.56 MPa under a saturated condition. In addition, its cohesion c dropped from 12.91 MPa under a natural-moisture-content condition to 10.23 MPa under a saturated condition, reducing by 20.8%. Similarly, its friction angle φ decreased from $37.65°$ to $35.12°$, reducing by 6.7%. This means that the hardness of the gangue decreased significantly after the gangue was soaked to saturation, and the gangue was more easily damaged when compressed.

3.3. Deformation Characteristics of LGBM

Figure 11 shows the axial strain—stress curves of the samples in each test case of the confined compression testing. Field monitoring datas from a number of mines (e.g., Pingdingshan twelfth coal mine [22], Zhai zhen coal mine [23], and Jining 3# coal mine [24]) have consistently shown that after LGBM is backfilled into the goaf, the roof experiences rapid subsidence (i.e., sagging), with a sagging rate of greater than 0.075 m/month (average value of the three Coal Mines). It is believed the subsidence reflects the compaction of the solid backfill materials are also in a rapid deformation stage with the same rate. When the roof sagging rate drops to less than 0.0075 m/month (average value of the three Coal Mines), the overburden strata is considered to be at a stable subsidence stage, so is the solid backfill materials. For this reason, in this work, the two key values 0.075 and 0.0075 were selected to divide the axial strain-axial stress curve of LGBM into three segments (rapid deformation stage, slow deformation stage, and stable deformation stage). The sagging rate was the slope of strain-stress curve, which was defined as the ratio of strain change due to stress change for each stress step. The maximum applied pressure in these tests was 16 MPa. When the load reached 16 MPa, the corresponding axial strain of the samples at that time was defined as the maximum axial strain. The relationship between the maximum axial strain and the height ratio of saturated gangue in the sample is shown in Figure 12.

Figures 11 and 12 show that as the load increased, the overall deformation of the samples in each test case could be divided into three segments: rapid deformation (segment I), slow deformation (segment II), and stable deformation (segment III). All of the samples were in the slow deformation stage when the applied axial stress was around 1.7 MPa. When the applied axial stress was around 12 MPa, the samples were in the stable deformation stage. At a given level of axial stress, the axial strain increased with the height ratio of saturated gangue in the sample. This is because the hardness of the gangue decreases after gangue is soaked in water. Gangue particles are more easily to be crushed under the compressive loading after the materials are saturated, which makes the saturated portion of the samples deformed more easily in the compression testing.

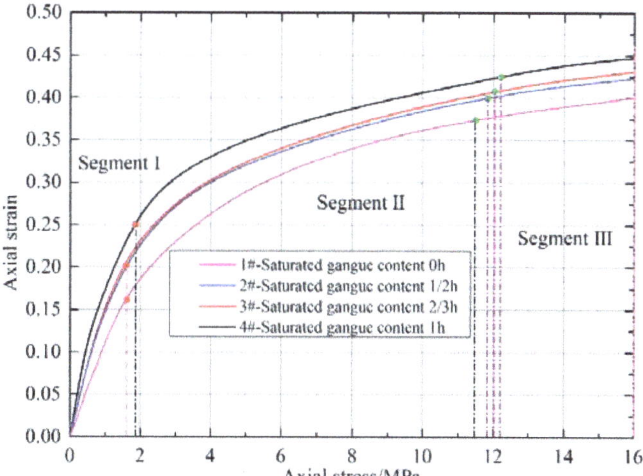

Figure 11. Axial strain-axial stress (ε-σ) curves.

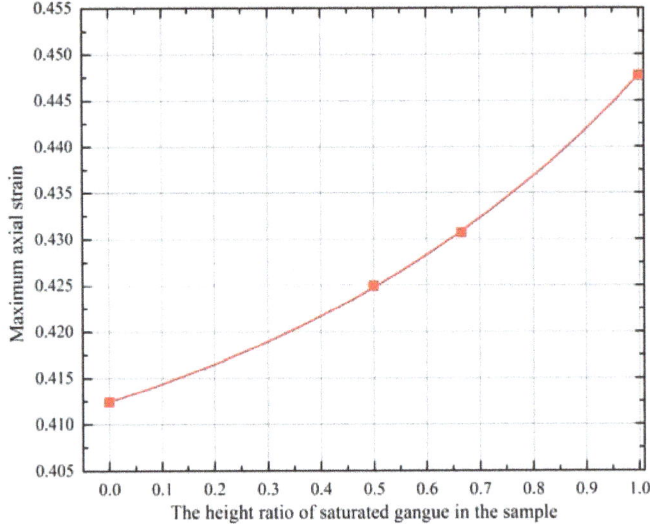

Figure 12. The relationship between the maximum axial strain and the height ratio of saturated gangue in the sample.

Therefore, the height ratio of saturated gangue in the samples was higher, the overall axial strain of the samples was greater. When the samples #1–#4 reached the slow deformation stage, their axial strain was 0.161, 0.2, 0.2, and 0.25, respectively. When the samples reached the stable deformation stage, the axial strain was 0.374, 0.398, 0.408, and 0.425, respectively. There was an exponential relationship between the maximum axial strain and the height ratio of saturated gangue in the sample. The maximum axial strain of the samples #1–#4 were 0.412, 0.425, 0.431, and 0.448, respectively. When compared with the maximum axial strain of sample #1, the maximum axial strain of samples #2–#4 increased by 3.2%, 4.6%, and 8.7%, respectively. Apparently, these changes were significant.

The above results indicated that the deformation resistance of LGBM used in SBCM would become significantly lower after it was soaked in water underground the mine. The higher the water level of LGBM, the lower the deformation resistance of the materials, leading to larger influence on the quality of backfill.

3.4. Crushing Characteristic of LGBM

(1) Quantitative index of particle crushing

In the loading process of gangue, the grain will be crushed. In order to study the impact of grain crushing on the strength and the deformation characteristics of LGBM, there is a need to quantitatively reflect the crushing degree of particle. Currently, the Marsal crushing ratio (B_g) [6], particle size control difference (B) [25], and relative crushing ratio (B_r) [26] are considered as three different quantitative indexes that are the most commonly used in geotechnical engineering. The Marsal crushing ratio (B_g) is defined as the sum of positive values of difference in particle content with different gradation before and after the test, which is expressed as a percentage. Its computational formula is, as follows:

$$B_g = \sum W_{ki} - W_{kf} \tag{1}$$

where, W_{ki}—particle content of a certain gradation in size before the test; and, W_{kf}—the corresponding particle content of the identical gradation in size after the test.

The Marsal crushing ratio is a multi-gradation difference-value summation method, which can reflect the full aspect of crushing condition of particle. Having the merits of simple calculation and clear physical definition, as well as high accuracy, which can meet the need of practical engineering, the Marsal crushing ratio (B_g) was used in this study as an essential quantitative index to describe the crushing condition of particle.

(2) Quantitative analysis of loose gangue crushing

After testing, the samples were allowed to dry naturally. Then, they were sieved to obtain the proportions of each particle size group and crushing ratio listed in Table 5. Figure 13 shows the relationship between the maximum axial strain and saturated gangue content, as well as the relationship between the crushing ratio and the saturated gangue content. Figure 14 shows the relationship between the maximum axial strain and the crushing ratio of each sample.

Table 5. Percentage of each particle size group and crushing ratio for each sample after testing.

		Percentage of Each Particle Size Group/%						
Range of Particle Diameters		0–10 mm	10–20 mm	20–30 mm	30–40 mm	40–50 mm	50–60 mm	Crushing Ratio B_g/%
	Before testing	0	20.00	20.00	20.00	20.00	20.00	0
	#1	43.11	21.43	13.28	9.39	10.79	1.98	44.56
Sample name	#2	47.50	20.68	12.66	8.53	8.02	2.32	48.47
	#3	48.70	20.62	12.90	10.22	4.95	2.65	49.28
	#4	48.41	18.67	13.50	9.38	10.04	0	58.41

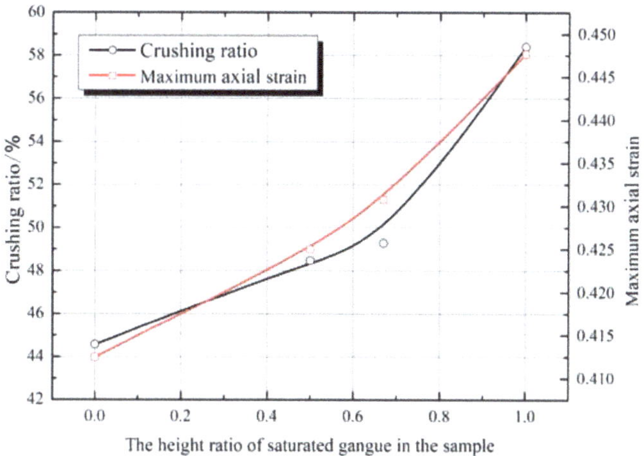

Figure 13. The relationship of the saturated gangue content with the crushing ratio or the maximum axial strain.

It can be obtained from Table 5 and Figures 13 and 14 that when the load reached 16 MPa, the particles of all loose gangue samples would be crushed significantly. The crushing ratios were all higher than 40%. The crushing ratio increased with the height ratio of saturated gangue in the sample. As for the samples that were not saturated in the water, the crushing ratio was 44.56%. In terms of the samples with the height ratio of saturated gangue of 0.5 h, 0.66 h, and 1 h, the crushing ratios were 48.47%, 49.28%, and 58.41%, respectively, rising by 8.8%, 10.6%, and 31.1%, respectively, when compared with the crushing ratio of the samples that were not saturated in the water. This can be explained by significantly decreased hardness of the saturated samples. When the samples were compressed under the same maximum axial stress of 16 MPa, the higher the height ratio of saturated gangue in the sample, the higher the crushing ratio and axial strain. In addition, there is a strong linear relationship between the axial strain and the crushing ratio, meaning that in the process of compression, particle crushing is the main cause of the deformation of the samples.

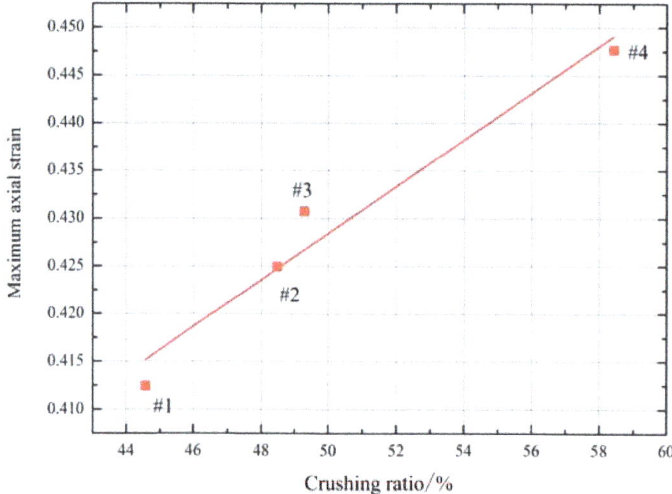

Figure 14. The relationship between the maximum axial strain and the crushing ratio.

4. Conclusions

This paper describes the experiments that were conducted to examine the mineral composition, mesostructure, and basic mechanical parameters of gangue rock. Then, details of the confined compression experiments that were carried out to study the effects of water soaked height on the deformation and the crushing characteristics of LGBM are presented. The conclusions are as follows:

(1) Sandstone-type "excavation gangue" is mainly composed of two minerals: quartz and kaolinite, with quartz and kaolinite accounting for 37% and 35% of the whole content, respectively. There are many small holes in the gangue material, which are the main reason why the material absorbs water and softens.

(2) The hardness of sandstone-type "excavation gangue" decreases significantly after the gangue is soaked in water. At a given load, the particles of saturated gangue are more easily crushed than gangue with natural moisture content.

(3) Under a given level of applied axial stress, the axial strain of gangue samples increased with the height ratio of saturated gangue in the sample. There was an exponential relationship between the maximum axial strain and the height ratio of saturated gangue in the sample. It had been shown that LGBM that is used to backfill the goaf in SBCM loses a certain amount of its ability to resist deformation after it is soaked in water existing in the mine. The higher the water level of the solid backfill materials was, the less likely that the solid backfill material can resist deformation, leading to a greater impact on the quality of backfill.

(4) When the loose gangue samples were compressed at 16 MPa, a higher saturated height ratio of the samples corresponded to a larger crushing ratio. The crushing ratios were all higher than 40%. In addition, the maximum axial strain was strongly correlated to the crushing ratio, meaning that particle crushing was the main cause of deformation of the gangue samples in the compression process.

There are some limitations in this study. Besides LGBM, the solid backfilling materials also include loess, aeolian sand, fly ash, and so on. This work solely focusses on the LGBM material and only single LGBM particle size distribution feature was studied. In the future studies, the effects of water soaked height on the deformation and the crushing characteristics of different solid backfilling materials would be strongly recommended. Moreover, the maximum pressure that was applied in this work was 16 MPa, which may not be representative of coal mining conditions at greater depth, thus it is recommended that the deformation characteristics of LGBM under greater loading conditions be studied in the future. In this work, the lateral deformation was monitored via two strain gauges, the average value was used together with the vertical displacement measurements to determinate the Poisson's ratio. The value may not be very reliable due to the limitation of the points, but this does not sacrifice the aim of the measurement, which is to gain the information of the key mechanical properties of gangue and how they are different to other rocks.

Author Contributions: J.L. and Y.H. conceived and designed the experiments; T.S. and G.K. performed the experiments; M.Q. and H.G. analyzed the data; L.G. contributed materials; J.L. wrote the paper; Y.H. and Z.C. modified the paper.

Acknowledgments: Financial support for this work provided by the Fundamental Research Funds for the Central Universities (2017BSCXA20) and Postgraduate Research & Practice Innovation Program of Jiangsu Province (KYCX17_1552) are gratefully acknowledged.

Conflicts of Interest: The authors declare no conflicts of interest.

References

1.   Miao, X.; Zhang, J.; Guo, G. *Waste Backfilling Method and Technology in Fully Mechanized Coal Mining*; China University of Mining & Technology press: Xuzhou, China, 2010.
2.   Miao, X. Progress of fully mechanized mining with solid backfilling technology. *J. China Coal Soc.* **2012**, *37*, 1247–1255.
3.   Zhang, Q.; Zhang, J.X.; Ju, F.; Tai, Y.; Li, M. Theoretical research on mass ratio in solid backfill coal mining. *Environ. Earth Sci.* **2016**, *75*, 586. [CrossRef]
4.   Li, M.; Zhang, J.; Huang, Y.; Zhou, N. Effects of particle size of crushed gangue backfill materials on surface subsidence and its application under buildings. *Environ. Earth Sci.* **2017**, *76*, 603. [CrossRef]
5.   Yilmaz, E.; Fall, M. *Paste Tailings Management*, 1st ed.; Springer International Publishing: Cham, Switzerland, 2017.
6.   Marsal, R.J. *Mechanical Properties of Rockfill Embankment Dam Engineering*; John Wiley and Sons Inc.: New York, NY, USA, 1973; pp. 109–200.
7.   Marsal, R.J. Large scale testing of rockfill materials. *J. Soil Mech. Found. Div.* **1967**, *93*, 27–43.
8.   Marsal, R.J. *Earth-Rock Dam Project*; Hydraulic press: Jiangsu, China, 1979.
9.   Kong, D.; Zhang, B.; Sun, X. Triaxial Tests on Particle Breakage Strain of Artificial Rock Fill Materials. *Chin. J. Geotech. Eng.* **2009**, *31*, 464–469.
10. Feda, J. Notes on the effect of grain crushing on the granular soil behavior. *Eng. Geol.* **2002**, *19*, 778–785.
11. Wang, B.; Jin, X.; Liu, H. Testing Study on Mechanical Properties of Slab Rockfill Dam Materials. *Chin. J. Rock Mech. Eng.* **2003**, *22*, 332–336.
12. Bagherzadeh-Khalkhali, A.; Mirghasemi, A.A. Nu-merical and experimental direct shear tests for coarse-grained soils. *Particuology* **2009**, *7*, 83–91. [CrossRef]
13. Kjaernsli, B.; Sande, A. Compressibility of some coarse-grained materials. In Proceedings of the 1st European Conference on Soil Mechanics and Foundation Engineering, Weisbaden, Germany, 15–18 October 1963; pp. 245–251.
14. Hall, E.B.; Gordon, B.B. Triaxial testing with large-scale high pressure equipment. *Lab. Shear Test. Soils* **1963**, *361*, 315–328.
15. Zhang, J. Study on Strata Movement Controlling by Raw Waste Backfilling with Fully-Mechanized Coal Winning Technology and Its Engineering Applications. *China Univ. Min. Technol.* **2008**, *33*, 183–186.

16. Huang, Y.; Li, J.; Teng, Y.; Dong, X.; Wang, X.; Kong, G.; Song, T. Numerical simulation study on macroscopic mechanical behaviors and micro-motion characteristics of gangues under triaxial compression. *Powder Technol.* **2017**, *320*, 668–684. [CrossRef]

17. Li, M.; Zhang, J.; Zhou, N.; Huang, Y. Effect of Particle Size on the Energy Evolution of Crushed Waste Rock in Coal Mines. *Rock Mech. Rock Eng.* **2017**, *50*, 1347–1354. [CrossRef]

18. Zhang, J.; Li, M.; Liu, Z.; Zhou, N. Fractal characteristics of crushed particles of coal gangue under compaction. *Powder Technol.* **2017**, *305*, 12–18. [CrossRef]

19. Zhou, N.; Han, X.; Zhang, J.; Li, M. Compressive deformation and energy dissipation of crushed coal gangue. *Powder Technol.* **2016**, *297*, 220–228. [CrossRef]

20. Li, M.; Zhang, J.; Gao, R. Compression Characteristics of Solid Wastes as Backfill Materials. *Adv. Mater. Sci. Eng.* **2016**, *2016*, 2496194. [CrossRef]

21. He, M.; Zhou, L.; Li, J. Experimental study on water absorption characteristics of deep well mudstone. *J. Rock Mech. Eng.* **2008**, *27*, 1113–1120.

22. Huang, Y.; Li, J.; Song, T.; Kong, G.; Li, M. Analysis on filling ratio and shield supporting pressure for overburden movement control in coal mining with compacted backfilling. *Energies* **2017**, *10*, 31. [CrossRef]

23. Zhang, Q.; Zhang, J.; Zhao, X.; Liu, Z.; Huang, Y. Industrial tests of waste rock direct backfilling underground in fully mechanized coal mining face backfilling. *Environ. Eng. Manag. J.* **2014**, *13*, 1291–1297.

24. Zhang, Q.; Zhang, J.X.; Kang, T.; Sun, Q.; Li, W.K. Mining pressure monitoring and analysis in fully mechanized backfilling coal mining face—A case study in Zhai Zhen Coal Mine. *J. Cent. South Univ.* **2015**, *22*, 1965–1972. [CrossRef]

25. Fu, Z.; Feng, J. *Concrete Faced Rock Fill Dam*; Huazhong University of Science and Technology Press: Wuhan, China, 1993.

26. Hardin, C.S. Crushing of soil particles. *J. Geotech. Eng.* **1985**, *111*, 1177–1192. [CrossRef]

 *processes*

Article

Key Parameters of Gob-Side Entry Retaining in A Gassy and Thin Coal Seam with Hard Roof

Shuai Yan [1,2] , **Tianxiao Liu** [1,2], **Jianbiao Bai** [1,2,*] **and Wenda Wu** [1,2]

1   Key Laboratory of Deep Coal Resource Mining (CUMT), Ministry of Education, Xuzhou 221116, China; yanshuai@cumt.edu.cn (S.Y.); ckliutx@163.com (T.L.); wwdcumt@126.com (W.W.)
2   School of Mines, China University of Mining & Technology, Xuzhou 221116, China
*   Correspondence: baijianbiao@cumt.edu.cn; Tel.: +86-516-8388-5602

Received: 4 April 2018; Accepted: 3 May 2018; Published: 7 May 2018

Abstract: Gob-side entry retaining (GER) employed in a thin coal seam (TCS) can increase economic benefits and coal recovery, as well as mitigate gas concentration in the gob. In accordance with the caving style of a limestone roof, the gas concentration and air pressure in the gob were analyzed, and a roof-cutting mechanical model of GER with a roadside backfill body (RBB) was proposed, to determine the key parameters of the GER-TCS, including the roof-cutting resistance and the width of the RBB. The results show that if the immediate roof height is greater than the seam height, the roof-cutting resistance and width of the RBB should meet the requirement of the immediate roof being totally cut along the gob, for which the optimal roof-cutting resistance and width of RBB were determined by analytical and numerical methods. The greater the RBB width, the greater its roof-cutting resistance. The relationship between the supporting strength of the RBB and the width of the RBB can be derived as a composite curve. The floor heave of GER increases with increasing RBB width. When the width of the RBB increased from 0.8 m to 1.2 m, the floor heave increased two-fold to 146.2 mm. GER was applied in a TCS with a limestone roof of 5 m thickness; the field-measured data verified the conclusions of the numerical model.

Keywords: gas concentration; gob-side entry retaining (GER); limestone roof; roof-cutting resistance; roadside backfill body (RBB)

---

1. Introduction

In China, thin coal seams (TCS), defined as less than 1.3 m thick, are rich in resources but challenging technically to mine. The mineable reserves of these seams account for 20.4% of the total coal resources in China but only 10.4% of the current total annual production [1]. The mining of TCS not only meets requirements of sustainable development but also has economic and social value on the gas extraction and the recovery rate of coal resources [2]. Entries in TCSs are excavated in mixed coal and rock mass, which decrease excavation speed of roadway and delay the mining plan and exacerbate environmental problems of gangue hills [3]. Gob-side entry retaining (GER), where a roadway of previous panel is retained without coal pillars by constructing a roadside backfill body (RBB) lagging behind the working face along the gob-side, was proposed to solve these problems, as shown in Figure 1.

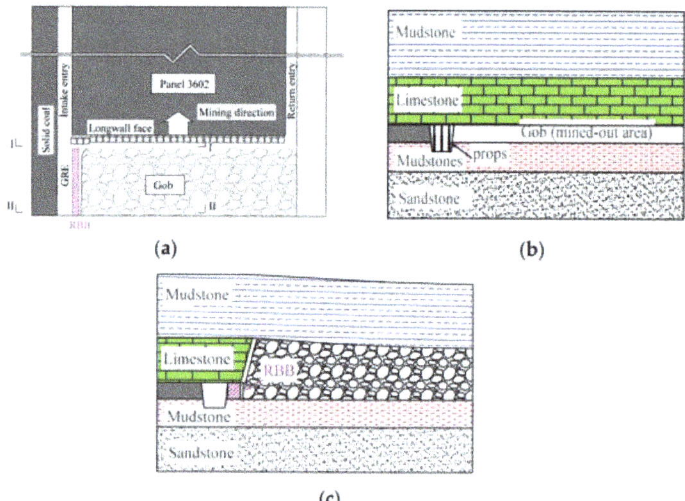

Figure 1. Schematic diagram of Gob-side entry retaining-thin coal seam-limestone roof (GER-TCS-LR):
(**a**) Plan view; (**b**) I-I section and (**c**) II-II section.

GER is an effective approach for sustainable and efficient exploitation of coal resources [4,5]. Through the analysis of the ground pressure and physical simulation, Lu [6] believed that the "given deformation" of roof subsidence in GER is proportional to mining height. Some researchers investigated the stress environment of GER in roof caving laws [7,8]. Based on the development of the Wilson model and the static analysis of structure, Whittaker and Woodron [9] proposed a mechanical model of separation rock, which can determine the parameters of roadway support. Hidalgo and Nordlund [10] compared the laboratory test result with the numerical simulation result, concluding that the laboratory test data can predict the excavation damage process of hard rock. Villaescusa et al. [11] used the method to analyze the performance quantization of the resin anchor of hard roof during underground mining. Sun et al. [12] determined the key parameters in GER formed by roof cutting and pressure releasing in TCS, such as the pre-split blasting height of roof, the pre-split blasting angle and the pattern of pre-split blasting holes.

In this paper, the GER technique was employed under conditions of a hard roof with strong ground pressure and quick-caving behaviors, the gas concentration and air pressure in the gob were analyzed and a roof-cutting mechanical model of GER with RBB was proposed to determine the key parameters of the GER-TCS, including the roof-cutting resistance and the width of the RBB. An alternative RBB with high-water, fast-setting material was constructed. Based on a mechanical model and a numerical model, the roof-cutting resistance of the RBB and RBB width were determined. A field case study in the Nantun coal mine, Shandong Province, China is also presented to validate the model results.

2. Materials and Methods

2.1. GER Techniques in a Thin Coal Seam (TCS) with a Limestone Roof (LR)

Based on the principle of GER with high-water, fast-setting material, the technical process of GER in a TCS with a limestone roof (LR) (GER-TCS-LR) is shown in Figure 1. The fracturing and caving of the roof occurs as it loses coal support in the gob and the load of the surrounding rock transfers to the solid coal that forms the abutment pressure. The bearing capacity of the shallow part of the gob-side coal body declines significantly when subjected to extensive coal mining, leading to decreasing pressure of GER. An abutment pressure zone and a decreasing pressure zone existing from deep to shallow, respectively, is the stress environment for GER.

Taking into account the complexity of the mining geology and the roof caving after coal mining, the inner area of the gob is simplified as a homogeneous and homogenous porous medium [13], the gob dimension in the simulation model is 200 m and 150 m in mining direction and dip direction, respectively. Because the roof boundary of the gob is uncertain and mining height is negligible compared with the gob length in the dip and strike direction, the gob is simplified as a two-dimensional plane. The density of the gas in the gob is assumed to be unchanged with time. The flow process is steady and constant and the gas in the gob is incompressible.

(1) Gas concentration in gob

As shown in Figure 2, the gas concentration in the gob under the Y-style ventilation mode of a mining face shows that the gas concentration around the gob-side entry in the dip direction of the gob is lower than that in the return entry under the influence of the airflow in the intake entry. The gas concentration in gob is high within the range of 25 m away from the return entry. Meanwhile, with the increase of gob depth in the strike direction, the concentration of gas increases. The direction of gas concentration gradient is approximately the same as that of air leakage.

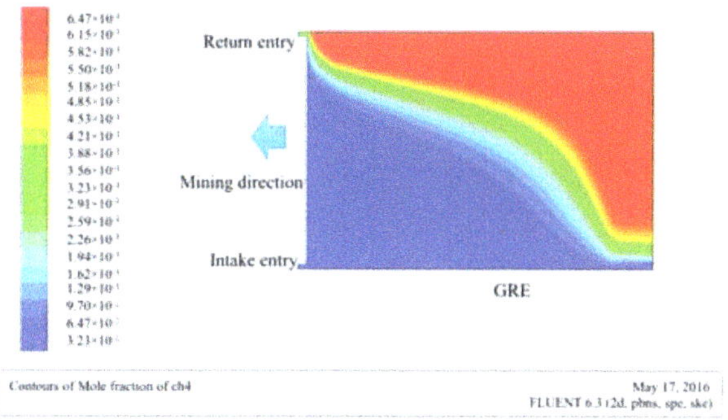

Figure 2. Contour of gas concentration in the gob of GER-TCS-LR.

(2) Contour of air pressure in the gob

Figure 3 shows the contour of air pressure in the gob under the Y-style ventilation mode, it can be seen that the air pressure of the return entry in the T-junction is the lowest and it is the highest in the intake area of the gob-side entry. The air pressure in the gob decreases gradually along the mining direction.

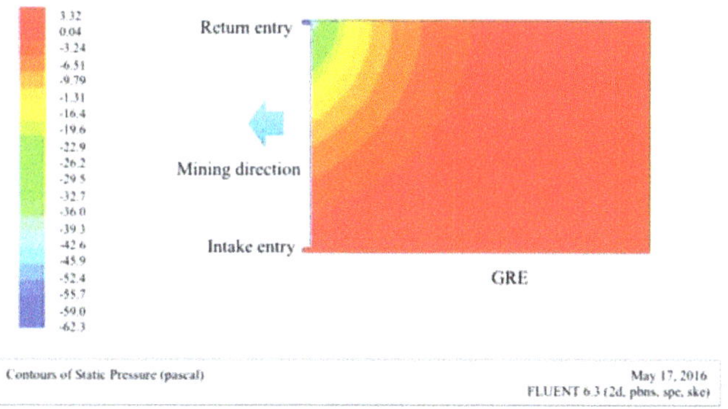

Figure 3. Contour of air pressure in the gob of GER-TCS-LR.

2.2. Mechanical Model for GER-TCS-LR

When the caved roof height is more than five times the mining height, the caved rocks can be full of gob. The caved roof at the gob edges often breaks into regular blocks that extrude and integrate mutually to form a "voussoir beam" structure. Using the method of roof load strip segmentation, a roof-cutting mechanical model of GER was proposed considering the actual size and proportion of the various rock seams [14,15], as shown in Figure 4. The mechanical model of the roof allows separation and misalignment in sedimentary rocks and applies to continuous materials and hinged rock blocks and not to loose rock mass.

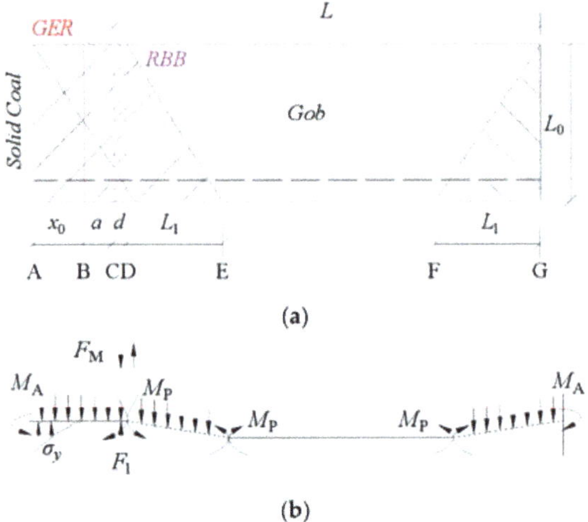

Figure 4. Roof-cutting mechanical model of GER: (**a**) Plan view: supported by three edges (periodic weighting) and (**b**) Section view: loading state of caved roof.

The assumptions in the model are as follows: (1) the intersection of the fracture zone and the plastic zone in the solid coal is the supporting point of the roof (clamped support or simple support); (2) the immediate roof is broken on the outside of the RBB under the effect of overlying rock, the immediate roof and the RBB; (3) when the immediate roof is broken down, M_P designates the ultimate bending moment of the facture point; and (4) the roof-cutting resistance of the RBB is simplified as a concentrated load. Using the block balance method, sections DE and AD (as shown in Figure 4a are derived according to Equations (1) and (2), respectively.

$$\begin{cases} \sum F_y = F_M - \gamma h_1 L_1 - q_1 L_1 = 0 \\ \sum M = 2M_P - \gamma h_1 L_1{}^2/2 - q_1 L_1{}^2/2 = 0 \end{cases} \tag{1}$$

$$\begin{aligned} \sum M = \; & M_A + F_1\left(a + \tfrac{d}{2} + x_0\right) + \int_0^{x_0} \sigma_y(x_0 - x)dx \\ & - M_P - \frac{\gamma h_1(a+d+x_0)^2}{2} - F_M(a + d + x_0) \\ & - \frac{q_1(a+d+x_0)^2}{2} = 0 \end{aligned} \tag{2}$$

Using Equation (1), $q_1 = 4M_P/L_1{}^2 - \gamma h_1$; combining this with Equations (2) and (3) is derived.

$$F_1 = \frac{M_P - M_A}{a + \frac{d}{2} + x_0} + \frac{4M_P}{L_1\left(a + \frac{d}{2} + x_0\right)}(a + d + x_0)$$
$$+ \frac{\left(4M_P - \gamma h_1 L_1{}^2\right)(a + d + x_0)^2}{2L_1{}^2\left(a + \frac{d}{2} + x_0\right)}$$
$$+ \frac{\gamma h_1 (a + d + x_0)^2}{2\left(a + \frac{d}{2} + x_0\right)} - \frac{\int_0^{x_0} \sigma_y (x_0 - x)dx}{a + \frac{d}{2} + x_0} \tag{3}$$

where F_1 is the roof-cutting resistance of the RBB, N/m; γ is the unit weight of the immediate roof, N/m^3; h_1 is the thickness of the immediate roof, m; a is the roadway width, m; d is the width of the RBB, m; q_1 is the load of soft rocks above the immediate roof, MPa; L_1 is the broken length of transverse caved blocks, which equals the periodic breakage length of roof, m; M_A is the bending moment of broken immediate roof, Nm; M_P is the ultimate bending moment when the immediate roof is breaking down, Nm; x_0 is the width of the limited equilibrium zone, m; σ_y is the bearing stress of the limited equilibrium area, MPa; and x_0, σ_y and L_1 can be expressed in the following general forms [16]:

$$x_0 = \frac{h_m A}{2 \tan \phi_0} \ln \frac{k\gamma_0 H + \frac{C_0}{\tan \phi_0}}{\frac{C_0}{\tan \phi_0} + \frac{P_c}{A}} \tag{4}$$

$$\sigma_y = \left(\frac{C_0}{\tan \phi_0} + \frac{P_c}{A}\right) e^{\frac{2\tan \phi_0}{h_m A} x} - \frac{C_0}{\tan \phi_0} \tag{5}$$

$$L_1 = h_1 \sqrt{\frac{R_t}{3q_1}} \tag{6}$$

where h_m is the mining height, m; A is the side abutment pressure coefficient; ϕ_0 is the internal friction angle of the coal seam, °; C_0 is the cohesion of the coal seam, MPa; k is the stress concentration factor; γ_0 is the unit weight of overlying strata, N/m^3; H is the mining depth, m; P_c is the support strength of solid coal, MPa; and R_t is the tensile strength of the immediate roof, MPa.

In Equation (3), the greater L_1 is, the smaller F_1 is, meaning there is an inversely proportional relationship between the roof-cutting resistance of the RBB and the periodic breakage length of the roof. The greater M_P is, the greater F_1 is, meaning the larger the resistibly flexural capacity of the roof, the larger the roof-cutting resistance needed.

When Point A is the simply supported end, $M_A = 0$ and Equation (3) was derived. It then follows that:

$$F_1 = \frac{\gamma h_1 L_1{}^2 + q_1 L_1{}^2}{4\left(a + \frac{d}{2} + x_0\right)} + \frac{\gamma h_1 L_1 + q_1 L_1}{a + \frac{d}{2} + x_0}(a + d + x_0)$$
$$+ \frac{(q_1 + \gamma h_1)(a + d + x_0)^2}{2\left(a + \frac{d}{2} + x_0\right)} - \frac{\int_0^{x_0} \sigma_y (x_0 - x)dx}{a + \frac{d}{2} + x_0} \tag{7}$$

Based on the characteristics of GER and the mechanical model of the immediate roof, the key parameters of GER in TCS are described below.

(1) Roof-cutting resistance of the RBB

When the thickness of the immediate roof is five times the mining height or more, the caved immediate roof can be full of a gob in TCS. However, when the main roof of mudstone and the immediate roof of hard limestone compose an "upper soft and lower hard" roof structure, the immediate roof cannot actively collapse. This requires a high roof-cutting resistance of the RBB to break down the immediate roof, while the main roof of weak strata collapses passively with immediate roof caving.

(2)   The width of the RBB

In Equation (7), the relationship between the roof-cutting resistance of the RBB and the RBB width is positive, that is, the roof-cutting resistance of the RBB increases with increasing RBB width. In view of economy and labor intensity of GER, a slender width of RBB is preferred. Therefore, the optimal width of RBB is a key parameter of GER-RRB-TCS-LR.

Traditional roadside support can be achieved with many techniques [9,17–19], including the wood stack, waste rock pillar, dense hydraulic props, concrete blocks and other artificial product supports of low intensity. The supporting resistance and deformability of traditional roadside support cannot adapt for the deformation of surrounding rock and is unable to isolate gob effectively, making it difficult to achieve high mechanization. Currently, the main roadside support material is high-water material which is made of sulphoaluminate cement clinker, gypsum, lime, compound retarder, suspension agent, compound accelerator and so on [19]. The uniaxial compressive strength (UCS) test of the standard specimens (diameter 50 mm and height 100 mm) with high-water quick-setting materials under water-to-cement ratio of 1.5:1 was conducted. The material is consolidated under the test temperature of 20 °C and the initial strength is 4.48 MPa after 2 h. The strain-stress curve of the standard specimens obtained by the laboratory test system MTS815 is shown in Figure 5. Point O, A, B and C mean the initial strength, ultimate strength, post-peak strength and residual strength, respectively.

The GER takes advantage of these combined impacts, including high-water material with rapidly increasing support resistance, the high support resistance of the RBB and the overlying load, which make the limestone roof cut off along the gob-side wall and reduce the load around the GER. Therefore, a dynamic disaster caused by a roof collapse of a large area can be avoided.

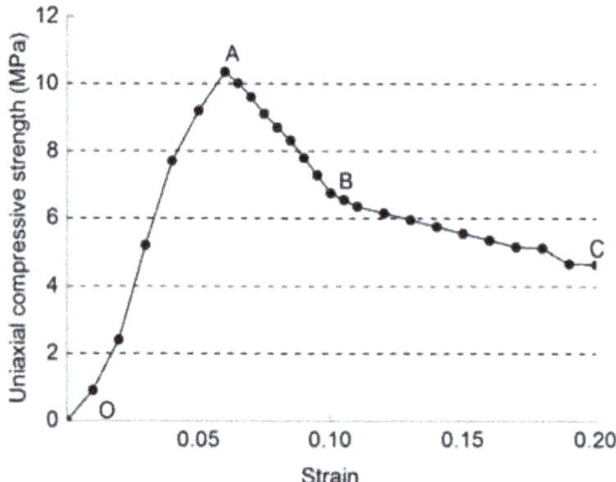

Figure 5. The stress-strain curve of the high-water material in the laboratory.

3. Determination of RBB Width

3.1. Mining and Geological Conditions of the Tested Roadway

The Nantun coal mine, which is located in the Yanzhou mining area of Shandong Province, eastern China (Figure 6), was selected for the case study. Panel 3602 in the Nantun coal mine mainly mines the No. 16-1 coal seam, buried at a depth of 575 m. The average thickness of the coal seam is 0.9 m, with an average inclination of 4°. The panel employs fully mechanized mining technology with a panel width of 164.5 m. The roof and floor lithology of panel 3602 is shown in Table 1.

Figure 6. Location of the test site.

Table 1. Roof and floor lithology of panel 3602.

Name	Lithology	Thickness (m)	Lithological Description
Main roof	Mudstone	6.4	Light grey, with mudstone and siltstone stripe
Immediate roof	Limestone	5.0	Dense hard, rich fracture
Coal bed	Coal	0.9	
Immediate floor	Aluminous mudstones	2.1	Light grey, water expand
Main floor	Sandstone	5.0	Grey with horizontal bedding

Rock mechanics tests and field observations of panel 3602 showed that for the high strength limestone roof, roof caving occurs suddenly when subjected to a small deformation. The main roof and immediate roof of panel 3602 are mudstone and limestone, respectively, that is, the typical "upper soft and lower hard" stratum, passively caved limestone roof. Therefore, the construction of the GER requires higher roof-cutting resistance or other measures to make the gob roof cave immediately with coal mining, to prevent a dynamic disaster or strong dynamic load to the surrounding rock of the GER.

3.2. Determination of the Roof-Cutting Resistance of the RBB

In accordance with the mining and geological conditions of panel 3602, where m = 0.9 m, A = 0.8, ϕ_0 = 21°, C_0 =1.4 MPa, k = 3.0, γ_0 = 2.5 × 104 N/m^3, H = 420 m, P_c = 0.1 MPa, γ = 2.7 × 104 N/m^3, h_1 = 5.0 m, q_1 = 0.5 MPa and R_t = 6.23 MPa, using the math software Mathcad, the above data were put into Equations (4) and (6) to give x_0 = 2.1 m and L_1 = 10.2 m.

The ratio of roof-cutting resistance of the RBB and the width of the RBB is defined as the RBB supporting strength, as described by Equation (8):

$$\sigma_F = \frac{F_1}{d} \tag{8}$$

When the average roadway width is 3.4 m, the relationship between the width of the RBB and the roof-cutting resistance of the RBB and supporting strength can be obtained from Equations (7) and (8), as shown in Figure 7.

From Figure 7, the relationship between the supporting strength of the RBB and the width of the RBB can be derived as a composite curve: $\sigma_F = 9.88 - 9.53 \ln(d - 0.1)$, variance is 0.978.

The derived curve of the supporting strength of the RBB and the RBB width shows that the required supporting strength of the RBB increases rapidly as the width of the RBB decreases, when the RBB width is between 0.1 m and 1.1 m. Otherwise, the required support strength of the RBB decreases slowly. Considering the cost and mechanical properties of the RBB, the width of the RBB should be greater than 1.1 m, while the roof-cutting resistance of the RBB should be greater than 8.465 MN/m and the supporting strength of the RBB should not be less than 7.696 MPa.

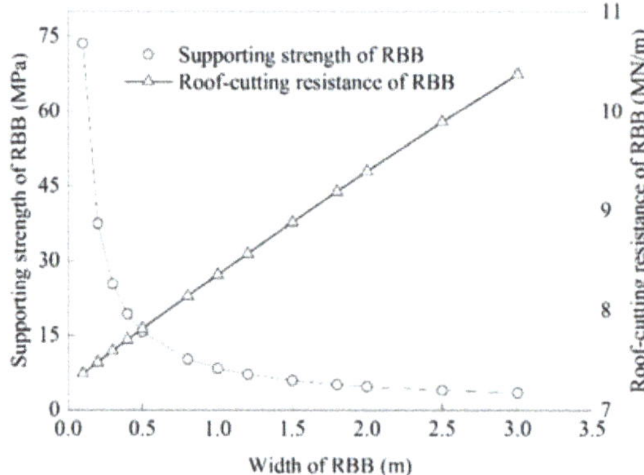

Figure 7. Roof-cutting resistance and support strength of the roadside backfill body (RBB) with respect to RBB width.

Using an example of an RBB with high-water, fast-setting material and considering a given safety coefficient k_0, the width of the RBB is d, the required supporting strength of the RBB is σ_F and the uniaxial compressive strength (UCS) of the high-water material (7 d age) is σ_0. Then:

$$\sigma_0 \geq k_0[\sigma_F] \tag{9}$$

If k_0 is defined as 1.2, the UCS of the high-water material (7 d age) is required to be greater than 9.24 MPa. When the water-to-cement ratio of high-water material is 1.5:1, its UCS is 10 MPa after 7 days [19], meeting the requirements of the RBB supporting strength.

The analytical results show that the roof-cutting resistance of the RBB should not be less than 8.465 MN/m when the width of the RBB is greater than 1.1 m and the water-to-cement ratio of the high-water material is 1.5:1.

3.3. Determination of the RBB Width

To determine an appropriate width of the RBB, a numerical simulation was used to analyze the vertical stress and the surrounding rock deformation of the GER with respect to different widths of RBB. Results of the numerical model were combined with the result of the theoretical calculations to determine the width of the RBB.

3.3.1. Numerical Model

Based on the mining and geological conditions of Panel 3602 in the Nantun coal mine, FLAC^{3D} [20] was used to construct a numerical model of plan-strain conditions, as shown in Figure 8. The model dimensions were 173.6 m × 200 m × 82 m, with one-half of panels 3604 and 3602 and the roadway system between them. At the top of the models, a vertical load was applied to simulate the overburden weight. The at-rest pressure coefficients along the x and y directions were both taken as 0.8 according to the in situ stresses of the No. 16-1 coal seam. The horizontal and bottom sides were roller constrained. The Mohr–Coulomb criterion was used to simulate the rock strata and the Cvisc creep model was used for the RBB of the GER.

Based on the laboratory results of the mechanical properties of rock and coal samples, the parameters employed in the model have been determined that elastic modulus, cohesion and tensile strength is 0.2 of the laboratory testing results and the Poisson ratio is 1.2 of the laboratory testing results, following the primary works of Perry [21] and Mohammad [22]. In addition, few trial model tests have been carried out and the mechanical parameters were adjusted in reference to the in situ measurement of the displacement and the width of yield zones developed around the roadways [23]. The final rock strata property parameters in the numerical model are shown in Tables 2 and 3. The modeling process was as follows: (i) calculation of the initial state induced by gravity; (ii) modeling the excavations of the headgate in panel 3602; (iii) retreating panel 3602; and (iv) modeling the retaining of the headgate in panel 3602 with respect to different RBB widths (0.8 m, 1.2 m, 1.6 m and 2.0 m).

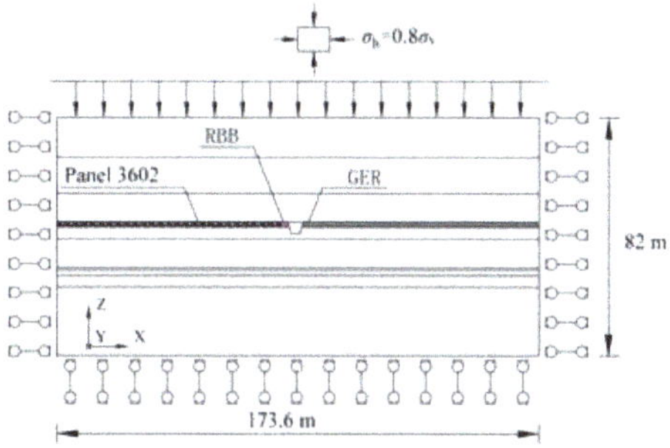

Figure 8. Numerical simulation model.

Table 2. Rock mechanical properties used in the numerical model.

Rock Strata	Thickness (m)	Denstity (kg/m³)	Bulk Modulus (GPa)	Shear Modulus (GPa)	Cohesion (MPa)	Friction Angle (°)
Underlying strata	12	2500	10.0	7.6	5.2	41
Siltstone	2.1	2550	8.8	6.3	4.0	38
Aluminum mudstone	1	1800	5.2	4.0	2.5	25
No. 16-2 coal seam	0.5	1400	4.3	2.7	1.6	23
Sandstone	5	2650	3.5	2.1	1.0	17
Aluminum mudstone	2.1	1700	4.2	2.4	2.2	25
No. 16-1 coal seam	0.9	1400	3.0	1.5	1.4	21
Limestone	5	2900	9.7	7.1	5.2	36
Mudstone	6.4	1800	4.3	2.2	1.5	28
Overlying strata	17	2500	8.9	6.2	3.2	32

Table 3. Mechanical parameters of RBB.

Properties	Values	Properties	Values
Bulk modulus (GPa)	2.9	K shear (GPa)	1.6
Shear modulus (GPa)	1.3	Kviscosity (GPa)	3.0
Cohesion (MPa)	3.2	Mshear (GPa)	2.2
Internal friction angle (°)	32	Mviscosity (GPa)	2.4

3.3.2. Results and Discussion

(1) Convergence of GER with respect to RBB width

The RBB width plays an important role in the convergence of surrounding rock in the GER, as shown in Figure 9. The roof-cutting effect of the RBB becomes stronger as the RBB width increases. When the width of the RBB is greater than 1.6 m, the roof subsidence changes little and the roof-cutting effect is weakened.

The floor heave of GER increases with increasing RBB width. When the width of the RBB increased from 0.8 m to 1.2 m, the floor heave increased two-fold to 146.2 mm, representing an increase of 68.3%.

The RBB convergence gradually decreases with the increasing width of the RBB. When the width of the RBB increased from 1.2 m to 1.6 m, the convergence of the RBB decreased from 281.1 mm to 237.7 mm, representing the convergence reduction of 15.4%. When the width of the RBB increased from 1.6 m to 2.0 m, the convergence of the RBB decreased 16.8 mm, a reduction of 15.4%.

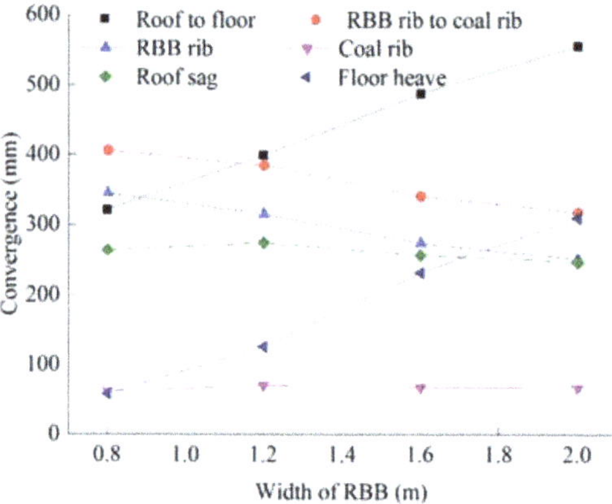

Figure 9. GER convergence with respect to the width of RBB.

(2) Stress distribution around the GER with respect to RBB width

Vertical stress distributions around the GER with respect to the different widths of RBB are shown in Figure 10.

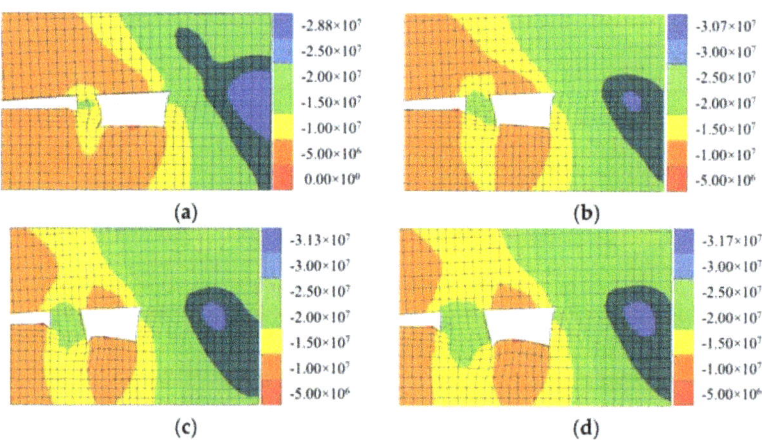

Figure 10. Vertical stress distribution with respect to RBB width (units: Pa): (**a**) RBB of 0.8 m wide; (**b**) RBB of 1.2 m wide; (**c**) RBB of 1.6 m wide; and (**d**) RBB of 2.0 m wide.

The results show that the vertical stress of the RBB gradually increases and the bearing capacity of the RBB becomes stronger with increasing RBB width. When the width of the RBB increased from 0.8 m to 1.2 m, the vertical stress of the RBB increased from 10.715 MPa to 13.25 MPa, an increase of 23.7%. When the width of the RBB increased to 1.6 m, the vertical stress of the RBB increased 6.8%

to 14.15 MPa, representing a relatively small increase. When the width of the RBB increased to 2.0 m, the vertical stress of the RBB remained stable.

With increasing RBB width, the vertical stress in the coal rib wall increases gradually and the peak stress is approximately 3.5–4.0 m away from the roadside. When the width of the RBB increased from 0.8 m to 1.2 m, the vertical stress of the coal rib wall increased from 28.86 MPa to 30.67 MPa, an increase of 6.3%. When the width of RBB was subsequently increased to 1.6 m, the vertical stress of the coal rib wall increased 2.1% to 31.3 MPa. When the RBB width increased to 2.0 m, the vertical stress in the solid coal rib wall remained stable.

Therefore, a headgate of panel 3602 is retained with the RBB width of 1.2 m; this could decrease the GER convergence and reduce RBB construction costs.

4. Field Test and Field Monitoring

According to the above analysis, the RBB is 1.2 m wide, 0.9 m high and has a 1.33 width-height ratio. High-water, fast-setting material with a water-cement ratio 1.5:1 was used to construct the RBB. Figure 11 shows the GER effect. In the 100 m ranges behind the working face, hydraulic props and a steel beam are used to temporarily support the roof of RBB. The roof in the gob nearby the GRE will collapse under the effect of RBB.

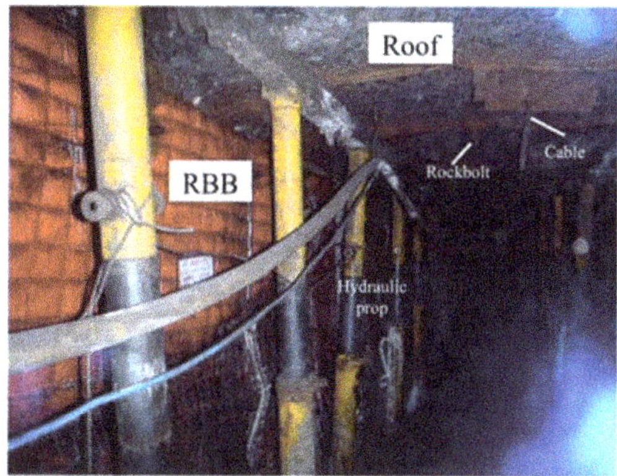

Figure 11. GER effect.

During construction of the GER of panel 3602, two monitoring stations were installed in the GER to record the convergence and RBB loading. The results in Figure 12 show that GER convergence behind the working face can be divided into three stages under the hard roof condition in a TCS (Stages I, II, III):

Stage I—in the range of 30 m behind the working face, the gob roof continuously subsides and the surrounding rock deformation increases after the RBB is constructed; in the area of 30 m behind the working face, the limestone roof of the GER in the gob side is cut off under the action of complicated forces, where the convergence almost accounts for the one-third of total convergence, and the convergence rate of surrounding rock reaches the maximum at this time.

Stage II—in the range from 30 to 110 m behind the working face, the convergence rate of surrounding rock gradually decreases. After the limestone roof collapses, the bearing structure composed of the coal seam support, caved rocks and RBB undertakes the load from the overlying rock, caved rocks compact the gob, and the surrounding rock deformation gradually increases as the convergence rate of the surrounding rock slows down.

Stage III—at a distance of 110 m behind the working face, the surrounding rock deformation of GER tends to be stable; the roof to floor convergence and RBB rib to coal rib convergence of the RBB are approximately 353 mm and 297 mm, respectively. The roof-to-floor convergence is always more than that of rib-to-rib in the mining period. It showed that the parameters of rock properties and RBB in the simulations were reasonable and reliable. Considering the requirement of GER cross-sections for ventilation, the section of the GRE in the mining period could meet the requirement. Field measurement results indicated that RBB width and roof-cutting resistance in the field were rational.

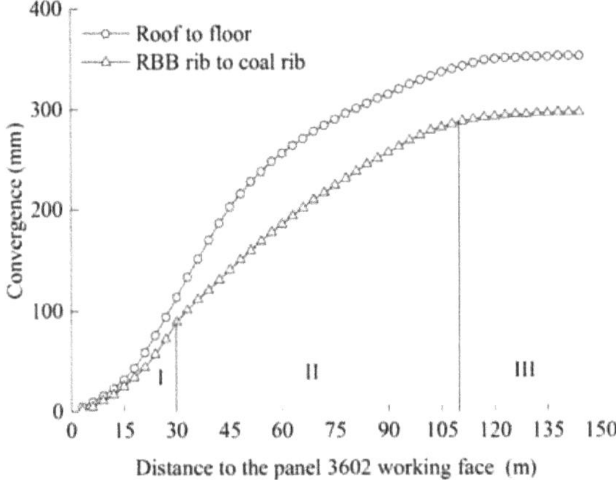

Figure 12. Convergence of GER-LR-TCS.

5. Conclusions

In accordance with the caving style of a limestone roof, using the Y-style ventilation mode, the air pressure of the return roadway in the T-junction is the lowest and it is the highest in the air inlet area of the gob-side entry. The gas concentration around the gob-side entry in the dip direction of the gob is mitigated.

The mechanics model of the GER roof and the RBB indicate that the hard roof collapsed passively under the interaction of the RBB and the overlying rock strata; the roof-cutting resistance of the RBB and its width play a key role in GER with limestone roof in a TCS. A roof-cutting mechanical model of GER with a roadside backfill body (RBB) was proposed to determine the roof-cutting resistance and the width of the RBB.

Based on the mechanical model, analytical results show that the greater the RBB width, the greater its roof-cutting resistance. Fitting algebraic relationships between the RBB's roof-cutting resistance and its width can determine an optimal roof-cutting resistance. The floor heave of GER increases with increasing RBB width. When the width of the RBB increased from 0.8 m to 1.2 m, the floor heave increased two-fold to 146.2 mm. Using numerical simulation to study the stress distributions and the deformation of GER with respect to RBB width, an appropriate RBB width can be determined.

Based on the geological conditions of panel 3602 in a TCS with a limestone roof, the roof-cutting resistance of the RBB was determined as 8.465 MN/m, and the width of the RBB and the water-to-cement ratio of high-water material were 1.2 m and 1.5:1, respectively. Field measurement indicated that these research results were successfully applied in engineering practice.

It is noted that this model was only tested on a TCS. Further study is necessary for medium-thick or thick seams. In addition, further research efforts to improve the high-water material are required to popularize the GER technique.

Author Contributions: S.Y. conceived the research and wrote the paper, T.L. analyzed the data and contributed numerical simulation. J.B. directed the paper structure. W.W. participated in collecting the data and language editing. All authors have read and approved the final manuscript.

Acknowledgments: This research was financially supported by the Fundamental Research Funds for the Central Universities (2014QNB29), National Natural Science Foundation of China (51604268 and 51574227), Independent Research Project of State Key Laboratory of Coal Resources and Safe Mining, CUMT (SKLCRSM18X08), and PAPD (SZBF2011-6-B35). Also, the authors are thankful to Davide Elmo and anonymous reviewers for their comments and suggestions.

Conflicts of Interest: The authors declare no conflict of interest.

References

1. Wang, D. *Study on Integrated Assessment and Regulation Policies of Coal Production Capacity in China*; China University of Mining and Technology Press: Xuzhou, China, 2013; pp. 12–16.
2. Huang, N.; Jiang, Y.; Li, B.; Liu, R. A numerical method for simulating fluid flow through 3-D fracture networks. *J. Nat. Gas Sci. Eng.* **2016**, *33*, 1271–1281. [CrossRef]
3. Deng, Y.; Wang, S. Feasibility analysis of gob-side entry retaining on a working face in a steep coal seam. *Int. J. Min. Sci. Technol.* **2014**, *24*, 499–503. [CrossRef]
4. Yang, H.; Cao, S.; Li, Y.; Sun, C.; Guo, P. Soft Roof Failure Mechanism and Supporting Method for Gob-Side Entry Retaining. *Minerals* **2015**, *5*, 707–722. [CrossRef]
5. Zhang, Z.Y.; Hideki, S.; Qian, D.Y.; Takashi, S. Application of the retained gob-side gateroad in a deep underground coalmine. *Int. J. Min. Reclam. Environ.* **2016**, *30*, 371–389. [CrossRef]
6. Lu, S. *Strata Behaviors of Entry without Chain Pillar*; China Coal Industry Press: Beijing, China, 1982; pp. 1–2.
7. Tan, Y.L.; Yu, F.H.; Ning, J.G.; Zhao, T.B. Design and construction of entry retaining wall along a gob side under hard roof stratum. *Int. J. Rock Mech. Min. Sci.* **2015**, *77*, 115–121. [CrossRef]
8. Huang, Y.L.; Li, J.M.; Song, T.Q.; Kong, G.Q.; Li, M. Analysis on Filling Ratio and Shield Supporting Pressure for Overburden Movement Control in Coal Mining with Compacted Backfilling. *Energies* **2017**, *10*, 31. [CrossRef]
9. Jiang, H.; Miao, X.; Zhang, J.; Liu, S. Gateside packwall design in solid backfill mining-A case study. *Int. J. Min. Sci. Technol.* **2016**, *26*, 261–265. [CrossRef]
10. Hidalgo, K.P.; Nordlund, E. Failure process analysis of spalling failure-Comparison of laboratory test and numerical modelling data. *Tunn. Undergr. Space Technol.* **2012**, *32*, 66–77. [CrossRef]
11. Villaescusa, E.; Varden, R.; Hassell, R. Quantifying the performance of resin anchored rock bolts in the Australian underground hard rock mining industry. *Int. J. Rock Mech. Min. Sci.* **2008**, *45*, 94–102. [CrossRef]
12. Sun, X.M.; Liu, X.; Liang, G.F.; Wang, D.; Jiang, Y.L. Key parameters of gob-side entry retaining formed by roof cut and pressure releasing in thin coal seams. *China J. Rock Mech. Eng.* **2014**, *33*, 1449–1456. (In Chinese)
13. Liu, R.; Jiang, Y.; Li, B.; Yu, L. Estimating permeability of porous media based on modified Hagen–Poiseuille flow in tortuous capillaries with variable lengths. *Microfluid. Nanofluid.* **2016**, *20*, 120. [CrossRef]
14. Chen, Y.; Bai, J.B.; Zhu, T.L.; Yan, S.; Zhao, S.H.; Li, X.C. Mechanisms of roadside support in gob-side entry retaining and its application. *Rock Soil Mech.* **2012**, *32*, 1427–1432.
15. Guo, Z.; Mou, W.; Huang, W.; Duan, H. Analysis on roadside support method with constant resistance yielding-supporting along the goaf under hard rocks. *Geotech. Geol. Eng.* **2016**, *34*, 827–834. [CrossRef]
16. Hou, C.J.; Ma, N.J. Stress in in-seam roadway sides and limit equilibrium zone. *J. China Coal Soc.* **1989**, *4*, 21–29.
17. Yang, D.W.; Ma, Z.G.; Qi, F.Z.; Gong, P.; Liu, D.P.; Zhao, G.Z.; Zhang, R.R. Optimization study on roof break direction of gob-side entry retaining by roof break and filling in thick-layer soft rock layer. *Geomech. Eng. Int. J.* **2017**, *13*, 195–215.
18. Zhang, Y.; Tang, J.; Xiao, D.; Sun, L.; Zhang, W. Spontaneous caving and gob-side entry retaining of thin seam with large inclined angle. *Int. J. Min. Sci. Technol.* **2014**, *24*, 441–445. [CrossRef]
19. Zhang, Z.Z.; Bai, J.B.; Chen, Y.; Yan, S. An innovative approach for gob-side entry retaining in highly gassy fully-mechanized longwall top-coal caving. *Int. J. Rock Mech. Min. Sci.* **2015**, *80*, 1–11. [CrossRef]
20. Itasca Consulting Group. *FLAC3D Instruction Manual*; Itasca Consulting Group: Minneapolis, MN, USA, 2007.
21. Perry, K.A.; Unrug, K.F.; Harris, K.W.; Raffaldi, M.J. Influence of roof/floor interface on coal pillar performance. In Proceedings of the 32th International Conference on Ground Control in Mining, Morgantown, WV, USA, 30 July–1 August 2012.

22. Mohammad, N.; Reddish, D.J.; Stace, L.R. The relation between in situ and laboratory rock properties used in numerical modelling. *Int. J. Rock Mech. Min. Sci.* **1997**, *34*, 289–297. [CrossRef]
23. Zhang, G.C.; Liang, S.J.; Tan, Y.L.; Xie, F.X. Numerical modeling for longwall pillar design: A case study from a typical longwall panel in China. *J. Geophys.* **2018**, *15*, 121–134. [CrossRef]

Article

Experimental Investigation of the Mechanical Behaviors of Grouted Sand with UF-OA Grouts

Yuhao Jin [1,2], Lijun Han [1,2], Qingbin Meng [2], Dan Ma [3], Guansheng Han [2,*], Furong Gao [1] and Shuai Wang [2]

1   School of Mechanics and Civil Engineering, China University of Mining and Technology, Xuzhou 221116, China; jinyuhao@cumt.edu.cn (Y.J.); hanlj@cumt.edu.cn (L.H.); furong_gao@126.com (F.G.)
2   State Key Laboratory for Geomechanics and Deep Underground Engineering, China University of Mining and Technology, Xuzhou 221116, China; mqb1985@cumt.edu.cn (Q.M.); shwzzh023@163.com (S.W.)
3   School of Resources & Safety Engineering, Central South University, Changsha 410083, China; dan.ma@csu.edu.cn
*   Correspondence: Han_GS@cumt.edu.cn; Tel.: +86-15162181307

Received: 29 March 2018; Accepted: 15 April 2018; Published: 19 April 2018

Abstract: A detailed understanding of the engineering properties for grouted sand is a key concern in foundation engineering projects containing sand layers. In this research, experiments of grouting with various grain sizes of sand specimens using a new type of improved chemical material-urea formaldehyde resin mixed with oxalate curing agent (UF-OA), which has rarely been used as grout in the reinforcement of soft foundations, were conducted on the basis of a self-developed grouting test system. After grouting tests, the effects on the mechanical behaviors of grouted sand specimens were investigated through uniaxial compression tests considering the grain size, the presence or absence of initial water in sand, and the curing time for grouted sand. Experimental results show that with the increase in the grain size and the presence of initial water in the sand specimen, the values of uniaxial compressive strength (UCS) and elastic moduli (E) of the grouted specimens decreased obviously, indicating that the increase of grain size and the presence of initial water have negative impacts on the mechanical behaviors of grouted sand; the peak strains (ε_c) were almost unchanged after 14 days of curing; no brittle failure behavior occurred in the grouted specimens, and desirable ductile failure characteristics were distinct after uniaxial compression. These mechanical behaviors were significantly improved after 14 days of curing. The micro-structural properties obtained by scanning electron microscopy (SEM) of the finer grouted sand indicate preferable filling performance to some extent, thereby validating the macroscopic mechanical behaviors.

Keywords: chemical grouts; grain size of sand; initial water contained in sand; grouted sand; macroscopic mechanical behaviors; microstructure characteristics; ductile failure

1. Introduction

Chemical grouting is a well-known reinforcement method in geotechnical engineering, and is used to reduce the permeability and increase the mechanical strength of the soil and rock mass. Grout flowing through the soil and fracture network within the rock mass are similar to the flow of fluid (or gas) through rock fracture networks [1–4]. Chemical grouts can penetrate tiny pores in some dense but weak sand layers, where conventional Portland cement grouts are likely to fail [5,6], although cement-based grouts are more widely used [7–9]. In recent years, many experimental results on the mechanical behaviors of grouted sand using chemical grouts have been achieved. For example, Anagnostopoulos, Papaliangas [10] investigated the mechanical behaviors of grouted sand and found that the epoxy resin grouts of low water content caused an increase in strength and a decrease in water permeability. Xing, Dang [11] conducted experiments to study the mechanical properties of chemically grouted sand and found that the dense polymer film–sand matrix after grouting also led to a significant decrease in water permeability and an increase in mechanical strength. Porcino, Ghionna [12] discussed the mechanical and hydraulic properties of sand injected with mineral-based grouts or silicate solution, and showed that both mechanical strength and initial shear modulus in the mineral-based grouted sand were higher than those of grouted sand using silicate solution. Dayakar, Raman [13] performed an experiment on the influence of cement grouts on the supporting capacity of sandy soil, and demonstrated that the strength of grouted specimens was improved with the increase of the cement content. Some researchers also investigated the strength properties of sodium silicate-grouted soils [14–18], and found that within a certain range, the mechanical strengths of the grouted soils decreased with the increase of water content in different types of grouts.

The above research results mainly focused on the influence of the different chemical and cement grouts (especially the amount of water in the grouts) on the mechanical parameters of the grouted sand; however, no variation of the curing time of grouted sand specimens was considered. Moreover, in terms of the grouting material, the improved chemical material UF-OA (a urea formaldehyde resin mixed with an oxalate curing agent that can be used to accelerate the curing process), characterized by high bonding strength, low production cost, sufficient raw materials and non-polluting nature [19], has rarely been used as grout for reinforcing soft foundations. Furthermore, the impact of the initial water contained in sand is also crucial to the mechanical mechanism, which has yet to be studied in chemical grouted sand up to now. To address these gaps in knowledge, a series of samples were prepared, and studies were conducted:

The gel time, concretion rate, and mechanical and rheological properties of UF-OA grouts with various volume ratios between A and B (A-volume of urea formaldehyde resin, and B-volume of oxalate curing agent) were determined to obtain a suitable volume ratio of A and B in the grouts.

The sand specimens with different grain sizes were prepared (defined as "large-grain sand", "medium-grain sand" and "small-grain sand", corresponding to different porosity), and the grouting apparatus was developed specifically for the grouting experiments in the study.

Through the tests of permeation grouting using the UF-OA chemical grouting material, uniaxial compression and scanning electron microscopy (SEM), the mechanical and microscopic characteristics of the grouted sand specimens were obtained, taking into account the different particle size (porosity), the existence (or absence) of the initial water (to facilitate the comparative study of the two, ω_{wc} of 0% and 5% were only selected in the study) in the sand specimens, as well as the different curing times of 3, 7, 14 and 28 days.

2. Experimental Work

2.1. Specimen Preparation

The results of tests of the physical parameters of the sand are shown in Table 1, including grain size, water content, average porosity and mass density. Based on the relative grain size, the sand specimens used in the experiment were divided into three major types (large-grain, medium-grain, and small-grain sand, which can be quantified as the porosity); the specimens were placed in a cylindrical grouting barrel with diameter of 150 mm and height of 200 mm and then subjected to grout injection with UF grouts during the grouting tests.

Table 1. Physical parameters of sand specimens before grouting.

Sand Category Used in the Tests	Grain Size (mm)	Water Content (%)	Average Initial Porosity	Average Mass Density (g/cm^3)
Large-grain	$0.5 < d < 0.85$	0	34	2.52
		5	30	2.61
Medium-grain	$0.25 < d < 0.42$	0	36	2.67
		5	32	2.74
Small-grain	$0.11 < d < 0.21$	0	40	2.72
		5	37	2.80

The parameter acquisition process is briefly described below:

Grain size:

The different grain sizes were gained by using a shaking screen, etc.

Water content:

$$\omega_{WC} = \frac{g_1 - g_2}{g_2} \times 100\% \tag{1}$$

where ω_{wc} refers to the water content of initial sand (before grouting), g_1 and g_2 denote the weight of the specimen with water and the dry specimen, respectively. It is worth noting that the water content in this paper was known (5%), and the amount of water needed to be added was calculated by using this formula in the process of sample making. A high-precision electronic balance and an oven are necessary.

Initial porosity:

We measured the volume of the material in the natural state V_0 (apparent volume), quality of the material m and density of the material ρ, and then used the formula for porosity expressed as:

$$\phi = (1 - \frac{m/V_0}{\rho}) \times 100\% \tag{2}$$

2.2. Chemical Reaction of Grout Components

When the urea formaldehyde resin was mixed with the oxalate curing agent to improve the property of the urea formaldehyde resin, the following reactions occurred, as shown in Equations (3)–(5) [20]. According to the reaction equation, in the addition reaction stage (Equation (3)), urea reacts with formaldehyde water, generating hydroxymethyl urea and dihydroxymethyl urea in the case of the weak base. Then, the oxalic acid reacts with the hydroxymethyl urea and dihydroxymethyl urea according to Equations (4) and (5), respectively, to produce acidic solution. In the acidic condition, various polymethylenes are produced to achieve solidification.

$$\underset{\overset{\displaystyle|}{NH_2}}{\overset{\displaystyle NH_2}{\overset{\displaystyle|}{CO}}} + HCHO \xrightarrow{pH7\sim9} \underset{\overset{\displaystyle|}{NH_2}}{\overset{\displaystyle NHCH_2OH}{\overset{\displaystyle|}{CO}}} + \underset{\overset{\displaystyle|}{NHCH_2OH}}{\overset{\displaystyle NHCH_2OH}{\overset{\displaystyle|}{CO}}} \tag{3}$$

$$\underset{\overset{\displaystyle|}{NH_2}}{\overset{\displaystyle NHCH_2OH}{\overset{\displaystyle|}{CO}}} \xrightarrow{-H_2O} \underset{\overset{\displaystyle|}{NH_2}}{\overset{\displaystyle N=CH_2}{\overset{\displaystyle|}{CO}}} \rightarrow \left[\underset{\overset{\displaystyle|}{NH_2}}{\overset{\displaystyle N-CH_2}{\overset{\displaystyle|}{CO}}} \right]_n \tag{4}$$

$$
\begin{array}{ccc}
\begin{array}{c} NHCH_2OH \\ | \\ CO \\ | \\ NHCH_2OH \end{array}
& \xrightarrow{-H_2O}
\begin{array}{c} N \;\; =CH_2 \\ | \\ CO \\ | \\ N \;\; =CH_2 \end{array}
& \rightarrow
\left[\begin{array}{c} N \;\; -CH_2 \\ | \\ CO \\ | \\ N \;\; =CH_2 \end{array}\right]_n
\end{array}
\tag{5}
$$

2.3. Properties of UF-OA Grouts

Through a series of physical and mechanical tests, the elastic modulus and Poisson's ratio of the grout gels with different volume ratios were obtained by mechanical test, as shown in Table 2.

Table 2. Elastic modulus and Poisson's ratio of the grout gel with different volume ratios obtained by using uniaxial compressive strength tests.

A:B	4:1	5:1	6:1	7:1
Elastic Modulus (MPa)	120	140	180	220
Poisson's Ratio	0.187	0.211	0.25	0.3

Moreover, the gel time and concretion rate with different volume ratios, viscosity variation with time, UCS variation with different curing time, and the ultimate compressive failure characteristics of the grout gel were determined: the gel time, concretion rate, and viscosity were tested by using a digital viscometer, as shown in Figure 1a. The uniaxial compressive strengths (UCSs) were obtained after the uniaxial compression tests, as displayed in Figure 1b. Figure 1c shows that the gel time and concretion rate both increased as the volume ratio of A and B increased. The values of gel time and concretion rate at A:B of 7 were 330 s and 99.2%, respectively. Similar to the above two parameters, the uniaxial compressive strengths at the same curing time were enhanced with the increasing A:B. In addition, the mechanical strengths also increased with the increase of curing time while keeping A:B unchanged, as shown in Figure 1e. The ultimate compressive failure of the UF-OA grouts, featuring the expansion and vertical splitting fracture of the gel, are as shown in Figure 1f. As shown in Figure 1d, the viscosity of the grouts increased with time but decreased with the increase in A:B. These findings indicated that the UF-OA grouts with an A:B of 7 would have better mechanical characteristics and flow properties than those with other volume ratios. Therefore, UF-OA grouts with a volume ratio of 7 were chosen in this study.

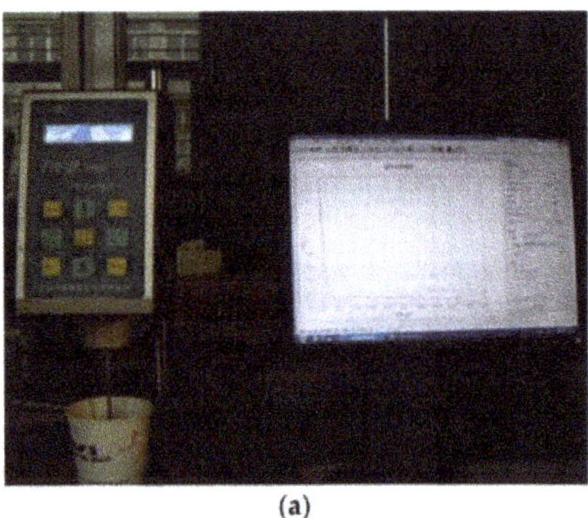

(a)

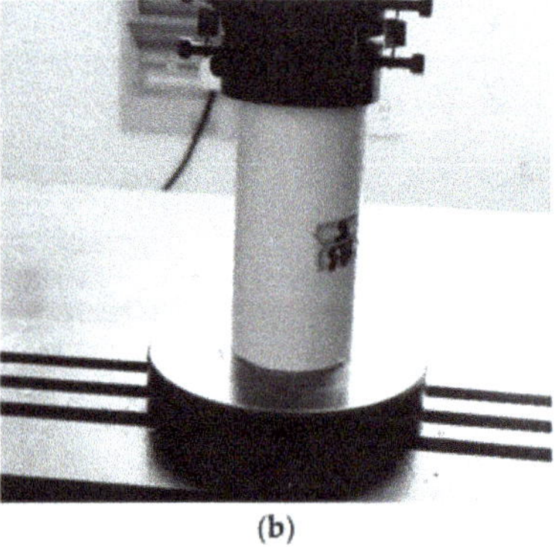

(b)

Figure 1. *Cont.*

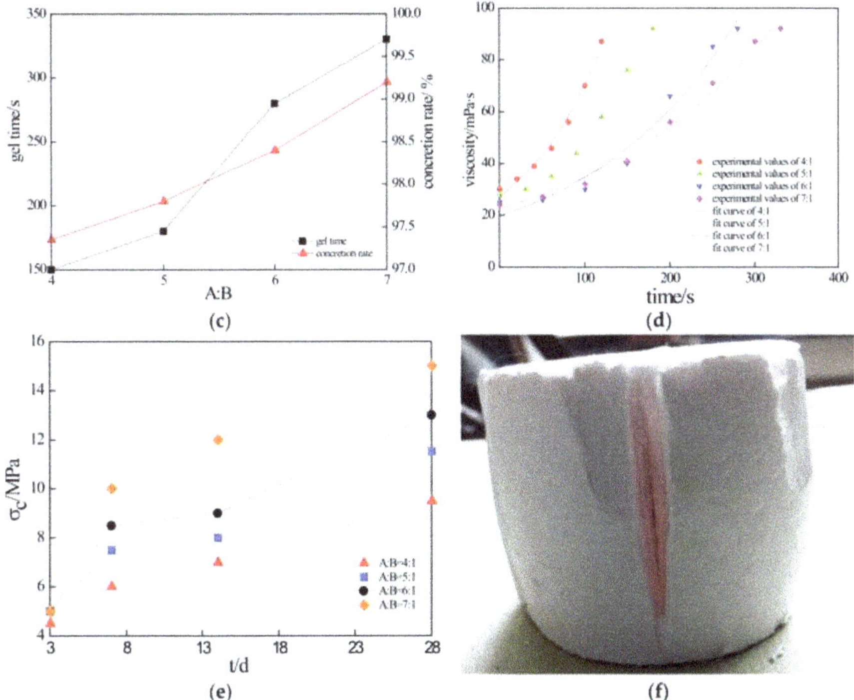

Figure 1. Rheological and mechanical properties of the UF-OA (urea formaldehyde resin mixed with oxalate curing agent) grouts, including (**a**) the digital viscometer; (**b**) specimen under compression; (**c**) gel time and concretion rate with different volume ratios; (**d**) viscosity variation with time; (**e**) uniaxial compressive strength (UCS) variation with different curing time; and (**f**) the ultimate compressive failure characteristics of grout gel.

2.4. Laboratory Procedure

The grouting test was conducted by injecting a measurable quantity of UF grouts into the sand specimen that was placed in the grouting barrel by using the self-designed grouting experimental setup, which is composed of a ground stress simulation system, a grouting test system (the cylindrical grouting barrel is assembled by two pieces of semi-cylindrical barrels for convenience in installation and disassembly), a grouting pressure gauge, a grout transmission system, and a grout supply system, as shown in Figure 2. Note that the testing machine can apply an axial pressure as well as a lateral restraining action (Poisson effect) on the specimen inside the grouting barrel via the load-bearing plate, which can be easily moved up and down in the grouting barrel to represent the ground stress in the real situation (Equation (6)). Moreover, the fixed axial pressure and the lateral restraint in the grouting apparatus can ensure that the porosity remains close to constant during the experiment.

$$\frac{F}{A} = \gamma H, \tag{6}$$

where F and A refer to the axial force provided by the testing machine and the area of load-bearing plate (0.02 m^2), respectively, γ denotes the average unit weight for sand layers (20 kN/m^3), and H is related to the ground depth. Here, H of 10 m was set up in the study, and the corresponding axial force F of 4 kN was calculated.

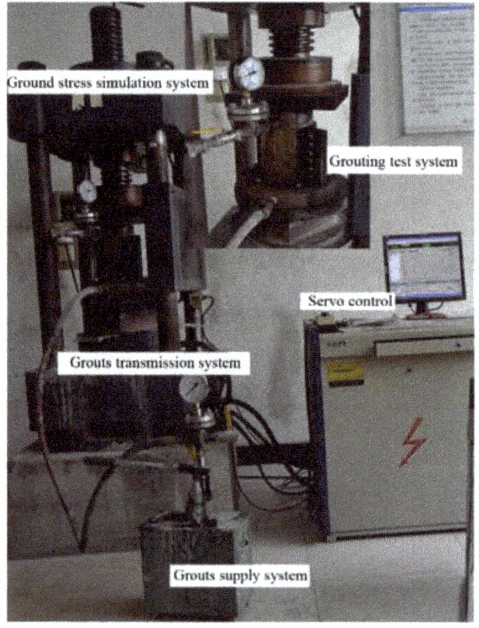

(**a**)

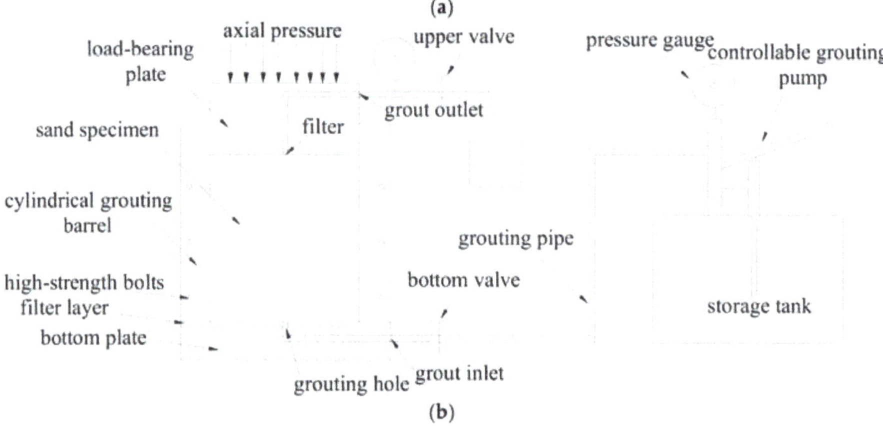

(**b**)

Figure 2. (**a**) Grouting experimental setup and (**b**) detailed schematic layout of the grouting test system.

After performing the grouting tests and curing for the designed times (3, 7, 14 and 28 days), referring to the uniaxial compression test standard [21], the grouted standard cylindrical sand specimens with the same height-to-diameter ratio of 2 (100 mm × 50 mm) for different grain size and initial water conditions were extracted from the grouting reinforced body (150 mm × 200 mm) in the grouting barrel, and then these grouted specimens were subjected to uniaxial compression tests using the CMT5305 electronic universal testing machine at a loading rate of 3×10^{-3} mm per second. Finally, after 28 days of curing, the microstructure scanning tests for the grouted specimens not subjected to the uniaxial compression tests were conducted using SEM.

To prevent the hydrofracturing of sand specimens from occurring, the grouting pressure during the grouting experiments must be maintained at a lower level (while the grouting pressures in the "hydrofracturing grouting" are larger and have larger serrated changes during the grouting experiments) using a hand grouting pump (which allows for easy adjustment of the grouting pressure) and a pressure gauge to ensure the penetration of the grouts in the sand samples. Note that each of the reported values for mechanical parameters in this study represent the average value of three specimens.

3. Results

3.1. Macroscopic Mechanical Parameters

To study the differences in mechanical behaviors among the multiple grouted specimens of different grain size and initial water conditions at 3, 7, 14 and 28 days of curing, different axial stress–strain curves of grouted sand specimens after the uniaxial compression tests are shown in Figure 3. Because the grouts in small-grain sand could not be sufficiently hardened in a short time, no corresponding effective experimental results were obtained on the small-grain grouted specimens after 3 days of curing, as shown in Figure 3a. More details are described in the Discussion.

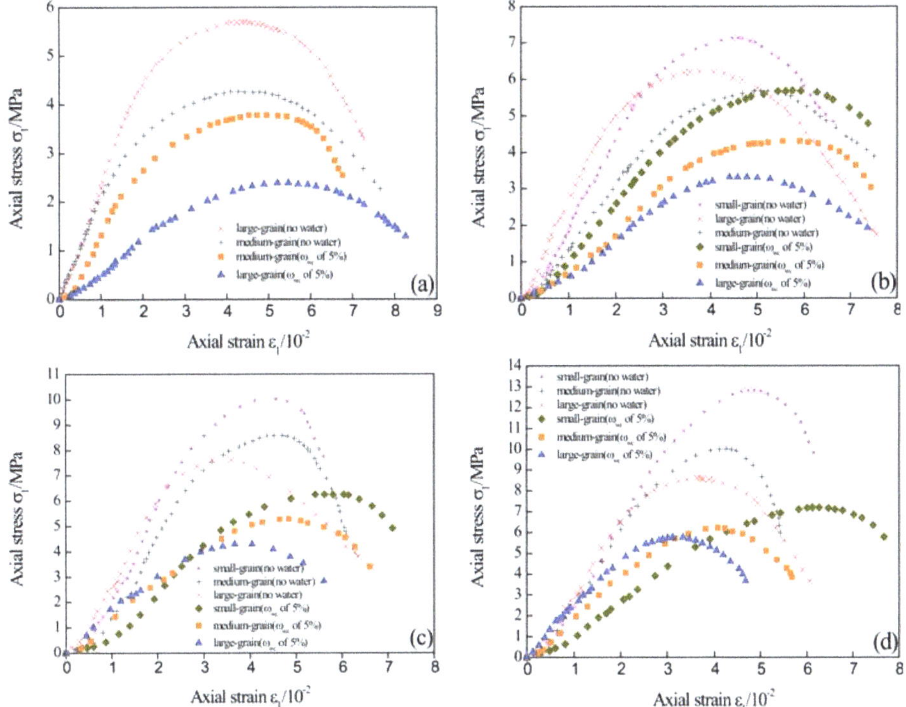

Figure 3. Axial stress–strain curves of representative grouted sand specimens with different grain size and water-contained condition of sand after curing treatment: (**a–d**) at 3, 7, 14 and 28 days of curing, respectively.

Figures 4 and 5 illustrate the evolution of the different mechanical parameters (namely, uniaxial compressive strength (UCS), elastic modulus (*E*), and peak axial strain (ε_c)) of the grouted sand specimens (see Table 3) with curing time (*t*) and porosity (*φ*), respectively. Figure 4a revealed that the average values of UCS for all of the grouted specimens in different grain sizes and initial water content increased distinctly with the increase in curing time (from 3 to 28 days). However, the slopes of curves decreased gradually after 14 days of curing (the slopes were quantified by linear fit, as shown in Figure 4a), this observation indicated that the mechanical behaviors for grouted specimens after 14 days of curing appeared to obtain satisfactory improvement. From the curing time of 3 (or 7) to 28 days, the average values of UCS of the grouted specimens not containing initial water increased by approximately 80% in the specimens with small grain size, compared to 76% and 51% with medium and large grain size, respectively, indicating that the curing time had positive effects on improving the mechanical characteristics of the sand after grouting, especially for the sand of smaller grain size. The evolution of UCS over curing time in grouted specimens containing initial water was consistent with that of the specimens without initial water. Additionally, as shown in Figure 5a, for the same curing time, the UCS curves all approximately increased with the increase in the porosity (corresponding to the grain size of sand) (Table 1); in other words, as the sand particle size decreased, the strength of the grouted sand increased, as confirmed by previous research results [22,23]. Zebovitz, Krizek [24] reported that this phenomenon occurs because the number of grain-to-grain contacts per unit volume of sand increases as the gradation of sand becomes finer or as the specific surfaces of the sand become larger; therefore, it is reasonable that the finer sand would exhibit higher strength than the coarser sand. Moreover, because of the presence of initial water in the specimen, the decrease in the UCS of grouted sand was also obviously observed. For example, the UCS of the small-grain grouted specimen without initial water (corresponding to the porosity of 40, see Figure 5) decreased from 12.83 MPa to 7.19 MPa (the UCS value of the small-grain grouted specimen containing initial water corresponding to the porosity of 37, see Figure 5) at the age of 28 days, a reduction of 44% compared to a reduction of 38% in a specimen of medium grain size and a reduction of 33% in a specimen of large grain size.

Figures 4b and 5b show the relationships of *E*–*t* and *E*–φ, respectively, where *E* is calculated as the average slope in the elastic deformation stage of the stress–strain curve. Overall, the average values of *E* appeared to be increasing with the increase in curing time and grain size (porosity), and this was similar to those of the UCS. For the grouted sand specimens with the same grain size, the average values of *E* in the specimens not containing the initial water were found to be significantly higher than those with the initial water. For example, they increased by approximately 66%, 55% and 57% after 28 days of curing in the specimens with small, medium, and large grain size, respectively, further demonstrating the adverse impact of initial water contained in sand on the mechanical behaviors of grouted sand.

The average values of peak axial strains of grouted specimens ranging from 3% to 6% are shown in Figure 4c. These results illustrated that the peak strains of different specimens were almost unchanged after 14 days of curing, revealing that the deformation characteristics of grouted specimens tended to be stable with the increase of curing time.

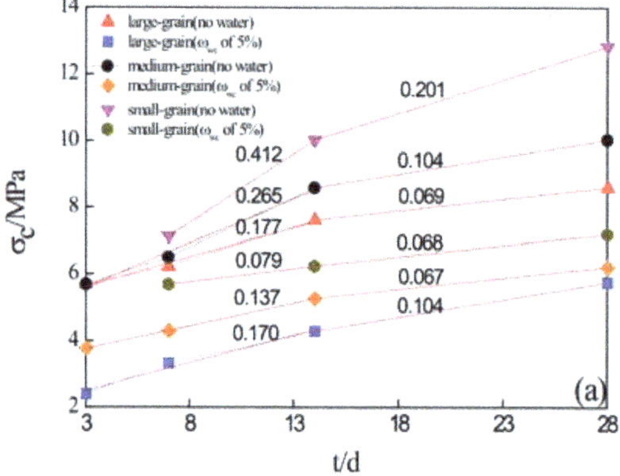

Figure 4. *Cont.*

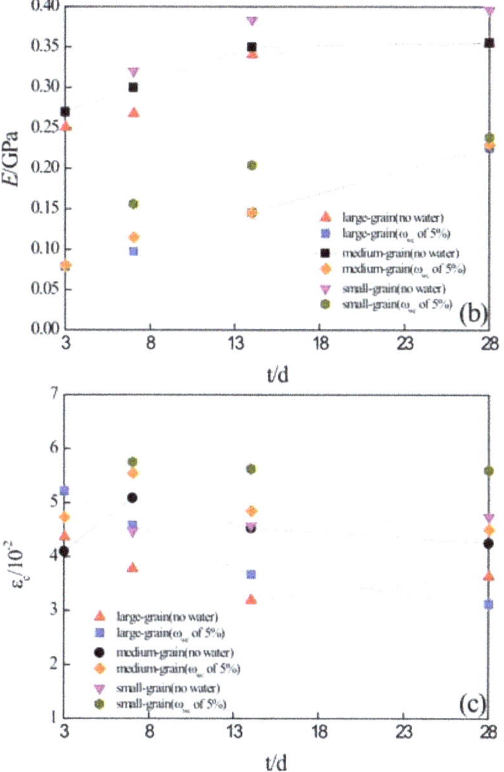

Figure 4. Different average values of mechanical parameters versus curing time. (a) σ_c, (b) E, (c) ε_c.

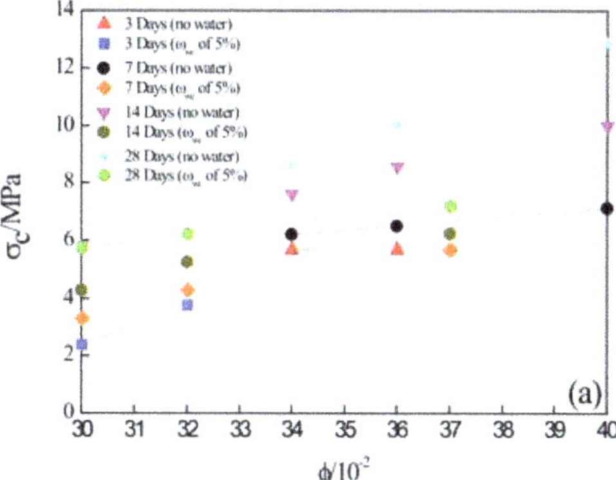

Figure 5. *Cont.*

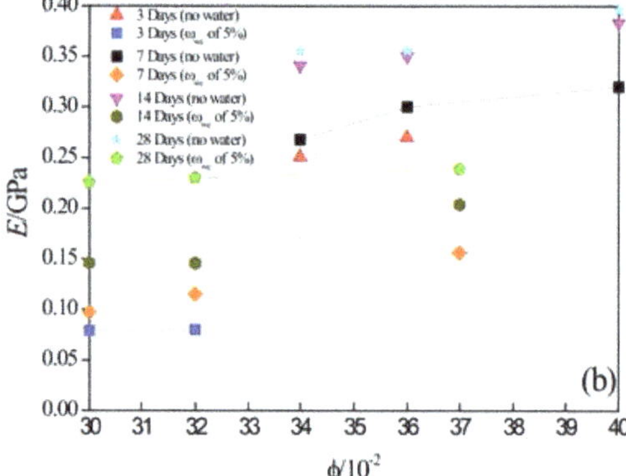

Figure 5. Different average values of mechanical parameters versus initial porosity. (**a**) σ_c, (**b**) E.

Table 3. Average values of mechanical parameters of grouted sand specimens.

Sand Types	ω_{wc} (%)	UCS (MPa)					E (GPa)					ε_c ($\times 10^{-2}$)					φ
		3 d	7 d	14 d	28 d	σ	3 d	7 d	14 d	28 d	σ	3 d	7 d	14 d	28 d	σ	
Large	0	5.70	6.21	7.62	8.59	1.34–3.12	0.25	0.27	0.34	0.35	0.14–0.29	4.36	3.78	3.19	3.63	0.04–0.09	34
	5	2.38	3.31	4.30	5.76	1.21–2.45	0.08	0.10	0.15	0.23	0.21–0.35	5.22	4.58	3.67	3.12	0.05–0.10	30
Medium	0	5.70	6.50	8.58	10.03	0.94–3.57	0.27	0.30	0.35	0.36	0.32–0.37	4.10	5.09	4.53	4.25	0.08–0.13	36
	5	3.78	4.30	5.28	6.22	0.89–2.63	0.08	0.11	0.15	0.23	0.19–0.54	4.73	5.55	4.85	4.5	0.03–0.05	32
Small	0	-	7.12	10.02	12.83	1.14–3.01	-	0.32	0.38	0.40	0.27–0.51	-	4.46	4.57	4.74	0.11–0.14	40
	5	-	5.69	6.24	7.19	1.25–1.85	-	0.16	0.20	0.24	0.38–0.43	-	5.75	5.62	5.60	0.09–0.15	37

Note: ω_{wc} refers water content of initial sand (before grouting); d represents days; σ means standard deviation; and φ represents porosity. For instance, UCS (E, ε_c) for 3 d means the average values of uniaxial compressive strength (elasticity modulus, peak axial strain) of the grouted sand specimens after 3 days of curing.

3.2. Microstructure Characteristics

According to Figure 6, which shows the SEM images of sections in grouted specimens (three pieces of specimens chosen with 28 days of curing), the preparation of SEM samples generally includes the following steps: determining the sampling site from the grouted samples, and then obtaining the SEM samples by thin section identification method. The abovementioned macroscopic mechanical results of the grouted sand (the finer the particle size of sand, the greater was the UCS of the grouted sand) can be explained from a microscopic perspective to a certain extent. For example, according to the magnified surfaces in the scale of 1:500 displayed in Figure 6(a₁–c₁), the finer the particle size of sand, the thicker and denser was the grout filling layer. In more detail, compared to the small-grain specimen magnified at the scale of 1:5000 (Figure 6(c₂)), micro-cracks that were not completely filled by UF grouts were found in the large- and medium-grain specimens because of the larger grain size of the sand, as shown in Figure 6(a₂,b₂). Moreover, referring to Figure 6(c₂), finer sand clearly possessed desirable filling performance because the UF grouts filled almost all of the voids and cracks between grains in many section surfaces; this observation was consistent with the observation results of other finer sand specimens after 7 and 14 days of curing. However, at present, it is difficult to draw reliable observations from qualitative SEM images. Therefore, the microstructural characteristics of grouted sand should be investigated and quantified in future research. Moreover, a study on the presence of such micro-cracks occurring in the grouted large-grain and medium-grain sand is needed in the next work because a dissipative phenomenon is associated to them [25–27]. Sometimes this dissipative behavior can be as a desired mechanical property, such as in absorbing seismic excitations, provided that it does not lower the resistance too much.

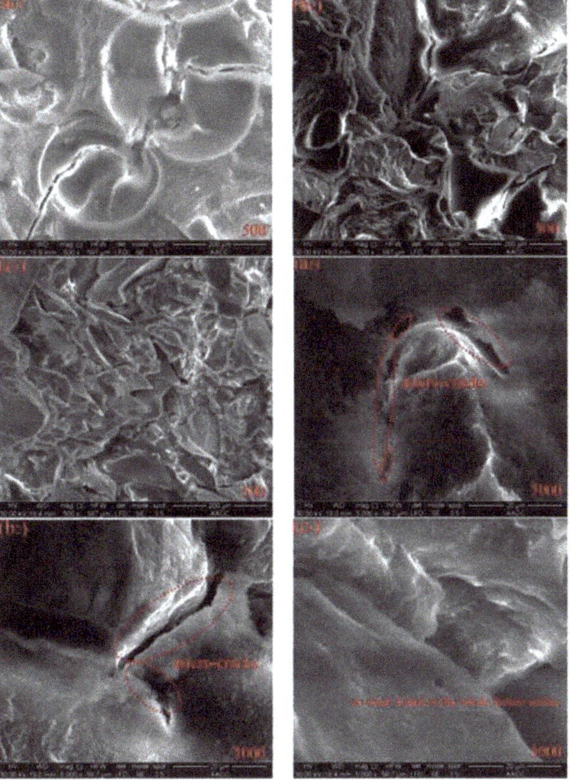

Figure 6. Representative SEM images (the magnifications are 500× (set 1) and 5000× (set 2)) of section surfaces in grouted sand specimens after 28 days of curing: (**a**) large-grain sand; (**b**) medium-grain sand; (**c**) small-grain sand.

4. Discussion

From the representative axial stress–strain curves shown in Figure 3, the ductile behaviors of various grouted sand specimens under uniaxial compression is analyzed below.

The stress–strain curves are divided into four stages: compaction stage, elastic deformation stage, yield stage, and post-peak stage. It was noticed that at the post-peak stage for all tested sand specimens, no obvious instant drop (brittle failures) occurred for the curves, demonstrating that the grouted sand specimen had the expected ductility with deformation up to a larger strain of approximately 7%, where the ultimate failure was observed; this property is advantageous to the delay of structural damage when disasters occur [28]. From the macroscopic and microscopic observations, the strong adhesion between the UF grouts and sand particles was found to essentially improve the mechanical and deformation characteristics of the grouted sand.

5. Conclusions

The finer grouted specimens were found to achieve higher strength, whereas the presence of initial water had a negative impact on the mechanical behaviors compared with the absence of water. These results are complemented and verified by the study of the microstructure through SEM, demonstrating that the grouts adhered better to the grains in the finer sand by almost filling all the cracks. Through the above study, the following conclusions can be drawn:

(1) The gel time, concretion rate, and mechanical and rheological properties of the improved chemical grouts (UF-OA grouts) with various volume ratios of A and B were investigated to obtain a suitable ratio (A:B of 7) in the grouts used in this study.

(2) With the increase in the grain size of the sand and the existence of initial water contained in sand, the values of the strength (UCS) and elastic moduli (E) for most of the grouted sand specimens decreased distinctly. Moreover, with the increase in the curing time, the UCS and E presented an increasing trend, and the mechanical behaviors of grouted sand after 14 days of curing were significantly improved. Moreover, the peak strains for the grouted specimens were found to remain constant with the increase of curing time after 14 days of curing.

(3) The microstructural characteristics indicate that the finer grouted sand was found to achieve higher mechanical strength via the better filling performance.

(4) Most of the grouted sand specimens under uniaxial compression at curing times of 3, 7, 14 and 28 days revealed desirable ductile failure characteristics.

Acknowledgments: The research described in this paper was financially supported by the National Natural Science Foundation of China (Grant Nos. 51704280 and 51574223), and the China Postdoctoral Science Foundation (Grant No. 2017T100420). The fourth author (Dan Ma) would like to thank the financial supported by State Key Laboratory for Geomechanics and Deep Underground Engineering, China University of Mining & Technology (SKLGDUEK1805) and Research Fund of State Key Laboratory of Coal Resources and Safe Mining, CUMT (SKLCRSM18KF024).

Author Contributions: Yuhao Jin, Lijun Han, Qingbin Meng and Guansheng Han conceived and designed the experiments; Yuhao Jin and Guansheng Han performed the experiments; Yuhao Jin and Dan Ma analyzed the data; Furong Gao and Shuai Wang contributed analysis tools; Yuhao Jin and Guansheng Han wrote the paper.

Conflicts of Interest: The authors declare no conflict of interest.

Notations

ω_{wc}	water content of initial sand (before grouting)
σ	standard deviation
φ	porosity
UF-OA	urea formaldehyde resin mixed with oxalate curing agent
σ_1	axial stress (MPa)
ε_1	axial strain (10^{-2})
UCS (σ_c)	uniaxial compressive strength (MPa)
E	elasticity modulus of grouted sand (GPa)
ε_c	peak axial strain (10^{-2})
t	curing time
SEM	scanning electron microscope

References

1. Liu, R.; Li, B.; Jiang, Y. A fractal model based on a new governing equation of fluid flow in fractures for characterizing hydraulic properties of rock fracture networks. *Comput. Geotech.* **2016**, *75*, 57–68. [CrossRef]

2. Liu, R.; Jiang, Y.; Li, B.; Wang, X. A fractal model for characterizing fluid flow in fractured rock masses based on randomly distributed rock fracture networks. *Comput. Geotech.* **2015**, *65*, 45–55. [CrossRef]

3. Huang, N.; Jiang, Y.; Li, B.; Liu, R. A numerical method for simulating fluid flow through 3-D fracture networks. *J. Nat. Gas Sci. Eng.* **2016**, *33*, 1271–1281. [CrossRef]

4. Liu, R.; Li, B.; Jiang, Y. Critical hydraulic gradient for nonlinear flow through rock fracture networks: The roles of aperture, surface roughness, and number of intersections. *Adv. Water Resour.* **2016**, *88*, 53–65. [CrossRef]

5. Avci, E. Performance of Novel Chemical Grout in Treating Sands. *J. Mater. Civ. Eng.* **2017**, *29*. [CrossRef]

6. Avci, E.; Mollamahmutoğlu, M. UCS Properties of Superfine Cement–Grouted Sand. *J. Mater. Civ. Eng.* **2016**, *28*. [CrossRef]

7. Mohammed, M.H.; Pusch, R.; Knutsson, S.; Hellström, G. Rheological Properties of Cement-Based Grouts Determined by Different Techniques. *Engineering* **2016**, *6*, 217–229. [CrossRef]
8. Tan, O.; Gungormus, G.; Zaimoglu, A.S. Effect of Bentonite, Fly Ash and Silica Fume cement injections on uniaxial compressive strength of granular bases. *KSCE J. Civ. Eng.* **2014**, *18*, 1650–1654. [CrossRef]
9. Bras, A.; Gião, R.; Lúcio, V.; Chastre, C. Development of an injectable grout for concrete repair and strengthening. *Cem. Concr. Comp.* **2013**, *37*, 185–195. [CrossRef]
10. Anagnostopoulos, C.A.; Papaliangas, T.; Manolopoulou, S.; Dimopoulos, T. Physical and mechanical properties of chemically grouted sand. *Tunnell. Undergr. Space Technol.* **2011**, *26*, 718–724. [CrossRef]
11. Xing, H.G.; Dang, Y.H.; Yang, X.G.; Zhou, J.W. Experimental study of physical and mechanical properties of chemically grouted sand and gravel. *Sens. Transducers* **2014**, *165*, 164–169.
12. Porcino, D.; Ghionna, V.N.; Granata, R.; Marcianò, V. Laboratory determination of mechanical and hydraulic properties of chemically grouted sands. *Geomech. Geoeng.* **2015**, *11*, 164–175. [CrossRef]
13. Dayakar, P.; Raman, K.V.; Raju, K.V.B. *Study on Permeation Grouting Using Cement Grout in Sandy Soil*; Abzena: Cambridge, UK, 2012; Volume 4, pp. 5–10.
14. Alaa, A.; Vipulanandan, C. Cohesive and Adhesive Properties of Silicate Grout on Grouted-Sand Behavior. *J. Geotechn. Geoenviron. Eng.* **1998**, *124*, 38–44.
15. Delfosse-Ribay, E.; Djeran-Maigre, I.; Cabrillac, R.; Gouvenot, D. Factors Affecting the Creep Behavior of Grouted Sand. *J. Geotech. Geoenviron. Eng.* **2006**, *132*, 488–500. [CrossRef]
16. Gonzalez, H.A.; Vipulanandan, C. Behavior of a Sodium Silicate Grouted Sand. In Proceedings of the Geo-Denver, Denver, CO, USA, 18–21 February 2007; pp. 1–10.
17. Hassanlourad, M.; Salehzadeh, H.; Shahnazari, H. Undrained triaxial shear behavior of grouted carbonate sands. *Int. J. Civ. Eng.* **2011**, *9*, 307–314.
18. Porcino, D.; Marcianò, V.; Granata, R. Static and dynamic properties of a lightly cemented silicate-grouted sand. *Can. Geotech. J.* **2011**, *49*, 1117–1133. [CrossRef]
19. Park, B. Properties of Urea-Formaldehyde Resin Adhesives with Different Formaldehyde to Urea Mole Ratios. *J. Adhes.* **2015**, *91*, 677–700.
20. Cao, X.X.; Zhang, Y.H.; Zhu, L.B.; Tan, H.Y.; Ji-You, G.U. Study on the Curing Characteristics and Synthesis Process of Modified Urea-formaldehyde Resin with Low Formaldehyde Release. In Proceedings of the 3rd International Conference and Exhibition on Biopolymers & Bioplastics, San Antonio, TX, USA, 12–14 September 2016.
21. Yang, S.Q.; Liu, X.R.; Jing, H.W. Experimental investigation on fracture coalescence behavior of red sandstone containing two unparallel fissures under uniaxial compression. *Int. J. Rock Mech. Min. Sci.* **2013**, *63*, 82–92. [CrossRef]
22. Schwarz, L.G.; Krizek, R.J. Hydrocarbon Residuals and Containment in Microfine Cement Grouted Sand. *J. Mater. Civ. Eng.* **2006**, *18*, 214–228. [CrossRef]
23. Markou, I.N.; Droudakis, A.I. Factors Affecting Engineering Properties of Microfine Cement Grouted Sands. *Geotech. Geol. Eng.* **2013**, *31*, 1041–1058. [CrossRef]
24. Zebovitz, S.; Krizek, R.J.; Atmatzidis, D.K. Injection of Fine Sands with Very Fine Cement Grout. *J. Geotech. Eng.* **1989**, *115*, 1717–1733. [CrossRef]
25. Scerrato, D.; Giorgio, I.; Della Corte, A.; Madeo, A.; Limam, A. A micro-structural model for dissipation phenomena in the concrete. *Int. J. Numer. Anal. Methods Geomech.* **2015**, *39*, 2037–2052. [CrossRef]
26. Giorgio, I.; Scerrato, D. Multi-scale concrete model with rate-dependent internal friction. *Eur. J. Environ. Civ. Eng.* **2016**, *21*, 821–839. [CrossRef]
27. Contrafatto, L.; Cuomo, M.; Fazio, F. An enriched finite element for crack opening and rebar slip in reinforced concrete members. *Int. J. Fract.* **2012**, *178*, 33–50. [CrossRef]
28. Christensen, R.M. Exploration of ductile, brittle failure characteristics through a two-parameter yield/failure criterion. *Mater. Sci. Eng. A* **2005**, *394*, 417–424. [CrossRef]

Article

Laboratory Investigation of Granite Permeability after High-Temperature Exposure

Lixin He [ID], Qian Yin and Hongwen Jing *

State Key Laboratory for Geomechanics & Deep Underground Engineering, China University of Mining and Technology, Xuzhou 221116, China; helixin_cumt@yeah.net (L.H.); Jeryin@foxmail.com (Q.Y.)
* Correspondence: hwjing@cumt.edu.cn; Tel.: +86-138-0520-9187

Received: 13 March 2018; Accepted: 17 April 2018; Published: 19 April 2018

Abstract: This study experimentally analysed the influence of temperature levels (200, 300, 400, 500, 600, and 800 °C) on the permeability of granite samples. At each temperature level, the applied confining pressure was in the range of 10–30 MPa, and the inlet hydraulic pressure varied below the corresponding confining pressure. The results are as follows: (i) With an increase in the temperature level, induced micro-fractures in the granites develop, and the decrement ratios of both the P-wave velocity and the density of the granite increase; (ii) The relationship between the volume flow rate and the pressure gradient is demonstrably linear and fits very well with Darcy's law. The equivalent permeability coefficient shows an increasing trend with the temperature, and it can be best described using the mathematical expression $K_0 = A \times 1.01^T$; (iii) For a given temperature level, as the confining pressure increases, the transmissivity shows a decrease, and the rate of its decrease diminishes gradually.

Keywords: scanning electron microscope (SEM) images; permeability; high temperature; Darcy's law; confining pressures

---

1. Introduction

Rock properties that are related to high-temperature exposure are involved in many fields, such as the disposal of nuclear wastes, underground coal gasification, and the exploitation and utilization of geothermal resources [1–4]. In the case of nuclear wastes disposal, the rock mass is subjected to a high-temperature environment and is affected by the temperature gradient. Numerous studies have demonstrated that both the physical and mechanical properties of rocks are affected by temperature exposure [5–7]. With an increase in temperature from 25 °C to 900 °C, the average mass loss rate of sandstone increases from 0 to 2.97% [8]. The thermal destruction can be significantly observed by scanning electron microscope (SEM) [8]. Yong et al. [9] quantify thermal cracking in granite as a function of thermal stresses. Additionally, researchers investigate thermal cracking in granite, which is monitored or quantified in different ways [10–12]. In the range of temperature exposure from 25 °C to 850 °C, the density and P-wave velocity of granite show decreases of 4.92% and 79.17%, respectively [13]. The tensile strength decreases from 12.8 MPa to 1.37 MPa at different rates [13]. The mechanical properties, including the peak strength and the elastic modulus of marble fluctuate when the temperature is lower than 400 °C, and gradually decrease when the temperature is greater than 400 °C [14].

Thermal damages result in changes in micro-characteristics, such as porosity, pore throat geometry, and the pore connectivity of rocks, which would directly affect the corresponding permeability [15]. Zoback et al. [16] investigate the influence of micro-fracturing on permeability generally. The permeability of sandstone after high-temperature exposure slowly decreases when the temperature is below 400 °C, and increases rapidly as the temperature increases from 400 °C to 600 °C [17]. The initial permeability of granite shows small changes when the temperature is below 300 °C, which is a slight increase from 300 °C to 500 °C, but a significant increase when the temperature exceeds 500 °C [18]. Although many studies have been reported, the effects of temperature and the confining pressure on the permeability have not been fully understood.

The purpose of this paper is to investigate the permeability of granites after high-temperature exposure. First, cylindrical granite samples were prepared and exposed to various temperature levels (200, 300, 400, 500, 600, and 800 °C). Next, a series of hydro-mechanical tests with respect to different inlet hydraulic pressures and increasing confining pressures from 10 MPa to 30 MPa were conducted. The equivalent permeability of granites as a function of temperature and confining pressure was analysed.

2. Experimental Methodology

2.1. Sample Preparation

The coarse granites for the experiment were taken from a single rock block in Rizhao, Shandong, China. The granites are fine-grained and composed mainly of quartz, feldspar, and calcite, containing abundant large phenocrysts and having an average density of approximately 2.68 g/cm³. In addition, the granite materials have no surface texture that is visible to the naked eye, and are gray in color in the natural state. This kind of granite has an initial uniaxial compressive strength of approximately 104.00 MPa. The cohesive force and the internal friction angle are 19.57 MPa and 49.17°, respectively. Based on the ISRM (International Society for Rock Mechanics) standard [19], a series of cylindrical samples with lengths of 100 mm and diameters of 50 mm were machined. Both ends of the samples were polished to achieve smooth surfaces for tests.

Then, by using a high-temperature furnace, the granite samples were exposed to high temperatures (200, 300, 400, 500, 600, and 800 °C). The heating rate was 10 °C/min until the set temperature level, and this temperature was maintained for 2 h to ensure that the samples heated evenly. When considering the heating rate of 10 °C/min, we assume that the heat travels fast in solids [20]. Finally, the samples were naturally cooled to room temperature. The samples after high-temperature exposure are shown in Figure 1.

Figure 1. Tested granite samples after high-temperature exposure.

2.2. Fluid Flow Test Procedure

The granite samples after high-temperature exposure were used to conduct fluid flow tests at room temperature using the LDY-50 permeability testing system (Haian county petroleum research instrument co. LTD, Nantong, China) (Figure 2). The maximum confining pressure of the system is 50 MPa. The system consists mainly of the following four units: (1) a water supplying system; (2) a triaxial cell clamping device; (3) a speed constant pressure pump; and, (4) a water measurement and collection system. Before testing, the samples were enclosed by a 3-mm-thick rubber jacket. Then, the samples were inserted into the triaxial cell clamping device with two porous stone platens over the ends to ensure an even distribution of the water pressure. Then, the water was fed using a fluid pump to the left side of the rock sample, flowed to the right side, and finally exited via a stainless steel tube.

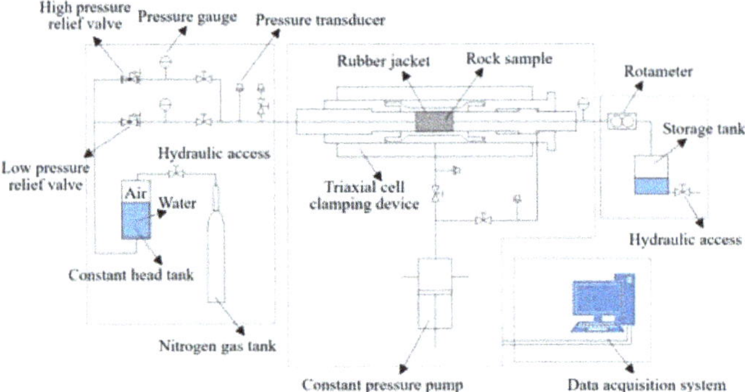

Figure 2. Schematic diagram of the rock permeability test system.

During the tests, for a given rock sample, the confining pressure p_c was increased from 10 MPa to 30 MPa at 5MPa intervals. For a given p_c, a wide range of inlet hydraulic pressures P_i lower than the designated p_c was adopted repeatedly. The hydraulic difference P_s was defined as the difference between the inlet hydraulic pressure P_i and the outlet hydraulic pressure P_o on the other end of the sample. In the experiment, P_s was continuously recorded using a differential pressure gauge, which had a resolution of 0.01 MPa. When the volume flow rates at the two sides of the samples were equal and stable for approximately ten minutes under a given P_s lower than p_c, the samples were considered to be water-saturated, and the fluid flow was considered in a steady state; the equivalent permeability coefficient could then be obtained. By this method, the effects of the confining pressure and high temperatures on granite permeability can be discussed.

3. Physical Properties of Granites after High-Temperature Exposure

The ultrasonic velocity is an effective index to characterize the overall damage in the thermally-cracked granite [21]. The velocities of the compressional waves in the tested granite samples were measured at room conditions using a PDS-SW sonic detector (Wuhan Geostar Scientific & Technological Co., Ltd., Wuhan City, Hubei Province, China). Physical properties of samples after high-temperature treatment, including P-wave (P-wave is one of the two main types of elastic body waves, called seismic waves in seismology, the first signal from an earthquake to arrive at a seismograph. It may be transmitted through gases, liquids, or solids.) velocity and density, were measured, as listed in Table 1.

Table 1. Physical properties of granite samples.

Samples	T (°C)	ρ (g/cm^3)	ρ' (g/cm^3)	$\Delta\rho$ (%)	v (km/s)	v' (km/s)	Δv (%)
CH-01#	200	2.68	2.67	0.17	4.70	4.50	4
CH-02#	300	2.68	2.67	0.33	4.94	4.10	17
CH-03#	400	2.67	2.66	0.41	4.97	3.70	26
CH-04#	500	2.68	2.66	0.64	4.54	3.08	32
CH-05#	600	2.67	2.65	0.79	4.70	2.60	45
CH-06#	800	2.67	2.64	1.08	4.70	1.70	64

Figure 3 presents the variations in density and P-wave velocity for the granite samples after high-temperature exposure, in which $\Delta\rho$ and Δv denote the decrement ratios of the density and P-wave velocity, respectively, and can be calculated, as follows:

$$\Delta\rho = \frac{\rho - \rho'}{\rho} \tag{1}$$

$$\Delta v = \frac{v - v'}{v} \tag{2}$$

in which ρ and ρ' are the densities of rock samples in natural state (25 °C) and after high-temperature exposure, respectively, and v and v' are the corresponding P-wave velocities.

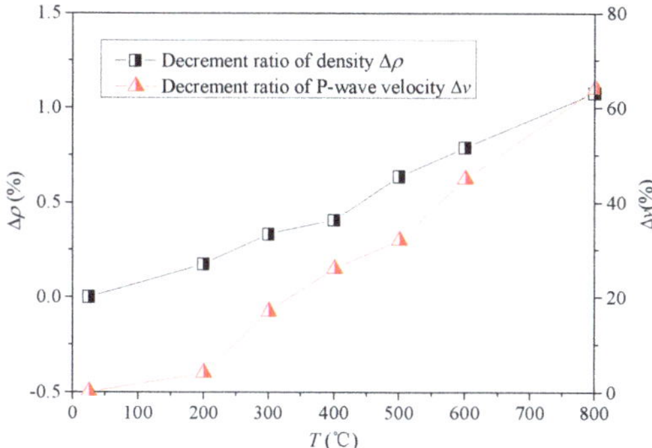

Figure 3. Effect of temperature on the decrement ratio of density and P-wave velocity of the tested granites.

From Figure 3, in the range of temperature (T) from 200 °C to 800 °C, $\Delta\rho$ increases from 0.17% to 1.08%, and Δv increases from 4.0% to 64.0%. There is a negative correlation between P-wave and temperature. A relatively weak decrease of velocity occurs at a low temperature level. However, it decreases significantly with a continuous increase in the temperature, especially after 500 °C. These results suggest that a great many new, thermally induced cracks were generated during the thermal treatment. These cracks form mainly because, with an increase in the heating temperature, the loss of interlayer water and bound water, as well as a transformation in the mineral composition, occurs in the granites, resulting in the degradation of the density and the P-wave velocity [20,22]. Through mercury injection testing, Zhang [23] studied the development of the internal porosity of granites with in heating temperature and found that the internal porosity shows an increase with a magnitude of 2.83 as T increases from 25 °C to 800 °C (Figure 4). Similar to Figure 3, the internal porosity experienced a slight increase from 100 °C to 500 °C. After that, the figures for porosity grow significantly above 500 °C, with a large increase of thermally induced cracks.

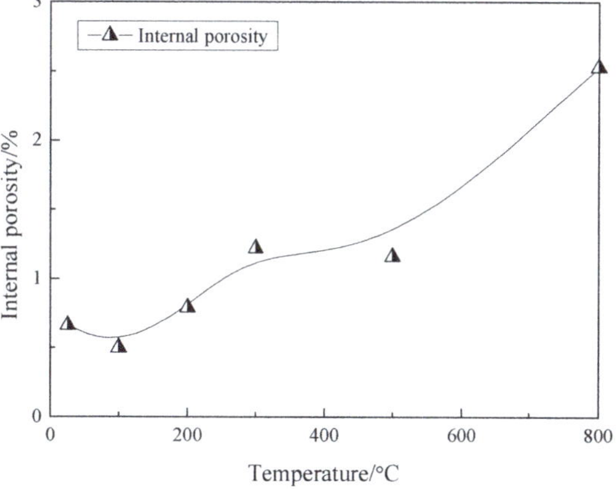

Figure 4. Variations of internal porosity of granites with temperature.

Then, scanning electron microscope (SEM) tests were conducted on the sample granites. Blocks with a size of approximately $8.0 \times 8.0 \times 5.0$ mm^3 were cut from the thermally damaged samples. To achieve a better comparison, SEM was also conducted on the samples at room temperature. Magnifications of 500, 1500, and 5000 were selected in the present study. During the scanning electron microscope observation, we first observed the fracture development from different angles in the sample, and then, the representative points were selected for scanning. Figure 5 shows the typical SEM images of the thermally damaged granites. For $T = 25\,^{\circ}$C, there exist almost no micro-fractures in the granite. In the range of 200–400 $^{\circ}$C, the quantity of the micro-cracks inside and across the aggregates increases slightly. However, as T increases from 400 $^{\circ}$C to 800 $^{\circ}$C, thermally induced cracks were developed along inter-crystalline boundaries and the surrounding crystals. Both of the cracks opening at the boundaries of aggregates and the number of internal pores in the rock increased greatly (Figure 5e). As analysed by Zhang et al. [24], thermally induced cracks are the result of strong bound water loss, dihydroxylation loss of constituent water, and solid mineral expansion in the temperature range of 100–500 $^{\circ}$C. Intra-granular cracks in feldspar and quartz crystal appeared successively at temperatures higher than the second threshold, 573 $^{\circ}$C [18]. During the heating process, physical properties present significant changes at the temperature of about 573 $^{\circ}$C, where the α-β transition occurs. Inner cracks extend quickly and the porosity increases, leading to a reduction of the density and an increase of the conductivity [25]. It is presumed that the increase in the inter-granular thermal stress significantly induced more inter-granular and more trans-granular cracks, for the anisotropic expansion that is linked to the α/β phase transformation of quartz, which is abundant in granite, at 573 $^{\circ}$C led to 5% volume growth in quartz [26,27]. Griffiths et al. [28] provided a Python-based open source tool for quantifying micro-cracks in the two-dimensional (2D) micro-crack density of granite samples using a newly developed algorithm and suggested that the continued evolution of physical properties at temperatures of 600 $^{\circ}$C and above is due to a widening of the existing micro-cracks rather than their formation or propagation. The existence of these pores and cracks doubtlessly affects the permeability of granites [29]. Therefore, it is of great significance to quantitatively evaluate the permeability characteristics of granites after high-temperature exposure.

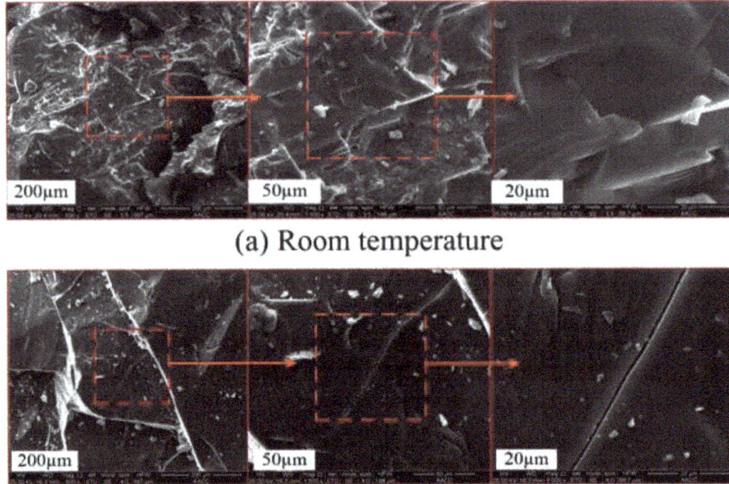

(a) Room temperature

(b) 200 °C

Figure 5. *Cont.*

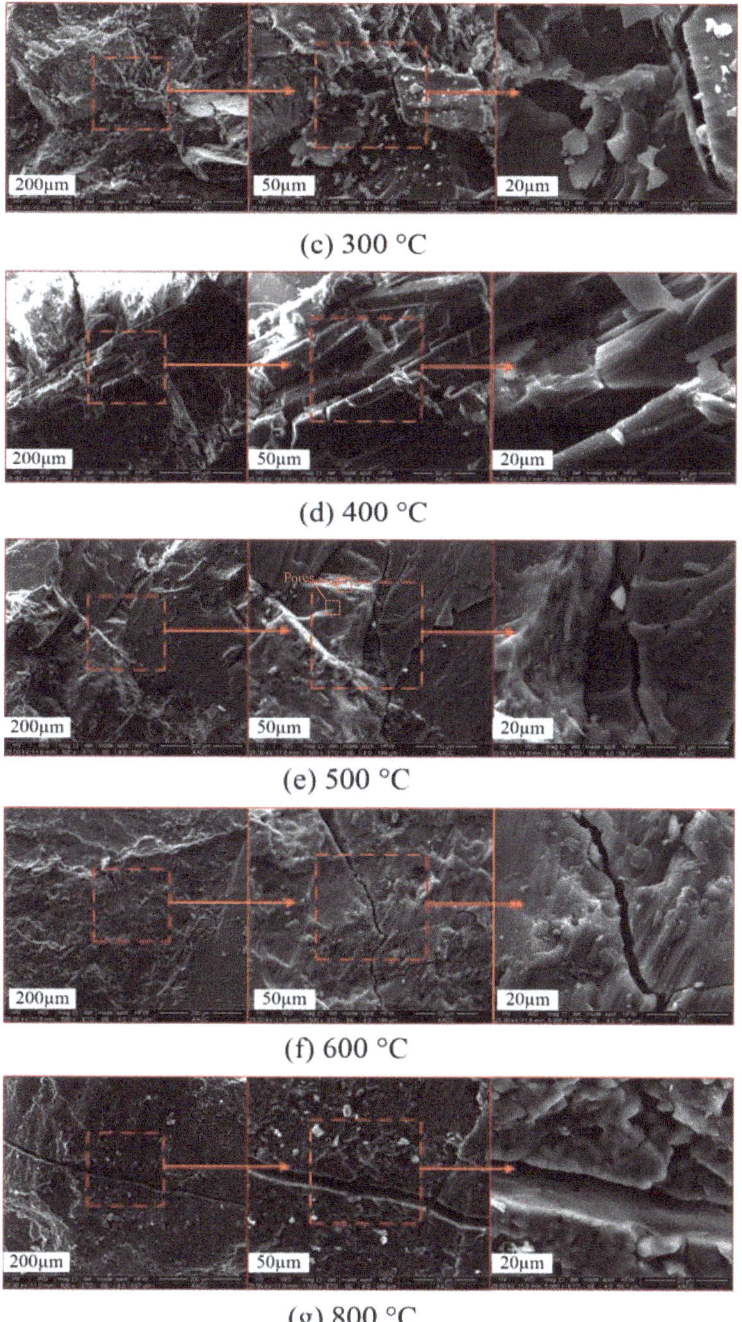

Figure 5. Scanning electron microscope (SEM) images of granites at room-temperature and after high-temperature exposure. (**a**) Room temperature; (**b**) 200 °C; (**c**) 300 °C; (**d**) 400 °; (**e**) 500 °C; (**f**) 600 °C; (**g**) 800 °C.

4. Fluid Flow Behaviours of Granites after High-Temperature Exposure

For all of the test cases, as the pressure gradient dP_s/dL increases, the volume flow rate Q flowing out of the sample shows an increasing trend. The relations between dP_s/dL and Q can be well described using a zero-intercept linear function, and are shown by the solid lines in Figure 6, in which L denotes the length of the sample. From Figure 6, the correlation coefficients or the R^2 values are all larger than 0.99, indicating that the linear Darcy's law fits the raw experimental data very well, and that the fluid flow is laminar. In addition, the increase in p_c does not change the linearity of the fluid flow through the samples, but the slopes of the dP_s/dL-Q fitting curves become steeper with the increase in p_c due to the closure of defects, indicating a higher flow resistance. However, for a given p_c, the slopes of the fitting curves show decreases with increases in T.

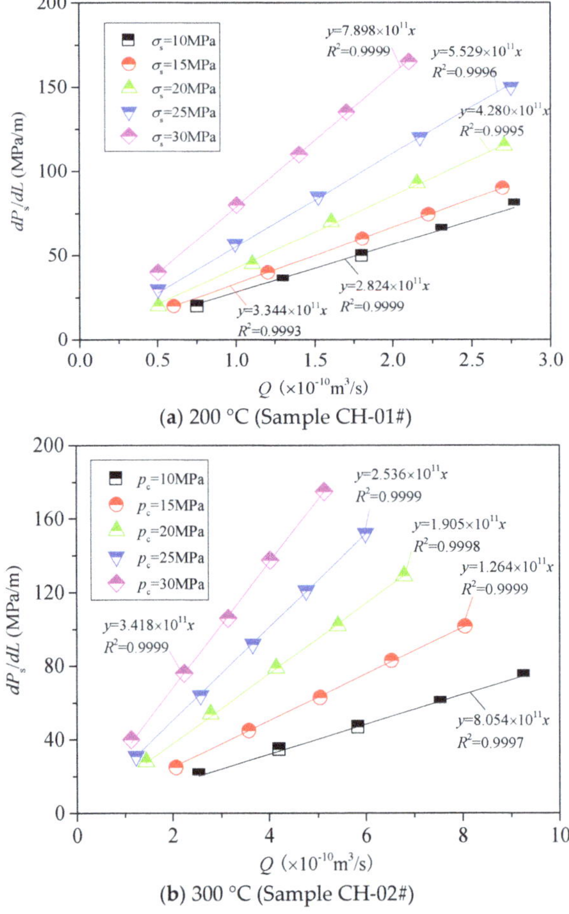

(**a**) 200 °C (Sample CH-01#)

(**b**) 300 °C (Sample CH-02#)

Figure 6. *Cont.*

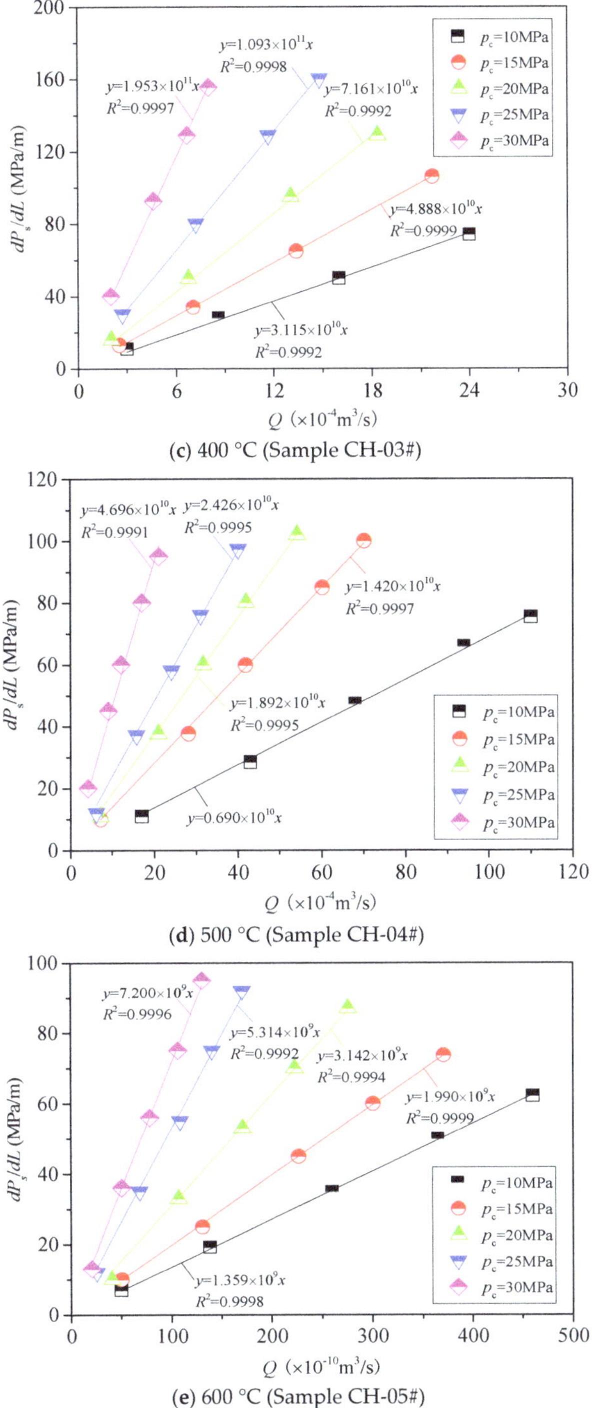

(**c**) 400 °C (Sample CH-03#)

(**d**) 500 °C (Sample CH-04#)

(**e**) 600 °C (Sample CH-05#)

Figure 6. *Cont.*

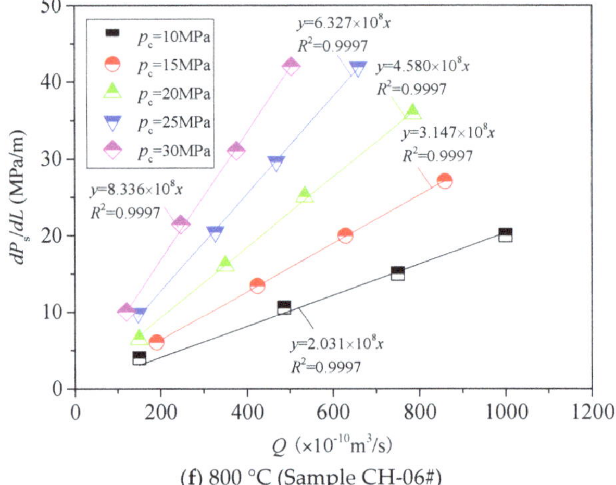

(**f**) 800 °C (Sample CH-06#)

Figure 6. Volume flow rate Q as a function of pressure gradient dP_s/dL of the granite samples after high-temperature exposure. (**a**) 200 °C (Sample CH-01#); (**b**) 300 °C (Sample CH-02#); (**c**) 400 °C (Sample CH-03#); (**d**) 500 °C (Sample CH-04#); (**e**) 600 °C (Sample CH-05#); (**f**) 800 °C (Sample CH-06#).

The equivalent permeability is one of the most important parameters to assess the hydraulic properties of fractured rock masses [30,31]. Using the linear Darcy's law, the equivalent permeability coefficient K_0 of the samples after high-temperature exposure can be calculated using Equation (3), and are listed in Table 2:

$$-\frac{dP_s}{dL} = \frac{\mu}{K_0 A_0} Q \tag{3}$$

in which μ is the dynamic viscosity, 10^{-3} Pa·s, and A_0 is the cross-sectional area of the sample, m². Here, the water was assumed to be viscous and incompressible.

Table 2. Equivalent permeability after different temperature under different confining pressure.

Temperature (°C)	Average Equivalent Permeability ($\times 10^{-18}$ m²)				
	p_c = 10 MPa	p_c = 15 MPa	p_c = 20 MPa	p_c = 25 MPa	p_c = 30 MPa
200	1.84	1.53	1.21	0.9	0.64
300	6.26	4.06	2.65	2.02	1.48
400	15.60	10.29	6.85	4.63	2.58
500	75.17	36.13	28.16	21.96	10.53
600	371.13	257.25	170.72	98.9	72.38
800	2336.46	1619.23	1130.04	799.18	608.17

Variations of K_0 as a function of T are displayed in Figure 7. For a given σ_s, as T increases, all K_0 exhibit nonlinear increasing trends, and the variation process can be divided into the following two stages.

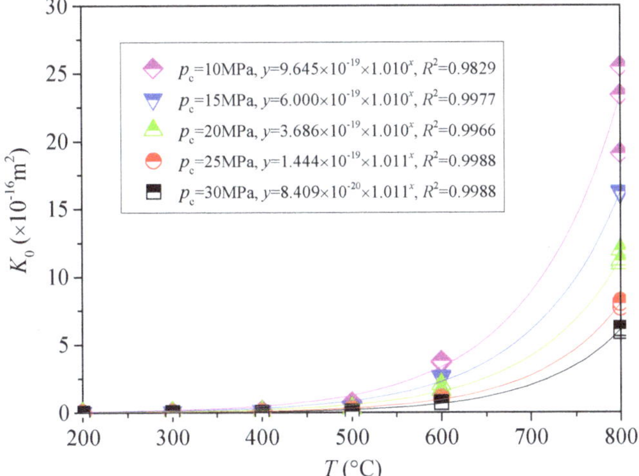

Figure 7. Variation in the equivalent permeability coefficients of granite samples with temperature.

In the range of 200–400 °C, K_0 increases slowly. For T = 200 °C, the K_0 values for the samples are 1.84×10^{-18} (p_c = 10 MPa), 1.53×10^{-18} (p_c = 15 MPa), 1.21×10^{-18} (p_c = 20 MPa), 8.90×10^{-19} (p_c = 25 MPa), and 6.43×10^{-19} m^2 (p_c = 30 MPa). However, for T = 400 °C, the K_0 values are 1.56×10^{-17} (p_c = 10 MPa), 1.03×10^{-17} (p_c = 15 MPa), 6.85×10^{-18} (p_c = 20 MPa), 4.62×10^{-18} (p_c = 25 MPa), and 2.58×10^{-18} m^2 (p_c = 30 MPa), showing increases by factors of 7.47, 5.74, 4.66, 4.14, and 3.02, respectively. From the SEM results in Figure 5, as T increases from 200 °C to 400 °C, several micro-cracks occur in the granites, but the cracks are not fully developed, which leads to a slight increase in the conductivity of the samples.

In the range of 400–800 °C, K_0 increases sharply. For T = 800 °C, the K_0 values for the samples are 2.34×10^{-15} (p_c = 10 MPa), 1.62×10^{-15} (p_c = 15 MPa), 1.13×10^{-15} (p_c = 20 MPa), 7.99×10^{-16} (p_c = 25 MPa), and 6.08×10^{-16} m^2 (p_c = 30 MPa), increasing by factors of 148.82, 156.37, 163.90, 171.82, and 234.36, respectively, over the values for T = 400 °C. The main reasons for the variations are as follows. At temperatures between 400 °C and 800 °C, thermal expansion and ion transformation occur in the granites. The samples undergo irreversible thermal damage. The interior cracks of the rocks extend quickly, and the porosity increases, resulting in a sharp increase in the conductivity.

From the experimental results that are discussed above, as T increases from 200 °C to 800 °C, variations of K_0 against T can be well described using the following exponential function:

$$K_0 = A \times 1.01^T \tag{4}$$

in which A is a fitting coefficient, m^2, representing the sensitivity of changes in the permeability coefficient to the confining pressure.

Chen et al. [18] studied the evolution of thermal damage and the permeability of the Beishan granite, and found that the permeability of granite samples after different temperature treatments and under a hydrostatic pressure of 5 MPa can be described using an exponential function (Figure 8). Generally, the evolution characteristics of permeability are consistent with the experimental results that are obtained here in this study. Then, the permeability in Chen et al.'s study as a function of temperature was evaluated using Equation (4). The comparison results are displayed in Figure 8. It shows that the fitting quality using Equation (4) is more suitable, with a correlation coefficient R^2 of 0.98.

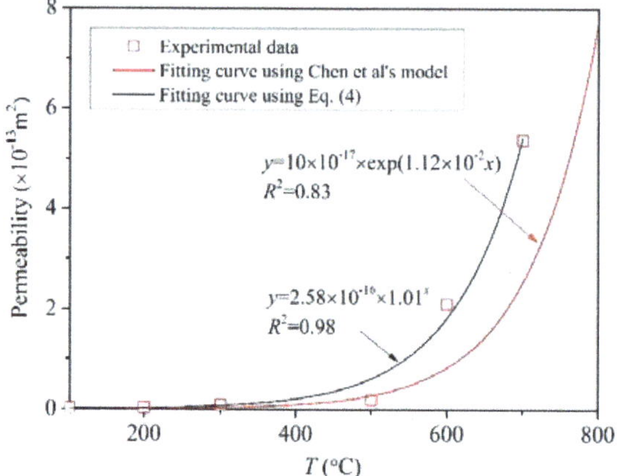

Figure 8. Evolution of permeability versus heat treatment temperature.

Figure 9 shows the variations of coefficient A with an increase in p_c. In the range of 10–30 MPa, A shows a decrease from 9.65×10^{-19} to 8.41×10^{-20} m^2, or by 91.28%. From Equation (4), for granite samples in the natural state ($T = 25\,^\circ$C), the K_0 values of the samples are 1.24×10^{-18} ($p_c = 10$ MPa), 7.69×10^{-19} ($p_c = 15$ MPa), 4.73×10^{-19} ($p_c = 20$ MPa), 1.90×10^{-19} ($p_c = 25$ MPa), and 1.10×10^{-19} m^2 ($p_c = 30$ MPa). In the range of p_c from 10 MPa to 30 MPa, K_0 shows a decrease of 91.06%, which is consistent with some previous results [32].

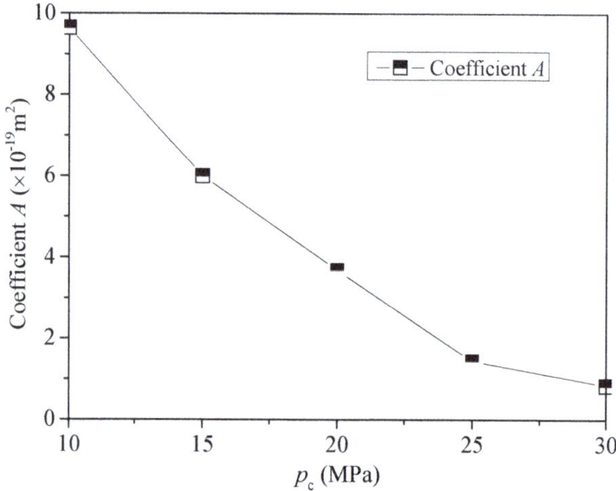

Figure 9. Variations in coefficient A with the confining pressure.

To evaluate the migration law for fluid in a fractured porous medium, the transmissivity (T_a) was defined [33–35].

$$-\frac{dP_s}{dL} = \frac{\mu}{T_a}Q \tag{5}$$

Combinations of Equations (3) and (5) yield the following Equation (6):

$$T_a = K_0 A_0 \tag{6}$$

in which, T_a is the transmissivity, m^4.

The variations in T_a as a function of p_c are displayed in Figure 10. As p_c increases, T_a shows a decrease. However, T_a shows an increase with the heating temperature, and the extent of increase gradually increases. For smaller p_c values (10, 15 and 20 MPa), T_a is sensitive to p_c due to pore/crack closure, especially for samples at higher temperature levels. However, for larger p_c values, the reduction rate for T_a gradually decreases, because the induced cracks that are due to thermal damage generally reach their residual crack apertures.

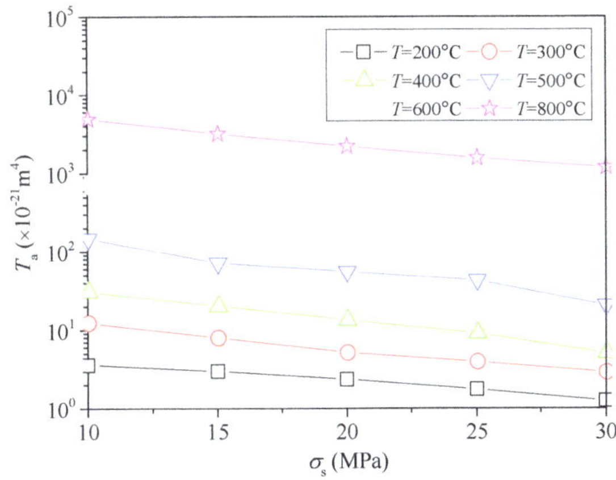

Figure 10. Transmissivity as a function of the confining pressure.

5. Conclusions

From the SEM results, thermal damages, such as micro-cracks and pores, initiate in the granites as a result of high-temperature exposure. As the temperature increases, both the density and the P-wave velocity show degradation.

The relationships between the volume flow rate and the pressure gradient in granite samples after high temperatures at various confining pressures can be well described using Darcy's law. The equivalent permeability coefficients of the samples show an increase with the temperature, and the variation process can be described using the mathematical expression $K_0 = A \times 1.01^T$.

As the confining pressure increases, the conductivity of the samples shows a decrease due to crack/pore closure, especially for samples after higher temperature treatment. In the range of confining pressure from 10 MPa to 30 MPa, the transmissivity shows a decrease of 65.12%–85.99%. These results are useful for research and applications that are involved in many areas, such as the disposal of nuclear wastes.

Acknowledgments: This work was supported by the National Key Basic Research and Development Program of China, China (No. 2017YFC0603001), National Natural Science Foundation of China, China (No. 51734009).

Author Contributions: Lixin He and Hongwen Jing conceived and designed the experiments; Lixin He performed the experiments; Qian Yin and Lixin He analyzed the data; Qian Yin contributed reagents/materials/analysis tools; Lixin He wrote the paper.

Conflicts of Interest: The authors declare no conflict of interest.

References

1. Wang, J.S.Y.; Mangold, D.C.; Tsang, C.F. Thermal impact of waste emplacement and surface cooling associated with geologic disposal of high-level nuclear waste. *Environ. Geol. Water Sci.* **1988**, *11*, 183–239. [CrossRef]
2. Nicholson, K. Environmental protection and the development of geothermal energy resources. *Environ. Geochem. Health* **1994**, *16*, 86–87. [CrossRef] [PubMed]

3. Luo, J.A.; Wang, L.G.; Tang, F.R.; Zheng, L. Variation in the temperature field of rocks overlying a high-temperature cavity during underground coal gasification. *Min. Sci. Technol.* **2011**, *21*, 709–713. [CrossRef]

4. Liu, R.; Li, B.; Jiang, Y.; Yu, L. A numerical approach for assessing effects of shear on equivalent permeability and nonlinear flow characteristics of 2-D fracture networks. *Adv. Water Resour.* **2018**, *111*, 289–300. [CrossRef]

5. Zhao, Y.S.; Wan, Z.J.; Feng, Z.J.; Yang, D.; Zhang, Y.; Qu, F. Triaxial compression system for rock testing under high temperature and high pressure. *Int. J. Rock Mech. Min. Sci.* **2012**, *52*, 132–138. [CrossRef]

6. Ozguven, A.; Ozcelik, Y. Effects of high temperature on physico-mechanical properties of Turkish natural building stones. *Eng. Geol.* **2014**, *183*, 127–136. [CrossRef]

7. Shao, S.; Ranjith, P.G.; Wasantha, P.L.P.; Chen, B.K. Experimental and numerical studies on the mechanical behaviour of Australian Strathbogie granite at high temperatures: An application to geothermal energy. *Geothermics* **2015**, *54*, 96–108. [CrossRef]

8. Sun, H.; Sun, Q.; Deng, W.; Zhang, W.; Lü, C. Temperature effect on microstructure and *p*-wave propagation in Linyi sandstone. *Appl. Therm. Eng.* **2017**, *115*, 913–922. [CrossRef]

9. Yong, C.; Wang, C.Y. Thermally induced acoustic emission in Westerly granite. *Geophys. Res. Lett.* **1980**, *7*, 1089–1092. [CrossRef]

10. Griffiths, L.; Lengliné, O.; Heap, M.J.; Baud, P.; Schmittbuhl, J. Thermal cracking in Westerly Granite monitored using direct wave velocity, coda wave interferometry and acoustic emissions. *J. Geophys. Res. Solid Earth* **2018**. [CrossRef]

11. Freire-Lista, D.M.; Fort, R.; Varas-Muriel, M.J. Thermal stress-induced microcracking in building granite. *Eng. Geol.* **2016**, *206*, 83–93. [CrossRef]

12. Wang, H.F.; Bonner, B.P.; Carlson, S.R.; Kowallis, B.; Heard, H.C. Thermal stress cracking in granite. *J. Geophys. Res.* **1989**, *94*, 1745–1758. [CrossRef]

13. Yin, T.B.; Li, X.B.; Cao, W.Z.; Xia, K.W. Effects of thermal treatment on tensile strength of Laurentian granite using Brazilian test. *Rock Mech. Rock Eng.* **2015**, *48*, 2213–2223. [CrossRef]

14. Zhang, L.Y.; Mao, X.B.; Lu, A.H. Experimental study on the mechanical properties of rocks at high temperature. *Sci. China* **2009**, *52*, 641–646. [CrossRef]

15. Chaki, S.; Takarli, M.; Agbodjan, W.P. Influence of thermal damage on physical properties of a granite rock: Porosity, permeability and ultrasonic wave evolutions. *Constr. Build. Mater.* **2008**, *22*, 1456–1464. [CrossRef]

16. Zoback, M.D.; Byerlee, J.D. The effect of microcrack Dilatancy on the permeability of Westerly granite. *J. Geophys. Res.* **1975**, *80*, 752–755. [CrossRef]

17. Ding, Q.L.; Ju, F.; Mao, X.B.; Ma, D.; Yu, B.Y. Experimental investigation of the mechanical behavior in unloading conditions of sandstone after high-temperature treatment. *Rock Mech. Rock Eng.* **2016**, *49*, 2641–2653. [CrossRef]

18. Chen, S.W.; Yang, C.H.; Wang, G.B. Evolution of thermal damage and permeability of Beishan granite. *Appl. Therm. Eng.* **2017**, *110*, 1533–1542. [CrossRef]

19. Verma, A.K.; Jha, M.K.; Maheshwar, S.; Singh, T.N.; Bajpai, R.K. Temperature-dependent thermophysical properties of Ganurgarh shales from Bhander group, India. *Environ. Earth Sci.* **2016**, *75*, 1–11. [CrossRef]

20. Yin, Q.; Ma, G.W.; Jing, H.W. Experimental study on mechanical properties of sandstone specimens containing a single hole after high-temperature exposure. *Géotech. Lett.* **2015**, *5*, 43–48. [CrossRef]

21. Hu, D.W.; Zhou, H.; Zhang, F.; Shao, J.F. Evolution of poroelastic properties and permeability in damaged sandstone. *Int. J. Rock Mech. Min. Sci.* **2010**, *47*, 962–973. [CrossRef]

22. Kumari, W.G.P.; Ranjith, P.G.; Perera, M.S.A.; Chen, B.K.; Abdulagatov, I.M. Temperature-dependent mechanical behaviour of Australian Strathbogie granite with different cooling treatments. *Eng. Geol.* **2017**, *229*, 31–44. [CrossRef]

23. Zhang, W. *Study on the Microscopic Mechanism of Rock Thermal Damage and the Evolution Characteristics of Macroscopic Physical and Mechanical Properties*; China University of Mining and Technology: Xuzhou, China, 2017.

24. Zhang, W.; Sun, Q.; Hao, S.; Geng, J.; Lv, C. Experimental study on the variation of physical and mechanical properties of rock after high temperature treatment. *Appl. Therm. Eng.* **2016**, *98*, 1297–1304. [CrossRef]

25. Pereira, A.H.A.; Miyaji, D.Y.; Cabrelon, M.D.; Medeiros, J.; Rodrigues, J.A. A study about the contribution of the α-β phase transition of quartz to thermal cycle damage of a refractory used in fluidized catalytic cracking units. *Cerâmica* **2014**, *60*, 449–456. [CrossRef]

26. Géraud, Y. Variations of connected porosity and inferred permeability in a thermally cracked granite. *Geophys. Res. Lett.* **1994**, *21*, 979–982. [CrossRef]
27. Xi, D.Y. Physical characteristics of mineral phase transition in the granite. *Acta Mineral. Sin.* **1994**, *14*, 223–227.
28. Griffiths, L.; Heap, M.J.; Baud, P.; Schmittbuhl, J. Quantification of microcrack characteristics and implications for stiffness and strength of granite. *Int. J. Rock Mech. Min. Sci.* **2017**, *100*, 138–150. [CrossRef]
29. David, C.; Menéndez, B.; Darot, M. Influence of stress-induced and thermal cracking on physical properties and microstructure of La Peyratte granite. *Int. J. Rock Mech. Min. Sci.* **1999**, *36*, 433–448. [CrossRef]
30. Liu, R.; Jiang, Y.; Li, B.; Wang, X. A fractal model for characterizing fluid flow in fractured rock masses based on randomly distributed rock fracture networks. *Comput. Geotech.* **2015**, *65*, 45–55. [CrossRef]
31. Liu, R.; Li, B.; Jiang, Y. A fractal model based on a new governing equation of fluid flow in fractures for characterizing hydraulic properties of rock fracture networks. *Comput. Geotech.* **2016**, *75*, 57–68. [CrossRef]
32. Takada, M.; Fujii, Y.; Jun-lchi, K. Study on the effect of confining pressure on permeability of rock in triaxial compression failure process. *J. MMIJ* **2012**, *127*, 151–157. [CrossRef]
33. Zhang, Z.; Nemcik, J. Fluid flow regimes and nonlinear flow characteristics in deformable rock fractures. *J. Hydrol.* **2013**, *477*, 139–151. [CrossRef]
34. Li, B.; Liu, R.; Jiang, Y. Influences of hydraulic gradient, surface roughness, intersecting angle, and scale effect on nonlinear flow behavior at single fracture intersections. *J. Hydrol.* **2016**, *538*, 440–453. [CrossRef]
35. Yin, Q.; Ma, G.W.; Jing, H.W.; Su, H.J.; Wang, Y.C.; Liu, R.C. Hydraulic properties of 3D rough-walled fractures during shearing: An experimental study. *J. Hydrol.* **2017**, *555*, 169–184. [CrossRef]

Review

Seepage Characteristics and Its Control Mechanism of Rock Mass in High-Steep Slopes

Hong Li [1] , **Hongyuan Tian [2] and Ke Ma [1],***

[1] School of Civil Engineering, Dalian University of Technology, Dalian 116024, China; hong.li@dlut.edu.cn
[2] College of Food Science and Engineering, Northwest Agriculture and Forestry University, Yangling 712100, China; shipinjiuye@126.com
* Correspondence: mark1983@dlut.edu.cn; Tel.: +86-186-009-52046

Received: 30 December 2018; Accepted: 24 January 2019; Published: 1 February 2019

Abstract: In Southwest China large-scale hydropower projects, the hydraulic conductivity and fracture aperture within the rock mass of a reservoir bank slope has dramatically undergone a time series of evolution during dam abutment excavation, reservoir impounding and fluctuation operation, and discharge atomization. Accordingly, seepage control measures by hydro-structures such as drainage or water insulation curtains should be guided by scientific foundation with a dynamic process covering life-cycle performance. In this paper, the up-to-date status of studying the evolution mechanism of seepage characteristics relating to fractured rock hydraulics from experimental samples to the engineering scale of the rock mass is reviewed for the first time. Then, the experimental findings and improved practice method on nonlinear seepage flow under intensive pressure drives are introduced. Finally, the scientific progress made in fractured rock seepage control theory and optimization of the design technology of high-steep slope engineering is outlined. The undertaken studies summarized herewith are expected to contribute to laying a foundation to guide the further development of effective geophysical means and integrated monitoring systems in hydropower station construction fields.

Keywords: hydro-power; high-steep slope; fractured rock; permeability; seepage control

---

1. Introduction

In Southwest China, from Sichuan to Yunnan province, large-scale hydro-power stations such as Xiaowan in Lantsang River, Jinping in Ya-lung (Nyag Chu in Tibetan) River (shown in Figure 1), Dagangshan in Dadu River, Xiluodu, Xiangjiaba, Wudongde and Baihetan successively in Chin-sha Chiang River have been constructed in the last 10 years. As shown in Table 1, taking four of them as examples, their complex rock slopes as dam abutments are high and steep.

Figure 1. A bird's eye view of Jinping I Hydropower Station by man-made satellite.

Table 1. Height of the dam abutment slopes for some hydro-power stations in southwest China [1].

Project Name	Xiaowan	Jinping	Dagangshan	Wudongde
Slope Natural Height (m)	700–800	>1000	>600	830~1036
Slope Excavation Height (m)	670	530	410	430

[1] After Zhou C.B., Presentation for End of State Key Fundamental Research 973 Project (No. 2011CB013500).

Slope stability is a major issue during design, construction. and operation of hydraulic and hydro-power engineering [1]. It is justified by the large quantity of research that the vast majority of landslides are related to water [2]. Distinguishing rock hydraulics [3] from classical poromechanics was recognized by humans by learning tragic lessons from world-shaking catastrophes such as the French Malpasset arch dam break in 1959 and the Italian Vajont tremendous left bank landslide in 1963 [4]. Hydraulic characteristics of rock discontinuities and the regularity of groundwater movement are the foundation and key to implement slope seepage analysis. In the case of saturated conditions, theoretical and experimental findings on the seepage in a single rock fracture are abundant. They reflect the influences of geometric configurations and stretch features of fracture walls on seepage flows in a fracture, the applicability of the cubic law and its modifications as well as stress sensitivity in fracture seepage [5–8]. Nevertheless, in the case of unsaturated conditions, there is not yet plenty of in-depth findings on water and air two-phase flow through a rock fracture. The theoretical models on multiphase flow in fractured rocks are mostly founded on soil–water characteristic curves in porous media [9–11]. More seriously, the study on flow regime as well as the hysteretic behavior of water and air two-phase flows through a rock fracture are rarely seen.

To study the seepage characteristics and seepage law of motion in fractured rock mass, equivalent continuum and discrete network approaches have been mainly adopted. Between them, the highlight of the equivalent continuum approach is to study the scale effect and anisotropy of hydraulic conductivity and specific storage coefficient so as to establish the equivalent permeability tensor of rock mass and discover the stress sensitive evolution regularity of seepage characteristics [12–14]. Although field testing methods represented by the cross-hole permeability test [15,16] have been made, progress to develop effective geophysical approaches such as microseismic monitoring is urgently needed. This facilitates understanding of how damage or failure occurring inside the rock mass influences its corresponding seepage characteristics while a dam abutment slope is being excavated or loaded [17,18]. Many researchers [19–21] studied the features of permeability evolution along the process of rock deformation as well as progressive failure. Their findings embodied the intensive scaling effect of permeability. Nevertheless, for unsaturated seepage properties of fractured rock mass, mechanisms of air and water two-phase flow as well as rainfall infiltration, study progresses only slowly.

Equivalent continuum, discrete network, and dual porosity methods are usually adopted to numerically simulate the seepage field of slope rock mass. Concerning the steady and transient seepage flow related problems, in order to solve the numerical stability problem caused by nonlinearity of flux spillage boundary, formulations of elliptical and parabolic Signorini's type variational inequality have been successively established [22]. Accordingly, incorporated with a drainage substructure method to numerically simulate its seepage flow, a precise simulation technique was proposed for analyzing large-scale drainage curtain structures [23]. Nowadays, the seepage control technique for slope rock mass has developed from the early stage of leakage prevention to the stage of taking drainage to be as important as leakproof. The rationality and effectiveness of slope seepage control design depends on not only correctly understanding the seepage control mechanism but also scientifically evaluating seepage control results. Fortunately, a number of researchers' achievements paved the way to evaluate the effect of seepage control in high-steep slopes [24–28]. It has been summarized in recent years that the mechanism comes down to process, state, parameter, and boundary, i.e., the four categories of rock mass seepage control [23]. Thus, laying a theoretical cornerstone for seepage control optimization design.

Therefore, it is badly needed in an intensive way to study the topics: evolution of seepage characteristics for intensively un-loading slope rock mass, unsaturated seepage and water-air two-phase flow motion features, rainfall infiltration mechanism as well as seepage control optimization design.

Encompassing three key aspects of the titled problem, i.e., seepage related properties, seepage motion regularities, and seepage control, it is the aim to systematically summarize the following latest findings in this review. These profound studies concern not only the macro- and meso-scopic mechanism, multiple scaling effect and anisotropic feature on the process evolution of the seepage characteristics of high-steep slope fractured rocks, but also the model solution and value selection of non-linear seepage flow parameters, as well as multi-purpose whole process dynamic inverse analysis of rock mass seepage fields.

2. Evolution Mechanism on Seepage Characteristics of High-Steep Slope Rock Mass

Natural evolution of deep valleys and large scale excavation of high-steep slopes both cause a strong unloading impact on the slope rock mass. Extensively developed discontinuous structures as well as rock damage play a dominant role in seepage flow properties. The following parts contain three aspects of the distinct findings: (1) the rapid laboratory technique on the permeability testing for low-permeable rocks and evolution models; (2) evolution mechanism and its scaling effect on the seepage characteristics of fractured rocks; (3) the multipurpose and whole process dynamic inverse analysis method on the seepage characteristics of fractured rock mass. They have been studied to deal with the challenge of predicting the performance of instantaneous seepage characteristics in the rock mass.

Utilizing a TAW-1000 deep water pore pressure tri-axial servo rock testing machine and a French tri-axial seepage rheological experimental system where the water pressure attains 70 MPa, the association between the seepage and deformation properties of fractured rock was studied [29]. Furthermore, acoustic emission signals were recorded from the whole stress–strain process of rock deformation and failure [30]. The inherent connection between deformation, permeability, and acoustic emission data was setup. In the meantime, sonic wave and radial testing using thick-wall cylindrical ring rock sample [31] was also attempted and developed, respectively. As well, simulators of various kinds of numerical methods such as Finite Element Method, Boundary Element Method, Distinct Element Method, Control Volume Finite Difference Method, and Particle Flow Code were adopted to analyze the experiments accordingly in both continuum and discontinuous, macro- and mesoscopic, parallel and large-scale high performance scientific computational ways. These studies obtained a large amount of fundamental systematic data sets on the relationship between hydraulic conductivities and the whole process of stress–strain behaviors, especially in (red) sandstone, limestone, granite, and marble. The fact that modelling computations displayed matching results with experimental data laid a solid foundation to further understand the evolution of seepage characteristics along with various loading and deformation states.

Blocked by high cost and long cycle, the existing rapid testing way of rock low permeability has become a technical difficulty. Through integrating an instantaneous pressure pulse device into the Temperature-Hydraulic-Mechanical coupling testing facility, shown in Figure 2, an experimental technique coupling water–air two-phase seepage flows and triaxial compression of low permeable rock samples was researched and developed. This system contains servo measuring of the axial, transverse surrounding and air– water partial pore pressures as well as automatic data acquisition and recording. As its range reaches 10^{-12}–10^{-22} m^2 wide with internationally equivalent level of precision, it was successfully applied in the Canadian URL (Underground Research Laboratory) deep underground laboratory [32,33].

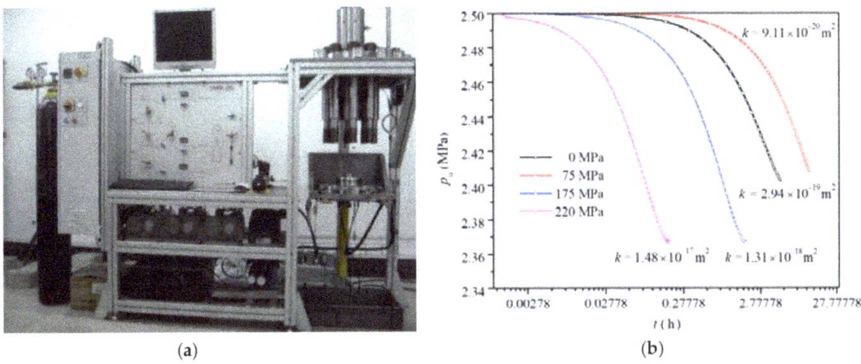

(a)   (b)

Figure 2. Coupled triaxial compression and two-phase flow testing system for low permeable rocks [34]: (a) Experimental facility by Research & Development; (b) Tested curves as an example.

Through analyzing sonic waveforms which penetrate different categories of rock samples under various conditions of moisture and saturation, the association between the features in time, frequency, time-frequency domains of acoustic signals and the pore and microcrack development, water-bearing condition of rocks was explained [35]. Through modeling acoustic emissions and energy dissipation of particle flows in rocks under water-mechanical coupling conditions, it was concluded that tensile cracks are always dominant under low confining pressure and shearing cracks are dominant when strain increases under high confining pressure [36]. Under an environment of particle flow discrete element software PFC2D, a fluid–solid coupling numerical simulation analysis of the particle flow discrete element is carried out by using embedded FISH (An imbedded programming language to enable users to define variables and functions). The influence of porosity dynamic distribution on permeability evolution characteristics under different osmotic pressures was explained by the micro-mechanism [37], as illustrated in Figure 3. The stress–strain curves and acoustic emission data of sandstone were obtained through a hydraulic coupling triaxial test and real-time Acoustic Emission monitoring. In the meantime, a hydraulic coupling bi-axial model of discrete element and particle flow was established to study the laws of acoustic emission and energy evolution in sandstone.

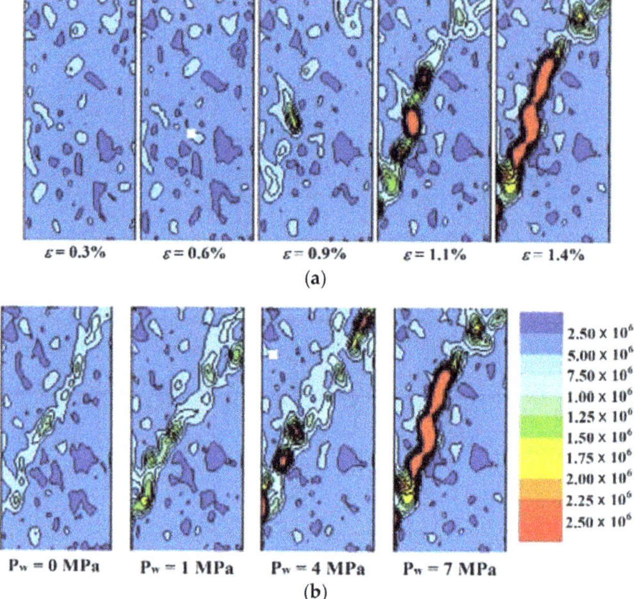

Figure 3. Contour of simulated progressive permeability distribution along with stress–strain process [37]: (**a**) under maintained pore pressure 7 MPa; (**b**) eventual peak stress states under various maintained pore pressures.

Coupled Hydraulic-Mechanical-Damage model and efficient solution of large-scale Finite Element equations for fully coupled problems were also studied [38]. The numerical solution to the two-part Hooke model (TPHM) was realized in TOUGH-FLAC3D. The conditions of its applicability were also defined. This method essentially reflects the nonlinear deforming behavior during the course of loading/unloading in a low stress stage so as to conform to the excavation unloading induced deforming features in the surrounding rocks. This model is more practical to design and analyze deformation characters of surrounding rocks in engineering with unloading disturbance [39]. Through the numerical simulation using the statistical meso-damage finite element method [40], as shown in Figure 4 [41], the effects of inhomogeneity, confining pressure, and joint density on the size effects of mechanical and seepage characteristics were discussed. A quantitative estimation method was proposed for the mechanical and hydraulic properties of rock mass with meso- to macro-levels of multi-scale fractures through numerous systematic simulations. The parameters of deformation, strength, fracture porosity, and permeability of the rock mass were represented in the numerical models. Although computational resources and cost are in most cases rather expensive and the applicability of cubic law as well as its restriction needs further validation, it can provide a reference for reasonably acquiring and extrapolating computational engineering parameters for practical high-steep slope rock mass.

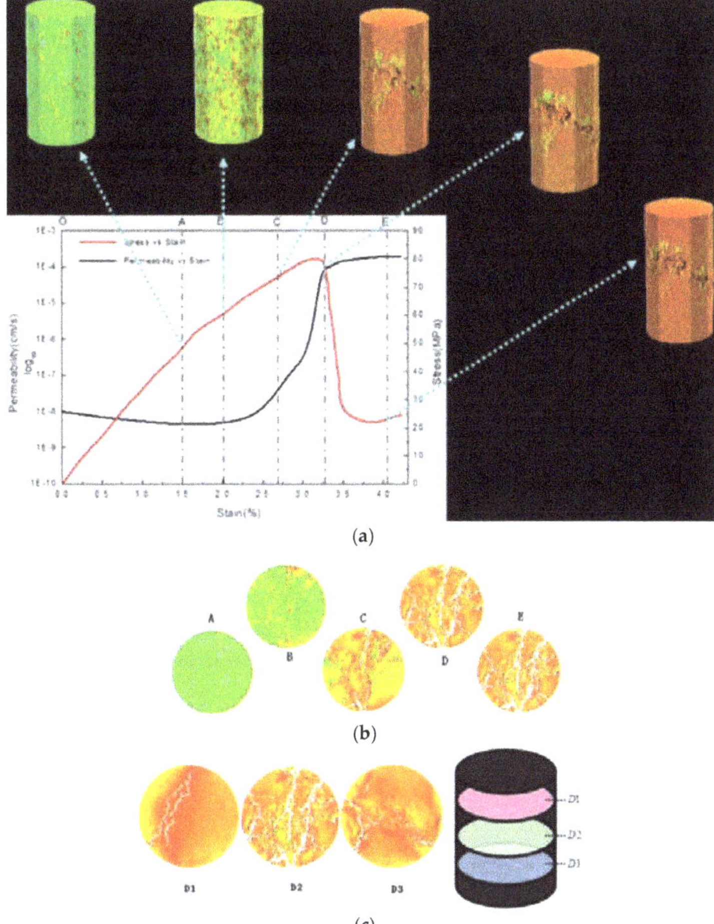

Figure 4. Illustration of high-performance fine modelling on a rock sample stress–strain process using statistical meso-damage Finite Element Method [41]: (**a**) Simulated permeability evolution along with stress–strain process; (**b**) Progress cracking states on the middle cross-section at five typical moments; (**c**) Cracking states on three typical cross-sections at the peak-stress moment.

In a recent study, an energy conservative microscopic and anisotropic damage mechanics model was established for brittle rocks [34]. The aforementioned study considered the normal stiffness recovery and sliding dilatancy of microcracks as follows [34]:

$$E = S^s : \Sigma + \frac{1}{4\pi} \oint_{l^2} \left[\beta(n) n \otimes n + \gamma(n) \overset{s}{\otimes} n \right] dS \tag{1}$$

where, E is the macropscopic strain tensor; S^s is the flexibility for rock matrix; Σ is the macroscopic stress tensor of inhomogeneous rock; β, γ are internal variables characterizing crack opening and sliding, respectively; l^2 is the unit spherical area, $\otimes$ represents the symmetrical tensor formed by cross multiplication from two vectors.

Also, considering crack initiation, propagation, changes in crack morphology, and volume fraction as well as the micro-structural feature of crack connectivity, a new damage-induced permeability evolution model was established [34]. This model in addition characterizes crack connectivity as an important microscopic structural feature as follows [34]:

$$k^{\text{Voigt}} = (1 - \varphi^c)k^s + \frac{1}{4\pi} \oint_{l2} k_0^c \frac{\beta^3}{\beta_0^2} \left(\frac{d}{d_0}\right)^{2\alpha - 2} (\delta - n \otimes n)dS \tag{2}$$

where, φ_c is the density function of cracks; n is the unit normal vector of a crack; δ is the two-order unit tensor; k^s is the permeability tensor of rock matrix; k_0^c is initial permeability of microcracks; β_0, β are the initial and current internal variables of microcracks; d^0, d are the initial and current density of crack distribution, respectively.

It was proved that the traditional multi-scale homogenization based model has a lower bound estimate nature and overcomes the limitation due to lack of describing the microcracks' connectivity. The number of parameters is less by 1 to 4 that of other models [42,43].

Compared with experimental tests on Swedish Äspö diorite, Lac du Bonnet, Senones and Beishan granite, the correctness of the above models was verified due to the consistent agreement between prediction and measurement [44]. Anisotropic damage of rock, structural plane dilatancy, and wear as well as statistical characteristics of fracture development is taken into account in this model. It reveals the macro–meso mechanism, multi-scale effect, and anisotropic characteristics of permeability evolution in rock mass, which was applied to study the excavation disturbance effect and permeability evolution law for Jinping I hydropower station left-bank slope.

Based on a multi-scale homogenization approach, the model of elasto-plastic damage and seepage characteristics evolution for fractured rock mass was established [45]. Among the scales of rock, structural plane and rock mass, macro-/meso-scopic mechanism, the multiscale effect and anisotropy in rock mass seepage characteristics evolution was revealed. The above mentioned models count not only the dominant function of fracture development features on rock mass permeability but also the contribution to rock mass seepage characteristics from rock damage caused by intensive disturbance and post-peak deformation behavior of structural planes. The advantage brought by converging between the rock mass structure model and the damage model created a condition for coupling mechanical and hydraulic analysis on rocks during implementing large-scale excavation. The findings were successfully applied in studying the influence of excavation relaxation on seepage characteristics evolution as well as field seepage control effects of the high-steep slopes and underground powerhouse group in Jinping I and Kala hydropower stations [46,47].

3. Nonlinearity in Seepage Flow of High-Steep Slope Rock Mass and Its Analysis

A series of theories and control technologies on seepage flow analysis under the condition of high osmotic pressure was established. It not only manifests the seepage motion characters in slopes under rainfall or atomization conditions, but also reveals the unsaturation, multi-phase coupling, and non-Darcy flow features of seepage flows in deformable rough-wall fractured rocks.

First, experiment and modeling on the seepage mechanism of a rough fracture under high osmotic pressure were undertaken. As shown in Figure 5, using high-precision 3D laser scanning, an accurate technology was proposed to reconstruct the initial aperture distribution of a structural plane.

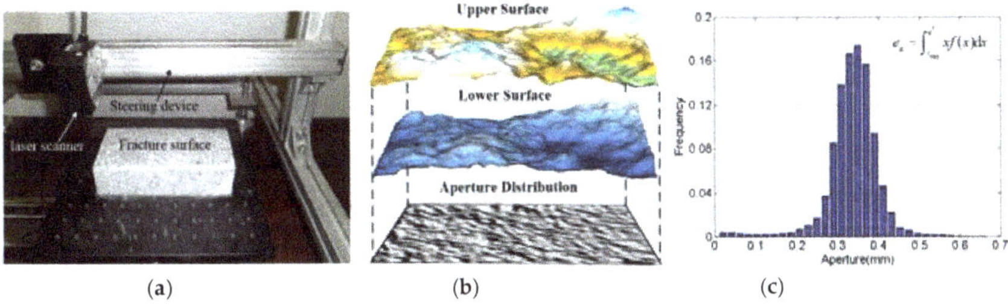

Figure 5. Morphology scanning and aperture reconstruction of a rough fracture: (**a**) Fracture surface morphology scanning [48]; (**b**) fracture aperture reconstruction and correction; (**c**) fracture aperture distribution.

In order to study the seepage characteristics of rough structural planes in a visual environment, a multi-phase percolating flow testing device which is capable of replicating a transparent rough crack by epoxy resin material was developed. At the same time, as shown in Figure 6, it provides an important means for visual observation of coarse-fracture multiphase flows, solute migration, and viscous fingering [49,50]. By laboratory experiments, the nonlinear coefficient for Forchheimer's law on the non-Darcy flow relationship between flow rate Q and pressure gradient P of a deformable rough fracture was determined. Furthermore, as shown in Figure 7, three mechanisms causing non-Darcy flow were revealed. They are namely inertia effect of water flow (a), pore swelling caused by dislocating rough fissure walls (b), and interaction between fluid–solid deformations (c). The physical parameter model for Forchheimer's law, critical Reynolds number model, and flow regime criterion were proposed and validated. Considering fracture morphology and fluid regime, the friction factor was given in a theoretical expression. Three flow zones covering fracture flow regimes were established. They are Darcy flow zone, Forchheimer flow zone, and complete turbulent zone [51,52]. It should be noted that the existing practice in fields had followed the linear seepage rule of Darcy flow over a long history. The newly developed permeability estimate method for high-pressure packer testing based on the above findings achieved favorable effects in the 450 m deep practice at the central region of Hainan Island.

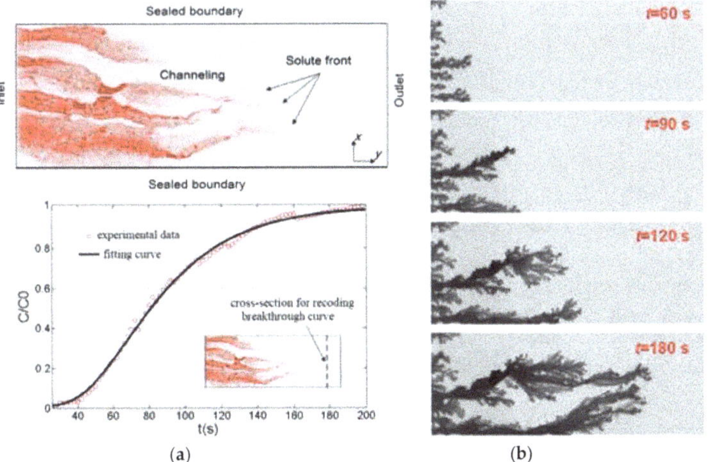

Figure 6. Experimental outcomes of solute migration and viscous fingering along a transparent rough fracture [52–56]: (**a**) Solute migration test and breakthrough curve; (**b**) developing process of viscous fingering.

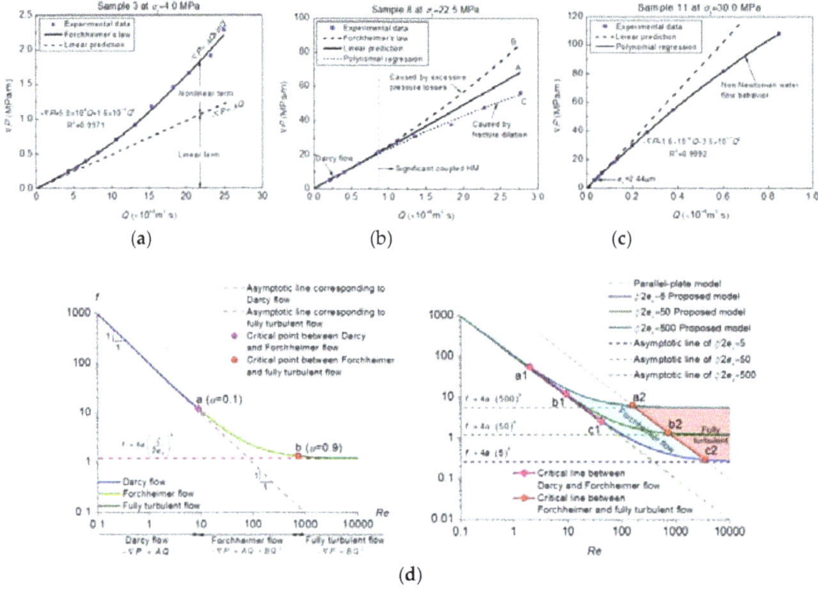

Figure 7. Experimental regularity and theoretical study of non-Darcy flows along a deformable rough rock fracture: (**a**) non-linearity from inertia effect of water flow (**b**) non-linearity from pore swelling due to dislocating rough fissure walls (**c**) non-linearity from coupling fluid–solid deformations [50]; (**d**) zone division for flow regimes [57].

Second, an analytical method for nonlinear seepage parameters of fractured rock mass was established. Based on a borehole high pressure test, aided by borehole video recording and sonic wave detection, it yielded a P–Q curve type, its feature and the intrinsic relationship between water injection pressure, rock mass integrity, and geostress level. An analytical model for high water pressure testing under nonlinear flow conditions was given. The model overcomes the disadvantage that the permeability coefficient obviously turns out to be smaller when conventional pressure test rules are applied to a high water pressure test which brings a huge safety risk to the engineering anti-percolation and drainage design. The test showed that the P–Q curve can be divided into three typical stages: Darcy flow segment (I), non-Darcy flow segment (II), and hydraulic splitting segment (III). As shown in Figure 8, it can also be divided into two basic types, A and B: Curve A-type only contains two sections I and II. The pressure rising curve and falling curve are basically coincident. The fracture irreversible deformation is not distinct; Curve B-type includes three sections I, II, and III. The rising and falling curves do not coincide and irreversible deformation or hydraulic cleavage occurs to the fracture [58]. Under the spherical diffusion flow assumption, an analytical model for nonlinear seepage parameters of rock mass based on nonlinear Izbash law and Forchheimer's law ($-\nabla P = AQ + BQ2$) under high pressure condition was established. The method to estimate the permeability coefficient of the rock mass in three different stages, namely linear, nonlinear, and hydraulic fracturing sections, was recommended [59,60]. It should be noted that the above model would soon be included in the revising test code after going through successful verifications in the field application.

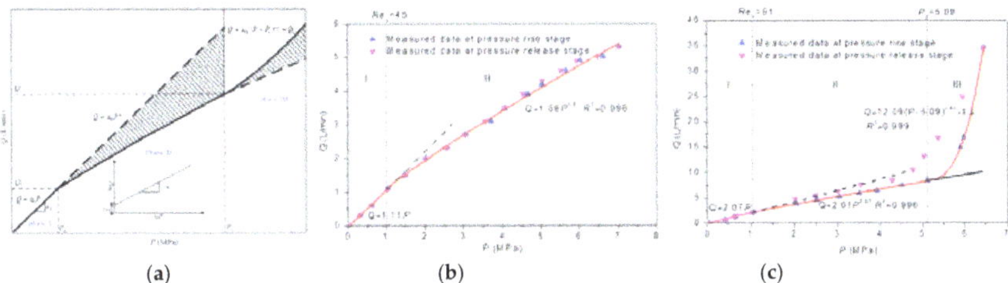

Figure 8. Types and features of P–Q curves in drilled borehole high-pressure Parker tests [59]: (**a**) Typical P–Q curve; (**b**) curve Type A; (**c**) curve Type B.

Third, as shown in Figure 9, a large-scale physical experiment was carried out to study the water coupling deformation features of a slope under strong atomization. The model scale is $8000 \times 2000 \times 4000$ mm^3, which realized water level fluctuation, simulated groundwater head varying in rock layers, and studied the evolution process of slope displacements under reservoir water fluctuation and rainfall conditions. The dip angle of the slope was adjusted to study the ultimately stable dip angle under different saturation. In addition, a contactable monitor system and a non-contactable digital speckle measurement system realized on-line continuous deformation measurement [61]. On the other hand, the model for the soil-water characteristic curve containing the deformation and hysteresis effect was established. It revealed the coupling mechanism between nonlinear deformation of soil swelling, its collapse by moisture, and unsaturated seepage flow. Based on these, as shown in Figure 10, the water-holding characteristic curve of deformable rock mass and the calculation method of unsaturated hydraulic conductivity were formed [62,63]. According to the proposed soil water characteristic curve model, a modified Mualem statistical model for the relative permeability coefficient of unsaturated soil, considering soil deformation effect, was established. The comparison with the experimental results showed that the model can estimate the relative permeability coefficient of the soil under deforming conditions, whose prediction ability is stronger than the Mualem statistical model. It provides a more reliable theoretical model for studying the seepage law of unsaturated soil and its numerical simulation of H–M coupling. In addition, according to the seepage characteristics of the slope rock mass in the atomized area, a coupled model of slope water–air two-phase flow motion considering rock mass deformation was established to explain the hysteretic effect of the wet front advancing process, gas migration, rock mass deformation, and drainage on the infiltration evolution process under heavy rainfall [64].

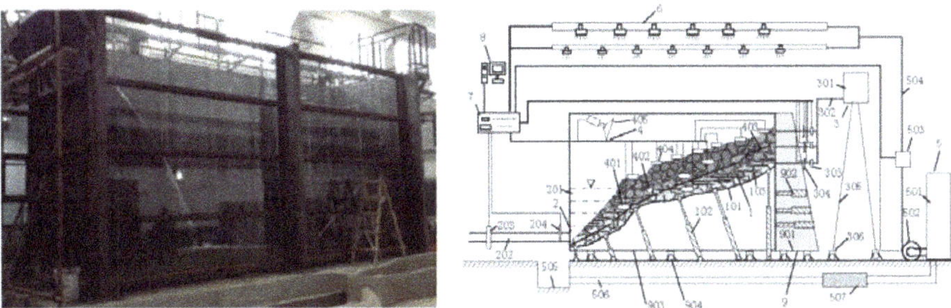

Figure 9. Large-scale physical experiment on deformation features of the reservoir bank under intensive atomization.

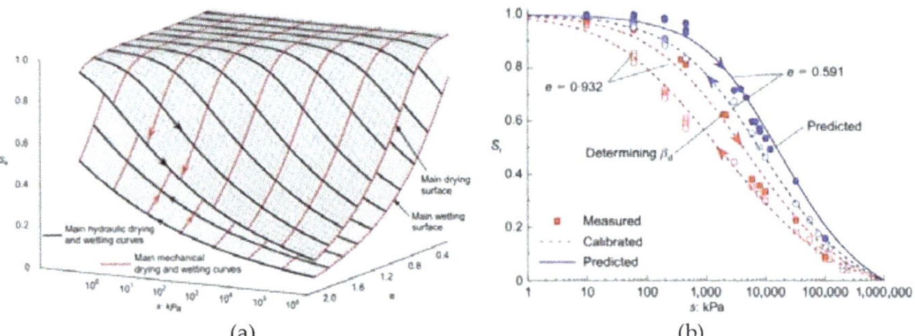

(a)   (b)

Figure 10. Water-holding characteristics curve of deformable soil/rockmass [62]: (**a**) Dominant moisture ab-/de-sorption contours; (**b**) Unsaturated characteristic curves containing dry/humid cycling effect on seepage and deformation.

Fourth, the thermo–hydro–mechanical coupling mechanism and modeling of unsaturated soils were established. A large number of studies have shown that the adhesion effect between soil particles in the unsaturated state has a significant impact on its deformation. Pore distribution as well as its evolution inside soil, play a decisive role on the soil-water characteristic curve. Based on the soil meso-structure and its evolution characters, an explicitly expressed factor which has definite mechanical meaning in characterizing bonding between unsaturated soil particles was proposed. Under the framework of the modified Cambridge model, an elastoplastic constitutive model of unsaturated soil was setup with a homogenized skeleton stress and particle bonding factor as basic variables. On the basis of the soil pore distributive evolution law during the deformation process, a soil-water characteristic curve model considering the deformation and hysteresis effect was expressed. As shown in Figure 11, the model of unsaturated seepage characteristics, the full-coupling model of elastoplastic deformation, and the capillary hysteresis of unsaturated soil were established under a thermodynamic framework, revealing the coupling mechanism of seepage flow on nonlinear deformation and unsaturation. Compared with the classic Barcelona (BBM) model in the field of unsaturated soil mechanics [65], the above fully coupled hydraulic model for unsaturated soil not only in a more fundamental way reflects two-way coupling characteristics of elastoplastic deformation and capillary hysteresis under the thermodynamic framework, but also has a more direct building method and clearer physical meaning. Only a single yield surface (BBMs has two) and fewer parameters (four less than BBM) are adopted. A mechanism of shear zone formation during a rain-induced landslide was revealed, laying a foundation for precisely simulating the hydraulic coupling process of the 300 m level of rockfill dam wall.

Lastly, the theory and method for seepage analysis of high-steep slope engineering were established. The seepage flow problem under high water head conditions involves strong physical and boundary nonlinearities. Conventional analysis methods generally have defects such as poor convergence and numerical instability. They are difficult to adapt to complex large engineering problems. The Signorini type unified complementary form for complex seepage flow boundary conditions such as infiltration, evaporation, and overflow were proposed. The Signorini parabolic variational inequality analysis method based on both continuum media and fracture network discrete models was established. The theory of seepage flow analysis in high-steep slope engineering was extended: from stationary/transit to saturated/unsaturated analyses; from linear Darcy to nonlinear non-Darcy flow analysis; from continuous medium to fracture network seepage flow analyses. In general, a high-efficiency numerical simulation method on the complex seepage flow process of high-steep slopes in hydropower projects with high water heads was solved under a unified framework.

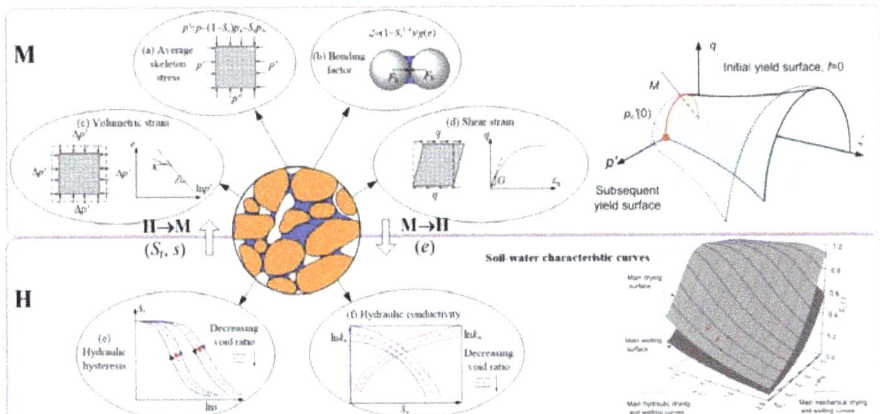

Figure 11. Fully coupling mechanism on elastoplastic deformation and capillary hysteresis of unsaturated soils [66].

4. Theory and Technology of Seepage Control in High-Steep Slope Engineering

This section mainly explains how to realize feedback of the seepage anomaly, the fine simulation of the seepage control structure, and the dynamic optimization design of leak-proof and drainage capability.

The standards of seepage control design in hydropower projects are mainly determined by hydraulic conductivity (q) of bedrock. Nevertheless, its hydrogeological significance and value acquisition under high water head conditions are not yet well understood. Thus a seepage control design criterion was proposed based on the maximum permeability of the rock mass under various pressures in a high-pressure packer permeability test. Based on the analytical model for non-linear seepage parameters of rock mass, this criterion has a rigorous theoretical foundation. It fits for laminar flow, turbulence, expansion, erosion, filling, and other P–Q curve types. Having been applied to the seepage control optimization design of the high-pressure diversion system in the Qiongzhong Pumped Storage Power Station, it was furthermore adopted by the energy industry standard 'The Rules for Drilling Packer Permeability Test for Hydropower Engineering' in China. A seepage control theory divided into medium property, boundary condition, initial state, and coupling process was proposed. Founded on seepage motion models, the theory broke through traditional empirical design methods such as engineering analogy, scheme comparison, and selection by quantifying leak-proof, drainage and filtration mechanisms. From the above, as shown in Figure 12, a leak-proof and drainage optimization design method on the basis of hydrogeological conditions, precise simulation of seepage control structure, and dynamic feedback was established. In the project design stage, it is beneficial to layout and optimize the seepage control system. In construction and initial impounding stages, it enables feeding abnormal seepage flow back and carrying out treatment. In the long-term operation stage, it realized seepage safety evaluation and regulation.

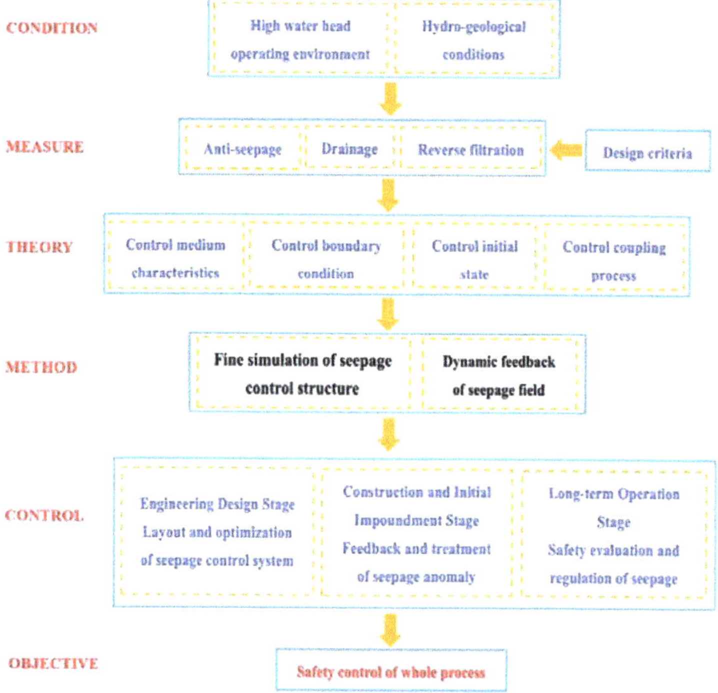

Figure 12. Optimization of the design method for seepage flow control of high-steep slope engineering [67].

As shown in Figure 13 [59,60], a multi-objective and whole-process dynamic inverse solution for analyzing seepage flow field was proposed based on orthogonal design, forwarding analysis of transient seepage, the BP neural network, and the non-inferior sequencing genetic algorithm. As to this approach: The fine precise simulation of the large drainage hole–curtain system and Signorini-type parabolic variational inequality method for transit seepage analysis are taken as the mean; the dynamic evolution of permeability and hydrogeological boundary conditions are taken as the parameters to be inversed; The optimal approximation of measured time series of data such as seepage pressure, water level and flow rate are taken as the objectives as follows [68],

$$\min f_1\left(K, \overline{\phi_L}\right) = \sum_{i=1}^{M} \frac{\left\|\phi_i\left(K, \overline{\phi_L}\right) - \phi_i^m\right\|_2^2}{\dim\phi_i}, \ \min f_2\left(K, \overline{\phi_L}\right) = \sum_{i=1}^{N} \frac{\left\|Q_i\left(K, \overline{\phi_L}\right) - Q_i^m\right\|_2^2}{\dim Q_i} \tag{3}$$

where $K, \overline{\phi_L}$ are the parameter sets of permeability coefficient and water-level boundary conditions to be inverted; Q_i, Q_i^m are the calculated and measured time series of data for flow amount at measuring point i, respectively; ϕ_i, ϕ_i^m are the calculated and measured time series of data of osmotic pressure at measuring point i, respectively; M and N are the numbers of measuring points for osmotic pressure and flow amount, respectively.

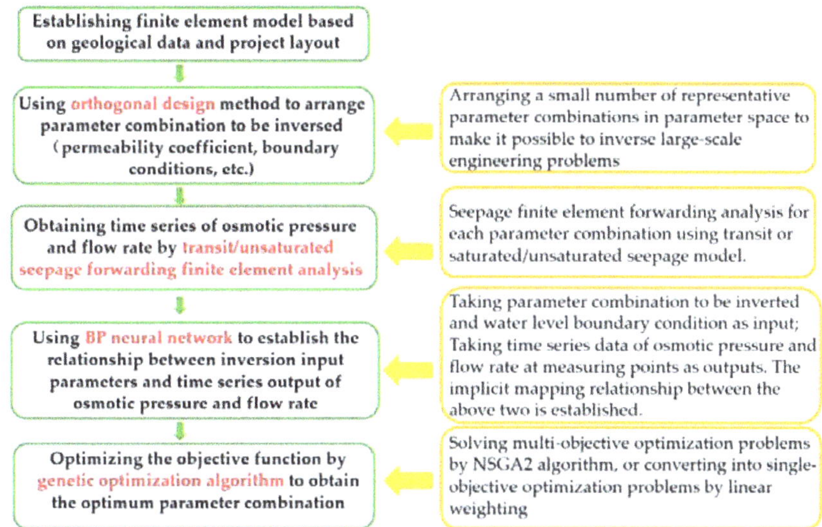

Figure 13. Flowchart for multi-objective, full process dynamic inverse analysis [69,70].

This method belongs to the positive feedback analysis method, which embodies the characteristics of multi-objective optimization and whole-process inversion. It realizes the uniqueness and reliability of inverse solution to seepage field and has good application to large-scale hydropower projects.

Based on multi-field coupling theory and precise and fine simulating technology for the engineering seepage control system, CoupledHM3D, a platform for both hydro-mechanical coupling analysis and seepage control optimization design, was developed by integrating the multi-scale model of seepage characteristics evolution, the unsaturated soil-water-mechanics fully coupling model, and the non-Darcy flow model of Geotechnical media. Considering seepage control, drainage, and reverse filtration, a global optimization design method for seepage control of water conservancy and hydropower engineering was developed. Thus a complete seepage control system including medium property control, boundary condition control, initial state control, and multi-field coupling control was realized.

The inverse analysis system has been successfully applied to dynamically analyze water gushing of a gallery in the dam foundation at the left bank in the Jinping hydropower project during the impoundment period. The Jinping first-class double-curvature arch dam on Yalong River has a maximum dam height of 305 m (shown in Figure 1) and complex geological conditions in the site area. Since the second stage of impoundment in June 2013, the water inflow of the drainage gallery in the base layer of the left bank (at elevation of 1595 m) had been large. The total water inflow exceeded 30 L/s and the water pressure inside the holes reached 0.3 MPa after closure of drainage holes at Pile No. 266 m.

Based on analyzing geology, water quality, monitoring and testing data, by means of multi-objective and whole-process inversion analysis of seepage field, the dynamic inversion analysis of the seepage tensor and hydrogeological boundary conditions of the rock mass at the left bank in Jinping was carried out in stages. As illustrated in Figure 14, three important conclusions were drawn as follows: (1) It was proved that water in the reservoir is the source of water inflow from the dam foundation; (2) The leakage passage of the gushing water was identified to be T2(6–8) 2–3z, i.e., namely two groups of predominant steeply dipping fractures along the river flow direction developed in marble. The geological origin of leakage passage and the mechanism of water inflow were also revealed. (3) It demonstrated the effectiveness of curtains, revealing the permeability characteristics as well as their strong anisotropy in the rock mass. A treatment measure was put forward to directly close and seal the drainage holes which fall in part with large water inflows.

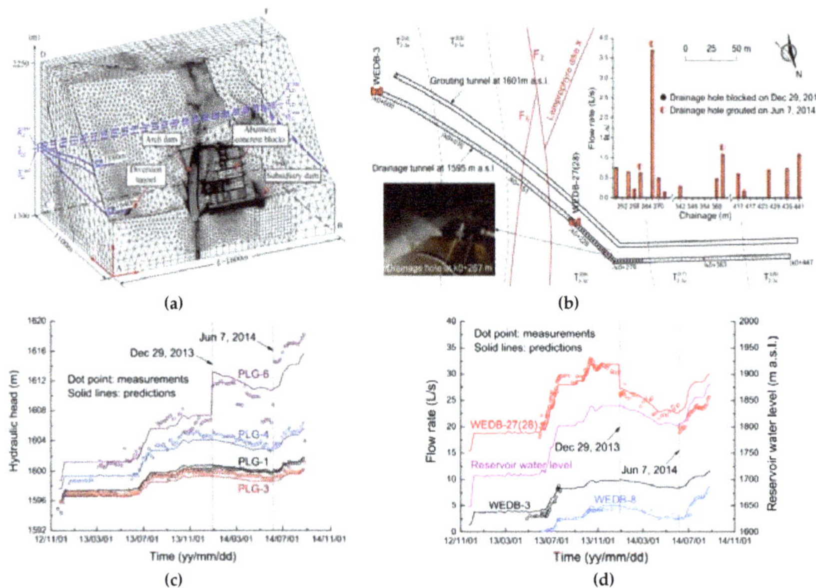

Figure 14. Inverse modelling and enforcement to water gushing from a drainage gallery at the left bank with altitude 1995 m in project Jinping [68]: (**a**) the inverse analysis numerical model; (**b**) the measured flow rates out of the drainage holes in the drainage tunnel at 1595 m above sea level on July 23, 2013, when the reservoir water level was about 1800 m above sea level; (**c**) comparison of the measured and calculated hydraulic heads at piezometers installed on the downstream side of grouting curtain; (**d**) comparison of the measured and calculated discharges through the measuring weirs.

5. Discussion and Conclusions

The up-to-date studies were reviewed on three main aspects from safety perspectives in Southwest China large-scale hydropower projects. They are the dynamic evolution of hydraulic properties of the slope rock mass fractured in different scales, regularities of non-Darcy flows under extreme pressure differences in lofty rock slopes, and field hydraulic controls based on systematical and feedback modelling analyses.

In this paper, the experimental tests on seepage characteristics of both rock samples and rough-walled fractures as well as the corresponding simulation analyses were summarized in a relatively systematic way. Not only the damage change defined by crack growth but also the evolution of seepage characteristics were predicted using the models established by researchers. Furthermore, the advances in studying the non-linear regularities of high-speed seepage flows under high pressure gradient and the unsaturated water-holding characteristics within deformable rock mass are introduced for the sake of broader application to hydraulic motion problems in high-steep slopes. Based on these latest findings, the significant achievements are also concisely elaborated in promoting seepage control theory such as a complete system covering medium property, boundary condition, initial state, and multi-field coupling as well as seepage control practice through inverse modeling analysis during construction of large-scale hydropower stations.

The hydraulic characteristics of fractured hard rocks of a reservoir bank slope undergo progressive evolution in the course of dam abutment excavation, reservoir impounding and fluctuation operation, and discharge atomization in hydropower engineering. Accordingly, seepage control measures by hydro-structures such as drainage or water insulation curtains should be guided by an innovative scientific foundation which concerns dynamic transformation of parameters, laws, and models during life-cycle performance.

Although applying hydraulic structures to seepage flowing control in a large-scale fractured rock mass resulted in a distinct achievement in Chinese hydropower engineering, a profound progress on fundamental fractured rock hydraulics is still on the way. It is expected that competent geophysical means and integrated data from effective monitoring systems will play a more valuable role in discovering and understanding the science intertwining complex flows and fractured rocks in engineering fields.

Author Contributions: Project administration, writing—original draft preparation, H.L.; methodology, formal analysis, H.T.; investigation, writing—review and editing, M.K.

Funding: This research was funded by National Basic Research Program of China, grant number 2011CB013503.

Acknowledgments: The authors would like to acknowledge the support by Ministry of Science and Technology of China for its funding to Project 'Seepage Characteristics and Seepage Control Mechanisms of Rock Mass in High-Steep Slopes' (No. 2011CB013503). Besides, a special acknowledgement is given to Professor Yi-Feng Chen in Wuhan University for his kind provision of the yet-not published Figures 6, 12 and 13.

Conflicts of Interest: The authors declare no conflict of interest.

References

1. Zhang, Y.T. *Rock Hydraulics and Engineering*; China Water & Power Press: Beijing, China, 2005; p. 227.
2. Sun, G.Z.; Yao, B.K. (Eds.) *Landslides in China—Selected Case Studies*; Science Press: Beijing, China, 1988.
3. Louis, C. Rock Hydraulics. In *Rock Mechanics*; Müller, L., Ed.; Springer: Vienna, Austria, 1974; pp. 299–387.
4. Zhang, Y.T. Analysis on several catastrophic failures of hydraulic projects in view of rock hydraulics. *J. Hydraul. Eng.* **2003**, *34*, 1–10.
5. Barton, N.; Bandis, S.; Bakhtar, K. Strength, deformation and conductivity coupling of rock joints. *Int. J. Rock Mech. Min. Sci.* **1985**, *22*, 121–140. [CrossRef]
6. Louis, C. *A Study of Groundwater Flow in Jointed Rock and Its Influence on Stability of Rock Mass*; Imperial College Rock Mechanics Report No.10; Imperial College of Science and Technology: London, UK, 1969.
7. Esaki, T.; Du, S.; Mitani, Y.; Ikusada, K.; Jing, L. Development of a shear-flow test apparatus and determination of coupled properties for a single rock joint. *Int. J. Rock Mech. Min. Sci.* **1999**, *36*, 641–650. [CrossRef]
8. Su, B.Y.; Zhan, M.L.; Zhao, J. The model test of the flow in smooth fracture and the study of its mechanism. *J. Hydraul. Eng.* **1994**, *5*, 19–24.
9. Reitsma, S.; Kueper, B.H. Laboratory measurement of capillary pressure-saturation relationships in a rock fracture. *Water Resour. Res.* **1994**, *30*, 865–878. [CrossRef]
10. Zhou, C.B.; Ye, Z.T.; Han, B. A preliminary study on unsaturated permeability of rock joints. *Chin. J. Geotech. Eng.* **1998**, *20*, 1–4.
11. Hu, Y.J.; Qian, R.; Su, B.Y. A numerical simulation method to determine unsaturated hydraulic parameters of fracture. *Chin. J. Geotech. Eng.* **2001**, *23*, 284–287.
12. Snow, D. Anisotropic permeability of fractured media. *Water Resour. Res.* **1969**, *5*, 1273–1289. [CrossRef]
13. Oda, M. Permeability tensor for discontinuous rock masses. *Geotechnique* **1985**, *35*, 483–495. [CrossRef]
14. Zhou, C.B.; Xiong, W.L. Study on coupling mechanism between permeability and deformation of geological discontinuity. *Hydrogeol. Eng. Geol.* **1998**, *20*, 1–4. (In Chinese)
15. Hsieh, P.A.; Neuman, S.P. Field determination of the three-dimensional hydraulic conductivity tensor of anisotropic media 1. Theory. *Water Resour. Res.* **1985**, *21*, 1655–1665. [CrossRef]
16. Hsieh, P.A.; Neuman, S.P.; Stiles, G.K.; Simpson, E.S. Field determination of the three-dimensional hydraulic conductivity tensor of anisotropic media 2. Methodology and application to fractured rocks. *Water Resour. Res.* **1985**, *21*, 1667–1676. [CrossRef]
17. Tang, C.A.; Hudson, J.A. *Rock Failure Mechanisms-Explained and Illustrated*, 1st ed.; CRC Press/Balkema: London, UK, 2010; pp. 1–319, ISBN 978-0-415-49851-7.

18. Xv, N.W.; Tang, C.A.; Sha, C.; Liang, Z.; Yang, J.; Zou, Y. Microseismic monitoring system establishment and its engineering applications to left bank slope of Jinping I Hydropower Station. *Chin. J. Rock Mech. Eng.* **2010**, *29*, 915–925.

19. Li, S.P.; Li, Y.S.; Wu, Z.Y. The permeability-strain equations relating to complete stress–strain path of rock. *Chin. J. Geotech. Eng.* **1995**, *17*, 13–19.

20. Peng, S.P.; Qv, H.L.; Luo, L.P.; Wang, L.; Duan, Y.E. An experimental study on the penetrability of sedimentary rock during the complete stress–strain path. *J. China Coal Soc.* **2000**, *25*, 113–116. (In Chinese)

21. Zheng, S.H.; Zhu, W.S. Theoretical analysis on a coupled seepage-damage model of fractured rock mass. *Chin. J. Rock Mech. Eng.* **2001**, *20*, 156–159.

22. Zheng, H.; Liu, D.F.; Lee, C.F.; Tham, L.G. A new formulation of Signorini's type for seepage problems with free surfaces. *Int. J. Numer. Meth. Eng.* **2005**, *64*, 1–16. [CrossRef]

23. Chen, Y.F.; Zhou, C.B.; Hu, R.; Li, D.Q.; Rong, G. Key issues on seepage flow analysis in large scale hydropower projects. *Chin. J. Geotech. Eng.* **2010**, *32*, 1448–1454.

24. Zhu, B.F. The analysis of the effect of draining holes in the seepage field by means of hybrid elements. *J. Hydraul. Eng.* **1982**, *9*, 32–41. (In Chinese)

25. Wang, L.; Liu, Z.; Zhang, Y.T. Analysis of seepage field near a drainage-hole's curtain. *J. Hydraul. Eng.* **1992**, *4*, 15–20. (In Chinese)

26. Zhu, Y.M.; Zhang, L.J. Solution to seepage field problem with the technique of improved drainage substructure. *Chin. J. Geotech. Eng.* **1997**, *19*, 69–76.

27. Wang, E.Z.; Wang, H.T.; Deng, X.D. Pipe to represent hole-numerical method for simulating single drainage hole in rock-masses. *Chin. J. Rock Mech. Eng.* **2001**, *20*, 346–349.

28. Chen, S.H.; Xv, Q.; Hu, J. Composite element method for seepage analysis of geotechnical structures with drainage hole array. *J. Hydrodyn. Ser. B* **2004**, *16*, 260–266.

29. Yv, J.; Li, H.; Chen, X.; Cai, Y.Y.; Wu, N.; Mu, K. Tri-axial experimental study of associated permeability-deformation of sandstone under hydro-mechanical coupling. *Chin. J. Rock Mech. Eng.* **2013**, *32*, 1203–1213.

30. Yv, J.; Li, H.; Chen, X.; Cai, Y.Y.; Mu, K.; Zhang, Y.Z.; Wu, N. Experimental study on seepage properties, acoustic emission of sandstone deformation during unloading of surrounding stress. *Chin. J. Rock Mech. Eng.* **2014**, *33*, 69–79.

31. Jiang, Z.B. Experiments and Mechanical Model Study on Rock Creep Properties under Multi-field Environment. Ph.D. Thesis, Dalian Maritime University, Dalian, China, November 2016.

32. Chen, Y.F.; Li, D.Q.; Jiang, Q.H.; Zhou, C. Micromechanical analysis of anisotropic damage and its influence on effective thermal conductivity in brittle rocks. *Rock Mech. Min. Sci.* **2012**, *50*, 102–116. [CrossRef]

33. Chen, Y.F.; Hu, S.H.; Zhou, C.B.; Jing, L. Micromechanical modeling of anisotropic damage-induced permeability variation in crystalline rocks. *Rock Mech. Rock Eng.* **2014**, *47*, 1775–1791. [CrossRef]

34. Chen, Y.F.; Hu, S.H.; Wei, K.; Hu, R.; Zhou, C.; Jing, L. Experimental characterization and micromechanical modeling of damage-induced permeability variation in Beishan granite. *Int. J. Rock Mech. Min. Sci.* **2014**, *71*, 64–76. [CrossRef]

35. Chen, X.; Yv, J.; Li, H.; Cai, Y.Y.; Zhang, Y.Z.; Mu, K. Experimental study of propagation features of acoustic waves in rocks varying in lithology and water content. *Rock Soil Mech.* **2013**, *34*, 2527–2533. (In Chinese)

36. Mu, K.; Yv, J.; Li, H.; Cai, Y.Y.; Chen, X. Particle flow simulation of acoustic emission and energy dissipation of sandstone under water-force coupling. *Rock Soil Mech.* **2015**, *36*, 1496–1504. (In Chinese)

37. Yv, J.; Mu, K.; Li, H.; Cai, Y.Y.; Zhang, Y.Z. Mesoscopic simulation analysis of porosity distribution of permeability evolution characteristics. *Eng. Mech.* **2014**, *31*, 124–131. (In Chinese)

38. Li, G.; Tang, C.A.; Li, L.C. Advances in research on coupled deformation and failure processes of water and rock. *Adv. Mech.* **2012**, *42*, 593–619. (In Chinese)

39. Li, L.C.; Liu, H.H.; Zhao, Y. Nonlinear elastic behavior analysis of rock based on Two-Part Hook model. *Chin. J. Rock Mech. Eng.* **2012**, *31*, 2119–2126.

40. Liang, Z.Z.; Zhang, Y.B.; Tang, S.B. Rock mass size effect and its parameter computation. *Chin. J. Rock Mech. Eng.* **2013**, *32*, 1157–1166.

41. Li, G.; Tang, C.; Li, L. Advances in rock deformation and failure process under water-rock coupling. *Adv. Mech.* **2012**, *42*, 593–619. (In Chinese)

42. Oda, M.; Takemura, T.; Aoki, T. Damage growth and permeability change in triaxial compression tests of Inada granite. *Mech. Mater.* **2002**, *34*, 313–331. [CrossRef]

43. Shao, J.F.; Zhou, H.; Chao, K.T. Coupling between anisotropic damage and permeability variation in brittle rocks. *Numer. Anal. Methods Geomech.* **2005**, *29*, 1231–1247. [CrossRef]

44. Hu, S.H.; Chen, Y.F.; Zhou, C.B. Experimental study and meso-mechanical analysis of seepage characteristics of Beishan granite. *Chin. J. Rock Mech. Eng.* **2014**, *33*, 2200–2209.

45. Chen, Y.F.; Zhou, C.B.; Sheng, Y.Q. Formulation of train-dependent hydraulic conductivity for a fractured rock mass. *Int. J. Rock Mech. Min. Sci.* **2007**, *44*, 981–996. [CrossRef]

46. Chen, Y.F.; Li, D.Q.; Rong, G. Micromechanical model of brittle rock damage and heat transfer characteristics. *Chin. J. Rock Mech. Eng.* **2011**, *30*, 1950–1969.

47. Chen, Y.F.; Zheng, H.K.; Wang, M.; Hong, J.M.; Zhou, C.B. Excavation-induced relaxation effects and hydraulic conductivity variation in the surrounding rocks of a large-scale underground powerhouse cavern system. *Tunn. Undergr. Space Technol.* **2015**, *49*, 253–267. [CrossRef]

48. Li, Y.; Chen, Y.; Zhou, C. Hydraulic properties of partially saturated rock fractures subjected to mechanical loading. *Eng. Geol.* **2014**, *179*, 24–31. [CrossRef]

49. Li, Y.; Chen, Y.F.; Zhou, C.B. Effective stress principle for partially saturated rock fractures. *Rock Mech. Rock Eng.* **2016**, *49*, 1091–1096. [CrossRef]

50. Chen, Y.; Zhou, J.; Hu, S.; Hu, R.; Zhou, C.B. Evaluation of Forchheimer equation coefficients for non-Darcy flow in deformable rough-walled fractures. *J. Hydrol.* **2015**, *529*, 993–1006. [CrossRef]

51. Zhou, J.; Hu, S.; Fang, S.; Chen, Y.F.; Zhou, C.B. Nonlinear flow behavior at low Reynolds numbers through rough-walled fractures subjected to normal compressive loading. *Int. J. Rock Mech. Min. Sci.* **2015**, *80*, 202–218. [CrossRef]

52. Chen, Y.F.; Fang, S.; Wu, D.-S.; Hu, R. Visualizing and quantifying the crossover from capillary fingering to viscous fingering in a rough fracture. *Water Resour. Res.* **2017**, *53*, 7756–7772. [CrossRef]

53. Hu, R.; Wu, D.-S.; Yang, Z.B.; Chen, Y.-F. Energy conversion reveals regime transition of imbibition in a rough fracture. *Geophys. Res. Lett.* **2018**, *45*, 8993–9002. [CrossRef]

54. Chen, Y.-F.; Wu, D.-S.; Fang, S.; Hu, R. Experimental study on two-phase flow in rough fracture phase diagram and localized flow channel. *Int. J. Heat Mass Transf.* **2018**, *122*, 1298–1307. [CrossRef]

55. Chen, Y.-F.; Guo, N.; Wu, D.-S.; Hu, R. Numerical investigation on immiscible displacement in 3D rough fracture Comparison with experiments and the role of viscous and capillary forces. *Adv. Water Resour.* **2018**, *118*, 39–48. [CrossRef]

56. Zhou, J.-Q.; Wang, L.; Chen, Y.-F.; Cardenas, M.B. Mass transfer between recirculation and main flow zones: Is physically-based parameterization possible? *Water Resour. Res.* **2019**, 55. [CrossRef]

57. Zhou, J.-Q.; Hu, S.-H.; Chen, Y.-F.; Wang, M.; Zhou, C.B. The friction factor in the Forchheimer equation for rock fractures. *Rock Mech. Rock. Eng.* **2016**, *49*, 3055–3068. [CrossRef]

58. Chen, Y.; Liu, M.; Hu, S.; Zhou, C.B. Non-Darcy's law-based analytical models for data interpretation of high-pressure packer tests in fractured rocks. *Eng. Geol.* **2015**, *199*, 91–106. [CrossRef]

59. Chen, Y.F.; Hu, S.H.; Hu, R.; Zhou, C.B. Estimating hydraulic conductivity of fractured rocks from high-pressure packer tests with an Izbash's law-based empirical model. *Water Resour. Res.* **2015**, *51*, 2096–2118. [CrossRef]

60. Liu, M.M.; Hu, S.H.; Chen, Y.F. Analysis of nonlinear seepage parameters of fractured rock mass based on high pressure water pressure test. *J. Hydraul. Eng.* **2016**, *47*, 752–762.

61. Lin, P.; Liu, X.; Zhou, W.; Wang, R.; Wang, S. Cracking, stability and slope reinforcement analysis relating to the Jinping dam based on a geomechanical model test. *Arab. J. Geosci.* **2015**, *8*, 4393–4410. [CrossRef]

62. Hu, R.; Chen, Y.F.; Zhou, C.B. A relative hydraulic conductivity model for unsaturated deformable soil. *Chin. J. Rock Mech. Geotech. Eng.* **2013**, *32*, 1279–1287.

63. Hu, R.; Chen, Y.F.; Liu, H.H.; Zhou, C.B. A water retention curve and unsaturated hydraulic conductivity model for deformable soils: Consideration of the change in pore-size distribution. *Géotechnique* **2013**, *63*, 1389–1405. [CrossRef]

64. Chen, Y.F. Effects of fluid flow, soil deformation and horizontal drains on soil slope stability subjected to rain infiltration. In *Landslides and Engineering Slopes: Proctecting Society through Improved Understanding*; Eberhardt, E., Froese, C., Turner, K., Lerouril, S., Eds.; Taylor & Francis Group, CRC Press: London, UK, 2012.

65. Alonso, E.E.; Gens, A.; Josa, A. A constitutive model for partially saturated soils. *Géotechnique* **1990**, *40*, 405–430. [CrossRef]

66. Hu, R.; Chen, Y.F.; Liu, H.-H.; Zhou, C.B. A coupled two-phase fluid flow and elastoplastic deformation model for unsaturated soils: Theory, implementation and application. *Int. J. Numer. Anal. Meth. Geomech.* **2016**, *40*, 1023–1058. [CrossRef]

67. Li, X.; Chen, Y.-F.; Hu, R.; Yang, Z. Towards an optimization design of seepage control: A case study in dam engineering. *Sci. China Technol. Sci.* **2017**, *60*, 1903–1916. [CrossRef]

68. Zhou, C.-B.; Liu, W.; Chen, Y.-F.; Hu, R.; Wei, K. Inverse modeling of leakage through a rock-fill dam foundation during its construction stage using transient flow model, neural network ad genetic algorithm. *Eng. Geol.* **2015**, *187*, 183–195. [CrossRef]

69. Chen, Y.-F.; Hong, J.-M.; Zheng, H.-K.; Li, Y.; Hu, R.; Zhou, C.-B. Evaluation of groundwater leakage into a drainage tunnel in Jinping-I arch dam foundation in Southwestern China: A case study. *Rock Mech. Rock Eng.* **2016**, *49*, 961–979. [CrossRef]

70. Hong, J.-M.; Chen, Y.-F.; Liu, M.M.; Zhou, C.-B. Inverse modelling of groundwater flow around a large-scale underground cavern system considering the excavation-induced hydraulic conductivity variation. *Comput. Geotech.* **2017**, *81*, 346–357. [CrossRef]

Technical Note

Experimental Study on the Shear-Flow Coupled Behavior of Tension Fractures Under Constant Normal Stiffness Boundary Conditions

Changsheng Wang [1,2], Yujing Jiang [1,2,*], Hengjie Luan [1], Jiankang Liu [1,2] and Satoshi Sugimoto [2]

[1] State Key Laboratory of Mining Disaster Prevention and Control Co-founded by Shandong Province and the Ministry of Science and Technology, Shandong University of Science and Technology, Qingdao 266510, China; cswang0635@163.com (C.W.); luanjie0330@126.com (H.L.); wodoyy@163.com (J.L.)

[2] School of Engineering, Nagasaki University, Nagasaki 852-8521, Japan; s-sugi@nagasaki-u.ac.jp

* Correspondence: jiang@nagasaki-u.ac.jp; Tel.: +81-095-819-2612

Received: 20 December 2018; Accepted: 16 January 2019; Published: 22 January 2019

Abstract: This study experimentally investigated the effects of fracture surface roughness, normal stiffness, and initial normal stress on the shear-flow behavior of rough-walled rock fractures. A series of shear-flow tests were performed on two rough fractures, under various constant normal stiffness (CNS) boundary conditions. The results showed that the CNS boundary conditions have a significant influence on the mechanical and hydraulic behaviors of fractures, during shearing. The peak shear stress shows an increasing trend with the increases in the initial normal stress and fracture roughness. The residual shear stress increases with increasing the surface roughness, normal stiffness, and initial normal stress. The dilation of fracture is restrained more significantly under high normal stiffness and initial normal stress conditions. The hydraulic tests show that the evolutions of transmissivity and hydraulic aperture exhibit a three-stage behavior, during the shear process—a slight decrease stage due to the shear contraction, a fast growth stage due to shear dilation, and a slow growth stage due to the reduction rate of the mechanical aperture increment. The transmissivity and hydraulic aperture decreased, gradually, as the normal stiffness and initial normal stress increase.

Keywords: rock fracture; shear-flow coupled test; constant normal stiffness conditions; transmissivity; hydraulic aperture

1. Introduction

Underground fracture rock masses consist of the intact rock matrix and various discontinuities. Discontinuities such as joints and fractures play a significant role in the hydro-mechanical behaviors of a host rock mass [1–7]. Hence, the analysis of the shear-flow behaviors of rock fractures is critical to many mass and energy transport engineering activities, such as nuclear waste disposal, oil, natural gas production, and geothermal energy extraction [8–12].

Many efforts have been made to take into account the influence of shear displacement on fluid flow in a single fracture [13–19]. Yeo et al. [13] carried out radial and unidirectional flow tests through single rough fractures. The results showed that with increasing shear displacement, the fracture aperture became heterogeneous and anisotropic, and the permeability, in the direction perpendicular to the shear, was larger than that along the shear direction. Esaki et al. [14] developed a laboratory technique for coupled shear-flow tests, to investigate the coupled effect of joint shear deformation, and dilatancy, on the hydraulic conductivity of rock joints. The results showed that the conductivity increased, rapidly, for the first 5 mm of shear displacement, and then gradually became a constant value, with continuous increase in the shear displacement. Li et al. [15] conducted a series of shear-flow coupling tests, to evaluate the influence of morphological properties of rock fractures, on their hydro-mechanical behavior. The results showed that the contact ratio drops rapidly at the very beginning of the shear and then kept a small value. However, the transmissivity changed inversely with the contact ratio change, during shearing. Javadi et al. [16] conducted coupled shear-flow tests on three granite specimens, to investigate the variation in the critical Reynolds number (Re_c), during the shear process, under different normal stresses. It was found that the Re_c was in the range of 0.001 to 25, as the shear displacement increased from 0 to 20 mm. Similar shear-flow tests were also conducted by Rong et al. [17]. They suggested the Re_c ranged from 1.5 to 13.0, as shear displacement increased from 0 to 10.9 mm.

Most of these shear-flow tests were conducted under constant normal load (CNL) boundary conditions, in which the normal load applying on the fracture surface was constant, during shearing. However, the CNL boundary condition was only appropriate for the non-reinforced rock slope or planar fractures, with no dilation during shearing [20–24]. For deep underground opening or rock anchor-reinforced slopes, the loads acting normal to the direction of shear were not constant, and the shear-induced dilation acted against the normal stiffness of the surround rock mass. The rock joints were subject to a variable normal load and the stiffness of the surrounding rock mass could significantly affect the shear behavior. In this case, during the shear process, the constant normal stiffness (CNS) boundary condition should be selected, rather than the CNL boundary condition. Some previous studies have experimentally investigated the shear behavior under CNS boundary conditions [20–22,25–27]. It was reported that the shear behavior was affected by the fracture surface roughness, normal stiffness, and initial normal stress. Note that the fracture aperture and its evolution during shear, under the CNS condition, was far more complicated than that under the CNL condition. On the one hand, shear-induced dilation could increase the fracture aperture. On the other hand, increasing the normal stress induced severe asperity degradation, which reduced joint dilation. However, few studies focused on the hydraulic behavior of fracture, during shear, under the CNS boundary condition. Li et al. [28] and Koyama et al. [29] investigated the effect of the boundary condition on the hydraulic behavior of the joint, during the shear process. The results showed that the hydraulic aperture under the CNL condition was larger than the CNS condition. Olsson et al. [30,31] investigated the effects of normal stiffness and initial normal stress on the hydraulic behavior of joint, during the shear process. The results showed that the transmissivity decreased with increasing normal stiffness and the initial normal stress. Sato et al. [32] studied the effect of roughness on the hydraulic behavior, under the CNS and the CNL condition. They expressed that the permeability of joints was larger for rougher fractures. However, the effects of surface roughness, normal stiffness, and initial normal stress on the shear-flow behavior were usually studied independently, and the corporate effects of these factors on the shear-flow behavior of fractures needed to be further investigated.

In this study, we conducted shear-flow tests on two types of fractures, with different surface roughness, under the CNS boundary conditions. For each type of fracture specimen, a series of hydraulic tests were performed for the different shear displacements (0 ~ 20 mm), under different normal stiffness (1.0, 2.0, 3.0 MPa/mm), and initial normal stresses (1.0, 2.0, 3.0 MPa). Finally, the effects of shear displacement, joint surface roughness, normal stiffness, and initial normal stress on mechanical and hydraulic characteristics of fractures were analyzed.

2. Experiment

2.1. Specimen Preparation

Two types of granite fractures (labeled as G1 and G3) were created using the Brazilian test. The fracture specimens were composed of upper and lower parts—200 mm in length, 100 mm in width, and 100 mm in height. In order to carry out repeated tests, under various boundary conditions, using specimens that had similar surface geometry, artificial replicas were manufactured by a mixture of plaster, water, and retardant, with a weight ratio of 1: 0.2: 0.005. Table 1 lists the physico-mechanical properties of these rock-like specimens.

Table 1. Physico-mechanical properties.

Physico-Mechanical Properties	Index	Unit	Value
Density	ρ	g/cm^3	2.066
Compressive strength	σ_c	MPa	38.5
Modulus of elasticity	E_s	GPa	28.7
Poisson's ratio	v	–	0.23
Tensile strength	σ_t	σ_t	2.5
Cohesion	c	MPa	5.3
Internal friction angle	φ	$^\circ$	60

Replica surface geometries were scanned using a high resolution 3D laser scanning profilometer system [25]. The scanning intervals in both X and Y axes were set to be 0.5 mm. Based on the scanned data, the surface topographies of the two replica of the tension fractures are shown in Figure 1a,b. To quantify the roughness of the rock joints, the Joint Roughness Coefficient (JRC) values of the replicas were evaluated using the replica profiles, which were obtained by the scanning data, along the length direction. The JRC was calculated according to the equations proposed by Tse and Cruden [33], written as:

$$Z_2 = \left[\frac{1}{(n-1)(\Delta x)^2} \sum_{i=1}^{n-1} (z_{i+1} - z_i)^2 \right]^{1/2} \tag{1}$$

$$\mathrm{JRC} = 32.2 + 32.47 \log Z_2 \tag{2}$$

where Z_2 is the root mean square slope of the profiles, based on the extracted data, z_i represent the coordinates of the fracture surface profile, n is the number of the data points, and Δx is the interval of the data points. The mean JRC values of the fractures G1 and G3 are 3.21 and 7.36, respectively.

2.2. Test Equipment and Procedures

The shear-flow coupled tests under the CNS conditions were carried out on the servo-controlled shear-flow test system, in the Nagasaki University, as shown in Figure 1c. The test system mainly consisted of three units—a hydraulic-servo actuator unit, a hydraulic testing unit, and a visualization unit. The hydraulic-servo actuator unit includes horizontal and vertical load jacks, to apply the shear and normal loads, through a servo-controlled hydraulic pump. The capacity of both the normal and the shear loads was 200 kN, with a precision of 99%. The shear and normal displacements were monitored by three linear variable differential transformers (LVDTs), with an accuracy of 0.001 mm. Two LVDTs were used to monitor the normal displacement. The capacities of the shear and normal LVDTs were 20 mm and 10 mm, respectively. In this system, the CNS boundary condition was achieved by detecting the signal of normal displacement, during shearing, and servo-controlling the normal load applied to the specimen.

The change in the normal stress, due to the application of normal stiffness (k_n) was calculated as follows [1]:

$$\Delta P_n = k_n \cdot \Delta \delta_n \tag{3}$$

$$P_n(\mathrm{t} + \Delta \mathrm{t}) = P_n(t) + \Delta P_n \tag{4}$$

where ΔP_n and $\Delta \delta_n$ are changes in the normal load and the normal displacement, respectively.

The shear-flow tests were conducted on the two sets of replicas, with their different roughness, under the various boundary conditions. The experimental cases and their corresponding boundary conditions are shown in Table 2. The flow tests were conducted, under a constant water head of 0.3 m, during shearing. The shear displacement increased from 0 to 20 mm, with an interval of 1 mm.

The shear rate was 0.5 mm/min. The outflow of the water was measured using an electrical balance that had a precision of 0.01 g. Numerous shear-flow tests were conducted using this testing system, showing a good sealing effect in these test studies [15,28,34,35].

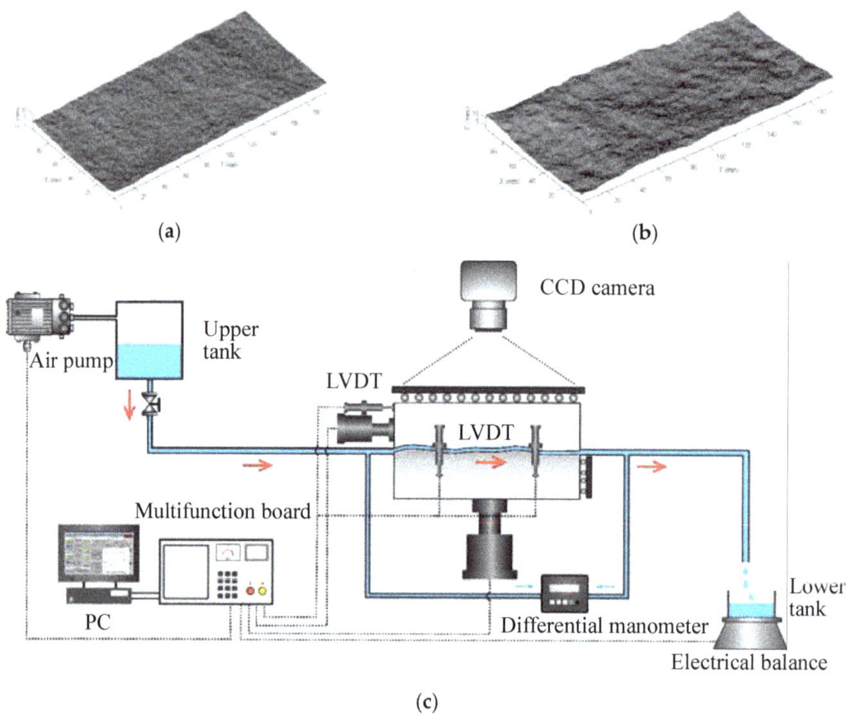

Figure 1. (**a**) Scanning graph of fracture G1. (**b**) Scanning graph of fracture G3. (**c**) Schematic view of the coupled shear-flow test system (Arrow represents water flow direction).

Table 2. Experimental cases and their corresponding boundary conditions.

Joint Specimens	Loading Cases	Roughness (JRC)	Normal Loading Conditions	
			Initial Normal Stress, σ_0 (MPa)	Normal Stiffness, k_n (MPa/mm)
	G1—1		1	1
	G1—2		1	2
G1	G1—3	3.21	1	3
	G1—4		2	2
	G1—5		3	2
	G3—1		1	1
	G3—2		1	2
G3	G3—3	7.36	1	3
	G3—4		2	2
	G3—5		3	2

3. Experimental Results

3.1. Effect of the Normal Stiffness on the Shear Characteristics

Figure 2 shows the results of the shear mechanical properties of the fractures G1 and G3, under different normal stiffness, with an initial normal stress of 1 MPa. In this study, the peck shear stress was defined as the point of shear stress, with a sharp change in curvature [20]. As shown in Figure 2a,b, in the pre-peak range, the shear stress increased linearly to the peak value, at the very beginning of shear. The test results indicate that the peak shear stress did not show obvious tendencies with respect to the normal stiffness. The peak shear stress values were similar, under various normal stiffness conditions. This is because the normal stress was approximately the same at the peak shear displacement (Figure 2e,f). In the post-peak range, the post shear strength was related to the fracture surface roughness and normal stiffness. For the relatively smooth fracture G1, when the normal stiffness was small (k_n = 1 MPa/mm), the shear strength abruptly increased to the peak, and then gradually decreased to the residual stress stage. With increase in the normal stiffness, the post peak strength decreased slightly, followed by a gradual increase. However, for the relatively rough fracture G3, as the normal stiffness increased, the post-peak strength grew, appreciably, with increasing shear displacement. This phenomenon showed that a higher value of the normal stiffness and larger roughness of the fracture surfaces indicate a more pronounced stress hardening behavior. It was also found that the residual shear stress showed an increasing trend, with increase in the value of the JRC and the normal stiffness.

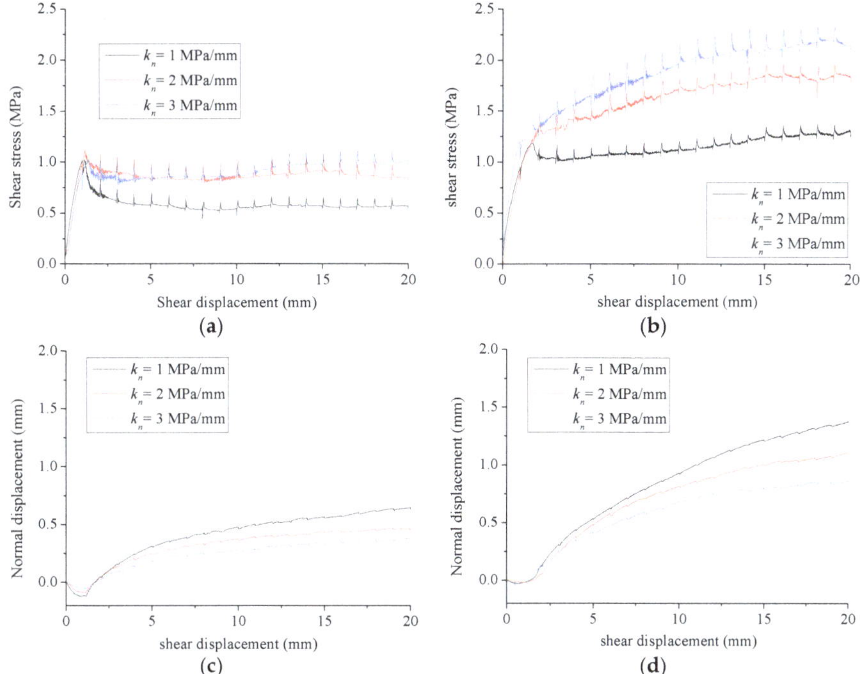

Figure 2. *Cont.*

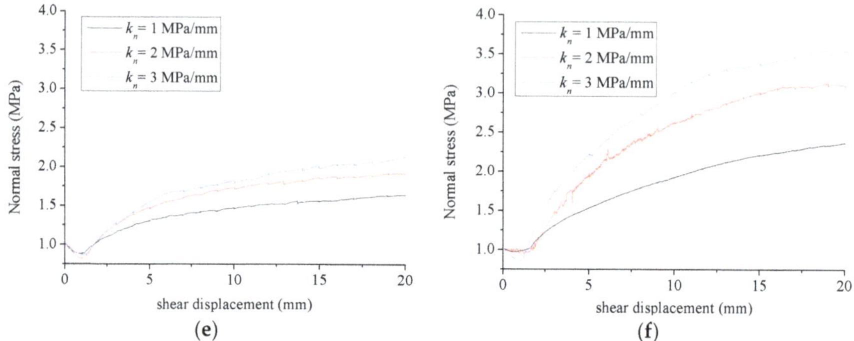

Figure 2. Shear behavior of rock fracture replicas under different normal stiffness. (**a**,**b**) Shear stress vs. shear displacement for the replicas G1 and G3; (**c**,**d**) Normal displacement vs. shear displacement for replicas G1 and G3; and (**e**,**f**) Normal stress vs. shear displacement for replicas G1 and G3.

The relationship between normal displacement and shear displacement are shown in Figure 2c,d. For all test cases, the normal displacement slightly decreased in the pre-peak stage and then increases, gradually, with increasing shear displacement. The normal displacement decrease was due to the deformation of the asperity and surface interlocking, at the beginning of the application of the shear load. With continuous increase in the shear displacement, the normal displacement increased due to shear-induced dilation. As the normal stiffness increased, the increment of normal displacement decreased, whereas the increasing rate of normal stress increased (Figure 2e,f). This indicates that dilation was restrained at the higher normal stiffness condition. Additionally, the rougher the replica surface, the larger dilation and normal stress that could be obtained.

3.2. Effect of the Initial Normal Stress on the Shear Characteristics

Figure 3 shows the shear mechanical properties of the fracture G1 and G3, under different initial normal stress, with a constant normal stiffness of 2 MPa/mm. As shown in Figure 3a,b, the shear stress was related to the initial normal stress and the replica fracture surface roughness. The peak shear stress and residual shear stress became larger, with an increase in the initial normal stress and JRC value. In the post-peak range, the test results indicated that the high initial normal stress condition increased the tendency toward the stress softening behavior. The transition from stress hardening to stress softening behavior was more obvious with an increase in the initial normal stress, for the rougher fracture G3.

Figure 3c,d show the relationship between normal displacement and shear displacement. The increasing rate of dilation decreased as the initial normal stress increased. For the rougher fracture G3, the dilation was larger than the fracture G1, under the same initial normal stress condition. Comparison with the dilation results, under different normal stiffness conditions, showed that the influence of the initial normal stress, on dilation, was more obvious than the normal stiffness. This was because the normal stress increased more significantly, according to Equations (3) and (4), and more significant failure of asperities was created, under higher normal stress conditions.

3.3. Evolution of Transmissivity during the Shear-Flow Tests

For a steady laminar flow through a single fracture, the flow rate is considered to be linearly proportional to the cubic of the fracture aperture, which is the cubic law [36–38], written as:

$$Q = -\frac{\Delta P}{L\mu}\frac{wb_h{}^3}{12} \tag{5}$$

where Q is volumetric flow rate, ΔP is the pressure drop along the flow direction, w is the fracture width, b_h is the hydraulic aperture, L is the fracture length over which the pressure drop takes place, and μ is the viscosity of the fluid. The transmissivity T equals to the term $wb_h{}^3/12$ in Equation (5), which is an important parameter to estimate the flow characteristics [36].

$$T = \frac{Q\mu}{\nabla Pw} = \frac{wb_h{}^3}{12} \tag{6}$$

Note that the fluid flow patch, in the shear direction, will progressively become short, during shear. Therefore, the transmissivity and the hydraulic aperture are calculated using the actual length [34].

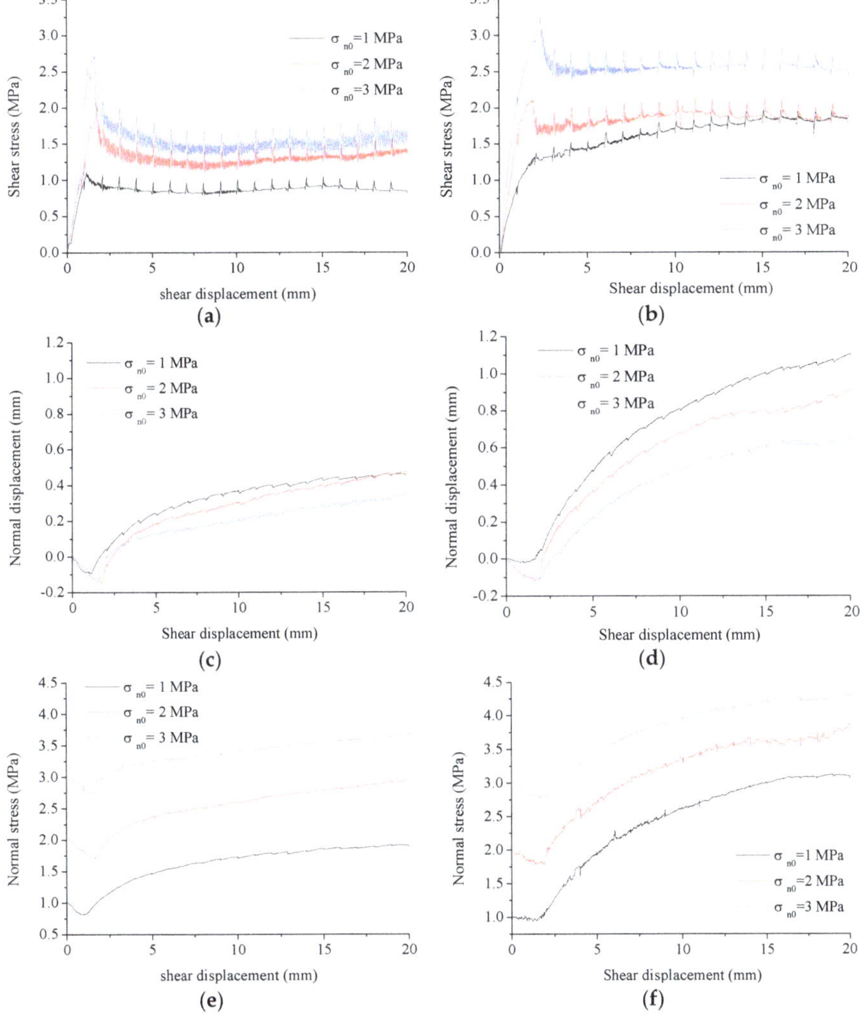

Figure 3. Shear behavior of the rock fracture replicas with different initial normal stresses. (**a,b**) Shear stress vs. shear displacement for the replicas G1 and G3; (**c,d**) Normal displacement vs. shear displacement for the replicas G1 and G3; and (**e,f**) Normal stress vs. shear displacement for the replicas G1 and G3.

The evolution transmissivities of the relicas G1 and G3, during shearing, under different boundary conditions, have been plotted in Figure 4. The changes in transmissivity exhibited three-stage behavior, during the shear process. For all test cases, the transmissivity experienced a slight descend at 0–1 mm shear displacement. Then the transmissivity increased rapidly, at a shear displacement of 1–4 mm. When the shear displacement exceeded 4 mm, the transmissivity gradually increased, until reaching a stable value. Similar behavior have also been reported in previous studies [14,15,28]. The transmissivity showed an increment of 2—3 orders of magnitude, as the shear displacement increased from 0 to 20 mm. For the rougher replica G3, the transmissivity increment was larger than the smooth replica G1. This is because the rougher joint created a relatively larger void space between the two surfaces of a fracture, during shearing. The transmissivity decreased with the increases in normal stiffness and the initial normal stress. This was because the rougher fracture created a relatively larger void space between the two surfaces of a fracture, during shearing. The transmissivity decreased with the increase in normal stiffness and the initial normal stress.

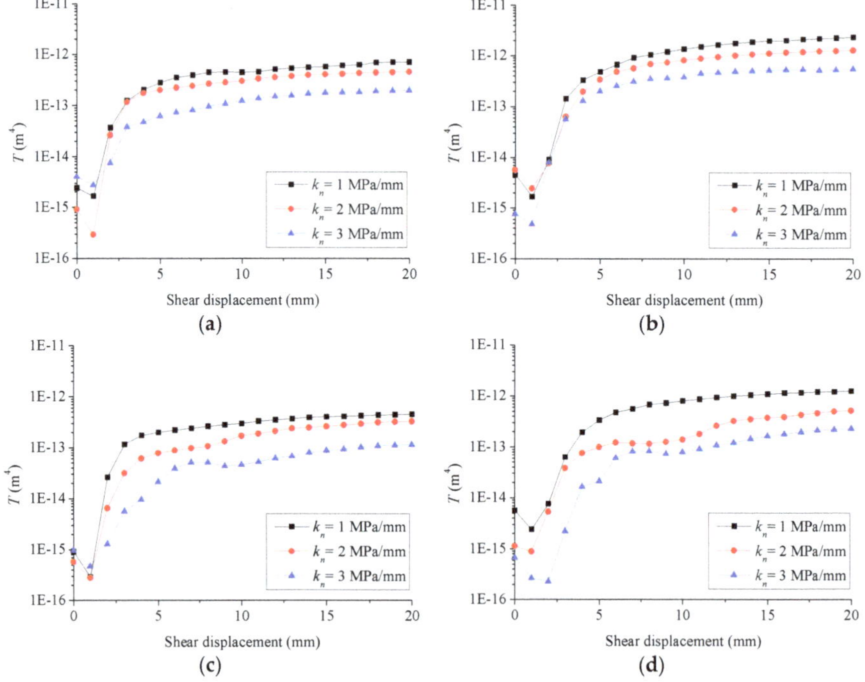

Figure 4. The evolutions of transmissivity of the replicas G1 and G3, during shearing, under different boundary conditions. (**a**) Transmissivity of replica G1, with different normal stiffness, under an initial normal stress of 1 MPa; (**b**) transmissivity of replica G3, with different normal stiffness, under an initial normal stress of 1 MPa; (**c**) transmissivity of replica G1, with different initial normal stresses, under a constant normal stiffness of 2 MPa/mm; (**d**) transmissivity of replica G3, with different initial normal stresses, under a constant normal stiffness of 2 MPa/mm.

3.4. Evolutions of the Mechanical Aperture and the Hydraulic Aperture

The evolution of the aperture, during shearing, was the key issue to study the fluid flow behaviors of fractures. Based on Equation (5), the hydraulic aperture at different shear displacements, can be calculated. The mean mechanical aperture (e_m) can be calculated based on the following equation [14,29]:

$$e_m = e_0 - \Delta b_n + \Delta d \tag{7}$$

where e_0 is the initial mechanical aperture, Δb_n is the change in mechanical aperture due to normal loading, and Δd is the change in mechanical aperture induced by shear dilation. The initial mechanical aperture at the small shear displacement (0–1 mm) was assumed to equal to the hydraulic aperture. Note that the normal stress changes with increasing the shear displacement, under CNS boundary conditions, therefore, the Δb_n should be revised, based on the corresponding normal stress.

Figure 5 shows the evolutions of hydraulic and mechanical aperture of the fractures G1 and G3, during shearing, under different boundary conditions. Similar with the changes in the transmissivity, the evolution of the hydraulic aperture also exhibited a three-stage behavior, during the shear process: (1) A declining stage due to the shear contraction of the fracture, at the beginning of shear; (2) A fast growth stage, in which the aperture increased rapidly, due to shear-induced dilation; (3) A slowly growth stage, during which the aperture gradually increased and approximately approached a constant value, due to the reduction of shear dilation. Both the hydraulic and mechanical aperture decreased, with increasing normal stiffness and initial normal stress. This indicated that the stiffness- and the initial-normal-stress-induced normal stress increase caused a larger degree of asperity damage, which decreased the mechanical aperture. The mean mechanical aperture was always larger than the hydraulic aperture. For a rougher fracture surface, the degree of hydraulic aperture deviated more significantly from the mechanical aperture than the smoother fracture surface.

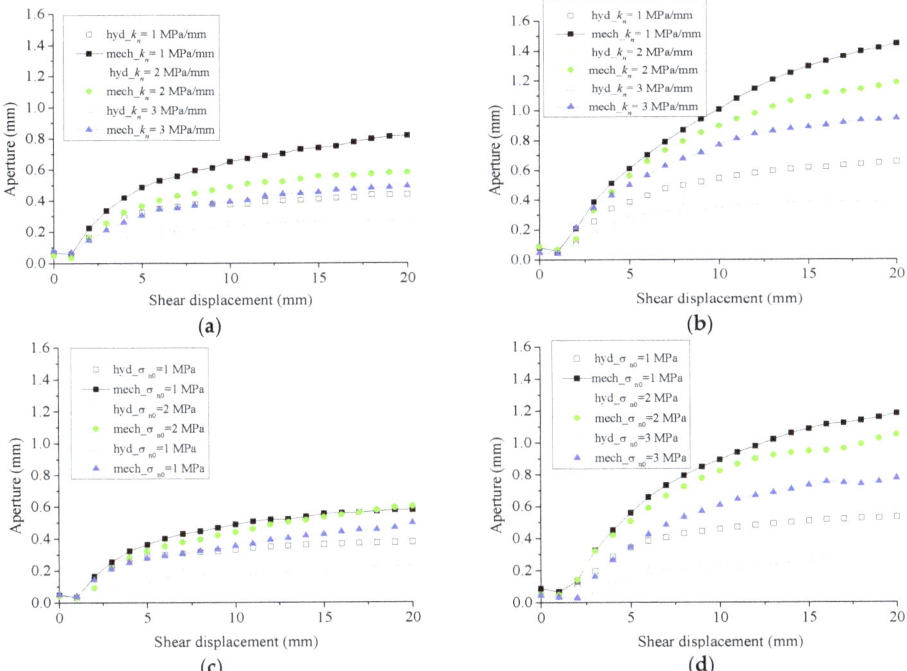

Figure 5. Evolution of the hydraulic and the mechanical apertures of the replicas G1 and G3, during shearing, under different boundary conditions. (**a**) Replica G1 with different normal stiffness, under an initial normal stress of 1 MPa. (**b**) Replica G3 with different normal stiffness, under an initial normal stress of 1 MPa. (**c**) Replica G1 with different initial normal stresses, under a constant normal stiffness of 2 MPa/mm. (**d**) Replica G3 with different initial normal stresses, under a constant normal stiffness of 2 MPa/mm.

4. Conclusions

A series of shear-flow tests for the rough fractures G1 and G3, under various CNS boundary conditions, were performed to investigate the effects of surface roughness, normal stiffness, and initial normal stress on shear-flow behavior. The peak shear stress showed an increasing trend with the increase in surface roughness and initial normal stress, while the peak shear stress did not show obvious tendencies, with respect to the normal stiffness. The residual shear stress was related to the surface roughness, normal stiffness, and initial normal stress. With increments of surface roughness, normal stiffness, and initial normal stress, the residual shear stress increased. The dilation of the fracture was significantly related to the roughness, normal stiffness and the initial normal stress. The dilation of the fracture in the shear process was restrained, more significantly, under higher normal stiffness and initial normal stress conditions.

The evolutions of transmissivity and hydraulic aperture exhibited a three-stage behavior, during the shear process. A slight decreasing stage occurred, first, due to the contraction of the fracture at the beginning of the shear. Then, the transmissivity increased rapidly, at a shear displacement of 1–4 mm, due to the shear-induced dilation. The transmissivity of the fractures increased by 2–3 orders of magnitude, during this stage. Finally, the transmissivity and the hydraulic aperture gradually increased and approximately approached constant values, due to the reduction of shear dilation. The transmissivity and hydraulic aperture decreased, gradually, as the normal stiffness and initial normal stress increased.

Author Contributions: C.W. and Y.J. conceived and designed the experiments. H.L. and S.S. performed the experiments. J.L. collected and analyzed the data. C.W. and Y.J. wrote and modified the paper.

Funding: This research was funded by National Natural Science Foundation of China (Grant No. 41772154), Natural Science Foundation of Shandong Province, China (Grant No. ZR2017MEE003), JSPS-NSFC Bilateral Joint Research Project (Grant No. 51611140122), Shandong University of Science and Technology Graduate Innovation Fund (NO.SDKDYC180303) and China Scholarship Council (CSC NO. 201608370099).

Conflicts of Interest: The authors declare no conflict of interest.

References

1. Jiang, Y.J.; Tanabashi, Y.; Xiao, J.; Nagaie. An improved shear-flow test apparatus and its application to deep underground construction. *Int. J. Rock Mech. Min. Sci.* **2004**, *41*, 170–175. [CrossRef]
2. Wu, X.Z.; Jiang, Y.J.; Guan, Z.C.; Wang, G. Estimating the support effect of energy-absorbing rock bolts based on the mechanical work transfer ability. *Int. J. Rock Mech. Min. Sci.* **2018**, *103*, 168–178. [CrossRef]
3. Barton, N.; Bandis, S.; Bakhtar, K. Strength, deformation and conductivity coupling of rock joints. *Int. J. Rock Mech. Min. Sci.* **1985**, *22*, 121–140. [CrossRef]
4. Makurat, A. The effect of shear displacement on the permeability of natural rough joints. Hydrogeology of rocks of low permeability. In Proceedings of the 17th International Congress on Hydrogeology, Tucson, AZ, USA, 7–12 January 1985; pp. 99–106.
5. Makurat, A.; Barton, N.; Rad, N.S.; Bandis, S. Joint conductivity variation due to normal and shear deformation. In Proceedings of the International Symposium on Rock Joints, Balkema, Loen, Norway, 4–6 June 1990; pp. 535–540.
6. Yang, S.Q.; Liu, X.R.; Jing, H.W. Experimental investigation on fracture coalescence behavior of red sandstone containing two unparallel fissures under uniaxial compression. *Int. J. Rock Mech. Min. Sci.* **2013**, *63*, 82–92. [CrossRef]
7. Yang, S.Q.; Yang, D.S.; Jing, H.W.; Li, Y.H.; Wang, S.Y. An experimental study of the fracture coalescence behaviour of brittle sandstone specimens containing three fissures. *Rock Mech. Rock Eng.* **2012**, *45*, 563–582. [CrossRef]
8. Mazumber, K.; Karnik, A.A.; Wolf, K.H.A.A. Swelling of coal in response to CO_2 sequestration for ECBM and its effect on fracture permeability. *SPE J.* **2006**, *11*, 390–398. [CrossRef]
9. Schmittbuhl, J.; Steyer, A.; Jouniaux, L.; Toussaint, R. Fracture morphology and viscous transport. *Int. J. Rock Mech. Min. Sci.* **2008**, *45*, 422–430. [CrossRef]

10. Leung, C.T.O.; Zimmerman, R.W. Estimating the hydraulic conductivity of two-dimensional fracture networks using network geometric properties. *Transp. Porous Media* **2012**, *93*, 777–797. [CrossRef]

11. Wang, C.G.; Zhai, P.C.; Chen, Z.W.; Liu, J.S.; Wang, L.S.; Xie, J. Experimental study of coal matrix-cleat interaction under constant volume boundary condition. *Int. J. Coal Geol.* **2017**, *181*, 124–132. [CrossRef]

12. Liu, R.; Li, B.; Jing, H.; Wei, W. Analytical solutions for water-gas flow through 3D rock fracture networks subjected to triaxial stresses. *Fractals* **2018**, *26*, 1850053. [CrossRef]

13. Yeo, I.W.; Freitas, M.H.D.; Zimmerman, R.W. Effect of shear displacement on the aperture and permeability of a rock fracture. *Int. J. Rock Mech. Min. Sci.* **1998**, *35*, 1051–1070. [CrossRef]

14. Esaki, T.; Du, S.; Mitani, Y.; Jing, L. Development of a shear-flow test apparatus and determination of coupled properties for a single rock joint. *Int. J. Rock Mech. Min. Sci.* **1999**, *36*, 641–650. [CrossRef]

15. Li, B.; Jiang, Y.J.; Koyama, T.; Jing, L.R. Experimental study of the hydro-mechanical behavior of rock joints using a parallel-plate model containing contact areas and artificial fractures. *Int. J. Rock Mech. Min. Sci.* **2008**, *45*, 362–375. [CrossRef]

16. Javadi, M.; Sharifzadeh, M.; Shahriar, K.; Mitani, Y. Critical Reynolds number for nonlinear flow through rough-walled fractures: The role of shear processes. *Water Resour. Res.* **2010**, *389*, 18–30. [CrossRef]

17. Frash, L.P.; Carey, J.W.; Lei, Z.; Rougier, E.; Ickes, T.; Viswanathan, H.S. High-stress triaxial direct-shear fracturing of Utica shale and in situ X-ray microtomography with permeability measurement. *J. Geophys. Res. Solid Earth* **2016**, *121*, 5493–5508. [CrossRef]

18. Frash, L.P.; Carey, J.W.; Ickes, T.; Porter, M.L.; Viswanathan, H.S. Permeability of fractures created by triaxial direct shear and simultaneous X-ray imaging. In Proceedings of the 51st U.S. Rock Mechanics/Geomechanics Symposium, San Francisco, CA, USA, 25–28 June 2017.

19. Rong, G.; Yang, J.; Cheng, L.; Zhou, C.B. Laboratory investigation of nonlinear flow characteristics in rough fractures during shear process. *J. Hydrol.* **2016**, *541*, 1385–1394. [CrossRef]

20. Jiang, Y.J.; Xiao, J.; Tanabashi, Y.; Mizokami, T. Development of an automated servo-controlled direct shear apparatus applying a constant normal stiffness condition. *Int. J. Rock Mech. Min. Sci.* **2004**, *41*, 275–286. [CrossRef]

21. Jiang, Y.J.; Li, B.; Tanabashi, Y. Estimating the relation between surface roughness and mechanical properties of rock joints. *Int. J. Rock Mech. Min. Sci.* **2006**, *43*, 837–846. [CrossRef]

22. Lee, Y.K.; Park, J.W.; Song, J.J. Model for the shear behavior of rock joints under CNL and CNS conditions. *Int. J. Rock Mech. Min. Sci.* **2014**, *70*, 252–263. [CrossRef]

23. Shrivastava, A.K.; Rao, K.S. Physical modeling of shear behavior of infilled rock joints under CNL and CNS boundary conditions. *Rock Mech. Rock Eng.* **2017**, *51*, 101–118. [CrossRef]

24. Li, Y.C.; Wu, W.; Li, B. An analytical model for two-order asperity degradation of rock joints under constant normal stiffness conditions. *Rock Mech. Rock Eng.* **2018**, *51*, 1431–1445. [CrossRef]

25. Indraratna, B.; Haque, A. Experimental study of shear behavior of rock joints under constant normal stiffness conditions. *Int. J. Rock Mech. Min. Sci.* **1997**, *34*, 3–4. [CrossRef]

26. Shrivastava, A.K.; Rao, K.S. Shear behaviour of rock joints under CNL and CNS Boundary Conditions. *Geotech. Geol. Eng.* **2015**, *33*, 1205–1220. [CrossRef]

27. Mirzaghorbanali, A.; Nemcik, J.; Aziz, N. Effects of cyclic loading on the shear behaviour of infilled rock joints under constant normal stiffness conditions. *Rock Mech. Rock Eng.* **2014**, *47*, 1373–1391. [CrossRef]

28. Li, B.; Jiang, Y.J.; Saho, R.; Tasaku, Y.; Tanabashi, Y. An investigation of hydromechanical behaviour and transportability of rock joints. *Asian Joint Symp. Geotech. Geo-Environ. Eng.* **2006**, 321–325.

29. Koyama, T.; Li, B.; Jiang, Y.J.; Jing, L.R. Numerical modelling of fluid flow tests in a rock fracture with a special algorithm for contact areas. *Comput. Geotech.* **2009**, *36*, 291–303. [CrossRef]

30. Olsson, R.; Lindblom, U. Direct Shear Tests Under Constant Normal Stiffness and Fluid Flow. In Proceedings of the 9th ISRM Congress, Paris, France, 25–28 August 1999.

31. Olsson, R.; Barton, N. An improved model for hydromechanical coupling during shearing of rock joints. *Int. J. Rock Mech. Min. Sci.* **2001**, *38*, 317–329. [CrossRef]

32. Saito, R.; Ohnishi, Y.; Nishiyama, S.; Yano, T.; Uehara, S. Hydraulic characteristics of single rough fracture under shear deformation. In Proceedings of the ISRM International Symposium (EUROCK), Brno, Czech Republic, 18–20 May 2005; pp. 503–509.

33. Tse, R.; Cruden, D.M. Estimating joint roughness coefficients. *Int. J. Rock Mech. Min. Sci. Geomech. Abstr.* **1979**, *16*, 303–307. [CrossRef]

34. Xiong, X.B.; Li, B.; Jiang, Y.J.; Koyama, T.; Zhang, C.H. Experimental and numerical study of the geometrical and hydraulic characteristics of a single rock fracture during shear. *Int. J. Rock Mech. Min. Sci.* **2011**, *48*, 1292–1302. [CrossRef]
35. Koyama, T.; Li, B.; Jiang, Y.J.; Jing, L.R. Coupled shear-flow tests for rock fractures with visualization of the fluid flow and their numerical simulations. *Int. J. Geotech. Eng.* **2012**, *2*, 215–227. [CrossRef]
36. Yin, Q.; Jing, H.W.; Ma, G.W.; Su, H.; Liu, R.C. Investigating the roles of included angle and loading condition on the critical hydraulic gradient of real rock fracture networks. *Rock Mech. Rock Eng.* **2018**, *51*, 3167–3177. [CrossRef]
37. Yin, Q.; Ma, G.W.; Jing, H.W.; Su, H.; Wang, Y.; Liu, R.C. Hydraulic properties of 3D rough-walled fractures during shearing: An experimental study. *J. Hydrol.* **2017**, *555*, 169–184. [CrossRef]
38. Liu, R.C.; Jing, H.W.; He, L.; Zhu, T.; Yu, L. An experimental study of the effect of fillings on hydraulic properties of single fractures. *Environ. Earth Sci.* **2017**, *76*, 684. [CrossRef]

Case Report

Numerical Simulation on the Dynamic Characteristics of a Tremendous Debris Flow in Sichuan, China

Yulong Chen [1,*], Zhenfeng Qiu [2,*], Bo Li [3,*] and Zongji Yang [4,*]

[1]   State Key Laboratory of Hydroscience and Engineering, Tsinghua University, Beijing 100084, China
[2]   Key Laboratory for Hydraulic and Waterway Engineering of Ministry of Education,
     Chongqing Jiaotong University, Chongqing 400074, China
[3]   Key Laboratory of Karst Environment and Geohazard, Ministry of Land and Resources, Guizhou University,
     Guiyang 550000, China
[4]   Institute of Mountain Hazards and Environment, Chinese Academy of Science, Chengdu 610041, China
*   Correspondence: chen_yl@tsinghua.edu.cn (Y.C.); qiuzhenfeng3012@163.com (Z.Q.);
     libo1512@163.com (B.L.); yzj@imde.ac.cn (Z.Y.)

Received: 26 June 2018; Accepted: 24 July 2018; Published: 1 August 2018

Abstract: The mega debris flow that occurred on 13 August 2010 in Zoumaling Valley in Mianzhu County, China has done great damage to the local inhabitants, as well as to the re-construction projects in the quake-hit areas. Moreover, it is of high possibility that a secondary disaster would reappear and result in worse consequences. In order to maximize risk reduction of this problem, the local government planned to construct seven debris-resisting barriers across each ditch for mitigation of debris flow hazards in the future. In this paper, the numerical simulation fields of flow velocity, pressure, and mud depth of the Zoumaling debris flow had been computed by using finite volume method software based on computational fluid dynamics (CFD). The Bingham fluid was chosen as the constitutive model of this debris flow. The debris flow geometry model was a 3D model. The initial conditions, boundary conditions, control equations, and parameters were determined and adjusted by the actual conditions and analyses. The flow field data obtained from numerical simulations were substituted into the finite element software ANSYS. Then the calculations of fluid-solid coupling action between the flow and dam had been done. All these results of simulations and analyses could be the guide and suggestion for the design and construction of prevention engineering of Zoumaling debris flow.

Keywords: debris flow; dynamic characteristics; numerical analysis; debris-resisting barriers

1. Introduction

Debris flows are a highly unpredictable hazard in areas of mountainous terrain and high runoff. The local geological configurations caused by orogenesis result in a large proportion of steep topography and mountainous areas where earthquakes occur frequently [1]. Meanwhile, torrential rains during the monsoon season are major factors that enhance the occurrence of debris flows, soil erosion, and landslides.

Computational fluid dynamics (CFD) numerical methods have been successfully used to analyze multiphase and multicomponent flows and to investigate fluid–structure interaction in various settings. From the standpoint of dynamic mechanics, Elverhøi et al. [2] performed a numerical model of submarine debris flows with Computational Fluid Dynamics X (CFX). Zakeri et al. [3] applied CFX to simulate the submarine debris flow impact on pipelines. Liu and Tian [4] employed CFX to analyze the impact forces of submarine landslides on free-span pipelines. In recent years, ANSYS CFX has become the most commonly adopted debris flow simulation model.

On 13 August 2010, high intensity heavy rain triggered a debris flow within the watershed, which was classified as a group-occurring debris flow, resulting in worse damages to human life and property than in previous decades [5]. Following this accident, the local government planned to construct seven debris-resisting barriers across each ditch for mitigation of debris flow hazards in the future, for which numerical analysis is an effective method to evaluate the performance of the barriers.

In this study, numerical simulation on the dynamic characteristics of the debris flow was conducted to evaluate the fields of flow velocity, pressure, and mud depth of the debris flow by using CFD software, ANSYS CFX 13.0 (ANSYS, Pittsburgh, PA, USA), which is based on the finite volume method. The Bingham fluid model was chosen as the constitutive model of this debris flow. The debris flow geometry was modeled in 3D. The initial conditions, boundary conditions, controlling equations, and parameters were determined and adjusted by the actual conditions and analyses. The flow field data obtained from numerical simulation were imported into the finite element software ANSYS. Then the calculations of coupled fluid-solid action between the flow and barriers was performed. The results of these simulations and analyses could be used as guidance and suggestions for the design and construction of prevention engineering of the debris flow.

2. Summary of the Study Area and Its Debris Flow Disaster

2.1. Study Area

The Mianyuan River, a branch of the Tuojiang River, is in Mianzhu County of Sichuan province (Figure 1). It is about 80 km northeast of the earthquake epicenter, with a basin area of approximately 400 km^2. The Yingxiu–Beichuan seismogenic fault of the Wenchuan earthquake goes through the middle part of the river basin. Outcrops within the river basin are locally covered by loose Quaternary materials (Figure 2).

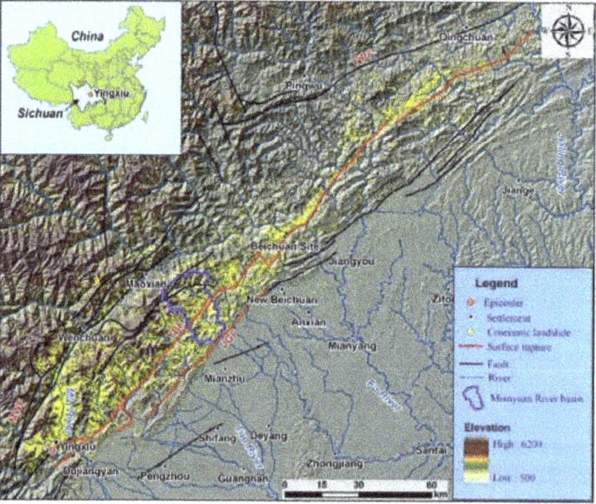

Figure 1. Location of the study area—the Mianyuan River basin. MWF: Miaoxian–Wenchuan Fault, YBF: Yingxiu–Beichuan Fault, JGF: Jiangyou–Guanxian Fault, QPF: Qingchuan–Pingwu Fault [5].

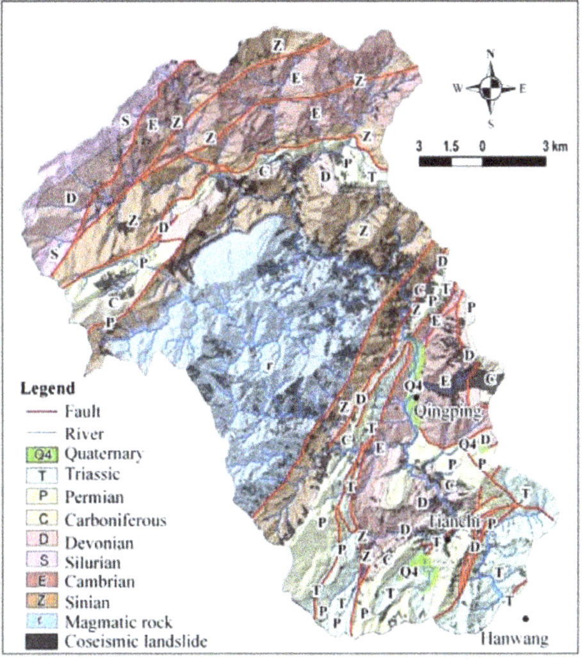

Figure 2. Geological setting of the Mianyuan River Basin [5].

2.2. Debris Flow Events in the Mianyuan River Basin

The co-seismic landslides triggered by the Wenchuan earthquake produced at least 4.0×10^8 m^3 of loose material in the Mianyuan River Basin. Loose materials are widespread on the hilly terrains and in gullies. Figure 3 shows the characteristics of long-term activity in the local watershed. Since numerous rainfall events, long cracks developed along the back edge of each gully resulting in an increasing source of potential landslide. These materials, as well as cracked slopes, are marginally stable under normal conditions, and can lose stability due to rainfall infiltration and become sources of deadly debris flows. The antecedent rainfall and rainfall intensity on 13 August 2010 were 82.6 mm/d and 37.4 mm/h, respectively.

A storm swept over the Mianyuan River Basin from 12 to 13 August 2010. The accumulative rainfall was about 200 mm and the duration was 10 h [7]. According to witnesses, seven debris flows occurred in Qingping Town at 23:45 on 12 August. Shortly afterward, another outbreak of debris flows occurred in other areas surrounding the town of Qingping. A depositional fan was formed and the Mianyuan River was blocked in a matter of seconds (Figure 4).

A field survey showed that the debris generated by this event buried the whole town, with a covering area of 1.4×10^6 m^2 (length of 4.3 km, width of 400–500 m) and an average thickness of 5 m (Figure 4b). In the event 379 houses were destroyed, which account for 20.9% of the total number of houses in Qingping Town. Furthermore, seven people were killed, seven were missing, and 33 were injured. The field investigation survey and remote sensing interpretation indicate that during the 12–13 August rainstorm, the total volume of deposited materials was 5.65×10^6 m^3.

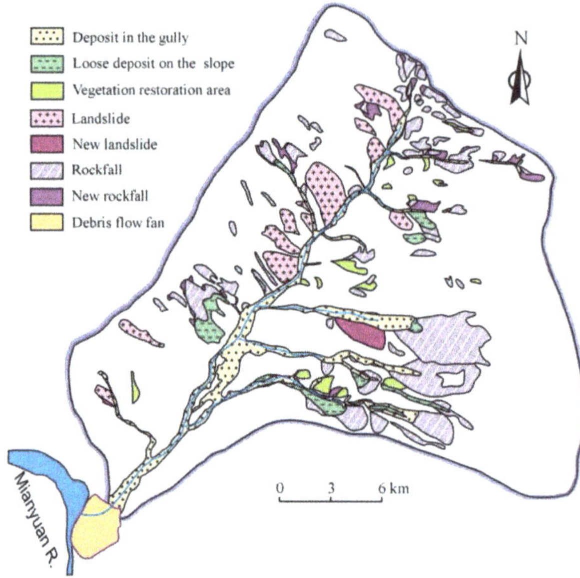

Figure 3. Evolution of the Zoumaling Gully after the Wenchuan earthquake [6].

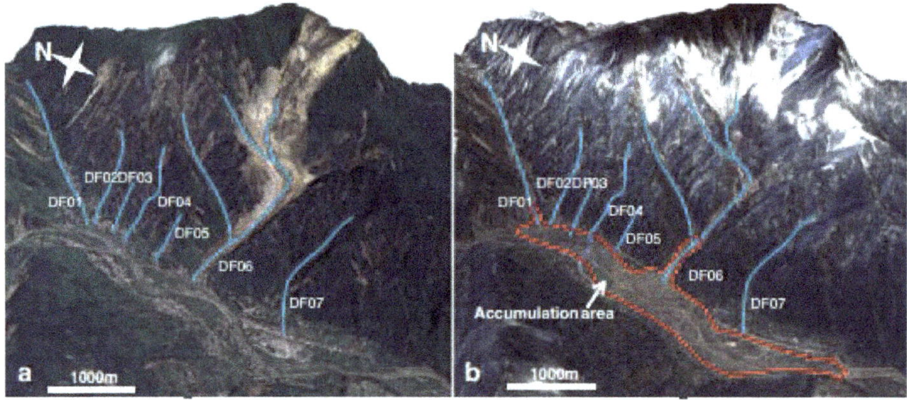

Figure 4. Images of the 13 August 2010 debris flow in Qingping Town: (**a**) aerial photo taken on 18 May 2008; and (**b**) a Worldview image from 13 December 2010.

According to the field survey, the Wenchuan earthquake has produced as much as 4.0×10^8 m^3 of loose material in this area. However, the Zoumaling debris flow has only transported about 5.65×10^6 m^3, which means the most remained resting on the hill slope nearby. If another flash flood breaks out, more debris flows would be more likely to take place. Hence, it is of great importance to understand the dynamic processes of the Zoumaling debris flow and examine the risk mitigation by the debris-resisting barriers.

The field investigation survey indicates that the slope is steep in the upstream gullies and flat in the downstream gullies, as well as alternating steep with flat in some local places. The width is narrow in the upstream gullies and broad in the downstream gullies, as well as alternating narrow with wide in some local places. The above geological condition is favorable for the construction of barriers. Due to the relatively short length of barriers and the relatively flat gully and broad valley upstream, the barriers have a relatively large reservoir capacity, playing a strong role in adjusting the peak flow rate of debris flow.

Due to the large amount of loose mass in the back edge of gully DF05-DF07 (see Figure 3), which may contribute to potential debris flow, multiple barriers were planned to be constructed in the three gullies to counteract erosion by reducing debris flow velocity, as well as to reinforce the foundations. Additionally, drain ditches were employed in the accumulation area to guide flows along designed paths from damaging residential houses and residents near the gully, for which the embankment wall is planned to be constructed uptown. A barrier should be constructed on a loose foundation within the watershed to prevent erosion from debris flow and to reduce the source contributing to debris flow. Barriers are installed upstream of the main gully to reinforce foundations and obstruct debris flows. Seven barriers will be constructed at a height of 8–12 m, with an upstream batter of 1:0.70, and a downstream batter of 1:20. The length of the axis of the barrier crest is 50–71 m. Barriers reserve coarse materials and fine materials pass, and they slow the velocity of debris flow. Thereby, sources reaching downstream are reduced, as is the unit weight, resulting in a smaller peak discharge. As a result, scour resistance walls and protection embankments are subjected to less pressure.

3. Modeling of Debris Flow

ANSYS CFX is a general purpose CFD program that includes a solver based on the finite volume (FV) method for unstructured grids, as well as pre- and post-processing tools for simulation definition and data extraction, respectively. The FV method uses the integral form of the conservation equations. With tetrahedra or hexahedra control volumes (CVs), unstructured grids are best adapted to the FV approach for complex 3D geometries [8]. In general, there are two types of multiphase flows: disperse flows and separated flows. The disperse flows consist of finite particles, such as drops or bubbles (the dispersed phase), distributed in a connected volume of another continuous phase (fluid), whereas the separated flows comprise two or more continuous streams of different fluids separated by interfaces [9].

3.1. Assumption

On the basis of CFX, the following assumptions are made for the numerical simulation of the debris flow: (1) debris flow fluid is a homogeneous single phase flow; (2) the gully and mountain models are stiff so that no deformation occurs during debris flow. Hence, erosion was not considered.

3.2. Construction of Model and Mesh

The 3D geological model gully for debris flow is regarded too complex to be directly modeled by ANSYS. The modeling process is combined with AutoCAD, Surfer, and Design Modeler, etc., following the rules of point to line, line to area, area to cubic, and bottom to top.

The first step was to create the contour line of the watershed where the Zoumaling gully debris flow occurred based on the topographically surveyed data. A plug-in was applied to input elevation points read from contour lines into Surfer. Then the plane range and division space of the model were selected in Surfer and a grid-formatted file was generated to create the terrain model by ANSYS. Finally, a terrain model was imported into Workbench to stretch the terrain to create an entity model of the watershed. The model was meshed with tetrahedrons for the terrain and hexahedrons for the fluid domain and barriers by CFD in Workbench. The number of nodes, cells, and areas are 180,203, 935,994, and 92,224, respectively, as shown in Figure 5. The safety factor of the barriers was calculated according to the Mohr-Coulomb strength criterion.

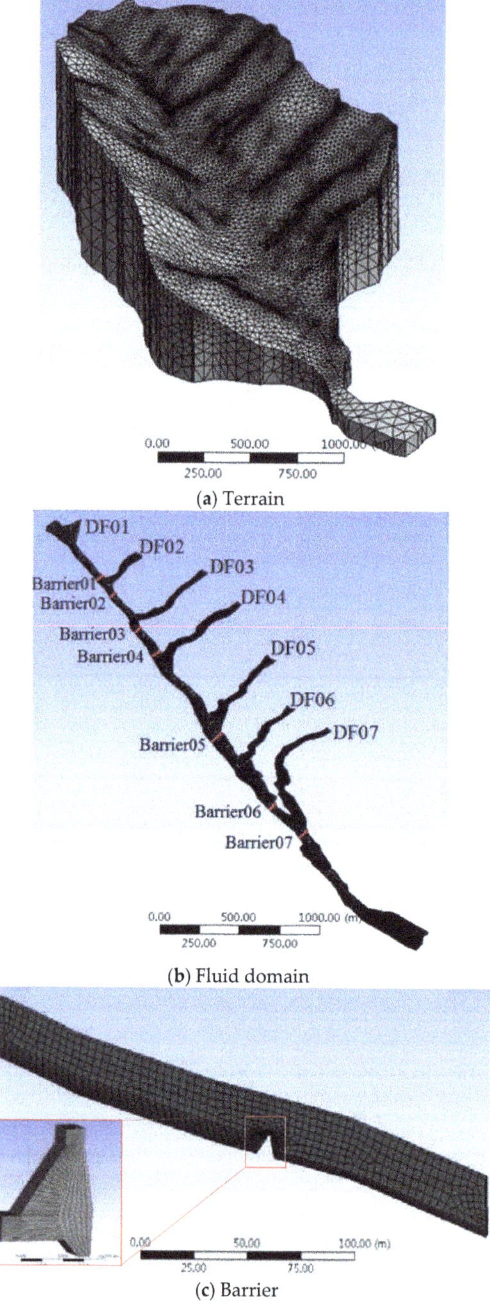

(**a**) Terrain

(**b**) Fluid domain

(**c**) Barrier

Figure 5. Mesh example with tetrahedron elements.

3.3. Preprocess

Dynamic parameters of debris flow associated with unit weight and flow rate at each gully should be determined in advance.

The unit weight of debris flow was found to be 18.00 KN/m^3 by field investigation of the slurry. The elastic modulus and Poisson's ratio are 3.15 × 10^7 kPa and 0.2, respectively. The coefficient of friction between the barriers and the debris flow is 0.35.

Referring to a 50-year flood discharge, the inlet flow discharges were 74.6 m^3/s for the main gully (DF01), 725.8 m^3/s for the DF02 ditch, 40.9 m^3/s for the DF03 ditch, 56.7 m^3/s for the DF04 ditch, 124.1 m^3/s for the DF05 ditch, 64.4 m^3/s for the DF06 ditch, and 57.0 m^3/s for the DF07 ditch, respectively.

The rheology of the debris flow follows the Bingham model:

$$\tau = \tau_y + \mu \frac{du}{dy} \tag{1}$$

where τ is the shear stress, τ_y is the yield stress, μ is the Bingham viscosity, and du/dy is the shear strain rate. For a Bingham fluid, viscosity that affects the maximum flow velocity was defined by the CFX Expression language. τ_y can be expressed as:

$$\tau_y = 0.098 \exp(B\varepsilon + 1.5) \tag{2}$$

$$\varepsilon = \frac{S_V - S_{V0}}{S_{Vm}} \tag{3}$$

$$S_V = \frac{S_{Vm}}{1 + \frac{1}{\lambda}} \tag{4}$$

$$S_{V0} = 1.26 S_{Vm}^{3.2} \tag{5}$$

where S_V is the volume concentration, S_{Vm} is the limit volume concentration, λ is the linear concentration, B is a constant. In this study, S_V, S_{Vm}, λ and B are 0.65, 0.56, 8.45, and 1.8.

Moreover, turbulence generated in the water was simulated using the k–ε model. Inherent is the dependency of the undrained shear strength on the shear strain rate.

4. Modeling Results

4.1. Debris Flow Simulation

The three-dimensional debris flow numerical model and computation conditions described previously were implemented to perform a debris flow simulation. Without considering barriers, the debris flow state, pressure nephogram of the riverbed, velocity vector, and shear stress on the riverbed wall are shown in Figure 6, Figure 7, Figure 8, Figure 9, respectively. The debris flow field data at the main gully in Zoumaling is shown in Table 1.

Table 1. Debris flow field data at the main gully in Zoumaling.

Position	Velocity (m/s)	Deposition Thickness (m)
Barrier01	2.29	1.49
Barrier02	5.69	0.82
Barrier03	2.41	1.29
Barrier04	3.64	0.82
Barrier05	5.69	0.58
Barrier06	7.83	0.84
Barrier07	6.92	1.83
Outlet	2.75	1.81

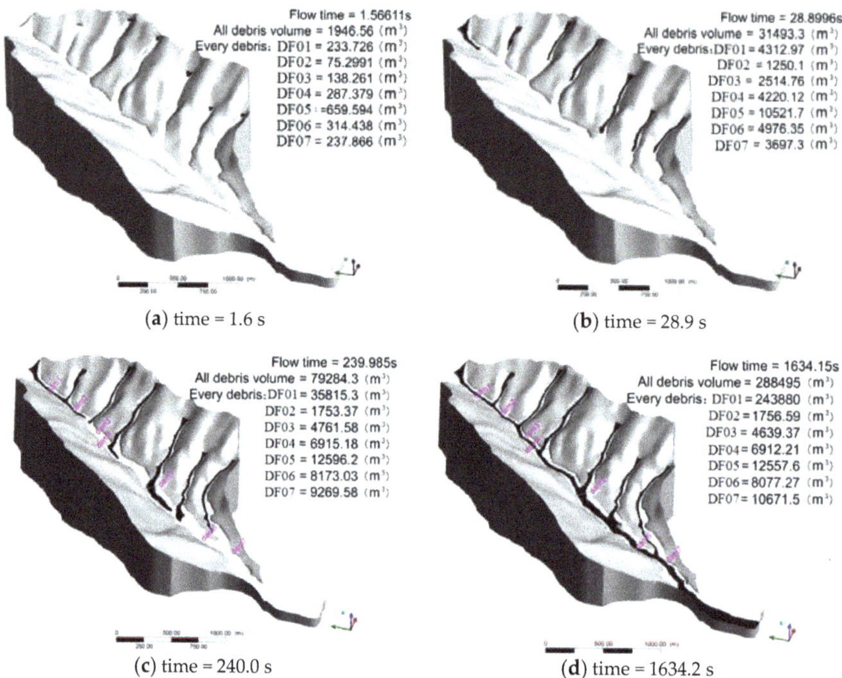

(**a**) time = 1.6 s (**b**) time = 28.9 s

(**c**) time = 240.0 s (**d**) time = 1634.2 s

Figure 6. Simulation of the debris flow state.

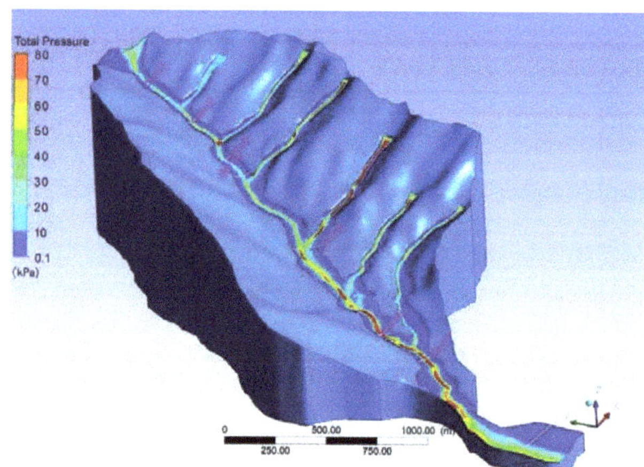

Figure 7. Debris flow pressure nephogram on the riverbed (time = 1634.2 s).

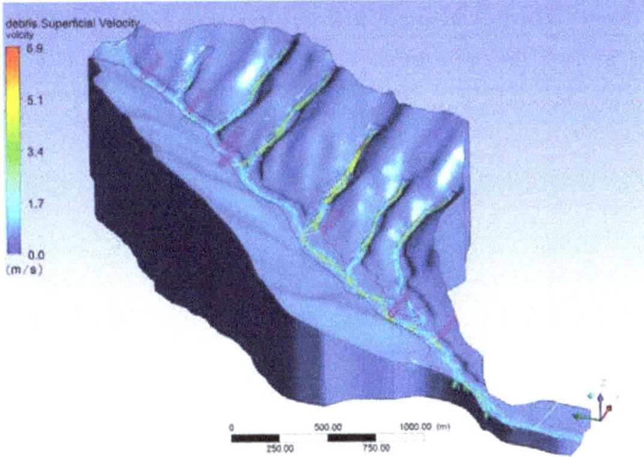

Figure 8. Debris flow velocity vector (time = 1634.2 s).

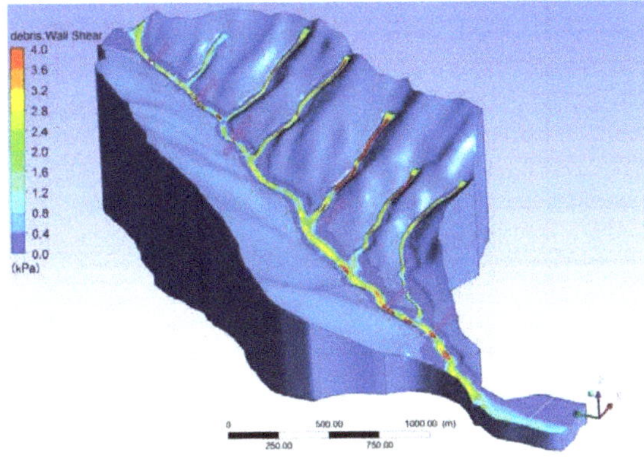

Figure 9. Debris flow shear stress on the riverbed wall (time = 1634.2 s).

This shows that the maximum pressure on riverbed was 149 kPa, occurring at the junction of gully DF02 and the main gully; maximum shear stress on the riverbed wall and velocity were 8.32 kPa and 13.57 m/s, respectively, occurring at gully DF05. This indicated that in gullies DF02 and DF05, the impact and erosion forces were so strong that the debris flow could greatly influence the stability of the gully bed and bank slope.

The debris flow velocity increased from upstream to downstream. A steeper or narrower riverbed resulted in a larger velocity of debris flow. The deposition thickness decreased with increasing debris flow velocity. A steep or wider riverbed resulted in a smaller deposition thickness of the debris flow.

4.2. Fluid-Solid Coupling Simulation of Debris Flow and Barriers

Debris flow field data at the main gully of Zoumaling was imported to ANSYS for fluid-solid coupling computation to yield flow pressure and characteristics on the barriers proposed to be constructed. Then we can obtain the stress and strain of the barrier in response to the debris flow. The debris-barrier interaction results are listed in Table 2. It can be seen that the stress and strain responses are similar for each barrier.

The deposition thickness of the debris flow in front of the barrier is shown in Figure 10. The deposition thickness increased with the debris-barrier interaction. When the debris flow reached the barrier, it was impeded by the obstruction of the barrier. Debris at the frontal portion ran up against the barrier. A deposition was formed by the debris that was stopped and trapped behind the barrier. The extent of the deposition developed when further debris was stopped behind the barrier. The debris flow subsequently rode on the plug and overtopped the barrier. Debris overflowing from the barrier launched into a ballistic flight, and resumed its travel on the runout trail after landing.

Table 2. Debris-barrier interaction.

Barrier	01	02	03	04	05	06	07
Max shear stress (Mpa)	1.22	1.29	1.05	0.898	0.332	0.456	0.534
Max von –mises stress (MPa)	2.36	2.49	2.04	1.74	0.634	0.857	1.02
Max equivalent elastic strain ($\times 10^{-5}$)	7.48	7.90	6.48	5.53	2.01	2.72	3.24
Max displacement at crest (mm)	0.334	0.368	0.388	0.333	0.12	0.193	0.229
Debris flow horizontal impact forces (kN)	3350	3690	4800	4110	2390	3870	5430
Debris flow vertical impact forces (kN)	9220	9470	12,600	11,400	55,500	8360	14,400
Barrier antioverturning counter moment (kN·m)	17,600	15,900	32,400	28,900	16,400	29,900	69,500
Safety factor (deposition filled)	2.12	2.01	2.45	2.87	2.63	2.34	1.94

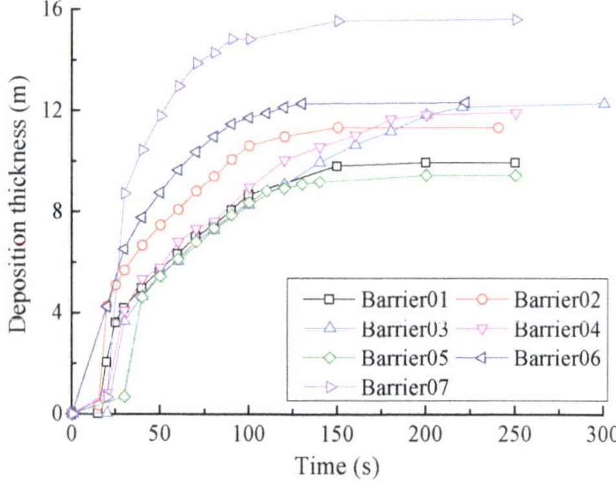

Figure 10. Deposition thickness at the front of the barrier.

Figure 11 shows the variation of safety factor of barrier with deposition thickness at the front of the barrier. The safety factor of a barrier decreased with the deposition thickness at the front of the barrier. When the deposition overtopped the barrier, the safety factor reached its lowest point.

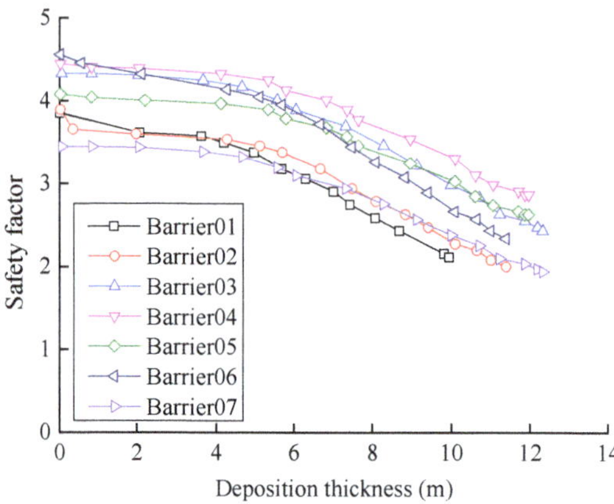

Figure 11. Variation of the safety factor of the barrier with deposition thickness at the front of the barrier.

The total pressure of debris flow on the barrier along the height is shown in Figure 12. It shows a similar distribution of total pressure of the debris flow on each barrier. Herein, we just present the results on Barrier 01. The total pressure distribution in front of Barrier 01 is shown in Figure 13. Accordingly, the total deformation, total stress, and safety factor of Barrier 01 are shown in Figure 14.

The maximum force on the barrier was 3360 kN in horizontal direction and 9220 kN in vertical direction. The friction force by the weight of the barrier itself and the vertical force on the it was 3220 kN, which was less than the maximum horizontal force. It meant that the barrier was prone to fail unless the soils along the flow path were protected from erosion. The lower deposition thickness led to a higher safety factor for the barriers.

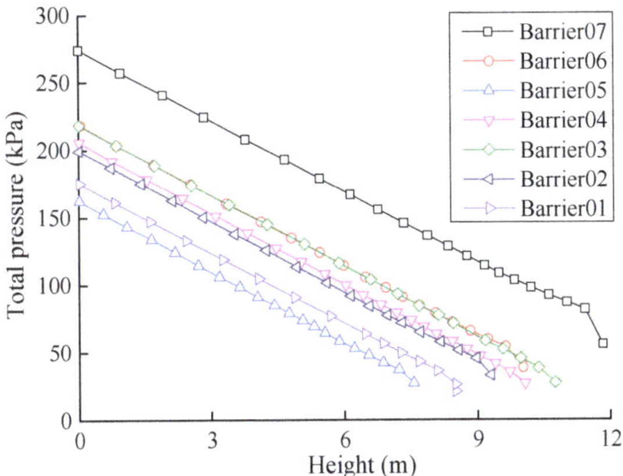

Figure 12. Total pressure with the height of the barrier.

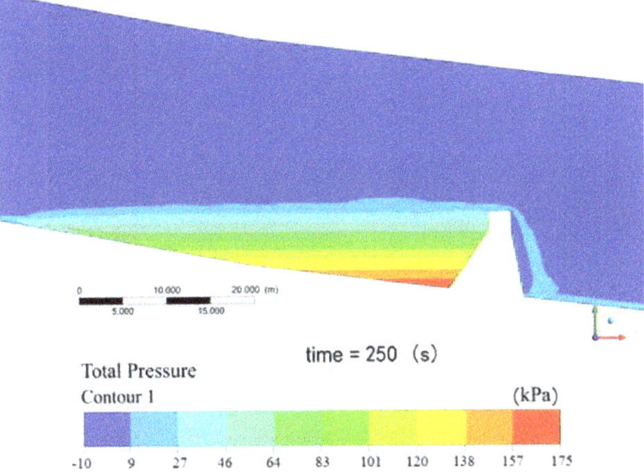

Figure 13. Total pressure distribution on Barrier 01.

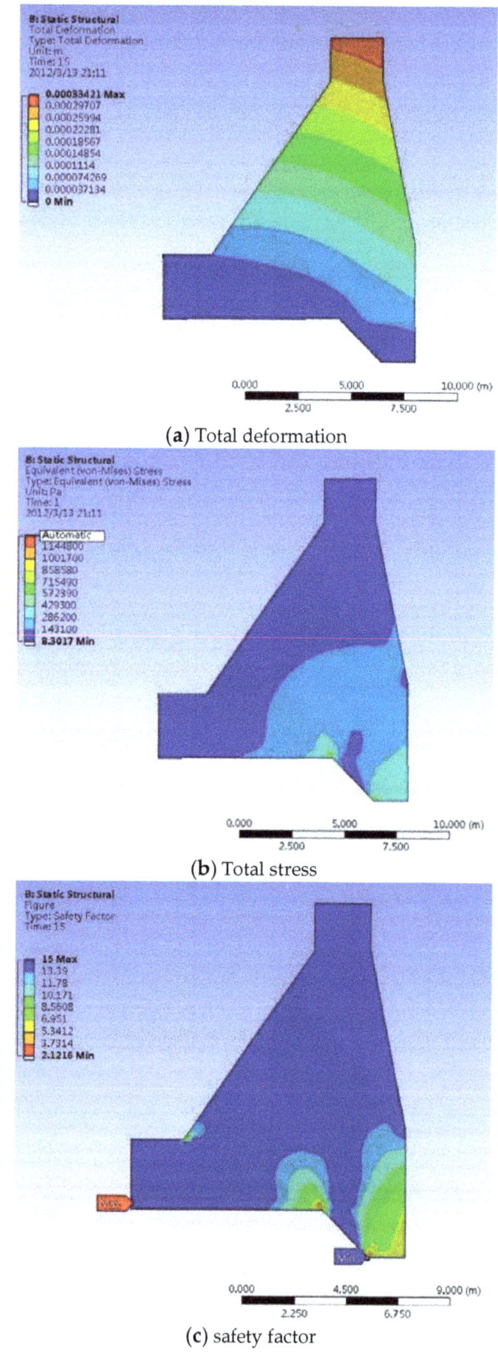

(**a**) Total deformation

(**b**) Total stress

(**c**) safety factor

Figure 14. Total deformation, total stress, and safety factor of Barrier 01.

5. Conclusions

On 13 August 2010, a debris flow occurred in the Zoumaling of Mianzhu County in the southwest part of China due to heavy rainfall. Meanwhile, it formed nearly 5.65×10^6 m^3 of deposits before being transformed into a debris flow. Based on numerical simulations of the formation process of the debris flow, the peak velocity of the debris flow, the maximum pressure and maximum shear stress on the riverbed wall were approximately 13.57 m/s, 149 kPa, and 8.32 kPa, respectively. Deposition thickness increased with the debris-barrier interaction. The safety factor of a barrier decreased with the deposition thickness at the front of the barrier. When the deposition overtopped the barrier, the safety factor reached its lowest point. The barrier was prone to fail unless the soils along the flow path were protected from erosion. The lower deposition thickness led to a higher safety factor for the barriers.

Based on the numerical analysis of this debris flow, it is suggested that the soils on the surface of gullies should be protected from erosion and soils on the riverbed should be removed so that the deposition is reduced.

Author Contributions: Y.C. performed the numerical simulations. Z.Q., B.L., and Z.Y. participated in the data treatment and the writing.

Funding: This work is supported by the China Postdoctoral Science Foundation under grant nos. 2017M620048 and 2018T110103.

Acknowledgments: The authors are grateful to Editors and the anonymous reviewers for their extensive and profound comments and suggestions, which substantially improved the quality of paper.

Conflicts of Interest: The authors declare no conflicts of interest.

References

1. Yang, Z.; Qiao, J.; Uchimura, T.; Wang, L.; Lei, X.; Huang, D. Unsaturated hydro-mechanical behaviour of rainfall-induced mass remobilization in post-earthquake landslides. *Eng. Geol.* **2017**, *222*, 102–110. [CrossRef]
2. Elverhøi, A.; Issler, D.; de Blasio, F.V.; Ilstad, T. Emerging insights into the dynamics of submarine debris flows. *Nat. Hazards Earth Syst. Sci.* **2005**, *5*, 633–648. [CrossRef]
3. Zakeri, A.; Høeg, K.; Nadim, F. Submarine debris flow impact on pipelines—Part II: Numerical analysis. *Coast. Eng.* **2009**, *56*, 1–10. [CrossRef]
4. Liu, J.; Tian, J. Impact forces of submarine landslides on free-span pipelines. In Proceedings of the 33rd International Conference on Ocean, Offshore and Arctic Engineering (OMAE 2014), San Francisco, CA, USA, 8–13 June 2014; p. V06BT04A037.
5. Huang, R.; Li, W. Post-earthquake landsliding and long-term impacts in the Wenchuan earthquake area, China. *Eng. Geol.* **2014**, *182*, 111–120. [CrossRef]
6. Zhang, Y.; Cheng, Y.; Yin, Y.; Lan, H.; Wang, J.; Fu, X. High-position debris flow: A long-term active geohazard after the Wenchuan earthquake. *Eng. Geol.* **2014**, *180*, 45–54. [CrossRef]
7. Xu, Q.; Zhang, S.; Li, W. The 13 August 2010 catastrophic debris flows after the 2008 Wenchuan earthquake, China. *Nat. Hazards Earth Syst. Sci.* **2012**, *12*, 201–216. [CrossRef]
8. Ferziger, J.H.; Perić, M. *Computational Methods for Fluid Dynamics*, 3rd ed.; Springer International Publishing: Berlin, Germany, 2002.
9. Brennen, C.E. *Fundamentals of Multiphase Flow*; Cambridge University Press: Cambridge, UK, 2005.

MDPI
St. Alban-Anlage 66
4052 Basel
Switzerland
Tel. +41 61 683 77 34
Fax +41 61 302 89 18
www.mdpi.com

Processes Editorial Office
E-mail: processes@mdpi.com
www.mdpi.com/journal/processes